W0257255

ENCYCLOPEDIA OF PHYSICS

EDITED BY

S. FLÜGGE

VOLUME XXVIII

SPECTROSCOPY II

WITH 223 FIGURES

SPRINGER-VERLAG BERLIN HEIDELBERG GMBH

1957

HANDBUCH DER PHYSIK

HERAUSGEGEBEN VON

S. FLÜGGE

BAND XXVIII

SPEKTROSKOPIE II

MIT 223 FIGUREN

SPRINGER-VERLAG BERLIN HEIDELBERG GMBH
1957

ISBN 978-3-642-45862-0   ISBN 978-3-642-45861-3 (eBook)
DOI 10.1007/978-3-642-45861-3

Inhaltsverzeichnis.

Microwave Spectroscopy.

By

Walter Gordy.

With 25 Figures.

A. Introduction.

1. The microwave region. Microwaves are Hertzian waves ranging in length from about thirty centimeters to a fraction of a millimeter, or in frequency from

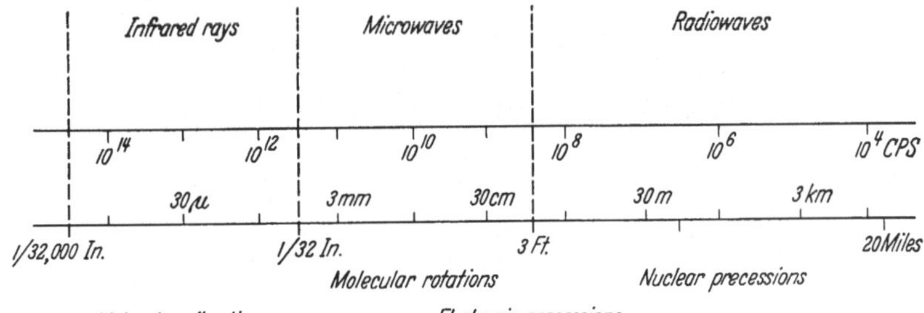

Fig. 1. A chart locating the microwave region within the spectrum.

about one billion[1] to five hundred billion cycles per second. The microwave region is thus the billion cycle range of the electromagnetic spectrum. Its low-frequency border is the upper megacycle range where ordinary radio tubes and circuits become ineffective and where the hollow cavity and waveguide components of the microwave region must be adopted. Its high-frequency border, the upper submillimeter wavelengths, is the region where tunable sources and detectors which characterize radio waves become ineffective and where optical and semi-optical infrared methods must be adopted. Fig. 1 is a chart which locates the microwave region within the spectrum and suggests the principal types of spectral transitions which occur within its bounds.

Even though microwaves were the first radio waves to be discovered by Hertz, they were the last radio waves to become useful to man. Although the klystron, the magnetron, the waveguide propagation were discovered before 1940, refined and versatile microwave

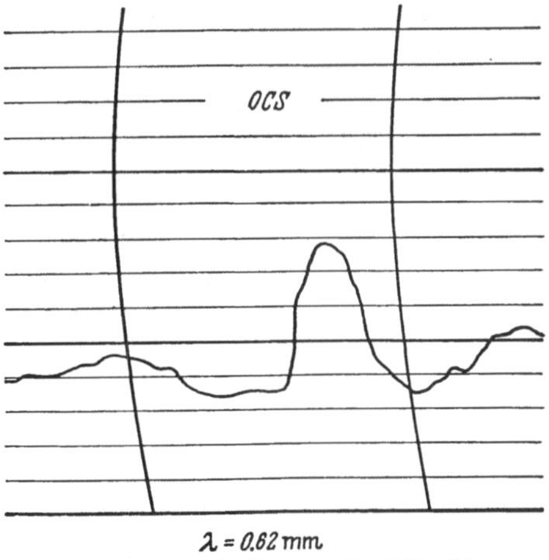

Fig. 2. Recording of the $J = 39 \rightarrow 40$ rotational frequency of carbonyl sulfide ($O^{16}C^{12}S^{32}$) at 486184.2 Mc/sec (wavelength = 0.617 mm), obtained with the 20th harmonic power of a 2 K 33 klystron. The line half-width is about 1 Mc/sec. (From M. Cowan and W. Gordy.)

---

[1] One billion means 10^9. The abbreviation cps means "cycles per second".

instruments and techniques were not developed until the need for them arose in wartime radar, 1940 to 1945. The large and important field of microwave spectroscopy has been developed almost wholly since 1945. The single pre-war spectral measurement which is generally classified as microwave is the low-resolution measurement of the inversion frequency of ammonia in the centimeter-wave region by CLEETON and WILLIAMS[1] in 1934. Not until 1954 were spectral measurements with microwave electronic (radio) methods[2] made to overlap those with infrared grating spectrometers[3]. As early as 1923, however, NICHOLLS and TEAR[4] in a sense closed this last gap in the electromagnetic spectrum by detecting energy from a spark gap generator similar to that employed by HERTZ. Measurements are now made with microwave elec-

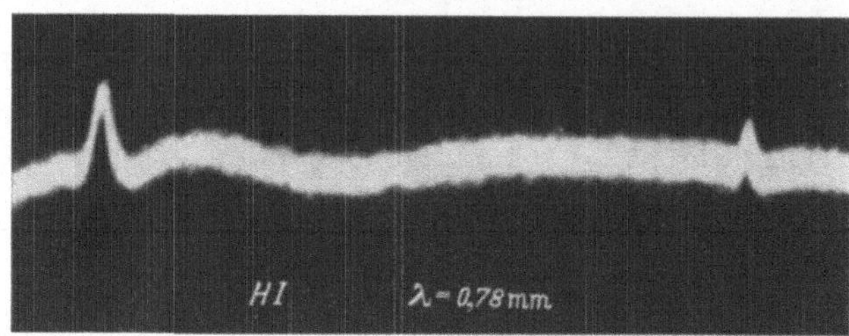

Fig. 3. Cathode ray display of two of the three hyperfine components of the $J=0 \to 1$ transition of HI at 0.778 mm wavelength. The line to the left is the $F=\frac{5}{2} \to \frac{7}{2}$ component at 385 385.11 Mc/sec, and the one to the right is the $F=\frac{5}{2} \to \frac{3}{2}$ component at 385 548.80 Mc/sec. [From M. COWAN and W. GORDY: Phys. Rev. **104**, 551 (1956).]

tronic methods down to 0.587 mm wavelength or 510 billion cps by our Duke University group[5]. Fig. 2 shows a spectral transition at 0.617 mm/sec detected with microwave methods. Spectral lines above 0.65 mm wavelength can now be displayed on the cathode ray scope (see Fig. 3). The sub-millimeter wave measurements[2,5] are made with the same high precision and resolution as those which characterize the centimeter region.

B. Microwave spectrometers.

2. Spectrometers for investigation of gases. Microwave spectrometers for the study of gases are usually designed for observation and measurement of sharp absorption lines—lines which are of the order of kilocycles or megacycles in width.

Various types of spectrometers for studying gases are in use [1], [2], [4], [5], [6], [16]. The particular design depends upon several factors such as the availability of components, the part of the microwave region for which the instrument is designed to operate, and the nature of the information which one wishes to gain from the measurements. Nevertheless, the various types have certain basic elements in common. They are: (1) a tunable microwave source; (2) a microwave detector; (3) a frequency meter; (4) an absorption cell; (5) a frequency modulator either for the source or for the absorption line; (6) an amplifier of the detected signal; and (7) a signal indicator. A diagram of a typical microwave spectrometer for gases is given in Fig. 4.

In the centimeter and lower frequency millimeter wave region the reflex klystron is usually employed as a microwave source. This convenient, versatile

[1] C. E. CLEETON and N. H. WILLIAMS: Phys. Rev. **45**, 234 (1934).

[2] C. A. BURRUS and W. GORDY: Phys. Rev. **93**, 897 (1954); **101**, 599 (1956).

[3] L. GENZEL and W. ECKHARDT: Z. Physik **139**, 592 (1954).

[4] E. F. NICHOLLS and J. D. TEAR: Phys. Rev. **21**, 378, 587 (1923). — Proc. Nat. Acad. Sci. U.S.A. **9**, 221 (1923).

[5] M. COWAN and W. GORDY: Phys. Rev. **104**, 551 (1956).

instrument is a highly stable, nearly monochromatic source of ample energy which can be mechanically tuned over a frequency range of about 20%, or elec-

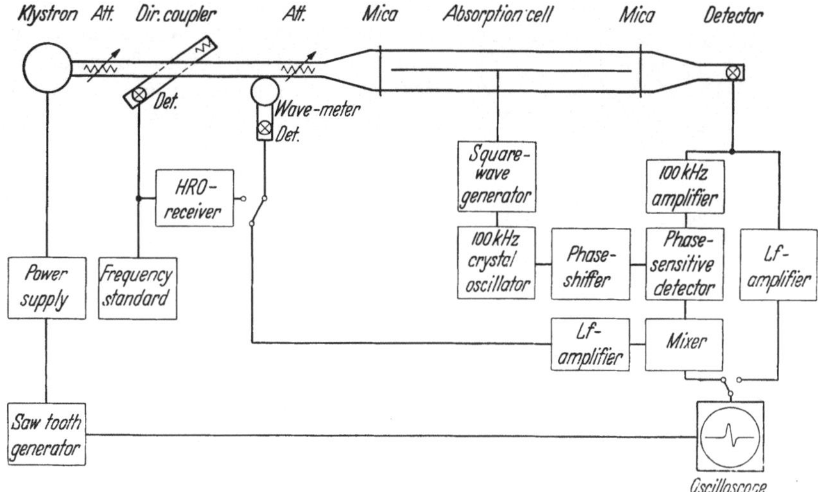

Fig. 4. Diagram of a STARK modulation spectrometer for gases. [From G. ERLANDSSON: Ark. Fysik 9, 399 (1955).]

trically swept at almost any desired rate over a range of approximately 50 mega-cycles. In the region above 60 kMc/sec (wavelengths below 5 mm) no klystrons

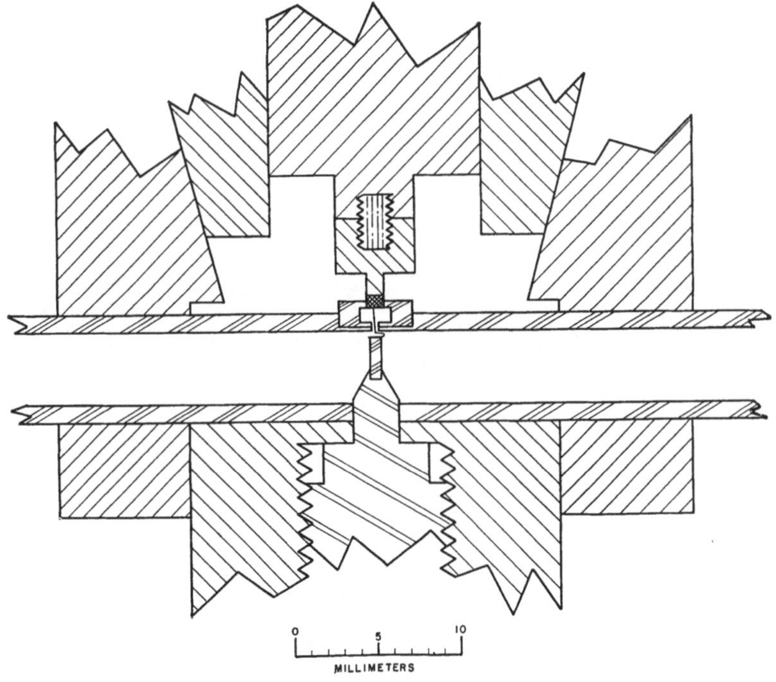

Fig. 5. Cross-section diagram of crystal multiplier employed for the lower millimeter and the sub-millimeter wave regions See also Fig. 6. [From KING and GORDY: Phys. Rev. 93, 407 (1954).]

are yet available. Silicon crystal harmonic generators driven by klystrons are generally employed. That shown in Fig. 5 employs a small crystal mounted directly in the waveguide with a differential screw mechanism for adjustment of the pressure of the tungsten "whisker". This design led to the first spectral measurements in the shorter millimeter[1] and sub-millimeter wave region[2].

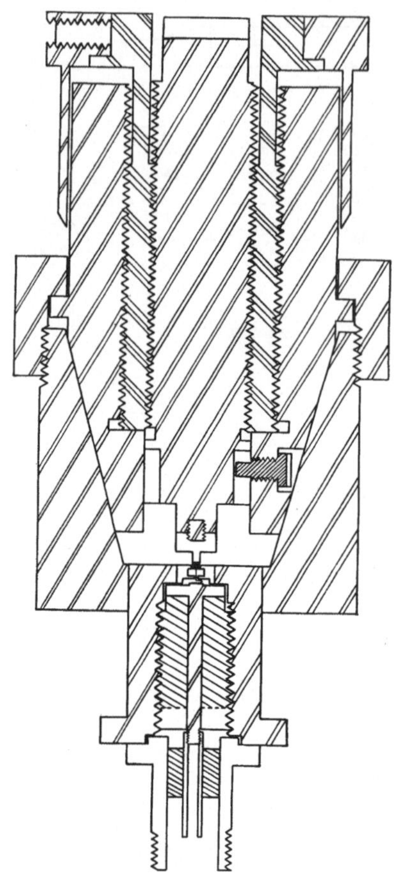

Silicon crystals are almost universally used as detectors in spectrometers for gases, whereas in spectrometers for study of paramagnetic resonance in solids bolometers are most often used. The difference arises because much larger amounts of power can be employed for paramagnetic resonance than can be used for gases without limitation of absorption through radiation saturation of the sample under investigation. Powers of the order of milliwatts generate excessive lowfrequency noise in crystals but not in thermal detectors. For the centimeter region

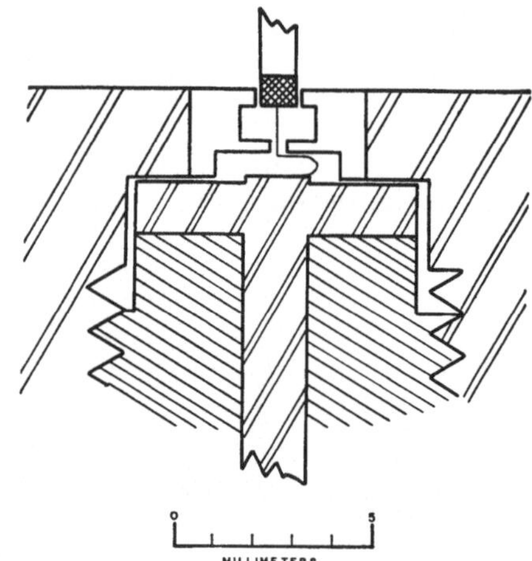

Fig. 6a and b. Cross-section diagram of crystal detector employed for the lower millimeter and the sub-millimeter wave regions. The diagram at the left shows details of the differential screw mechanism used for the critical adjustment of the pressure of the crystal against the tungsten "whisker" in both the multiplier (Fig. 5) and the detector. An enlargement of the waveguide section which contains the crystal and "whisker" is given on the right. [From KING and GORDY: Phys. Rev. **93**, 407 (1954).]

commercially available, coaxially mounted silicon crystals such as the 1N26 are used. For the 3 to 10 mm wave region the similar but smaller 1N53 crystal can be used. For the 1 to 3 mm region and for the sub-millimeter region small silicon crystals mounted directly in the millimeter waveguide with a differential screw mechanism, for adjustment of the pressure of the "whisker" have proved more satisfactory[1]. Fig. 6 shows a cross section of such a mount used at Duke University for measurements in the 0.6 to 4 mm wave region. This design makes possible easy removal of the crystal for repolishing of its surface and easy removal of the tungsten "whisker" for repointing, which is accomplished by electrolytic etching.

---

[1] W. C. KING and W. GORDY: Phys. Rev. **90**, 319 (1953). — C. A. BURRUS and W. GORDY: Phys. Rev. **92**, 274 (1953).

[2] C. A. BURRUS and W. GORDY: Phys. Rev. **93**, 897 (1954); **101**, 599 (1956). — M. COWAN and W. GORDY: Phys. Rev. **104**, 551 (1956).

Two devices for frequency measurement are used in combination in most microwave spectrometers for gases. One of these is a simple, tunable, resonant cavity, accurate only to about 5 to 25 megacycles. This is used in the search for spectral lines and for identification of the more accurate markers used in the final measurement of the spectral lines. The second frequency meter is a more elaborate device which provides sharp, standard frequency markers, accurate to a few parts in 10^8. Such markers are customarily produced by multiplication of standard low frequencies to the desired microwave region. The low frequency signals, usually in the megacycle range, are generated by a crystal-controlled oscillator. The crystal which controls the frequency of the oscillator is kept at a constant temperature. In the United States, the master oscillator is continuously monitored with the standard frequencies broadcast by the National Bureau of Standards. The multiplier[1] in its lower frequency stages consists of conventional radio tube circuits; in its intermediate stages, of a double cavity klystron multiplier; and in its final stages, of a crystal multiplier circuit. In many systems the klystron multiplier is omitted[2]. Such frequency standards usually provide markers separated by 30 to 90 megacycles. Markers spaced closer than 30 Mc are difficult to identify with the cavity wave meter. Either a calibrated radio receiver or a beat-frequency oscillator is employed for interpolation between the markers.

When the STARK effect is not employed, the gas to be studied is most often placed in a simple metallic waveguide cell sealed at either end with thin mica windows. STARK cells are most often constructed with a metallic conducting strip placed in the center of the waveguide cell perpendicular to the electric vector of the microwave radiation. The strip is supported at either side by a dielectric material such as Teflon.

In searching for absorption lines one tunes the source oscillator so as to sweep its frequency over that of the spectral line. The absorption line is detected as a decrement in the received power when the oscillator passes over its characteristic frequency. In the simple video system the source is swept over the line several times a second, and the sweep rate is synchronized with a cathode-ray trace which displays the response of the amplifier. Fig. 3 shows absorption lines displayed in this way. The amplifier is designed to amplify the FOURIER components of the spectral line for the chosen rate of sweep. This system, although convenient, is not particularly sensitive for two reasons. First, the video amplifier for the FOURIER components of such a signal must be in the audio region, where the crystal noise generated by microwave power of the order of milliwatts is high. Second, for passing the signal components with such rapid sweep, an amplifier with a wide noise bandwidth, of the order of kilocycles, must be used. The required amplifier bandwidth decreases with decreasing rate of sweep over the spectral line. When high sensitivity is required, one must sweep so slowly that a cathode ray tracing of the line cannot be made, and an automatic, pen-and-ink recorder is used to trace out the signal.

To avoid amplification of low-frequency noise of the detector crystal and of the tubes it is necessary to amplify the signal at high audio or radio frequencies. Some form of intensity modulation of the signal power at the amplifier frequency is necessary. The intensity modulation can be achieved through STARK effect modulation of the spectral line itself[3], or through frequency modulation of the

---

[1] R. R. UNTERBERGER and W. V. SMITH: Rev. Sci. Instrum. **19**, 580 (1948).
[2] W. E. GOOD and D. K. COLES: Phys. Rev. **71**, 383 (1947).
[3] R. H. HUGHES and E. B. WILSON: Phys. Rev. **71**, 562 (1947).

source power[1]. In the first method the STARK components are moved into and out of the frequency of the microwave source. The microwave power is thus modulated in intensity at the rate of the STARK modulation. An amplifier tuned to the modulation frequency will amplify the fundamental component of the signal power. In the second method, a rapid frequency modulation is superimposed upon the slower sweep of the source oscillator. The microwave frequency is then moved rapidly into and out of the regions of greater absorption as it sweeps slowly over the line. An intensity modulation of the detected power is thus produced which allows amplification of the signal at the modulation frequency. In both systems the modulation amplifier is followed by a second detector, usually a phase-sensitive detector, which detects the absorption line signal superimposed upon the "carrier wave" or the modulation frequency. Of the two, the STARK modulation method is preferable for the centimeter and upper millimeter wave range, where ample microwave power is available. Source modulation is used in the shorter millimeter and sub-millimeter region, where the power is severly limited. The recording of Fig. 2 was obtained with source modulation.

Sine wave and square wave STARK modulation are both commonly used. A sine wave modulator is simple to construct, and the sensitivity achieved with it is comparable to that achieved in square wave modulation. With square wave modulation the STARK components can be resolved and measured. Suitable square wave modulators are described by HEDRICK[2] and by SHARBAUGH[3].

3. Spectrometers for study of paramagnetic resonance. Fig. 7 is a diagram of a typical spectrometer for the study of magnetic resonance of electrons in solids.

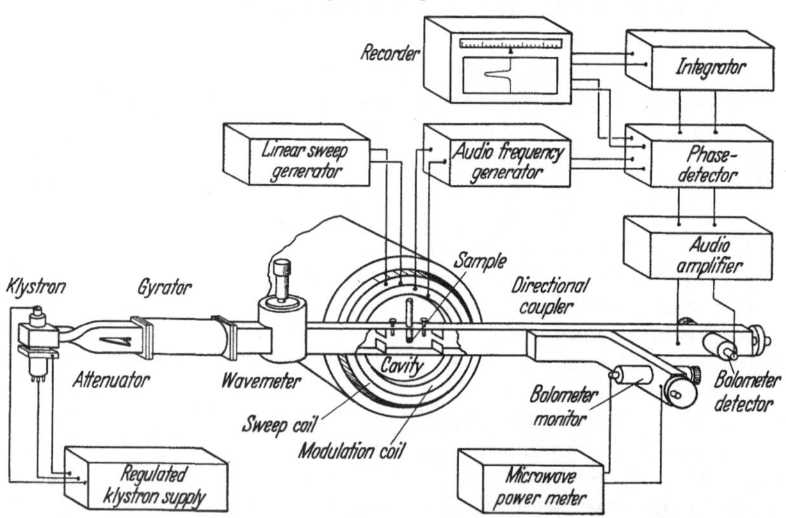

Fig. 7. Diagram of a typical microwave spectrometer for measurement of electron-magnetic resonance in solids and liquids. (From J.E. WERTZ [22].)

It can also be used for observation of magnetic resonance of liquids and gases. Some of its components are common to spectrometers designed for observation of rotational spectra of gases. A basic difference in design results from the fact that the frequency of a magnetic resonance is determined by an imposed magnetic

---

[1] W. GORDY and M. KESSLER: Phys. Rev. **72**, 644 (1947). — W. D. HERSHBERGER: J. Appl. Phys. **19**, 411 (1948).

[2] L. C. HEDRICK: Rev. Sci. Instrum. **20**, 781 (1949); **22**, 537 (1951).

[3] A. H. SHARBAUGH: Rev. Sci. Instrum. **21**, 120 (1950).

field and can be tuned to any desired microwave frequency by variation of this field. This circumstance makes unneccessary the sweeping of the microwave oscillator and thus makes possible the use of fixed-frequency resonant cavities for absorption cells and fixed, sharply tuned source oscillators and receivers. Also, the possibility of using many milliwatts of power without saturating the paramagnetic resonance of most samples makes simple, convenient bolometer detection as sensitive as superheterodyne detection employing a crystal mixer. Such amounts of microwave power do not generate extra low-frequency noise (above the JOHNSON noise) in the bolometer as they do in the crystal mixer or detector. The greater power required for saturation in paramagnetic resonance over that in rotational spectra is caused primarily by the much smaller coupling dipole effective in the paramagnetic resonance transitions. The magnetic dipole responsible for the absorption of radiation in paramagnetic resonance is of the order of a BOHR magneton, whereas the electric dipole moment responsible for the absorption in rotational spectra is of the order of a DEBYE unit. The latter is 100 times as strong as the former, and its coupling to the radiation field is $(100)^2$, or 10000, times more effective.

As in the detection of rotational spectra, some form of modulation of the received microwave power is needed for detection of the resonance. In magnetic resonance this modulation is most easily achieved with an alternating component superimposed upon the magnetic field which determines the resonance frequency. This modulates the magnetic resonance frequency and thereby the intensity of absorption of microwave power at the fixed microwave frequency. Usually the range of the frequency modulation is small as compared with the line width. With this small modulation and the customary phase-sensitive detector, the detected shape of the magnetic resonance appears as the first derivative of the actual absorption contour when the receiver is tuned to the fundamental of the modulating frequency, and as the second derivative when the receiver is tuned to the second harmonic of the modulating frequency. It is not easy to achieve the necessary modulation amplitude with iron-core magnets at high frequencies, nor will the usual bolometer respond to modulating frequencies much in excess of a kilocycle. Hence the modulating frequency is usually in the low audio range.

The phase-lock-in detector and automatic recorder are used for high sensitivity, as they are in spectrometers for gases, but cathode ray scopes can be used to display the stronger magnetic resonances. The resonance is swept out by variation of the magnetic field. The magnetic resonance is thus a plot of intensity of absorption versus magnetic field strength, rather than a plot of intensity of absorption versus frequency as in ordinary spectroscopy.

C. Microwave molecular spectra.

4. Inversion spectra. The first and most thoroughly studied microwave spectrum is the inversion spectrum of ammonia[1-3]. Furthermore, ammonia including its various isotopic forms seems to be unique in exhibiting a microwave inversion spectrum, despite the fact that inversion is theoretically conceivable, given unlimited time, in any non-planar molecule. Inversion may be described as the reflection of all the nuclei of a non-planar molecule at its center of mass. In this way a new equilibrium configuration of the molecule is obtained which

[1] C. E. CLEETON and N. H. WILLIAMS: Phys. Rev. **45**, 234 (1934).
[2] B. BLEANEY and R. P. PENROSE: Nature, Lond. **157**, 339 (1946).
[3] W. E. GOOD: Phys. Rev. **70**, 213 (1946).

in non-planar molecules cannot be obtained from the original one through any succession of simple rotations. To obtain the new equilibrium configuration, the molecule must pass through a potential barrier, a feat which is described quantum mechanically as "barrier tunneling". In practically all non-planar molecules the barrier is so great that inversion is not achieved because the inversion time, if not infinite, is too long to observe. For NH_3, however, the inversion frequency in the ground vibrational state falls in the microwave region at about 24 kMc (kilomegacyles per second), and in ND_3 it falls near 16 kMc. For molecules of this symmetry the inversion consists simply of a turning inside out of the molecule. It may be regarded as a form of molecular vibration.

In inversion, a pyramidal molecule like ammonia can be treated simply as a one-dimensional oscillator of reduced mass μ, oscillating in a double minima potential field like that in Fig. 8. There are equivalent vibrational energy levels on either side of the hill, and the oscillator has equal probability of existing in either. If the barrier is not infinite, the oscillator will "resonate" between these two states with a "resonant" frequency inversely related to the height of the barrier.

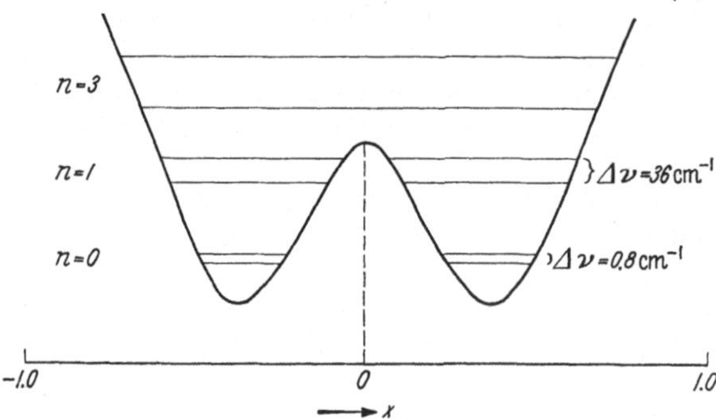

Fig. 8. Double minima of the potential curve for ammonia which give rise to the inversion splitting of the vibrational levels. The NH_3 absorption in the 1.25 cm wave region results from transition between the two components of the ground vibrational state ($n = 0$). The diagram is not drawn to scale.

The wave functions of the equivalent vibrational states on either side of the hill overlap to some extent. The resulting interaction leads to a splitting of the vibrational levels into doublets of separation ΔE_{inv}, as shown in Fig. 8. The splitting varies inversely as the height of the barrier and hence increases with the vibrational quantum number for the symmetrical bending mode in which the N vibrates along the symmetry axis perpendicular to the plane of the three hydrogens. The two inversion sub-levels of the vibrational state have opposite symmetry. The dipole matrix elements connecting the states do not vanish, and therefore transition between them giving rise to an absorption frequency $\nu_{\mathrm{inv}} = \Delta E_{\mathrm{inv}}/h$ is possible.

The inversion splitting ΔE_{inv} is related[1] to ΔE_v, the over-all separation of the vibrational levels considered, and to A, the area under the potential barrier, by:

$$\Delta E_{\mathrm{inv}} = \Delta E_v/\pi A^2 \tag{4.1}$$

or

$$\nu_{\mathrm{inv}} = \nu_v/\pi A^2 \tag{4.2}$$

where

$$A = \exp\left[\left(\frac{2\pi}{h}\right) \int\limits_0^{x_1} [2\mu(V - E_v)]^{\frac{1}{2}}\, dx\right] \tag{4.3}$$

and μ = reduced mass;

x = inversion coordinate;

V = potential energy;

$x = 0$ at potential maximum and $x = x_1$ at $V = E_v$.

[1] D. M. DENNISON and G. E. UHLENBECK: Phys. Rev. 41, 313 (1932).

The inversion frequency is lowest therefore for the ground vibrational state and increases with the vibrational quantum number. Since A increases exponentially as μ, the inversion frequency decreases rapidly with increasing μ. For these reasons the inversion frequency for the next lightest molecule similar to ammonia, PH_3, the expected inversion frequency for the ground vibrational state is only a fraction of a megacycle. Something like a year would be required for AsH_3 to execute one inversion; some centuries would be required for SbH_3 to do so.

As a result of interaction of vibration and rotation, the inversion splitting varies slightly for the different rotational states of the molecule. This leads to the fine structure of the inversion transition. Some sixty-six components of this fine structure have been measured for $N^{14} H_3$. The frequency of these in Mc/sec fit approximately the relation[1]:

$$\nu = 23\,787 - 151.3\,J(J+1) + 211.0\,K^2 + 0.5503\,J^2(J+1)^2 - \\ - 1.531\,J(J+1)\,K^2 + 1.055\,K^4, \tag{4.4}$$

where J and K are the usual rotational quantum numbers for the symmetric-top molecule.

The inversion frequency of NH_3 was originally observed by CLEETON and WILLIAMS[2] in 1934. Its fine structure was observed by BLEANEY and PENROSE[3] and independently by GOOD[4] in 1946. The inversion spectrum of ND_3 has been observed by LYONS, RUEGER, NUCKOLLS and KESSLER[5], and of mixed form ND_2H and NH_2D by WEISS and STRANDBERG[6].

5. Pure rotational spectra. The principal class of microwave molecular spectra arises from transitions between quantized rotational energies of molecules in singlet electronic ground states. Although pure rotational spectra of a few light molecules have been resolved with optical techniques in the far infrared region and although the rotational spectra of many heavy molecules originate in the longer wave radio region, the greater part of all measurements of pure rotational spectra has been made at microwave frequencies. The microwave region appears to be the optimum one for the study of rotational spectra. The theory of rotational spectra, which was developed by infrared spectroscopists long before the microwave region was opened up, is generally adequate, even though the great resolution of microwave spectroscopy makes it necessary to extend it to include fine or hyperfine structure normally unresolvable in the infrared region. This structure may arise from externally imposed electric or magnetic fields or from such weak internal interactions of the molecule as those of molecular electrons with magnetic dipole or electric quadrupole moments of the nuclei. The perturbations giving rise to fine and hyperfine structure are treated in Sects. 10 and 11.

For discussion of their rotational spectra molecules are classified as linear, symmetric-top, and asymmetric rotors. In a linear rotor, two of the principal moments of inertia are zero; in a symmetric-top, two principal moments of inertia are equal, but none is zero; and in an asymmetric rotor, none of the principal moments of inertia is equal to another, and none is zero. Because the spherical-top rotor, in which all three principal moments of inertia are equal, has no observable rotational spectrum, it will not be treated here.

---

[1] J. W. SIMMONS and W. GORDY: Phys. Rev. **73**, 713 (1948).

[2] C. E. CLEETON and N. H. WILLIAMS: Phys. Rev. **45**, 234 (1934).

[3] B. BLEANEY and R. P. PENROSE: Nature, Lond. **157**, 339 (1946).

[4] W. E. GOOD: Phys. Rev. **70**, 213 (1946).

[5] H. LYONS, L. T. RUEGER, R. G. NUCKOLLS, and M. KESSLER: Phys. Rev. **81**, 630 (1951).

[6] M. T. WEISS and M. W. P. STRANDBERG: Phys. Rev. **83**, 567 (1951).

α) *Diatomic and linear polyatomic molecules.* Diatomic or linear polyatomic molecules having no unbalanced electronic angular momenta can be treated in the first approximation as rigid rotors. The wave equation for such a rotor is:

$$\frac{1}{\sin\vartheta}\cdot\frac{\partial}{\partial\vartheta}\left(\sin\vartheta\,\frac{\partial\psi}{\partial\vartheta}\right)+\frac{1}{\sin^2\vartheta}\cdot\frac{\partial^2\psi}{\partial\varphi^2}+\frac{8\pi^2\,I\,E}{h^2}=0 \tag{5.1}$$

where ϑ is the polar angle with reference to the space-fixed Z axis, φ is the azimuthal angle, h is PLANCK'S constant, I is the moment of inertia, and E is the rotational energy. Solutions of this equation which have physical significance are possible only for discrete values, E_J, of the rotational energy,

$$E_J=\frac{h^2}{8\pi^2\,I}\,J(J+1)\,, \tag{5.2}$$

where J, the rotational quantum number, can have only the values: $J=0, 1, 2, \ldots$. Complete solution of Eq. (5.1) yields the rotational wave functions:

$$\psi_J=N_{JM}\,e^{iM\varphi}\,P_J^{|M|}(\cos\vartheta)\,, \tag{5.3}$$

where N_{JM} is a normalizing factor determined by $\int\psi_J\,\psi_J^*\,d\tau=1$ to be

$$N_{JM}=\frac{1}{\sqrt{2\pi}}\left[\frac{(2J+1)}{2}\cdot\frac{(J-|M|)!}{(J+|M|)!}\right]^{\frac{1}{2}}. \tag{5.4}$$

$P_J^{|M|}(\cos\vartheta)$ is the associated LEGENDRE function. M, the "magnetic" quantum number, can take the integral values:

$$M=J, J-1, J-2, \ldots -J. \tag{5.5}$$

With the BOHR postulate,

$$h\nu=E_1-E_2\,, \tag{5.6}$$

and the selection rule for absorption of radiation through rotational transition, $J\to J+1$, the rotational frequencies of a "rigid" linear molecule are found from Eq. (5.2) to be

$$\nu_r=2B(J+1)\,, \tag{5.7}$$

where $B=h/(8\pi^2 I)$ is the spectral constant in frequency units and J is the quantum number of the lower of the two energy levels between which the transition occurs. From Eq. (5.7) it is obvious that the rotational spectrum of a rigid linear molecule is a series of equally spaced lines of separation $2B$ and with the lowest frequency having the value $2B$.

If B is sufficiently large (I, sufficiently small), the rotational spectrum of a linear molecule may originate in the infrared region and therefore not have a microwave rotational spectrum. Actually, however, all but the lighter diatomic hydrides such as HF or HCl have their rotational spectra originating at frequencies which can now be reached with microwave or radiofrequency methods. Rotational lines of simple molecules such as DBr, TCl and HI have already been measured with microwave methods (see Table 1), and the first rotational lines of DCl and HBr fall in the workable microwave region.

If an observable transition between the rotational levels J and J' is to occur, all the matrix elements of the electric dipole moment μ along space-fixed axes,

$$\mu_J^{J'}=\int\psi_J\,(\mu_f)\,\psi_{J'}\,d\tau\,, \tag{5.8}$$

must not vanish. This will occur only when $\mu\neq0$ or when the molecule has a permanent dipole moment. From the wave function of Eq. (5.3) substituted in Eq. (5.8) it is found that even if $\mu \neq 0$, all the matrix elements vanish unless $J = J' \pm 1$. This latter condition is the basis for the selection rule already stated.

Although the simple formula (5.7) represents a good first approximation to the rotational spectra of diatomic and linear polyatomic molecules, it is inadequate for refined measurements. In addition to the nuclear and external field perturbations treated in later sections, effects of centrifugal distortion and vibrational motions are easily detected in microwave measurements. Theoretically, such effects can be included in the original HAMILTONian used in obtaining the wave equation for the rotational motion. It is simpler, however, to treat them as perturbations upon the rigid rotor states already found.

For a particular vibrational state the rotational levels and frequencies of the non-rigid diatomic or linear polyatomic molecule with no unbalanced electronic momentum are given with sufficient accuracy for microwave measurement by the equations:

$$E_J = c\left[B_v J(J+1) - D_v J^2 (J+1)^2\right] \tag{5.9}$$

and

$$\nu_r = 2 B_v (J+1) - 4 D_v (J+1)^3, \tag{5.10}$$

where B_v is the "effective" spectral constant, $h/(8\pi^2 I_v)$, for the particular vibrational state for which the rotational spectrum is observed, and where D_v is the centrifugal stretching constant for that state.

In terms of the constants B_e and D_e for the hypothetical vibrationless state, B_v and D_v for diatomic molecules are:

$$B_v = B_e - \alpha\left(v + \tfrac{1}{2}\right), \tag{5.11}$$

$$D_v = D_e + \beta\left(v + \tfrac{1}{2}\right), \tag{5.12}$$

where α and β are interaction constants which are very small in comparison with B and D respectively, and where v is the vibrational quantum number. For linear polyatomic molecules there is more than one vibrational mode, and the above equations must be written in the more general form:

$$B_v = B_e - \sum_i \alpha_i\left(v_i + \frac{d_i}{2}\right), \tag{5.13}$$

$$D_v = D_e + \sum_i \beta_i\left(v_i + \frac{d_i}{2}\right), \tag{5.14}$$

where the summation is taken over all the fundamental modes of vibration. The subscript i refers to the i-th mode, and d_i represents the degeneracy of that mode.

There is an additional complication for degenerate bending modes. A splitting of the rotational lines into doublets, known as l-type doublets, arises from the interaction between rotation and vibration. This splitting increases with J according to the equation,

$$\Delta \nu = 2 q_l (J+1), \tag{5.15}$$

where q_l, the interaction constant, is of the order of $2 B^2/\omega$, in which ω is the frequency of the bending vibration. In applying Eq. (5.10) one can simply employ the mean of the doublet frequencies.

When it is possible to make measurements of rotational transitions in all the fundamental vibrational modes, one can find the interaction and stretching constants and the equilibrium value B_e from microwave measurements alone. Since D is itself a small, second-order constant in comparison with B, it is usually possible to neglect its variations with the vibrational state.

When the B values are measured for a diatomic molecule, the corresponding moments of inertia and internuclear distances can be immediately calculated. In terms of the moments of inertia and two atomic masses, m_1 and m_2, the internuclear distance r is:

$$r = \left[I\left(\frac{1}{m_1} + \frac{1}{m_2}\right)\right]^{\frac{1}{2}}. \tag{5.16}$$

The effective internuclear distance varies slightly for different vibrational states. Because of the zero point vibrational energy, the effective value r_0 for the ground state differs slightly from the value r_e for the hypothetical, vibrationless state.

Table 1. *Spectral constants for some hydrogen halides obtained from millimeter wave spectroscopy.*

Molecule	B_0 in Mc/sec	eQq coupling for Cl, Br, or I in Mc/sec	Bond length r_0 in Å	References
DCl^{35}	161656.09	-67.3	1.28125	g
DCl^{37}	161182.93	-53.1	1.28124	g
TCl^{35}	111075.76	-67.0	1.28003	a
TCl^{37}	110601.51	-53.0	1.28002	a
DBr^{79}	127358.2	533	1.42136	b
DBr^{81}	127280.0	445	1.42136	b
TBr^{79}	86252.24	530	1.42012	a, c
TBr^{81}	86174.33	443	1.42011	a, c
DI^{127}	97537.2	-1823	1.6165	d, e
HI^{127}	192658.6	-1831	1.60904	f

a) C. A. BURRUS and W. GORDY, B. BENJAMIN and R. LIVINGSTON: Phys. Rev. **97**, 1661 (1955).
b) W. GORDY and C. A. BURRUS: Phys. Rev. **93**, 419 (1954).
c) A. E. NETHERCOT and B. ROSENBLUM: Phys. Rev. **97**, 84 (1955).
d) J. A. KLEIN and A. H. NETHERCOT: Phys. Rev. **91**, 1018 (1953).
e) C. A. BURRUS and W. GORDY: Phys. Rev. **92**, 1437 (1953).
f) M. COWAN and W. GORDY: Phys. Rev. **104**, 551 (1956).
g) From M. COWAN and W. GORDY.

Table 1 lists spectral constants and internuclear distances evaluated for hydrogen halides from microwave spectroscopy. In the evaluation of these constants the observed rotational transitions were corrected for any nuclear coupling effects, as described in Sect. 10 and Sect. 11.

A linear molecule of n atoms has in general $(n-1)$ independent internuclear dimensions. To evaluate the complete structure of a linear molecule having more than two atoms from its rotational constants alone, one must make measurements on different isotopic species of the molecule and assume that the internuclear distances are not altered by the isotopic substitution. Because of differences in zero point vibrational energies of the different isotopic species, the latter assumption is not exactly true. Nevertheless, internuclear distances evaluated with different isotopic combinations, when more observables than unknown parameters are available, usually agree within 0.1%. To evaluate the complete structure of a linear molecule with n unknown structural parameters, one must make measurements on at least n different isotopic combinations of the molecule, with only one isotopic substitution per atom. Only one substitution per atom leads to additional independent structural equations. However, additional substitutions are useful as a check on the evaluations and in the estimation of the error caused by the zero point vibrations. Simplified methods for such calculations with discussion of the inherent errors are available[1] [2], [4], [6]. Results on some

---

[1] J. KRAITCHMAN: Amer. J. Phys. **21**, 17 (1953).

representative linear molecules are given in Table 2. Comprehensive tabulations of spectral constants for linear molecules are given elsewhere [3], [4], [6].

Table 2. *Structures of some linear molecules measured with microwave spectroscopy.*

XYZ	d_{XY} in Å	d_{YZ} in Å	XYZ	d_{XY} in Å	d_{YZ} in Å
HCN	1.064	1.156	NNO	1.126	1.191
ClCN	1.629	1.163	OCS	1.1637	1.5584
BrCN	1.790	1.159	OCSe	1.1588	1.7090
ICN	1.995	1.159	TeCS	1.904	1.557

Remarks: Values taken from [4].

β) *Symmetric-top molecules.* Non-linear molecules which have axial symmetry about an internal axis such that two principal moments of inertia are equal are commonly designated as symmetric tops or symmetric rotors. When the two smaller principal moments of inertia are equal $(I_a = I_b < I_c)$, they are oblate symmetric tops; when the two larger ones are equal $(I_a < I_b = I_c)$, they are prolate symmetric tops. Solutions of the wave equation for a rigid prolate symmetric-top molecule[1] which have physical significance require that the rotational energies have only the values:

$$E_{J,K} = \frac{h^2}{8\pi^2}\left[\frac{J(J+1)}{I_b} + \left(\frac{1}{I_a} - \frac{1}{I_b}\right)K^2\right] \tag{5.17}$$

$$= h\left[BJ(J+1) + (A-B)K^2\right], \tag{5.18}$$

where $A = h/(8\pi^2 I_a)$ and $B = h/(8\pi^2 I_b)$. Solutions for the oblate top requires an equation of the same kind, with the subscript a replaced by c. The quantum number J refers to the total angular momentum of the molecule, and as for a linear molecule it can take only integral values, including zero, whereas K is a quantum number which measures in units of $h/(2\pi)$ the component of the total momentum $J(J+1)h/(2\pi)$ which is directed along the symmetry axis, i.e. the a axis for the prolate top under discussion.

In all but the accidentally symmetric top the permanent dipole moment will lie along the symmetry axis. Under these conditions the selection rules for absorption of radiation through rotation are:

$$J \rightarrow J+1, \quad K \rightarrow K.$$

These rules, like those for linear molecules, are based upon the non-vanishing of the dipole moment matrix elements with reference to axes fixed in space. Application of these rules with the BOHR postulate to Eq. (5.18) gives for the rotational lines the values:

$$\nu = 2B(J+1), \tag{5.19}$$

exactly as obtained for the linear molecule. Since only B or I_b appears in this formula, it clearly applies equally well to both prolate and oblate tops. Furthermore, the spectrum of a rigid linear rotator should obviously be the same as that of a completely rigid symmetric top, since the former can be regarded as a special case of the latter for which $I_a = 0$ and $I_b = I_c$. Nevertheless, no molecule is completely rigid; differences in centrifugal stretching cause distinguishable differences in the rotational spectra of the actual linear and symmetric-top molecules.

The effects of centrifugal stretching on the rotational energies of the symmetric top have been calculated with perturbation theory by SLAWSKY and DENNISON[2],

---

[1] D. M. DENNISON: Phys. Rev. **28**, 318 (1926); Rev. Mod. Phys. **3**, 280 (1931).
[2] Z. I. SLAWSKY and D. M. DENNISON: J. Chem. Phys. **7**, 509 (1939).

who obtained an equation for the energy of the non-rigid symmetric rotor:

$$E_{JK} = h\left[B J(J+1) + (A-B) K^2 - D_J J^2 (J+1)^2 - D_{JK} J(J+1) K^2 - D_K K^4\right], \quad (5.20)$$

where the new quantities D_J, D_{JK}, and D_K are centrifugal stretching constants. Used with the selection rules, $J \to J+1$ and $K \to K$, this equation leads to the formula:

$$\nu = 2B(J+1) - 4D_J(J+1)^3 - 2D_{JK}(J+1) K^2 \quad (5.21)$$

for the rotational absorption frequencies. As before, J is the rotational quantum number for the lower of the two levels. Formula (5.21) differs from the corresponding one for the non-rigid linear molecule, Eq. (5.10), only in the additional stretching term, $-2D_{JK}(J+1) K^2$. Since this last term involves K^2, it leads

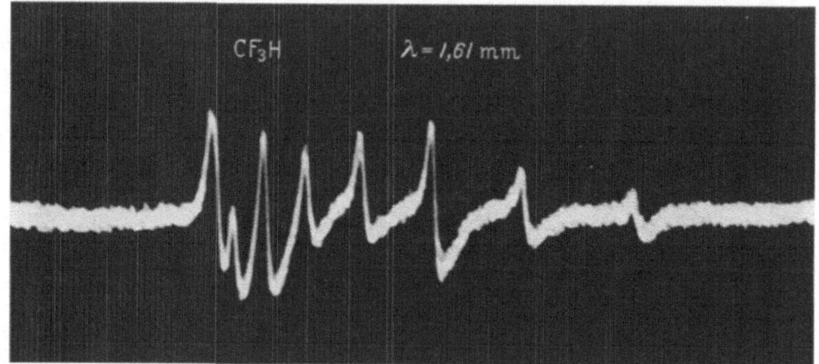

Fig. 9. Cathode-ray display of a rotational transition in a symmetric top molecule: the $J = 8 \to 9$ transition of CF$_3$H at 1.61 mm wavelength. The components corresponding to different values of K from 0 to 8 are separated by centrifugal distortions. There is no nuclear hyperfine structure. The total spread is 21 Mc/sec. [From BURRUS and GORDY: J. Chem. Phys. (in press).]

to a splitting of a rotational transition of a symmetric-top molecule into $(J+1)$ components which are closely spaced but usually resolvable. Fig. 9 illustrates this splitting with the $J = 8 \to 9$ transition of fluoroform at 1.61 mm wavelength. A rotational transition of a symmetric-top molecule is both split into components and displaced by centrifugal stretching, whereas a rotational transition of a linear molecule is only displaced. In the centimeter region the splitting and the displacements are very small, usually of the order of kilocycles or megacycles, but in the region of one millimeter wavelength they may be of the order of tens to hundreds of megacycles. Tabulations of the constants are available [3], [4], [6].

In the analysis of spectra the variations of the stretching constants D_J and D_{JK} with vibrational state are customarily neglected. It is seldom possible to obtain sufficient data even for evaluation of these effects upon B and for determination of B_e for the simpler symmetric tops. Formula (5.13) for the linear molecule applies for the symmetric top also. A complication due to interaction between rotation and vibration is encountered for degenerate bending modes of symmetric-top molecules. This condition, similar to l-type doubling in linear molecules, is treated by NIELSEN[1].

To obtain the molecular structure of a symmetric-top molecule one must apply the principle of isotopic substitution as described for linear polyatomic molecules. Since only one observable parameter involving the molecular dimensions can be obtained for each isotopic species, the number of isotopic species

---

[1] H. H. NIELSEN: Phys. Rev. **77**, 130 (1950).

employed must equal the total number of unknown parameters of the molecule. Table 3 gives some structures which have recently been evaluated with millimeter-wave spectroscopy. Structural parameters evaluated through isotopic substitution are usually limited by effects of zero point vibrations to four significant figures. For tabulation of values, see [4], [6].

Table 3. *Spectral constants of hydrides and deuterides of nitrogen, phosphorus, arsenic, and antimony obtained from millimeter wave spectroscopy.*

Molecule	B_0 in Mc/sec	eQq coupling for N, As, or Sb in Mc/sec	Bond angle	Bond length in Å	Ref.
$N^{14}D_3$	154162.7	−4.10	107°	1.0144	d
PH_3	133478.3		93°27′	1.4206	a
PD_3	69470.41		93°10′	1.4166	a
$As^{75}H_3$	112468.46	− 160.1	91°50′	1.5192	b
$As^{75}D_3$	57477.15	− 165.9	91°30′	1.5145	b
$Sb^{121}H_3$	88031.92	458.7	91°18′	1.7073	c
$Sb^{121}D_3$	44693.29	465.4	90°56′	1.7021	c
$Sb^{123}H_3$	88015.54	586.0	91°18′	1.7073	c
$Sb^{123}D_3$	44677.13	592.8	90°56′	1.7021	c

a) C. A. BURRUS, A. W. JACHE, and W. GORDY: Phys. Rev. **95**, 706 (1954).
b) G. S. BLEVINS, A. W. JACHE, and W. GORDY: Phys. Rev. **97**, 684 (1955).
c) A. W. JACHE, G. S. BLEVINS, and W. GORDY: Phys. Rev. **97**, 680 (1955).
d) G. ERLANDSSON and W. GORDY: Phys. Rev. (inpress).

Remark: The mixed asymmetric species such as PD_2H, etc., have transitions falling in the cm-wave region. These spectra have been studied by C. C. LOOMIS and M. W. P. STRANDBERG: Phys. Rev. **81**, 798 (1951).

γ) *Asymmetric-top molecules.* The spectra of asymmetric rotors are considerably more complex than those of linear or symmetric-top molecules. Except for low J values it is impossible to express the characteristic rotational energies of an asymmetric rotor in closed formulae, as can be done for the symmetric top. There is no quantized component of the angular momentum about an internal axis as there is for the latter, although the total angular momentum is still quantized according to the relation $P_{\text{total}}^2 = J(J+1) h^2/(2\pi)^2$ and has, as before, quantized components $P_z = M h/(2\pi)$ along a space-fixed axis.

Several theoretical treatments of the asymmetric rotor have been made[1], all of which are too involved for discussion here. In most of them it is assumed that the wave functions can be represented as an expansion of an orthogonal set of functions such as those for the symmetric rotor. The secular equations are then obtained and solved for the unknown coefficients in the expanded wave functions and for the allowed rotational energies. The secular determinant is broken down into subdeterminants, the order of which increases with J. They can be solved completely for only relatively low values of J.

The characteristic energies of the rigid asymmetric rotor are commonly expressed in terms of the so called reduced energies $E(\varkappa)$, which are a function of the asymmetry parameter,

$$\varkappa = \frac{2B - A - C}{A - C}. \tag{5.22}$$

Here A, B, and C are the spectral constants $h/(8\pi^2 I_a)$, $h/(8\pi^2 I_b)$, and $h/(8\pi^2 I_c)$, where h is PLANCK's constant and I_a, I_b, and I_c are the principal moments of inertia. Conventionally, the inertial axes are chosen so that $I_a < I_b < I_c$, and hence $A > B > C$. The prolate and oblate symmetric rotors may be regarded

---

[1] For discussion of basic theory and references, see reviews by D. M. DENNISON: Rev. Mod. Phys. **3**, 280 (1931); and by H. H. NIELSEN: Rev. Mod. Phys. **23**, 90 (1951).

as limiting cases of the asymmetric rotor for which $\varkappa$ equals -1 and $+1$ respectively. The most asymmetric rotor corresponds to $\varkappa = 0$. In terms of $E(\varkappa)$ the characteristic values of the total energy are:

$$E_J = \left(\frac{A+C}{2}\right) J(J+1) + \left(\frac{A-C}{2}\right) E(\varkappa). \qquad (5.23)$$

Explicit values[1] of $E(\varkappa)$ are:

$J_{K_{-1}K_1}$	$E(\varkappa)$	$J_{K_{-1}K_1}$	$E(\varkappa)$
0_{00}	0	3_{13}	$2\left[\varkappa - (\varkappa^2 + 15)^{\frac{1}{2}}\right]$
1_{10}	$\varkappa + 1$	3_{03}	$5\varkappa - 3 - 2(4\varkappa^2 + 6\varkappa + 6)^{\frac{1}{2}}$
1_{11}	0	4_{40}	$-$
1_{01}	$\varkappa - 1$	4_{41}	$5\varkappa + 5 + 2(4\varkappa^2 - 10\varkappa + 22)^{\frac{1}{2}}$
2_{20}	$2\left[\varkappa + (\varkappa^2 + 3)^{\frac{1}{2}}\right]$	4_{31}	$10\varkappa + 2(9\varkappa^2 + 7)^{\frac{1}{2}}$
2_{21}	$\varkappa + 3$	4_{32}	$5\varkappa - 5 + 2(4\varkappa^2 + 10\varkappa + 22)^{\frac{1}{2}}$
2_{11}	$4\varkappa$	4_{22}	$-$
2_{12}	$\varkappa - 3$	4_{23}	$5\varkappa + 5 - 2(4\varkappa^2 - 10\varkappa + 22)^{\frac{1}{2}}$
2_{02}	$2\left[\varkappa - (\varkappa^2 + 3)^{\frac{1}{2}}\right]$	4_{13}	$10\varkappa - 2(9\varkappa^2 + 7)^{\frac{1}{2}}$
3_{30}	$5\varkappa + 3 + 2(4\varkappa^2 - 6\varkappa + 6)^{\frac{1}{2}}$	4_{14}	$5\varkappa - 5 - 2(4\varkappa^2 + 10\varkappa + 22)^{\frac{1}{2}}$
3_{31}	$2\left[\varkappa + (\varkappa^2 + 15)^{\frac{1}{2}}\right]$	4_{04}	$-$
3_{21}	$5\varkappa - 3 + 2(4\varkappa^2 + 6\varkappa + 6)^{\frac{1}{2}}$	5_{42}	$10\varkappa + 6(\varkappa^2 + 3)^{\frac{1}{2}}$
3_{22}	$4\varkappa$	5_{24}	$10\varkappa - 6(\varkappa^2 + 3)^{\frac{1}{2}}$
3_{12}	$5\varkappa + 3 - 2(4\varkappa^2 - 6\varkappa + 6)^{\frac{1}{2}}$		

The subscripts K_{-1} and K_1 are the notation of KING, HAINER, and CROSS for indicating the sequence of the sub-levels. They are not quantum numbers since only J and its component M are good quantum numbers for the asymmetric rotor. They have the significance that in the two limiting cases K_{-1} transforms into the corresponding K for the prolate symmetric top ($\varkappa = -1$) and K_1 transforms into the K for the oblate symmetric top ($\varkappa = +1$).

The selection rules for J as determined by the non-vanishing property of the matrix elements of the molecular dipole moment are:

$$\varDelta J = 0 \quad \text{or} \quad \pm 1.$$

Each of these changes can give rise to absorption of radiation. In addition to these, there are restrictions on the changes in the subscripts K_{-1} and K_1. The permitted changes which can give rise to dipole absorption of radiation are:

	$\varDelta K_{-1}$	$\varDelta K_1$
$\mu_a \neq 0$ (Dipole component along axis of least moment of inertia.)	$0, \pm 2 \ldots$	$\pm 1, \pm 3 \ldots$
$\mu_b \neq 0$ (Dipole component along axis of intermediate moment of inertia.)	$\pm 1, \pm 3 \ldots$	$\pm 1, \pm 3 \ldots$
$\mu_c \neq 0$ (Dipole component along axis of greatest moment of inertia.)	$\pm 1, \pm 3 \ldots$	$0, \pm 2 \ldots$

The restrictions on the changes in K_{-1}, and K_1 arise from the symmetry properties of the momental ellipsoid and are sometimes called the symmetry selec-

---

[1] G. W. KING, R. M. HAINER, and P. C. CROSS: J. Chem. Phys. **11**, 27 (1943).

tion rules. If the dipole moment of the molecule lies wholly along one of the principal inertial axes, only the transitions indicated for a non-vanishing of the component of μ along that axis can occur, since the components of μ along the other two inertial axes would then be zero. If the dipole moment has components along all three inertial axes, all transitions indicated above can occur. Although any changes in K_{-1} and K_1 permitted by the symmetry selection rules can occur, transitions corresponding to large changes in either subscript are very weak. CROSS, HAINER, and KING[1] give useful tables for calculation of line intensities.

KING, HAINER, and CROSS[2] have published numerical tables from which the reduced energies corresponding to various degrees of asymmetry can be readily obtained to a useful approximation for J values up to 12. The development of machines for rapid computation make the problem of the asymmetric rotor a less formidable one.

In some asymmetric-top molecules, primarily those having at least one small principal moment of inertia, the effects of centrifugal distortion upon the microwave rotational lines are great. These effects come about because in asymmetric-top molecules, unlike linear and symmetric-top molecules, transitions between levels of rather high rotational energy can give rise to microwave absorption lines. NIELSEN[3], STRANDBERG and collaborators[4], KIVILSON and WILSON[5] among others, have treated these effects.

A few of the simpler structures of asymmetric-top molecules calculated from microwave rotational spectra are given in Table 4. More comprehensive tables are given elsewhere [3], [4], [6].

Table 4. *Structure of some nonlinear triatomic molecules evaluated from microwave spectroscopy.*

Molecule	Angle	Bond length in Å	References
H_2S	$92° 6'$	1.323	a
H_2Se	$91°$	1.460	b
SO_2	$119° 33'$	1.433	c
O_3	$116° 49'$	1.278	d

a) C. A. BURRUS and W. GORDY: Phys. Rev. **92**, 274 (1953).

b) A. W. JACHE, P. W. MOSER, and W. GORDY: J. Chem. Phys. **25**, 209 (1956).

c) M. H. SIRVETZ: J. Chem. Phys. **19**, 938 (1951). — G. F. CRABLE and W. V. SMITH: J. Chem. Phys. **19**, 502 (1951).

d) R. H. HUGHES: J. Chem. Phys. **21**, 959 (1953). — R. TRAMBARULO, S. N. GHOSH, C. A. BURRUS, and W. GORDY: J. Chem. Phys. **21**, 851 (1953).

Because of their influence on atmospheric propagation of microwaves, the microwave absorption lines of the asymmetric rotor H_2O are of primary interest. Only two of significant strength are predicted to fall in the millimeter and centimeter range, the $5_{2,3} \rightarrow 6_{1,6}$ line and the $2_{2,0} \rightarrow 3_{1,3}$ line. These have both been measured, the former[6] at 22235.22 Mc/sec ($\lambda = 1.35$ cm) and the latter[7] at 183311.04 Mc/sec ($\lambda = 1.63$ mm). Other strong lines are predicted to occur between 0.7 and 1.0 mm wavelength. Several microwave lines of D_2O and HDO

[1] P. C. CROSS, R. M. HAINER, and G. W. KING: J. Chem. Phys. **12**, 210 (1944).

[2] G. W. KING, R. M. HAINER, and P. C. CROSS: J. Chem. Phys. **11**, 27 (1943).

[3] H. H. NIELSEN: Rev. Mod. Phys. **23**, 90 (1951).

[4] R. B. LAWRENCE and M. W. P. STRANDBERG: Phys. Rev. **83**, 363 (1951). — R. E. HILLGER and M. W. P. STRANDBERG: Phys. Rev. **83**, 575 (1951).

[5] D. KIVILSON and E. B. WILSON jr.: J. Chem. Phys. **20**, 1575 (1952).

[6] G. E. BECKER and S. H. AUTLER: Phys. Rev. **70**, 300 (1946). — C. H. TOWNES and R. F. MERRITT: Phys. Rev. **70**, 558 (1946).

[7] W. C. KING and W. GORDY: Phys. Rev. **93**, 407 (1954).

have been observed, but they do not affect atmospheric propagation significantly because of the relatively low abundance of D. Microwave propagation in the upper atmosphere may be influenced somewhat by the absorption of the ozone layer. Strong, low J transitions of O_3 have been measured in the millimeter region[1], the $2_{0,2} \rightarrow 1_{1,1}$ line at 42832.62 Mc/sec, the $2_{0,2} \rightarrow 2_{1,1}$ line at 96228.8 Mc/sec, the $4_{0,4} \rightarrow 4_{1,3}$ line at 101736.83 Mc/sec, and the $0_{0,0} \rightarrow 1_{1,1}$ line at 118364.3 Mc/sec. Many higher J transitions of this asymmetric rotor have also been observed in the centimeter region[2]. The structure of this interesting molecule was first correctly assigned from its microwave spectrum.

6. Molecules with electronic moments. Practically all stable molecules have singlet Σ electronic ground states with no resultant electronic angular momenta. The discussion of the previous section applies to this predominant class. A few stable molecules, notably O_2, NO, NO_2, and ClO_2, all of which have been studied with microwave spectroscopy, have resultant electronic angular momenta in the electronic ground state.

The electronic angular momentum of a molecule may arise from an unbalanced electronic spin, from an orbital motion, or from a combination of the two. These causes are best treated separately.

α) *Spin multiplets.* If a molecule has only a resultant spin moment, as have O_2, NO_2, and ClO_2, with no resultant orbital momentum, the resultant spin moment S will be coupled, but not strongly, to the molecular rotational axis, Hund's case (b), through interaction of the spin magnetic moment with the weak magnetic moment generated by the molecular rotation. This weak interaction gives rise to a splitting of each rotational level into $(2S+1)$ components, commonly designated as the spin multiplets. Because the molecular rotational moment increases with rate of rotation, the magnitude of the splitting of the spin multiplets increases with the rotational quantum number.

In molecules with electronic spin, the total angular momentum (exclusive of nuclear spins), designated as J, is the sum of the electronic spin S and the angular momentum corresponding to the end-over-end rotation of the molecule, designated as N. The quantum number J can therefore have only the values:

$$J = N + S, N + S - 1, \ldots |N - S|,$$

with

$$N = 0, 1, 2 \ldots .$$

In its ground state O_2 has a spin moment corresponding to $S = 1$. For each value of N, there are then three values of J. Each rotational level is a triplet. Transitions between these triplets (changes in J with no changes in N) give rise to a group of closely spaced absorption lines in the 4 to 6 mm region and to a single component in the 2.5 mm region. The change in energy arises essentially from a flipping of the electron spin vector in the molecular rotational magnetic field with no change in the rotational momentum vector N. The microwave absorption arises from interaction of the electronic spin magnetic moment with the magnetic vector of the microwave radiation.

The theory of the splitting of the spin multiplets in O_2 was originally worked out by Schlapp[3], and the formulae have recently been refined by Miller and

[1] R. Trambarulo, S. N. Ghosh, C. A. Burrus, and W. Gordy: J. Chem. Phys. **21**, 851 (1953).

[2] R. H. Hughes: J. Chem. Phys. **21**, 959 (1953).

[3] R. Schlapp: Phys. Rev. **51**, 342 (1937).

Townes[1] and by Mizushima[2] after measurement of the fine structure[3] of the millimeter wave spectrum revealed that the Schlapp theory was inadequate.

The unresolved O_2 absorption in the 4—6 mm wave region was first observed by Beringer[4] and later re-examined by Strandberg, Meng, and Ingersoll[5]. The individual components of the fine structure were first resolved and measured by Burkhalter *et al.*[3] with a magnetic modulation spectrograph. Later Anderson *et al.*[6] observed the single component in the 2.5 mm wave region. Beringer and Castle[7] have investigated the magnetic resonance of O_2 (transitions between the Zeeman components of the individual fine structure levels) in the region of 3 cm wavelength. The O_2 absorption strongly affects propagation of millimeter waves through the atmosphere.

Nitrogen dioxide has an unbalanced spin moment, with $S = \frac{1}{2}$. It also has an electric dipole moment and observable microwave rotational spectrum. In the absence of nuclear interactions, each of its rotational levels should be split into a doublet corresponding to the two orientations, $M_S = \frac{1}{2}$ and $-\frac{1}{2}$, of S along N. This should lead to an observable doubling of all rotational lines with a doublet spacing which would increase with N. A rotational transition of this molecule has been observed by McAfee[8] to have six components, which apparently arise from the three components of the N^{14} nuclear magnetic hyperfine structure superimposed upon the doublet structure. In this molecule the spin doublets and nuclear hyperfine structure are of the same order of magnitude. Nitrogen dioxide provides an example of very weak coupling between the unbalanced electronic spin moment and the rotational moment.

Castle and Beringer[9] have observed magnetic resonance of NO_2 at 9360 Mc per sec (magnetic field of 3350 gauss) and found it to conform to the Paschen-Back formula:

$$E_H = g_S\, \beta\, M_S\, H + A\, M_S\, M_I + B_S M_{\,S} M_J, \qquad (6.1)$$

in which the first term, $g_S\, \beta\, M_S\, H$, represents the interaction of the essentially free electronic spin moment with the imposed field; the second term, $A\, M_S M_I$, represents the interaction of the nuclear magnetic moment of N^{14} and the electronic spin moment, both of which are resolved along H in the Paschen-Back case; and the third term, $B_S\, M_S\, M_J$, represents the interaction of the electronic spin moment with the components of the rotational moment which are resolved along H. The third term, which might be called the fine structure term, is actually smaller than the second term, which represents the hyperfine structure. Castle and Beringer found that $A = 132$ Mc/sec and that $B \approx A/2$. The resonance frequencies are obtained from Eq. (6.1) with the selection rules $\Delta M_S = 1$, $\Delta M_I = 0$, and $\Delta M_J = 0$.

A rotational transition of ClO_2 has been observed by Schawlow and Sanders[10] to have a doublet fine structure corresponding to the two orientations of S,

[1] S. L. Miller and C. H. Townes: Phys. Rev. **90**, 537 (1953).

[2] M. Mizushima: Phys. Rev. **91**, 222 (1953). See also M. Mizushima and R. M. Hill: Phys. Rev. **93**, 745 (1954).

[3] J. H. Burkhalter, R. S. Anderson, W. V. Smith and W. Gordy: Phys. Rev. **79**, 651 (1950).

[4] R. Beringer: Phys. Rev. **70**, 53 (1946).

[5] M. W. P. Strandberg, C. Y. Meng and J. G. Ingersoll: Phys. Rev. **75**, 1524 (1949).

[6] R. S. Anderson, C. M. Johnson and W. Gordy: Phys. Rev. **83**, 1061 (1951).

[7] R. Beringer and J. G. Castle: Phys. Rev. **78**, 581 (1950).

[8] K. B. McAfee: Phys. Rev. **78**, 340 (1950); **82**, 97 (1951).

[9] J. G. Castle and R. Beringer: Phys. Rev. **80**, 114 (1950).

[10] A. L. Schawlow and T. M. Sanders, quoted by C. H. Townes and A. L. Schawlow [6].

$M_S = \frac{1}{2}$ and $-\frac{1}{2}$, in the molecular rotational magnetic field. Each component of this doublet has a nuclear hyperfine structure of four closely spaced components corresponding to the $(2I+1)$ orientations of the Cl nucleus, both stable isotopes of which have $I = \frac{3}{2}$ and comparable nuclear magnetic moments.

β) Molecules with electronic orbital momentum. When the electrons of a diatomic molecule have a resultant electronic orbital momentum, this will be coupled strongly to the internuclear axis by the strong electric field of the molecule. The orbit will precess about the bond direction so rapidly that the orbital component normal to the bond axis will be essentially nullified; a quantized component along the bond axis will be left. The quantum number for this component is conventionally designated as λ. If the molecule also has an uncancelled spin moment, this usually will be coupled to the internuclear axis also through the interaction of the spin magnetic moment directed along λ or the bond axis. With this type of coupling [HUND's case (a)], only the component of S which lies along the bond axis, designated as Σ, will be defined since again the component normal to this axis will be, in the vector model picture, cancelled by the rapid precession of S about λ. The components λ and Σ both lie along the bond axis and are added to give the total or resultant electronic angular momentum, Ω, in units of $h/2\pi$. The quantum number Ω can take the values:

$$\Omega = \lambda + \Sigma,\ \lambda + \Sigma - 1,\ \dots |\lambda - \Sigma|.$$

The total angular momentum of the molecule is now the sum of Ω and N; the latter again represents the momentum for end-over-end rotation. The quantum number representing the total angular momentum (exclusive of nuclear spin) is therefore:

$$J = N + \Omega,\ N + \Omega - 1,\ \dots |N - \Omega|.$$

Like the non-linear, symmetric-top molecule already discussed, this type of diatomic molecule has a component of its total angular momentum quantized along an internal "symmetry" axis. If electron decoupling effects are neglected, its molecular axis executes a motion like that of the symmetric-top molecule and its quantum mechanical treatment takes a form similar to that for the symmetric top. In the first approximation, its rotational levels are given by:

$$E = BJ(J+1) + (A - B)\Omega^2, \tag{6.2}$$

which is seen to have the same form as Eq. (5.18). Here, as before, $B = \dfrac{h}{8\pi^2 I}$, where I is the moment of inertia of the molecule and A is a constant which depends on the strength of the spin-orbit coupling. If one applies the BOHR postulate [Eq. (5.6)] with the selection rules for rotational absorption (changes in N but no change in Ω), $J \to J+1$, $\Omega \to \Omega$, the rotational frequencies are seen to be:

$$\nu = 2B(J+1), \tag{6.3}$$

as for the rigid symmetric-top or linear molecule without electronic angular momentum. There is, nevertheless, a significant difference in the quantum number J, which leads to a big displacement in the rotational spectrum. Here J does not necessarily take integral values but increases by integral increments beginning with Ω. The lowest rotational frequency is not a $0 \to 1$ transition as before, but a $\Omega \to \Omega + 1$ transition. For the lowest electronic state of NO, the $^2\pi_{\frac{1}{2}}$ state, $\Omega = \frac{1}{2}$ and the first rotational line is a $J = \frac{1}{2} \to \frac{3}{2}$ transition which occurs at approximately $3B$ or 150 kMc/sec. For the $^2\pi_{\frac{3}{2}}$ state which lies only 124 cm^{-1} above the ground state, $\Omega = \frac{3}{2}$, and in this state the lowest rotational frequency corresponds to $J = \frac{3}{2} \to \frac{5}{2}$ and has the value of $5B$ or 250 kMc/sec. If the molecule had no electronic angular momentum, the lowest rotational frequency would have the value $2B$.

Because of the interaction of the rotational and electronic motions, the B of Eq. (6.3) is not exactly constant and in more refined treatments[1] must be replaced by B_{eff}, which to a good approximation is:

$$B_{\text{eff}} = B\left(1 \pm \frac{B}{\lambda A}\right), \tag{6.4}$$

where A is the constant which measures the spin-orbit interaction and where the plus sign is used when $\Omega = \lambda + S$ and the minus sign is used when $\Omega = \lambda - S$. For NO, $A = 3720$ kMc. The first rotational line of the $^2\pi_{\frac{1}{2}}$ state of $N^{14}O^{16}$, which gives $3B_{\text{eff}}$, has been measured[2] at 150373 Mc/sec, after corrections for λ doubling and hyperfine structure. With $\lambda = 1$ and $A = 3720$ kMc (from optical spectroscopy), this leads to $B_0 = 50818.0$ Mc/sec, and yields the interatomic distance $r_0 = 1.1540$ Å.

The observed rotational transition of $N^{14}O^{16}$ was found to have a doublet fine structure (λ doublet) discussed below and a hyperfine structure caused by the N^{14} nuclear interaction, which has been analyzed by MIZUSHIMA[3]. The ZEEMAN effect of the hyperfine structure has been measured and analyzed by MIZUSHIMA, COX, and GORDY[4].

Electronic magnetic resonance of NO in the lowest rotational level ($J = \frac{3}{2}$) of the $^2\pi_{\frac{3}{2}}$ state has been investigated by BERINGER and CASTLE[5]. The coupling was found to conform closely to HUND's case (a). If nuclear effects are neglected, the g factor observed agrees with the value, $3/J(J+1)$, calculated from the simple vector model for HUND's case (a). Nuclear interactions cause some slight decoupling effects which are accounted for by MARGENAU and HENRY[6].

There is another important difference in the spectra of molecules with and without electronic orbital momentum. In the first approximation the levels for λ and $-\lambda$ will have the same energy, i.e., the ground state will be doubly degenerate. The interaction between end-over-end rotation and electronic motions breaks down this degeneracy. The rotational levels are split into doublets called λ doublets. The splitting arises from a slight decoupling of L with the molecular axis; and, although small for low rotational states (except when $\Omega = 0$) it increases rapidly with the molecular rotation, approximately quadratically with increase of the rotational quantum number for most electronic states. The nature of the λ doubling depends upon the type of electronic ground state and particularly upon the degree of coupling between the various vectors of angular momentum. Generally, the splitting decreases rapidly with increase of λ or of Ω. It is usually negligible for $\lambda > 1$, or when Ω is large. Several cases have been treated by VAN VLECK, MULLIKEN and CHRISTY[7] and others. A survey of the various cases is given by HERZBERG [9]. The λ doublets of the $J = \frac{1}{2} \to \frac{3}{2}$ rotational transition of NO in the $^2\pi_{\frac{1}{2}}$ state have been found[2] to be separated by 355.1 Mc/sec.

[1] J. H. VAN VLECK: Phys. Rev. 33, 469 (1929). — R. S. MULLIKEN and H. CHRISTY: Phys. Rev. 38, 87 (1931).

[2] C. A. BURRUS and W. GORDY: Phys. Rev. 92, 1437 (1953). See also J. J. GALLAGHER, F. D. BEDARD and C. M. JOHNSON: Phys. Rev. 93, 729 (1954).

[3] M. MIZUSHIMA: Phys. Rev. 94, 569 (1954). See also C. C. LIN and M. MIZUSHIMA: Phys. Rev. 100, 1726 (1955).

[4] M. MIZUSHIMA, J. T. COX and W. GORDY: Phys. Rev. 98, 1034 (1955).

[5] R. BERINGER and J. G. CASTLE: Phys. Rev. 75, 1963 (1949).

[6] H. MARGENAU and H. HENRY: Phys. Rev. 78, 587 (1950).

[7] J. H. VAN VLECK: Phys. Rev. 33, 469 (1929). — R. S. MULLIKEN and H. CHRISTY: Phys. Rev. 38, 87 (1931).

Although NO is the only stable diatomic molecule which has electronic orbital momentum, numerous unstable radicals such as OH, CH, and BH have unbalanced electronic orbital momentum. Furthermore, in light radicals like these, the λ doubling is sufficiently large for highly populated rotational states that direct transitions between λ doublets should be observable in the microwave region. Direct transitions between λ doublets of the OH and OD free radicals have already been observed in the microwave region[1]. Microwave spectroscopy offers a means for very precise measurement of the various coupling constants for such radicals. Electronic magnetic resonance, as exemplified by BERINGER'S work on O_2, NO, and NO_2, should prove the simplest and perhaps the most effective method for study of such radicals at microwave frequencies. Fig. 10 shows the λ doublet separation as a function of J for several radicals calculated from optical spectroscopy by MULLIKEN and CHRISTY. Note that all have separations corresponding to microwave quanta.

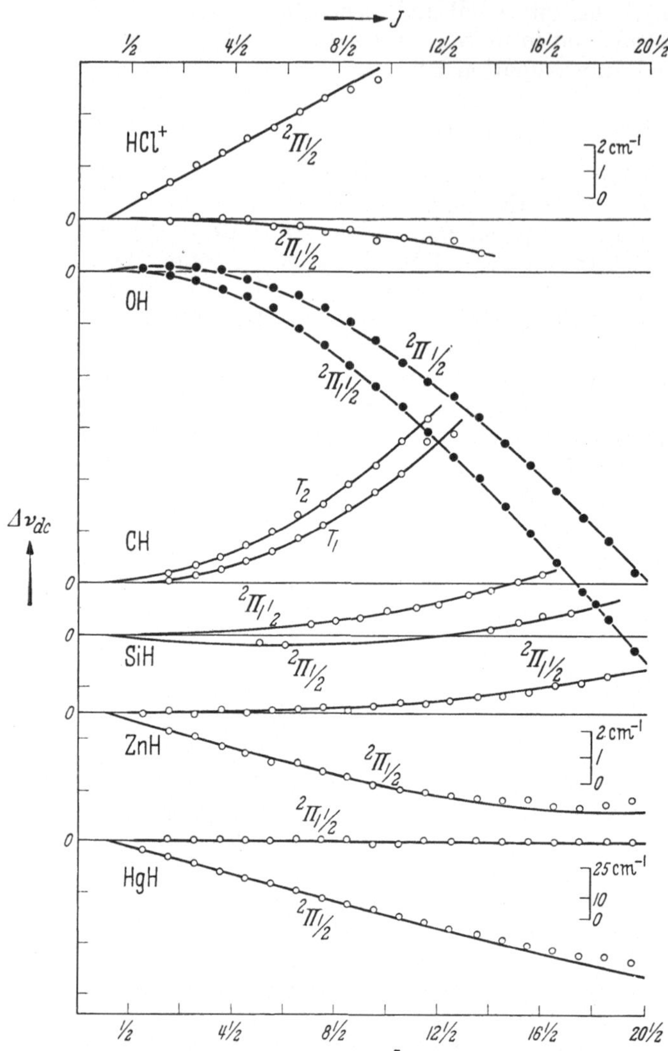

Fig. 10. Predicted separations of λ doublets as a function of J for some light diatomic radicals. Note that these all have separations for low J (populated states) corresponding to microwave $h\nu$. The scale unit, 2 cm^{-1} = 60 kMc/sec. [From MULLIKEN and CHRISTY: Phys. Rev. **38**, 87 (1931).]

7. Line shapes and intensities. Microwave spectra are almost always observed in absorption. Emission lines cannot be observed except under specialized experimental conditions such as the molecular beam techniques[2] in which molecules in the upper and lower states can sometimes be separated in space, or from extended astronomical sources such as the observation of the 21 cm emission line of hydrogen in radio astronomy[3]. We shall discuss here absorption lines only.

---

[1] T. M. SANDERS, A. L. SCHAWLOW, Q. C. DOUSMANIS and C. H. TOWNES: Phys. Rev. **89**, 1158 (1953). — G. C. DOUSMANIS, T. M. SANDERS and C. H. TOWNES: Phys. Rev. **100**, 1735 (1955).

[2] J. P. GORDON, H. J. ZEIGER and C. H. TOWNES: Phys. Rev. **95**, 282 (1954); **99**, 1264 (1955). — See also N. G. BASSOV and A. M. PROKHOROV: Proc. Acad. Sci. (U.S.S.R.) **101**, 47 (1945); — J. Exptl. Theoret. Phys. **27**, 431 (1954).

[3] H. I. EWEN and E. M. PURCELL: Phys. Rev. **83**, 881 (1951). — C. A. MULLER and J. H. OORT: Nature, Lond. **168**, 357 (1951).

An excellent treatment of coherent, spontaneous radiation in the microwave region is given by DICKE[1], with suggested methods for exciting "super-radiant" states (states having abnormally large spontaneous radiation rates).

Microwave absorption lines are usually observed as relatively sharp resonances. The "intensity" of such a line is often given in terms of the absorption coefficient at the peak or maximum of absorption. The absorption coefficient α_ν for any frequency is defined as the ratio of the rate of loss of power with length to the total power passing through the substance:

$$\alpha_\nu = -\frac{1}{P}\frac{dP}{dx}. \tag{7.1}$$

The integrated intensity is:

$$I_{\text{int}} = \int \alpha_\nu \, d\nu \tag{7.2}$$
$$\approx \alpha_{\max}(\Delta\nu),$$

where $\Delta\nu$ is the width of the line measured between half-power points and where $\alpha_{\max}$ is the α_ν at the frequency for maximum absorption.

Experimentally, α_ν is obtained by measurement of the loss in relative power on passage of the microwave radiation through the absorption cell. If α_c represents the coefficient of loss for the cell itself exclusive of the gas and if α_ν represents the absorption coefficient of the gas in the cell, then

$$\frac{P_r}{P_i} = e^{-(\alpha_c + \alpha_\nu)l}, \tag{7.3}$$

and

$$\alpha_\nu = -\alpha_c - \frac{1}{l}\log_e\frac{P_r}{P_i}, \tag{7.4}$$

where P_r is the received power at the output of the cell, P_i is the power input to the cell, and l is the effective cell length.

The absorption coefficient for the microwave absorption of a substance can be theoretically derived from the EINSTEIN transition coefficients [4]. For a quantum transition $J \to J'$ it can be expressed as:

$$\alpha_\nu = \frac{8\pi^3 \nu^2 N_J}{c\,kT}(J\,|\mu_f|\,J')\,S(\nu, \nu_0), \tag{7.5}$$

where
$\qquad N_J =$ number of particles per cc in the lower state J;
$\qquad k =$ BOLTZMANN's constant;
$\qquad T =$ absolute temperature;
$\qquad c =$ velocity of light;
$\qquad \nu =$ frequency at which α_ν is measured;
$\qquad \nu_0 =$ resonant frequency at maximum absorption;
$(J\,|\mu_f|\,J') =$ matrix elements of the dipole moment connecting the states J and J';
$\quad S(\nu, \nu_0) =$ the line shape function.

The population factor N_J depends upon a number of factors and is different for different types of spectra. The dipole matrix elements depend upon the nature of the quantum states involved in the transitions. Descriptions of the various factors in Eq. (7.5) are given elsewhere [2], [4], [6]. The shape function is briefly described below.

Under varying circumstances many factors may influence the width and shape of microwave absorption lines. In most microwave experiments, however, line

---

[1] R. H. DICKE: Phys. Rev. 93, 99 (1954).

widths and shapes are determined primarily by the interaction of the absorbing particles with their close neighbors. Natural line widths are far too small to be measured at microwave frequencies. Although DOPPLER broadening can be easily detected at very low pressures, only in special applications such as the stabilization of oscillators with spectral lines does it become a significant broadening factor; for these purposes it can be largely eliminated through molecular beam techniques. In gases at the pressures most often employed in microwave spectral measurements, 10 to 10^{-2} mm of Hg pressure, the molecules are too far apart to interact significantly except during collisions. Under these conditions the lines are said to be pressure broadened. In paramagnetic resonance of solids, spin-lattice interactions and dipole-dipole coupling between neighboring particles are mainly responsible for the broadening, although exchange interaction often influences significantly the shapes and widths of the resonances.

In the pressure region of most interest, 10 to 10^{-2} mm of Hg, the VAN VLECK-WEISSKOPF[1] line shape function,

$$S(\nu, \nu_0) = \frac{\nu}{\pi \nu_0} \left[\frac{\Delta \nu}{(\nu_0 - \nu)^2 + (\Delta \nu)^2} + \frac{\Delta \nu}{(\nu_0 + \nu)^2 + (\Delta \nu)^2} \right] \tag{7.6}$$

has been found to agree satisfactorily with experimental observations. In this expression, ν_0 is the peak resonant frequency, ν the variable frequency, and

$$\Delta \nu = \frac{1}{2 \pi \tau}, \tag{7.7}$$

where τ is the mean time between collisions. In the derivation it is assumed that τ is long as compared with the collision duration. The assumption, which holds well at microwave, but not at optical frequencies, that the collision duration is short as compared with the period of oscillation of the absorbed radiation is also made.

In most microwave measurements of spectral lines of gases, the pressure is reduced until the lines are relatively sharp and $\nu \approx \nu_0$. Under these conditions the last term of Eq. (7.6) is negligible in comparison with the first, and Eq. (7.6) reduces to the LORENTZian shape function

$$S(\nu, \nu_0) = \frac{1}{\pi} \left[\frac{\Delta \nu}{(\nu_0 - \nu)^2 + (\Delta \nu)^2} \right]. \tag{7.8}$$

The maximum value of S in Eq. (7.8) occurs when $\nu = \nu_0$ and is $1/(\pi \Delta \nu)$. The value of S is reduced to half this maximum value when $|\nu_0 - \nu| = \Delta \nu$. Thus $\Delta \nu$ measures the frequency span from the peak to the half-power point, and $2 \Delta \nu$ represents the width of such a line as measured between the half-power points.

If the maximum value, $1/(\pi \Delta \nu)$, is substituted for $S(\nu, \nu_0)$ in Eq. (7.5), the maximum value of the absorption coefficient $\alpha_{\max}$ is obtained. In the pressure range 10 to 10^{-2} mm of Hg, the line breadth parameter $\Delta \nu$ has been found to vary directly as the pressure. Under these conditions it is readily seen that the peak absorption or $\alpha_{\max}$ does not depend upon pressure. As the pressure is lowered over this wide range, the peak intensity remains essentially unchanged whereas the line becomes sharper. For this reason relatively sharp lines of gases are more easily detectable with conventional microwave spectrometers than are broad ones. This is a fortunate circumstance for if the position of the line ν_0 is to be measured accurately, its width must not be broad.

8. STARK effect. The electric fields of a few thousand volts per centimeter ordinarily applied in microwave rotational spectra produce splittings of the

---

[1] J. H. VAN VLECK and V. F. WEISSKOPF: Rev. Mod. Phys. **17**, 227 (1945).

rotational levels which are small as compared with the rotational energies. Under these conditions the STARK effect can be treated satisfactorily with the first or second order perturbation theory.

Let us assume that a rotating molecule has an electric dipole moment μ. The classical energy of this dipole in an electric field is $-\mu \mathcal{E} \cos \vartheta$, where ϑ is the angle between μ and $\mathcal{E}$. In the usual situation $\mathcal{E}$ is fixed in space, and μ is a direction fixed in the molecule. The angle ϑ, and hence the energy $-\mu \mathcal{E} \cos \vartheta$, varies with the rotation of the molecule. However, for each rotational state $\cos \vartheta$ will have a mean value which can be used for calculation of the STARK energy for that state. The first order rotational energy is given by the average of $-\mu \cdot \mathcal{E}$ over each rotational state. If ψ_J^0 represents the normalized rotational wave function of the field-free rotor in state J, the average is obtained by evaluation of the integral over all the coordinates:

$$\begin{aligned} E_\mathcal{E} &= -\int \psi_J^0 \, (\mu \cdot \mathcal{E}) \, \psi_J^{0*} \, d\tau \\ &= -\mu \, \mathcal{E} \int \psi_J^0 \cos \vartheta \, \psi_J^{0*} \, d\vartheta . \end{aligned} \right\} \qquad (8.1)$$

The last integral represents the direction cosine matrix elements which have been evaluated [4] for various classes of molecular rotors.

For the linear rotor the first-order STARK effect is zero. This may be readily seen by substitution of the wave function of Eq. (5.3) in Eq. (8.1) and by evaluation of the integral. This is true also for the symmetric-top molecule when $K=0$. In the vector model $\boldsymbol{J}$ precesses about the field $\mathcal{E}$, and only the component of μ along $\boldsymbol{J}$ has a non-vanishing interaction with $\mathcal{E}$.

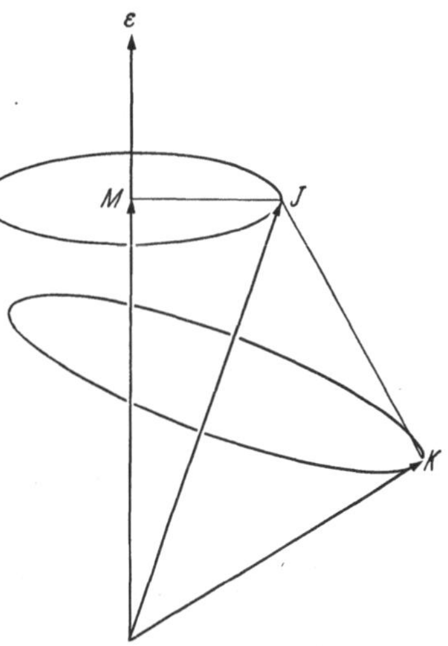

Fig. 11. Vector model for the symmetric-top molecule without nuclear coupling.

In the linear molecule μ lies along the molecular axis and in the first-order approximation has no component along $\boldsymbol{J}$ and hence none along $\mathcal{E}$.

When $K \neq 0$, the symmetric-top molecule has a first-order STARK effect which can be found from the direction cosine matrix elements as indicated above. It can be derived also from the vector model of Fig. 11. In this model μ is directed along the symmetry axis and therefore lies along $\boldsymbol{K}$, the quantized component of internal momentum. Also, $\boldsymbol{J}$ precesses rapidly about $\mathcal{E}$ while $\boldsymbol{K}$ precesses much more rapidly about $\boldsymbol{J}$. The component of μ normal to J is averaged out by the precession of K about J, but there is a non-vanishing component along J which is seen to be:

$$\mu_J = \mu \, \frac{K}{\sqrt{J(J+1)}} , \qquad (8.2)$$

since $K/\sqrt{J(J+1)}$ is the cosine of the angle between $\boldsymbol{J}$ and $\boldsymbol{K}$. Likewise, μ_J will have a non-vanishing component along $\mathcal{E}$:

$$\mu_\mathcal{E} = \mu_J \cos \vartheta = \mu_J \, \frac{M}{\sqrt{J(J+1)}} = \frac{\mu M K}{J(J+1)} . \qquad (8.3)$$

The first-order STARK energy $-\mu_\mathcal{E}\,\mathcal{E}$ therefore has the characteristic values:

$$E_{J,K,M} = -\mathcal{E}\mu \left(\frac{MK}{J(J+1)} \right) . \qquad (8.4)$$

This is seen to be zero when $K=0$. Here M, the "magnetic" quantum number, measures the component of J along $\mathcal{E}$ and can take the values:

$$M = J, J - 1, \ldots - J.$$

The selection rules for M are:

$$M \to M \quad \text{and} \quad M \to M \pm 1.$$

The first, $\Delta M = 0$, gives the π components; the latter $\Delta M = \pm 1$, gives the σ components. For absorption of radiation through electric dipole coupling the $\Delta M = 0$ components are observed when the electric field is parallel to the electric vector of the microwave radiation. These are the components usually observed in microwave spectroscopy. The $\Delta M = \pm 1$ components are circularly polarized and are observed in electric dipole absorption only when $\mathcal{E}$ is perpendicular to the electric vector of the microwave radiation.

Although there is no first-order STARK effect for linear molecules and for the $K = 0$ levels of symmetric tops, there is a second-order effect which can ordinarily be observed with a few thousand volts/cm for low J values. Physically, the second-order effect can be attributed to perturbations of the rotational states in such a manner as to give rise to a component of μ along $\boldsymbol{J}$. The second-order effect is obtained by an averaging of $\boldsymbol{\mu} \cdot \mathcal{E}$ over the first-order "perturbed" wave function $\psi_J^{(1)}$, which in terms of the zero-th order or unperturbed function $\psi_J^{(0)}$ can be expressed as:

$$E_{\mathcal{E}}^{(2)} = (\mu\mathcal{E})^2 \sum_{\substack{J \\ J \neq J'}} \frac{[\int \psi_J^{(0)} (\cos \vartheta)\, \psi_{J'}^{*,(0)}\, d\tau]^2}{E_J^{(0)} - E_{J'}^{(0)}}. \tag{8.5}$$

The integral in this expression represents the direction cosine matrix elements already mentioned. Although the summation is extended over all functions (except for $J = J'$), orthogonality of the wave functions causes all terms to vanish except when $J = J \pm 1$. The denominator $E_J^0 - E_{J'}^{(0)}$ represents the differences in the rotational energy of the states J and J'. Summation yields the characteristic second-order energies:

$$E_{\substack{J, M \\ J \neq 0}}^{(2)} = - \frac{\mu^2 \mathcal{E}^2}{2hB} \left[\frac{3M^2 - J(J+1)}{J(J+1)(2J-1)(2J+3)} \right]. \tag{8.6}$$

For the special case $J = 0$ not included in this formula, the levels are not split but are displaced by the amount:

$$E_{00} = - \frac{\mu^2 \mathcal{E}^2}{6hB}. \tag{8.7}$$

In these expressions B is the rotational constant already defined. The displacements of the rotational frequencies are obtained from these formulae by application of the BOHR postulate and application of the selection rules as already described. The formulae apply both for the linear and the symmetric tops when $K = 0$. The second-order effect can usually be neglected in comparison with the first-order effect when $K \neq 0$.

In a symmetric-top molecule there are $(2J+1)$, $(M \to M)$ first-order STARK components for each value of K excluding zero. Since the K components of the zero-field lines are usually closely spaced, the first-order STARK components are seldom resolvable. For this reason the second-order, $K = 0$ components are the ones most often used for dipole moment measurement. Because of the double degeneracy in the second-order levels, there are only $(J+1)$ components for the $(K = 0)$ lines. There are likewise $(J+1)$ components for each $M \to M$, $J \to J+1$ rotational line of a linear molecule.

The relative intensity formulae for the STARK components are the same as those for the ZEEMAN effect and will be given in Sect. 9.

No simple closed formulae can be given for the STARK effect in asymmetric rotors. However, GOLDEN and WILSON[1] have expressed the STARK energies of the asymmetric rotor in terms of the line-intensity factors, numerically tabulated for certain asymmetries and for J up to 12 by CROSS, HAINER, and KING[2]. The relationship between line intensity and STARK splitting arises from their mutual dependence on matrix elements of the dipole moment.

Treatments of the STARK effect in molecules with nuclear coupling are generally complex. Discussions of them are available [4], [6].

In addition to its use for identification of spectral transitions and for measurement of molecular dipole moments, the STARK effect often aids in the detection of spectral lines in the STARK modulation spectrometer. The ease with which the STARK splitting is resolved in rotational spectra is one of the notable advantages of microwave spectroscopy over optical spectroscopy. Fig. 12 shows the spectacular splitting of the OCS rotational line originally obtained by DAKIN, GOOD, and COLES[3] in the first observation of this effect in rotational spectra.

9. ZEEMAN effect in molecular spectra. Most stable molecules are "nonmagnetic" in the sense that they have singlet electronic ground states with no primary electronic magnetic moment.

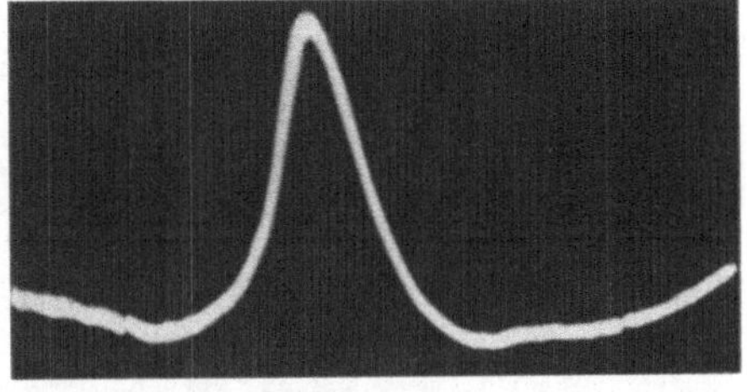

a) zero field

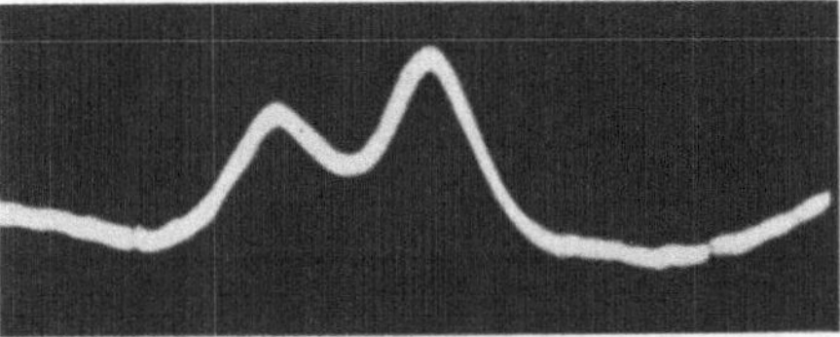

b) 750 volts/cm

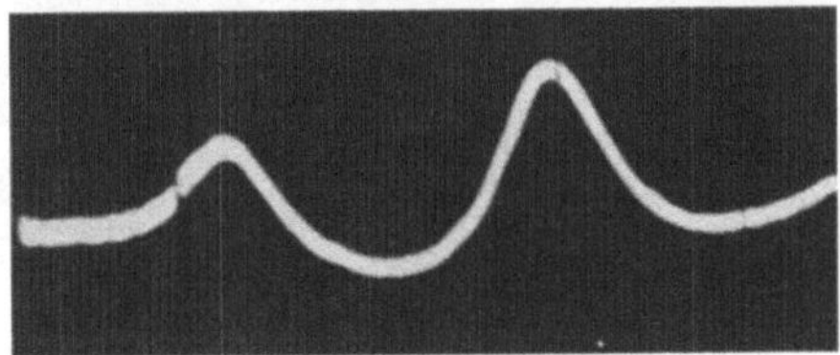

c) 1070 volts/cm

Fig. 12 a—c. STARK splitting of the $J = 1 \rightarrow 2$ rotational transition of OCS. [From DAKIN, GOOD and COLES: Phys. Rev. **70**, 560 (1946).]

Nevertheless, all molecules have small magnetic moments, of the order of a nuclear magneton, which arise from molecular rotation, from nuclear magnetic moments, or from slight intermixing of the ground electronic state with excited "magnetic" states. These small moments give rise to ZEEMAN splitting of rotational lines which can usually be resolved in the microwave region with fields of the order of 10 kilogauss. Fig. 13 illustrates the resolution of the ZEEMAN effect which can be achieved with such "nonmagnetic" molecules.

In a linear molecule without nuclear magnetic coupling and without uncancelled electronic moments, the molecular magnetic moment is generated by the molecular rotation. For this case the molecular rotational moment lies along J, and its magnitude is proportional to the magnitude of the rotational momentum or to $\sqrt{J(J+1)}$. In terms of the molecular rotational g_J factor which can be here

[1] S. GOLDEN and E. B. WILSON jr.: J. Chem. Phys. **16**, 669 (1948).

[2] P. C. CROSS, R. M. HAINER and G. W. KING: J. Chem. Phys. **12**, 210 (1944).

[3] T. W. DAKIN, W. E. GOOD and D. K. COLES: Phys. Rev. **70**, 560 (1946).

regarded as a proportionality constant, the rotational moment μ_J can be expressed as:

$$\mu_J = g_J \beta_I \sqrt{J(J+1)}. \tag{9.1}$$

In this expression β_I represents the nuclear magneton, and g_J is in nuclear magneton units. It is evident that μ_J will have a component along $\boldsymbol{H}$ of

$$\left.\begin{aligned}\mu_H &= \mu_J \cos(J, H) \\ &= \mu_J \frac{M}{\sqrt{J(J+1)}} = g_J \beta_I M.\end{aligned}\right\} \tag{9.2}$$

The characteristic ZEEMAN energies, the eigenvalues of $-\,\mu \cdot H$ or $-\mu_H H$, are readily seen to be:

$$E_H = -g_J \beta_I M H. \tag{9.3}$$

From this formula the splitting of the $J \to J'$ rotational lines is found to be:

$$\left.\begin{aligned}\Delta\nu = [-g_{J'}\Delta M + \\ + (g_J - g_{J'})M]\beta_I H/h.\end{aligned}\right\} \tag{9.4}$$

The selection rules of dipole radiation permit $\Delta M = 0,\ \pm 1$. For electric dipole coupling to the microwave field as in ordinary rotational spectra, the $\Delta M = 0$ component is observed when the electric vector of the microwave radiation is parallel to the imposed magnetic field $\boldsymbol{H}$, and the $\Delta M = \pm 1$ components are observed when the electric vector is perpendicular to $\boldsymbol{H}$.

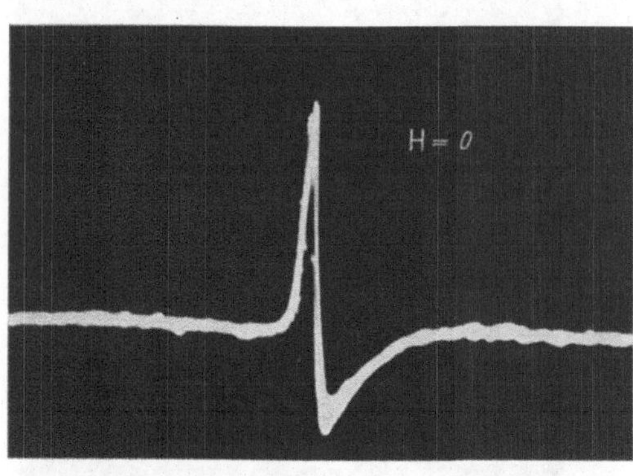

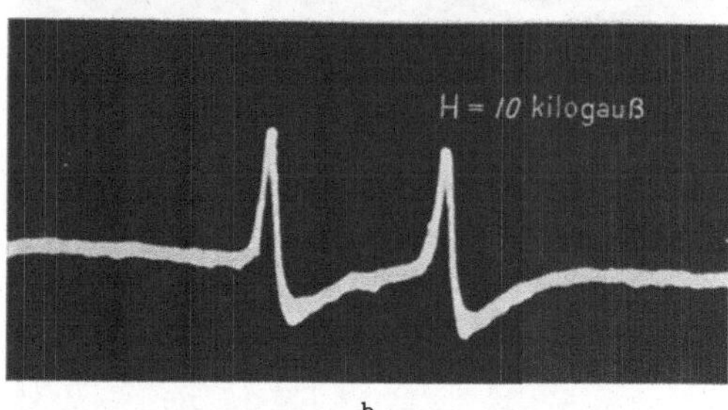

Fig. 13a and b. ZEEMAN splitting ($\Delta M = \pm 1$ components) of the $J = 0 \to 1$ rotational transition of CO. [From Cox and GORDY: Phys. Rev. **101**, 1298 (1956).]

When, as is usually found true for linear molecules, the g_J factor does not vary significantly with rotational state, one can set $g_{J'} = g_J$ and can obtain the simpler formula:

$$\Delta\nu = -g_J \beta_I H \Delta M/h. \tag{9.5}$$

From this formula it is seen that the ZEEMAN effect most commonly observed for linear molecules consists of a single undisplaced π component ($\Delta M = 0$) and doublet σ components ($\Delta M = \pm 1$) of total separation $2g_J \beta_I H/h$.

The rotation about the symmetry axis in symmetric-top molecules gives rise to a rotational magnetic moment directed along this axis and of magnitude proportional to the angular momentum $Kh/(2\pi)$. In terms of the corresponding g factor, g_K, this component can be expressed as:

$$\mu_K = g_K \beta_I K. \tag{9.6}$$

Similarly, the component of μ perpendicular to the symmetry axis can be expressed as:

$$\mu_\perp = g_\perp \beta_I [J(J+1) - K^2]^{\frac{1}{2}}, \tag{9.7}$$

in which $g_\perp$ is the component of g normal to the symmetry axis. One can obtain μ_J by resolving μ_K and $\mu_\perp$ along J which from the vector model (Fig. 11) is seen to be:

$$\mu_J = \frac{\mu_K\, K}{\sqrt{J(J+1)}} + \frac{\mu_\perp\,[J(J+1)-K^2]^{\frac{1}{2}}}{\sqrt{J(J+1)}}$$
$$= \frac{g_K\,\beta_I\, K^2}{\sqrt{J(J+1)}} + \frac{g_\perp\,\beta_I\,[J(J+1)-K^2]}{\sqrt{J(J+1)}}. \qquad (9.8)$$

Furthermore, $\cos(J,H) = \dfrac{M}{\sqrt{J(J+1)}}$, and therefore characteristic values of the ZEEMAN energy, $-\mu_J \cdot H$, are:

$$E_H = -\frac{\mu_J\, H\, M}{\sqrt{J(J+1)}} = -\left[g_\perp + (g_K - g_\perp)\left(\frac{K^2}{J(J+1)}\right)\right]\beta_I\, M\, H. \qquad (9.9)$$

Usually $g_\perp$ und g_K can be treated as constants independent of the rotational state. When this is done, the displacement of a rotational line ($J \to J+1$, $K \to K$ transition), $M \to M'$ component is:

$$\Delta\nu = -\left[g_\perp\,(M'-M) + \frac{(g_K-g_\perp)\,K^2\,M'}{(J+1)\,(J+2)} - \frac{(g_K-g_\perp)\,K^2\,M}{J(J+1)}\right]\frac{\beta_I\, H}{h}. \qquad (9.10)$$

When $K=0$ this formula reduces to that [Eq. (9.5)] for a linear molecule. By measurement on $K=0$ lines one can obtain $g_\perp$ without a knowledge of g_K. This value can then be employed in evaluation of g_K from lines for which K is not zero.

Relative intensities within a given STARK or ZEEMAN multiplet may be determined with the formulae:

$$\left.\begin{aligned} I_{M\to M} &= P\,M^2 \\ I_{M\to M\pm1} &= (P/4)\,(J \mp M)\,(J \pm M + 1) \end{aligned}\right\}\ J \to J, \qquad (9.11)$$

$$\left.\begin{aligned} I_{M\to M} &= Q\,[(J+1)^2 - M^2] \\ I_{M\to M\pm1} &= (Q/4)\,(J \pm M + 1)\,(J \pm M + 2) \end{aligned}\right\}\ J \to J+1, \qquad (9.12)$$

$$\left.\begin{aligned} I_{M\to M} &= Q\,(J^2 - M^2) \\ I_{M\to M\pm1} &= (Q/4)\,(J \mp M)\,(J \mp M - 1) \end{aligned}\right\}\ J \to J-1, \qquad (9.13)$$

in which P and Q are parameters independent of M. The same intensity rules apply either to the STARK or ZEEMAN effect of nuclear hyperfine structure when J is replaced by F and M is replaced by M_F.

Helpful discussions of the ZEEMAN effect are given by JEN[1], who did much of the early work on ZEEMAN effect at microwave frequencies. Useful theory

Table 5. *Some representative g factors obtained from microwave ZEEMAN effect in rotational spectra.*

Molecule	g factor in n.m.		Ref.
CO		$\lvert g_J\rvert = 0.268$	a
N$_2$O		$\lvert g_J\rvert = 0.086$	b
OCS		$g_J = -0.029$	b
CH$_3$F	$\lvert g_\perp\rvert = 0.0624$	$g - g_K = 0.549$	a
CH$_3$CCH	$g_\perp \approx 0$	$g_K = 0.298$	a
SO$_2$		$g_J = 0.084$ for	b
		$(7_{2,6} \to 8_{1,7}$ transition)	

a) J. T. COX and W. GORDY: Phys. Rev. **101**, 1298 (1956).
b) C. K. JEN: Physica, Haag **17**, 379 (1951).

[1] C. K. JEN: Phys. Rev. **74**, 1396 (1948); **76**, 471 (1949); **81**, 197 (1951). — Physica, Haag **17**, 379 (1951). — Amer. J. Phys. **22**, 553 (1954).

is developed by ESHBACH and STRANDBERG[1]. A number of applications to linear and symmetric-top molecules are made by COX and GORDY[2]. Asymmetric rotors are treated by STRANDBERG [5] and by TOWNES and SCHAWLOW [6]. The ZEEMAN effect of nuclear hyperfine structure in rotational spectra has been used in some cases for measurement of magnetic moments of nuclei [4]. Treatments of ZEEMAN and PASCHEN-BACK effects in the hyperfine structure are available elsewhere [4], [6]. HERZBERG [9] discusses its effect in diatomic molecules with electronic magnetic moments. Some illustrative examples of g values obtained for various types of molecules are given in Table 5.

10. Nuclear quadrupole coupling in molecules. Atomic nuclei which have spin values greater than $\frac{1}{2}$ generally have electric quadrupole moments which can interact with the electric field gradients of the molecule to couple the nuclear spin moment to the molecular axis. The resulting interaction gives rise to a splitting of the rotational lines, a nuclear quadrupole hyperfine structure in rotational spectra, which is ordinarily resolvable in the microwave region. From this nuclear quadrupole hyperfine structure, nuclear spins and quadrupole moments as well as significant information about the chemical bond can be obtained [4], [6]. Nuclei which have $I=0$ or $\frac{1}{2}$ have spherical symmetry and no electric quadrupole moment.

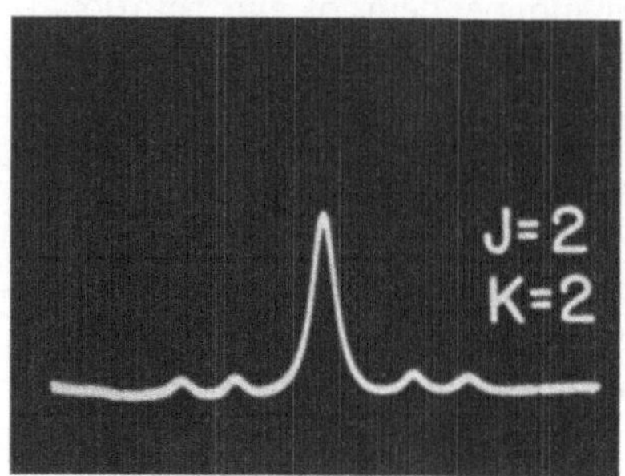

Fig. 14. Fine structure component ($J=2$, $K=2$) of NH$_3$ inversion spectrum showing N^{14} nuclear quadrupole hyperfine structure. [From W.E. GOOD: Phys. Rev. **70**, 213 (1946).]

Nuclear quadrupole hyperfine structure in microwave spectra was first detected by GOOD in the inversion spectrum of ammonia[3]. It has since been observed in the microwave rotational spectra of scores of molecules. In an inversion spectrum ($\Delta J=0$) the hyperfine structure is beautifully symmetric. Fig. 14 shows the N^{14} nuclear quadrupole hyperfine structure of the $J=2$, $K=2$ ammonia line as originally detected by GOOD.

The basic theory for nuclear quadrupole interaction with extra-nuclear electrons has been developed by CASIMIR[4]. The HAMILTONian which describes the interaction is:

$$H = \sum_i e\, Q_i \left\langle\!\left\langle \frac{\partial^2 V}{\partial Z^2} \right\rangle_{\!i}\!\right\rangle_{\mathrm{Av}} \left[\frac{3(\boldsymbol{J}\cdot\boldsymbol{I}_i)^2 + \frac{3}{2}(\boldsymbol{J}\cdot\boldsymbol{I}_i) - \boldsymbol{J}^2\,\boldsymbol{I}_i^2}{2J(2J-1)\,I_i(2I_i-1)} \right]. \tag{10.1}$$

where e is the unit of electronic charge, Q_i is the electric quadrupole moment of the i-th coupling nucleus, $\boldsymbol{I}_i$ is the vector of the spin momentum of the i-th nucleus, and $\boldsymbol{J}$ is the angular momentum of the molecule exclusive of nuclear spin. The quantity $\left\langle\!\left\langle \frac{\partial^2 V}{\partial Z^2} \right\rangle_{\!i}\!\right\rangle_{\mathrm{Av}}$ is the average of the second derivative of the potential at the nucleus arising from the electrons and other nuclei of the molecule. The reference axis is the spacefixed Z axis along which J has its maximum projection $M_J=J$. Theoretically, all extra-nuclear charges contribute to this quantity, but in practice only the electronic charge within a small sphere of the order of an angstrom in radius around the coupling nucleus

[1] J. R. ESHBACH and M. W. P. STRANDBERG: Phys. Rev. **85**, 24 (1952).
[2] J. T. COX and W. GORDY: Phys. Rev. **101**, 1298 (1956).
[3] W. E. GOOD: Phys. Rev. **70**, 213 (1946).
[4] H. B. G. CASIMIR: On the Interaction between Atomic Nuclei and Electrons. Haarlem: Teyler's Tweede Genootschap, E. F. Bohn 1936.

usually need be included. The quantity in the brackets is independent of molecular type and applies also to free atoms when J represents the total angular momentum of the atom exclusive of the nuclear spin.

For an atom or for a single coupling nucleus in a molecule, the first-order nuclear quadrupole perturbation energies can be expressed in relatively simple closed formulas. Characteristic values of the expression in brackets of Eq. (10.1) can be simply evaluated from the vector model (Fig. 15). In the vector model I and J combine to form a resultant F, about which they both precess. The total momentum quantum number, now F, can take the values: $F = J + I, J + I - 1 \ldots |J - I|$. The magnitudes of F, J, and I are $\sqrt{F(F+1)}$, $\sqrt{J(J+1)}$, and $\sqrt{I(I+1)}$ respectively. Therefore $J^2 I^2 = J(J+1) \times I(I+1)$. From the law of the cosine applied to Fig. 15 it is found that:

$$J \cdot I = \tfrac{1}{2}[F(F+1) - J(J+1) - I(I+1)]. \quad (10.2)$$

With these values substituted in Eq. (10.1) the expression for the characteristic quadrupole energies of a single coupling nucleus is obtained as:

$$E_Q = e\, Q \left\langle \frac{\partial^2 V}{\partial Z^2} \right\rangle_{\mathrm{Av}} \times \\ \times \frac{\tfrac{3}{4} C(C+1) - I(I+1)\, J(J+1)}{2J(2J-1)\, I(2I-1)}, \quad (10.3)$$

where

$$C = F(F+1) - J(J+1) - I(I+1).$$

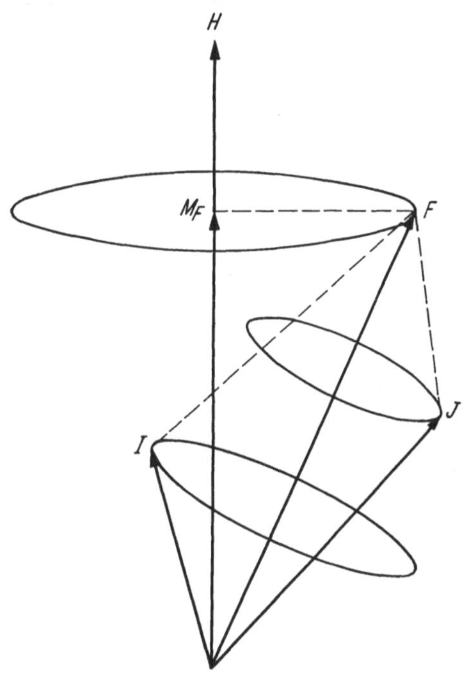

Fig. 15. Vector model for weak field case of atom with nuclear coupling.

The only factor of Eq. (10.3) which varies for atoms and molecules or for different molecular types is the average field gradient which can be expressed as:

$$\left\langle \frac{\partial^2 V}{\partial Z^2} \right\rangle_{\mathrm{Av}} = \left\langle \sum_k e_k \left[\frac{3\cos^2 \vartheta_k - 1}{r_k^3} \right] \right\rangle_{\mathrm{Av}}, \quad (10.4)$$

where r_k is the radius vector between the nucleus and the k-th extra-nuclear charge and where ϑ_k is the angle between r_k and Z. The summation is taken over all extra-nuclear charges of the atom or molecule, and the average is taken over the electronic or molecular states of the charged particles. In the observation of pure rotational spectra of molecules the electronic and vibrational states do not change with the rotational transition. One can therefore treat the field gradients with reference to molecular-fixed axes as characteristic molecular constants, independent of the rotational state. To find a formula which describes the rotational hyperfine structure we need only to express the field gradient in molecular-fixed coordinates and take the average over the particular rotational state, with $M_J = J$. The molecular coupling constants can then be obtained from the observed hyperfine structure, and these constants in turn can be interpreted in terms of the electronic structure of the molecule.

In a linear molecule the reference axis is chosen as the molecular or bond axis. Following convention, we designate this as the z axis, and we designate the gradient at the coupling nucleus with reference to this axis as q_{zz}. The average of q_{zz} for a particular rotational state with $M_J = J$ gives the result:

$$\left\langle \frac{\partial^2 V}{\partial Z^2} \right\rangle_{\mathrm{Av}} = -\frac{J}{2J+3}\, q_{zz}. \tag{10.5}$$

Details of this transformation are given by Ramsey [19]. Eq. (10.5) with Eq. (10.3) yields the characteristic quadrupole coupling energies for a single coupling nucleus in a linear molecule:

$$\begin{aligned}
E_Q &= -e\, Q\, q_{zz} \left(\frac{\tfrac{3}{4} C(C+1) - J(J+1)\, I(I+1)}{2(2J-1)(2J+3)\, I(2I-1)} \right) \\
&= -\chi_{zz}\, Y(F).
\end{aligned} \tag{10.6}$$

For brevity we designate the coupling constant $e\, Q\, q_{zz} \equiv \chi_{zz}$. Numerical values of the function:

$$Y(F) \equiv \frac{\tfrac{3}{4} C(C+1) - J(J+1)\, I(I+1)}{2(2J-1)(2J+3)\, I(2I-1)} \tag{10.7}$$

are tabulated in the appendix of [4] for J values up to 20 and for spin values up to $\tfrac{9}{2}$.

If a symmetric-top molecule has only a single coupling nucleus, this will (to preserve the symmetry) be located on the symmetry axis. The molecular reference axis, the z axis, is now chosen along the symmetry axis, which is not normal to J except when $K=0$. We can obtain the formula for quadrupole coupling by expressing the coupling constant q_{zz} along the symmetry axis and by obtaining the average of $\left\langle \frac{\partial^2 V}{\partial Z^2} \right\rangle_{\mathrm{Av}}$ over the wave function $\psi_{J,K,M_J=J}$ for the symmetric rotor. The process gives:

$$\left\langle \frac{\partial^2 V}{\partial Z^2} \right\rangle_{\mathrm{Av}} = \frac{J}{2J+3} \left(\frac{3K^2}{J(J+1)} - 1 \right) q_{zz} \tag{10.8}$$

and leads to the formula:

$$E_Q = \chi_{zz} \left(\frac{3K^2}{J(J+1)} - 1 \right) Y(F), \tag{10.9}$$

in which $\chi_{zz} = e\, Q\, q_{zz}$ and $Y(F)$ is the function defined by Eq. (10.7).

For an asymmetric-top molecule it is most convenient to take as the reference axes for the quadrupole coupling the three principal inertial axes conventionally designated a, b, and c, so chosen that $I_a < I_b < I_c$. There are thus three coupling constants for the asymmetric molecule, but because Laplace's equation ($\nabla^2 V = 0$) applies here, only two of them are independent. With reference to the a, b, and c coordinates fixed in the molecule, relevant coupling constants are:

$$q_{aa} = \frac{\partial^2 V}{\partial a^2}, \quad q_{bb} = \frac{\partial^2 V}{\partial b^2}, \quad q_{cc} = \frac{\partial^2 V}{\partial c^2},$$

and

$$\chi_{aa} = e\, Q\, q_{aa}, \quad \chi_{bb} = e\, Q\, q_{bb}, \quad \chi_{cc} = e\, Q\, q_{cc}.$$

Bragg and Golden[1] have evaluated $\left\langle \frac{\partial^2 V}{\partial Z^2} \right\rangle_{\mathrm{Av}}$ and have obtained an expression for the quadrupole coupling energies of an asymmetric rotor in terms of the couplings χ_{aa}, χ_{bb}, and χ_{cc}, the reduced energies $E(\varkappa)$, and the inertial asymmetry parameter $\varkappa$ (Sect. 5). The resulting formula for the characteristic coupling

---

[1] J. K. Bragg and S. Golden: Phys. Rev. **75**, 735 (1949).

energies of the asymmetric rotor is:

$$E_Q = \frac{Y(F)}{J(J+1)} \left\{ \chi_{aa} \left[J(J+1) + E(\varkappa) - (\varkappa+1)\left(\frac{\partial E(\varkappa)}{\partial \varkappa}\right) \right] + \\ + 2\chi_{bb}\frac{\partial E(\varkappa)}{\partial \varkappa} + \chi_{cc}\left[J(J+1) - E(\varkappa) + (\varkappa-1)\left(\frac{\partial E(\varkappa)}{\partial \varkappa}\right) \right] \right\}. \quad (10.10)$$

Again $Y(F)$ is the function defined by Eq. (10.7) and tabulated in [4]. Expressions for the reduced energies $E(\varkappa)$ for certain low J values are given in Sect. 5. Approximate values of $E(\varkappa)$ for J values up to 12 can be obtained from tables of KING, HAINER, and CROSS[1]. A second formula derived by BRAGG[2] develops the coupling energies in terms of line intensity factors tabulated by CROSS, HAINER, and KING[3].

The above formulas derived from first-order perturbation theory hold only when E_Q is small as compared with the rotational energy. Formulas which give the quadrupole coupling energy for a single coupling nucleus to second order have been worked out by BARDEEN and TOWNES[4] for the linear and symmetric-top molecule and by BRAGG[2] for the asymmetric rotor.

In general, the quadrupole hyperfine structure of rotational spectra is complex when there are two or more coupling nuclei. Linear molecules with two coupling nuclei have been treated by BARDEEN and TOWNES[5], and symmetric-top molecules with three symmetrically placed, identical coupling nuclei have been treated by BERSOHN[6] and by MIZUSHIMA and ITO[7].

In the calculation of the quadrupole hyperfine structure of a given rotational transition, the energies E_Q are added to the energies of the unperturbed rotor. The energy difference of the particular levels between which the transitions occur determines the frequency of the particular component of the multiplet according to the BOHR relation. In addition to those given in Sect. 5, which apply to the unperturbed rotor, the selection rules $\Delta F = 0$, ± 1 now apply.

The relative intensities of the components of a hyperfine multiplet of a particular rotational transition are given by the formulas:

$$
\begin{aligned}
I_{\mp} &= \left(\frac{1}{F}\right) \cdot Q(F) \cdot Q(F-1) & \Delta F &= \mp 1 \\[4pt]
I_0 &= \frac{(2F+1)}{F(F+1)} \cdot P(F) \cdot Q(F) & \Delta F &= 0 \\[4pt]
I_{\pm} &= \left(\frac{1}{F}\right) \cdot P(F) \cdot P(F-1) & \Delta F &= \pm 1
\end{aligned}
\quad \Bigg\} \quad \Delta J = \pm 1
$$

$$
\begin{aligned}
I_0 &= \frac{(2F+1)}{F(F+1)} \cdot R^2(F) & \Delta F &= 0 \\[4pt]
I_{\pm} &= \left(\frac{1}{F}\right) \cdot P(F) \cdot Q(F-1) & \Delta F &= \pm 1
\end{aligned}
\quad \Bigg\} \quad \Delta J = 0
\qquad (10.11)
$$

where

$$
\begin{aligned}
P(F) &= (F+J)(F+J+1) - I(I+1), \\
Q(F) &= I(I+1) - (F-J)(F-J+1), \\
R(F) &= F(F+1) + J(J+1) - I(I+1),
\end{aligned}
$$

[1] W. G. KING, R. M. HAINER and P. C. CROSS: J. Chem. Phys. **11**, 27 (1943).
[2] J. K. BRAGG: Phys. Rev. **74**, 533 (1948).
[3] P. C. CROSS, R. M. HAINER and W. G. KING: J. Chem. Phys. **12**, 210 (1944).
[4] J. BARDEEN and C. H. TOWNES: Phys. Rev. **73**, 627, 1204 (1948).
[5] J. BARDEEN and C. H. TOWNES: Phys. Rev. **73**, 97 (1948).
[6] R. BERSOHN: J. Chem. Phys. **18**, 1124 (1950).
[7] M. MIZUSHIMA and T. ITO: J. Chem. Phys. **19**, 737 (1951).

in which J and F are the larger values for the two levels involved. Numerical tabulation of the relative intensities is available [4], [6].

The coupling constants obtained from the analysis of the quadrupole hyperfine structure are of the form eQq, where e is the electronic charge, Q is the quadrupole moment of the coupling nucleus, and q is a quantity which depends upon the electronic structure of the molecule. When q can be calculated, the nuclear quadrupole moment can be obtained from the coupling constants. In practice, however, it is not possible to calculate q with very good accuracy because of the incompleteness of our knowledge of the electronic structures of even the simplest molecules. Nevertheless, useful estimates of the quadrupole moments of many nuclei have been obtained in this manner, moments which were not previously known from any other source. Since the evaluation of nuclear moments from microwave spectra is treated in detail in Vol. XXXVIII and elsewhere [4], [6], the subject will not be developed here. If Q is known from other sources, the coupling constants yield directly the quantity q, which can give information about the electronic structure of the molecule. From experiments in atomic beam resonance the quadrupole moments of several nuclei, including the halogens Cl, Br, and I, are now accurately known [19], [21].

The potential V at the nucleus arising from a charge e at a distance r is e/r. The second derivative of this quantity with respect to the z axis is:

$$\frac{\partial^2}{\partial z^2}\left(\frac{e}{r}\right) = \frac{\partial^2}{\partial z^2}\left[\frac{e}{(x^2 + y^2 + z^2)^{\frac{1}{2}}}\right] = \frac{e\,(3\cos^2\vartheta - 1)}{r^3}, \tag{10.12}$$

where ϑ is the angle between r and the z axis. Let us now assume that the charge e is that of the i-th molecular electron in an orbital described by the wave function ψ_i, and let us choose z as the molecular or bond axis on which the nucleus is located. A molecular electron cannot of course be considered as at a fixed distance from a nucleus. Furthermore, q_{zz} arises from all the extra-nuclear electrons. Therefore, to obtain the total q_{zz} one must average the above quantity over the wave function for each electron and sum the results. Thus q_{zz} can be expressed as:

$$q_{zz} = \sum_i e \int \psi_i \left[\frac{3\cos^2\vartheta_i - 1}{r^3}\right] \psi_i^* \, d\tau, \tag{10.13}$$

if ψ_i is properly normalized. The integral is taken over the volume occupied by the electron considered, and the summation is taken over all the electrons of the molecule. Theoretically, the charges of other nuclei in the molecule must be included in the calculation of q; but, because of the inverse cube variation of q with r, their effects can be neglected. Also as a result of the inverse cube variation with r, only the electronic charge within a small distance, of the order of one Ångstrom, contributes significantly to q_{zz}. One can therefore neglect all electronic charge which is essentially localized in atomic orbitals of atoms of the molecule other than the one with the coupling nucleus. Furthermore, all electronic density in s orbitals and all electrons of completely filled shells about the coupling nucleus, because of their spherical symmetry, make no contribution to q_{zz} except for small effects caused by distortions. Thus in practice the summation in Eq. (10.13) need include only non-s electrons of the valence shell of the coupling atom. Nevertheless, the calculation of q_{zz} for an atom in a molecule is a formidable problem which has been solved in accurate manner only for the simple hydrogen molecule[1].

---

[1] A. Nordsieck: Phys. Rev. **58**, 310 (1940).

Since the valence shell electrons primarily determine q_{zz}, it is evident that the quadrupole coupling will depend in a sensitive manner on the number and type of bonds formed by the atom. The molecular quadrupole coupling, when Q or the atomic coupling is known, then gives an additional observable parameter which can be used with dipole moments, bond energies, or other measurable bond properties for finding information about the chemical bond. The atomic coupling is now known for the halogens Cl, Br, and I from atomic beam experiments. TOWNES and DAILEY[1] made the first attempts at relating quadrupole coupling of these halogens to chemical bond properties before the atomic couplings were known. The large amount of a hybridization of the halogen orbital which they predicted ($\sim 20\%$ for Cl) does not appear to be supported by later results. In such molecules as BrCl or ICl where there are two observable q values and only one chemical bond involved, a fairly detailed description of the chemical bond can be obtained from the quadrupole couplings alone. It has been shown[2] that a consistent calculation of the observed q_{Cl} and q_{Br} in BrCl requires an ionic character of the BrCl bond of 6 to 10%, with negligible bond hybridization effects in both atoms. Similarly, an ionic character of 23 to 24% is obtained[3] for the ICl bond with no evident effects of hybridization of the bonding orbitals. In H_2S, however, the quadrupole coupling of S^{33} indicates about 10 to 12% s character in the bonding orbitals [8]. In this asymmetric molecule there are two observable coupling constants, and hence more information is obtained than in the symmetric-top molecule. There is also evidence from quadrupole coupling data for significant s character, 8 to 10%, in the bonding orbitals As, Sb, and Bi [4], [6], [8]. Simplified equations for expressing the coupling constants in terms of chemical bond properties for molecules of certain classes are available [4], [6], [8].

11. Nuclear magnetic coupling in molecules. Practically all stable molecules have singlet electronic ground states with no uncancelled electronic magnetic moment. As mentioned in the discussion of ZEEMAN effect (Sect. 9), such molecules nevertheless have small molecular magnetic fields, of the order of a few gauss, generated mainly by the molecular rotation. Effects of the interaction of the molecular rotational magnetic moment with the nuclear moments of the molecule are extremely small, but nevertheless they have been detected for a few molecules [4], [6]. Such magnetic interactions were first detected in the inversion spectrum of $N^{14}H_3$ by SIMMONS and GORDY[3], who noted that the N^{14} nuclear quadrupole coupling did not exactly account for the splitting of the various fine structure lines and that additional displacements by the N^{14} nuclear magnetic moment are present. Since the $(2I+1)$ degeneracy of the N^{14} nucleus in NH_3 is lifted by the N^{14} quadrupole interaction, the N^{14} magnetic interaction does not lead to additional components but to further displacement of the quadrupole components. However, an additional splitting of certain of these components was later detected by GOOD[4] et al. and was accounted for by VAN VLECK[5] in terms of magnetic interaction involving the three hydrogen nuclei.

When the molecular magnetic moment is generated by the rotation of the molecule, it is reasonable to assume that for linear molecules the molecular

[1] C. H. TOWNES and B. P. DAILEY: J. Chem. Phys. **17**, 782 (1949).
[2] W. GORDY: Disc. Faraday Soc. **19**, 14 (1955). See also W. GORDY: J. Chem. Phys. **19**, 792 (1951); **22**, 1417 (1954).
[3] J. W. SIMMONS and W. GORDY: Phys. Rev. **73**, 713 (1948).
[4] W. E. GOOD, D. K. COLES, G. R. GUNTHER-MOHR, A. L. SCHAWLOW and C. H. TOWNES: Phys. Rev. **83**, 880 (1951).
[5] J. H. VAN VLECK: Phys. Rev. **83**, 880 (1951).

magnetic field will be proportional to $\boldsymbol{J}$. The HAMILTONian for the nuclear magnetic interaction is then of the form $c\,\boldsymbol{I}\cdot\boldsymbol{J}$. Since magnetic interactions in molecules in $^1\Sigma$ states are always very small compared with the rotational energy, the first-order perturbation theory or the vector model treatment can be used to derive the characteristic interaction energies, E_H. For a linear molecule,

$$E_H = \tfrac{1}{2}c\left[F(F+1) - J(J+1) - I(I+1)\right], \tag{11.1}$$

where c is the coupling constant.

From perturbation theory HENDERSON and VAN VLECK[1] have derived the formula:

$$E_H = \left[a + \frac{b\,K^2}{J(J+1)}\right]\left[F(F+1) - J(J+1) - I(I+1)\right] \tag{11.2}$$

for the characteristic nuclear magnetic interaction in symmetric-top molecules. From a treatment similar to that used for the ZEEMAN effect in symmetric tops (Sect. 9), this formula can be derived simply from the vector model [4]. It accounts completely for the N^{14} nuclear magnetic interaction found for NH_3.

Molecules which have an unbalanced electronic momentum usually have a corresponding electron magnetic moment of the order of a BOHR magneton. In such molecules the nuclear magnetic interaction will be very large, often comparable to the rotational energy. The hyperfine structure will depend upon the relative degree of coupling among the molecular, electronic, and nuclear rotational momentum vectors. There are relatively few stable molecules for which the total electronic angular momentum is not zero. Practically all of these are special cases which require special treatment. Some interesting examples which have been studied with microwave spectroscopy are $N^{14}O^{16}$ in the $^2\pi_{\frac{3}{2}}$ state[2] and in the $^2\pi_{\frac{1}{2}}$ state[3] and $O^{16}O^{17}$ in the $^3\Sigma$ state[4]. TOWNES and SCHAWLOW [6] discuss a number of different coupling cases and summarize the observed coupling constants for molecules with and without electronic angular momentum. TOWNES[5] *et al.* discuss the nuclear magnetic coupling in terms of chemical bond properties

D. Microwave atomic spectra.

In the range of 0.03 to 16 cm^{-1} (1 to 500 kilo megacycles) now covered by microwave spectroscopy, direct transitions between fine or hyperfine levels of many atoms are theoretically observable. Because of the low intensity of microwave transitions of atoms and the difficulty of obtaining large concentrations of free atoms in the observation cell, only a few such microwave spectra have been observed. These, however, include the celebrated LAMB shift of the hydrogen atom, the indirect detection of spectral transitions in the unique "atom" positronium, and the detection of the extremely important atomic hydrogen emission line in interstellar space.

A method of great promise for the detection of microwave transitions in atoms is the hybrid optical and radiofrequency method introduced by BITTER[6]

[1] R. S. HENDERSON and J. H. VAN VLECK: Phys. Rev. **74**, 106 (1948).

[2] R. BERINGER and J. G. CASTLE jr.: Phys. Rev. **78**, 581 (1950). — H. MARGENAU and A. HENRY: Phys. Rev. **78**, 587 (1950).

[3] C. A. BURRUS and W. GORDY: Phys. Rev. **92**, 1437 (1953). — J. J. GALLAGHER, F. D. BEDARD and C. M. JOHNSON: Phys. Rev. **93**, 729 (1954). — M. MIZUSHIMA: Phys. Rev. **94**, 569 (1954).

[4] S. L. MILLER, A. JAVAN and C. H. TOWNES: Phys. Rev. **82**, 454 (1951). — S. L. MILLER and C. H. TOWNES: **90**, 537 (1953).

[5] C. H. TOWNES, G. C. DOUSMANIS, R. L. WHITE and R. F. SCHWARZ: Disc. Faraday Soc. **19**, 56 (1955).

[6] F. BITTER: Phys. Rev. **76**, 833 (1949).

and by KASTLER and BROSSEL[1]. In this method atoms excited by light of a given polarization are shifted by the absorption of radio or microwave quanta to other magnetic sub-states from which they emit light of a different polarization. The ZEEMAN effect is employed in the experiments, and the radio waves cause transitions between ZEEMAN levels of the optically excited states. The absorption of the microwave or radiofrequency quanta is detected through its alteration of the relative intensities of the components of the optically polarized light.

Perhaps the simplest and most widely applicable microwave method for study of atoms is the electronic magnetic resonance (paramagnetic resonance). This method is used extensively for the study of solids (Part. E, p. 43). Its application to atoms with different electronic ground states will be described in later paragraphs.

12. Fine structure. In dealing with microwave spectra of atoms one is concerned primarily with the electrons in incompletely filled shells. The inner closed shells form spherically symmetric clouds with no resultant moments and give rise to no microwave spectra. Except for their screening of the electrons in the uncompleted shells from the nuclear charge, the closed shells can generally be ignored in the calculation of the fine or hyperfine structure or the ZEEMAN splitting which gives rise to microwave absorption.

The orbital momentum of a single atomic electron is represented by the vector l which according to quantum theory has the magnitude $\sqrt{l(l+1)}\,(h/2\pi)$, where l is an integer, the orbital quantum number. A magnetic moment directed along l but in the opposite direction to it and of magnitude $\mu_l = \beta\sqrt{l(l+1)}$ is generated by the orbital motion. Here β, the BOHR magneton, equals $\dfrac{e\,h}{4\pi\,m\,c}$ or 0.92712×10^{-20} ergs per gauss. The electron spin momentum is represented by the vector s which has the magnitude $\sqrt{s(s+1)}\,(h/2\pi)$, where $s=\tfrac{1}{2}$ is the electron spin quantum number. Associated with the spin motion is a magnetic moment directed along the spin axis in an opposite direction to s, of magnitude $2\beta\sqrt{s(s+1)}$. Because of the interaction of the spin and orbital magnetic moments, the vectors l and s combine to form a resultant j, of magnitude $\sqrt{j(j+1)}\,(h/2\pi)$, where the quantum number j can have the values $(l+\tfrac{1}{2})$ and $(l-\tfrac{1}{2})$, corresponding to the two allowed orientations of s with l. The interaction energy is different for the two orientations of s with respect to l, and this difference gives rise to the well known doublet fine structure of electronic levels of atoms which have only a single valence electron. For hydrogen-like atoms, quantum theory predicts the characteristic energies [11], [12]:

$$E = -h\,c\left[\frac{R\,Z^2}{n^2} + \frac{R\,\alpha^2\,Z^4}{n^3}\left(\frac{1}{j+\tfrac{1}{2}} - \frac{3}{4\,n}\right) + \text{higher order terms}\right], \qquad (12.1)$$

where n is the total quantum number, Z is the atomic number, α is the fine structure constant, and R is the RYDBERG constant. This formula gives for the separation of the fine structure doublet (difference in the levels for $j=l+\tfrac{1}{2}$ and $j=l-\tfrac{1}{2}$, $n=n$):

$$\Delta\nu = \frac{R\,\alpha^2\,Z^4}{n^3\,l(l+1)}. \qquad (12.2)$$

For heavier atoms, effects of nuclear screening by inner, closed shells require that n or Z, or both, be replaced by effective values which can be estimated from optical spectra [11].

---

[1] A. KASTLER and J. BROSSEL: C. R. Acad. Sci., Paris **229**, 1213 (1949). — A. KASTLER: J. Phys. Radium **11**, 255 (1950).

There is of course no orbital magnetic moment for electrons in s orbitals. For this reason the ground states of atoms such as H, Li, Na, etc., with only a single electron in the unfilled shell, have no fine structure splitting. In the absence of an external field their ground states, which are $^2S_{\frac{1}{2}}$ states, are doubly degenerate.

When more than one electron is in the valence shell of an atom, its fine and hyperfine structure and even its magnetic resonance (ZEEMAN splitting) can be complicated. For the lighter elements in which the RUSSELL-SAUNDERS coupling predominates, the gross aspects of the microwave spectra are readily interpreted with vector models, although the finer details revealed by the high resolution microwave methods are not so easily explained.

In the RUSSELL-SAUNDERS scheme which applies closely for the ligher elements, the orbital moments of the individual electrons are coupled to form a resultant orbital momentum L. The individual spins, $s_1, s_2, \ldots$, are likewise coupled to form a resultant or total electronic spin S. The spin-orbit interaction now couples S and L to form a resultant angular momentum J, in complete analogy to the interaction of the spin moment of a single electron with its orbital moment. Fine structure arises from changes in the relative orientation of S relative to L which lead to different values of the total angular momentum J. The quantum number J takes the values:

$$J = L + S, \quad L + S - 1, \ldots |L - S|.$$

The fine structure will therefore have $(2S+1)$ components corresponding to the allowed values of J for a given value of L. This structure is no longer restricted to a doublet, as it is for atoms with a single valence electron, but may be a singlet, doublet, triplet, or higher multiplet depending upon the value of the resultant spin S.

Spacings of the fine structure of numerous atoms for ground and certain excited states have been measured in optical spectra and tabulated in various places[1]. Methods for calculating or estimating the fine structure spacing are described in all standard texts on atomic spectra. The doublet fine structure separation of the ground state of boron is 16 cm^{-1}, for example.

Matrix elements of the electronic magnetic dipole moment of the atom are non-vanishing for a transition between fine structure levels, and hence such transitions can give rise to observable microwave absorption spectra when the spin-orbit interaction is of the correct magnitude. Although the fine structure of the ground states of practically all atoms is too widely spaced to be observable in the microwave region, the splitting varies inversely with the cube of the quantum number, n_{eff}, and therefore the fine structure intervals of certain excited states always fall in the microwave region. Nevertheless, it seems very difficult, if not impossible, to obtain sufficient population to make direct microwave measurements on any excited atomic states except metastable ones.

$\alpha)$ *The* LAMB *shift*[2]. The DIRAC quantum theory applied to hydrogen-like atoms predicts that states with the same total quantum number n and the same total angular momentum quantum number j are degenerate. This degeneracy is evident from Eq. (12.1). The theory leads to Eq. (12.2) for the calculation of the fine structure splitting. Formula (12.2) is independent of j and indicates a doublet splitting which depends upon the orbital quantum number l and upon the total quantum number n. According to Eqs. (12.1) and (12.2) the $2^2P_{\frac{1}{2}}$ state

---

[1] R. BACHER and S. GOUDSMIT: Atomic Energy States. New York: McGraw-Hill Book Co. 1932. — C. E. MOORE: Atomic Energy Levels, National Bureau of Standards Circular 467, U.S. Govt. Printing Office, Washington, D.C., 1949.

[2] Cf. H. A. BETHE and E. E. SALPETER, Vol. XXXV, especially p. 189—193, and the contribution of L. WILETS in Vol. XXVII of this Encyclopedia.

of hydrogen should be degenerate with the $2\,^2S_{\frac{1}{2}}$ state, but the $2\,^2P_{\frac{3}{2}}$ state should lie above the $2\,^2P_{\frac{1}{2}}$ state by about 0.365 cm^{-1}.

By an ingenious microwave experiment employing atomic beam methods, LAMB and RETHERFORD[1] proved that, in contradiction to the previously accepted DIRAC theory of the hydrogen-like atom, the $2\,^2S_{\frac{1}{2}}$ level of H is actually 1057.777 Mc/sec above the $2\,^2P_{\frac{1}{2}}$ level, although the separation of the $2\,^2P_{\frac{1}{2}}$ and the $2\,^2P$ levels was found to be in agreement with the DIRAC theory. Likewise LAMB and SKINNER[2] found that the $2\,^2S_{\frac{1}{2}}$ level of ionized helium is 14020 Mc/sec above the $2\,^2P_{\frac{1}{2}}$ level. The shift of the $2\,^2S_{\frac{1}{2}}$ level above the $2\,^2P_{\frac{1}{2}}$ level, now called the LAMB shift, has been accounted for by BETHE[3] and others[4] in terms of the self-interaction of the electron with its radiation field.

β) *Positronium.* The strange and short-lived ($\sim 10^{-8}$ sec) atom positronium[5], made of a positron and an electron, has been studied in the microwave region by WEINSTEIN, DEUTSCH, and BROWN[6]. The difference in energy between the triplet state ($J=1$) and the singlet state ($J=0$) is a millimeter wave quanta of 203 kMc frequency. This frequency interval was measured indirectly with lower microwave frequencies by use of the ZEEMAN displacement of the levels. The triplet state is much more stable than the singlet state because triple X-ray quanta must be emitted in the annihilation of the triplet state in order to conserve the momentum, whereas only a double quanta is required in the annihilation of the positronium atom when in the singlet state. WEINSTEIN, DEUTSCH, and BROWN therefore used microwave energy to induce transitions from the triplet to the singlet state and observed the resonance by the increased rate of emission of the pairs of X-ray quanta. The observed frequency (203.38 $\pm$ 0.04) kMc is in good agreement with the calculated value, 203.37 kMc.

13. Transitions within hyperfine multiplets. If the nucleus of an atom has a non-zero spin momentum I, it will have an associated magnetic moment which can interact with the resultant electronic magnetic moment and thus couple I and J to form the resultant F, of magnitude $\sqrt{F(F+1)}\,(h/2\pi)$. The total angular momentum quantum number F can take the values:

$$F = J + I, \quad J + I - 1, \dots |J - I|.$$

The different orientations of I relative to J correspond to different interaction energies. This splitting of the levels for a given J is known as the magnetic hyperfine structure. The magnetic interaction is proportional to the cosine of the angle between I and J and has the characteristic values:

$$E_I = \frac{A}{2}\left[F(F+1) - I(I+1) - J(J+1)\right], \tag{13.1}$$

where A is the interaction constant. The selection rules for magnetic dipole transitions allow $\Delta F = 0$, ± 1. The transition $F \rightarrow F+1$ corresponds to the absorption of energy of frequency:

$$\nu = \frac{A(F+1)}{h}. \tag{13.2}$$

If A is of the proper magnitude, ν will fall in the microwave region.

[1] W. E. LAMB and R. C. RETHERFORD: Phys. Rev. **72**, 241 (1947); **75**, 1325 (1949); **79**, 549 (1950).

[2] W. E. LAMB and M. SKINNER: Phys. Rev. **78**, 539 (1950).

[3] H. A. BETHE: Phys. Rev. **72**, 339 (1947).

[4] N. M. KROLL and W. E. LAMB: Phys. Rev. **75**, 388 (1949). — J. B. FRENCH and V. F. WEISSKOPF: Phys. Rev. **75**, 1240 (1949).

[5] M. DEUTSCH: Phys. Rev. **82**, 455 (1951). — Cf. vol. XXXIV of this Encyclopedia.

[6] R. WEINSTEIN, M. DEUTSCH and S. C. BROWN: Phys. Rev. **94**, 758 (1954).

The interaction constant A depends upon the type of orbital of the interacting electron as well as upon the values of the nuclear moments. For non-s orbitals (p, d, f, etc.) which have no density $\psi_0\psi_0^*$ at the nucleus, the interaction is a dipole-dipole type between the s orbital magnetic moment and the orbital and spin magnetic moments of the electron. This dipole-dipole interaction, when averaged over the wave function for the electron in hydrogen-like orbitals —if small relativistic effects are neglected—gives [19], [20]:

$$A = \frac{h\,c\,g_I\,R\,\alpha^2\,Z^3}{n^3\,(l+\tfrac{1}{2})\,j\,(j+1)\,\varrho} \qquad \text{hydrogen-like non-}s\text{ orbitals}, \qquad (13.3)$$

where g_I is the nuclear g factor, R is the RYDBERG constant, α the fine structure constant, and ϱ is the ratio of the mass of the proton to that of the electron. Eq. (13.3) holds only for non-penetrating, hydrogen-like orbitals. For penetrating orbitals various methods of approximating A have been developed [11], [12], [19], [20]. The most common approach is to divide the orbitals into inner and outer parts to which different amounts of nuclear screening and hence different Z_{eff} apply. When the fine structure splitting $\Delta\nu$ for the atomic state is measured, it can be used for calculation of A [11], [12], [19], [20].

For an s electron the dipole-dipole interaction of the above type averages to zero because of the spherical symmetry of the s orbital. However, the s orbital, unlike higher orbitals with $l \neq 0$, has a non-vanishing density $\psi_0\psi_0^*$ at the nucleus which gives rise to an interaction constant [19], [20]:

$$A = \frac{16}{3}\,\beta\,\beta_I\,g_I(\psi_0\psi_0^*) = \frac{8}{3}\,\frac{h\,c\,g_I\,R\,\alpha^2\,Z^3}{n^3} \qquad \text{hydrogen-like } s \text{ orbitals}. \qquad (13.4)$$

Examples of microwave observations of hyperfine intervals are provided in the measurements of Cs^{133} at 9192.6 Mc/sec by ROBERTS, BEERS, and HILL[1] and of Na^{23} at 1772 Mc/sec by SHIMODA and NISHIKAWA[2]. The famous emission line of interstellar hydrogen[3] at 1420 Mc/sec ($\lambda = 21$ cm) is of this origin. Since $L=0$ in the ground state of this atom, the absorption of microwaves results essentially from a reorientation of the nuclear spin magnetic moment in the magnetic field of the electron spin moment.

If a nucleus of an atom has a spin value greater than $\tfrac{1}{2}$, it will in general have an electric quadrupole moment which can interact with the non-spherical electronic charge of non-S states to give an additional displacement of the levels of the magnetic hyperfine structure. The nuclear quadrupole splitting is given by the theory of CASIMIR, described briefly in Sect. 10. Although large in some atoms, the quadrupole interactions are often much smaller than the accompanying magnetic interactions. When this is true, formula (10.3) can be used to compute the quadrupole interactions.

14. Magnetic resonance of free atoms. In non-singlet S states ($L=0$ but $S \neq 0$) the spin degeneracy can be lifted by an imposed magnetic field, and transitions between the separated ZEEMAN levels can be observed by microwave absorption when the field is of a few kilogauss. The resonance frequency is given by:

$$\nu = \frac{g_S\,\beta\,H}{h} \qquad (14.1)$$

[1] A. ROBERTS, Y. BEERS and A. G. HILL: Phys. Rev. **70**, 112 (1946).
[2] K. SHIMODA and T. NISHIKAWA: J. Phys. Soc. Japan **6**, 512 (1951).
[3] H. I. EWEN and E. M. PURCELL: Phys. Rev. **83**, 881 (1951). — C. A. MULLER and J. H. OORT: Nature, Lond. **168**, 357 (1952).

where β is the Bohr magneton and g_S (in the absence of nuclear interaction) is the effective g factor for the electron spin, which is very close to 2.0023 for S states. Interactions of the nuclear magnetic moment (when $I \neq 0$) with the electron spin magnetic moment give rise to a hyperfine structure superimposed upon the magnetic resonance. If the hyperfine splitting is small as compared with the Zeeman splitting, the electron spin and nuclear spin vectors both precess about the direction of the applied field, and the Back-Goudsmit effect (strong-field case) is observed. The energy is then:

$$E_H = (g_S \beta H M_S + g_I \beta_I H M_I) + A M_I M_S, \tag{14.2}$$

where M_S and M_I are the electron spin and nuclear spin magnetic quantum numbers respectively, where g_I is the nuclear g factor, and where β_I is the nuclear magneton. The coupling constant A is the same as that in Eq. (13.3). In the usual magnetic resonance experiments M_S flips over in the field, but M_I remains unchanged $(M_S - 1 \to M_S;\ M_I \to M_I)$. The resonance frequency is then:

$$h \nu = g_S \beta H + A M_I. \tag{14.3}$$

Since M_I can have any of the values: $M_I = I,\ I - 1, \ldots -I$, all of which are equally probable, there will be $(2I + 1)$ equally spaced (separation A) and equally intense components.

In paramagnetic resonance measurements the frequency is usually left fixed and the magnetic field is swept over the various components. According to Eq. (14.3), the resonant value of the field is:

$$H = \frac{h \nu - A M_I}{g_S \beta}. \tag{14.4}$$

It is evident that the splitting $A M_I$ is equivalent to an additional magnetic field ΔH of value $A M_I / g_S \beta$ added to H, or in magnetic field units the hyperfine components are separated by $\dfrac{A}{g_S \beta}$. Numerically, A (in Mc/sec) equals $1.40\, g_S$ (ΔH) (in gauss) for the strong-field case.

When, as is often true, the hyperfine structure is comparable to the Zeeman energy, the difficult intermediate-field case is encountered for which one must solve the secular equation to obtain the energy eigenvalues. For the important case $J = \frac{1}{2}$, however, the Breit-Rabi formula[1] $[19]$

$$E_H = -\frac{\Delta E}{2(2I + 1)} - g_I \beta_I H M_F \pm \frac{\Delta E}{2} \sqrt{1 + \frac{4 M_F}{2I + 1} x + x^2}, \tag{14.5}$$

in which

$$\Delta E = \frac{A}{2}(2I + 1), \quad \text{and} \quad x = \frac{-g_J \beta H + g_I \beta_I H}{\Delta E},$$

which gives the splitting for any field value is applicable. Atoms with one valence electron—H, Li, Na, etc.—generally have $^2S_{\frac{1}{2}}$ ground states and hence $J = S = \frac{1}{2}$ and $g_J = g_S$. (The symbols in the S state indicates $L = 0$ and should not be confused with the spin quantum number S.) The separation of the hyperfine components in these atoms is large: 1420.405 Mc/sec for H, 803.5 Mc/sec for Li[7], and 1771.6 Mc/sec for Na[23].

The electron spin resonance of atomic hydrogen in the $^2S_{\frac{1}{2}}$ ground state has been measured by Beringer and Heald[2] with microwave magnetic resonance

---

[1] G. Breit and I. Rabi: Phys. Rev. **38**, 2082 (1931).
[2] R. Beringer and M. A. Heald: Phys. Rev. **95**, 1474 (1954).

at 9500 Mc/sec. They obtained a value of $g_S = 2.002296 \pm 0.000006$ for the g factor of the electron spin, in good agreement with the atomic beam results of Koenig, Prodell, and Kusch[1]. Interestingly, Livingston, Zeldes, and Taylor[2] measured at 23 033 Mc/sec the magnetic resonance of H atoms produced by gamma irradiation of solids ($HClO_4$), H_2SO_4, and H_2PO_4 at liquid air temperature and found that $g = 2.0022$ to 2.0025. The hyperfine doublet separation calculated with the Breit-Rabi formula (14.5) from these solid-state measurements gives $\Delta \nu$ as varying from 1407.1 to 1423.4 Mc/sec for the various solids as compared with the value of 1420.41 Mc/sec from atomic beam measurements. Thus the resonance of non-bonded H atoms trapped in solids (at least in some solids) is close to that of free atoms.

Atoms with five electrons in the valence shell—N, As, Sb, etc.—generally have $^4S_{\frac{3}{2}}$ ground states. The spin multiplets of these ground states are degenerate. Like the $^2S_{\frac{1}{2}}$ states, these ground states can be studied at microwave frequencies through their electron spin resonance. There is an important difference, however, in their magnetic hyperfine structure. In the $^2S_{\frac{1}{2}}$ state the coupling of the electron spin magnetic moment to the nuclear spin magnetic moment is large because of the non-vanishing density $\psi_0 \psi_0^*$ of the S wave function at the nucleus. In the $^4S_{\frac{3}{2}}$ state the coupling to the nucleus is very small. In fact, before a small nuclear splitting of the electron spin resonance was observed in the electron spin resonance of N by Heald and Beringer[3] and in P by Dehmelt[4], one might have predicted from the following arguments that the nuclear coupling in $^4S_{\frac{3}{2}}$ states would have been unobservably small. In the formation of the $^4S_{\frac{3}{2}}$ state two of the five electrons are expected to pair off in the s orbital to form a closed sub-shell. These electrons, like those in closed sub-valence shells, would not be detected by spin resonance. Since $L = 0$ and $J = S = \frac{3}{2}$, we know that the remaining three electrons must have their spin pointing in the same direction. According to the Pauli exclusion principle this can be true only if they occupy separate p orbitals (p_x, p_y, and p_z corresponding to $m_l = 1, 0$, and -1). The observable electron spins would have no interaction with the nucleus. The density of the P wave function at the nucleus is zero, and the resultant dipole-dipole interaction of the electrons outside the nucleus would in an S state average zero. Thus the observed splittings in N^{14} (triplet with a coupling constant[3] $A = (10.45 \pm 0.02)$ Mc/sec or 3.73 gauss for the strongfield case) and in P^{31} (doublet with a separation[4] of (56.2 ± 1.5) Mc/sec or 20 gauss for the strong-field case) are of interest because they indicate some type of configuration interaction, possibly a small contribution of s wave function to the orbitals occupied by the unpaired spins. A similar and larger effect has been found in the spin resonance of Mn^{2+} ions $^6S_{\frac{5}{2}}$ ground state) in crystals (see Sect. 19). The hyperfine doublet observed for P^{31} is shown in Fig. 16.

The g factors observed for atomic nitrogen[3] (2.0021) and for atomic phosphorus[4] (2.0019) are close to that for atomic hydrogen. Atoms in non-S ground states, $L \neq 0$ (P, D, F, etc. states), will have a fine structure if $S \neq 0$, also a hyperfine structure if $I \neq 0$. The magnetic resonance frequencies of those which obey the Russell-Saunders coupling scheme and which have no nuclear interaction can be shown by the vector model treatment to be:

$$\nu = \frac{g_J \beta H}{h}, \tag{14.6}$$

[1] S. Koenig, A. G. Prodell and P. Kusch: Phys. Rev. **88**, 191 (1952).

[2] R. Livingston, H. Zeldes and E. H. Taylor: Phys. Rev. **94**, 725 (1954). — Disc. Faraday Soc. **19**, 166 (1955).

[3] M. A. Heald and R. Beringer: Phys. Rev. **96**, 645 (1954).

[4] H. G. Dehmelt: Phys. Rev. **99**, 527 (1955).

where g_J is the Landé g factor,

$$g_J = \frac{g_S}{2}\left(1 + \frac{J(J+1) + S(S+1) - L(L+1)}{2J(J+1)}\right).\qquad(14.7)$$

Eq. (14.6) is seen to reduce to Eq. (14.1) when $L = 0$.

When $I \neq 0$ and the nuclear magnetic interaction is large as compared with the Zeeman interaction, the vector model treatment can again be applied. The

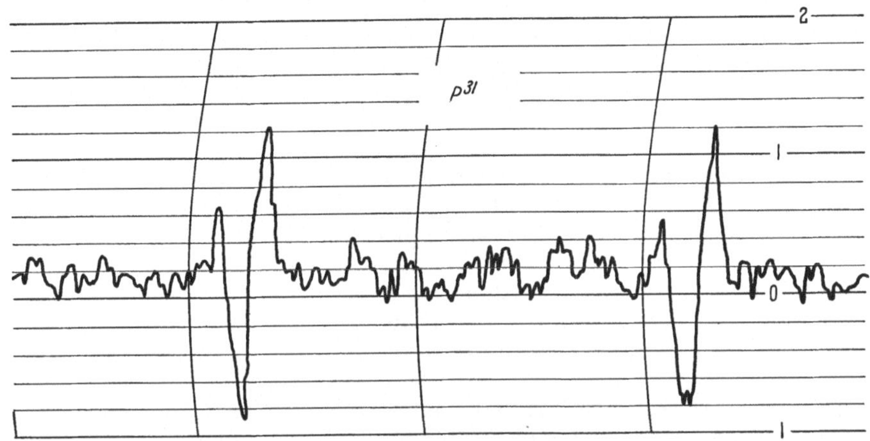

Fig. 16. Doublet splitting by P³¹ nucleus ($I = \frac{1}{2}$) in the microwave magnetic resonance of atomic phosphorus. [From H. Dehmelt: Phys. Rev. 99, 527 (1955).]

magnetic resonance frequency is thus found to be:

$$\nu = \frac{g_F \beta H}{h}\qquad(14.8)$$

where

$$g_F = g_J\left(\frac{F(F+1) + J(J+1) - I(I+1)}{2F(F+1)}\right) - g_I\left(\frac{F(F+1) - J(J+1) + I(I+1)}{2F(F+1)}\right).\qquad(14.9)$$

When the nuclear interactions are small as compared with the Zeeman splitting, the Back-Goudsmit effect as already described for S states will be observed. The complex intermediate-field case is likely to be encountered for atoms in non-S states. When $J = \frac{1}{2}$, the splitting for this case is given by the Breit-Rabi formula (14.5).

Nuclear quadrupole interactions might also be observed in magnetic resonance experiments on atoms with $L \neq 0$ and $I > \frac{1}{2}$, but no such cases have yet been studied with microwave magnetic resonance.

E. Electronic magnetic resonance in solids.

15. Introduction. Electronic magnetic resonance, first observed by Zavoisky[1] in Russia in 1945, has become an effective means for investigation of the solid state and for measurement of nuclear moments. Electronic magnetic resonance in gaseous molecules and radicals is discussed in Sect. 6 and that in free atoms in Sect. 14. Its occurrence in solids will be treated here.

Quantum mechanically, electronic magnetic resonance represents direct transitions between Zeeman components of a particular electronic state, usually the lowest one. The transitions are induced through a coupling of the magnetic

[1] E. Zavoisky: J. Phys. USSR. 9, 211 (1945); 10, 197 (1946).

dipole moment of the resonating electrons to the magnetic component of the microwave radiation. Classically, the electronic momentum vector executes a precessional motion about the direction of the applied field H at the frequency (when in resonance) of the microwave radiation. If the effective g factor for the ZEEMAN splitting is represented by g_{eff}, the characteristic energy is:

$$E_H = g_{\text{eff}}\,\beta\,H\,M. \tag{15.1}$$

where β is the BOHR magneton, H is the magnetic field strength, and M is the magnetic quantum number which measures the component of the electronic angular momentum along H. The resonance frequency is:

$$\nu = \frac{g_{\text{eff}}\,\beta\,H}{h}. \tag{15.2}$$

The frequencies of magnetic resonance are thus determined by the applied magnetic field. Hence, theoretically at least, they can occur anywhere within a wide range of microwave and radio frequencies. However, the resonances are stronger and are more easily observable at microwave than at lower radio frequencies. It is in the microwave region that most measurements have been made. Because of the practical difficulties of obtaining the necessarily high fields (~ 100 kilogauss), no observations have been made in the submillimeter or infrared region.

Eq. (15.1) and (15.2) are like those for free atoms (Sect. 14). However, the g_{eff} here is not the LANDÉ g factor of Eq. (14.7), which applies to free atoms with RUSSELL-SAUNDERS coupling. The coupling of the orbital and electronic spin vectors of electrons within solids is usually broken down, either partly or completely, by the strong electric fields within solids. For this reason the effective g factor, appropriately called the spectroscopic splitting factor, is often near the free electron spin value and seldom near the free atom value except for atoms in S states. In organic free radicals and in conductors or semi-conductors the unpaired electrons which give rise to the resonance can seldom be considered as in localized atomic orbitals of particular atoms but must be treated as in molecular orbitals which envelop more than one atom, and sometimes many different atoms. In paramagnetic salts of the transition elements the electrons at resonance are usually localized on particular atoms or ions with the orbitals of the latter strongly modified by the internal fields.

16. Characteristic energies. In the analysis of paramagnetic resonance each ion and even each crystal is a problem in itself. No generally applicable theory can be given. Nevertheless, the work of ABRAGAM and PRYCE [*13*] provides a generalized procedure for analysis of the resonances in crystals of transition elements, one which can also be applied to electron spin resonances in other solids.

The procedure is to find the "effective" spin wave function ψ and a spin-HAMILTONian operator $\mathscr{H}$ for the particular level or sub-level which is split by the magnetic field to produce the resonance. Usually this sub-level is the lowest level or multiplet of the magnetic element situated within the solid. The spin-HAMILTONian operator applied to the effective spin wave function yields the energy eigenvalues E,

$$\mathscr{H}\psi = E\,\psi, \tag{16.1}$$

from which the resonance frequencies can be computed with the selection rules and the BOHR relation (5.6). The correct spin-HAMILTONian is the one whose energy eigen values thus obtained will account for the observed resonance including its fine and hyperfine structure. The form of the Hamiltonian employed is obtained or guessed from what is already known about the paramagnetic element or ion and about the crystalline fields in which it is situated. The various parameters in the general form are then evaluated by comparison of the predicted with the observed spectrum.

The effective electronic spin S' employed with the spin-Hamiltonian is not necessarily the actual spin S of the free ion. It is determined by setting $(2S'+1)$ as equal to the multiplicity of the state (excluding nuclear spin degeneracies) which gives rise to the resonance even though this multiplicity may actually result from a complicated mixture of spin and orbital states. The simplification which this innovation makes will become apparent from illustrations given in later sections. Since the effective spin is always employed in applications of the spin-Hamiltonian whether it happens to coincide with the actual spin S or not, we shall for brevity often drop the prime and use S for the effective spin as well as for the free ion spin whenever the meaning is clear.

The spin-Hamiltonian which is sufficiently general to encompass the effective spin multiplets which give rise to magnetic resonance in most transition group crystals is [13], [15]:

$$\mathcal{H} = \beta\left(g_z H_z S_z + g_x H_x S_x + g_y H_y S_y\right) + D\left\{S_z^2 - \tfrac{1}{3}S(S+1)\right\} + \\ + E(S_x^2 - S_y^2) + A_z S_z I_z + A_x S_x I_x + A_y S_y I_y + \\ + P\left\{I_z^2 - \tfrac{1}{3}I(I+1)\right\} + P'(I_x^2 - I_y^2) - g_N \beta_N \boldsymbol{H} \cdot \boldsymbol{I}, \tag{16.2}$$

in which the principal axes x, y, and z are assumed to be the same in all terms, an assumption found to hold for all cases studied [15]. The first term in parentheses represents the splitting of the $(2S'+1)$ spin multiplets by the applied field. In this term g_x, g_y, and g_z are the values of the spectroscopic splitting factor g along the principal axes. For any arbitrary orientation of $\boldsymbol{H}$ with the principal axes,

$$g = \sqrt{\left(g_x^2 \cos^2(\boldsymbol{H}, x) + g_y^2 \cos(\boldsymbol{H}, y) + g_z^2 \cos^2(\boldsymbol{H}, z)\right)}, \tag{16.3}$$

where $\cos(\boldsymbol{H}, z)$, etc., represents the cosines of the angle between $\boldsymbol{H}$ and the respective axis. The term $D\left(S_z^2 - \tfrac{1}{3}S(S+1)\right)$ represents a possible initial splitting (with $H=0$) of the spin multiplet by an axially symmetric (trigonal or tetragonal) component of the crystalline field; and the term $E(S_x^2 - S_y^2)$ represents possible initial splitting by crystalline field components of still lower symmetry (rhombic). The remaining terms account for nuclear interactions: the terms $A_z S_z I_z$, etc. represent the nuclear magnetic interactions, and the terms $P\left(I_z^2 - \tfrac{1}{3}I(I+1)\right)$ and $P'(I_x^2 - I_y^2)$ represent, respectively, the nuclear quadrupole interactions with the axially symmetric component and with the rhombic component of the crystalline electric field. Finally, the term $g_I \beta_I \boldsymbol{H} \cdot \boldsymbol{I}$ takes account of the interaction of the nuclear magnetic moment with the externally imposed $\boldsymbol{H}$. The last term is very small and is usually neglected.

The various constants—g_x, g_y, g_z, A, D, E, P, and I—can be regarded as observable parameters to be evaluated from the experimental data. When thus evaluated they obviously reveal the symmetries of the crystalline fields about the paramagnetic element. The constant A is of the same nature as the nuclear magnetic coupling constant for free atoms and can be used to calculate the nuclear moments as described in Sect. 13. The constant P is related to the coupling constant eQq for free atoms and molecules (Sect. 10) by:

$$P = \frac{3eQq}{4I(2I-1)}. \tag{16.4}$$

When the crystalline field which perturbs the effective spin states under consideration has only an axially symmetric component, as is very often found to be true, and when in addition the very small term $g_I \beta_I \boldsymbol{H} \cdot \boldsymbol{I}$ is neglected, the spin-HAMILTONian takes the simpler but still rather general form [13]:

$$\mathscr{H} = \beta \left\{ g_{\parallel} H_z S_z + g_{\perp}(H_x S_x + H_y S_y) \right\} + D\left(S_z^2 - \tfrac{1}{3}S(S+1)\right) + \\ + A\,S_z I_z + B(S_x I_x + S_y I_y) + P\{I_z - \tfrac{1}{3}I(I+1)\}. \tag{16.5}$$

In this transformation $g_z = g_{\parallel}$, $g_x = g_y = g_{\perp}$ and $A_z = A$, $A_x = A_y = B$. The spectroscopic splitting factor ϑ for any orientation of H with the symmetry axis z becomes:

$$g = \sqrt{g_{\parallel}^2 \cos^2 \vartheta + g_{\perp}^2 \sin^2 \vartheta} \tag{16.6}$$

where $g_{\parallel}$ and $g_{\perp}$ are the values for H parallel and perpendicular to this axis.

The calculation of energy eigenvalues from the spin-HAMILTONian is usually involved. Special cases, however, with particular, favorably chosen orientations of the crystal in the magnetic field are often solvable without difficulty (Sects. 19 to 22). From these special cases one can usually gain most of the information obtainable from the more laborious general solution.

The general procedure in finding the energy eigenvalues is to set up and solve the secular equation obtained by equating to zero the determinant of the coefficients a_i in the set of equations:

$$\sum a_i \left[(i|\mathscr{H}|j) - E_i \delta_{ij} \right] = 0, \tag{16.7}$$

in wich E_i represents the energy eigenvalue of the spin state i, δ_{ij} represents the KRONICKER delta which is unity for $i = j$ and zero for $i \neq j$, and $(i|\mathscr{H}|j) = \int \psi_i \mathscr{H} \psi_i^* \, d\tau$ represents the matrix elements of the spin Hamiltonian operator.

Since all quantities in $\mathscr{H}$, except the spin operators, can be considered constants and factored from the integrals, the known matrix elements of the spin operators with reference to space-fixed axes x, y, and z can be employed to obtain the desired matrix elements of $\mathscr{H}$. In the representation M_S, M_I, with the z axis chosen as the axis of quantization, the non-vanishing matrix elements of the spin operators are:

$$(M_S, M_I | S_z | M_S, M_I) = M_S, \tag{16.8}$$

$$(M_S, M_I | S_x | M_S \pm 1, M_I) = \tfrac{1}{2}\left[S(S+1) - M_S(M_S \pm 1)\right]^{\frac{1}{2}}, \tag{16.9}$$

$$(M_S, M_I | S_y | M_S \pm 1, M_I) = \pm \frac{i}{2}\left[S(S+1) - M_S(M_S \pm 1)\right]^{\frac{1}{2}} \tag{16.10}$$

and

$$(M_S, M_I | I_z | M_S, M_I) = M_I, \tag{16.11}$$

$$(M_S, M_I | I_x | M_S, M_I \pm 1) = \tfrac{1}{2}\left[I(I+1) - M_I(M_I \pm 1)\right]^{\frac{1}{2}}, \tag{16.12}$$

$$(M_S, M_I | I_y | M_S, M_I \pm 1) = \pm \frac{i}{2}\left[I(I+1) - M_I(M_I \pm 1)\right]^{\frac{1}{2}}. \tag{16.13}$$

Since the electronic and nuclear spin functions are separable, the non-vanishing elements of the products, $S_z I_z$, etc., are readily obtainable. For example, the non-vanishing elements of $S_z I_z$ are:

$$(M_S, M_I | S_z I_z | M_S, M_I) = M_S M_I, \tag{16.14}$$

and those for $S_x I_x$ are:

$$(M_S, M_I | S_x I_x | M_S \pm 1, M_I \pm 1) = \tfrac{1}{4}\left[S(S+1) - M_S(M_S \pm 1)\right]^{\frac{1}{2}} \times \\ \times \left[I(I+1) - M_I(M_I \pm 1)\right]^{\frac{1}{2}}. \tag{16.15}$$

The matrix elements of the squared operators, S_x^2, etc., are obtained from Eqs. (16.8) to (16.13) by means of the product rule:

$$(M_S, M_I | S_x^2 | M_S', M_I') = \sum (M_S, M_I | S_x | M_S'', M_I'') (M_S'', M_I'' | S_x | M_S', M_I'), \qquad (16.16)$$

and the Hermitian property of the matrices. Thus, with z again the axis of quantization, the non-vanishing matrix elements of S_z^2 are:

$$(M_S, M_I | S_z^2 | M_S, M_I) = M_S^2; \qquad (16.17)$$

and the non-vanishing ones of S_x^2 and S_y^2 are:

$$(M_S, M_I | S_x^2 | M_S, M_I) = (M_S, M_I | S_y^2 | M_S, M_I) = \tfrac{1}{2}[S(S+1) - M_S^2], \qquad (16.18)$$

$$\left.\begin{aligned}(M_S, M_I | S_x^2 | M_S \pm 2, M_I) &= -(M_S, M_I | S_y^2 | M_S \pm 2, M_I)\\ &= \tfrac{1}{4}[S(S+1) - M_S(M_S \pm 1)]^{\frac{1}{2}}[S(S+1) - (M_S \pm 1)(M_S \pm 2)]^{\frac{1}{2}}.\end{aligned}\right\} \quad (16.19)$$

The matrix elements of I_z^2, I_x^2, and I_y^2 are analogous to these. It should be remembered that:

$$M_S = S, S-1, \ldots -S \quad \text{and} \quad M_I = I, I-1, \ldots -I.$$

With the above matrix elements of the spin operators, the matrix elements of $\mathscr{H}$ can be obtained, and the secular equation can be set up and solved for the allowed energies. Because $F_z = S_z + I_z$ commutes with $\mathscr{H}$, all matrix elements not diagonal in $M_F = M_S + M_I$ vanish, and the secular determinent breaks up into sub-determinents, each corresponding to a different value of the sum $M_S + M_I$. In some applications, the terms which are responsible for the off-diagonal elements of $\mathscr{H}$ are zero, or negligible. The energy eigenvalues are then given simply by the diagonal matrix elements of $\mathscr{H}$:

$$E_H = (M_S, M_I | \mathscr{H} | M_S, M_I). \qquad (16.20)$$

Helpful illustrative examples of the solution for the energy eigenvalues are given by Bowers and Owen [15].

17. Line breadths in solids. In solids the factors which broaden electron resonances are so potent that under ordinary conditions it is often impossible to observe resonances because of their great breadths. Even when the resonances are observable, it is seldom possible to resolve the fine and hyperfine structure without special effort to reduce the line widths. The principal factors which contribute to width of electron magnetic resonances in solids are: (α) spin-lattice relaxation which reduces the lifetime in the energy states, (β dipole-dipole interactions between neighboring magnetic elements, and (γ) exchange interactions between the unpaired electrons of neighboring magnetic elements. In powdered samples the anisotropy in the spectroscopic splitting factor g produces a broadening which can only be avoided by use of single crystals. Unresolved line structure such as nuclear hyperfine structure causes an apparent broadening when the individual components are unresolved.

α) *Spin lattice relaxation.* To produce an observable resonance the microwave radiation induces transitions between the Zeeman levels, or the effective spin states separated by the magnetic field. At the same time such transitions are constantly being induced by the lattice vibrations. The two processes are, so to speak, competing in the business of making the electron-spin vectors flip over in the magnetic field. At ordinary temperatures and with the usual microwave power employed, the lattice vibrations are far the more potent of the two.

The effects of the microwave power as a broadening factor are, in fact, negligible except under very special conditions where other broadening factors can be virtually eliminated. The spin-lattice relaxation time τ is a measure of the time required for the exchange of quanta between the spin system and the lattice. According to the uncertainty principle, the spread in the energy states is a function of the lifetime in the states. If ΔE represents the spread and Δt the time in the state, the uncertainty principle requires:

$$\Delta E \cdot \Delta t \approx \frac{h}{(2\pi)}. \tag{17.1}$$

When other broadening factors are negligible, $\Delta E/h$ is a measure of the line width $\Delta \nu$. Eq. (17.1) therefore indicates that $\Delta \nu \approx 1/(2\pi\Delta t)$. When the spin lattice relaxation is the principal broadening process, then $\Delta t \approx \tau$ and $\Delta \nu \approx 1/(2\pi\tau)$, where τ is the spin lattice relaxation time. The line breadth factor arising from spin lattice relaxation is therefore inversely proportional to the relaxation time. The spin lattice relaxation time τ increases rapidly with decrease in temperature, and spin-lattice relaxation although very important at room temperature becomes a negligible line breadth factor at the temperature of liquid helium. Also, τ depends upon the separation of the ground electronic level from the first excited level and increases rapidly with increase in this separation.

β) *Dipole-dipole broadening.* Magnetic dipole-dipole interactions between the effective spin magnetic moments are significant to broadening when the magnetic elements are concentrated in the sample. The broadening arises from spread of the effective magnetic field acting on a given magnetic element caused by the magnetic fields of the closely neighboring dipoles. Assuming that the magnetic moments of all the electrons are directed along $\boldsymbol{H}$, the field contributed at an electron i by a neighbor j at r_{ij} distance away is $\mu_j(1 - 3\cos^2\vartheta_{ij})r_{ij}^{-3}$, where ϑ_{ij} is the angle of r_{ij} with H, and μ_j is the dipole moment of j. Since the various neighbors of a given electron will have different values of r and ϑ and since these quantities vary in time, there will be a spread in the internal field which is added to the externally applied H in the determination of the resonant frequency.

The contribution of this process to the line width has been calculated by VAN VLECK[1]. If a gaussian curve is assumed as the line shape, his formula for the half-width (in terms of H) of the dipole-dipole broadened resonance is:

$$\Delta H = 1.18\sqrt{\langle \Delta H^2 \rangle_{\text{Av}}}, \tag{17.2}$$

where $\langle \Delta H^2 \rangle_{\text{Av}}$, the mean square moment of the line width, is:

$$\langle \Delta H^2 \rangle_{\text{Av}} = \tfrac{3}{4} g^2 \beta^2 S(S+1) \sum_j (1 - 3\cos^2\vartheta_{ij})^2 r_{ij}^{-6}. \tag{17.3}$$

The quantity $g^2\beta^2 S(S+1)$ is the square of the electron spin magnetic moment. It is assumed that the interaction is between like paramagnetic elements. Because of the rapid decrease in dipole-dipole broadening with r, only relatively close neighbors, those within a radius of about ten angstroms, need be included in the summation. For the same reason this type of broadening decreases rapidly with increasing separation of the magnetic elements and can be reduced to insignificance by dilution with diamagnetic substances. Although GAUSSIAN shape was assumed for formula (17.2), VAN VLECK has shown that a dipole-dipole broadened line has a more blunted peak than has the GAUSSIAN curve.

---

[1] J. H. VAN VLECK: Phys. Rev. **74**, 1168 (1948).

γ) Exchange interaction. Exchange of the paramagnetic electrons between different magnetic elements or ions can either broaden or sharpen the resonance, depending upon the nature of the exchange interaction. As was shown by van Vleck[1], isotropic exchange interaction between like elements—those which have the same g factors or the same Zeeman energies in a common field—tends to decrease the width of the resonance and to give it a peaked shape. Such exchange narrowing is analogous to the narrowing of nuclear resonances by the tumbling motions in liquids and can be attributed to a smoothing out of the slowly varying internal fields by the rapid exchange motions of the electrons. If the magnetic elements are unlike—have different g factors or Zeeman frequencies—the exchange will tend to average the two frequencies and thus to broaden the resonance. Exchange interactions also scramble the nuclear interactions of the elements involved and can make the hyperfine structure more complex, or can average out the nuclear interactions so that no such structure is observable.

Exchange interactions, like direct dipole-dipole interactions, can be decreased or eliminated entirely by separation of the magnetic elements through dilution with a suitable diamagnetic substance.

18. Paramagnetism of the transition elements. In the transition elements the filling of the outer or valence shell begins before the inner shells are completely filled with electrons. The valence electrons form bonds, sometimes covalent but usually ionic, to leave the partially filled shells on the bonded atom or stranded ion. The looseness of the unfelled shell permits parallel alignment of some of the electrons in the unfilled shells without conflict with the Pauli exclusion principle. According to Hund's rule, the electron will tend to form the maximum total spin consistent with the exclusion principle, which requires that no two electrons have all four quantum numbers the same. Thus the maximum spin will generally occur when the uncompleted shell is half filled. When the shell lacks only one electron for completion, the spin will be $\frac{1}{2}$, just as for a shell with only a single electron. In the iron group elements (atomic numbers 22 to 29, Ti to Cu) the uncompleted shell which gives rise to the paramagnetism is the $3d$ shell; in the palladium group (atomic numbers 40 to 47, Zr to Ag), the $4d$ shell; in the rare earths (atomic numbers 58 to 70, Ce to Yb), the $4f$ shell; in the platinum group (atomic numbers 72 to 79, Hf to Au), the $5d$ shell; and in the actinide group (atomic number 90 and up, Th to end of table), the $5f$ or $6d$ shells.

The influence of the internal fields on the paramagnetism or paramagnetic resonance of the transition elements depends sensitively upon the particular shell or orbital—whether $3d$, $4f$, etc.—in which the unpaired electrons are found. In the rare earths, for example, the paramagnetic $4f$ electrons are inner shell electrons relatively well shielded from the crystalline fields within the solids, whereas the $3d$ shell of the ions of the iron group and the $4d$ shell of those of the palladium group are outer shells, directly exposed to these fields.

In a discussion of the effects of internal crystalline fields on the magnetic properties of the transition elements it is convenient to define a further classification based on the strength of the crystalline field interactions relative to the strength of the electronic couplings within the paramagnetic element.

(a) In the weak field case the crystalline field interactions are much less than the spin orbit interactions and thus are not sufficiently strong to prevent L and S from forming a resultant J, as in the free atom. The effect of the crystalline field is to lift the J degeneracy either wholly or partly. This case approximates

---

[1] J. H. van Vleck: Phys. Rev. **74**, 1168 (1948).

that of the free atom. The rare earth salts are representative of the weak field case.

(b) In the intermediate field case the internal crystalline interactions are sufficiently strong to prevent L and S from forming a resultant—are greater than the spin orbit interactions—but are not sufficient to prevent the l vectors from forming a resultant L and the s vectors from forming a resultant S. The ions of this group can be treated as "free ions" which are subjected to strong perturbing forces. The ionically bound elements of the iron group are representative of this case.

(c) In the strong field case the internal fields are sufficiently strong to prevent the formation of resultants L or S. It is no longer possible to treat the ion as a free ion perturbed by the crystalline field since its spectroscopic ground state is altered completely by the crystalline field interaction, that is, the crystalline field interactions are greater than the COULOMB interactions between the electrons in the unfilled shell. The palladium group and the covalently bonded complexes of the iron group are representative of this case.

The above classifications were recognized, and many of the magnetic properties of compounds of transitions elements were learned from the earlier measurements of magnetic susceptibilities and from the theoretical work of VAN VLECK and others before magnetic resonances were detected. This earlier information is essential to the understanding of paramagnetic resonance, but no review of it can be given here. The book on susceptibilities by VAN VLECK [17], despite its age, remains perhaps the most valuable source for this information. The less comprehensive but later book by GORTER [18] is also very useful.

The basic theory now generally used for interpretation of paramagnetic resonance compounds of the transition elements was developed in a valuable paper by ABRAGAM and PRYCE [13], at Oxford. The interpretation and application of this theory has been carried out mainly by the experimental group at Oxford, led by BLEANEY. The recent papers by BLEANEY and STEVENS (Part I) [14] and by BOWERS and OWEN (Part II) [15] provide comprehensive reviews of the essential theory and results. In the interest of brevity we shall here make reference to these reviews rather than to cite the numerous original sources.

19. The iron group (Atomic numbers 22 to 29). The elements of the iron group generally form bonds in solids to leave "semi-free" paramagnetic ions. The ions of these elements with their spectroscopic ground states and other pertinent information are listed in Table 6. They constitute the group most extensively investigated through microwave magnetic resonance [14], [15].

To understand the basis for the spin HAMILTONian and the nature of the electronic resonance of the iron group, one needs to examine the gross effects of the crystalline field on the $3d$ shell. The crystalline electric field cannot interact directly with the electronic spin. Its principal effect upon the ions of the iron group is to break down the $L \cdot S$ coupling and to lift the $(2L+1)$ orbital degeneracy, either partly or completely. The crystalline field splitting is essentially an internal STARK splitting of the orbital levels. The L degeneracy will not be completely lifted, except in a field of very low symmetry (rhombic field). The field about the paramagnetic ions in the iron group salts is predominately cubic. Most of these salts are of the form $M^{n+}(X)_6$, where M^{n+} is the magnetic ion and X is a negative ion or the negative end of an electric dipole. An example is $Cu^{2+}(H_2O)_6$ of the salt, $K_2Cu(SO_4)_2 \cdot 6H_2O$. The Cu^{2+} is at the center of an octahedron of which the O of the H_2O forms the six corners.

Table 6. *Some properties of the iron group.*

Ion	Configuration	Spectroscopic state of free ion	Spin orbit coupling constant λ (cm^{-1})	Degeneracy of orbital levels in cubic field
Ti^{3+}	$3d^1$	$^2D_{\frac{3}{2}}$	154	2, 3*
V^{3+}	$3d^2$	$^3F_2$	105	1, 3, 3*
V^{2+}, Cr^{3+}	$3d^3$	$^4F_{\frac{3}{2}}$	55 (V^{2+}), 87 (Cr^{3+})	1*, 3, 3
Cr^{2+}, Mn^{3+}	$3d^4$	$^5D_0$	59 (Cr^{2+})	2*, 3
M^{2+}, Fe^{3+}	$3d^5$	$^6S_{\frac{5}{2}}$		1
Fe^{2+}	$3d^6$	$^5D_4$	-100	2, 3*
Co^{2+}	$3d^7$	$^4F_{\frac{9}{2}}$	-180	1, 3, 3*
Ni^{2+}	$3d^8$	$^3F_4$	-355	1*, 3, 3
Cu^{2+}	$3d^9$	$^2D_{\frac{5}{2}}$	-852	2*, 3

Remark: * indicates lowest level.

From group theory BETHE[1] has shown how the L degeneracy will be lifted for the various iron group ions in fields of different symmetry. These are listed in Table 6, with an asterisk indicating the lowest level in the cubic field. These are in addition to spin degeneracies. The separation of the non-degenerate orbital levels by the predominately cubic field in the iron group salts is great, of the order of 10^4 cm^{-1}. Further splitting of the remaining degeneracies is usually produced by weaker fields of lower symmetry superimposed upon the cubic fields. The latter fields are often axially symmetric, trigonal or tetragonal.

BOWERS and OWEN [15] give satisfying physical descriptions of the splitting of the d orbitals by a cubic field and of the quenching, or partial quenching, of the orbital moments. The description is simplest for the case of only one d electron but can be extended to other cases. Let us suppose that there is only one electron in the d shell. In the free ion the five d orbitals are degenerate, and the single electron will occupy all five equally. In the usual cubic crystal the magnetic ion will be at the center of an octahedron which has six negative charges at its corners. In such crystals the single d electron will not occupy all five of the orbitals equally but will favor those orbitals which most nearly avoid the six negative corners of the octahedron. If the coordinate system of the electron is chosen as the center of the magnetic ion M with the maximum lobes of two of the five d orbitals (the $d_{(x^2-y^2)}$ and the $d_{(2z^2-x^2-y^2)}$) pointing in the direction of the negative X elements, the maximum lobes of the other three orbitals (the d_{xy}, d_{xz}, and d_{yz}) bisect the various coordinate axes in the three coordinate planes and thus point in between the negative X elements. The electron will prefer the latter three equivalent orbitals (designated as the $d\varepsilon$ triplet) over the former two (designated as the $d\gamma$ doublet) which approach the negative X group more closely. In the iron group salts the $d\varepsilon$ triplet lies below the $d\gamma$ doublet by about 10^4 cm^{-1}, and hence at room temperature the single electron will occupy the $d\varepsilon$ level essentially all the time. It is evident that its orbital motion is more restricted than in the free ion. If an effective l' is defined which will give the $(2l'+1=3)$ degeneracy of the $d\varepsilon$ level, the effective l' will equal one rather than two, as it does for the free ion. Thus the orbital momentum has been effectively lowered or partially quenched by the crystalline field.

Now let us consider the effect of adding other electrons in the d orbitals of the ion in the cubic field. Because of the COULOMB repulsion, the electrons will go

---

[1] H. A. BETHE: Ann. Physik **3**, 133 (1929).

into different orbitals as long as empty orbitals of sufficiently low energy are available. According to HUND's rule, the spin of the electrons will tend to become aligned to form a maximum total spin that is consistent with the exclusion principle. If there are three d electrons, as in Cr^{3+} ions, these will go into separate $d\varepsilon$ levels, and their spins will become aligned to form a total $S = \frac{3}{2}$, as for the free ion. Because the three $d\varepsilon$ levels are symmetrical about the origin, the orbital moments of the three electrons will balance to zero; and thus the effective L' will equal zero, in contrast to the free d^3 ion for which L equals three. The orbital moment is thus effectively quenched by the cubic field. The Cr^{3+} and V^{2+} ions in cubic field behave somewhat as if the ions were in an S state. The observed g factors are isotropic and are close to the free spin value. The spin degeneracy is lifted by interaction with the higher $d\gamma$ state in most salts by the order of 10^{-1} cm^{-1}. Nevertheless, it is of advantage to consider as a first approximation that the ions are in a pseudo S state.

When the $3d$ shell is more than half filled, it is simpler to treat the ions as having vacancies or positively charged holes in the otherwise symmetrically filled d shell. Let us consider the $3d^9$ case (Cu^{2+}, for example) which lacks only one electron of having a completed $3d$ shell. The "positive hole" will be "attracted" by the negative X ions and will therefore go into the doubly degenerate $d\gamma$ orbital level. The ground level will then be an orbital doublet with $S = \frac{1}{2}$. The orbital doublet and triplet of the d^1 case are thus inverted for the d^9 case.

Often the combined effects of the crystalline field (cubic plus a weaker and usually axially symmetric component) and the spin orbit interactions leave an orbital singlet as the ground level. This is illustrated by Fig. 17. The higher levels nearest it are usually more than 100 cm^{-1} above, so that at helium temperatures the orbital singlet will be the only significantly populated level and hence the one which is observed in paramagnetic resonance at low temperature. In a purely cubic field, a singlet ground orbital level occurs for d^3 or d^8 (Cr^{3+} or Ni^{2+}) and for d^5, which is a special case because in the corresponding free ion L is also zero. There the observed resonance will arise from a orbital singlet even at room temperature. Most of these singlet orbital levels have spin degeneracies which can be lifted by a magnetic field to given observable resonance.

Although the crystalline field cannot interact directly with the electron spin, it can decompose the spin degeneracy through indirect interactions such as through the spin-orbit coupling. Usually the splitting of the spin levels by the internal fields, when it occurs, is smaller than the customary ZEEMAN splitting and gives rise to a fine structure of the magnetic resonance.

When the lowest orbital level is not a singlet in the predominant cubic field, perturbations of the higher order often have the effect of mixing the lower orbital and spin states, causing a ground manifold having a magnetic moment comprised of both orbital and spin contributions. For this ground manifold ABRAGAM and PRYCE [13] define a fictitious spin S' by setting $(2S' + 1)$ equal to the total multiplicity of the ground manifold even though this manifold may be a complicated mixture of orbital and spin states. The fictitious spin S' will have an associated "spin" magnetic moment and a spectroscopic splitting factor g, but these quantities are obviously not the same as those for the free electron or for the free ion. This definition of S' allows one to set up a spin HAMILTONIAN like that of Eq. (16.2), the energy eigenvalues of which will give the magnetic resonance spectrum.

When the number of electrons is even, the components $M_{S'}$ will be integral, and the levels will usually be resolved by the crystalline field into a singlet corresponding to $M_{S'} = 0$ and into doublets corresponding to $M_{S'} = \pm 1$, etc. If the separation of these components is greater than the $h\nu$ used for observation, one could not detect the usual magnetic resonance although when the ± 1 doublet happens to lie lowest, one might, in sufficiently asymmetric fields, observe weak resonances corresponding to the first forbidden ($\Delta M_{S'} = \pm 2$) transitions.

If the number of electrons of the ion is odd, KRAMERS' theory[1] shows that all levels must have even degeneracies whatever the nature of the field. For these, combined effects of the cubic and weaker components of other fields acting on the ion are to resolve the levels into a series of doublets often separated by the order of 100 cm^{-1} or more. The effective (fictitious!) spin S' for the KRAMERS doublet is $\frac{1}{2}$. Its degeneracy can be lifted by the externally imposed H, and transitions can be observed between the two levels corresponding to $M_{S'} = \frac{1}{2}$ and $-\frac{1}{2}$. Crystals with Ti^{3+}, Co^{2+}, and Cu^{2+} are examples. Fig. 17 shows the

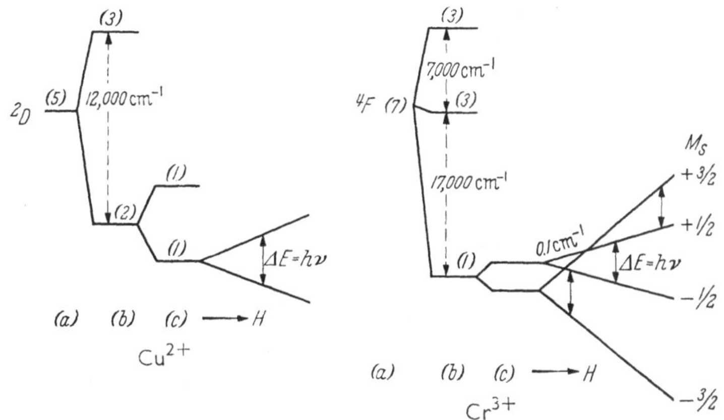

Fig. 17. Diagram (not to scale) showing: (a) levels of free ions Cu^{2+} and Cr^{3+}, (b) splitting of these by crystalline electric field of cubic symmetry, and (c) splitting by an additional component of lower symmetry. Finally the splitting of the lower remaining levels by an imposed magnetic field is shown with the transitions indicated which give rise to microwave magnetic resonance. [From J.H.E. GRIFFITHS: Disc. Faraday Soc. **19**, 3790 (1955).]

levels of Cu^{2+} (neglecting nuclear effects) in a cubic plus a weaker tetragonal field. An axial field of higher symmetry (trigonal) does not resolve the two Cu^{2+} doublets, but in crystals of this symmetry (CuSiF$_6 \cdot 6$H$_2$O, for example) the doublets are separated by the JAHN-TELLER effect (here a tunneling motion of the H$_2$O molecules in the Cu^{3+}(H$_2$O)$_6$ complex).

Table 7 comprises the effective spin values S' which have been found for the iron group with the actual spins of the corresponding free ions. In the same table are given typical g values which have been found.

A rather complete survey of the paramagnetic resonance work on salts of the iron group is made in the review by BOWERS and OWEN [15]. A few illustrative examples of different types of ions will be given here.

Single crystals of the two titanium salts, CsTi(SO$_4$)$_2 \cdot 12$H$_2$O and KTi(C$_2$O$_4$)$_2 \cdot 2$H$_2$O, have been studied. In them the cubic field leaves an orbital triplet lowest in the Ti^{3+} ion (state $3d^1$), but this triplet is split further into three KRAMERS doublets by components of the field of lower symmetry. Observations were made on the lowest of the doublets. Interpretation of the results was possible with $S = \frac{1}{2}$ and with the spin HAMILTONIAN:

$$\mathcal{H} = g_{\parallel} H \beta S_z + g_{\perp} \beta (H_x S_x + H_y S_y). \tag{19.1}$$

---

[1] H. A. KRAMERS: Proc. Amsterdam Acad. Sci. **33**, 959 (1930).

Table 7. *Actual and effective spin values and range of g values found for ions of the iron group for which resonance has been observed. (From* Bowers *and* Owen *[15].)*

No. of $3d$ electrons	Ions	S	S'	g	r.m.s. g
1	Ti^{3+}	$\frac{1}{2}$	$\frac{1}{2}$	1.1–1.3	1.2
	V^{2+}, Mn^{6+}		$\frac{1}{2}$	2.0 (a)	2.0
3	V^{2+}, Cr^{3+}	$\frac{3}{2}$	$\frac{3}{2}$	2.0	2.0
4	Cr^{2+}	2	2	1.9_5–2.0	2.0
5	Mn^{2+}, Fe^{3+}	$\frac{5}{2}$	$\frac{5}{2}$	2.0	2.0
			$\frac{1}{2}$ (b)	0.7–2.6	1.9
6 ,	Fe^{2+}	2	$\frac{1}{2}$ (c)	0–9	5
7	Co^{2+}	$\frac{3}{2}$	$\frac{3}{2}$ (d)	2.3	2.3
			$\frac{1}{2}$	1.4–7	4.5
8	Ni^{2+}	1	1	2.2–2.3	2.2_5
9	Cu^{2+}	$\frac{1}{2}$	$\frac{1}{2}$	2.0–2.5	2.2

Remarks: (a) Complexes almost certainly not octahedral; (b) $M(CN)_6$ complex; (c) FeF_6 complex; (d) tetrahedral $CoCl_4$ complex.

No nuclear effects were detected. The frequencies are given by $\nu = g\beta \dfrac{H}{h}$, where $g^2 = g_\parallel^2 \cos^2\vartheta + g_\perp^2 \sin^2\vartheta$, where ϑ is the angle of H with the symmetry axes of the crystalline field. For $CsTi(SO_4)_2 \cdot 12H_2O$, $g_\parallel = 1.25$ and $g_\perp = 1.14$; for $KTi(C_2O_4)_2 \cdot 2H_2O$, $g_\parallel = 1.86$ and $g_\perp = 1.96$ [15].

In Fe^{2+} the number of electrons is even, and the ground state is $3d^6$, $^5D_4$. Only one crystal, FeF_2, diluted with ZnF_2, has been observed and analyzed in detail. The observations were made at low temperature to avoid broadening by rapid spin-lattice relaxation. The results were interpreted in terms of the spin Hamiltonian:

$$\mathcal{H} = g_z \beta H_z + \Delta S_x, \tag{19.2}$$

with the effective spin $S = \frac{1}{2}$ and with the perpendicular components of g assumed to be zero. The characteristic energies obtained from this spin Hamiltonian are:

$$E = M_S \sqrt{(g_z \beta H \cos\vartheta)^2 + \Delta^2}, \tag{19.3}$$

where $M_S = \pm\frac{1}{2}$ and where ϑ is the angle between H and the crystalline axes. Here Δ represents a small initial splitting of the lowest doublet within which the transition occurs. This splitting indicates a small rhombic component of the field which can lift all degeneracy in the even ions (no Kramers' degeneracy). The observed value of Δ for the salt studied is 0.224 at 90° K and 0.203 at 20° K. The value obtained for g_z is 8.97 at both temperatures. No nuclear effects were observed.

The V^{2+} ion, $3d^3$, $^4F_{\frac{3}{2}}$ ground state, has been studied in detail in diluted vanadium ammonium sulfate. An orbital singlet with four fold spin degeneracy ($S = \frac{3}{2}$) is the ground level in a cubic field. This quartet is further resolved into a pair of Kramers doublets by components of lower symmetry, but this splitting is very small in the crystal studied. One must consider the quartet as the ground manifold with the effective spin equal to the actual spin, $\frac{3}{2}$. The appropriate spin Hamiltonian is:

$$\mathcal{H} = g\beta(H_z S_z + H_x S_x + H_y S_y) + D\left(S_z^2 - \tfrac{5}{4}\right) + E\left(S_x^2 - S_y^2\right) + A\left(S_z I_z + S_x I_x + S_y I_y\right). \tag{19.4}$$

The term in E is small in the ammonium sulfate, and the nuclear effects can be readily analyzed with the good approximation $g\beta H \gg E, A$, if H is large (of high microwave frequency) and is chosen parallel to z. The axis of quantization of both S and I is then z. The off-diagonal matrix elements of $\mathscr{H}$ are thus eliminated. The energy eigenvalues for this case are given simply by the diagonal elements, $(M_S, M_I | \mathscr{H} | M_S, M_I)$, as:

$$E_\| = g\beta H M_S + D(M_I^2 - \tfrac{5}{4}) + A M_S M_I, \tag{19.5}$$

from which the frequencies are readily obtained. The most abundant isotope V^{51} has $I = \tfrac{7}{2}$. The values obtained for the Zn diluted ammonium sulfate are: $g = 1.951$, $D = 0.158$ cm^{-1}, $E = 0.049$: for V^{51}, $A = 0.0088$ cm^{-1}. Since the fine

structure constant is much larger than A, each fine structure component (for V^{51}) has an eight-component hyperfine structure. This structure is illustrated in Fig. 18. The V^{2+} paramagnetic resonance of the isotope of low abundance, V^{50}, was used for the first determination[1] of its nuclear spin. For this isotope the hyperfine structure consists of 13 components of equal intensity and of approximately equal spacing. Here the spin is 6.

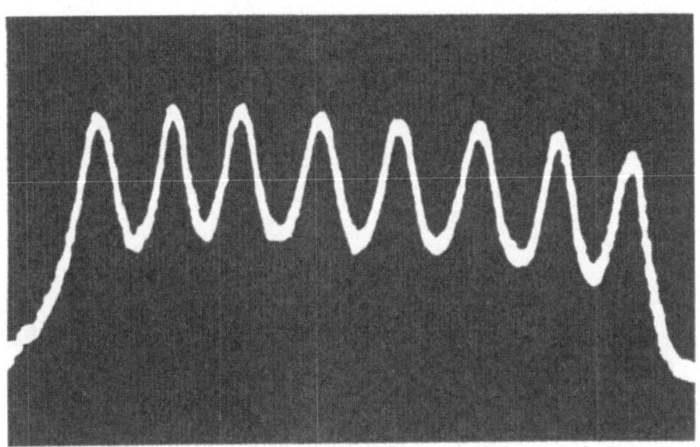

Fig. 18. Hyperfine structure caused by V^{51} ($I = \tfrac{7}{2}$) nuclear interactions in the electron magnetic resonance of vanadous ammonium sulfate. [From BLEANEY, INGRAM and SCOVIL: Proc. Phys. Soc. A 64, 601 (1951).]

Ions which have S states must be treated differently from those which have orbital angular momentum. Since the crystalline field cannot interact directly with the electronic spin, the splitting of the spin levels arises from weak, indirect interactions of more than one type. Because the non-s orbitals have no density at the nucleus, the $3d^5$ S state should have no coupling to the nucleus unless the magnetic shell has some s orbital contribution. Thus in the first approximation the resonance of an ion such as Mn^{2+} within solids should have only a single isotropic component at $g\beta H/h$, where g is the free electron spin. Actually, the Mn^{2+} salts have been found to have a hyperfine structure of considerable proportions and those with non-cubic fields, a significant fine structure. ABRAGAM and PRYCE [13] have accounted for the essential features of this structure. Their general spin-HAMILTONian for ions of this type (with $S = \tfrac{5}{2}$) is:

$$\mathscr{H} = g\beta \mathbf{H} \cdot \mathbf{S} + D\left(S_z^2 - \tfrac{1}{3} S(S+1)\right) + E(S_x^2 - S_y^2) + \tfrac{a}{6}(S_1^4 + S_2^4 + S_3^4) + \left.\vphantom{\tfrac{a}{6}}\right\} \tag{19.6}$$
$$+ A(S_z I_z) + B(S_x I_x + S_y I_y).$$

The only term here not included in Eq. (16.2) is $\tfrac{a}{6}(S_1^4 + S_2^4 + S_3^4)$, which represents the interaction of the ion with the cubic field, where S_1, S_2, and S_3 are the spin components along the three axes of the cubic field. This term is very small and is usually neglected. Also the term in E is often negligible. The term in D represents the interaction of the axially symmetric component of the field and gives rise to an observable fine structure in fields with significant trigonal or

---

[1] C. KIKUCHI, M. H. SIRVETZ and V. W. COHEN: Phys. Rev. **92**, 109 (1953).

tetrahedral components. The axially symmetric field produces an ellipsoidal distortion of the spherical distribution of the electronic charge and thus alters the mutual spin-spin interaction of the five electrons. To a good approximation, the terms in $a/6$ and E can be neglected. For the strong-field case, the first-order energies for an orientation, ϑ, of H with the crystalline axis are:

$$E_H = g\beta H M_S + D\left(M_S^2 - \tfrac{35}{12}\right)(3\cos^2\vartheta - 1) + A M_S M_I. \qquad (19.7)$$

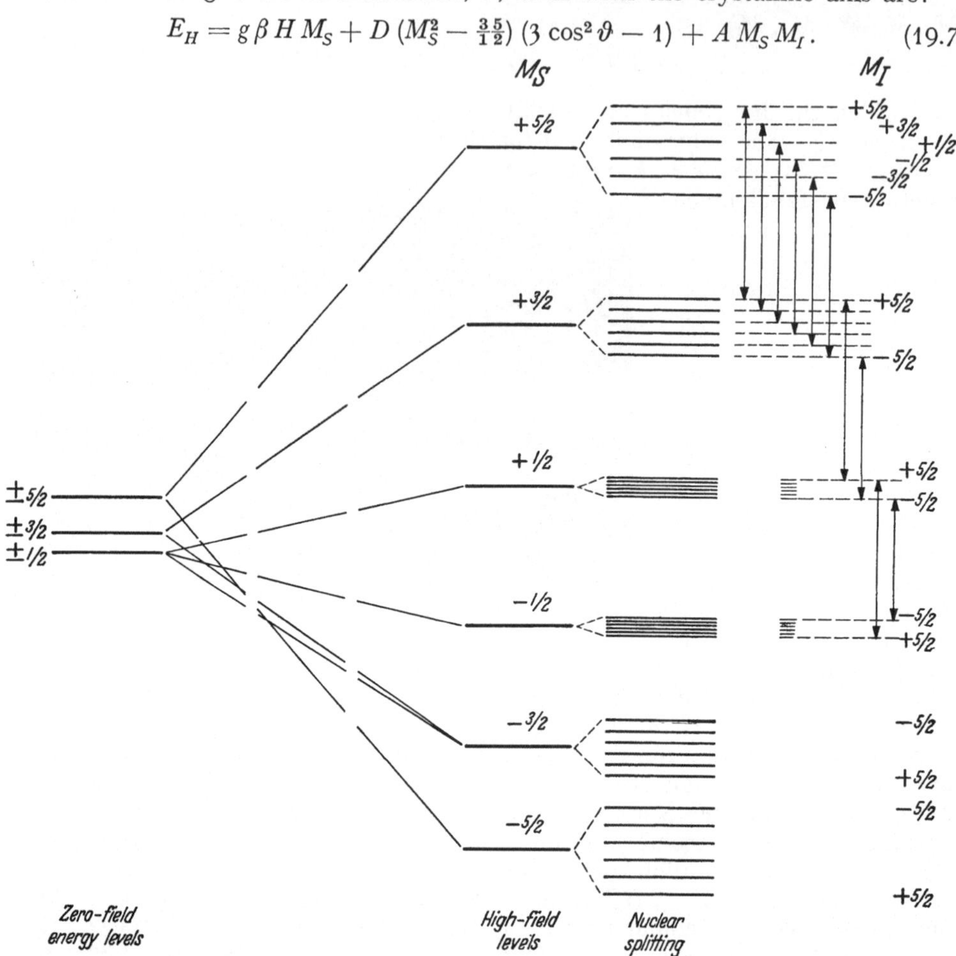

Fig. 19. Energy level diagram for Mn^{2+} ions, showing crystalline field splitting of electron spin levels in zero magnetic field, the splitting of the spin levels in a strong magnetic field, and finally the splittings caused by nuclear interactions. Arrows indicate some typical transitions. [From BLEANEY and INGRAM: Proc. Roy. Soc. Lond., Ser. A **205**, 336 (1951).]

The frequencies can be obtained from this formula with the selection rules $\Delta M_S = +1$, $\Delta M_I = 0$. Both S and I for Mn^{2+} are $\tfrac{5}{2}$. Fig. 19, from BLEANY and INGRAM, is a diagram of the splitting of the levels for a typical case. Complications arise because the fine and hyperfine interactions are often of the same order of magnitude.

Many Mn crystals have essentially cubic fields with no significant component of lower symmetry. For these the term in D is out, and the energy formula becomes simply:

$$E_H = g\beta H M_S + A M_S M_I, \qquad (19.8)$$

and the frequencies:

$$h\nu = (g\beta H + A M_I). \qquad (19.9)$$

It is found, however, that the first-order formula does not exactly fit the observed frequencies. The spacings between the components are not quite equal but increase toward the higher-field side. The formula:

$$h\nu = g\beta H + A M_I + \frac{A^2}{2h\nu}[I(I+1) - M_I^2] + \frac{A^2}{2h\nu}(2M_S + 1), \qquad (19.10)$$

derived from second-order perturbation theory accounts well for the hyperfine structure. In Eq. (19.10) M_S is the value for the lower level involved in the transition.

In solids the Mn^{2+} hyperfine structure has been found to have a total spread ranging from about 300 to 500 gauss (A, 60 to 100 gauss). In dilute liquid water solutions, the Mn^{2+} hyperfine structure has a total spread near the upper limits of the solid state values, or 475 gauss. ABRAGAM and PRYCE have accounted for this large hyperfine structure by postulating that the ground state is not simply $3d^5$ but is an admixture, $3s\, 3d''\, 4s$. Because of the large nuclear coupling of the $3s$ orbital, only a small $3s$ contribution ($\sim 0.1\%$) to the magnetic shell is required to account for the large splitting observed. This admixture is similar to that found in the gaseous free atoms N and P (Sect. 14), and since it persists for the ions in solution it seems to be a property of the free ion, i.e., the admixture is not induced by the crystalline fields. HERSHBERGER and LEIFER[1] have studied a number of solid phosphors containing Mn, in which the Mn^{2+} resonance exhibited both the fine and hyperfine structure. VAN WIERINGEN[2] measured the hyperfine structure in a number of solid solutions with essentially cubic fields (fine structure negligible). Table 8 summarizes the results which he obtained. The nuclear magnetic coupling constant A is seen to vary from 59 to 98 in the dilute solutions, with the largest g factor for the smallest A. VAN WIERINGEN attributes this decrease in A from its highest value (presumably that for the free ion) to covalent character in the Mn bonding. For a given solution which is relatively concentrated there is also a decrease in A with further concentration of the magnetic ions. This is attributed to exchange interaction between the magnetic ions. The variation in g recorded in Table 8 is slightly out of the experimental error, and VAN WIERINGEN attributes this to a slight departure of the Mn^{2+} ions from the true S state, i.e., there is some orbital contribution to the magnetic moment.

Table 8. *Nuclear coupling constant and g factor for* Mn^{2+} *ions in various phosphors.*
[From J. S. VAN WIERINGEN, Discussion Faraday Soc. **19**, 118 (1955).]

Compound	Approximate (atomic Mn) %	A (oersted) (± 1)	g (± 0.0025)
$KMgF_3$	0.1	98	2.004
CaF_2	0.1	99	2.004
$CsCaF_3$	0.1	97	2.004
CaO	unknown	91	2.004
$2MgO \cdot \frac{3}{2}Al_2O_3$ *	0.1	87	2.004
$6MgO \cdot As_2O_5$ *	0.1	87	2.004
$MgAl_2O_4$ *	0.1	87	2.004
$Zn\ 0.1 \cdot 1Al_2O_3$	0.01	81	2.004
ZnS	0.005	69	2.004
CdS	0.1	69	2.004
ZnSe	0.05	65	2.006
CdTe	0.01	59	2.008

* In these phosphors most of the Mn is tetravalent. The paramagnetic resonance absorption (caused by Mn^{2+}) was accordingly weak.

[1] W. O. HERSHBERGER and H. N. LEIFER: Phys. Rev. **88**, 714 (1952).
[2] J. S. VAN WIERINGEN: Disc. Faraday Soc. **19**, 118 (1955).

20. Palladium and platinum groups (Atomic numbers 40 to 47 and 72 to 79). Paramagnetic salts of the palladium and platinum groups generally represent the strong internal electric field case. The magnetic electrons are in $4d$ and $5d$ shells, respectively. Because of the projections of the $4d$ and $5d$ orbitals over those of the $3d$ shell of the iron group, they are more subjected to effects of the crystalline field. Many of the group form covalently bonded complexes which are not paramagnetic. Some, however, form magnetic salts, a few of which have been investigated. Observation of the resonance often requires magnetically dilute samples because of a strong exchange interaction which tends to align the spins of closely spaced magnetic ions in an antiparallel manner. In addition to the reviews already mentioned [14] to [16], paramagnetic resonance in the palladium and plutonium groups is treated in a valuable paper by Griffiths, Owen, and Ward[1].

Salts of $4d$ and $5d$ groups tend to have octahedral complexes with the magnetic ion at the center, as do those of the $3d$ shell. Just as for the $3d$ iron group, the crystalline field is predominantly cubic, with weaker components of lower symmetry superimposed. The cubic field splits the d orbitals into the $d\varepsilon$ triplet and the $d\gamma$ doublet, as already described. A difference arises, however, in the filling of these levels. Let us suppose that each of the three $d\varepsilon$ levels already has an electron. A fourth electron will now share a lower $d\varepsilon$ orbital rather than go into the higher $d\gamma$ doublet. This condition arises because the separation of the $d\varepsilon$ and $d\gamma$ levels in the strong-field case is greater than the repulsion energy of two electrons in the d orbital. According to the Pauli principle, two electrons in the same orbital must have opposing spins. Thus for a d^4 ion of the strong cubic field, $S=1$ rather than 2, as for the weak-field or free-ion case. Similarly, additional electrons go into the $d\varepsilon$ state until all three of its orbitals have a pair of electrons. The d^6 ions in the strong cubic field thus have a closed $d\varepsilon$ sub-shell with $S=0$ and $L=0$, and they are not paramagnetic. The $d\varepsilon$ sub-state in these salts is in many respects like a p shell of the free atom.

Except for possible differences in the ground state arising from the stronger crystalline field interaction, the magnetic resonance spectra of the $4d$ and $5d$ ions are generally similar to those of the $3d$ ions. The approach to interpretation of their resonance spectra, like those for the $3d$ ions, is to find an effective spin S' and a spin-Hamiltonian for the ground state which predicts correctly the g factor and the fine and hyperfine structure of the resonance. The general form of the spin-Hamiltonian depends upon the symmetries of the crystalline field. When there is no covalent character in the bonding to the magnetic ion, it can take the same form as for the iron group. For example, the resonance of Mo^{5+}, $4d^1$, has been observed as an impurity in single crystals of $K_3[InCl_6] \cdot 2H_2O$. Except for nuclear effects, its observed resonance could be interpreted with the spin-Hamiltonian, Eq. (19.1), for the $3d^1$ ions Ti^{3+} and Mn^{6+}.

An example of the departure in behavior of the $5d$ from the $3d$ shell is Ir^{4+}, which has five unpaired $5d$ electrons, $5d^5$. The effective spin for this ion in various salts is found to be $\frac{1}{2}$, whereas the corresponding spin of the free ion is $\frac{5}{2}$ according to Hund's rule. For the corresponding $3d^5$ ion, Fe^{3+}, the effective spin is $\frac{5}{2}$, the same as that expected for the free ion. Furthermore, considerable covalent character is found for the bonding to Ir in the irridium salts, and this leads to a further difference from the $3d$ group. The effective spin for Ir^{4+}, $S'=\frac{1}{2}$, can be regarded as arising from a single electron hole in the $d\varepsilon$ sub-shell, which has three orbitals and behaves somewhat like a p shell. Let us consider the

---

[1] J. H. E. Griffiths, J. Owen and I. M. Ward: Proc. Roy. Soc. Lond., Ser. A **219**, 526 (1953).

$[IrCl_6]^{2-}$ and $[IrBr_6]^{2-}$ complexes, which have been studied in detail by GRIF-FITHS and OWEN[1]. If possible distortions and covalent character are neglected, an Ir^{4+} ion is at the center of an octahedron, the six corners of which are formed by the Cl^- or Br^- ions. The three $d\varepsilon$ orbitals are the equivalent $5d_{xy}$, $5d_{xz}$, and $5d_{yz}$ orbitals, which have their lobes pointing between the negative ions. The field is cubic, and, if nuclear interactions are neglected, the spin-HAMILTONian would be simply:

$$\mathscr{H} = \beta (g_x H_x S_x + g_y H_y S_y + g_z H_z S_z). \qquad (20.1)$$

However, both stable isotopes of Ir have a non-zero spin, $I = \frac{3}{2}$, and in the resonance of these complexes hyperfine structure arising not only from Ir but from halogen nuclei has been observed. This interesting observation made by GRIFFITHS and OWEN[1] proves that there is some covalent π bond character in the bonding and provides a method for measuring its amount.

The approximate spin-HAMIL-TONian[1] for this complex, including nuclear magnetic interactions, is

$$\mathscr{H} = \beta (g_x H_x S_x + g_y H_y S_y + g_z H_z S_z) + A_x S_x I_x + A_y S_y I_y + A_z S_z I_z + A_{1'} S_x [(I_x)_1 + (I_x)_4] + A_{2'} S_y [(I_y)_2 + (I_y)_5] + A_{3'} S_z [(I_z)_3 + (I_z)_6], \qquad (20.2)$$

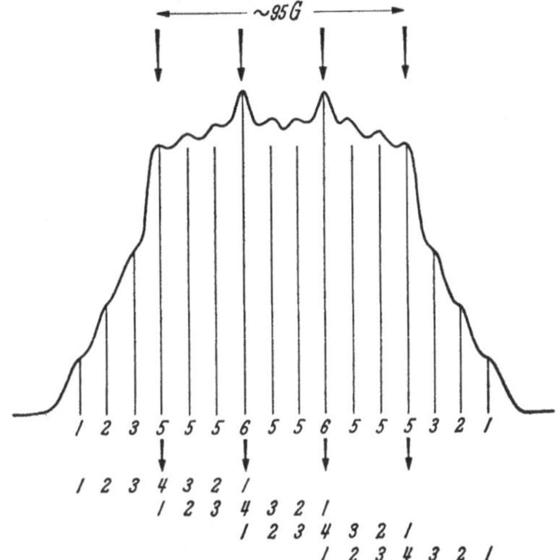

Fig. 20. Hyperfine structure caused by both Cl and Ir nuclei in the electron-magnetic resonance spectrum of $(NH_4)_2[Ir\,Cl_6]$ with H parallel to a ClIrCl axis. For this orientation the hyperfine structure arises only from the two Cl and the Ir nuclei which are on this axis. The positions of the four Ir components $(I = \frac{3}{2})$ are marked by the arrows above. Each of these components is further split into seven components by the two equally coupling Cl nuclei $(I_1 + I_2 = 3)$. The bars below give a complete reconstruction of the spectrum which consists of 4 sets of 7 components, neglecting small differences in coupling of the Cl^{37} and Cl^{35} isotopes. [From GRIFFITHS and OWEN: Proc. Roy. Soc. Lond., Ser. A **226**, 96 (1954).]

where the effective spin is $S' = \frac{1}{2}$. The terms A_x, A_y, and A_z represent the interaction of the Ir nucleus $(I = \frac{3}{2})$ with the electron spin. The terms in $A_{1'}$, $A_{2'}$, and $A_{3'}$ represent the interactions of the electron spin with the six halogen nuclei. For Cl and Br all stable isotopes have $I = \frac{3}{2}$. Hence for their complexes, I_1, I_2, etc., equal $\frac{3}{2}$. At any arbitrary orientation of such a complex in the magnetic field, the hyperfine structure of the resonance is very complex. However for a strong H parallel to any one of the cubic axes, hyperfine structure is observed for only the two halogens which lie on that axis. For H parallel to z and for the strong-field (BACK-GOUDSMIT) case,

$$E_{\parallel} = g_z \beta H M_S + A_z M_S M_I + A_{3'} M_S [(M_I)_3 + (M_I)_6], \qquad (20.3)$$

and

$$h\nu_{\parallel} = g_z \beta H + A_z M_I + A_{3'} [(M_I)_3 + (M_I)_6], \qquad (20.4)$$

where for an Ir complex with Cl or Br, M_I, $(M_I)_3$, and $(M_I)_6$ can take independently the values $\frac{3}{2}$, $\frac{1}{2}$, $-\frac{1}{2}$, $-\frac{3}{2}$.

Fig. 20 gives for a parallel orientation of H with x, y, or z the hyperfine structure for the $[IrCl_6]^{2-}$ complex in $(NH_4)_2[IrCl_6]$ diluted with the analogous salt

[1] J. H. E. GRIFFITHS and J. OWEN: Proc. Roy. Soc. Lond., Ser. A **226**, 96 (1954).

of platinum. The ratio of the observed coupling for a given halogen to the atomic coupling constant of the free atom gives a measure of the time the electron hole remains on the halogen and hence of the amount of π character in the bonding. If the overlap charge is neglected, the π character of each of the six bonds will be $\frac{1}{6}(1-\beta^2)$ for each Ir-halogen bond, where β^2 is a measure of the time the vacancy or hole is in the $d\varepsilon$ shell of the Ir ion and where $\beta^2 = 1$ for purely ionic bonding. From the hyperfine structure GRIFFITHS and OWEN thus obtained $\beta^2 = 0.74$ for both $[IrCl_6]^{2-}$ and $[IrBr_6]^{2-}$. The covalent character of the bonding also influences the g factor[1], and independent estimates from the effects on g give $\beta^2 = 0.66$, in fair agreement.

Table 9 gives the effective spins and g factors which have been found for the ions in the various crystals which have been investigated. Just as for the iron group, the g values are much nearer to the value for the free electron spin than to the LANDÉ g factor of the free ion.

Table 9. *Actual and effective spin values and range of g values found for ions of the palladium and platinum groups for which resonance has been observed.*
(From BOWERS and OWEN [*15*].)

Configuration	Ion	S	S'	g	r.m.s. g
$4d^1$	Mo^{5+}	$\frac{1}{2}$	$\frac{1}{2}$	1.94–1.96	1.95
$4d^3$	Mo^{3+}	$\frac{3}{2}$	$\frac{3}{2}$	1.90–2.0	1.95
$4d^5$	Ru^{3+}	$\frac{5}{2}$	$\frac{1}{2}$	1.0–3.24	1.9
$4d^9$	Ag^{2+}	$\frac{1}{2}$	$\frac{1}{2}$	2.04–2.18	2.1
$5d^3$	Re^{4+}	$\frac{3}{2}$	$\frac{3}{2}$		$\sim$1.8
$5d^5$	Ir^{4+}	$\frac{5}{2}$	$\frac{1}{2}$	0.75–2.20	1.8

21. The rare earths (Atomic numbers 58 to 70). Crystals of the rare earths typify the weak-field case of crystalline field splitting. The magnetic electrons are in the $4f$ shell, partially screened by the valence shell electrons from neighboring negative ions. Spin orbit coupling is stronger and crystalline field interaction weaker than for the d electron groups. Also, unlike the iron, palladium, or platinum groups, the salts of the rare earths do not have predominantly cubic crystalline fields but instead are often found to have axially symmetric fields. The spin orbit coupling is not broken down by the crystalline field, and the g factors found are usually very anisotropic. The $(2J+1)$ degeneracy of the free ion, however, is lifted either partly or completely by the crystalline fields. The crystalline field splitting of the J levels is usually of the order of 10^2 cm^{-1}, whereas the spin orbit coupling in the $4f$ shell is of the order of 10^3 cm^{-1}. When the ion is in an axially symmetric field—as it commonly is in the rare earth crystals—quantized components of $\boldsymbol{J}$ are resolved along this axis.

Because the magnetic moment which gives rise to the ZEEMAN splitting in the rare earth ions—except Eu^{2+} and Gd^{3+} which have S ground states—is coupled strongly to the lattice, the spin-lattice relaxation process is very efficient (Sect. 17). The lifetime in the magnetic states for this reason is very short and the resonance too broad for detection except at low temperatures where the lattice vibrations are frozen out. At liquid helium temperatures, where the measurements are usually made, only the lowest level of J_z is populated. Often the level is a KRAMERS' doublet corresponding to $M_J = \pm\frac{1}{2}$. In this case a single line corresponding to $M_J = \frac{1}{2} \rightarrow -\frac{1}{2}$ is observed, if nuclear splitting is neglected. To avoid the direct, dipole-dipole interactions between the magnetic ions which broaden the

---

[1] J. H. E. GRIFFITHS and J. OWEN: Proc. Roy. Soc. Lond., Ser. A **226**, 96 (1954).

lines so much in concentrated samples that the fine and hyperfine structure is not resolvable, the measurements are usually made in magnetically diluted samples. Furthermore, $g_{\parallel}$ and $g_{\perp}$ are usually very different in magnitude so that single crystals are also required for good resolution.

The principal rare earth crystals for which magnetic resonance has been investigated [14], [15] are the ethyl sulfates and double nitrates, both of which have crystalline fields of trigonal symmetry. The general formula for the ethyl sulfate is $M'''(C_2H_5SO_4)_3 \cdot 9H_2O$, in which M''' signifies the trivalent rare earth ion. The double nitrates are represented by $X_3''M'''(NO_3)_{12} \cdot 24H_2O$ in which X'' signifies a divalent cation and M''', the trivalent rare earth ion. Theory of the crystalline field splitting of the J levels in these crystals has been developed, mainly by ELLIOTT and STEVENS[1]. Except for the $4f^7$ ions Eu^{2+} and Gd^{3+} which have S states, the trigonal fields split the J_z levels of those rare earth ions which have odd numbers of electrons into a series of doublets spaced by the order of 100 cm^{-1} apart. According to KRAMERS' theorem this doublet degeneracy can only be lifted by an external field.

In cerium and samarium ethyl sulfates the doublet corresponding to $J_z = \pm\frac{1}{2}$ is lowest, although in the former the $J_z = \pm\frac{5}{2}$ lies only a few wave numbers above [14]. The lowest levels of neodymium sulfate and erbium ethyl sulfate are admixtures of $J_z = \pm\frac{5}{2}$ and $J_z = \pm\frac{7}{2}$ states. Thus $\Delta M = \pm 1$ transitions corresponding to magnetic dipole absorption are possible in the ground level of each of these salts. Although transitions $-\frac{5}{2} \to \frac{5}{2}$ or $-\frac{7}{2} \to \frac{7}{2}$ would violate the selection rules, the transitions between the $\pm\frac{5}{2}$ and $\pm\frac{7}{2}$ states are allowed.

In these salts the lowest doublet can be treated as a spin doublet with an effective spin $S' = \frac{1}{2}$, obtained by setting $(2S'+1)$ equal to the degeneracy of the level, Sect. 16. In a magnetic field the degeneracy is removed, and a resonance is obtained as a transition, $M = -\frac{1}{2} \to \frac{1}{2}$, between the two components of the "spin" doublet. Since there are no remaining degeneracies, there can be no fine structure of the usual type. The spin HAMILTONian which applies to this class

$$\mathcal{H} = \beta\left\{g_{\parallel}H_z S_z + g_{\perp}(H_x S_x + H_y S_y)\right\} + A S_z I_z + \left. + B(S_x I_x + S_y I_y) + P\left\{I_z^2 - \tfrac{1}{2}I(I+1)\right\}, \right\} \tag{21.1}$$

is obtained from Eq. (16.5) if the fine structure term, $D\left(S_z^2 - \tfrac{1}{3}S(S+1)\right)$, is dropped.

All stable isotopes of Ce are even-even and thus have $I = 0$. For the Ce^{3+} salts, for which the ground level is $J_z = \pm\frac{1}{2}$, the spin HAMILTONian reduces to:

$$\mathcal{H} = \beta\left\{g_{\parallel}H_z S_z + g_{\perp}(H_x S_x + H_y S_y)\right\}, \tag{21.2}$$

with effective spin $S' = \frac{1}{2}$ which has the energy eigen values:

$$E_H = g\beta H M_{S'}, \tag{21.3}$$

and a resonance frequency:

$$\nu = g\beta \frac{H}{h}, \tag{21.4}$$

where

$$g^2 = g_{\parallel}^2 \cos^2\vartheta + g_{\perp}^2 \sin^2\vartheta, \tag{21.5}$$

where ϑ is the angle between H and the crystalline axes. For this transition in cerium ethyl sulfate $g_{\parallel} = 0.955$ and $g_{\perp} = 2.185$ have been observed [14]. The same equations are applicable to the lower $\pm\frac{1}{2}$ doublets for isotopes of the other elements of this class which have $I = 0$.

[1] R. J. ELLIOTT and K. W. H. STEVENS: Proc. Phys. Soc. Lond. A **64**, 205, 932 (1951); **65**, 370 (1952). — Proc. Roy. Soc. Lond., Ser. A **215**, 437 (1952); **219**, 387 (1953).

When $I \neq 0$ and both the nuclear magnetic and nuclear quadrupole interactions are sufficiently large to be observable, as is true for the ethyl sulfate salts of Er^{167}, all terms in Eq. (21.1) must be employed, and calculation of the allowed energies is somewhat involved. However, if H is chosen along z and if the strong field case is assumed, the off-diagonal elements are avoided. The eigenvalue of S_z is M_S and that of I_z is M_I, and the energy eigenvalues, given by the diagonal elements $(M_S, M_I \mid \mathcal{H} \mid M_S, M_I)$, are:

$$E_\parallel = g_\parallel \beta H M_S + A M_S M_I + P [M_I^2 - \tfrac{1}{3} I(I+1)], \qquad (21.6)$$

from which the stronger components for this orientation can be obtained with the selection rules $M_S \rightarrow M_S + 1$, $M_I \rightarrow M_I$. They are:

$$h \nu_\parallel = g_\parallel \beta H + A M_I. \qquad (21.7)$$

For Er^{167}, $I = \tfrac{7}{2}$, and hence $M_I = \tfrac{7}{2}, \tfrac{5}{2}, \ldots - \tfrac{7}{2}$. The spectrum for the orientation $H \parallel z$ consists of a hyperfine structure with eight components equally spaced and equally intense. Non-zero changes in M_I can give rise to weak satellites of these lines. The spectra for salts of Nd^{143} and Nd^{145}, both of which have $I = \tfrac{7}{2}$, are similar. The solution for the perpendicular orientation is more involved, and the resonance pattern is more complex.

It is apparent that the nuclear spins can be learned from a count of the $(2I + 1)$ components of the hyperfine structure, and furthermore that the coupling constants A and P, from which the nuclear magnetic moment and nuclear quadrupole moments can in theory be computed, can be obtained from the spacings of the lines. The nuclear spins of Nd^{143}, Nd^{145}, Sm^{147}, Sm^{149}, and Er^{167} together with estimates of their magnetic and nuclear quadrupole moments were first obtained from this method [14].

When the number of electrons in the magnetic ions is even, it is possible for all the $(2J + 1)$ degeneracy to be lifted by the internal field, although the strictly trigonal field would still leave a series of doublets, as in the odd electron ions. Resonances have been found in ethyl sulfates of praseodymium and terbium. The lowest doublet in these was found to have a small initial splitting possibly caused by the JAHN-TELLER effect or by deviations from trigonal symmetry in the crystal. The spin HAMILTONian found to describe the resonance (excluding nuclear quadrupole effects) is:

$$\mathcal{H} = g_\parallel \beta H_z S_z + A S_z I_z + \Delta S_x \qquad (21.8)$$

with: for Pr^{3+}: $g_\parallel = 1.69$, $A = 0.083$ cm^{-1}, and $\Delta \sim 0.04$ cm^{-1}; and for Tr^{3+}: $g_\parallel = 17.72$, $A = 0.209$ cm^{-1} and $\Delta = 0.387$ cm^{-1}.

No resonance could be found in the ethyl sulfates of dysprosium and ytterbium even at low temperatures, and it is believed that the ground level is a $J_z = \pm \tfrac{3}{2}$ doublet (no transition allowed between its two components), with possible admixtures of other J_z values which differ by six [14].

The Gd^{3+} and Eu^{2+} ions have $^8S_{\frac{7}{2}}$ ground states, and in the first-order approximation their resonance is that of a free electron spin. Resonances of their salts are observable at room temperature with a g factor near 2.00 and almost isotropic. The spin degeneracy is 8. Various distortions of the ion can lift this spin degeneracy to produce a fine structure. These splittings, although small, are rather complex and will not be discussed here. BLEANEY and STEVENS[14] give the spin HAMILTONian which described the various distortions in gadolinium ethyl sulfate. Two stable isotopes, Gd^{155} and Gd^{157}, have $I \neq 0$, but no hyperfine structure was observed for them. Evidently there is no admixture involving s orbitals as there is in Mn^{2+} ions.

Table 10 lists the ground states and effective spin values of the rare earth ions for which resonances have been observed.

Table 10. *Effective spin values and other properties of rare earth ions for which resonance has been observed.*
(From Bowers and Owen [15].)

No. of of $4f$ electrons	Ion	Free ion ground state	S'	Remarks
1	Ce^{3+}	$^2F_{\frac{5}{2}}$	$\frac{1}{2}$	one or two Kramers doublets
2	Pr^{3+}	$^3H_4$	$\frac{1}{2}$	split doublet
3	Nd^{3+}	$^4I_{\frac{9}{2}}$	$\frac{1}{2}$	Kramers doublet
5	Sm^{3+}	$^6H_{\frac{5}{2}}$	$\frac{1}{2}$	Kramers doublet
7	Eu^{2+}, Gd^{3+}	$^8S_{\frac{7}{2}}$	$\frac{7}{2}$	
8	Tb^{3+}	$^7F_6$	$\frac{1}{2}$	split doublet
11	Er^{3+}	$^4I_{\frac{15}{2}}$	$\frac{1}{2}$	Kramers doublet

22. Actinide series (Atomic numbers 90 to 100). Less is known about the electronic configurations of the actinide than of those of the other transition groups, and paramagnetic resonance studies of them are only beginning. Nevertheless, these microwave studies offer promise of clearing up many of the uncertainties which exist and of providing knowledge of these important elements. The name "actinide group" originates from analogy with the rare earth series. Seaborg[1] in 1949 postulated that an incomplete $5f$ shell analogous to the rare earch $4f$ shell begins with actinium ($N = 89$), which is analogous to lanthanum at the beginning of the rare earth or lanthanide group. In 1952 Dawson[2] interpreted available data from magnetic susceptibility to indicate that the magnetic electrons of the actinide group are in the $6d$ shell when there are one, two, or perhaps three unpaired electron spins and in the $5f$ shell when there are more than three. Bleaney[3] in 1955 summarized the paramagnetic evidence along with the results which have been made on paramagnetic resonance. Table 11

Table 11. *Number of unpaired electrons in various actinide ions.*
[From B. Bleaney, Disc. Faraday Soc. **19**, 112 (1955).]

n	0	1	2	3	4	5	6	7
Th	Th^{4+}	(Th^{3+})	(Th^{2+})					
Pa	Pa^5	Pa^{4+}	(Pa^{3+})					
U	UO_2^{2+}	UO_2^+	U^{4+}	U^{3+}				
Np		NpO_2^{2+}	NpO_2^+	Np^{4+}	Np^{3+}			
Pu			PuO_2^{2+}	PuO_2^+	Pu^{4+}	Pu^{3+}		
Am				AmO_2^{2+}	AmO_2^+	(Am^{4+})	Am^{3+}	$? Am^{2+}$
Cm								Cm^{3+}
Bk							Bk^{3+}	Bk^{4+}
Cf						Cf^{3+}		
99					99^{3+}			
100				100^{3+}				

Remarks: Number n of unpaired electron spins in the actinide ions. () indicates observed only in the solid, not in solution. ? doubtful.

For the ions above curium the number of unpaired spins shown is based on the assumption that the electrons are in the $5f$-shell, so that the number n of unpaired spins is $(14-n')$, where n' is the number of magnetic electrons.

[1] G. Seaborg: Nucleonics **5**, 16 (1949).
[2] J. K. Dawson: Nucleonics **10**, 39 (1952).
[3] B. Bleaney: Disc. Faraday Soc. **19**, 112 (1955).

shows the number of unpaired electron spins in various ions of the series as summarized by BLEANEY. With reservations, those with $n=1$ and 2 may be regarded as $6d$, those with $n=4$ to 7 as $5f$ electrons. Unlike the rare earth elements, the actinide group tends to form very stable ionic complexes with oxygen such as $(UO_2)^{2+}$, which is known to be linear in structure. The complex $(UO_2)^{2+}$ is also non-magnetic, and it is believed to form dative covalent bonds such as to leave no unpaired spins in the complex[1]. The similar complexes of neptunium and plutonium $(NpO_2)^{2+}$ and $(PuO_2)^{2+}$ have one and two unpaired electrons, respectively (see Table 11). Paramagnetic resonance in both these complexes has been observed. That of $(PuO_2)^{2+}$ in $NaPuO_2(C_2H_3O_2)_3$, diluted with the diamagnetic analogue $NaUO_2(C_2H_3O_2)_3$, has been observed by HUTCHISON and LEWIS[2] at $4°$ K. They found $g_{\parallel} \approx 5.9$ and $g_{\perp} \approx 0$, measured with respect to the linear axis of $(PuO_2)^{2+}$, and interpreted their results to indicate that the unpaired electrons are $5f$ electrons. They also observed a doublet hyperfine structure to confirm the spin $\frac{1}{2}$ for the Pu nucleus. The $(NpO_2)^{2+}$ resonance has been observed by BLEANEY, LLEWELLYN, PRYCE, and HALL[3] in diluted single crystals in $Rb(NpO_2)(NO_3)_3$. Very large nuclear magnetic and nuclear quadrupole effects were detected from which the nuclear moments were evaluated. The ground level is KRAMERS doublet, and the observed spectrum could be interpreted in terms of the spin-HAMILTONian:

$$\mathscr{H} = g_{\parallel}\beta H_z S_z + g_{\perp}\beta(H_x S_x + H_y S_y) + A S_z I_z + \left. + B(S_x I_x + S_y I_y) + P[I_z^2 - \tfrac{1}{3}I(I+1)]. \right\} \tag{22.1}$$

With an effective spin $S' = \frac{1}{2}$ and $I = \frac{5}{2}$ (for Np^{237}), the experimentally determined parameters are[3]: $g_{\parallel} = 3.413$, $g_{\perp} = 0.21$, $A = 0.166$ cm^{-1}, $B = 0.0175$ cm^{-1}, and $P = -0.030$ cm^{-1}. This value of P corresponds to $eQq = -12080$ Mc/sec, which is an extremely large nuclear quadrupole coupling. From this coupling EISENSTEIN and PRYCE[4] estimate $Q \approx 15\times10^{-24}$ cm^2 for Np^{237}.

Resonance of powdered UF_3 by GHOSH, GORDY, and HILL[5] yields: $g_{\parallel} \approx 2.8$ and $g_{\perp} \approx 2.1$. The theoretical work of O'BRIEN[6] shows that these values are most nearly in harmony with the $5f^3$ configuration for the unpaired electrons of the U^{3+} ion.

The magnetic resonance results so far obtained tend to support the $5f$ configuration[1] for ions with only one, two, or three unpaired spins. Thus it appears that the magnetic electrons may be $5f$ electrons for the entire series, as postulated by SEABORG, but the electron resonance results are not yet sufficiently complete to be decisive.

23. Conductors and semi-conductors. The valence electrons which give rise to observable microwave resonances in conductors and semi-conductors are not localized on individual ions or groups as are the magnetic electrons of salts of the transition elements but must be regarded as in molecular-type orbitals which spread over many atoms. The atoms of the conducting solid together form energy bands, the outermost (conduction bands) of which are only partially filled with electrons. The bands below the conducting ones are completely filled

---

[1] B. BLEANEY: Disc. Faraday Soc. **19**, 112 (1955).

[2] C. A. HUTCHISON and W. B. LEWIS: Phys. Rev. **95**, 1096 (1954).

[3] B. BLEANEY, P. M. LLEWELLYN, M. H. L. PRYCE and G. R. HALL: Phil. Mag. **45** a 992, b 991 (1954).

[4] J. C. EISENSTEIN and M. H. L. PRYCE: Proc. Roy. Soc. Lond., Ser. A **229**, 20 (1955).

[5] S. N. GHOSH, W. GORDY and D. G. HILL: Phys. Rev. **96**, 36 (1954).

[6] M. C. M. O'BRIEN: Proc. Phys. Soc. Lond. A **68**, 351 (1955).

and are non-magnetic, according to the requirements of the exclusion principle. The outer, partially filled bands, are responsible not only for normal conduction but also for the cyclotron resonance motions and for the magnetic susceptibility which gives rise to observable electron spin resonance. The looseness of the unfilled bands allows the electron to occupy separate spatial orbitals so that the spins can flip over in an imposed magnetic field and absorb radiation without violation of the exclusion principle.

α) *Electron spin resonance.* The primary difficulty in observation of electron spin resonance in conductors and semiconductors is the shielding properties of the conducting electrons which tend to prevent penetration of the microwave power into the metal (the well known skin effect). GRISWOLD, KIP, and KITTEL[1] circumvented the skin effect by grinding the conductors into granules, ~ 10 microns or less in diameter, and suspending them in a non-conducting medium (paraffin wax). They were then able to observe electron spin resonance at 9240 Mc per sec in metallic sodium. They found a g value between 1.998 and 2.004, which with the limits of error is that of the free electron. Thus it must be concluded that the spin orbit interaction is extremely small, as would be expected since the conducting electrons are in delocalized orbitals which are constructed from combinations of atomic s orbitals of the sodium. However, the line width observed (78 gauss) is much greater than would be expected for the free electron model,

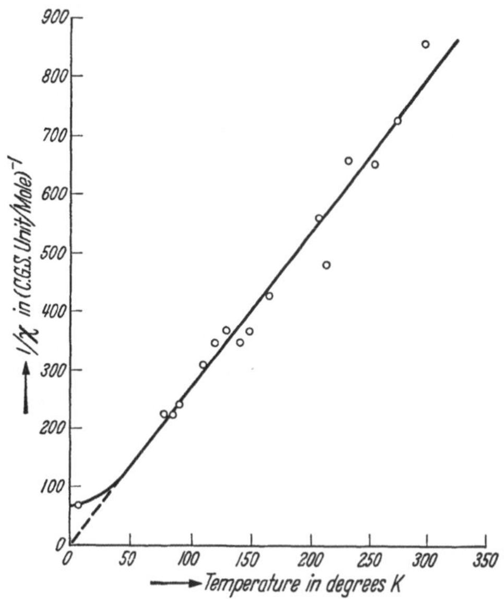

Fig. 21. Plot of reciprocal of paramagnetic susceptibility of Si semi-conductor as a function of temperature. The susceptibility at the various temperatures was obtained from strengths of the electron spin resonance at 9000 Mc/sec. [From PORTIS, KIP, KITTEL and BRATTAIN: Phys. Rev. **90**, 988 (1953.)]

in which the spin-lattice relaxation time would be long and the spin-spin broadening (including interactions with the nuclear spin) would be averaged out by the motions of the electrons in the conducting band. GRISWOLD, KIP, and KITTEL suggest that the dominant broadening mechanism here may be the random phase and frequency modulation of the microwave field as observed from coordinates moving with a conducting electron diffusing about in the eddy current field of the metal.

PORTIS, KIP, KITTEL, and BRATTAIN[2] applied the fine particle method described above to semiconducting silicon and germanium crystals with small amounts of phosphorus added ($\sim 2 \times 10^{18}$ P atoms per cm³). They observed a strong spin resonance of the conducting electrons with $g = 2.001 \pm 0.001$, independent of T between 4 and 300° K. Again the line width appeared to be anomalous. The reciprocal of the magnetic susceptibility of the conducting electrons as determined by the strength of the resonance is plotted as a function of temperature in Fig. 21. Note the point where the curve departs from the straight line expected for LANGEVIN or non-degenerate paramagnetism and follows that for PAULI or

---

[1] T. F. GRISWOLD, A. F. KIP and C. KITTEL: Phys. Rev. **15**, 951 (1952).
[2] A. M. PORTIS, A. F. KIP, C. KITTEL and W. H. BRATTAIN: Phys. Rev. **90**, 988 (1953).

degenerate paramagnetism. A degeneracy temperature, $T_0 = 38°$ K, was obtained from these measurements.

Nuclear hyperfine structure of As^{75} and of P^{31} has been observed by FLETCHER et al.[1] at low temperature, 4° K, in the electron spin resonance at 24000 Mc/sec of specially treated single crystals of silicon which contained the P or As atoms as impurities. To produce the hyperfine structure the samples were plastically deformed by compression at 1000° C. Four equally spaced, equally intense components corresponding to the $(2I + 1)$ orientations of the As^{75} nucleus $(I = \frac{3}{2})$ with component spacing of 73 gauss were observed for the As doped samples. Two components separated by 42 gauss were observed for P^{31}, for which $I = \frac{1}{2}$. No hyperfine structure could be observed for either As or P in similar samples which had not been deformed, and none could be observed for B or Al with these elements as impurities with the sample deformed or not. Thus it seems reasonable that P or As may have a free valency in the deformed samples. The unpaired electron must be essentially localized on a single P or As atom since a more complex hyperfine structure or no resolvable structure would be observed if the unpaired spin were to shift from one coupling atom to another at a rate fast as compared with the lifetime in the ZEEMAN states. Nothing is said about orientation dependence of the structure, but in the deformed samples all angles of the deformed axis must be present. Hence one might expect that any anisotropic terms in the nuclear coupling would mainly broaden the components and that the isotropic s-orbital coupling would be responsible for their separation. It is of interest that the 42 gauss splitting of the P^{31} doublet is about twice that found by DEHMELT (Sect. 14) in the similar resonance (strong field case) of gaseous free atoms of P^{31}. In the latter, the splitting must arise from s-orbital contribution to the ground configuration. Only a small $3s$ contribution would be required to account for the splitting in either case. Atoms of P or As in the deformed samples may employ four of their electrons and the s and p orbitals (or hybrids of them) mainly for bonding (to behave like Si), and thus the fifth or unpaired electron would be forced into the higher d or s orbitals or into a hybrid of these with perhaps small contributions of the lower s and p. Such an atomic orbital would presumably be lower than the conductivity band only for the deformed samples under the conditions of the observation.

β) *Cyclotron resonance.* Cyclotron resonance in solids, sometimes called diamagnetic resonance, was first detected in 1953 by DRESSELHAUS, KIP, and KITTEL[2] in germanium semi-conductor at 9050 Mc/sec and $T = 4°$ K. It has since been detected at other frequencies in other semi-conductors or conductors including silicon crystals[3] and bismuth metal[4]. Such resonances have been observed for both p and n type semiconductors—resonances of holes as well as of electrons. Fig. 22 shows the resonances for Si semi-conductor at 24 kMc/sec. Microwave cyclotron resonance of electrons in gaseous discharge has also been observed[5].

Cyclotron resonance is different from electron spin resonance in that it arises from the spiraling orbital motions of the conducting electron in a static magnetic

[1] R. C. FLETCHER, W. A. YAGER, G. L. PEARSON, A. N. HOLDEN, W. T. READ and F. R. MERRITT: Phys. Rev. **94**, 1392 (1954).

[2] G. DRESSELHAUS, A. F. KIP and C. KITTEL: Phys. Rev. **92**, 827 (1953).

[3] G. DRESSELHAUS, A. F. KIP and C. KITTEL: Phys. Rev. **98**, 368 (1955).

[4] J. K. GALT, W. A. YAGER, F. R. MERRITT, B. B. CETLIN and H. W. DAIL: Phys. Rev. **100**, 748 (1955).

[5] R. N. DEXTER and B. LAX: Phys. Rev. **100**, 1216 (1955). — J. W. BENSON: J. Appl. Phys. **23**, 757 (1952). — R. V. JONES, W. DOBROWOLSKY, W. B. KUNKEL and C. D. JEFFRIES: Phys. Rev. **99**, 646 (1955). — M. GILDEN and L. GOLDSTEIN: Phys. Rev. **100**, 1233 (1955).

field and not from spin motions. The name comes from the close analogy with the motions of electrons or ions in an ordinary cyclotron. Coupling to the radiation field occurs through the electric rather than the magnetic component of the radiation. The transitions are thus electric dipole transitions and not magnetic dipole transitions as for electron spin resonance. Transition probabilities for cyclotron resonance are therefore orders of magnitude greater than for spin resonance (although this should not be taken to mean that the former are necessarily the easier to observe). If magnetic resonance of any kind is to be detected, the static magnetic field must be arranged normal to the particular coupling vector of the radiation field. Thus a different orientation of H relative to the absorption cavity is required for observation of the two kinds of resonances.

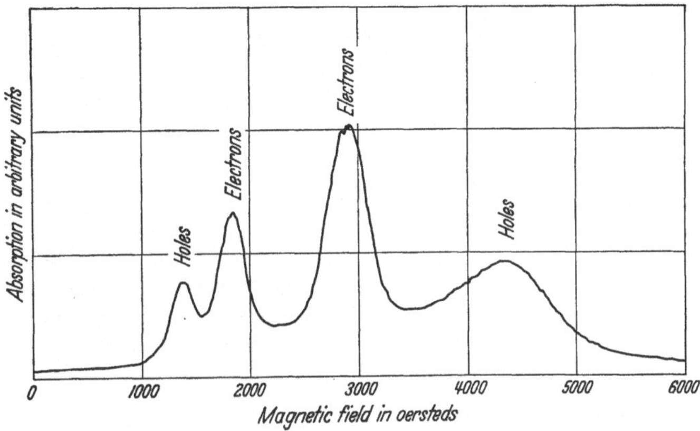

Fig. 22. Cyclotron resonances at 24 kMc/sec of electrons and holes in Si semi-conductor. [From DRESSELHAUS, KIP and KITTEL: Phys. Rev. 98, 373 (1955).]

Classically, cyclotron resonance occurs when the frequency of rotation of the electron in the magnetic field is equal to the frequency of the microwave radiation. This frequency is easily found to be:

$$\nu = \frac{eH}{2\pi m^* c},\qquad(23.1)$$

by analogy with the frequency in an ordinary cyclotron. However, m^* is not the actual mass of the free electron but is its effective mass in the conduction band. The use of an effective (fictitious) mass here is somewhat like the use of an effective (fictitious) spin moment in the spin-HAMILTONian for calculation of spin resonance frequencies (Sect. 16).

Measurement of cyclotron resonant frequencies in solids provides a direct means of evaluation of effective masses m^* of the conduction electrons. These masses depend upon the nature of the energy surfaces within the solid and are related to a number of other properties of the conductor. Resonances were observed[1] at 9050 Mc/sec in n-type germanium crystals at 370 gauss. This corresponds to an effective mass $m^* = 0.11\,m$ where m is the normal mass of the electron. Two resonances were observed[1] at 9050 Mc/sec in p-type germanium crystals, one at 125 gauss with 50 gauss half-width and one at 970 gauss with 100 gauss half-width. They gave $m^* = 0.04\,m$ and $m^* = 0.30\,m$, respectively. The positions of the resonances were found to be independent of orientation of the crystal in the field. This observation was later confirmed by LAX *et al.*[2] who

[1] G. DRESSELHAUS, A. F. KIP and C. KITTEL: Phys. Rev. 92, 827 (1953).

[2] B. LAX, H. J. ZEIGER, R. N. DEXTER and E. S. ROSENBLUM: Phys. Rev. 93, 1418 (1954). — R. N. DEXTER, H. J. ZEIGER and B. LAX: Phys. Rev. 95, 557 (1954).

showed, however, that the resonance for n-type germanium crystals is orientation dependent.

Theoretical discussions of possible cyclotron resonance were made by DORFMANN[1], by DINGLE[2], and by SHOCKLEY[3] before such resonances were detected. The second paper by DRESSELHAUS, KIP, and KITTEL[4] gives a theoretical treatment as well as more complete experimental results on Si and Ge semi-conductors. The theory of the resonance in metallic Bi is developed by LAX et al.[5].

24. Ferromagnetic and anti-ferromagnetic resonance. Electron spin resonance in ferromagnetic substances was first observed by GRIFFITHS[6] after the possibility of such resonance was theoretically predicted by LANDAU and LIFSHITZ[7]. It has since been observed by YAGER and BOZORTH[8] and others.

The observed g factor as calculated with Eq. (15.2) depends upon the shape of the specimen and is often several times greater than that for the free electron spin. KITTEL[9] has shown that if correction is made for the strong magnetization of the specimen, a g factor very close to that for the free electron spin is always obtained. The magnetization and particular formula for the resonance frequency depend on the shape of the sample. For a thin plane surface with the applied H and the magnetic component of the radiation both parallel to the plane of the specimen, the resonance frequency can be expressed as:

Table 12. *Comparison of representative values of spectroscopic splitting factor (g) and magnetomechanical ratio (g′) for ferro magnetic substances.*
[From C. KITTEL: Phys. Rev. **76**, 743 (1949).]

	Microwave resonance g	Gyromagnetic experiments, g'
Iron	2.12–2.17	1.93
Cobalt . . .	2.22	1.87
Nickel . . .	2.19–2.42	1.92
Magnetite . .	2.20	1.93
HEUSLER alloy	2.01	2.00
Permalloy . .	2.07–2.14	1.91

$$\nu = \frac{g\beta\sqrt{BH}}{h} \qquad (24.1)$$

where $\boldsymbol{B} = \boldsymbol{H} + 4\pi\,\boldsymbol{M}$ is the magnetic induction and $\boldsymbol{M}$ is the magnetization.

Some spectroscopic splitting factors, g, obtained from microwave resonances with Eq. (24.1) are compared in Table 12 with the gyromagnetic ratios, g', obtained in the usual gyromagnetic experiments. The small differences in g and g' are attributed by KITTEL to spin-lattice interactions.

In both ferromagnetic and anti-ferromagnetic substances there is a strong exchange interaction between neighboring magnetic elements. In ferromagnetic substances the exchange integral is positive, and the exchange forces tend to align the spins parallel. This is the quantum mechanical basis for the WEISS field postulated to align the spins of ferromagnets. In anti-ferromagnetic substances the exchange integral is negative, and hence the exchange forces tend to align the electronic spins on alternate elements in an anti-parallel manner. If a given spin flips over (changes orientation) without a compensating change in orientation by a neighboring spin, there is an increase in the energy of the spin system which must be supplied from outside itself. When this energy, which is a measure of the exchange interaction, is of the magnitude of a microwave quanta, microwave radiation will be absorbed. This resonant absorption

[1] J. DORFMANN: Dokl. Acad. Sci. USSR. **81**, 795 (1951).
[2] R. B. DINGLE: Proc. Roy. Soc. Lond., Ser. A **212**, 38 (1952).
[3] P. W. SHOCKLEY: Phys. Rev. **90**, 491 (1953).
[4] G. DRESSELHAUS, A. F. KIP and C. KITTEL: Phys. Rev. **98**, 368 (1955).
[5] B. LAX, K. J. BUTTON, H. J. ZEIGER and L. M. ROTH: Phys. Rev. **102**, 715 (1956).
[6] J. H. E. GRIFFITHS: Nature, Lond. **158**, 670 (1946).
[7] L. LANDAU and E. LIFSHITZ: Phys. Z. Sowjet. **8**, 153 (1935).
[8] W. A. YAGER and R. M. BOZORTH: Phys. Rev. **72**, 80 (1947).
[9] C. KITTEL: Phys. Rev. **73**, 155 (1948); **76**, 743 (1949). — J. Phys. Radium **12**, 291 (1951).

of radiation by the internally coupled, anti-parallel electron spin system is called anti-ferromagnetic resonance. The exchange forces can be considered equivalent to an effective internal magnetic field H_i acting on a given electron spin moment. This H_i is independent of any externally applied field, and therefore anti-ferromagnetic resonance is observable without the application of an external H. The latter can be used, however, for displacing further the initially split spin levels and thus moving the resonance frequency to a more desirable observation frequency. The exchange interaction in most antiferromagnetic substances is strong, and the zero field resonances are expected to fall in the shorter millimeter or sub-millimeter wave region. It has, however, been observed[1] at radiofrequencies for $CuCl_2 \cdot 2H_2O$ at very low temperatures. For MnF_2 antiferromagnetic resonance[2] has been observed in the 125 kMc/sec region at liquid air temperature.

The theory of anti-ferromagnetic resonance[3] is developed by KITTEL and KEFFER, who show that for cubic or axially symmetric crystals at 0° K the resonance frequency in an externally applied field H should be:

$$h\nu = g\beta \left[H_0 \pm \sqrt{H_A(2H_E + H_A)} \right], \qquad (24.2)$$

where H_E is the effective exchange field (WEISS field) and H_A is the anisotropic field. For the most common anti-ferromagnetic substances, which have curie temperatures of the order of 100° K, KEFFER and KITTEL estimate $H_E \sim 10^6$ gauss, $H_A \sim 10^3$ gauss, and $\nu \sim 150$ kMc/sec with $H_0 = 0$.

25. Free radicals. Chemical groups with an odd number of electrons have an unpaired spin moment which can give rise to electron spin resonance. Such radicals with a free valence are generally unstable, but many can be chemically prepared which are stable for weeks, months, or even longer. The more stable radicals are usually large organic species which are stabilized by quantum mechanical resonance. Bi-radicals having two unpaired spins and a triplet electronic ground state can be produced with an even number of electrons[4].

Simple radicals such as atomic hydrogen, CH_3, $(C_2H_4)^+$, C_2H_5, etc., are apparently too reactive and too mobile to remain long enough in liquid solution for detectable quantities to be built up easily by irradiation or by chemical means. However, in solids subjected to ionizing irradiation at liquid air temperature and in many instances at room temperature such radicals have been detected from their electron spin resonance and identified with the proton hyperfine structure of this resonance. Table 13 includes simple ones which contain only C and H, which are identified with the patterns of their microwave resonance. Fig. 23 shows the proton hyperfine structure of a more complex radical produced by X-irradiation of solid acetyl valine. This 18-component structure may be regarded as a set of 9 doublets arising from equal coupling of the electron spin to 8 proton spins with a coupling of about half as much to a ninth. Surprisingly, certain of these radicals are stable even at room temperature, in many solids for weeks or months. Although the simplest radical of all, the mobile hydrogen atom, has not been detected in solids at room temperature, it has been observed in at least three gamma-irradiated solids at 77° K by LIVINGSTON, ZELDES, and

[1] N. J. POULIS, J. VAN DEN HANDEL, J. UBBINK, J. A. POULIS and C. J. GORTER: Phys. Rev. **82**, 552 (1951). — J. UBBINK, N. J. POULIS, A. N. GERRITSEN and C. J. GORTER: Physica, Haag **18**, 361 (1951).

[2] F. M. JOHNSON and A. H. NETHERCOT jr.: Bull. Amer. Phys. Soc. **1**, 283 (1956).

[3] C. KITTEL: Phys. Rev. **82**, 565 (1951). — F. KEFFER and C. KITTEL: Phys. Rev. **85**, 329 (1952). — F. KEFFER: Phys. Rev. **87**, 608 (1952). See also T. NAGAMIYA: Progr. Theoret. Phys. **6**, 350 (1951).

[4] Observation of spin resonance in bi-radicals has been reported by H. S. JARRETT, G. J. SLOAN and W. R. VAUGHAN: J. Chem. Phys. **25**, 697 (1956).

Table 13. *Some hydrocarbon radicals produced in solids by irradiation.*

Number of components	Relative intensity (theoretical)	Multiplet spread in gauss	Probable radical	Some solids in which the multiplet has been produced by X-irradiation
2	1, 1	504	H	H_2SO_4, H_3PO_4, $HClO_4$ [1]
3	1, 2, 1	30—45	$(CH_2)^+$ or (RCH_2) Where R has no coupling nuclei	acetic acid [2], acetone [2], acetanilide [3], glycine [4], methyl alcohol [3], sodium methoxide [3], guanalitic acid [5], acetamide [3]
4	1, 3, 3, 1	70—80	CH_3 or $(RCH_3)^+$	propionic acid [2], palmitic acid [2], zinc dimethyl [6]
5	1, 4, 6, 4, 1	90—110	$(C_2H_4)^+$	alanine [2], valine [2], glutamic acid [2], mercury dimethyl [6], ethyl alcohol [3], lactic acid [2], alanyl alanine [7], propionamide [3]
6	1, 5, 10, 10, 5, 1	110—130	C_2H_5	mercury diethyl [6], acetyl alanine [7]
7	1, 6, 15, 20, 15, 6, 1	150—160	$(C_2H_6)^+$ or $RC(CH_3)_2$	valine [2], isoleucine [8]

[1] R. LIVINGSTON, H. ZELDES and E. H. TAYLOR: Phys. Rev. **94**, 725 (1954). — Disc. Faraday Soc. **19**, 166 (1955).

[2] W. GORDY, W. B. ARD and H. SHIELDS: Proc. Nat. Acad. U.S.A. **41**, 983 (1955); **41**, 996 (1955).

[3] C. F. LUCK and W. GORDY: J. Amer. Chem. Soc. **78**, 3240 (1956).

[4] J. COMBRISSON and J. UEBERSFELD: C. R. Acad. Sci. Paris **258**, 1397 (1954).

[5] W. GORDY and H. SHIELDS: Bull. Amer. Phys. Soc. **1**, 267 (1956).

[6] C. G. McCORMICK and W. GORDY: J. Amer. Chem. Soc. **78**, 3243 (1956).

[7] W. GORDY and C. G. McCORMICK: Bull. Amer. Phys. Soc. **1**, 200 (1956).

[8] W. GORDY and H. SHIELDS: Bull. Amer. Phys. Soc. **1**, 267 (1956).

TAYLOR (see Sect. 14). Its doublet hyperfine structure was found to have a spacing of 504 gauss for the strong-field case, almost exactly that for free H atoms in the gas (Sect. 14). The other radicals listed in Table 13 have been postulated to account for numerous resonances detected in a variety of X-irradiated substances by our Duke University group. Their identification from their theoretically predicted proton hyperfine patterns is not as certain as that for the H atom because the corresponding patterns for the gaseous state have not yet been observed. Other radicals produced by ionizing radiation will be treated in Sect. 26.

There are three features which characterize spin resonance in organic free radicals. (1) The g factor is always very nearly isotropic and very close to the free electron spin value of 2.0023. In a recent survey WURTZ [22] lists some 40 organic free radicals which have isotropic g factors measured to 4 or 5 significant figures. These all fall between 2.001 and 2.0064. (2) When the radicals are concentrated, the resonances are usually sharp and without structure. Some have line widths of less than a gauss. (3) In the highly diluted solutions (solid or liquid) nuclear hyperfine structure arising from more than one nucleus is often observed.

All three of these features could be attributed directly or indirectly to the delocalization of the orbital wave function associated with the unpaired spin. The migration of the unpaired electron over several atoms tends to average out the orbital moment and to leave the electron spin free to precess about the applied field. Exchange interaction between the radicals in the concentrated samples (which depends upon the spreading of the odd electron wave function) gives rise to motional narrowing (exchange narrowing) and the averaging out of the nuclear interactions (see Sect. 17) to leave only a single, sharp resonance, as mentioned under (2). By dilution of the radical in a suitable, non-magnetic substance—liquid or solid—the exchange interaction can be broken down. Radicals produced by irradiation are conveniently diluted when produced. Dilution also eliminates the spin-spin broadening which would occur in the concentrated

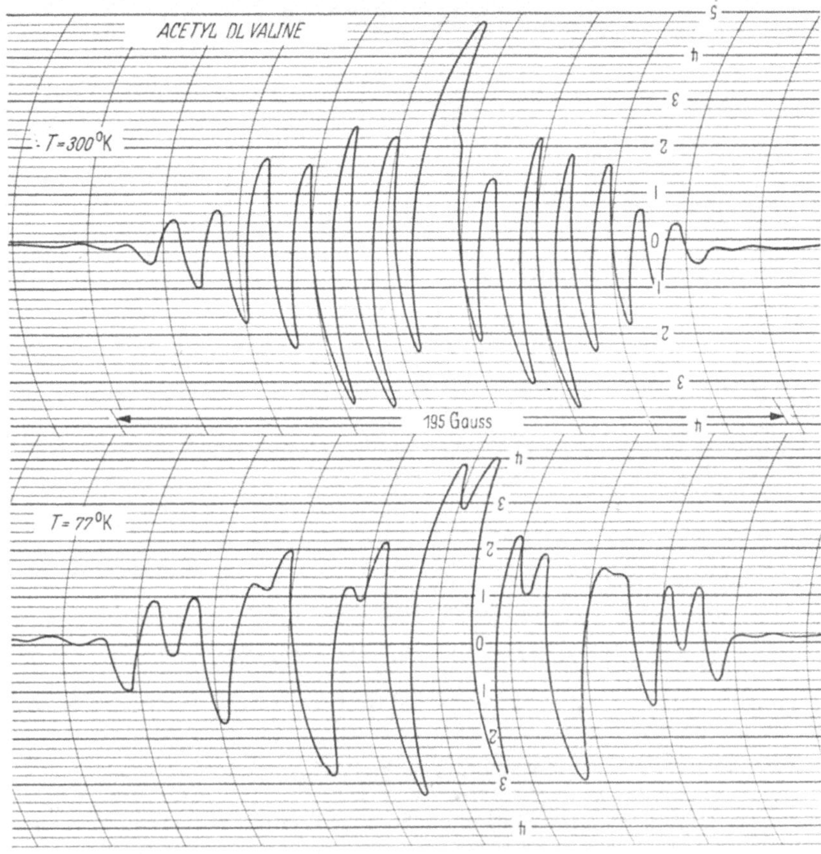

Fig. 23. Proton hyperfine structure of the electron-spin resonance of X-irradiated acetyl valine in the polycrystalline form (From G. McCormick and W. Gordy.)

solids without exchange interaction. In the samples thus diluted, the observed hyperfine structure gives evidence of the spreading of the orbital wave function associated with the free spin.

Fig. 24 shows the hyperfine structure observed for the triphenyl methyl radical in dilute liquid solution. Since the only abundant isotopes with non-zero spin in this radical are the hydrogens and since a single hydrogen ($I = \frac{1}{2}$) can give only a doublet, this complex pattern proves that the electron is not localized but that it migrates around the carbon rings and spends some time in the vicinity of the several H atoms. The tumbling motions in the liquid solution for which this observation was made would average out any anisotropic $I \cdot S$ interactions. Thus the unpaired electron seems to be some of the time in the s orbitals of the various hydrogens.

Relatively simple theory will acount very closely for the electron spin resonance frequencies (including the hyperfine structure) of most organic free radicals so far investigated. The effective spin of the odd radicals is $S = \frac{1}{2}$. There is no fine structure, and the g factors and nuclear hyperfine structure are isotropic or nearly so. The spin Hamiltonian (Sect. 16) for the usual organic free radical is:

$$\mathscr{H} = g\beta \, \boldsymbol{H} \cdot \boldsymbol{S} + \sum_i A_i \, (\boldsymbol{S} \cdot \boldsymbol{I}_i), \tag{25.1}$$

which has the energy eigen values:

$$E = g\beta H M_S + \sum_i A_i M_S M_{I_i}. \tag{25.2}$$

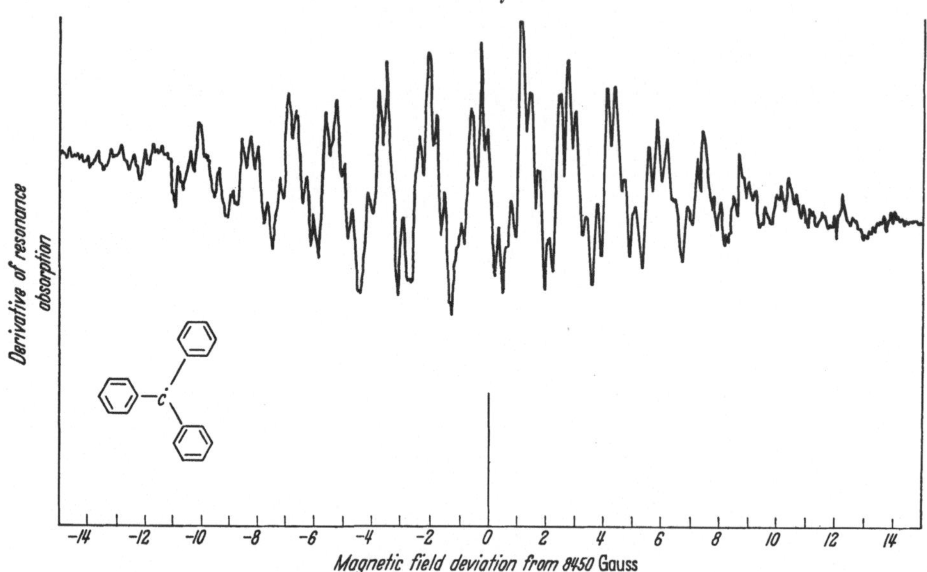

Fig. 24. Proton hyperfine structure of the electron-spin resonance of the triphenyl methyl radical in dilute solution in benzene. [From Jarrett: J. Chem. Phys. 21, 761 (1953).]

For the transitions $(M_S \to M_{S'} \pm 1)$ $(M_I \to M_I)$ for the strong field (Back-Goudsmit) case it yields the frequencies

$$h\nu = g\beta H + \sum_i A_i M_{I_i}. \tag{25.3}$$

The subscript i refers to a particular nucleus with coupling to the electron spin. The summation is taken over all such nuclei in the radical. In general there will be $\sum_i (2I_i + 1)$ components corresponding to all values of $M_{I_i} = I_i, I_i - 1, \ldots -I_i$.

When the coupling constant for all nuclei is the same ($A = A_1 = A_2 = A_3 \ldots$) Eq. (25.3) can be written:

$$h\nu = g\beta H + A \sum_i M_{I_i}. \tag{25.4}$$

If in addition the spins of the equally coupling nuclei are also identical ($I_1 = I_2 = I_3 \ldots$ as well as $A_1 = A_2 = A_3 \ldots$), one can take the total spin $T = nI$, where n is the number of coupling nuclei, and define $M_T = T, T-1, T-2, \ldots -T$ For these conditions, which hold for symmetrical radicals such as those in Table 13, Eq. (25.3) becomes:

$$h\nu = g\beta H + A M_T. \tag{25.5}$$

and the hyperfine structure consists of $(2T + 1)$ or $(2nI + 1)$ components, equally spaced but not equally intense. In Eq. (25.5) A is measured by the component separation and is the magnetic coupling constant for the individual nucleus as defined in Sect. 13.

The relative intensities of the components of a BACK-GOUDSMIT hyperfine structure (strong field case) arising from one or more nuclei such as those described above can be calculated from simple probability theory on the assumption that all permitted orientations of the nuclei (all values of M_I and M_{I_i} but not M_T) are equally probable. This assumption is good to a very high approximation because the BOLTZMANN factor which gives the difference in population of the nuclear orientation levels in the applied field is small indeed. The strengths of the different lines are proportional to the number of different combinations of the M_{I_i} values which give the same value for the nuclear interaction field on the electron spin. For example, if there are two equally coupling hydrogens ($I = \frac{1}{2}$), there will be four equally probable combinations: ↑↑, ↓↑, ↑↓, and ↓↓. The two center combinations give the same (zero) interaction with S and lead to a central component of weight 2. The outside components corrsponding to ↑↑ and ↓↓ can be formed in only one way each and have a weight of 1. The relative intensities of the resulting triplet will be 1, 2, 1. If, on the other hand, the three components are joined by a single nucleus with $I = 1$ (such as deuterium), the three allowed orientations giving $M_I = 1$, 0, and 1 are equally probable, and the three components are equally intense. From similar reasoning, the four equally coupled nuclei (with $I = \frac{1}{2}$ and $T = 2$) would give a symmetrical quintet hyperfine structure with relative intensities 1, 4, 6, 4, 1. Two equally coupled nuclei (with $I = 1$ and $T = 2$) would also give a symmetrical quintet hyperfine structure, but with the different relative intensities 1, 2, 3, 2, 1. Relative intensities for certain other combinations are given in Table 13.

Although anisotropy in the g factor of organic free radicals is very small, perhaps immeasurably small for most of them, measurements made in the shorter millimeter wave region with its accompanying high fields permit determination of the anisotropy for many radicals[1]. The resonant condition, $\nu = g\,\beta\,\dfrac{H}{h}$, shows that the ΔH caused by a small difference ($g_1 - g_2$) is proportional to the observation frequency. Hence a separation of the resonances for an orientation difference in g will be 10 times greater at 100 kMc/sec (3 mm wavelength) than at 10 kMc/sec (3 cm wavelength). Single crystals of the three organic radicals listed in Table 14 were measured at 75 kMc/sec ($\lambda = 4$ mm). The first two radicals were found to be axially symmetric, with the axis of symmetry probably along the NN bond of the diphenyl picryl hydrazyl and along the NO bond in the di-p-anisyl nitrogen oxide. No hyperfine structure is observable in the concentrated form. The HAMILTONian which describes the resonance in each of these crystals is:

$$\mathcal{H} = \beta\,[g_{\parallel} H_z S_z + g_{\perp}(H_x S_x + H_y S_y)], \qquad (25.6)$$

with $S = \frac{1}{2}$, and with $g_{\parallel}$ and $g_{\perp}$ as listed. The unpronouncable radical given in the last row was found to have lower symmetry. Its resonance is described by the spin HAMILTONian:

$$\mathcal{H} = \beta\,(g_z H_z S_z + g_x H_x S_x + g_y H_y S_y), \qquad (25.7)$$

with $S = \frac{1}{2}$ and with the values of g_x, g_y, and g_z as listed.

Earlier experiments on diphenyl-picryl hydrazyl[2,3] in the centimeter wave region revealed an anisotropy which, it was suggested[3], might arise from the

---

[1] A. VAN ROGGEN, L. VAN ROGGEN and W. GORDY: Bull. Amer. Phys. Soc. 1, 215 (1956). — Phys. Rev. 105, 50 (1957).

[2] C. A. HUTCHISON jr., R. C. PASTOR and A. G. KOWALSKY: J. Chem. Phys. 20, 534 (1952).

[3] C. KIKUCHI and W. V. COHEN: Phys. Rev. 93, 394 (1954).

Table 14. *Spectroscopic splitting factors, g, for some single crystals of organic free radicals.* [From VAN ROGGEN, VAN ROGGEN and GORDY: Phys. Rev. **105**, 50 (1957).]

Radical	g factor
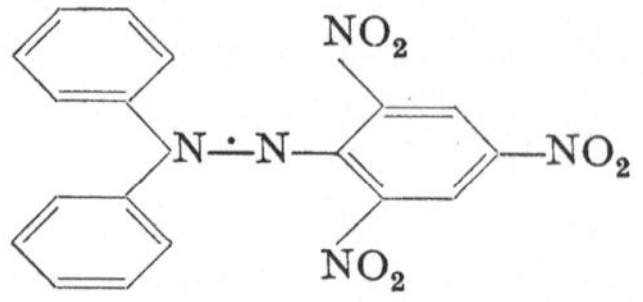 1.1-diphenyl-2-picryl hydrazyl	$g_\parallel = 2.0035$ $g_\perp = 2.0043$
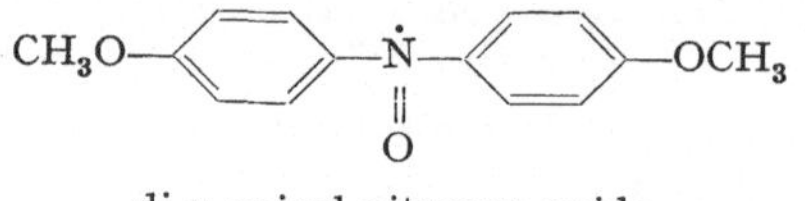 di-p-anisyl nitrogen oxide	$g_\parallel = 2.0095$ $g_\perp = 2.0035$
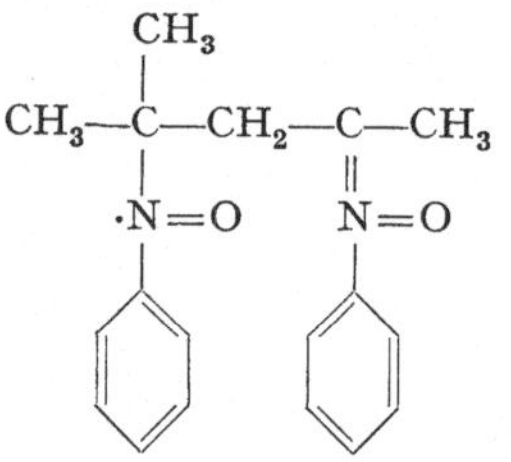 2-(phenyl nitrogen oxide)-2-methyl pentane-4-one-oxime-N-phenyl ether	$g_x = 2.0042$ $g_y = 2.0064$ $g_z = 2.0083$

diamagnetism of the attached ring system. However, the degree of the anisotropy seems too large for it to be attributed entirely to this source. The same is true for the other radicals of Table 14. The anisotropy in all three is believed to arise primarily from a residual spin orbit coupling[1]. In diphenylpicryl hydrazyl the N^{14} hyperfine structure measured in dilute solutions[2] shows that the unpaired spin is equally coupled to the two N atoms with no evidence of coupling to other nuclei of the radical. From this result one can conclude that the unpaired electron has its greatest probable density on the $-N-N-$ group. Likewise, from the N^{14} hyperfine structure[1] of diluted samples of the other two radicals it is concluded that the odd electron is primarily on an $-N=O$ group. Thus the evidence for some spin orbit coupling is consistent with the nuclear hyperfine structure in indicating that the odd electron is confined largely to a two-atom group in these large radicals. This does not preclude, however, an exchange interaction in the pure crystal nor some migration of the electron in the isolated radical.

26. Resonance induced by irradiation. Magnetic resonance of the resulting unpaired electronic spins provides a powerful means for study of the effects of ionizing radiation on solid matter. The great potentialities of this method are only now becoming evident with the preliminary investigation of a variety of substances—inorganic crystals, organo-metallic compounds, glasses, plastics, organic acids, alcohols, sugars and cellulose, amino acids, peptides, nucleic acids, vitamins, hormones, and proteins. Among the more promising applications of electron spin resonance is the study of primary effects of ionizing radiations on biological systems. These investigations, though only beginning, are already

[1] A. VAN ROGGEN, L. VAN ROGGEN and W. GORDY: Phys. Rev. **105**, 50 (1957).
[2] C. A. HUTCHISON jr., R. C. PASTOR and A. G. KOWALSKI: J. Chem. Phys. **20**, 534 (1952).

too numerous for discussion here. We must therefore limit our consideration to a few illustrative examples from which some tentative conclusions can be drawn. A summary of results and references to the earlier studies are given by WURTZ [22]. References to some of the later papers are given in Table 13.

Despite the complex interactions in the solid state, characteristic microwave magnetic resonance patterns of numerous organic free radicals are obtainable in solids which have been subjected to ionizing radiations[1]. Some hydrocarbon radicals believed to be identified by the proton hyperfine structure of their electron spin resonance are listed in Table 13. Characteristic hyperfine patterns of these radicals have been observed in a variety of types of irradiated chemicals and biochemicals.

α) *The cage effect in solids.* The various results obtained in our Duke laboratory indicate that caging is an important factor[1,2,3] in determination of ultimate effects of radiation on molecules of many solids. When a unit of a molecular solid is ionized it often breaks up, but the parts are unable to escape immediately as in a gas (or to a lesser extent in a liquid). They are trapped, so to speak, in the vacancy or cage occupied by the parent molecule where they react among themselves and sometimes with molecules which form the cage until the most stable new assembly is formed. If the substance is an ideal nonconductor so that no electron can come in to take the place of the one originally removed by irradiation, the assembly will still contain a radical (an unpaired spin), but this radical may be quite different from the one originally made by the ionization. For example, the proton hyperfine structure indicates that the radical formed by irradiation of frozen CH_3HgCH_3 is $(C_2H_4)^+$. It is difficult to understand how such a radical could be formed as a direct product of irradiation. The same pattern is observed—and incidentally the same radical $(C_2H_4)^+$—in X-irradiated alanine (CH_3CHNH_2COOH), which if ionized could break up and reform the more stable assembly $(C_2H_4)^+$, NH_3, and CO_2; but it is difficult to understand how a radical giving the symmetrical quintet hyperfine structure characteristic of four equally coupled protons could be formed as a primary product of the irradiation of alanine. Many similar instances could be cited. There are, of course, exceptions to the cage effect described here. Small parts or radicals such as H atoms might escape rapidly through the cage wall or lattice, and macromolecules such as the proteins described below do not seem to fall apart when ionized—even the simpler ones mentioned here probably would not do so at low temperatures.

β) *Proteins.* Varied and complex patterns are obtained for irradiated amino acids and peptides. See Table 13 and Fig. 23 for some examples. However, for numerous proteins[1] constructed from these smaller units only two patterns, either separately or in combination, are obtained. One of these is characteristic of irradiated cystine or cysteine and is believed to result from an electron vacancy on the sulfur. The other is a doublet similar to that found to occur in numerous hydrogen-bridged complexes when X-irradiated. It possibly arises from an unpaired electron localized on an O and interacting with a single proton to which it is bonded through a normal O—H bond or a hydrogen bond O————H, probably the latter. If this is true, the proteins do not break up when ionized, as do many of their simpler constituents, but an electron vacancy created at a given point in a protein can migrate out to the point where it has greatest stability.

---

[1] W. GORDY, W. B. ARD and H. SHIELDS: Proc. Nat. Acad. Sci. U.S.A. **41**, 983 (1955); **41**, 996 (1955).

[2] W. GORDY and G. McCORMICK: J. Amer. Chem. Soc. **78**, 3243 (1956).

[3] C. F. LUCK and W. GORDY: J. Amer. Chem. Soc. **78**, 3240 (1956).

If the doublet found in many proteins actually arises from a dipole-dipole interaction of the electron spin localized on an atom X with the moment of a single proton, the doublet separation should have—if the strong field case is assumed—the orientation dependence:

$$H = \frac{M_I \mu_I \beta_I}{I} \left\langle \frac{1}{r^3} \right\rangle_{Av} (3 \cos^2 \vartheta - 1)_{Av}, \qquad (26.1)$$

where ϑ is the angle between the applied field and the X—H axis, and r is the radius vector separating the electron and proton. In the strong field case both S and I are resolved along the field H. In our case certainly, S would precess about H, but possibly I would not. Even so, Eq. (26.1) would always give the correct ΔH for the $\parallel$ and $\perp$ orientations, and it is seen that for the $\parallel$ orientation the doublet spacing should be twice that for the $\perp$ orientation. This relation was tested on silk for which X-ray and infrared data have shown that the hydrogen bridges are at approximately right angles to the strands. Thus, if the odd electron of the irradiated silk is on the O, interacting with the bridging H as shown in B,

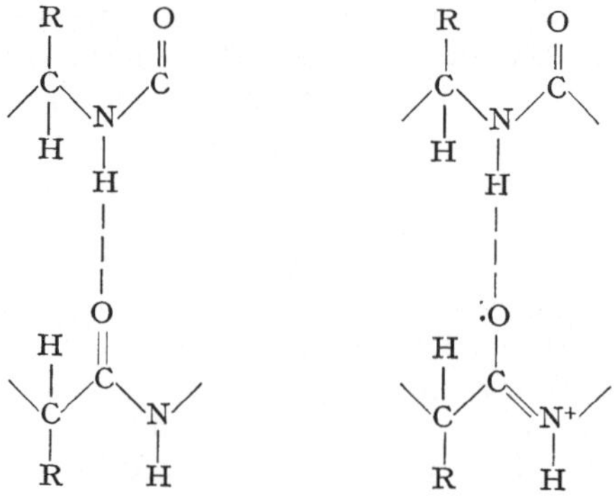

the applied field when parallel to the silk strands should have an approximately perpendicular orientation with all O————H axes, $\vartheta = 90°$, and the doublet structure should be well resolved. With the applied field perpendicular to the silk strands, the O————H axes would have various orientations (within the plane perpendicular to the silk strands), and thus the doublet structure would not be well resolved. The observations, Fig. 25, conform to this model. If this is indeed the correct explanation of the doublet resonance found in several proteins, the resonance should be useful in finding the direction of hydrogen bridges in certain biological substances. For example, the orientation dependence for the doublet found in irradiated feather quill indicates that the hydrogen bonds are along the quill as they should be for the α keratin with the axis of the α helix along the axis of the quill.

A second possible interpretation of the doublet in silk is that an R group is knocked off to leave an odd electron mainly localized on a C to which a single H is attached.

$\gamma)$ *Color centers in the alkali halides.* The first observation of electron magnetic resonance induced by ionizing radiations was that by HUTCHISON[1] on color centers in the alkali halides. The resonance was found to be rather broad with

---

[1] C. A. HUTCHISON jr.: Phys. Rev. **75**, 1769 (1949). — C. A. HUTCHISON jr. and G. A. NOBLE: Phys. Rev. **87**, 1125 (1952). See also E. E. SCHNEIDER and T. S. ENGLAND: Physica, Haag **17**, 221 (1951).

$g = 2.00$ and with no resolvable structure. Table 15 summarizes the results of F center resonance in NaCl, KCl, NaBr, and KBr. The widths are much too great to be attributed to the usual broadening processes. Spin-spin broadening is

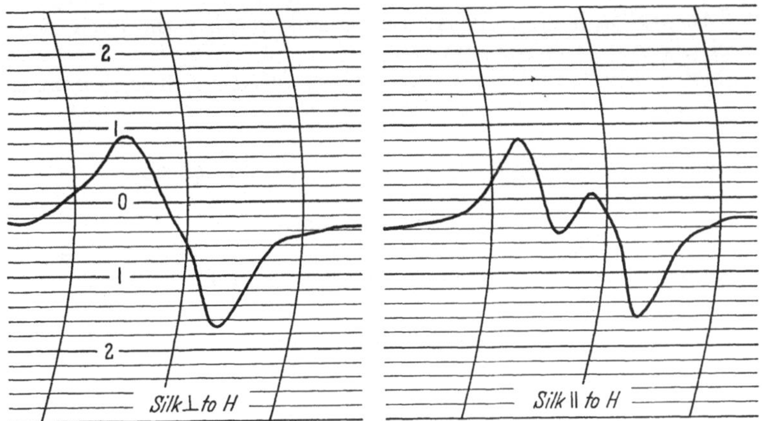

Fig. 25. Electron-spin resonance in X-irradiated silk for **H** parallel and for **H** perpendicular to the silk strands. [From GORDY and SHIELDS: Bull. Amer. Phys. Soc. **1**, 199 (1956).]

essentially negligible in these systems for which the magnetic elements have low concentration. Furthermore, the g factors, very near that for free electron spin, indicate that the spin-orbit coupling is very slight and hence that the broadening

Table 15. *Summary of experimental results on F-center resonance.*
[From KIP, KITTEL, LEVY and PORTIS: Phys. Rev. **91**, 1070 (1953).]

Crystal	g	Observed full width between points of maximum dispersion (oersteds)	Equivalent full width between points of half-maximum absorption (oersteds)
KCl	1.995	61	54
K⁴¹Cl	1.995	40	36
KBr	1.980	162	146
NaCl	1.987	180	162

by way of spin lattice relaxation is likewise small. KIP, KITTEL, LEVY, and PORTIS[1] attribute the abnormal widths to unresolved hyperfine structure arising from interaction of F center electrons with nuclei of adjacent ions.

A 19-component hyperfine structure superimposed on a central broad peak has been reported by SCHNEIDER[2] for resonance of X-irradiated LiF. JEN and LORD, after reporting their inability to detect the structure for the resonance[3], later reported[4] a hyperfine structure of 19 lines with component spacing of 14 gauss. There are yet some differences to be cleared up between the two experiments. The existence and perhaps the nature of this hyperfine structure may depend on some unrecognized pretreatment of the crystal or upon the nature of the irradiation.

COHEN[5] has X-irradiated and examined the alkali halide resonances at the temperature of liquid air and has found a complex and widely spaced hyperfine

[1] A. F. KIP, C. KITTEL, R. A. LEVY and A. M. PORTIS: Phys. Rev. **91**, 1066 (1953).
[2] E. E. SCHNEIDER: Phys. Rev. **93**, 919 (1954).
[3] C. K. JEN and N. W. LORD: Phys. Rev. **96**, 1150 (1954).
[4] N. W. LORD and C. K. JEN: Bull. Amer. Phys. Soc. **1**, 12 (1956).
[5] M. H. COHEN: Phys. Rev. **101**, 1432 (1956).

structure which he attributes to V center resonances arising from Hal_2^- radicals in which the unpaired electron, or hole, spends equal time on the two halogen atoms. H. Shields and the author have checked the observation of a complex and widely spaced structure for LiF observed under these conditions but have not finished the theoretical analysis to ascertain whether the pattern fits the V center model (F_2^- ion) proposed by Cohen.

Although the various results on F and V centers are still somewhat tentative, with many details yet to be worked out and checked, it is already evident that microwave magnetic resonance will eventually give a rather detailed knowledge of F and V centers in the alkali halides and other similar solids.

Acknowledgment.

I am indebted to my wife, Vida Miller Gordy, and to one of my students, Harvey N. Rexroad, for valuable assistance in the preparation and correction of this manuscript.

General references.

[1] Gordy, W.: Microwave Spectroscopy, Rev. Mod. Phys. **20**, 668—717 (1948).
[2] Coles, D. K.: Microwave Spectroscopy, Adv. Electronics **2**, 299—362 (1950).
[3] Maier, W.: Die Mikrowellenspektren molekularer Gase und ihre Auswertung, Ergebn. exakt Naturw. **24**, 275—370 (1951).
[4] Gordy, W., R. F. Trambarulo and W. V. Smith: Microwave Spectroscopy. New York: Wiley & Sons 1953.
[5] Strandberg, M. W. P.: Microwave Spectroscopy. London: Methuen & Co. 1954.
[6] Townes, C. H., and A. L. Schawlow: Microwave Spectroscopy. New York: McGraw-Hill 1955.
[7] Wilson jr., E. B.: Microwave Spectroscopy of Gases. Annual Rev. Phys. Chem. **2**, 151—176 (1951).
[8] Gordy, W.: Microwave and Radiofrequency Spectroscopy. Chap. II in Chemical Applications of Spectroscopy, edit. W. West. New York: Interscience Publ. 1956.
[9] Herzberg, G.: Molecular Spectra and Molecular Structure, 2nd edit. New York: Van Nostrand 1950.
[10] Herzberg, G.: Infrared and Raman Spectra of Polyatomic Molecules. New York: Van Nostrand 1945.
[11] White, H. E.: Introduction to Atomic Spectra. New York: McGraw-Hill 1934.
[12] Condon, E. U., and G. H. Shortley: Theory of Atomic Spectra. New York: Macmillan & Co. 1935.
[13] Abragam, A., and M. H. L. Pryce: Theory of Nuclear Hyperfine Structure of Paramagnetic Resonance Spectra in Crystals. Proc. Roy. Soc. Lond. **205**, 135—153 (1951).
[14] Bleaney, B., and K. W. H. Stevens: Paramagnetic Resonance. I. Rep. Progr. Phys. **16**, 109—157 (1953).
[15] Bowers, K. D., and J. Owen: Paramagnetic Resonance. II. Rep. Progr. Phys. **18**, 304—373 (1955).
[16] Ingram, D. J. E.: Spectroscopy at Radio and Microwave Frequencies. London: Butterworth & Co. 1956.
[17] Vleck, J. H. van: The Theory of Electric and Magnetic Susceptibilities. Oxford: Clarendon Press 1932.
[18] Gorter, C. J.: Paramagnetic Relaxation. Amsterdam: Elsevier 1947.
[19] Ramsey, N. F.: Nuclear Moments. New York: Wiley & Sons 1953.
[20] Kopfermann, H.: Kernmomente. Leipzig: Akademische Verlagsgesellschaft 1940, and Edwards Bros., Ann. Arbor, Michigan.
[21] Kopfermann, H.: Kernmomente, 2. Aufl. Frankfurt: Akademische Verlagsgesellschaft 1956.
[22] Wertz, J. E.: Nuclear and Electronic Spin Magnetic Resonance. Chem. Rev. **55**, 829—955 (1955).

Kontinuierliche Spektren.

Von

W. Finkelnburg und Th. Peters.

Mit 85 Figuren.

1. Einleitung und Übersicht über das gesamte Erscheinungsgebiet. Im Gegensatz zu den diskontinuierlichen Linien- und Bandenspektren, die Übergängen zwischen stationären, im allgemeinen recht scharf definierten Energiezuständen von Atomen und Molekülen entsprechen, beruhen die echten kontinuierlichen Spektren auf Energiezustandsänderungen, bei denen das emittierende oder absorbierende System im Anfangs- oder Endzustand, oder in beiden, sich nicht in einem stationären, sondern in einem freien Zustand befindet. Solche „freie" Zustände sind beispielsweise der ionisierte Zustand eines Atoms oder der dissoziierte Zustand eines Moleküls, weil diese Zustände wegen der beliebigen Beträge an kinetischer Energie, die die Systeme Ion plus Elektron oder Atom plus Atom besitzen können, nicht scharf definiert sind, sondern mehr oder weniger ausgedehnte kontinuierliche Energiebereiche darstellen. Kontinuierliche Spektren können aber weiter auch durch Übergänge zwischen an sich stationären Energiezuständen von Atomen oder Molekülen entstehen, wenn letztere durch innere Störungen oder solche ihrer Umgebung sehr stark verbreitert werden. Von diesen echten kontinuierlichen Spektren scharf zu unterscheiden sind die *unechten*, nur scheinbar kontinuierlichen Spektren, die in Wirklichkeit nicht aufgelöste Linien- oder Bandenspektren besonders mehratomiger Moleküle sind, die nur infolge ungenügender Auflösung des Spektralapparates kontinuierlich erscheinen. Wenn wir von diesen unechten kontinuierlichen Spektren absehen, haben wir also zwei verschiedene Gruppen kontinuierlicher Spektren zu unterscheiden.

α) *Primärkontinuierliche Spektren*, d.h. solche, deren Träger im Anfangs- oder Endzustand oder in beiden nicht-stationäre Systeme darstellen. Diese primärkontinuierlichen Spektren deuten also auf *Vorgänge* in dem emittierenden oder absorbierenden System hin. Beispiele sind die durch Lichtabsorption bewirkte Ionisierung eines Atoms oder Dissoziation eines Moleküls (Anfangszustand stationär, d.h. gebunden, Endzustand frei) sowie die Umkehrvorgänge der Wiedervereinigung von Ion und Elektron zum Atom bzw. die Rekombination zweier Atome zum Molekül (Anfangszustand frei, Endzustand gebunden), schließlich die Abbremsung freier schneller Elektronen unter Emission (Anfangs- und Endzustand frei).

β) *Kontinuierliche Spektren*, die durch gegenseitige Überlagerungen mehr oder weniger stark verbreiterter Linien oder Molekülbanden zustande kommen. Hier handelt es sich primär um Übergänge zwischen stationären Zuständen von Atomen oder Molekülen, bei denen durch innere Störungen im System (Autoionisation bzw. Prädissoziation) oder durch die störende Einwirkung der Umgebung erst die primär diskontinuierlichen Spektren kontinuierlich werden. Diese inneren oder äußeren Störungen bewirken dann, daß bei dem emittierenden oder absorbierenden System der Unterschied zwischen gebundenem diskreten oder freiem kontinuierlichen Energiezustand verschwindet. Auch bei ihnen handelt es sich daher um echte kontinuierliche Spektren. Beispiele für diesen Typ waren das kontinuierliche Spektrum eines 1000 Atm-Eisen-Lichtbogens oder -Funkens sowie das von Hochdruckfunken oder Höchstdruckbögen in Wasserstoff bzw. Wasserdampf. Zu diesem Kontinuatyp kann man auch die Spektren vieler Flüssigkeiten und Festkörper zählen, bei denen die absorbierenden oder emittierenden Elektronen durch die fest gepackte Umgebung so stark gestört sind, daß der Unterschied zwischen dem gebundenen stationären und dem freien kontinuierlichen Zustand verschwindet. Den Extremfall der kontinuierlichen Strahlung solcher Plasmen oder Festkörper hoher Temperatur stellt die im Grenzfall schwarze Strahlung Planckscher Intensitätsverteilung dar, die in erster Näherung von der Sonne und den Sternen, in schlechterer Näherung aber auch von manchen hocherhitzten Festkörpern emittiert wird.

Die Bedeutung der kontinuierlichen Spektren liegt wie die aller Spektren in erster Linie darin, daß sie Schlüsse auf den Zustand ihrer Träger, hier also insbesondere auf wichtige Elementarvorgänge erlaubt. Ionisierung und Wiedervereinigung, Moleküldissoziation und Molekülrekombination sind ebenso wie die Bremsung oder Beschleunigung von Ladungsträgern Primärprozesse, deren Auftreten und relative Häufigkeit aus der Beobachtung und genaueren Untersuchung der kontinuierlichen Spektren erschlossen werden kann. Das Interesse der Astrophysik an der Kontinuaforschung ist damit klar. Der Photochemiker kann aus dem Absorptionsspektrum eines Stoffes sofort ersehen, ob bei Absorption von Licht einer bestimmten Wellenlänge primär Anregung, Dissoziation oder Ionisation der ihn interessierenden Moleküle erfolgt und kennt damit den Primärprozeß seiner photochemischen Reaktion. Für den Lichtquellentechniker schließlich sind, um noch ein weiteres Beispiel zu nennen, die kontinuierlichen Spektren von oft entscheidender Bedeutung, weil der lichttechnische Wirkungsgrad einer Lichtquelle wesentlich davon abhängen kann, ob im Spektralbereich der erwünschten Ausstrahlung nur einzelne Linien oder ein den gesamten Spektralbereich überdeckendes kontinuierliches Spektrum emittiert werden kann. Gasentladungslichtquellen mit intensivem kontinuierlichem Spektrum wie der Xe-Hochdruckbogen oder der Quecksilber-Höchstdruckbogen gewinnen aus diesem Grunde immer mehr an Bedeutung.

Das Schrifttum über kontinuierliche Spektren umfaßt mehrere tausend Nummern, so daß hier nur einige der wichtigsten und vorwiegend neueren Arbeiten zitiert werden können. Bis zum Jahre 1937 einschließlich ist die gesamte Literatur (über 1700 Nummern) in einer Monographie von Finkelnburg [149] angeführt und eingehend besprochen. Die gesamte Literatur über astrophysikalisch interessierende Kontinua ausschließlich der Molekülkontinua bis in die neueste Zeit hat Unsöld [449] ausführlich behandelt.

In dem vorliegenden Beitrag kann nur auf die Gebiete der Kontinuaforschung näher eingegangen werden, auf denen seit 1938 grundsätzliche Fortschritte erzielt worden sind, wie das bei der Theorie der Elektronenkontinua und ihrer experimentellen Prüfung mittels der Hochtemperaturplasmen der Fall ist. Bei den übrigen Gebieten, bei denen wie in der Molekülkontinua-Forschung im wesentlichen neuere *Einzel*ergebnisse vorliegen, muß bezüglich der Theorie und aller älteren Untersuchungen auf [149] verwiesen werden, während hier aus Raumgründen nur ein knapper Überblick über den heutigen Stand der Kenntnis mit Angabe der wichtigsten neueren Literatur gegeben werden kann.

2. Meßmethodische Vorbemerkungen. α) *Trennung von Linien und Kontinuum.* Bevor wir in die eigentliche Behandlung der kontinuierlichen Spektren ein-

treten, sind ein paar meßtechnische Vorbemerkungen erforderlich. Hat man es
mit rein kontinuierlichen Spektren in Absorption oder Emission zu tun, so ist
die Messung der Intensitätsverteilung kein grundsätzlich schwieriges Problem.
Anders ist es in den häufigen Fällen, wo Linien bzw. Banden und kontinuierliche
Spektren einander überlagert auftreten. Hier muß sichergestellt werden, daß die
gemessene Intensitätsverteilung des Kontinuums nicht durch die Flügel der
Atomlinien verfälscht erscheint.

Solange die Spektrallinien in der Wellenlängenskala hinreichend weit voneinander entfernt sind, ihre Linienflügel sich also nicht mehr überlagern, bietet
dies keine besonderen Schwierigkeiten. Es gibt aber auch Spektren (wie z.B.
bei „späten" Spektraltypen der Sterne), bei denen die Linien so dicht liegen, daß
das „wahre" Kontinuum völlig überdeckt ist. Hier kann nur indirekt über die
Theorie der Linienbreiten auf die Intensität des eigentlichen Kontinuums geschlossen werden (s. Auswertung des Sonnenspektrums in Ziff. 24).

Beim Vergleich zwischen Linienintensität und Intensität des Kontinuums
bei bestimmten λ-Werten ist aber darauf zu achten, daß verschiedenartige
Fehlerquellen die wahre Linienkontur und in geringem Umfang natürlich auch
die Verteilung des Kontinuums verzerren. Einmal wird durch *Streulicht* Strahlung
aus anderen Wellenlängenbereichen an die betreffende Stelle λ getragen. Sodann
bestimmen *Beugungserscheinungen*, die *endliche Spaltbreite* und *geometrisch-
optische Bildfehler* überhaupt erst ein endliches *Trennungsvermögen* des Spektrographen. Als Grundlage für die „Entzerrung" der gemessenen Intensitätsverteilung muß das sog. *Apparateprofil* oder die *Apparatefunktion* — d.i. das vom
Spektralapparat gelieferte Profil einer ideal scharfen Linie — empirisch bestimmt
werden. Diese Funktion geht dann in die verschiedenen mathematischen Ent-
zerrungsverfahren ein. Im allgemeinen sind die Intensitätsschwankungen des
Kontinuums in einem nicht zu großen Wellenlängenbereich so gering, daß die
Entzerrungsprobleme praktisch nur bei Linien eine Rolle spielen. Liegen diese
aber wie in dem bereits erwähnten Beispiel sehr dicht, so wird auch das nur
indirekt meßbare Kontinuum von der vorangehenden Linienentzerrung beeinflußt. Der Einfluß insbesondere der *Spaltbreiten* auf das kontinuierliche Spektrum
ist unter anderem von MICHARD [*312*], CRINO [*107*], GREAVES [*188*], RÖSSLER [*387*]
und LORD [*286*] behandelt worden.

β) Eichung. Zur Bestimmung der *Energieverteilung* und der *Absolutintensität*
im Spektrum dienen Eichlampen bzw. Normalstrahler oder mit ihnen geeichte
Strahlungsempfänger. Als Normalstrahler kommen die Wolframbandlampe de-
finierter Farbtemperatur [*174*], das Hg-UV-Normal [*261*], (s. Ziff. 15), der positive Krater eines Graphitbogens [*307*], [*354*], [*139*] (Farbtemperatur $= 3995°$ K)
und neuerdings auch die Xenon-Hochdrucklampe [*203*] (Farbtemperatur $=
5242°$ K) in Frage. Als Strahlungsempfänger (z.B. bei Monochromatoren) können
geeichte Thermoelemente oder Photozellen Verwendung finden. Durch Vergleich der zu messenden Strahlung mit den geeichten Strahlern werden somit die
selektiven Materialeigenschaften der Spektrographen eliminiert. Bei den photographischen Verfahren ist unbedingt erforderlich, das Spektrum des Normal-
strahlers zusammen mit dem zu messenden auf *einer* Platte aufzunehmen (gleiche
Entwicklungsbedingungen), wobei die Normalstrahlung gleichzeitig durch Verwendung von Stufenfiltern, Graukeilen, rotierenden Sektorblenden oder durch
aufeinanderfolgende Belichtung mit verschiedenen Blendeneinstellungen (Raum-
winkelbegrenzung) zur Herstellung von Schwärzungsmarken auf der Platte
dienen kann. Ein Beispiel für diese Arbeitsweise zeigt Fig. 43 in Ziff. 15, wo als
geeichter Strahler der positive Krater des Graphitbogens als Normalstrahler
diente.

3. Grundbegriffe der Strahlungstheorie. Bevor wir auf die eigentliche Theorie der Elektronenkontinua eingehen, geben wir zunächst eine kurze Darstellung der für die Behandlung der Kontinua notwendigen Begriffe aus der Strahlungstheorie.

α) *Intensität und Strahlungsdichte.* Als Strahlungsintensität I_ν sei diejenige Energiemenge bezeichnet, die pro Raumwinkeleinheit und im Frequenzbereich zwischen ν und $\nu + d\nu$ in der Sekunde durch eine senkrecht zur Richtung des Strahlenbündels stehende Einheitsfläche strömt. Die Intensität der Gesamtstrahlung ist dann

$$I = \int\limits_0^\infty I_\nu \, d\nu. \tag{3.1}$$

Der Zusammenhang mit der Strahlungsdichte u_ν (Strahlungsenergie/cm^3) ist allgemein gegeben durch

$$u_\nu = \frac{1}{c} \int I_\nu \, d\omega; \quad d\omega = \text{Raumwinkelelement}. \tag{3.2}$$

Für isotrope Strahlung gilt also

$$u_\nu = \frac{4\pi}{c} I_\nu. \tag{3.3}$$

Die Gesamtstrahlungsdichte wird entsprechend Gl. (3.1) zu

$$u = \int\limits_0^\infty u_\nu \, d\nu. \tag{3.4}$$

In der Literatur wird die Strahlungsintensität bzw. die Strahlungsdichte wahlweise auf ein Frequenzintervall 1 oder auf ein Wellenlängenintervall 1 bezogen. Die Umrechnung ergibt sich aus den Beziehungen

$$I_\lambda = \frac{c}{\lambda^2} I_\nu \quad \text{bzw.} \quad u_\lambda = \frac{c}{\lambda^2} u_\nu. \tag{3.5}$$

β) *Emission und Absorption.* Bezeichnen wir mit ε_ν die pro sec und Volumeneinheit in den Raumwinkel 1 und im Frequenzbereich 1 emittierte Energie, so ist die gesamte Ausstrahlung des Volumenelements dV pro sec gegeben durch

$$S_V = dV \int\limits_0^\infty\!\!\int\limits_\omega \varepsilon_\nu \, d\nu \, d\omega. \tag{3.6}$$

Der *Emissionskoeffizient* ε_ν ist im allgemeinen eine Funktion der Zustandsgrößen der Materie, der Frequenz ν und eventuell abhängig von der Strahlungsrichtung. Bei isotroper Ausstrahlung gilt

$$S_V = dV \, 4\pi \int\limits_0^\infty \varepsilon_\nu \, d\nu. \tag{3.7}$$

Der Intensitätsverlust, den ein Strahlenbündel der Intensität I_ν durch Absorption erfährt, wenn es eine materielle Schicht der Dicke l durchläuft, wird durch die Formel

$$\frac{dI_\nu}{dl} = -\varkappa_\nu I \tag{3.8}$$

beschrieben. Dabei ist der *Absorptionskoeffizient* $\varkappa$ [cm^{-1}] wiederum eine Funktion der Frequenz ν, der Materialeigenschaften und eventuell der Richtung von dl.

Gelegentlich kommen in der Literatur noch weitere Definitionen für Absorptionskoeffizienten vor: Der Massenabsorptionskoeffizient $\varkappa_{\nu,M}$ [cm^2 g^{-1}] wird definiert durch die Gleichung

$$\frac{dI_\nu}{dl} = -\varkappa_{\nu,M}\,\varrho\,I_\nu, \tag{3.9}$$

in der ϱ die Dichte bedeutet. Der atomare Absorptionskoeffizient $\varkappa_{\nu,\,\mathrm{at}}$ [cm^{-2}] ist definiert durch

$$\frac{dI_\nu}{dl} = -\varkappa_{\nu,\,\mathrm{at}}\,N\,I_\nu \tag{3.10}$$

mit $N =$ Anzahl der betreffenden Atome/cm^3. $\varkappa_{\nu,\,\mathrm{at}}$ [cm^2] kann direkt als *Wirkungsquerschnitt* eines Atoms für den betreffenden Absorptionsprozeß bezeichnet werden. Entsprechend bedeutet $\varkappa_{\nu,M}$ die Summe der Wirkungsquerschnitte pro Gramm Materie, und $\varkappa_\nu$ selbst die pro cm^3. Die Integration von Gl. (3.8) liefert

$$I_\nu = I_{\nu,0}\,e^{-\int \varkappa_\nu\,dl} = I_{\nu,0}\,e^{-\tau_\nu}, \tag{3.11}$$

wo $I_{\nu,0}$ die einfallende Strahlungsintensität ist. Die dimensionslose Größe $\tau_\nu = \int \varkappa_\nu\,dl$ wird als *optische Dicke* der Schicht bezeichnet.

$\gamma)$ *Strahlung im thermischen Gleichgewicht (Hohlraumstrahlung)*[1]. Steht die Strahlung mit ihrer Umgebung im Temperaturgleichgewicht (Hohlraumstrahlung der Temperatur T), so ist das Strahlungsfeld isotrop, und es gilt das KIRCHHOFF-sche Gesetz

$$\frac{\varepsilon_\nu}{\varkappa_\nu} = I_\nu(\nu, T) = B_\nu(\nu, T), \tag{3.12}$$

wobei

$$B_\nu = \frac{2h\nu^3}{c^2}\,\frac{1}{e^{h\nu/kT} - 1} \quad\text{bzw.}\quad B_\lambda = \frac{2hc^2}{\lambda^5}\,\frac{1}{e^{hc/\lambda kT} - 1} \tag{3.13}$$

die PLANCK-Funktion ist. Die Strahlungsdichte der Hohlraumstrahlung ist nach Gl. (3.3) dann gegeben durch

$$u_\nu = \frac{8\pi h\nu^3}{c^3}\,\frac{1}{e^{h\nu/kT} - 1}. \tag{3.14}$$

Bei sehr kleinen Frequenzen $h\nu/kT \ll 1$ ergibt sich aus (3.13) das RAYLEIGH-JEAN-sche Gesetz

$$B_\nu = \frac{2\nu^2}{c^2}\,kT \quad\text{bzw.}\quad u_\nu = \frac{8\pi\nu^2}{c^3}\,kT \tag{3.15}$$

für den klassischen Grenzfall, in dem h verschwindet, und für große Frequenzen $h\nu/kT \gg 1$ das WIENsche Strahlungsgesetz

$$B_\nu = \frac{2h\nu^3}{c^2}\,e^{-h\nu/kT} \quad\text{bzw.}\quad u_\nu = \frac{8\pi h\nu^3}{c^3}\,e^{-h\nu/kT}.$$

Nach der Definition von I_ν ist die gesamte Ausstrahlung pro cm^2 und sec in dem ganzen Halbraum

$$\pi I_\nu = 2\pi \int\limits_0^{\pi/2} I_\nu \cos\vartheta \sin\vartheta\,d\vartheta.$$

ϑ definiert die Richtung eines Raumwinkelelements. Bei der Hohlraumstrahlung ergibt sich daraus mit $I_\nu = B_\nu$ nach Gl. (3.13) und nach Integration über alle Frequenzen das STEFAN-BOLTZMANNsche Gesetz

$$\pi I_\nu = \frac{2\pi^5 k^4}{15 c^2 h^3}\,T^4 = \sigma T^4, \quad\text{mit}\quad \sigma = 5{,}6716 \cdot 10^{-5}\ \mathrm{cgs.} \tag{3.16}$$

---

[1] Über Temperaturstrahlung vgl. den Artikel von G. A. W. RUTGERS in Bd. XXVI dieses Handbuches.

δ) *Temperaturdefinitionen.* Kontinuierliche Spektren werden hinsichtlich ihrer *Energieverteilung* oder ihrer *Absolutintensität* oft mit den entsprechenden Eigenschaften der Hohlraumstrahlung in Beziehung gesetzt, wobei verschiedenartige „Temperatur"-Definitionen Verwendung finden.

1. Wird die Energieverteilung des beobachteten kontinuierlichen Spektrums in einem gewissen Wellenlängenbereich λ_1 bis λ_2 hinsichtlich ihres *relativen Verlaufs* einer PLANCK-Kurve angepaßt, so erhält man die sog. „*Farbtemperatur* T_F im Bereich λ_1 bis λ_2".

Rückt der Wellenlängenbereich immer enger zusammen, bis zu einer mittleren Wellenlänge $\bar{\lambda}$, so erfaßt man praktisch die Tangente $\left(\dfrac{dI_\lambda}{d\lambda}\right)_{\bar{\lambda}}$ und spricht dann von der „*Farbtemperatur* T_F oder *Gradationstemperatur* bei $\bar{\lambda}$".

2. Bezieht man sich bei dem Vergleich mit der PLANCK-Funktion auf die *absolute* Strahlungsintensität in einem engen Bereich (λ_1, λ_2), so bekommt man die sog. „*Strahlungstemperatur*". Vergleicht man bei einer bestimmten Wellenlänge $\bar{\lambda}$, so spricht man von der „*schwarzen Temperatur*" oder auch „*Strahlungstemperatur* T_S bei $\bar{\lambda}$".

3. Deutet man die gemessene Gesamtstrahlung im Sinne des STEFAN-BOLTZMANNschen Gesetzes, so erhält man die „effektive Temperatur T_e".

Alle diese Temperaturangaben haben aber lediglich den Wert einer Beschreibung der zugrunde liegenden Messungen und daher keinen besonderen physikalischen Sinn.

Beispielsweise habe das im thermischen Gleichgewicht befindliche Plasma eines elektrischen Lichtbogens eine Temperatur von $T = 10000°$ K. Damit ist gesagt, daß die Geschwindigkeitsverteilung der Atome, Ionen und Elektronen, der Ionisationszustand und die Besetzung der Anregungszustände der Atome und Ionen genau dieser Temperatur entspricht. Die Energiedichte der Strahlung wird aber im allgemeinen nicht dieser Temperatur entsprechen. „Meist" ist sie erheblich kleiner als diejenige der Hohlraumstrahlung von $T = 10000°$ K. Die effektive Temperatur wird beispielsweise nur den Wert von $3000°$ K besitzen, die Strahlungstemperatur T_S kann alle möglichen Werte $< 10000°$ K annehmen, und die Farbtemperatur T_F, welche die relative Verteilung im kontinuierlichen Spektrum beschreibt (und, wie wir sehen werden, im wesentlichen von der Termanordnung der Atome abhängt), kann ebensogut 5000 wie $50000°$ K betragen.

ε) EINSTEINs *Übergangswahrscheinlichkeiten.* Nach EINSTEIN läßt sich die Wechselwirkung der Strahlung mit der Materie durch folgende *Übergangsprozesse* und *Übergangswahrscheinlichkeiten* (A_{nm}, B_{mn}, B_{nm}) beschreiben:

1. Spontane Emission $(m \to n)$.

Ein angeregtes Atom im Zustand E_m geht unter Emission eines Lichtquants der Energie

$$h\nu_{nm} = E_m - E_n \tag{3.17}$$

spontan in den energieärmeren Zustand E_n über. Befinden sich N_m Atome im angeregten Zustand E_m, so ist die Anzahl der Quantensprünge pro sec:

$$A_{nm} N_m. \tag{3.18}$$

2. Absorption $(n \to m)$.

Die Anzahl der Übergänge pro sec von E_n nach E_m ist proportional der Zahl der Atome (N_n) im Zustand E_n und der Energiedichte der absorbierten Strahlung u_ν, also

$$B_{mn} N_n u_\nu. \tag{3.19}$$

3. Induzierte oder erzwungene Emission.

Durch ein Lichtquant der Frequenz $\nu = \nu_{nm}$ wird ein Atom im angeregten Zustand E_m veranlaßt, ein dem ankommenden hinsichtlich Frequenz und Richtung genau gleiches Lichtquant zu emittieren. Dieser Prozeß entspricht der Ausstrahlung eines klassischen Oszillators beim Mitschwingen in einem Strahlungsfeld. Die Anzahl dieser Prozesse pro sec ist

$$B_{nm} N_m u_\nu. \tag{3.20}$$

Im thermischen Gleichgewicht muß die Zahl der Übergänge $m \to n$ pro sec der Anzahl der umgekehrten Quantensprünge gleich sein. Es gilt also

$$(A_{nm} + B_{nm} u_\nu)\, N_m = B_{mn} u_\nu N_n. \tag{3.21}$$

Nun ist nach dem BOLTZMANNschen Gesetz

$$\frac{N_m}{N_n} = \frac{g_m}{g_n}\, e^{-\frac{E_m - E_n}{kT}} = \frac{g_m}{g_n}\, e^{-\frac{h\nu}{kT}} \tag{3.22}$$

(g_m und g_n sind die statistischen Gewichte der betreffenden Zustände).

Durch Einsetzen von (3.22) in (3.21) erhält man für die Energiedichte der *Hohlraumstrahlung*

$$u_\nu = \frac{A_{nm}}{g_n/g_m\, B_{mn}\, e^{h\nu/kT} - B_{nm}} \tag{3.23}$$

und daraus durch den Vergleich mit der PLANCKschen Strahlungsformel (3.14) die folgenden Beziehungen zwischen den drei *Übergangswahrscheinlichkeiten*:

$$\left. \begin{aligned} &g_m B_{nm} = g_n B_{mn}, \\ &A_{nm} = \frac{8\pi h \nu^3}{c^3} B_{nm} = \frac{8\pi h \nu^3}{c^3} \frac{g_n}{g_m} B_{mn}. \end{aligned} \right\} \tag{3.24}$$

Der Zusammenhang zwischen den *Übergangswahrscheinlichkeiten* und den *Emissions- bzw. Absorptionskoeffizienten* ist aus dem Vorangehenden ohne weiteres ersichtlich.

Die pro sec, cm³ und Raumwinkeleinheit im Frequenzintervall $d\nu$ *emittierte* Energie (spontane Emission) ergibt sich zu

$$\varepsilon_\nu = \frac{A_{nm}}{4\pi} N_m h\nu. \tag{3.25}$$

Die unter denselben Bedingungen *absorbierte* Energie ist nach Gl. (3.8) unter Berücksichtigung von (3.3) in der Form

$$\varkappa_\nu I_\nu = \varkappa_\nu \frac{c}{4\pi} u_\nu = \frac{1}{4\pi} B_{mn} h\nu N_n u_\nu \tag{3.26}$$

zu schreiben. Demnach ist

$$\varkappa_\nu = \frac{h\nu}{c} B_{mn} N_n. \tag{3.27}$$

Berücksichtigt man ferner noch wie in (3.21) die *erzwungene Emission*, die gelegentlich auch als *negative Absorption bezeichnet wird*, so folgt schließlich unter Verwendung der Beziehung (3.24) wieder der KIRCHHOFFsche Satz in der Form

$$\varepsilon_\nu = \varkappa_\nu (1 - e^{-h\nu/kT})\, B_\nu = \varkappa_\nu' B_\nu. \tag{3.28}$$

Der Faktor $(1 - e^{-h\nu/kT})$ ist also kennzeichnend für die erzwungene Emission. $\varkappa_\nu'$ bezeichnet man als *effektiven* Absorptionskoeffizienten.

A. Atom- und Elektronenkontinua.

I. Theorie der Elektronenkontinua.

4. Anschauliche Bedeutung der freien und gebundenen Zustände der Elektronen.
In den folgenden Kapiteln werden die kontinuierlichen Spektren behandelt, die
durch Energiezustandsänderungen von Elektronen zustande kommen. Da den
stationären Bohrschen „Ellipsenbahnen" der im Atom oder Molekül gebundenen
Elektronen einzelne diskrete Energiezustände entsprechen, eine kontinuierliche
Mannigfaltigkeit von Energiewerten aber nur freien, vom Atom oder Molekül
losgelösten Elektronen (auf „Hyperbelbahnen") zur Verfügung steht, zeigt das
Auftreten eines kontinuierlichen Spektrums Zustandsänderungen an, bei denen
die Elektronen im Anfangs- oder Endzustand des Überganges oder in beiden
frei waren. Kontinuierliche Elektronenspektren treten infolgedessen bei drei
verschiedenen Vorgängen auf:

1. Bei der Absorption und Emission von Strahlung durch freie Elektronen,
wenn diese auch im Endzustand frei bleiben. Anschaulich bedeutet dies Übergänge zwischen den Hyperbelbahnen.

2. Bei dem Übergang freier Elektronen in diskrete Atom- oder Molekülzustände unter Ausstrahlung der Bindungsenergie plus ihrer kinetischen Energie
(Strahlungsrekombination). Anschauliche Bedeutung: Übergänge von hyperbolischen in elliptische Bahnen.

3. Bei der Umkehrung dieses Vorganges, nämlich bei der Absorption von
Strahlung durch gebundene Atom- oder Molekülelektronen, wobei diese aus dem
Atom- oder Molekülverband befreit werden und im Endzustand freie Elektronen
mit kinetischer Energie darstellen (Photoionisation).

Wir besprechen zunächst die allen drei Prozessen gemeinsame Theorie in an-
schaulicher Weise, wobei wir mit dem letzten Vorgang beginnen.

Wird ein normales Atom durch Energiezufuhr angeregt, so gelangt das Leucht-
elektron dadurch in seine verschieden angeregten Zustände. Diesen entsprechen
diskrete Energiewerte, und bei Übergängen zwischen ihnen werden die Spektrallinien des Atoms emittiert oder absorbiert. Führen wir dem Atom aber einen
Energiebetrag zu, der die Bindungsenergie des Leuchtelektrons übersteigt, so
wird das Atom ionisiert. Das Elektron gelangt dadurch in einen freien Zustand,
und der kontinuierlichen Mannigfaltigkeit von Beträgen kinetischer Energie, die
das freie Elektron besitzen kann, entspricht sein kontinuierlicher Energiebereich
im Termschema. Erfolgt die Anregung und Ionisierung des Atoms speziell durch
Lichtabsorption, so werden beim Übergang in die angeregten Zustände die
einzelnen Spektrallinien absorbiert, an deren Seriengrenze sich nach kurzen Wellen
zu ein kontinuierliches Absorptionsspektrum anschließt, welches anzeigt, daß
das Elektron bei der Abtrennung (Photoionisation) kontinuierlich variable Beträge kinetischer Energie mitnehmen kann.

Die Umkehrung der Photoionisation stellt die Rekombination eines freien
Elektrons mit einem positiven Atomion zum neutralen Atom dar. Bei Übergängen aus angeregten Atomzuständen in tiefere Zustände werden die diskreten
Spektrallinien emittiert. Geht dagegen ein freies Elektron in einen diskreten,
gebundenen Zustand über, so wird außer der Bindungsenergie des Elektrons in
dem betreffenden Zustand noch seine kinetische Energie ausgestrahlt. Beobachtet wird also ein kontinuierliches Emissionsspektrum, das sich genau wie das
Absorptionsgrenzkontinuum nach kurzen Wellen an eine Atomseriengrenze anschließt. Fig. 1 zeigt diese Vorgänge im Termschema. Den Übergängen A_L
entsprechen die Absorptionslinien, den Übergängen A_k das an ihre kurzwellige

Grenze sich anschließende Absorptionskontinuum. In Emission zeigt Fig. 1 die Verhältnisse an zwei Serien mit verschiedenem Endterm: E_L sind die den Linien entsprechenden Übergänge, während die Übergänge E_k dem Einfang freier Elektronen mit kinetischer Energie in zwei verschiedene stationäre Zustände entsprechen und den Anlaß zur Emission von Seriengrenzkontinua darstellen.

Außer diesen Atomgrenzkontinua in Absorption und Emission haben wir noch den unter 1. genannten Fall der Elektronenkontinua als Folge der Kombination zweier freier Zustände im kontinuierlichen Energiebereich, angedeutet durch den obersten Doppelpfeil in Fig. 1. Ein solcher frei-frei-Übergang, bei dem Anfangs- und Endzustand im Kontinuum liegen, bedeutet, daß freie Elektronen im Feld positiver Ionen einen Teil ihrer kinetischen Energie ausstrahlen oder umgekehrt durch Absorption kinetische Energie gewinnen. Klassisch entspricht der frei-frei-Strahlung die Ausstrahlung eines in einem elektrischen Feld ungleichförmig sich bewegenden Ladungsträgers. Nach der Bohrschen Theorie handelt es sich um Übergänge zwischen Hyperbelbahnen verschiedener Energie, da ja den stationären Zuständen Kreis- und Ellipsenbahnen, den freien Zuständen dagegen Parabel- und Hyperbelbahnen entsprechen.

Der Spektralbereich, in dem diese Kontinua auftreten, hängt von der Größe der Energieänderung ab. Handelt es sich bei den Übergängen in Fig. 1 um solche des Leuchtelektrons eines Atoms, so liegen die Spektren im Ultrarot, Sichtbaren oder Ultraviolett. Haben wir es dagegen bei einem Übergang A_k mit der Photoionisation eines der inneren Elektronen eines Vielelektronenatoms zu tun, dessen Ionisierungsenergie einige tausend Volt beträgt, so liegt das entsprechende Absorptionskontinuum im Röntgengebiet. Entsprechend ist es bei der frei-frei-Strahlung. Haben wir als Anfangszustand Elektronen mit vielen tausend Volt Geschwindigkeit (Kathodenstrahlen), die im Feld der Antikathodenatome etwa

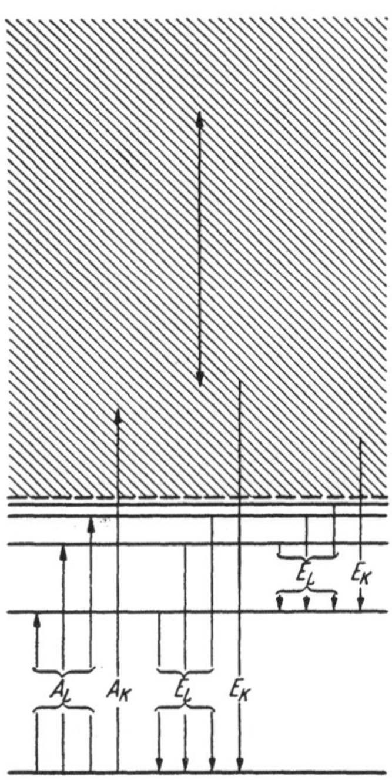

Fig. 1. Schematische Darstellung der den Atomspektren entsprechenden Übergänge. A_L und A_K: Absorptionslinien und Grenzkontinuum in Absorption. E_L und E_K: Emissionslinien und Grenzkontinuum in Emission. Oberster Doppelpfeil: frei-frei-Übergang.

einer Röntgenröhre abgebremst werden, und dabei ihre Energie ausstrahlen, so liegt diese frei-frei-Strahlung im Röntgengebiet (kontinuierliches Röntgenbremsspektrum), und der obere Endpunkt unseres Pfeils in Fig. 1 wäre im gleichen Maßstab etwa 10 m höher im Kontinuum zu zeichnen. Werden umgekehrt Elektronen bis zu einigen Volt Geschwindigkeit in einer Gasentladung abgebremst, so erhalten wir ein Bremskontinuum, welches sich vom Radiofrequenzbereich bis ins Sichtbare erstreckt.

Wir haben bei dieser anschaulichen Darstellung der Verhältnisse einen schrittweisen Übergang von der Elektronenanregung zur Ionisation gemacht und werden in diesem Sinne auch weiterhin ein Atom und das aus ihm durch Ionisation entstehende System Ion + Elektron als das gleiche physikalische System betrachten. Das ist auch die in der Wellenmechanik übliche Auffassung, in der die Schrödinger-Gleichung das gesamte Verhalten des Atoms beschreibt, wobei als Nullpunkt der Energiezählung die des Systems Ion + Elektron bei kinetischer Energie Null gewählt wird. Den diskreten Energiezuständen des gebundenen

Atomelektrons entsprechen dann die negativen diskreten Eigenwerte der Energie, den freien Elektronen der positive Energiebereich. In diesem Sinne gehören also auch die kontinuierlichen Eigenwertbereiche zum Atomtermschema. Fig. 2 zeigt ein solches für ein Mehrelektronenatom. E_{00} bezeichnet den Grundzustand des Atoms E_{10}, E_{20} usw. die Anregungszustände des „äußersten" am leichtesten abtrennbaren Atomelektrons, an die sich von der Ionisierungsgrenze dieses Elektrons E_{G0} an die kontinuierlichen Zustände E_{k0} dieses Elektrons anschließen. Irgendein Zustand E_{k0} kann nun gleichzeitig wieder als Grundzustand des einfach geladenen positiven Ions aufgefaßt werden, das bei Anregung eines zweiten

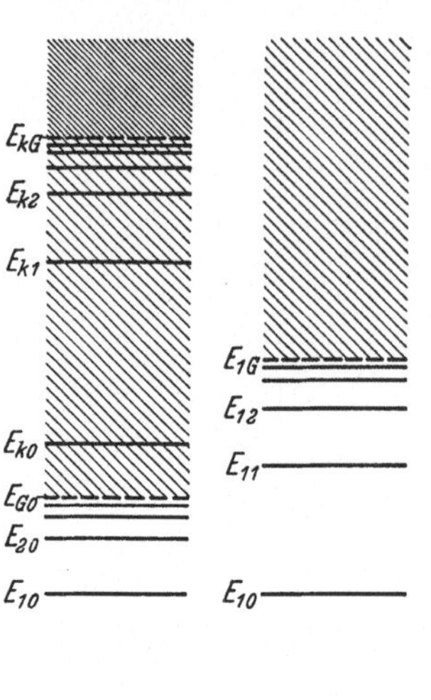

Fig. 2. Termschema eines Mehrelektronen-Atoms, zur Veranschaulichung von Doppelanregung und strahlungslosen Übergängen.

Elektrons die diskreten Energiezustände E_{k1}, E_{k2} usw. annehmen kann. An die Ionisierungsgrenze dieses zweiten Elektrons E_{kG} schließt sich dessen kontinuierliches Energiegebiet E_{kk} an, in dem nun wieder die Energiezustände des doppelt positiv geladenen Atomions liegen usw. Das vollständige Termschema des Atoms ist aber noch viel komplizierter. Die Anregung des zweiten Elektrons braucht nämlich nicht, wie bisher angenommen, erst zu erfolgen, wenn das erste bereits abgetrennt ist; sie kann vielmehr bereits stattfinden, wenn das erste nur angeregt ist. So ist z.B. im Zustand E_{10} das erste Elektron im ersten angeregten Zustand, das zweite im tiefsten Zustand. In der Folge möge nun das Elektron 2 angeregt werden (Zustände E_{11}, E_{12}, E_{13} ...), bis es bei E_{1G} ionisiert werden möge, wobei in diesem Fall ein freies Elektron und ein angeregtes Ion entstehen. Lassen wir nun auch das erste Elektron alle möglichen Anregungsstufen durchlaufen, so ergibt sich schon bei nur zwei Leuchtelektronen ein äußerst verwickeltes Termschema, in dem diskrete Energiezustände und kontinuierliche Energiebereiche sich vielfach überlagern. Das bedeutet aber, daß bei einer bestimmten Energie zwei oder mehr verschiedene Elektronenanordnungen des Systems möglich sind. Wir werden in Ziff. 12 auf die aus der Überlagerung diskreter und kontinuierlicher Elektronenzustände folgenden Effekte kurz eingehen.

5. Klassisch-korrespondenzmäßige Theorie. Den einfachsten Zugang zur Theorie kontinuierlicher Strahlung bietet nach wie vor die *klassisch-korrespondenzmäßige* Theorie von Kramers [257], die eine weitgehende Bestätigung durch die Quantenmechanik erfahren hat. Kramers geht aus von der klassischen Strahlung eines freien Elektrons, das sich ungleichförmig im Felde eines positiven Kerns der Ladung Ze bewegt, behandelt also zunächst nur die frei-frei-Strahlung (Bremsstrahlung). Bei nichtrelativistischer Rechnung $v/c \ll 1$ folgt unter der Annahme, daß die ausgestrahlte Energie zu vernachlässigen ist gegenüber der kinetischen Energie des Elektrons, daß letzteres eine *Hyperbelbahn* beschreiben muß.

Nach der klassischen Elektrodynamik strahlt eine beschleunigte Ladung in der Zeit dt die Energie

$$dR = \frac{2e^2}{3c^3}\dot{v}^2\,dt \tag{5.1}$$

aus, wo der Punkt wie üblich die zeitliche Differentiation andeutet. Bezeichnen wir den Ablenkungswinkel in Fig. 3 mit $\pi - 2\varphi_0$, so ist (Rutherfordsche Streuformel)

$$\tan \varphi_0 = \frac{m\,p\,v^2}{Z\,e} \tag{5.2}$$

mit p als Stoßparameter.

Nach Einführung von Polarkoordinaten (r, φ) und Berücksichtigung des Flächensatzes

$$r^2\,\dot\varphi = p\,v = \text{const}, \tag{5.3}$$

ergibt sich für das Quadrat der Beschleunigung

$$\dot v^2 = \frac{Z^2\,e^4}{m^2\,r^4}. \tag{5.4}$$

Die gesamte ausgestrahlte Energie ist somit unter Verwendung von Gl. (5.3)

$$R = \frac{2Z^2\,e^6}{c^3\,m^2} \int\limits_{-\infty}^{+\infty} \frac{dt}{r^4} = \frac{2Z^2\,e^6}{3c^3\,m^2\,p\,v} \int\limits_{\varphi_0}^{2\pi-\varphi_0} \frac{d\varphi}{r^2}. \tag{5.5}$$

Zwischen den Koordinaten der Hyperbelbahnen besteht ferner die Beziehung

$$\frac{1}{r} = \frac{1 - \varepsilon \cos \varphi}{p \tan \varphi_0} \tag{5.6}$$

mit $\varepsilon = \dfrac{1}{\cos \varphi_0}$ als Exzentrizität.

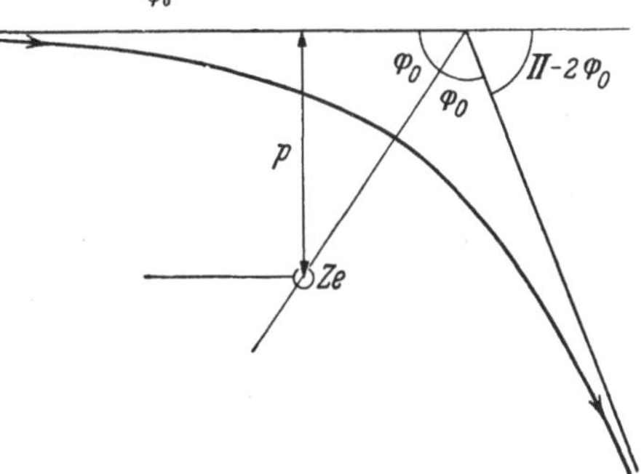

Fig. 3. Ablenkung eines Elektrons mit dem Stoßparameter p und der Geschwindigkeit v durch ein Ion mit der Ladung $Z\,e$.

Führen wir Gl. (5.6) unter Berücksichtigung von Gl. (5.2) in Gl. (5.5) ein, so folgt

$$R = \frac{2Z^4\,e^{10}}{3c^3\,m^4\,p^5\,v^5} \int\limits_{\varphi_0}^{2\pi-\varphi_0} (1 - \varepsilon \cos \varphi)^2\,d\varphi, \tag{5.7}$$

d. h.

$$R = \frac{2Z^4\,e^{10}}{3c^3\,m^4\,p^5\,v^5} \left[(2\pi - 2\varphi_0)\left(1 + \frac{1}{2}\,\frac{1}{\cos^2 \varphi_0}\right) + 3 \tan \varphi_0 \right]. \tag{5.8}$$

Dieser Ausdruck läßt sich leicht berechnen für die Grenzfälle $\tan \varphi_0 \ll 1$ (also φ_0 sehr klein) und $\tan \varphi_0 \gg 1$ (also $\varphi_0 \approx 90°$, fast geradlinige Bahn).

Im ersten Fall wird das Elektron um nahezu 180° abgelenkt. Die Bahn kann besonders in Kernnähe als *Parabel* ($\varepsilon = 1$) angesehen werden. Mit

$$\tan \varphi_0 = \frac{m\,p\,v^2}{Z\,e^2} \ll 1 \tag{5.9}$$

wird der Klammerausdruck $= 3\pi$ und für die Gesamtstrahlung gilt

$$R = \frac{2\pi\,Z^4\,e^{10}}{c^3\,m^4\,p^5\,v^5}. \tag{5.10}$$

Die Verteilung der Gesamtstrahlung auf die verschiedenen Frequenzen findet man aus dem zeitlichen Verlauf der Beschleunigung nach Richtung und Größe beim Durchlaufen der gesamten Bahn. Diese Bahnbeschleunigung teilt KRAMERS auf in je eine Komponente senkrecht und parallel zur Bahnachse. Zur Normierung der Rechnung wird $p\,v = 1$ und $\dfrac{Z\,e^2}{m} = 1$ gesetzt. Der kürzeste Abstand Elektron-

Kern sei $\frac{1}{2}$. Drückt man noch die trigonometrischen Funktionen rational aus, indem man $\tan \varphi/2 = -z$ setzt, so ist

$$\sin \varphi = \frac{2z}{1+z^2}; \qquad \cos \varphi = -\frac{1-z^2}{1+z^2}; \qquad r = \frac{1+z^2}{2}. \tag{5.11}$$

Für die Beschleunigungskomponenten ergeben sich dann die Ausdrücke

$$\dot{v}_\perp = -\frac{\sin \varphi}{r^2} = \frac{8z}{(1+z^2)^3}; \qquad \dot{v}_\| = -\frac{\cos \varphi}{r^2} = \frac{4(1-z^2)}{(1+z^2)^3}. \tag{5.12}$$

Zur Berechnung der Verteilung der Energie auf die einzelnen Frequenzintervalle ersetzen wir diese Funktionen durch Fourier-Integrale:

$$\dot{v}_\perp(t) = \int_0^\infty \varphi(\gamma) \sin \gamma t \, d\gamma; \qquad \dot{v}_\|(t) = \int_0^\infty \psi(\gamma) \cos \gamma t \, d\gamma. \tag{5.13}$$

Dann gilt auch

$$\varphi(\gamma) = \frac{1}{\pi} \int_{-\infty}^{+\infty} \dot{v}_\perp(\tau) \sin \gamma \tau \, d\tau; \qquad \psi(\gamma) = \frac{1}{\pi} \int_{-\infty}^{+\infty} \dot{v}_\|(\tau) \cos \gamma \tau \, d\tau. \tag{5.14}$$

In den letzten Integralen ist die Zeit mit τ bezeichnet. Führt man nun die Variable z aus Gl. (5.12) ein, so erhält man nach Umformung

$$\varphi(\gamma) = \gamma \, i^{\frac{4}{3}} \cdot 3^{-\frac{1}{2}} H_{\frac{1}{3}}^{(1)}(i\gamma/3); \qquad \psi(\gamma) = \gamma \, i^{\frac{5}{3}} \cdot 3^{-\frac{1}{2}} H_{\frac{2}{3}}^{(1)}(i\gamma/3), \tag{5.15}$$

wobei $H_n^{(1)}(x)$ die Hankel-Funktion erster Art der Ordnung n ist.

Für die totale emittierte Energie gilt nach Gl. (5.10) zusammen mit den eingeführten Normierungen

$$R = \frac{2e^2}{3c^3} \int_{-\infty}^{+\infty} (\dot{v}_\perp^2 + \dot{v}_\|^2) \, dt = \frac{2\pi e^2}{c^3}. \tag{5.16}$$

Andererseits ist

$$\int_{-\infty}^{+\infty} (\dot{v}_\perp^2 + \dot{v}_\|^2) \, dt = \pi \int_0^\infty [\varphi^2(\gamma) + \psi^2(\gamma)] \, d\gamma \tag{5.17}$$

und hat nach Gl. (5.16) den Wert 3π.

Setzen wir also den relativen Anteil der Strahlungsenergie im Frequenzintervall $d\gamma$ gleich $P(\gamma)$, so ist

$$P(\gamma) \, d\gamma = \tfrac{1}{3}[\varphi^2(\gamma) + \psi^2(\gamma)] \, d\gamma \quad \text{mit} \quad \int_0^\infty P(\gamma) \, d\gamma = 1. \tag{5.18}$$

Diese Verteilungsfunktion läßt sich mit Hilfe der Gl. (5.15) berechnen. Für spätere Zwecke merken wir an, daß

$$\int_0^\infty \frac{P(\gamma)}{\gamma} \, d\gamma = \frac{4}{\pi\sqrt{3}} \tag{5.19}$$

ist. Aus den Normierungsbedingungen erhalten wir weiterhin den Zusammenhang zwischen der Frequenz ν und der normierten Frequenz γ:

$$2\pi\nu = \gamma \, \frac{Z^2 e^4}{m^2 p^3 v^3}. \tag{5.20}$$

Der zweite Grenzfall der fast *geradlinigen Bahn* mit der Bedingung

$$\tan \varphi_0 = \frac{m\,p\,v^2}{Z\,e^2} \gg 1 \quad \left(\varphi_0 \approx \frac{\pi}{2}\right) \tag{5.21}$$

wird ganz analog behandelt. Für die gesamte Ausstrahlung erhält man dann

$$R = \frac{\pi}{3}\,\frac{Z^2\,e^6}{c^3\,m^2\,p^3\,v} \tag{5.22}$$

und für die entsprechenden Funktionen, die aus der FOURIER-Integraldarstellung entstehen:

$$\varphi'(\gamma) = -\gamma\,H_1^{(1)}(i\,\gamma); \quad \psi'(\gamma) = i\,\gamma\,H_0^{(1)}(i\,\gamma). \tag{5.23}$$

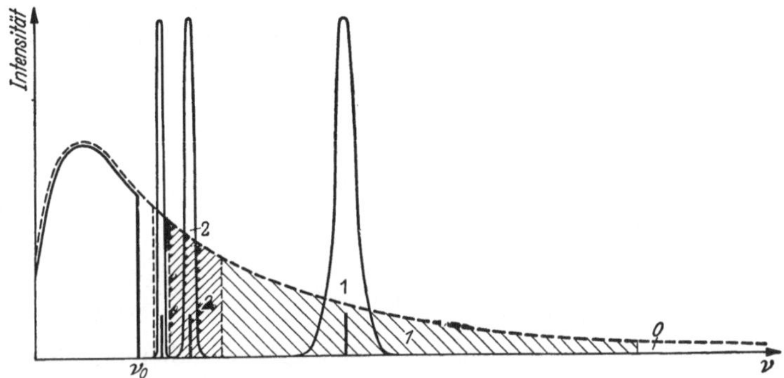

Fig. 4. Korrespondenzmäßig berechneter Verlauf des durch Übergänge freier Elektronen bestimmter Geschwindigkeit entstehenden Spektrums. (Nach KRAMERS [*257*].)

Die Verteilungsfunktion lautet in diesem Falle

$$P'(\gamma)\,d\gamma = 2\left(\varphi'^2(\gamma) + \psi'^2(\gamma)\right) d\gamma \quad \text{mit} \quad \int\limits_0^\infty P'(\gamma)\,d\gamma = 1 \tag{5.24}$$

und die Verknüpfung zwischen v und γ:

$$\gamma = 2\pi v\,\frac{p}{v}. \tag{5.25}$$

Nach einem Stoß zwischen Elektron und Kern kann das Elektron entweder noch frei oder gebunden sein. Das statistische Ergebnis einer großen Anzahl von Zusammenstößen mit freien Elektronen derselben Geschwindigkeit v führt daher zur Emission eines kontinuierlichen Spektrums (frei-frei-Strahlung) im Frequenzbereich v bis v_0 (wobei v_0 durch $h v_0 = \frac{1}{2} m v^2$ gegeben ist) und eines anschließenden „Linienspektrums" (Fig. 4).

Da es sich hier um ein Gedankenexperiment handelt, dürfen diese Linien nicht mit wirklichen Spektrallinien identifiziert werden. Wir werden sehen, daß aus diesen Linien, wenn wir alle möglichen Elektronengeschwindigkeiten zulassen, das betreffende Seriengrenzkontinuum entsteht.

Das Korrespondenzprinzip läßt nun erwarten, daß die Intensitätsverteilung dieses Spektrums aus der klassischen Theorie abgeschätzt werden kann. Zumindest die Verteilung der frei-frei-Strahlung wird von dieser Theorie sicher richtig wiedergegeben werden.

Nehmen wir den parabolischen Fall an, und bezeichnen mit $q(v)\,dv$ die Wahrscheinlichkeit dafür, daß ein Stoß der betrachteten Art Anlaß gibt zu einer Strahlung im Frequenzintervall dv, so führt das Korrespondenzprinzip zu der

Annahme, daß in erster Näherung zusammen mit den Gln. (5.10) und (5.20)

$$q(\nu)\,h\nu\,d\nu = \frac{2\pi\,Z^4\,e^{10}}{c^3\,m^4\,p^5\,v^5}\,P(\gamma)\,\frac{d\gamma}{d\nu}\,d\nu,$$
$$= \frac{4\pi^2\,Z^2\,e^6}{c^3\,m^2\,p^2\,v^2}\,P\left(2\pi\nu\,\frac{p^3\,v^3\,m^2}{Z^2\,e^4}\right)d\nu \qquad\qquad (5.26)$$

ist. Für den *hyperbolischen* Fall gilt entsprechend mit Gl. (5.22) und (5.25):

$$q(\nu)\,h\nu\,d\nu = \frac{2\pi^2\,Z^2\,e^6}{3\,c^3\,m^2\,p^2\,v^2}\,P'\left(2\pi\nu\,\frac{p}{v}\right)d\nu. \qquad (5.27)$$

Diese beiden Ausdrücke führen zu einer vollständigen Theorie der kontinuierlichen Spektren einschließlich der Röntgenspektren.

KRAMERS versucht weiterhin, die Gln. (5.26) und (5.27) auch auf Einfangprozesse anzuwenden, auf die die klassische Theorie im Prinzip nicht mehr anwendbar ist.

Wird ein freies Elektron der Geschwindigkeit v in einem diskreten Zustand der Energie

$$E_n = \frac{2\pi^2\,Z^2\,e^4\,m}{h^2\,n^2}$$

gebunden, so wird Strahlung der Energie

$$h\nu = E_n + \frac{m\,v^2}{2}$$

emittiert. KRAMERS nimmt nun an, daß das korrespondierende Frequenzintervall geschrieben werden kann in der Form:

$$\Delta\nu \approx \frac{2\pi^2\,Z^2\,e^4\,m}{h^3}\left[\frac{1}{(n-\tfrac{1}{2})^2} - \frac{1}{(n+\tfrac{1}{2})^2}\right] \approx \frac{4\pi^2\,Z^2\,e^4\,m}{h^3\,n^3}. \qquad (5.28)$$

Es sei ferner q_n die Wahrscheinlichkeit für den Einfang in den n-ten Quantenzustand, dann ist nach Gl. (5.26)

$$q_n\,h\nu = \frac{4\pi^2\,Z^2\,e^6}{c^3\,m^2\,p^2\,v^2}\,P\left(2\pi\nu\,\frac{p^3\,v^3\,m^2}{Z^2\,e^4}\right)\Delta\nu,$$
$$= \frac{16\pi^4\,Z^4\,e^{10}}{c^3\,m\,h^3\,p^2\,v^2\,n^3}\,P\left(2\pi\nu\,\frac{p^3\,v^3\,m^2}{Z^2\,e^4}\right). \qquad\qquad (5.29)$$

Die Funktion $P(\gamma)$ ist im Bereich $\Delta\nu$ als hinreichend konstant angenommen worden. Die durch Gl. (5.29) dargestellte Energie konzentriert sich also nach KRAMERS in den in Fig. 4 angedeuteten Linien.

Bislang haben wir nur stoßende Elektronen mit festem Stoßparameter p und gleicher Anfangsgeschwindigkeit v betrachtet.

Wir berechnen nun die Strahlungsintensitäten, die pro Sekunde im Frequenzintervall $d\nu$ von einem Flächenelement $d\sigma$ einer unendlich dünnen Schicht, die A Ionen bzw. Kerne der Ladung Ze pro cm² enthält, in den gesamten Raum ausgestrahlt wird, wenn S Elektronen der Geschwindigkeit v pro Sekunde die Fläche von 1 cm² durchsetzen. Dazu muß zunächst (5.29) über alle vorkommenden Stoßparameter p integriert werden. Es gilt daher

$$i_\nu\,d\sigma\,d\nu = S\,A\,d\sigma\,d\nu \int_0^\infty q(\nu)\,h\nu\,2\pi\,p\,dp. \qquad (5.30)$$

Diese pro cm² emittierte Strahlung geht in den Emissionskoeffizienten ε_ν über, wenn statt der Zahl A der Ionen pro cm² die Zahl N_i der Ionen pro cm³ eingesetzt wird, und wenn ferner die Anzahl S der Elektronen, die pro Sekunde die Flächeneinheit durchsetzen, gleich der in dem Quader von 1 cm Grundfläche und der Länge v enthaltenden Elektronen $S = v N_{e,v}$ ist. Der Index v soll dabei betonen, daß Elektronen bestimmter Geschwindigkeit v gemeint sind. Schließlich soll die Strahlungsintensität noch durch den Faktor $1/4\pi$ auf die Raumwinkeleinheit bezogen werden. Damit wird

$$\varepsilon_\nu = N_{e,v} N_i \frac{v}{4\pi} \int\limits_0^\infty q(\nu)\, h\nu\, 2\pi\, p\, dp. \qquad (5.31)$$

Setzen wir hier den Ausdruck (5.26) für den *parabolischen* Fall ein, so ist

$$\varepsilon_\nu = N_{e,v} N_i \int\limits_0^\infty \frac{2\pi^2 Z^2 e^6}{c^3 m^2 p v}\, P\left(2\pi\nu\, \frac{p^3 v^3 m^2}{Z^2 e^4}\right) dp. \qquad (5.32)$$

Aus Gl. (5.20) erhalten wir ferner bei fester Frequenz ν

$$dp = \frac{Z^2 e^4}{3 \cdot 2\pi\nu\, m^2 p^2 v^3}\, d\gamma \qquad (5.33)$$

und somit für den Emissionskoeffizienten die Form

$$\varepsilon_\nu = \frac{N_{e,v} N_i\, 2\pi^2 Z^2 e^6}{3 c^3 m^2 v} \int\limits_0^\infty \frac{P(\gamma)}{\gamma}\, d\gamma, \qquad (5.34)$$

die unter Berücksichtigung von Gl. (5.19) übergeht in

$$\varepsilon_\nu = \frac{8\pi Z^2 e^6 N_{e,v} N_i}{3\sqrt{3}\, m^2 c^3 v}. \qquad (5.35)$$

Wie wir sehen, ist ε_ν *unabhängig* von der Frequenz ν.

Wir werden diese Gleichung sofort auf die uns im wesentlichen interessierenden kontinuierlichen Spektren im optischen Bereich anwenden. In Ziff. 18 wird sie den Ausgangspunkt auch für die Berechnung der kontinuierlichen Röntgenstrahlung bilden.

In einem ionisierten Gas, dessen freie Elektronen eine thermische Geschwindigkeitsverteilung der Elektronentemperatur T_e besitzen, wird kontinuierliche Strahlung emittiert, die sowohl von den frei-frei-Übergängen (Bremskontinuum) als auch von den Einfangprozessen (Seriengrenzkontinua) herrührt.

Behandeln wir zunächst die frei-frei-Strahlung allein: Strahlung im Frequenzintervall zwischen ν und $\nu + d\nu$ kann nur emittiert werden von freien Elektronen, die eine kinetische Energie

$$\frac{m v^2}{2} \geq h\nu \qquad (5.36)$$

haben. Im Sinne des MAXWELL*schen Verteilungsgesetzes* bedeutet $N_{e,v}$ die Anzahl der Elektronen im Geschwindigkeitsintervall dv:

$$N_{e,v} = N_e \frac{4 v^2}{\sqrt{\pi}}\, \frac{e^{-\frac{m v^2}{2 k T_e}}}{\left(\frac{2 k T_e}{m}\right)^{\frac{3}{2}}}\, dv. \qquad (5.37)$$

Setzen wir dieses in Gl. (5.35) ein und integrieren über alle Elektronen, deren kinetische Energie der Bedingung (5.36) genügt, so folgt für den *Emissionskoeffizienten der frei-frei-Strahlung* der Ausdruck

$$\varepsilon_\nu = \frac{32\pi^2 Z^2 e^6}{3\sqrt{3}\, c^3 (2\pi m)^{\frac{3}{2}}} \frac{N_e N_i}{(kT_e)^{\frac{1}{2}}}\, e^{-h\nu/kT}.\tag{5.38}$$

Wir müssen an dieser Stelle kurz noch auf eine bislang nicht erwähnte Unkorrektheit eingehen, die gerade bei der Theorie der Bremsstrahlung von Bedeutung ist. In den Gln. (5.32) bzw. (5.34) wurde über alle Werte von p bzw. γ integriert (von 0 bis ∞), obgleich der Integrand nur Gültigkeit für *parabolische* Bahnen mit dem begrenzten p-Bereich nach Gl. (5.9)

$$p \ll \frac{Z e^2}{m v^2}\tag{5.39}$$

besitzt. Um auch den *hyperbolischen* Fall mit

$$p \gg \frac{Z e^2}{m v^2}\tag{5.40}$$

nach Gl. (5.21) berücksichtigen zu können, teilt Kramers das Integral der Gl. (5.31) auf in zwei Anteile. Im ersten gelte die Bedingung (5.39) (Parabelbahnen), im zweiten die Bedingung (5.40) (Hyperbelbahnen). Dieses Verfahren ist natürlich nicht ganz korrekt, weil beide Möglichkeiten nur Grenzfälle darstellen. Es zeigt sich aber, daß an allen bisher angeschriebenen Emissionsformeln nur eine Korrekturfunktion $g(\nu)$ anzubringen ist, die mit Ausnahme sehr kleiner Frequenzen praktisch den Wert 1 hat. Das heißt, daß bei nicht zu kleinen Frequenzen der Hauptteil der frei-frei-Strahlung von den stark abgelenkten Elektronen (parabelähnliche Bahnen) herrührt. Für $\nu \to 0$ nimmt die Funktion $g(\nu)$ allerdings große Werte an; es überwiegen also mehr und mehr die Beiträge der fast geradlinigen Hyperbelbahnen. Nun können wiederum nicht beliebig große Stoßparameter zugelassen werden. Wir dürfen überhaupt nur mit Zweierstößen rechnen, solange p kleiner als der halbe mittlere Abstand zweier benachbarter Ionen ist. Der maximale Stoßparameter ist also gegeben durch

$$p_{\text{max}} = \frac{1}{2}\left(\frac{4\pi}{3}\, N_i\right)^{-\frac{1}{3}}.$$

Eine Rechnung von Elwert [*135*] und Burkhardt, Elwert und Unsöld [*76*] zeigt, daß die Funktion g für sehr kleine Frequenzen den Wert

$$g = \frac{\sqrt{3}}{\pi}\ln\frac{kT_e}{1{,}44\, Z e^2 N_i^{\frac{1}{3}}}\tag{5.41}$$

annimmt.

Wenn wir wegen $h\nu/kT \ll 1$ in Gl. (5.38) $e^{-h\nu/kT} \approx 1$ setzen und noch mit dem Korrekturfaktor g nach Gl. (5.41) multiplizieren, so erhalten wir für den Emissionskoeffizienten der frei-frei-Strahlung bei sehr kleinen Frequenzen

$$\varepsilon_\nu = \frac{32\pi Z^2 e^6 N_i N_e}{3 c^3 (2\pi m)^{\frac{3}{2}} (kT_e)^{\frac{1}{2}}}\ln\frac{kT_e}{1{,}44\, Z e^2 N_i^{\frac{1}{3}}},\tag{5.42}$$

eine Formel, welche die Grundlage für die Theorie der *radiofrequenten Sonnenstrahlung* bildet (s. Ziff. 24).

Setzt man in Gl. (5.35) die Verteilung der Elektronen nach Maxwell [Gl. (5.37)] ein, integriert aber im Gegensatz zur bisherigen Rechnung über den gesamten Geschwindigkeitsbereich, so ist

$$\begin{aligned}
\varepsilon_\nu &= \frac{32\,\pi\,Z^2\,e^6\,N_i\,N_e}{3\,\sqrt{3}\,m^2\,c^3\,\pi^{\frac12}\left(\dfrac{2kT_e}{m}\right)^{\frac32}} \int_0^\infty v\,e^{-\frac{mv^2}{2kT_e}}\,dv, \\[2mm]
&= \frac{32\,\pi^2\,Z^2\,e^6\,N_i\,N_e}{3\,\sqrt{3}\,c^3\,(2\pi\,m)^{\frac32}\,(kT_e)^{\frac12}}\,.
\end{aligned}\qquad (5.43)$$

Wir haben damit das Verteilungsgesetz der gesamten kontinuierlichen Emission gewonnen, einschließlich der diskret-kontinuierlichen Strahlung. Wenn wir nämlich Elektronen beliebig kleiner Geschwindigkeit bei der Berechnung des Emissionskoeffizienten ε_ν an der Stelle ν der Frequenzskala zulassen, so bedeutet dies, daß auch Rekombinationsprozesse zur Emission beitragen. Wie aus der Art der Ableitung hervorgeht, haben wir uns allerdings der „klassischen Verteilung" nach Gl. (5.26) bzw. Fig. 4 bedient, also auf die Quantelung der Energiezustände im Atom keine Rücksicht genommen. Dieses Verfahren (UNSÖLD [443]) wird in erster Näherung zulässig sein, wenn die Energiezustände im Atomtermschema hinreichend dicht liegen, so daß sie näherungsweise als „*Termkontinuum*" anzusehen sind. Die Forderung ist am besten realisiert für

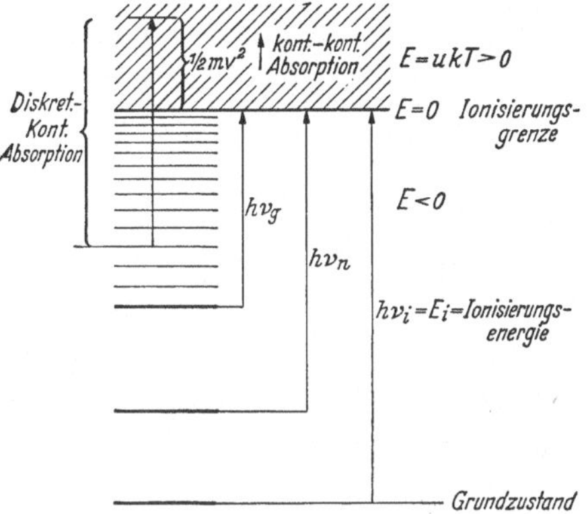

Fig. 5. Termschema der Atome des „Modellgases."

Terme, die in der Nähe der Ionisierungsgrenze liegen. Die Termschemata der Atome enthalten aber auch eine große Anzahl relativ weit voneinander getrennter Einzelterme, auf welche die vorstehende Überlegung nicht mehr anwendbar ist und deren zugehörige Grenzkontinua daher einzeln zu berücksichtigen sind. Die frequenzunabhängige Emission nach Gl. (5.43) wird daher nicht im ganzen Spektralbereich Gültigkeit haben, worauf zuerst HAHN und FINKELNBURG [195] aufmerksam machten.

Zur Veranschaulichung der Theorie betrachten wir ein aus „Modellatomen" bestehendes Gas, das sich im thermischen Gleichgewicht befindet, für das also $T_e = T$ sei. Die Modellatome bzw. die gleichwertigen Systeme Ion + Elektron sollen das in Fig. 5 dargestellte Termschema besitzen (s. auch MAECKER und PETERS [295]). Auf den Grundterm folgt ein einzelner Anregungszustand und schließlich oberhalb der Energie $E_g = h\nu_g$ (von der Ionisierungsgrenze aus gerechnet) eine relativ dichte Termfolge bis zur Ionisierungsgrenze, die als *Termkontinuum* angesehen werden soll. Das entsprechende Emissionskontinuum zeigt Fig. 6, in der als Abszisse die Größe $u = h\nu/kT$ und als Ordinate $\log\varepsilon_\nu$ aufgetragen ist. Der nach Gl. (5.38) mit $\exp(-h\nu/kT)$ abfallenden frei-frei-Strahlung überlagern sich die Emissionskontinua. Bis zur Frequenz ν_g — also bis zum Abszissenwert $u_g = h\nu_g/kT$ — folgen diese einzelnen Grenzkontinua wegen der dichten Termfolge so rasch aufeinander, daß Gl. (5.43) anwendbar ist, die einen *frequenzunabhängigen* Strahlungsverlauf ergibt. Da unterhalb der Energie $h\nu_g$ im Termschema zunächst keine weiteren Terme mehr folgen, kommen auch im Spektrum für $u > u_g$ zunächst keine neuen Grenzkontinua hinzu: Der Emissionskoeffizient muß daher absinken.

Eine dem bisherigen Rechengang ganz analoge Überlegung auf der Grundlage von Gl. (5.29) ergibt für den Verlauf des Emissionskoeffizienten eines einzelnen

Grenzkontinuums (n-ter Quantenzustand) die Form

$$\varepsilon_{\nu,n} = \frac{128\,\pi^4 Z^4 e^{10}\,m\,N_i\,N_e}{3\sqrt{3}\,c^3\,h^2\,(2\pi\,m\,kT)^{\frac{3}{2}}}\,\frac{1}{n^3}\,e^{h\,(\nu_n-\nu)/kT}. \tag{5.44}$$

Jedes Seriengrenzkontinuum fällt also auf der kurzwelligen Seite der Seriengrenze ν_n genau wie das Bremskontinuum nach (5.38) mit $\exp(-h\nu/kT)$ ab. Den gleichen relativen Abfall zeigt folglich auch das Gesamtkontinuum nach der Grenze ν_g (bzw. u_g in der Fig. 6). Die Rechnung ergibt für den Bereich $\nu_g \lesssim \nu < \nu_n$:

$$\varepsilon_\nu = \frac{32\,\pi^2 Z^2 e^6\,N_i\,N_e}{3\sqrt{3}\,c^3\,(2\pi\,m)^{\frac{3}{2}}\,(kT)^{\frac{1}{2}}}\,e^{h\,(\nu_g-\nu)/kT}. \tag{5.45}$$

Beim Fortschreiten nach höheren Frequenzen hin tritt bei $u = u_n$ das Seriengrenzkontinuum des *einzelnen* angeregten Zustandes additiv nach Gl. (5.44) hinzu und schließlich von $u = u_i = E_i/kT$ ab auch das Hauptseriengrenzkontinuum.

In einem wirklichen Gas, welches aus neutralen Atomen und Ladungsträgern besteht (Plasma), tritt noch ein Effekt hinzu, der sich durch das Übergreifen der Seriengrenzkontinua zu kleineren Frequenzen hin bemerkbar macht. Wie wir besonders in Ziff. 15 feststellen werden, beginnen die Seriengrenzkontinua nie an der theoretischen Grenze, sondern meist schon ein beträchtliches Stück vorher. Dies rührt daher, daß die Ionisierungsgrenze auf Grund der von den Ladungsträgern erzeugten Mikrofelder und der daraus resultierenden Verbreiterung der obersten Terme um einen Betrag ΔE_i herabgesetzt wird. Die ausgezeichneten Energiewerte $h\nu_g, h\nu_n, \dots$ im Termschema sind demnach von dieser effektiven Ionisierungsgrenze $(E_i - \Delta E_i)$ ab zu rechnen. In Fig. 6 ist der Effekt des Übergreifens durch gestrichelte Linien angedeutet. Da die Abszisse in u-Einheiten angegeben ist, werden die Kanten gerade um $\Delta u = \Delta E_i/kT$ vorgezogen.

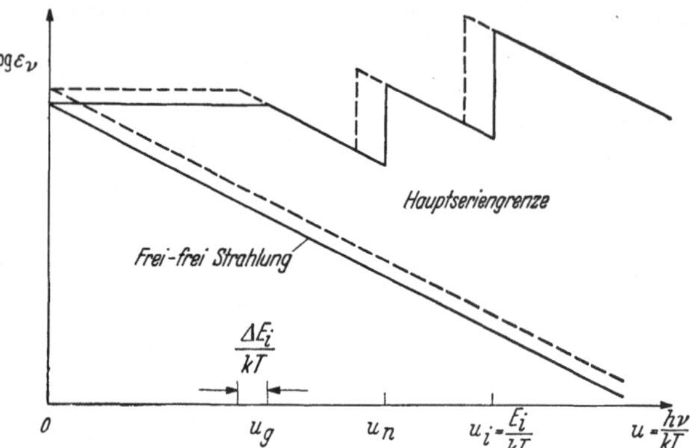

Fig. 6. Kontinuierliche Emission des „Modellgases" (Termschema Fig. 5). Das frei-frei-Kontinuum ist gesondert eingezeichnet. Bei Erniedrigung der Ionisierungsenergie um $(-\Delta E_i)$ durch Mikrofelder der Ladungsträger im Plasma werden die Seriengrenzen in der u-Skala gerade um $\Delta u = \Delta E_i/kT$ vorgezogen. [Übergreifen des Kontinuums zu längeren Wellenlängen und Erhöhung vor der Grenzfrequenz und im frei-frei-Anteil (gestrichelte Linien).]

Wenn wir die bislang aufgestellten Formeln nicht nur auf Wasserstoff bzw. wasserstoffähnliche Atome, sondern näherungsweise auch auf alle anderen Atome anwenden wollen, müssen wir berücksichtigen, daß die Elektronenbahnen teilweise auch in die Elektronenhülle eintauchen werden und somit einem stärkeren Feld als dem der Ionenladung Ze ausgesetzt sind. Dieser Abweichung vom Wasserstoffmodell wird Rechnung getragen durch Einführung eines mittleren Faktors $\overline{(Z+s)^2}$ an Stelle von Z^2.

Der Faktor darf nicht verwechselt werden mit dem in der Atomtheorie, z.B. auch der Röntgenstrahlung auftretenden $(Z-s)^2$, wo s die Abschirmungszahl ist. Der Unterschied liegt in der verschiedenen Bedeutung von Z. Im Röntgengebiet ist Z die Kernladungszahl, weil schnelle Elektronen die Atomhülle praktisch ungehindert durchdringen und erst in unmittelbarer Nähe des Kerns abgelenkt werden. Der abschirmende Einfluß der kernnahen Hüllenelektronen (K-Schale) wird durch Angabe der Abschirmungszahl s in dem Faktor $(Z-s)$ berücksichtigt. Im optischen Bereich haben wir es hingegen mit Strahlung relativ langsamer Elektronen zu tun, die nur in den äußeren Teil der Elektronenhülle des Atoms eindringen können. Hier bedeutet Z also die Ionenladungszahl, und beim tieferen Eindringen eines Elektrons in die Hülle verstärkt sich das wirksame Feld, was durch den Korrekturfaktor $(Z + s)$ zum Ausdruck kommt.

Der Zahlenwert für $\overline{(Z + s)^2}$ darf nun nicht aus den Termdefekten gegen Wasserstoff entnommen werden, weil es in diesem Fall für die Energieberechnung auf die Mittelung über $1/r$ ($r =$ Radiusvektor der Elektronenbahn) ankommt, während zur Bestimmung der Übergangswahrscheinlichkeiten nach Gl. (7.4) über r^2 zu mitteln ist. Hier bekommen also die kernfernen Gebiete relativ großes Gewicht und $\overline{(Z + s)^2}$ wird kleiner. Man wird also auf jeden Fall sagen können, daß dieser Faktor für das Kontinuum zwischen Z^2 und dem aus dem Termdefekt nach der Formel

$$\overline{(Z + s)^2} = n^2 \frac{E_i - E_n}{E_{iH}} \tag{5.46}$$

($n =$ Hauptquantenzahl, $E_i =$ Ionisierungsenergie der betreffenden Atomart, $E_n =$ Anregungsenergie, $E_{iH} =$ Ionisierungsenergie des Wasserstoffes)

folgenden Wert liegen sollte.

Das frequenzunabhängige Verteilungsgesetz (5.43) wird gelegentlich in einer anderen Schreibweise verwandt. In einem ionisierten Gas im thermischen Gleichgewicht ist nach SAHA das Ionisationsgleichgewicht gegeben durch

$$\frac{N_e N_i}{N_0} = \frac{2 U_1}{U_0} \frac{(2 \pi m k T)^{\frac{3}{2}}}{h^3} e^{-(E_i - \Delta E_i)/kT}, \tag{5.47}$$

wo U_1 bzw. $U_0 =$ Zustandssumme des Ions bzw. des Atoms bedeuten. Ersetzt man nun das Produkt $N_e N_i$ in Gl. (5.43) mit Hilfe der SAHA-Gleichung (5.47),

Tabelle 1.

Art des Kontinuums	Emissionskoeffizient $\left[\dfrac{\text{erg}}{\text{sec cm}^3\, \Delta\nu\, \text{ster}}\right]$ *	Absorptionskoeffizient [cm^{-1}]
Frei-frei-Kontinuum	$\varepsilon_\nu = C\,\overline{(Z+s)^2}\,\dfrac{N_e N_i}{(kT)^{\frac{1}{2}}}\,e^{-h\nu/kT}$ (5.38a)	$\varkappa_\nu = C\,\dfrac{c^2}{2h}\,\overline{(Z+s)^2}\,\dfrac{N_e N_i}{(kT)^{\frac{1}{2}}}\,\dfrac{1}{\nu^3}$ (5.38 b
Frei-frei-Kontinuum im Grenzfall sehr kleiner Frequenzen (Radiofrequenzstrahlung)	$\varepsilon_\nu = C\,\dfrac{\sqrt{3}}{\pi}\,Z^2\,\dfrac{N_e N_i}{(kT)^{\frac{1}{2}}}\,\ln\!\left(\dfrac{kT}{1,44\,Z\,e^2\,N_i^{\frac{1}{3}}}\right)$ (5.42a)	$\varkappa_\nu = C\,\dfrac{\sqrt{3}\,c^2}{2\pi}\,Z^2\,\dfrac{N_e N_i}{(kT)^{\frac{3}{2}}}\,\dfrac{1}{\nu^2}\,\ln\dfrac{kT}{1,44\,Z\,e^2\,N_i^{\frac{1}{3}}}$ (5.42b)
Seriengrenzkontinuum des n-ten Quantenzustandes (wasserstoffähnlich)	$\varepsilon_{\nu,n} = C\,\dfrac{4\pi^2 e^4 m}{h^2 n^3}\,(Z+s)^4\,\dfrac{N_e N_i}{(kT)^{\frac{3}{2}}}\,e^{\frac{h}{kT}(\nu_n - \nu)}$ (5.44a)	$\varkappa_{\nu,n} = C\,\dfrac{2\pi^2 e^4 m c^2}{h^3 n^3}\,(Z+s)^4\,\dfrac{N_e N_i}{(kT)^{\frac{3}{2}}}\,\dfrac{1}{\nu^3}\,e^{\frac{h\nu_n}{kT}}$ (5.44b)
Gesamtkontinuum. Überlagerung der frei-frei-Strahlung und der Seriengrenzkontinua sehr eng benachbarter Terme	$\varepsilon_\nu = C\,\overline{(Z+s)^2}\,\dfrac{N_e N_i}{(kT)^{\frac{1}{2}}}$ (5.43a) frequenzunabhängig!	$\varkappa_\nu = C\,\dfrac{c^2}{2h}\,\overline{(Z+s)^2}\,\dfrac{N_i N_e}{(kT)^{\frac{1}{2}}}\,\dfrac{1}{\nu^3}\,e^{h\nu/kT}$ (5.43b)

* ster = Steradiant = Raumwinkeleinheit

$$C = \frac{32\,\pi^2 e^6}{3\sqrt{3}\,c^3\,(2\pi m)^{\frac{3}{2}}} = 6{,}36 \cdot 10^{-47}\ \text{cgs-Einh.}$$

so erhält man

$$\varepsilon_\nu = \frac{64\,\pi^2\,\overline{(Z+s)^2}\,e^6}{3\,\sqrt{3}\,h^3\,c^3}\,\frac{U_1}{U_0}\,N_0\,kT\,e^{-(E_i-\Delta E_i)/kT}.\tag{5.48}$$

Wird nun hierin der Partialdruck der Neutralgasteilchen N_0kT näherungsweise durch den Gesamtgasdruck p ersetzt, *so gilt der daraus entstehende Ausdruck nur für ein schwach ionisiertes Gas*, eine Einschränkung, die in der Literatur nicht immer beachtet wird.

Zum Schluß geben wir in Tabelle 1 noch eine kurze Zusammenstellung der wichtigsten, aus der KRAMERS-UNSÖLDschen Theorie folgenden Ergebnisse, wobei wir auch die Absorptionskoeffizienten $\varkappa_\nu$ mit anschreiben wollen. Befindet sich das strahlende System im thermischen Gleichgewicht, so gilt der KIRCHHOFFsche Satz (3.12). Berücksichtigen wir auch die erzwungene Emission, so ist nach Gl. (3.28)

$$\varepsilon_\nu = \varkappa_\nu\,(1 - e^{-h\nu/kT})\,B_\nu = \varkappa'_\nu\,B_\nu,$$

woraus bei gegebenem Emissionskoeffizienten ε_ν der zugehörige Absorptionskoeffizient $\varkappa_\nu$ zu berechnen ist. Die Ergebnisse der Rechnungen von UNSÖLD [449] speziell für Wasserstoff findet man in Ziff. 15.

6. Schwarze Strahlung von Gasen. Als Anwendungsbeispiel für die soeben entwickelten Vorstellungen über kontinuierliche Strahlung von Gasen soll die z.B. von FINKELNBURG [151] angeschnittene Frage behandelt werden, unter welchen Bedingungen ein hoch erhitztes, mehr oder minder ionisiertes Gas eine Strahlung emittiert, die näherungsweise der eines schwarzen Strahlers von der Gastemperatur T entspricht.

Betrachten wir eine homogene Gasschicht der Dicke l, so gilt, falls nur Absorption im Spiel ist, die Gl. (3.8). Da im vorliegenden Fall Strahlung sowohl absorbiert als auch wieder emittiert wird, müssen wir ansetzen

$$\frac{dI_\nu}{dl} = \varepsilon_\nu - \varkappa'_\nu I_\nu.\tag{6.1}$$

Unter Berücksichtigung des KIRCHHOFFschen Satzes (3.28) wird daraus

$$\frac{dI_\nu}{dl} = \varkappa'_\nu\,(B_\nu - I_\nu).\tag{6.2}$$

Die Integration liefert mit $I_\nu = 0$ für $l = 0$

$$I_\nu = B_\nu\,(1 - e^{-\varkappa'_\nu l}).\tag{6.3}$$

Ist die optische Dicke $\tau_\nu = \varkappa'_\nu l \ll 1$, so führt die Reihenentwicklung auf

$$I_\nu = \varkappa'_\nu\,l\,B_\nu \quad (\varkappa'_\nu l \ll 1).\tag{6.4}$$

Die Intensität wächst also proportional mit der Schichtdicke an. Für große optische Tiefen $\tau_\nu \gg 1$ gilt andererseits

$$I_\nu = B_\nu \quad (\varkappa'_\nu l \gg 1).\tag{6.5}$$

In diesem Fall sendet das Gas also eine seiner Temperatur entsprechende schwarze Strahlung aus. Wie wir sehen, kommt es nicht allein auf den *Absorptionskoeffizienten*, sondern ebenso auf die *Schichtdicke* des Gases an. Im allgemeinen sind Gasspektren ja Linienspektren mit kontinuierlichem Untergrund. Der Absorptionskoeffizient weist deshalb eine starke Abhängigkeit von der Wellenlänge auf; dicht beieinander liegende Spektralgebiete unterscheiden sich im Absorptionskoeffizienten oft um viele Größenordnungen. Trotzdem kann sich ein solches Gas nach Gl. (6.5) wie ein schwarzer Körper verhalten, wenn es genügend ausgedehnt ist. Strahlung der Wellenlänge einer intensiven Linie (etwa einer Resonanzlinie) wird beispielsweise bereits von 1 mm Schichtdicke vollständig absorbiert, während Strahlung einer benachbarten Wellenlänge, für die der Absorptionskoeffizient 10^8mal kleiner sein möge, zu vollständigen Absorption eine Schichtdicke von 100 km benötigt. Derartige Ausdehnungen strahlender und absorbierender Materie kommen nur in kosmischen Systemen vor (s. Ziff. 24 und 25).

Wir wollen für das folgende die Linienstrahlung unberücksichtigt lassen und fragen, unter welchen Bedingungen ein im Laboratorium erzeugtes strahlendes Gas, das naturgemäß nur wenige Millimeter oder Zentimeter Ausdehnung haben kann, eine so starke *kontinuierliche* Absorption aufweist, daß es näherungsweise als schwarzer Strahler der Gastemperatur T angesehen werden kann. Die Gasschicht habe eine Ausdehnung von $l = 2$ mm — eine Schichtdicke, die vielfach bei side-on-Beobachtungen von Hochdruckbögen vorliegt. Ferner sei gefordert, daß

$$I_\nu \geq 0,9\, B_\nu \tag{6.6}$$

ist, der *Absorptionsgrad* $\alpha = 1 - e^{-\varkappa'_\nu l}$ also mindestens 90% beträgt. Folglich muß mit $\alpha = 0,9$

$$\tau_\nu = \varkappa'_\nu\, l \geq \ln 10 = 2,3$$

sein, und mit $l = 0,2$ cm gilt

$$\varkappa'_\nu = \varkappa_\nu (1 - e^{-h\nu/kT}) \geq 11,5. \tag{6.7}$$

Zur weiteren Berechnung benutzen wir den Absorptionskoeffizienten des Gesamtkontinuums nach Tabelle 1, Gl. (5.43 b). Die Bedingung (6.7) lautet dann

$$\frac{16\,\pi^2\,\overline{(Z+s)^2}\,e^6\,N_e\,N_i}{3\sqrt{3}\,c\,h\,(2\pi m)^{\frac{3}{2}}\,(kT)^{\frac{1}{2}}}\,\frac{e^{h\nu/kT}-1}{\nu^3} \geq 11,5$$

oder, wenn wir näherungsweise $\overline{(Z+s)^2} = 1$ setzen:

$$4,32\,\frac{N_e\,N_i}{(kT)^{\frac{1}{2}}}\,\frac{e^{h\nu/kT}-1}{\nu^3} \geq 11,5. \tag{6.8}$$

Betrachten wir des praktischen Interesses wegen etwa die Absorption im Sichtbaren bei $\lambda = 5000$ Å ($\nu = 6 \cdot 10^{14}$ Hz), so folgt aus (6.8)

$$N_e\,N_i \geq 5,75 \cdot 10^{44}\,\frac{(kT)^{\frac{1}{2}}}{e^{2,88\cdot 10^4/T}-1}. \tag{6.9}$$

Mit Hilfe der SAHA-Gleichung (6.47) für die betreffende Gasart mit der Ionisierungsenergie E_i und unter weiterer Berücksichtigung der Gasgleichung

$$p = (N_0 + N_i + N_e)\,kT$$

kann nun das Druck-Temperaturgebiet ermittelt werden, in dem das Gas die eingangs gestellte Forderung (6.6) erfüllt. Fig. 7 gibt eine graphische Darstellung der numerischen Rechnungen für gasförmiges Caesium, Quecksilber, Xenon und Wasserstoff. Unter diesen hat Cs die geringste Ionisierungsspannung (3,87 V). Hg und X unterscheiden sich zwar in ihren Ionisierungsspannungen, aber die Differenz (12,08 gegenüber 10,38 V) wird durch die recht unterschiedlichen Zustandssummen, die in die SAHA-Gleichungen eingehen, wieder wettgemacht, so daß für beide Plasmen praktisch nur *eine* Grenzkurve für $I_\nu = 0,9\,B_\nu$ resultiert.

Bei kleinen Temperaturen sind relativ große Drucke erforderlich, um schwarze Strahlung der betreffenden Temperatur zu erhalten, weil verhältnismäßig wenig freie Elektronen für Rekombinationsprozesse zur Verfügung stehen. Nach Durchlaufen eines Minimums steigen die Kurven wieder an, weil die Gasdichte mit wachsender Temperatur bei konstantem Druck abnimmt. Da wir nur jeweils eine Ionisierungsstufe berücksichtigt haben, schmiegen sich diese ansteigenden Kurvenstücke einer Einhüllenden an, die sich aus Gl. (6.9) berechnen läßt, wenn

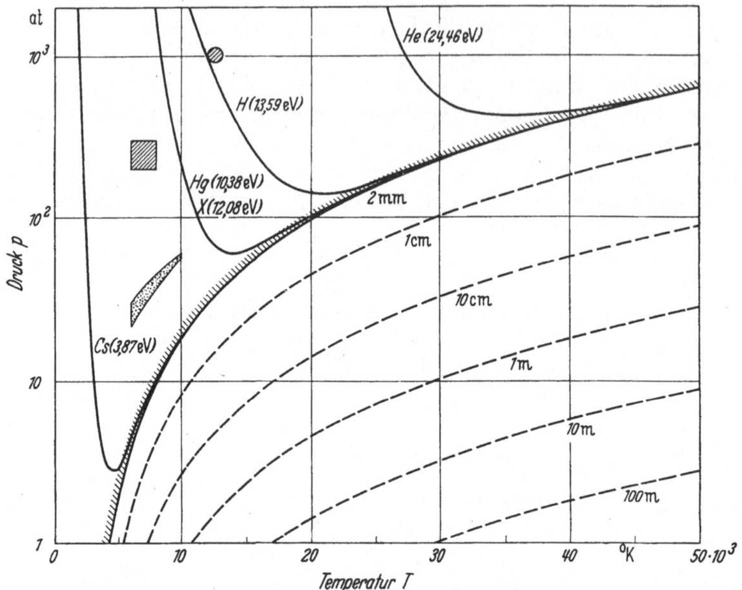

Fig. 7. Druck-Temperaturgebiete zur Emission schwarzer Strahlung von Gasen ($\lambda = 5000$ Å). Oberhalb der eingezeichneten Kurven strahlen die betreffenden Gase bei 2 mm Schichtdicke im sichtbaren Spektralgebiet praktisch schwarz ($I_\nu \geq 0,9\ B_\nu$). Wird jeweils nur die erste Ionisationsstufe berücksichtigt, so ergibt sich für alle Gase bei konstanter Schichtdicke eine einheitliche Grenzkurve, die dem Ionisationsgrad $x = 1$ entspricht. Diese Grenzkurven sind für verschiedene Schichtdicken des Plasmas gestrichelt eingetragen. ▨ Xe-Hochdrucklampen; ▨ Hg-Höchstdrucklampen; ◉ wasserstabilisierter Hochdruckbogen.

vollständige Ionisierung der Gase vorliegt. In diesem Fall werden nämlich die atomaren Unterschiede der einzelnen Gasarten unwesentlich. Bei vollständiger Ionisation können wir

$$N_i = N_e = \frac{p}{2kT}$$

setzen, und daher gilt für die Einhüllende nach Gl. (6.9)

$$p(T) = 4{,}8 \cdot 10^{22} \frac{(kT)^{\frac{5}{4}}}{\left(e^{2,88 \cdot 10^4/T} - 1\right)^{\frac{1}{2}}} \quad [\text{dyn cm}^{-2}], \left.\begin{array}{c} \\ \\ \end{array}\right\}$$

$$(I_\nu = 0{,}9\ B_\nu, \quad l = 0{,}2\ \text{cm}). \qquad\qquad (6.10)$$

In Fig. 7 sind ebenfalls die entsprechenden Grenzkurven für größere Schichtdicken gestrichelt eingezeichnet. Wie man sieht, ist schwarze Strahlung bei kleinen Drucken und hohen Temperaturen nur dann zu erwarten, wenn Gasausdehnungen kosmischer Größenordnungen (Sterne) vorliegen. Für die Zwecke der Lichttechnik, für welche schwarze Strahlung von Sonnentemperatur (5 bis 6000° K) erwünscht ist, würde ein Cs-Bogen von Atmosphärendruck oder wenig darüber am zweckmäßigsten sein (Mohler [327]).

Zum Vergleich mit der Theorie haben wir die Betriebswerte der bisher bekannten Hochdruckbögen eingetragen. Die Hg-Höchstdrucklampen (ELENBAAS [132]) kommen dem Schwarzstrahlergebiet für Hg sehr nahe, insbesondere aber sind die Xenon-Lampen von LARCHÉ [273], deren kathodische Plasmakugel bei einer Temperatur von etwa 10000° K eine Strahlung mit einer Leuchtdichte von über 10^6 cd/cm² emittiert, näherungsweise schon als schwarze Strahler zu betrachten. Auch im wasserstabilisierten Hochdruckbogen wurde von PETERS [355] — allerdings nicht bei stationärem Betrieb — schwarze Strahlung von etwa 12 bis 13 000° K erzielt, woran wesentlich das sog. Elektronenaffinitätsspektrum des negativen Wasserstoffions H⁻ (s. Ziff. 15 und 24) beteiligt ist. Über schwarze Strahlung von Funkenplasma wird Ziff. 15 berichten.

7. Wellenmechanische Theorie. α) *Theorie der Bremsstrahlung.* Die wellenmechanische Theorie der Bremsung freier Elektronen an Atomkernen ist von OPPENHEIMER [352], SUGIURA [413], [414], GAUNT [170], [171] und in besonders geschlossener und vollständiger Form von SOMMERFELD [409], SCHERZER [415] und MAUE [304] gegeben worden. Diese Autoren gehen teilweise auf durchaus verschiedenen Wegen vor, gelangen aber zu grundsätzlich gleichen Endformeln, die sich, wie erwähnt, von den KRAMERSschen nur wenig unterscheiden. Den Vergleich mit dem KRAMERSschen Resultat hat MAUE [304] sehr klar durchgeführt. Es zeigt sich, daß die in den klassischen Gleichungen auftretende Funktion $g(\nu) \approx 1$ zu setzen ist, um beste Übereinstimmung zu erreichen. Für langsame Elektronen und im langwelligen Spektralgebiet ist die Übereinstimmung ausgezeichnet, doch betragen die Abweichungen auch im kurzwelligen Gebiet nur bis zu 20%. Genaue Einzelheiten des Rechenganges findet man in SOMMERFELDS ,,Atombau und Spektrallinien", Bd. 2.

β) *Theorie der diskret-kontinuierlichen Elektronenkontinua.* Dem Vorgang von UNSÖLD folgend, hatten wir die klassisch korrespondenzmäßigen Rechnungen von KRAMERS näherungsweise auch auf diskret-kontinuierliche Übergänge angewandt. Das wesentlichste Ergebnis war die *Frequenzunabhängigkeit* der Emission, sofern die Terme im Atomtermschema hinreichend dicht liegen und als wasserstoffähnlich zu behandeln sind. Sobald diese Voraussetzungen fallen, wenn also einzelne Seriengrenzkontinua isoliert liegender Terme betrachtet werden, die auch nicht mehr näherungsweise wasserstoffähnlich sind, ist die Anwendung quantenmechanischer Methoden unumgänglich. Das gilt insbesondere für die Hauptseriengrenzkontinua.

In der SCHRÖDINGER-Gleichung

$$\frac{\hbar^2}{2m}\,\Delta\psi_a + \left(E_a - U(\mathfrak{r})\right)\psi_a = 0 \tag{7.1}$$

mit dem Potentialfeld $U(\mathfrak{r})$ des Atoms oder Ions, entspricht jedem Zustand der Energie E_a eine überall eindeutige und endliche Eigenfunktion $\psi_a(\mathfrak{r})$. Dabei sind für $E < 0$ solche eindeutigen und endlichen Lösungen nur für bestimmte Eigenwerte der Energie E_a möglich, so daß die E_a eine diskrete Folge mit einer Häufungsstelle bei $E = 0$ bilden. Für $E > 0$ dagegen ist eine kontinuierliche Folge von Energiewerten möglich. Wir haben von diesen Tatsachen schon des öfteren bei der Darstellung eines Atomtermschemas Gebrauch gemacht.

Die physikalische Bedeutung der ψ-Funktion ist die, daß $\psi_a \cdot \psi_a^* \, dV$ die Wahrscheinlichkeit dafür angibt, im Energiezustand E_a das Elektron in dem Volumelement dV anzutreffen. Entsprechend lautet die Normierungsbedingung $\int \psi_a \cdot \psi_a^* \, dV = 1$.

Die Berechnung der Intensität von Atomlinien wie der kontinuierlichen Emission bzw. Absorption eines Atoms setzt die Kenntnis dieser Eigenfunktionen

ψ voraus. Nach der klassischen Elektrodynamik ist die mittlere Aussstrahlung (erg/sec) eines harmonischen Oszillators der Frequenz ν, dessen elektrisches Moment $\mathfrak{M} = e \mathfrak{r}$ ist, gegeben durch

$$I = \frac{2}{3 c^3} \overline{\left(\frac{d^2 \mathfrak{M}}{d t^2} \right)^2}. \tag{7.2}$$

Für das einem Quantensprung $E_m \to E_n$ zugeordnete elektrische Dipolmoment $\mathfrak{M}_{mn}$ ergibt sich aus der Quantenmechanik

$$\mathfrak{M}_{mn} = e^{2 \pi i \nu_{mn} t} \, e \int \psi_m \mathfrak{r} \, \psi_n^* \, dV. \tag{7.3}$$

Setzen wir den hieraus nach zweimaliger Differentiation und Bildung des zeitlichen Mittelwertes folgenden Ausdruck wegen der Korrespondenzbeziehung zwischen klassischen und Quantenvorgängen in Gl. (7.2) ein, so folgt

$$I = \frac{64 \pi^4 e^2}{3 c^3} \, \nu_{mn}^4 \, \left| \int \psi_m \mathfrak{r} \, \psi_n^* \, dV \right|^2. \tag{7.4}$$

Bei den diskret-kontinuierlichen Übergängen treten in Gl. (7.4) die Eigenfunktionen gebundener Zustände, die wir mit ψ^i bezeichnen, und die Eigenfunktionen ψ^f freier (kontinuierlicher) Zustände auf. Rechnen wir in Absorption (Photoionisation), so gilt nach BATES [38] für den Absorptionsquerschnitt = Absorptionskoeffizient/Atom [s. Gl. (3.10)]

$$\varkappa_{\nu, \text{at}} = \frac{32 \pi^4 m^2 e^2}{3 h^3 c} \cdot \frac{1}{g_i} \sum_i \sum_f v \, \nu \left| \int \psi^{i*} \left(\sum_j \mathfrak{r}_j \right) \psi^f dV \right|^2. \tag{7.5}$$

g_i ist das statistische Gewicht des Ausgangszustandes, ν die Frequenz des eingestrahlten Lichtes, v die Geschwindigkeit des aus dem Atomverband austretenden Elektrons und $\mathfrak{r}_j$ der Ortsvektor des j-ten Elektrons. Die Summationen $\sum_i$ und $\sum_f$ gehen über alle Ausgangs- und Endzustände. Frequenz ν und Geschwindigkeit v sind bekanntlich verknüpft durch

$$h \nu = E_i + \tfrac{1}{2} m v^2, \tag{7.6}$$

wobei E_i die Ionisierungsenergie aus dem i-ten Anfangszustand bedeutet.

Die exakte Berechnung der Gl. (7.5) bietet allerdings außerordentliche Schwierigkeiten, so daß man gezwungen ist, auf Näherungslösungen zurückzugehen. Bei den wichtigen Zentralkraftfeldern kann durch Einführung von Kugelkoordinaten (r, ϑ, φ) die ψ-Funktion separiert werden, und zwar folgt für den gebundenen Zustand der Ausdruck

$$\psi^i = \left(\frac{(2l + 1)}{4 \pi} \cdot \frac{(l - m)!}{(l + m)!} \right)^{\frac{1}{2}} \cdot R_{n, l}(r) \cdot P_l^m (\cos \vartheta) \cdot e^{i m \varphi} \tag{7.7}$$

und für die freien Zustände

$$\psi^f = \sum_{l=0}^{\infty} (2l + 1) \, i^l e^{i \eta_{\varepsilon, l}} \cdot R_{\varepsilon, l}(r) \cdot P_l^0 (\cos \vartheta). \tag{7.8}$$

Dabei ist n die Hauptquantenzahl, l die Azimutal- (Bahndrehimpuls-) Quantenzahl, m die magnetische Quantenzahl und ε die kinetische Energie des befreiten Elektrons. Die Normierungsbedingungen für die radialen Anteile der Eigenfunktionen lauten für gebundene Zustände

$$\int_0^{\infty} R_{n, l}^2 \, r^2 \, dr = 1 \tag{7.9}$$

und für freie

$$R \approx \frac{1}{k\,r} \sin\left(k\,r - \frac{\pi}{2}\,l + \eta_{\varepsilon,\,l} + \alpha \cdot \ln 2k\,r\right) \tag{7.10}$$

mit

$$k = \frac{m\,v}{\hbar}$$

und

$$\alpha = \frac{Z\,e^2}{\hbar\,v} \quad (Z = \text{Ionenladungszahl}).$$

Für die Ionisation gelten folgende Auswahlregeln:

a) die Multiplizität ändert sich nur um ± 1,

b) der gesamte Bahndrehimpuls L um 0, $\pm 1, \ldots, \pm 1$.

Nach Integration über die Raumwinkel ϑ und φ und Berücksichtigung der Auswahlregeln erhält man, wenn noch

$$\varkappa_{v,\,\text{at}} = \sum k_{v,\,\text{at}} \quad \text{(Summation über alle Ionisationsstufen)}$$

gesetzt wird

$$k_{v,\,\text{at}} = \frac{32\,\pi^4\,m^2\,e^2}{3\,h^3\,c}\,\nu\,v\,4\pi\,\zeta_p\left(C_{l-1}\left|\int\limits_0^\infty R_{n,\,l}^i R_{\varepsilon,\,l-1}^f r^3\,dr\right|^2 + C_{l+1} \times \left. \times \left|\int\limits_0^\infty R_{n,\,l}^i R_{\varepsilon,\,l+1}^f r^3\,dr\right|^2\right).\right. \tag{7.11}$$

Im allgemeinen interessiert man sich nur für die tiefste Ionisationsstufe. Die Funktion ζ_p beschreibt die Korrektur des Absorptionsquerschnittes durch den Einfluß der inneren noch gebundenen Elektronen. Da deren Störung gewöhnlich klein bleibt, erhält ζ_p näherungsweise den Wert 1. Die komplizierten Ausdrücke für C_{l-1} und C_{l+1} sind in einer Arbeit von BATES [38] tabuliert.

Die wesentliche Aufgabe besteht nun darin, die in Gl. (7.11) auftretenden Integrale, vor allem die Eigenfunktionen im Integranden zu berechnen. Dazu wird oft die Methode des *self consistent field* von HARTREE [201], [202] verwandt, die aber vom Standpunkt der Quantenmechanik aus nicht ganz korrekt ist, weil hierbei die *Austausch- (Resonanz-) Effekte* (s. FOCK [154]) nicht berücksichtigt sind. BATES und MASSEY [35] haben die Bedeutung dieser Resonanzeffekte sowie der Polarisierbarkeit für die Berechnung der Eigenfunktionen hervorgehoben.

Eine möglichst genaue Bestimmung der ψ-Funktionen ist deshalb so wichtig, weil der Integrand in Gl. (7.11) meist oszilliert und geringe Fehler den Gesamtwert des Integrals schon stark beeinflussen. Möglicherweise können sich sogar positive und negative Anteile des Integranden bei der Integration praktisch gegenseitig aufheben (BATES [38]), so daß eine zuverlässige Angabe über die Größe des Absorptionskoeffizienten unmöglich wird.

Insgesamt zeigen die quantenmechanischen Rechnungen, daß in der Frequenzabhängigkeit des Absorptionskoeffizienten recht erhebliche Abweichungen vom wasserstoffähnlichen Verhalten ($\varkappa_\nu \sim \nu^{-3}$) auftreten können. In einigen Fällen steigt der Absorptionsquerschnitt, von der Hauptseriengrenze an gerechnet, mit wachsender Frequenz sogar an (z.B. Neon). Dagegen ergibt sich z.B. bei Natrium ein sehr steiler Abfall hinter der Grenze. Bei Kalium wiederum fällt der Absorptionsquerschnitt zunächst ab, und steigt nach Durchlaufen eines Minimums wieder an (s. Fig. 13). Auf jeden Fall haben die Absorptionsquerschnitte der neutralen Atome und positiven Ionen einen endlichen Wert an der Seriengrenze. Ein völlig anderes Verhalten ergeben die Rechnungen über negative Ionen: Hier geht der Querschnitt an der Seriengrenze gegen Null (Massey [*300*]) und durchläuft bei höheren Frequenzen ein Maximum (vgl. H⁻ in Ziff. 9 und 15).

Eine genauere Diskussion der bis 1946 durchgeführten Rechnungen und der Zuverlässigkeit der so ermittelten Eigenfunktionen findet man bei Bates [*38*]. Über die Ergebnisse der Theorie werden wir im Zusammenhang mit den experimentellen Resultaten in Ziff. 9 berichten.

II. Erzeugung und Beobachtung der Elektronenkontinua in Absorption und Emission.

a) Seriengrenzkontinua in Absorption und Photoionisation.

8. Allgemeines. Wir behandeln als erste Gruppe die in Absorption an die Atomlinienserien sich anschließenden Grenzkontinua und die mit diesen verwandten Erscheinungen, also die Absorption kontinuierlicher Strahlung durch gebundene Atom- und Molekülelektronen. Der Anfangszustand der absorbierenden Elektronen ist also jeweils ein diskreter Energiezustand, während der Endzustand im kontinuierlichen Energiebereich (vgl. Fig. 1) liegt, d.h. die Elektronen nach der Absorption frei sind.

Am einfachsten liegen die Verhältnisse beim Atom. Absorption von Spektrallinien ergibt Anregung des absorbierenden Leuchtelektrons in höhere diskrete Zustände, Absorption einer Wellenlänge des an die Grenze einer Linienserie nach kurzen Wellen sich anschließenden Grenzkontinuums Ionisation des Atoms, d.h. Abtrennung des absorbierenden Elektrons. Diese Loslösung von Elektronen unter Absorption der zur Abtrennung erforderlichen Energie bezeichnet man als *Photoionisation* oder allgemeiner als *Photoeffekt*, wenn die Befreiung des Elektrons nicht von einem bestimmten Atom oder Molekül, sondern aus einem Atomkomplex, etwa einem Metallgitter, erfolgt.

Die Bedeutung der Untersuchung der Absorptionsgrenzkontinua liegt nun zum Teil darin, daß die Ausbeute der Photoionisation durch Licht einer bestimmten Wellenlänge der Absorptionsintensität eben dieser Wellenlänge proportional ist. Die Ermittlung der Wellenlängenabhängigkeit des kontinuierlichen Absorptionskoeffizienten eines bestimmten Gases ergibt also direkt die Ausbeute der Photoionisation bei Einstrahlung dieser Wellenlängen und damit nach Gl. (7.6) ein Maß für die Geschwindigkeitsverteilung der auftretenden Photoelektronen.

Die Lage dieser Grenzkontinua im Spektrum hängt von der Bindungsfestigkeit, d.h. von der Ionisierungsenergie E_G der absorbierenden Elektronen ab, die nach $E_G = h\nu_G$ die Lage der langwelligen Grenze des Kontinuums bestimmt. Jedes Atom besitzt folglich außer dem an die Resonanzserie sich anschließenden kurzwelligsten Grenzkontinuum, das der Photoionisation des im Grundzustand befindlichen Leuchtelektrons entspricht, noch eine große Zahl langwelligerer Grenzkontinua, die sich an die langwelligeren Linienserien anschließen und der Photoionisation *angeregter* Elektronen entsprechen. Voraussetzung für ihre Beobachtung ist natürlich, daß genügend Elektronen in den betreffenden Zuständen vorhanden sind, und das ist im allgemeinen nur bei sehr hoher Temperatur, besonders also in den Sternatmosphären, der Fall. Im Laboratorium wird man daher praktisch in Absorption nur die zum Grundzustand der Atome gehörenden Grenzkontinua beobachten (Ziff. 9). Diese Überlegungen gelten offensichtlich in gleicher Weise für *Moleküle*, auf deren Absorptionsgrenzkontinua wir in Ziff. 10 kurz eingehen.

Die Absorption kontinuierlicher Strahlung braucht aber nicht nur durch die äußersten Elektronen (Valenzelektronen) zu erfolgen, sondern genügend kurzwellige Strahlung, etwa Röntgenstrahlung, kann auch durch sehr fest gebundene innere Elektronen von Atomen höherer Ordnungszahl absorbiert werden, wobei diese dann mit kinetischer Energie den Atomverband verlassen. So beträgt z.B. die Ionisierungsspannung eines der beiden innersten K-Elektronen des Silberatoms rund $E_G = 25\,000$ V, und tatsächlich beobachtet man bei Durchstrahlung einer dünnen Silberschicht mit kontinuierlichen Röntgenstrahlen ein Absorptionskontinuum, das sich von der diesem Energiewert entsprechenden Grenze $\lambda_G = 0{,}485$ Å aus nach kurzen Wellen zu erstreckt. In Ziff. 11 gehen wir kurz auf diese *Röntgenabsorptionskontinua* ein.

In diesem Abschnitt werden endlich zwei nur locker mit der Photoionisation zusammenhängende Sonderfälle behandelt, die *Autoionisation* in Ziff. 12 und das COMPTON-*Kontinuum* in Ziff. 13. Die Autoionisationsspektren entstehen durch Absorptionsübergänge von Elektronen in die in Ziff. 4 bereits erwähnten halbstationären Zustände, die durch Überlagerung diskreter und kontinuierlicher Energiezustände gleicher Energie entstehen (Fig. 2). Das COMPTONsche sog. Streukontinuum aber entsteht dadurch, daß monochromatische Lichtquanten hoher Energie, also harte Röntgenstrahlen bestimmter Wellenlänge, nach Absorption der zur Abtrennung eines Atomelektrons erforderlichen Energie wieder ausgestrahlt quantenhaft gestreut werden. Da die durch diesen Prozeß frei werdenden Elektronen verschiedene Beträge kinetischer Energie miterhalten, entsteht aus der primär scharfen Röntgenlinie ein *Streukontinuum*, dessen Intensitätsverteilung den Ablauf des COMPTON-Effektes im einzelnen zu untersuchen gestattet.

9. Spezielle Atom-Grenzkontinua. Zur quantitativen Beschreibung eines Absorptionskontinuums benutzt man vielfach den Begriff der „*differentiellen Oszillatorenstärke*". Der Zusammenhang mit dem Absorptionsquerschnitt = Absorptionskoeffizient pro Atom ist gegeben durch die Gleichung

$$\varkappa_{\nu,\,\mathrm{at}} = \frac{\pi\,e^2}{m\,c}\,\frac{df}{d\nu} = \frac{\pi\,e^2}{m\,c^2\,R}\,\frac{df}{dE} \tag{9.1}$$

mit $E = \nu/Rc$ (Energie des befreiten Elektrons in Einheiten der Ionisierungsenergie des Wasserstoffatoms) und R als RYDBERG-Zahl.

Für die kontinuierliche Gesamtabsorption ergibt sich weiterhin

$$\int\limits_{\nu_g}^{\infty} \varkappa_\nu\,d\nu = \frac{\pi\,e^2}{m\,c}\int\limits_{\nu_g}^{\infty}\frac{df}{d\nu}\,d\nu = \frac{\pi\,e^2}{m\,c}\,f; \tag{9.2}$$

f, abgekürzt vielfach durch $\int f_{\mathrm{kont}}$, bezeichnet man als die *Oszillatorenstärke* des betreffenden Kontinuums.

α) *Neutrale Atome.* Wasserstoff. Der Absorptionsquerschnitt (Absorptionskoeffizient pro Atom) im n-ten Quantenzustand ergibt sich aus der Theorie (KRAMERS [257], SUGIURA [413], GAUNT [170], [171], MENZEL und PEKERIS [309]) zu

$$\varkappa_{\nu,\,\mathrm{at}} = \frac{64\,\pi^4}{3\sqrt{3}}\,\frac{m\,e^{10}}{c\,h^6\,n^5}\,\frac{1}{\nu^3}\,g, \tag{9.3}$$

wobei g sich nur wenig von dem Wert 1 unterscheidet.

Die der Ionisation aus dem zwei- und dreiquantigen Zustand entsprechenden BALMER- und PASCHEN-Absorptionsgrenzkontinua sind in Sternspektren zu beobachten (s. Fig. 62).

Helium. Quantentheoretische Berechnungen der Absorption vom *Helium-grundzustand* 1^1S aus bei $\lambda \lesssim 504$ Å sind nach verschiedenen Verfahren von Wheeler [471], Vinti [452], [453], Körwien [256] und Su-Shu-Huang [227] durchgeführt worden. Für die Gesamtoszillatorenstärke f erhält Wheeler 1,58, Vinti 1,55, Körwien 1,56 und Huang 1,50 (für den Wasserstoffgrundzustand gilt z.B. $f = 0{,}437$). Da nach Kuhn [264], Thomas und Reiche [373] die Summe der f-Werte aller von einem Zustand aus möglichen Übergänge gleich ist der Zahl n der für den Übergang in Betracht kommenden Elektronen (f-Summensatz), so gibt das obige f-Resultat für das Heliumkontinuum mit $n = 2$ an, daß etwa 78% der gesamten Absorption normaler Heliumatome zur Ionisation führen. Einen Vergleich der Ergebnisse von Wheeler und Vinti für die Absorptionsin-

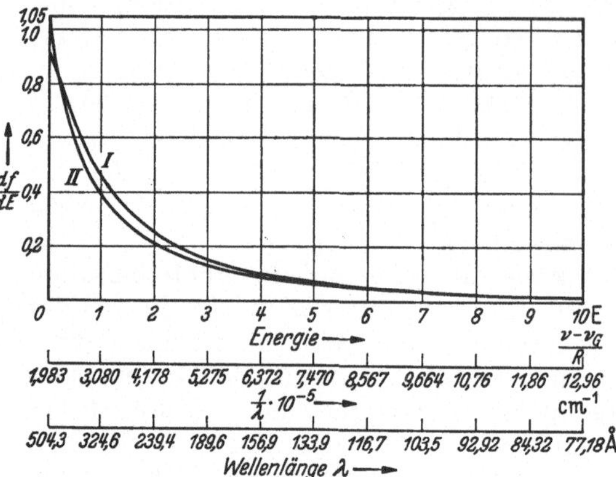

Fig. 8. Theoretischer Verlauf von df/dE im Hauptseriengrenz-kontinuum des He, aufgetragen gegen $\frac{v-v_G}{R}$, v und λ. (I nach Wheeler [471], II nach Vinti [453].)

tensität im Grenzkontinuum, die von 504 Å bis hinunter zu 10 Å berechnet worden ist, gibt Fig. 8. Körwiens Ergebnisse stimmen in dieser Darstellung fast vollkommen mit denen von Vinti (Kurve II) überein. Der Abfall nach kurzen Wellen erfolgt etwa $\sim 1/v^2$, also viel langsamer, als der Formel für wasserstoffähnliche Atome entspricht.

Die Grenzkontinua der scharfen und diffusen *Singulett*- und *Triplett*-Nebenserien sind zuerst von Goldberg [184], [185] und neuerdings teilweise noch einmal in anderer Weise von Su-Shu-Huang [227] berechnet worden. Die einzelnen Ergebnisse für die Absorptionskoeffizienten sind in Fig. 9 als Funktion von $\log v/R$ aufgetragen. Die Kontinua der höheren Heliumterme mit $n > 2$ können ebenso wie die frei-frei-Übergänge wasserstoffähnlich berechnet werden.

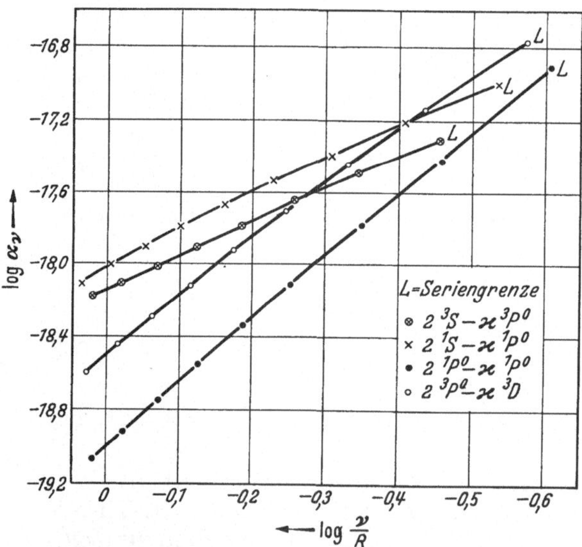

Fig. 9. Absorptionskoeffizienten der Grenzkontinua der Helium-Serien $2S - nP$ und $2P - nD$. (Nach Goldberg [184].)

Alkalimetalle. Für einen Vergleich von Theorie und Experiment eignen sich am besten die Absorptionskontinua der Alkalien, weil diese infolge der geringen Ionisierungsspannungen der Atome in einem spektroskopisch noch einigermaßen zugänglichen Spektralbereich liegen (erste Beobachtung durch Wood [480], [481] und Holtsmark [216]). Dafür besteht bei den Alkalien aber die Schwierigkeit, daß stets ein gewisser Bruchteil von Molekülen vorhanden ist, deren Absorption sich von der der Atome nicht immer einwandfrei trennen läßt. So erhielt Bott [69] an der Hauptseriengrenze des Natriums einen zu hohen Wert für den Absorptionsquerschnitt, weil der Beitrag der Molekülabsorption nicht eliminiert werden konnte.

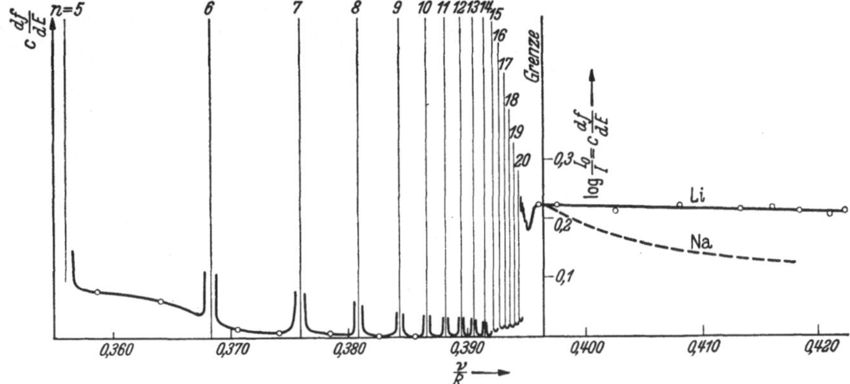

Fig. 10. Gemessener Verlauf von df/dE für Lithium und Natrium. (Nach Trumpy [439].)

Die Messung geschieht im Prinzip nach Gl. (3.10). Dazu muß die Anzahl der absorbierenden Atome über die gesamte Länge des Absorptionsrohres, also der Dampfdruck und die Temperatur, bekannt sein. Wegen gewisser Unsicherheiten in der Dampfdruckbestimmung konnten die ersten von Harrison [199] und Trumpy [437], [438], [439] durchgeführten Messungen nur relative Werte für die Absorptionsquerschnitte liefern (vgl. Fig. 10). Die neueren Meßmethoden sind von Ditchburn, Tunstead und Yates [116] eingehend beschrieben. Ditchburn, Jutsum und Marr [120] benutzen für die Absorptionsmessungen am *Natrium* ein etwa 2 m langes Nickelabsorptionsrohr, das teilweise auf etwa 400 bis 500° C aufgeheizt wird. Als Strahlungsquellen dienen Wasserstoffentladungen. Bei diesen Untersuchungen konnte der auf Molekülabsorption zurückgehende Anteil an der Gesamtabsorption durch Messung bei verschiedenen Dampfdrucken eliminiert werden. In derselben Arbeit wird auch ein allgemeiner Vergleich zwischen den bisher durchgeführten Rechnungen und Beobachtungen an den Alkalien angestellt. Für *Lithium* (Fig. 11) hat Trumpy bereits festgestellt, daß die Absorption an der Seriengrenze ein Maximum hat und auf der kurzwelligen Seite $\sim\lambda^{3,5}$ abfällt. Demgegenüber erhält Hargreaves [198] auf Grund seiner Rechnungen ein schwaches Maximum bei 1950 Å, welches aber von dem Experiment nicht bestätigt wird (Tunstead [440]).

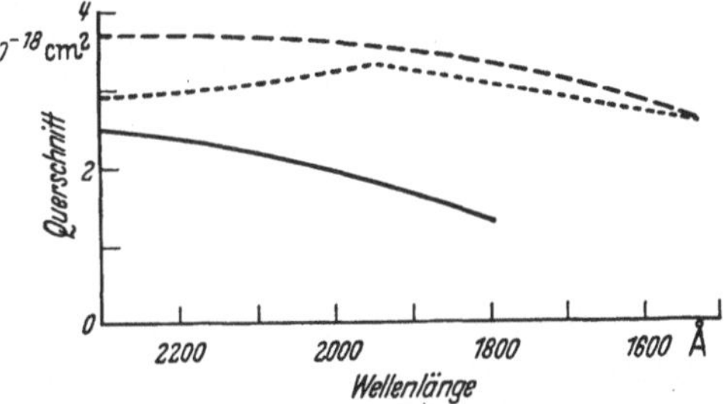

Fig. 11. Absorption von Lithium-Dampf. Meßergebnisse von Tunstead (1953) ———, theoretische Kurve von Trumpy (1931) — — — und Hargreaves (1929) - - - -. (Nach Ditchburn, Jutsum und Marr [120].)

Die Messungen am *Natrium* (Ditchburn, Jutsum und Marr [120]) und *Kalium* (Ditchburn, Tunstead und Yates [116]) stehen in relativ guter Übereinstimmung mit den Rechnungen von Bates [39] und Seaton [398]. Auffallend ist das Minimum auf der kurzwelligen Seite der Seriengrenze (Fig. 12 und 13).

Das gleiche Ergebnis haben die von Mohler und Boeckner [319] und Braddick und Ditchburn [70] durchgeführten Untersuchungen am *Caesium* erbracht (Fig. 15). Der bisher gemessene Verlauf des Absorptionsquerschnittes des *Rubidiums* in Fig. 14 deutet ebenfalls auf die Möglichkeit eines Minimums hin.

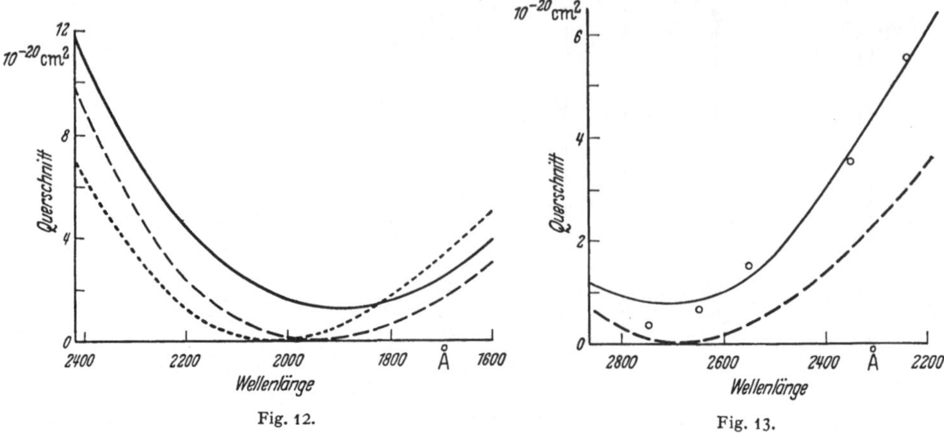

Fig. 12. Absorption von Natrium-Dampf. Meßergebnisse von Ditchburn, Jutsum und Marr ———. Berechnete Kurven von Seaton (1951) — — —, - - - -. (Nach Ditchburn, Jutsum und Marr [120].)

Fig. 13. Absorption von Kalium-Dampf. Meßergebnisse von Ditchburn und Mitarbeitern (1943) ———. Die Kreise geben die Beobachtungen von Mohler und Boeckner (1929) wieder. Berechneter Verlauf von Bates (1947) — — —. (Nach Ditchburn, Jutsum und Marr [120].)

Allgemein ist die Übereinstimmung zwischen Theorie und Experiment bei den Alkalien recht gut, mit Ausnahme der Minima. An diesen Stellen wird, wie erwähnt, die Rechnung unsicher, weil die beiden Integrale in Gl. (7.11) sich gegen-

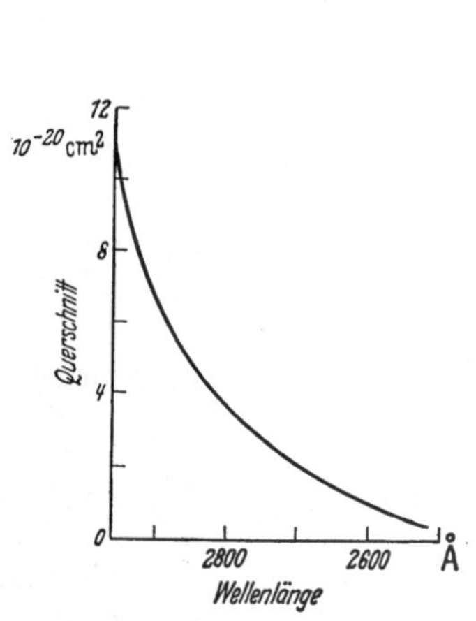

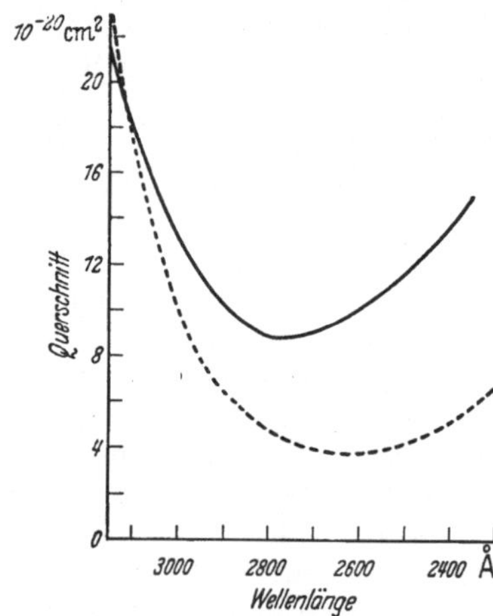

Fig. 14. Absorption von Rubidium nach Mohler und Boeckner (1929). (Nach Ditchburn, Jutsum und Marr [120].)

Fig. 15. Absorption von Caesium nach Braddick und Ditchburn (1935) ——— und von Mohler und Boeckner (1929) - - - -. (Nach Ditchburn, Jutsum und Marr [120].)

seitig kompensieren und daher geringe Fehler in der Bestimmung der Eigenfunktionen bedeutenden Einfluß gewinnen.

Erdalkalimetalle. Die bei ihrer Untersuchung des Na-Absorptionskontinuums verwendete Meßtechnik für das Vakuum-UV konnte von Ditchburn und Marr [118] auch zur Bestimmung des *Magnesium*-Absorptionskontinuums ausgenutzt werden. Wie Fig. 16 zeigt, fällt der Absorptionsquerschnitt sehr rasch (etwa mit λ^{15}) nach der Seriengrenze ab.

Während aber für Mg keine quantenmechanischen Rechnungen vorliegen, haben BATES [38] das *Beryllium-* und BATES und MASSEY [34] das Calcium-Absorptionskontinuum theoretisch behandelt. Es ergab sich für Be ein Querschnitt von $8{,}2 \cdot 10^{-18}$ cm² an der Grenze

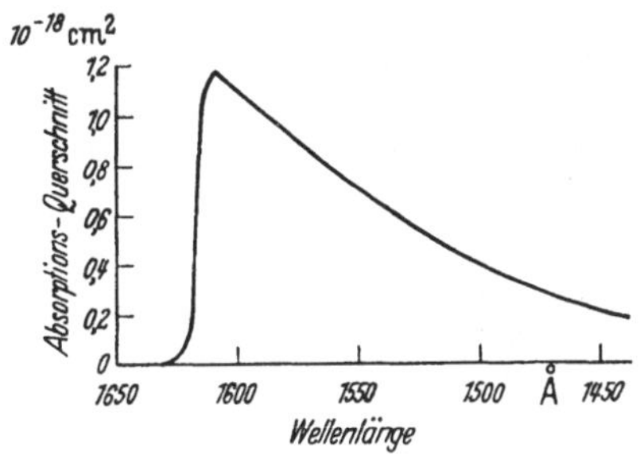

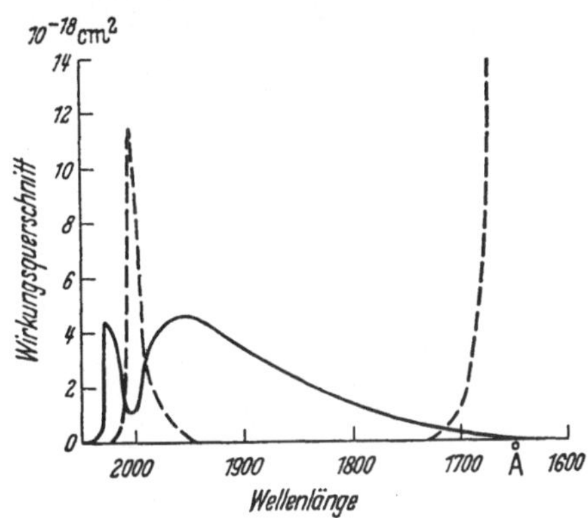

Fig. 16. Wirkungsquerschnitt für Photoionisation von Magnesium. (Nach DITCHBURN und MARR [118].)

Fig. 17. Trennung des Thallium-Kontinuums von den Autoionisationslinien nach MARR [296]. ——— Kontinuum; — — — Autoionisationslinien.

mit einem schnellen Absinken auf der kurzwelligen Seite, und für Ca ein Grenzwert von $25 \cdot 10^{-18}$ cm² und ein entsprechender Abfall mit λ^3.

Von den Elementen der *III. Gruppe* des Periodischen Systems ist nur *Thallium* experimentell von MARR [296] untersucht worden. Seine Arbeit ist auch insofern besonders interessant, als das Hauptserien-Grenzkontinuum von einigen *Autoionisations*-Linien überlagert ist, deren Intensität bei der Ermittlung des wahren Kontinuumverlaufes von der gemessenen Gesamtintensität in dem betreffenden Wellenlängenbereich abgezogen werden muß (Fig. 17). Auf die Bedeutung der Autoionisation gehen wir in Ziff. 12 kurz ein.

Schließlich ist zu erwähnen, daß BATES [32] die Absorptionsquerschnitte der Elemente B, C, N, O, F, Ne für den Grundzustand über einen weiten Frequenzbereich quantenmechanisch berechnet hat. Wie man der Fig. 18 entnimmt, kommen nacheinander recht verschiedene Verteilungsfunktionen vor. Die Messungen von LEE und WEISSLER [279], [280] am Neon im Bereich zwischen 230 und 800 Å stehen in guter Übereinstimmung mit den Rechnungen von BATES und SEATON [400]. Weitere Arbeiten sind in Tabelle 2 angegeben.

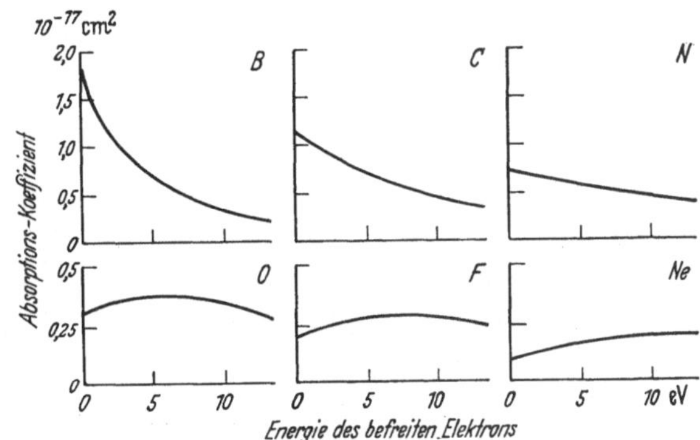

Fig. 18. Verlauf der kontinuierlichen Absorptionskoeffizienten (2p-Elektronen) der Atome Bor bis Neon als Funktion der Energie des befreiten Elektrons nach BATES [32].

β) *Positive Ionen.* Der He⁺-Absorptionsquerschnitt kann ebenso wie der des Wasserstoffes wieder völlig exakt mit $Z = 2$ berechnet werden. Für das Kontinuum des *Grundzustandes* gilt folglich nach Gl. (9.3)

$$\varkappa_\nu = \frac{64\,\pi^4}{3\,\sqrt{3}}\,\frac{m\,e^{10}\,Z^4}{c^4\,h^3}\,\lambda^3 . \tag{9.4}$$

Tabelle 2. *Arbeiten über Absorptionsquerschnitte.*

Neutrale Atome	Experimentell	Theoretisch
H (exakt) He	Balmer-Grenzkontinuum in Emission, s. Ziff. 15 —	Kramers [257] 1923 Sugiura [413] 1929 Gaunt [170], [171] 1930 Menzel und Pekeris [309] 1936 Vinti [452], [453] 1932 Wheeler [471] 1933 Körwien [256] 1934 Goldberg [184], [185] 1939 Huang [227] 1948
Li Na K Rb Cs	Trumpy [437], [438], [439] 1929 Tunstead [440] 1953 Trumpy [437], [438], [439], 1928 Ditchburn und Jutsum [117] 1950 Ditchburn, Jutsum und Marr [120] 1953 Ditchburn [115] 1928 Mohler und Boeckner [319] 1929 Ditchburn, Tunstead und Yates [116] 1943 Lawrence und Edlefsen [276] 1929 Mohler und Boeckner [319] 1929 Mohler und Boeckner [319] 1929 Braddick und Ditchburn [70] 1935	Hargraeves [198] 1929 Trumpy [437], [438], [439] 1927 Trumpy [437], [438], [439] 1931 Rudkjøbing [391] 1940 Seaton [398] 1951 Phillips [358] 1932 Bates [39] 1946 — —
Be Mg Ca	— Ditchburn und Marr [118] 1953 —	Bates [38] 1946 — Bates und Massey [34] 1941
O	—	Bates, Buckingham, Massey und Unwin [33] 1939 Yamanouchi und Kotani [491] 1940
B, C, N, O, F, Ne	Ne: Lee und Weissler [279], [280] 1953	Bates [32] 1939 Bates und Seaton [40] 1949 Seaton [400], [397], [399] 1951—1954
Tl	Marr [296] 1954	—

Positive Ionen	Experimentell	Theoretisch
He+ K+ Ca+ Mg+ Si+ C+, N+, O+, F+, Ne+, Na+ K+	— — — — — — —	Exakt wasserstoffähnlich zu berechnen Seaton [398] 1950 Bates und Massey [35] 1941 Green [189] 1949 Green und Weber [190] 1950 Green, Weber und Krawitz [191] 1951 Biermann und Lübeck [64] 1948 Biermann [62] 1947 Biermann und Lübeck [64] 1948 Bates [38] 1946 Seaton [398] 1950

Tabelle 2. (Fortsetzung.)

Negative Ionen	Experimentell	Theoretisch
H^-	In Emission s. Ziff. 15	WILDT [475], [476] 1939 JEN [244] 1933 BATES und MASSEY [301] 1940 CHANDRASEKHAR [90] 1945 CHANDRASEKHAR und BREEN [91] 1946 STAVER [429] 1949
Na^-	—	HUANG [226] 1945
O^-	—	BATES und MASSEY [36] 1943 BATES [37] 1946
Cl^-	—	MASSEY und SMITH [299] 1936

Außer dem Heliumion sind die Ionen K^+, Ca^+, Mg^+, Si^+ und die Gruppe C^+, N^+, O^+, F^+, Ne^+, Na^+ (Fig. 19) theoretisch behandelt worden (Autoren s. Tabelle 2).

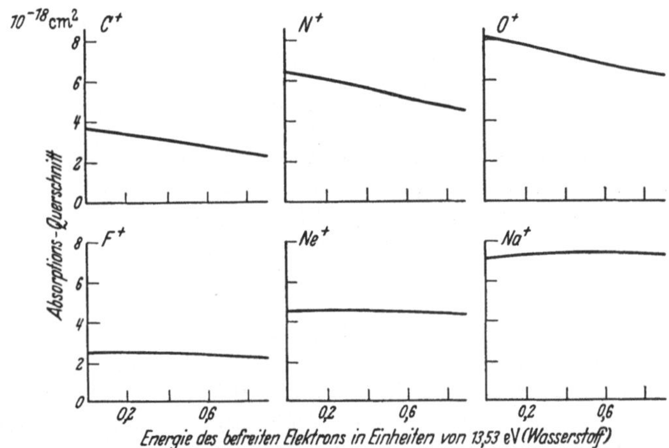

Fig. 19. Absorptionsquerschnitte der positiven Ionen C^+, N^+, O^+, F^+, Ne^+, Na^+ nach BATES [38]. Ionisierungsenergien in eV:

C^+ 24,26	F^+ 34,91
N^+ 29,44	Ne^+ 40,86
O^+ 34,94	Na^+ 47,14

γ) Negative Ionen. Die ersten quantenmechanischen Rechnungen über den H^--Absorptionsquerschnitt (auch frei-frei-Absorption) stammen von JEN [244], BATES und MASSEY [301]. Später haben sich besonders CHANDRASEKHAR [90] und CHANDRASEKHAR und BREEN [91] eingehend mit der Absorption des H^--Ions beschäftigt. Auf ihre speziellen Rechenmethoden können wir nicht im einzelnen eingehen. Eine kurze Darstellung der Theorie und tabellierte Werte findet man bei UNSÖLD [449]. Fig. 20 gibt ein Ergebnis von CHANDRASEKHAR wieder, das wahrscheinlich typisch für alle negativen Ionen ist. Aus der Ionisierungsenergie des H^--Ions im Grundzustand $(1s)^2\,{}^1S$ von 0,75 eV folgt, daß sich die diskret-kontinuierliche Absorption von der entsprechenden Wellenlänge $\lambda = 16550$ Å nach kürzeren Wellen erstreckt. An der Grenze beginnt die Absorption mit Null, erreicht bei $\lambda \sim 8500$ Å ein Maximum und nimmt dann nach kurzen Wellen wieder ab. Die frei-frei-Absorption nimmt von kurzen nach langen Wellen auch über die Grenze $(\lambda = 16550$ Å$)$ hinaus stetig zu.

Über die Rolle des H⁻-Ions in der Gasentladungsphysik und Astrophysik wird in den Ziff. 15 und 24 berichtet.

Weitere Arbeiten über negative Ionen (Na^-, O^-, Cl^-) findet man wiederum in Tabelle 2.

Wegen der besonderen Bedeutung der in dieser Ziffer aufgeführten Literatur für die Physik der kontinuierlichen Spektren, stellen wir die wichtigsten Arbeiten noch einmal in Form einer Tabelle zusammen.

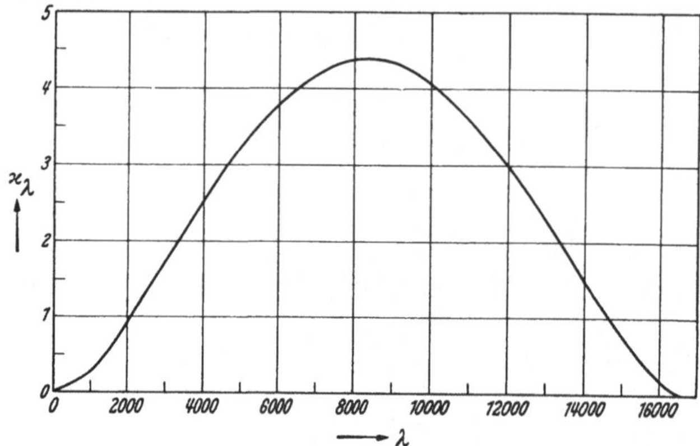

Fig. 20. Absorptionsquerschnitt (in 10^{-17} cm²) pro H⁻-Ion, berechnet von Chandrasekhar [90].

10. Ionisationskontinua von Molekülen. Eine Elektronenemission unter Absorption kontinuierlicher Strahlung muß grundsätzlich auch bei Molekülen möglich sein. Infolge der Überlagerung der Elektronenzustände der Moleküle durch die Schwingungs- und Rotationszustände entsprechen den Linienserien der Atome bei den Molekülen Rydberg-*Serien* von Bandensystemen, die einer Konvergenzstelle zustreben, an die sich nach kurzen Wellen hin ein die Ionisation anzeigendes Absorptionskontinuum anschließt. Die Verhältnisse werden aber dadurch recht kompliziert, daß im allgemeinen infolge der Anregung bzw. Ionisation der Bindungscharakter der Moleküle sich ändert und daher nach dem Franck-Condon-*Prinzip* (Ziff. 28) in vielen Fällen eine Photodissoziation wahrscheinlicher werden kann als eine Photoionisation.

Mit Sicherheit nachgewiesen sind *Molekül-Ionisationskontinua* z.B. von Price [361] bis [367] für eine Reihe mehratomiger Moleküle. Bei den mehratomigen Molekülen liegen die Verhältnisse insofern einfacher als bei zweiatomigen, als bei Anregung bzw. Photoionisierung eines nichtbindenden Elektrons die Atomanordnung des Moleküls sich nicht ändert und daher Schwingungsspektren nur in sehr beschränktem Umfange auftreten können. So sind denn von Price bei C_2H_2, C_2H_4, C_6H_6, H_2CO, $(CH_3)_2CO$, CH_3HCO, CH_3OH, CS_2, SO_2, H_2S und vielen anderen mehratomigen Molekülen Rydberg-Serien gefunden worden, deren intensivste im allgemeinen zur Grenze des Molekülions gehen und durch Extrapolation eine sehr genaue Bestimmung der betreffenden Ionisationsenergie gestatten. Bei einer Reihe dieser Moleküle (H_2O, H_2S, C_2H_2, C_6H_6, Alkylhalogenide, HCOOH und verwandte Moleküle) sind auch ausgeprägte Photoionisationskontinua gefunden worden. Neuerdings wurde von Watanabe, Marmo und Inn [461] das Ionisationskontinuum des NO-Moleküls im Spektralbereich zwischen 1070 und 1345 Å durchgemessen. Aus der Kontinuumsgrenze bei 1345 Å folgt eine Ionisierungsenergie von 9,2 eV.

11. Röntgenabsorptionskontinua. Wir haben bisher nur die Photoionisation der am leichtesten abtrennbaren Elektronen (Leuchtelektronen) von Atomen und Molekülen behandelt. Auch innere Elektronen abgeschlossener Schalen können aber unter Absorption kontinuierlicher Strahlung das Atom verlassen, und zwar wieder unter Mitnahme kinetischer Energie. Entsprechend der größeren Abtrennenergie dieser inneren Elektronen liegen ihre Ionisationskontinua im wesentlichen im Röntgengebiet, wo sie als Absorptionskanten der K-, L-, M- usw. Kante bezeichnet werden. Sie erstrecken sich, wie alle Grenzkontinua, von der durch die Ionisierungsenergie $E_G = h\nu_G$ gegebenen langwelligen Grenze mit abnehmender Absorptionsintensität nach kurzen Wellen zu. Fig. 21 zeigt die Verhältnisse an Hand eines Atomtermschemas, in dem jetzt im Gegensatz zu unserer bisherigen Darstellung (Fig. 1 und 2) auch die vollbesetzten Elektronenschalen

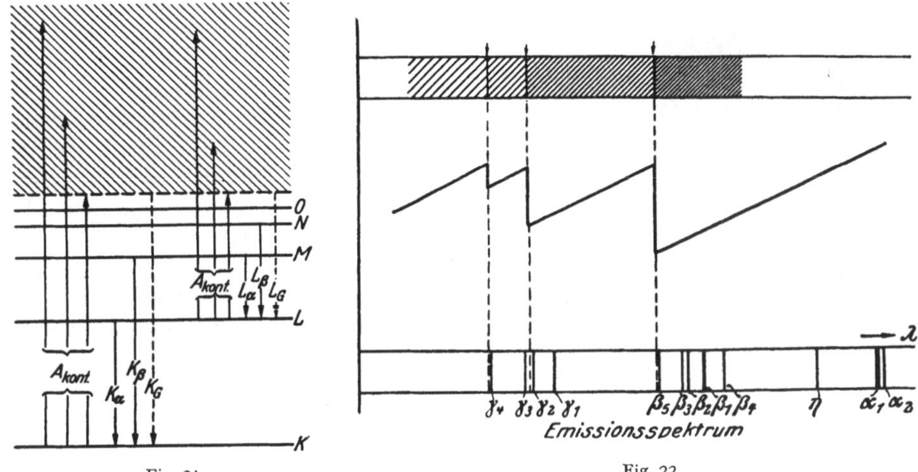

<table>
<tr><td>Fig. 21.</td><td>Fig. 22.</td></tr>
</table>

Fig. 21. Röntgen-Übergangsschema. Zusammenhang der K- und L-Absorptionskontinua mit den Emissionslinien der K- und L-Serie.

Fig. 22. Schematische Darstellung des Zusammenhanges zwischen Absorptionskontinuum, Absorptionsverlauf und Emissionslinien bei der L-Serie.

eingezeichnet sind, und zwar die K-Schale um den Betrag der Bindungsenergie eines K-Elektrons unterhalb der Ionisierungsgrenze. A_K stellt dann die Ionisation eines K-Elektrons unter Absorption kontinuierlicher Röntgenstrahlung dar. Die dadurch in der K-Schale entstandene Lücke wird durch ein Elektron einer höheren Schale unter Emission des frei werdenden Energiebetrages (charakteristische Röntgen-Linienstrahlung) ausgefüllt. Die K-Emissionsserie liegt also, wie Fig. 21 erkennen läßt, auf der langwelligen Seite der K-Absorptionskante. Entsprechend liegt die L-Emissionsserie auf der langwelligen Seite des durch die Photoabsorption eines L-Elektrons entstehenden L-Absorptionskontinuums. Entsprechend der Vielfachheit der L-, M- usw. Energiezustände gibt es drei L-Kanten, fünf M-Kanten usw.

Fig. 22 zeigt den Zusammenhang der L-Absorptionskanten mit den L-Emissionslinien; die Absorption ist dabei nach oben aufgetragen. Als Beispiel gibt Fig. 23 eine Aufnahmenserie von WAGNER [458] wieder, der gleichzeitig mit DE BROGLIE [72] diese Röntgenabsorptionskontinua erstmalig beobachtet und gedeutet hat.

Nimmt man ein kontinuierliches Röntgenspektrum mit einer photographischen (Bromsilber-) Platte auf, so erhält man auf der Platte Schwärzungs-Diskontinuitäten an den Stellen der K-Absorptionskante des Silbers und des Broms, und zwar zeigt sich die verstärkte Absorption durch vergrößerte Schwärzung an. Auf den Aufnahmen a, c, e und g von Fig. 23 erkennt man deutlich die Ag-Kante;

die verstärkte Schwärzung nimmt von der Kante nach kurzen Wellen (links) zu mit der Absorption ab. Die Aufnahmen b, d und f sind nun dadurch gewonnen worden, daß in den Weg des Röntgenstrahls eine absorbierende Schicht aus Cadmium, Silber bzw. Palladium eingeschaltet wurde. Durch sie wurden aus dem Röntgenspektrum die auf der kurzwelligen Seite der Metallabsorptionskanten liegenden Gebiete herausabsorbiert; die photographische Platte erscheint an diesen Stellen also weniger stark geschwärzt. Aufnahme d zeigt, daß die Ag-Kante, wie zu erwarten, genau mit der einen Kante der Bromsilberschicht übereinstimmt.

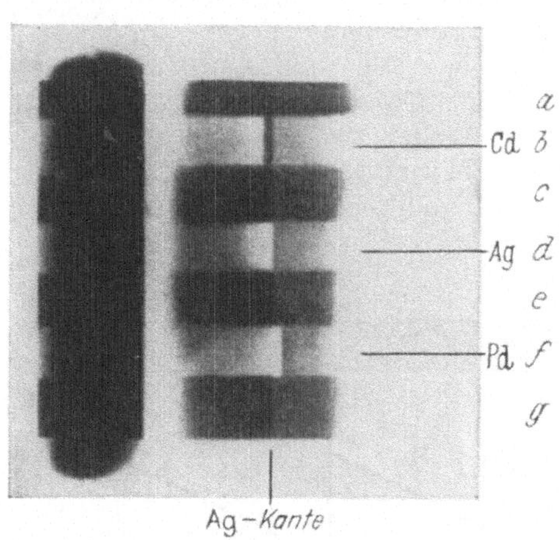

Fig. 23. Aufnahme der Absorptionskanten von Cd, Ag und Pd auf Bromsilberplatte von Wagner [458].

Über die Abhängigkeit der Absorptionsstärke von der Wellenlänge und der Ordnungszahl des absorbierenden Elementes liegt ein umfangreiches Material vor, auf das wir an dieser Stelle nicht näher eingehen können (s. Band XXXI des Handbuches).

In grober Näherung kann wiederum die korrespondenzmäßige Theorie von Kramers herangezogen werden. Dann gilt

$$\varkappa_\lambda = C\,Z^4\,\lambda^3,$$

wo Z die Ordnungszahl und C eine für die K-, L-Kante jeweils charakteristische Konstante ist. Im Gegensatz zu den „optischen" Elektronen hängt die Photoionisierungswahrscheinlichkeit eines inneren Elektrons also in erster Näherung nur von der Kernladungszahl Z und nicht mehr von der Struktur der äußeren Elektronenhülle ab, die die großen Unterschiede der in Ziff. 9 behandelten Absorptionsgrenzkontinua bedingt.

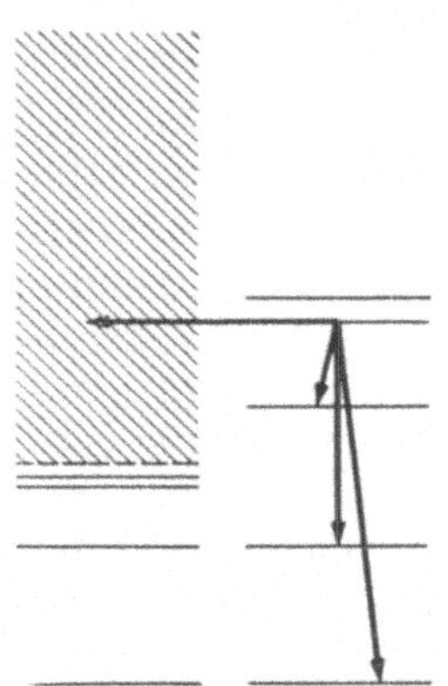

Fig. 24. Termschema zur Veranschaulichung der Konkurrenz zwischen strahlungslosen und Strahlungsübergängen bei Autoionisation.

12. Autoionisationsspektren und Auger-Effekt. Im Anschluß an die Photoionisationskontinua müssen wir uns kurz mit den spektralen Erscheinungen befassen, die die strahlungslose Ionisation von Atomen und Molekülen begleiten. In Ziff. 4 wurde schon gezeigt, daß bei Mehrelektronensystemen nicht selten Energieresonanz zwischen einem stationären (etwa doppelt angeregten) und einem freien, ionisierten Zustand vorkommt. Ist der Übergang zwischen diesen beiden Zuständen nach den Auswahlregeln (Shenstone [405]) erlaubt, so kann aus dem stationären Zustand ein strahlungsloser Übergang in den ionisierten Zustand erfolgen, wobei das zweite vorher angeregte Elektron in den Grundzustand übergeht. Man bezeichnet diesen von selbst ohne Einwirkung von außen verlaufenden Vorgang als *Autoionisation*, gelegentlich auch als *Präionisation*. Für einen solchen doppelt angeregten, über der Ionisationsgrenze liegenden Atomoder Molekülzustand steht also der strahlungslose Übergang in den ionisierten Zustand in Konkurrenz mit den unter Strahlungsemission verlaufenden Übergängen zu tiefer gelegenen Zuständen (vgl. Fig. 24). Wentzel [468] und

FUES [*166*] haben zuerst wellenmechanisch die relative Wahrscheinlichkeit der mit Strahlung verbundenen und der strahlungslosen Übergänge von einem bestimmten autoionisierenden Term aus berechnet. Infolge der großen Wahrscheinlichkeit strahlungsloser Übergänge autoionisierender Zustände ist deren

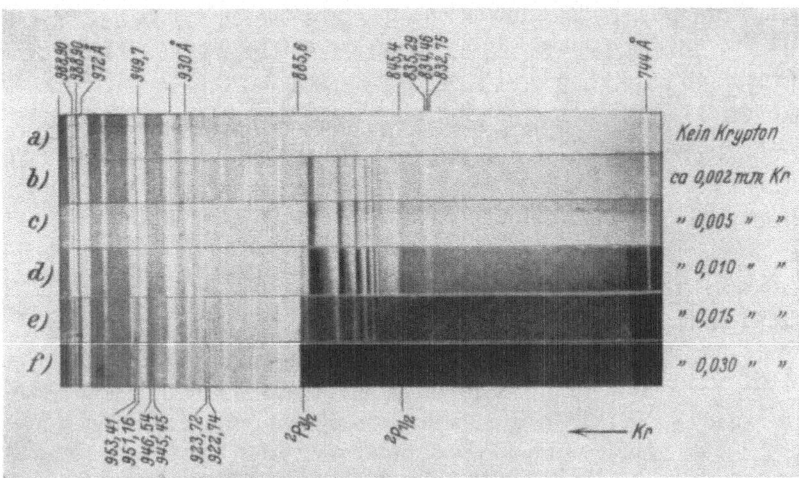

Fig. 25. Absorptionsspektrum des Kryptons mit Nachweis der Autoionisation auf der kurzwelligen Seite der ^2P-Grenze. (Nach BEUTLER [*58*].)

Lebensdauer oft um mehrere Größenordnungen kleiner als die gewöhnlicher, nur unter Strahlungsemission bzw. -absorption kombinierender Zustände, und nach der HEISENBERGschen Unschärferelation macht sich diese verkürzte Lebens-

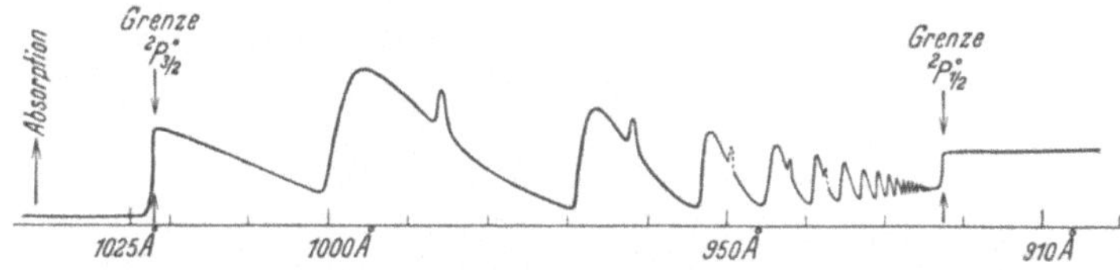

Fig. 26. Graphische Darstellung der bandartigen Linienverbreiterungen infolge von Autoionisation im Absorptionsspektrum des Xenons. (Nach BEUTLER [*58*].)

dauer des oberen Zustandes durch eine vergrößerte Breite der entsprechenden Linien bemerkbar. Von den neueren theoretischen Arbeiten erwähnen wir die Rechnung von WU [*485*] über einige doppelt angeregte Heliumzustände. Die Ergebnisse (Lebensdauern von 10^{-14} bis 10^{-15} sec) stimmen mit denen von BRANSDEN und DALGARNO [*71*] gut überein. WU und OUROM [*486*] führten ähnliche Betrachtungen am Beryllium durch. Sie erhielten jedoch erheblich zu große Werte im Vergleich zu den Beobachtungen von KRÜGER und PASCHEN [*263*].

Umfangreiche Messungen über Autoionisationslinien der Elemente K, Rb, Cs, Zn, Cd, Hg, Tl, A, Kr, X sind von BEUTLER und Mitarbeitern [*51*] bis [*56*], [*58*] durchgeführt worden. Von besonderem Interesse ist BEUTLERs Untersuchung [*58*] über die asymmetrische Verbreiterung autoionisierender Zustände der Edelgase, die zu einem bandartigen Aussehen der oberhalb der Ionisierungsgrenze liegenden Absorptionslinien Anlaß gibt (s. Fig. 25 und 26). Die schon im Zusammenhang mit dem Tl-Grenzkontinuum erwähnte Arbeit von MARR [*296*] zeigt, in Übereinstimmung mit den Ergebnissen von BEUTLER und DEMETER [*56*], drei Autoionisationslinien bei 2007, 1610 und 1490 Å.

Während aber eine beträchtliche Anzahl von Publikationen über die Identifizierung solcher Linien vorliegt (s. auch Garton [169]), sind nur wenige Halbwertsbreiten wirklich vermessen worden. Die Beobachtungen deuten allgemein darauf hin, daß recht verschiedene Lebensdauern vorliegen können. Allen [6] und Shenstone [407] untersuchten Autoionisationslinien im Cu-Spektrum, deren Breite eine Lebensdauer der angeregten Zustände von 10^{-12} sec ergab. Im Bogenspektrum des Silbers fand Shenstone [406] drei Linien, die er als Autoionisationslinien deutete. Aus der breitesten Linie ergab sich eine Lebensdauer von 10^{-14} sec. Beutler und Jünger [59], [60] ist es auch gelungen, die Autoionisation von Molekülen bei Untersuchung des kurzwelligen ultravioletten Absorptionsspektrums des H_2 aufzufinden. Anzeichen für Präionisation in den Absorptionsspektren des CO und H_2O fand auch Henning [204].

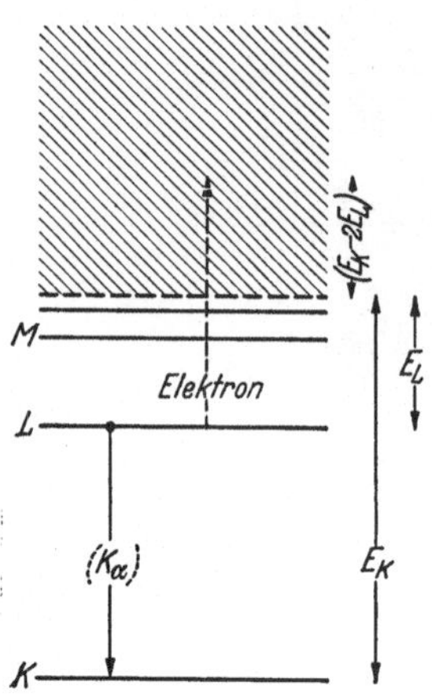

Fig. 27. Röntgen-Termschema zur Veranschaulichung des Auger-Effektes.

Auf die bei Molekülen sehr häufige, der Autoionisation verwandte Erscheinung der Prädissoziation kommen wir im Zusammenhang mit den Molekülkontinua in Ziff. 29 noch zu sprechen.

Die der Autoionisation entsprechende Erscheinung im Röntgengebiet bezeichnet man nach ihrem Entdecker als Auger-*Effekt*. Eine z.B. in der K-Schale eines Atoms durch Ionisation entstandene Elektronenlücke wird durch Übergang eines Elektrons aus einer der äußeren Schalen ausgefüllt. Die dabei frei werdende Energie kann entweder in der üblichen Weise in Form einer Linie der K-Serie ausgestrahlt werden; sie kann aber auch zur Emission eines Elektrons der gleichen Schale verwandt werden. Ist E_K in Fig. 27 die Bindungsenergie eines K-Elektrons, E_L die eines L-Elektrons, so kann die Energie $E_K - E_L$ auf ein Elektron der L-Schale übertragen werden, das nach Überwindung seiner Bindungsenergie E_L das Atom mit der kinetischen Energie

$$E_{\text{kin}} = E_K - 2E_L$$

verläßt. Es besteht hier also Energieresonanz und damit die Möglichkeit zu strahlungslosen Übergängen zwischen einem in der K-Schale ionisierten und einem in der L-Schale doppelt ionisierten Atom, da ja durch Übergang eines L-Elektrons in die K-Lücke und die Emission eines zweiten L-Elektrons in der L-Schale zwei Lücken entstehen. Die Emission solcher Elektronen mit der kinetischen Energie $E_K - 2E_L$ wurde nun von Auger [18] mittels der Wilson-Kammer beobachtet. Neuere Arbeiten über die Auger-Übergänge und die Breiten von Röntgenenergie-Termen der Elemente von $73 \leq Z \leq 92$ wurden von Cooper [104] durchgeführt.

13. Das Comptonsche Streukontinuum. Als Sonderfall der bei der Photoionisation von Atomen und Molekülen auftretenden Kontinua haben wir noch das Comptonsche *Streukontinuum* zu behandeln. Im allgemeinen wird die Photoionisation ja durch ein Absorptionskontinuum angezeigt. Es entsteht nach Ziff. 8 dadurch, daß Lichtquanten verschiedener Energie von den Atomen absorbiert werden, wobei die abgetrennten Elektronen den ihre Ionisierungsspannung übersteigenden Energiebetrag als kinetische Energie mitnehmen. Beim Compton-Effekt dagegen absorbieren die Atome zunächst monochromatische Lichtquanten großer Energie (Röntgenstrahlen *bestimmter* Wellenlänge). Von dieser absorbierten Energie wird ein aus Energie- und Impulssatz folgender

Betrag zur Abtrennung eines Atomelektrons unter Mitgabe kinetischer Energie aufgewandt, während der Rest in Form von Röntgenquanten geringerer Energie, also größerer Wellenlänge, wieder emittiert wird. Den Gesamtvorgang bezeichnet man als Quantenstreuung.

Wentzel [469] wies zuerst darauf hin, daß der Compton-Effekt quantentheoretisch in üblicher Weise als Übergang eines Elektrons aus dem gebundenen in den ionisierten, freien Zustand aufgefaßt werden kann, und berechnete die Wahrscheinlichkeit solcher Übergänge. Da der auf das Elektron übertragene Energiebetrag sich elementar aus den Erhaltungssätzen von Energie und Impuls berechnen läßt, hängt er und damit der Abstand der Streulinie von der Primärlinie vom Streuwinkel ab. Bei Rückwärtsstreuung des Lichtquants wird der größtmögliche Betrag an kinetischer Energie auf das Elektron übertragen, und die Energie des gestreuten Quants ist entsprechend die kleinstmögliche. Im umgekehrten Fall sehr kleiner Streuwinkel fallen Primärlinie und Streustrahlung fast zusammen.

Wir betrachten nun die unter einem bestimmten Winkel gestreute Strahlung. Besäßen die die Energie übernehmenden Elektronen im Anfang alle gleiche Energie und gleichen Impuls, erfolgte die Compton-Streuung also an

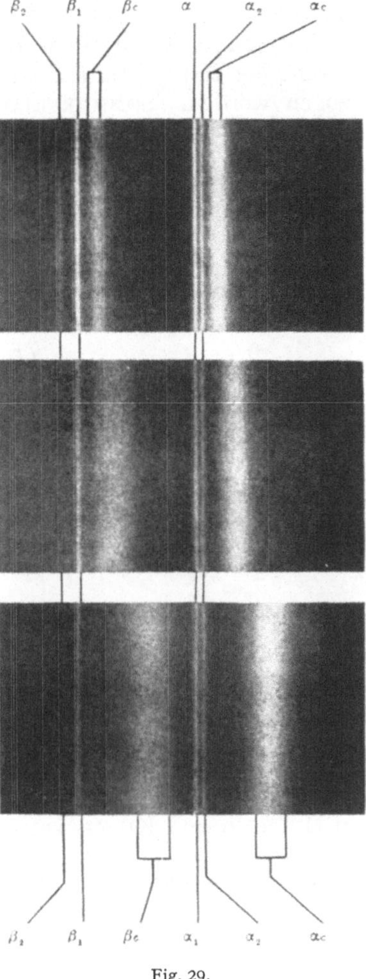

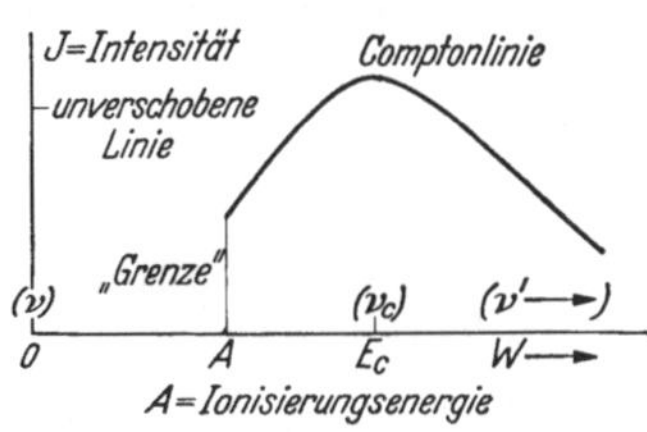

Fig. 28.   Fig. 29.

Fig. 28. Theoretischer Intensitätsverlauf der an einzelnen Atomelektronen gestreuten primär linienhaften Strahlung (Compton-Kontinuum). (Nach Franz [164].)

Fig. 29. Aufnahme des Comptonschen Streukontinuums von DuMond [124]; Molybdän-K-Strahlung, gestreut an Graphit, aufgenommen mit dem Vielkristall-Spektrographen von DuMond und Kirkpatrick. Die Aufnahmen zeigen die Abhängigkeit der Breite des Streukontinuums vom Streuwinkel Θ und stellen den ersten direkten Nachweis der statistischen Verteilung der Impulse der streuenden Atomelektronen dar. Von oben nach unten: $63° < \Theta < 64°$, $89°30' < \Theta < 90°30'$, $156°12' < \Theta < 156°42'$ (20 mA; 50 kV).

freien, ruhenden Elektronen, so müßte die in eine bestimmte Richtung gestreute Strahlung nach dem Energie- und Impulssatz auch nach der Streuung streng monochromatisch sein, also eine scharfe Streulinie darstellen. Beim Compton-Effekt an Atomelektronen ist nun zwar deren Anfangsenergie konstant, nicht dagegen ihr Impuls, der nach der Bohrschen Vorstellung um einen bestimmten Betrag streut. Schon Wentzel [469] hat darauf hingewiesen, daß dieser Streuung des Anfangsimpulses der Elektronen eine gewisse Breite des Streuspektrums entsprechen muß, daß also ein Streukontinuum auftreten muß. Dabei ist der Abstand des Intensitätsmaximums von der Primärlinie durch den Streuwinkel, die Breite und Intensitätsverteilung des Kontinuums dagegen außerdem durch die

Streuung des Anfangsimpulses der Elektronen bestimmt (s. Fig. 28). Den experimentellen Nachweis dieses Kontinuums und die Messung seiner Intensitätsverteilung verdanken wir den schönen Präzisionsarbeiten von DuMond und seinen Mitarbeitern [124], [126]. Fig. 29 zeigt ein Beispiel ihrer Aufnahmen. Die Untersuchung des Compton-Kontinuums bietet also die Möglichkeit zur Messung der Impulsbreite der Atomelektronen und ist hierzu auch zuerst herangezogen worden. Sommerfeld [408], [410] hat dann darauf aufmerksam gemacht, daß das Streukontinuum, dessen Intensitätsverteilung für H-Atome als Streuatome etwa die Form

$$I = \frac{I_{\max}}{(1 + x^2)^3} \tag{13.1}$$

besitzt ($x =$ Abstand des Maximums von der Primärlinie), eine kurzwellige Grenze besitzen muß. Das folgt aus der Energiegleichung

$$h\nu_0 - E_G = h\nu + \frac{m}{2}v^2, \tag{13.2}$$

wenn man die kinetische Energie des freien Elektrons Null setzt. Hierin bedeutet ν_0 die Frequenz des primären, und ν die des gestreuten Lichtquants und E_G die Bindungsenergie des Elektrons im Atom. Der Wellenlängenabstand der kurzwelligen Grenze des Kontinuums von der Primärlinie ergibt sich dann aus Gl. (13.2) zu

$$\Delta\lambda = \lambda_0 - \lambda = \frac{\lambda^2 E_G}{hc}. \tag{13.3}$$

Auf der kurzwelligen Seite dieser Grenze müssen sich an das Kontinuum diskrete Linien anschließen, die dadurch entstehen, daß bei dem Streuprozeß das Atomelektron nicht abgetrennt, sondern nur auf ein höheres Atomniveau gehoben, angeregt wird (atomarer Raman-Effekt). Rechnungen über die Intensitätsverteilung von Kontinuum und Streulinien sind von Burkhardt [75], Schnaidt [418] und Franz [164] sowie von Hicks [213] ausgeführt worden; quantitative experimentelle Untersuchungen liegen vor von Kappeler [247] sowie besonders von DuMond und Kirkpatrick [126].

b) Seriengrenzkontinua in Emission und Elektronenrekombination. Elektronen-Affinitätsspektren.

14. Allgemeines. Wie wir an Hand der Fig. 1 (S. 87) sahen, entstehen Seriengrenzkontinua in Emission bei der Rekombination von Elektronen mit positiven Ionen zu neutralen Atomen bzw. zu Ionen geringerer positiver Ladung (Umkehrvorgang zur Photoionisation). Dabei kann ein Elektron von einem Ion in den verschiedenen Anregungszuständen *„eingefangen"* werden, wobei die kinetische Energie des Elektrons, vermehrt um die Bindungsenergie E_g des Elektrons in dem betreffenden Zustand, in Form eines Lichtquants der Energie

$$h\nu = \frac{mv^2}{2} + E_g \tag{14.1}$$

emittiert wird. Beim Einfangen einer großen Zahl von Elektronen mit verschiedenen kinetischen Energien wird also ein kontinuierliches Spektrum emittiert, das sich von der Grenze $\nu_g = E_g/h$ des betreffenden Zustandes aus nach kurzen Wellen zu erstreckt, und dessen Intensitätsverteilung von der Geschwindigkeitsverteilung der rekombinierenden Elektronen abhängt. Da Rekombinationen in alle möglichen Atomzustände erfolgen, entstehen gerade so viel einzelne Seriengrenzkontinua wie Atomzustände vorhanden sind. Das gesamte kontinuierliche

Spektrum besteht also aus einer Überlagerung aller dieser einzelnen Grenzkontinua, und der frei-frei-Strahlung, deren Theorie wir in Ziff. 19 behandeln werden.

Wie die Gl. (5.43) zeigt, sind intensive Emissionsgrenzkontinua besonders in Systemen mit hoher Elektronen- und Ionendichte bei kleiner Elektronengeschwindigkeit bzw. Elektronentemperatur zu finden. Entsprechend bedient sich die neuere Forschung als Quellen kontinuierlicher Elektronenspektren vorwiegend der elektrischen Bogen- und Funkenentladungen, in denen die höchsten Elektronenkonzentrationen zu erreichen sind. Wir werden uns daher zunächst in Ziff. 15 eingehend mit den Emissionskontinua dieser Entladungstypen beschäftigen, zumal hierbei, wegen des zumindest in den Bogenentladungen vorherrschenden thermischen Gleichgewichtes, die besten Möglichkeiten zum quantitativen Vergleich von Theorie und Beobachtung vorliegen.

In den Wasserstoff- bzw. Wasserdampf-Lichtbögen wurde erstmals auch ein Elektronenaffinitätsspektrum, nämlich das des negativen Wasserstoffions (H^-) gefunden, welches besonders astrophysikalische Bedeutung besitzt (s. Ziff. 24). Elektronenaffinitätskontinua in Emission sind bei allen elektronegativen Atomen nach der Gleichung

$$X + \text{Elektron} \rightarrow X^- + h\,\nu$$

zu erwarten. Außer dem H^--Kontinuum sind aber bislang noch keine weiteren Elektronen-Affinitätskontinua mit Sicherheit nachgewiesen worden. (Für die Rechnungen über Absorptionsquerschnitte der negativen Ionen vgl. Ziff. 9.)

Im Anschluß an die Behandlung der Emissionskontinua der Funkenentladungen und Stoßwellen, die inzwischen schon weitgehend quantitativ erfaßbar sind, werden in Ziff. 16 die Kontinua der Niederdruckentladungen behandelt. Der quantitative Vergleich zwischen Theorie und Experiment ist hierbei aber wegen des Fehlens thermischen Gleichgewichtes viel schwieriger durchzuführen als bei den Bogenentladungen. Dementsprechend sind die meisten Beobachtungen bei Niederdruckentladungen nur qualitativer Art.

Zu den Elektronengrenzkontinua gehören im Prinzip auch diejenigen, die bei der Strahlungsrekombination von Molekülionen und Elektronen entstehen (Umkehrvorgang der Photoionisierung von Molekülen). Die Bedingungen für ihre Beobachtung sind dieselben wie für die Atomgrenzkontinua, nämlich hohe Konzentrationen von Molekülionen und Elektronen bei kleiner Elektronengeschwindigkeit. Stabile Molekülionen existieren aber nur unter Verhältnissen, bei denen auch schon in beträchtlichem Umfang Dissoziationen und darauffolgend Atomrekombinationen stattfinden, so daß solche Molekülkontinua nur gleichzeitig mit anderen Molekülkontinua und Atomgrenzkontinua auftreten, was unter weiterer Berücksichtigung der an sich schon komplizierten Struktur der Molekülspektren und ihrer zu erwartenden geringen Intensität die Beobachtung außerordentlich erschweren muß. Solche die Elektronenrekombination anzeigenden Molekül-Grenzkontinua sind daher auch noch nicht nachgewiesen worden.

Schließlich sei erwähnt, daß Röntgen-Grenzkontinua in Emission, die bei dem Einfang freier Elektronen in innere Elektronenschalen emittiert werden müßten, nicht beobachtet werden, weil die inneren Elektronenschalen normalerweise vollbesetzt sind, so daß die Wahrscheinlichkeit dieser Prozesse außerordentlich klein ist.

15. Kontinuierliche Strahlung von Bogenentladungen, Funkenentladungen und Stoßwellen. Nachdem die eingangs dargestellte korrespondenzmäßige KRAMERsche Theorie von EDDINGTON [*129*] und MILNE [*315*] auf astrophysikalische Probleme angewandt worden war, hat FINKELNBURG [*144*], [*145*], [*146*] in einer

Reihe von Arbeiten darauf hingewiesen, daß die Kontinua stromstarker Gasentladungen ebenfalls als Elektronenbrems- und Rekombinationsstrahlung gedeutet werden können. Unsöld [443] hat dann diese Vorstellungen zu einer Theorie erweitert, die unter anderem die Frequenzunabhängigkeit des Kontinuums für nicht zu große Frequenzen ergab, und hat einen quantitativen Vergleich mit den Kontinua einiger Gasentladungslichtquellen versucht. Dieser Vergleich, der im wesentlichen an Quecksilber-Hochdrucklampen (Elenbaas [132]) unternommen wurde, fiel seinerzeit noch nicht sehr befriedigend aus. Die berechnete Ausstrahlung blieb um nahezu eine Größenordnung unter der beobachteten. Diese Diskrepanz ist aber vor kurzem von Göing, Meier und Meinen [183] aufgeklärt worden; sie beruhte auf einer zu niedrigen Temperaturangabe für die untersuchten Hg-Lichtbögen.

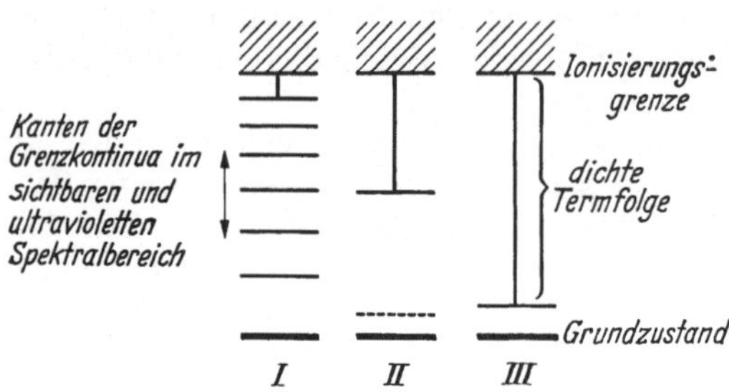

Fig. 30. Einteilung der Termschemata der Atome in drei Typen. (Nach Maecker und Peters [295].)

Der elektrische Lichtbogen mit lokalem thermischen Gleichgewicht im Plasma ist inzwischen zum idealen Untersuchungsobjekt für die Spektroskopie und damit auch für die Kontinuaforschung entwickelt worden. Die beiden nach Gl. (5.43) für die Emission wesentlichen Größen, Elektronenkonzentration und Temperatur, können einerseits spektroskopisch sehr genau ermittelt, andererseits durch äußere Bedingungen, wie Stromstärke, Gasdruck usw. in gewünschter Weise variiert werden. Quantitative Messungen der Kontinua solcher Bögen stimmen, mit Ausnahme der Metallkontinua bei hohen Frequenzen, so gut mit den Unsöldschen Rechnungen überein, daß die Finkelnburg-Unsöldsche Auffassung über die Entstehung der Kontinua in Gasentladungen als völlig gesichert angesehen werden kann. Die Verhältnisse haben sich insofern sogar umgekehrt, als man beginnt, aus Absolutintensität und Energieverteilung der kontinuierlichen Strahlung die Temperatur und Elektronenkonzentration von Plasmen zu bestimmen.

Die Frequenzunabhängigkeit des Gesamtkontinuums nach Gl. (5.43) gilt, der Voraussetzung der Theorie gemäß, nur bis zu einer aus dem Termschema des betreffenden Elementes zu entnehmenden Grenzfrequenz ν_g, die so zu wählen ist, daß die Termenfolge zwischen der Ionisierungsgrenze und der Energie $h\nu_g$ als hinreichend dicht angesehen werden kann. Nach höheren Frequenzen hin fällt dann $\varepsilon_\nu \sim \exp(-h\nu/kT)$ bzw. $\varkappa_\nu \sim 1/\nu^3$ ab, bis sich ein neues Rekombinationskontinuum überlagert usw. Vergleicht man verschiedene Atomarten hinsichtlich ihrer Strahlungseigenschaften miteinander, so kann man grob schematisch eine Einteilung in drei Typen vornehmen (Fig. 30).

Zum ersten Typ sollen die Elemente zählen, die eine ziemlich aufgelockerte Termfolge besitzen, bei denen eine frequenzunabhängige Emission also nur im Ultraroten zu erwarten ist, während im übrigen Spektralbereich ein ziemlich zerklüftetes „Gebirge" von überlagerten Einzelkontinua erscheint. Zu diesen Elementen gehören H, He, die Gruppen I, II (z. B. Hg) und ein Teil der III. Gruppe des periodischen Systems. Bei dem zweiten Typ von Elementen besteht das Termschema aus einer praktisch dichten Termfolge der angeregten Elektronenzustände, die durch eine breite termlose Energielücke vom Grundterm getrennt sind. Auf eine konstante Emission vom fernen Ultrarot bis etwa 4000 Å folgt hier der beschriebene Abfall bis zur Hauptseriengrenzfrequenz. Diese Elementengruppe umfaßt im wesentlichen die Edelgase außer Helium, die Halogene, dann aber auch schon die jeweils vor den Halogenen stehenden Elemente der IV., V. und VI. Gruppe (z.B. O, N, C), bei denen im Gegensatz zu den Edelgasen und Halogenen allerdings noch einige Anregungszustände in der Nähe des Grundzustandes liegen. Schließlich wollen wir in einem dritten Typ alle diejenigen Atomarten zusammenfassen, die sich durch eine hohe Multiplizität der Terme auszeichnen. Die Termfolge ist hierbei bis in die Nähe des Grundzustandes als dicht zu betrachten, so daß die Frequenzunabhängigkeit der kontinuierlichen Emission durch das gesamte Spektrum hindurch fast bis zur Hauptseriengrenzfrequenz gelten sollte. Hierher gehören die Elemente der Gruppen IIIa bis VIIa und die sonstigen Übergangselemente (Eisen- und Platinmetalle).

α) *Emissionskontinua von Elementen des Typs I.* Eine genaue Berechnung der Überlagerung der einzelnen Rekombinationskontinua im gesamten Spektralbereich ist praktisch nur beim Wasserstoff möglich. Nach der Rechnung von Unsöld [*449*] ergibt sich, wenn man nur die Seriengrenzkontinua der vier untersten Wasserstoffterme (Lyman-, Balmer-, Paschen- und Brackett-Kontinuum) getrennt aufsummiert und alle höheren Grenzkontinua als verschmiert betrachtet, zunächst für den Absorptionskoeffizienten pro cm

$$\varkappa_\nu = \frac{64\,\pi^4}{3\,\sqrt{3}} \cdot \frac{m\,e^{10}}{c\,h^6} \cdot \frac{1}{\nu^3} \left\{ \sum_{u_1}^{u_4} g_n \frac{e^{u_n}}{n^3} + \frac{e^{u_5}}{2\,u_1} \right\} N_{\mathrm{H}}\, e^{-E_i/kT} \,, \tag{15.1}$$

$$u_n = \frac{R\,h\,c}{n^2\,kT}\,; \qquad R = \frac{2\,\pi^2\,m\,e^4}{c\,h^3} = \text{Rydberg-Konstante},$$

$$u_1 = \frac{E_i}{kT}\,; \qquad N_{\mathrm{H}} = \text{Anzahl der H-Atome/cm}^3,$$

$$g_n = \text{Gaunt-Faktoren nach Menzel und Pekeris [}309\text{]}$$

oder nach Umformung

$$\varkappa_\nu = \frac{16\,\pi^2}{3\,\sqrt{3}} \frac{e^6}{c\,h^4} \frac{e^{u_5}}{\nu^3} \left\{ 1 + 2\,u_1 \sum_{u_1}^{u_4} \frac{g_n}{n^3} e^{u_n - u_5} \right\} N_{\mathrm{H}}\, kT\, e^{-E_i/kT} \,. \tag{15.2}$$

Schließlich folgt für den Emissionskoeffizienten bei optisch dünner Schicht nach Gl. (3.28) mit $u = \dfrac{h\,\nu}{kT}$

$$\varepsilon_\nu = \frac{32\,\pi^2}{3\,\sqrt{3}} \frac{e^6}{c^3\,h^3} e^{u_5 - u} \left\{ 1 + 2\,u_1 \sum_{u_1}^{u_4} \frac{g_n}{n^3} e^{u_n - u_5} \right\} N_{\mathrm{H}}\, kT\, e^{-E_i/kT} \tag{15.3}$$

bzw. nach Umformung mit der Saha-Gleichung für Wasserstoff

$$\varepsilon_\nu = \frac{32\,\pi^2}{3\,\sqrt{3}} \frac{e^6}{c^3\,(2\,\pi\,m)^{\frac{3}{2}}} \frac{N_e\,N_i}{(kT)^{\frac{1}{2}}} e^{u_5 - u} \left\{ 1 + 2\,u_1 \sum_{u_1}^{u_4} \frac{g_n}{n^3} e^{u_n - u_5} \,. \right\} \tag{15.4}$$

Für $u < u_4$ geht diese Form in Gl. (5.45) über und für $u \leqq u_5$ wieder in die frequenzunabhängige Gl. (5.43).

Bei genügend hohen Drucken und nicht zu hohen Temperaturen liefert auch das Elektronenaffinitätsspektrum des Wasserstoffes (H⁻-Kontinuum) einen beträchtlichen Beitrag zur kontinuierlichen Emission (s. Ziff. 9). An das neutrale Wasserstoffatom kann sich ein Elektron anlagern, dessen Bindungsenergie dann 0,75 eV beträgt. Außerdem können frei-frei-Übergänge von Elektronen im Felde des neutralen Wasserstoffatoms stattfinden. Der aus diesen Prozessen sich ergebende Absorptionskoeffizient und seine quantenmechanische Berechnung wurde bereits in Ziff. 9 behandelt. Diese Rechnungen bilden die Grundlage für

das Verständnis des Sonnenkontinuums (s. Ziff. 24) und für den Nachweis des
H⁻-Kontinuums in Gasentladungen. Es gilt

$$\varkappa_\nu(\mathrm{H}^-) = x(\nu,\,T)\,N_\mathrm{H}\,p_e,\qquad (15.5)$$

wobei $x(\nu,\,T)$ eine bei Chandrasekhar tabulierte Funktion der Frequenz und
der Temperatur ist. Vitense [454] hat nach dieser Gleichung das Druck-Temperaturgebiet berechnet, in dem das H⁻-Kontinuum in Wasserstoffplasmen

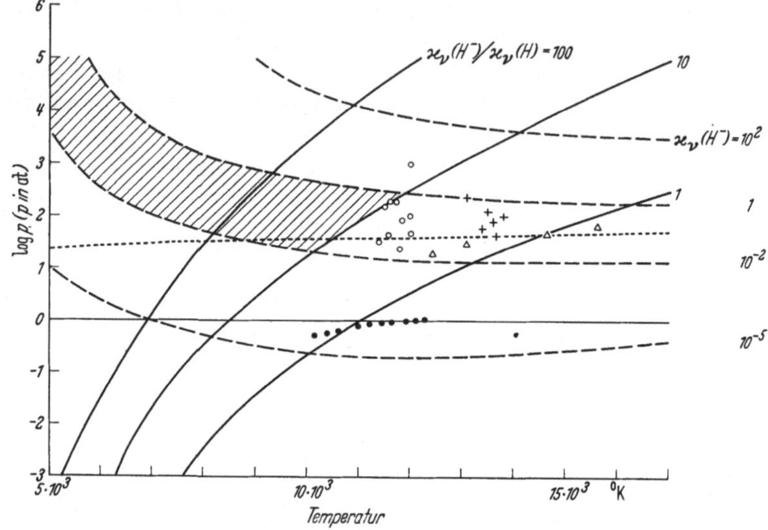

Fig. 31. Die Linien $\varkappa_\nu(\mathrm{H}^-)/\varkappa_\nu(\mathrm{H})=$ const (ausgezogen) und $\varkappa_\nu(\mathrm{H}^-)=$ const (gestrichelt) sind als Funktion von Druck
(p in Atm; logarithmische Skala) und Temperatur T für reinen Wasserstoff bei $\lambda=5060$ Å aufgetragen. Der Temperatur-
Druck-Bereich, in dem das H⁻-Kontinuum im Laboratorium gut beobachtbar sein sollte, ist schraffiert; bei niedrigen
Temperaturen dürften sich Molekülkontinua in zunehmendem Maße störend bemerkbar machen. Die punktierte Linie
gibt die Druck-Temperatur-Kurve einer ebenen Stoßwelle für den Anfangsdruck $p_1=\frac{1}{20}$ Atm wieder. (Nach Vitense
[454].) Die eingezeichneten Meßpunkte stellen die Zustandsgrößen des Wasserstoffplasmas in der Säulenachse der bislang
spektroskopisch untersuchten Wasserstoff-Lichtbögen und -Funken dar: △ Wasserstoffgleitfunken nach Fuchs [165];
Wasserstoffbögen nach Lochte-Holtgreven und Nissen [285]; + Wasserstoffbögen nach Nissen [350]. Ferner
sind eingetragen die Zustandswerte ○ des Wasserdampf-Hochdruckplasmas nach Peters [355].

nachweisbar sein sollte. Damit der Nachweis eindeutig ist, sollte die Intensität
des H⁻-Kontinuums möglichst größer als die des H-Kontinuums, als etwa

$$\frac{I_\nu(\mathrm{H}^-)}{I_\nu(\mathrm{H})}\sim\frac{\varkappa_\nu(\mathrm{H}^-)}{\varkappa_\nu(\mathrm{H})}\geq 10$$ sein, andererseits darf $\varkappa_\nu(\mathrm{H}^-)\cdot l$ nicht wesentlich größer als 1

sein, da man sonst nahezu schwarze Strahlung erhält. Das Gebiet, in dem beide Bedingungen erfüllt sind, ist in Fig. 31 durch Strichelung markiert. Man hat daran
gedacht, mit Hilfe einer Detonationswelle in den erwünschten Druck-Temperatur-
bereich zu gelangen. Bei einem festen Ausgangsdruck von $p_1=\frac{1}{20}$ Atm würde
man bei verschiedener Stärke der Detonation die einzelnen Punkte der eingezeichneten punktierten Kurve erreichen. Inzwischen sind aber von Lochte-
Holtgreven und Mitarbeitern erfolgreiche Versuche unternommen worden,
günstige Bedingungen in einem Wasserstoff-Lichtbogen bzw. -Funken zu reali-
sieren. Zuerst hat Fuchs [165] in einer Kondensatorentladung durch ein ab-
geschlossenes Wasserstoffvolumen Temperaturen von 12500 bis 15500° K bei
einem Druck von etwa 30 Atm erreicht. Die Entladungsdauer war durch
einen Widerstand im Entladungskreis so vergrößert worden, daß zumindest ein
nur „langsam veränderliches" thermisches Gleichgewicht angenommen werden
konnte. Sodann wurde von Lochte-Holtgreven und Nissen [285] ein Wasser-
stoffbogen entwickelt, der in einer wassergekühlten Quarzkammer bei Drucken

zwischen 0,4 und 1 Atm und Temperaturen zwischen 10000 und 12000° K brannte. In beiden Fällen wurde ein Kontinuum gefunden, welches im Mittel doppelt so stark war, als der Rekombination und den frei-frei-Übergängen im Felde von Wasserstoffionen (H-Kontinuum) allein entsprach. Nach Fig. 31 ist $\frac{\varkappa_\nu(\mathrm{H}^-)}{\varkappa_\nu(\mathrm{H})} \approx 1$ gerade das theoretisch erwartete Verhältnis. Zu etwas höheren Verhältniszahlen kam dann NISSEN [350] in einem wirbelstabilisierten Hochdruckbogen bei Drucken von 40 bis 140 Atm und Temperaturen zwischen 13000 und 14000° K, und schließlich wurden Intensitätsverhältnisse der Größenordnung 10

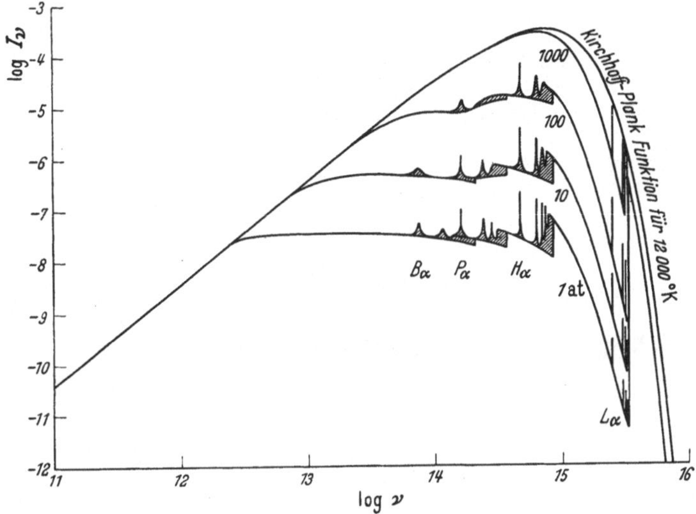

Fig. 32. Kontinuierliche Emission eines Wasserdampf-Plasmas von 2 mm Schichtdicke bei 12000° K und bei $p = 1$, 10, 100, 1000 Atm (berechnet). Die Wasserstofflinien sind nur angedeutet. (Nach PETERS [355].)

und darüber von PETERS [355] mit dem wasserstabilisierten Hochdruckbogen bei Drucken bis zu 1000 Atm und Temperaturen von etwa 12000° K erreicht. Alle diese Zustandspunkte sind in Fig. 31 eingetragen. Bei höheren Drucken werden die Verhältnisse für die Erzeugung des H⁻-Kontinuums noch etwas günstiger, als im Bilde angegeben ist, wenn man die wegen der hohen Elektronendichte auftretende relativ starke Erniedrigung der Ionisierungsspannung berücksichtigt. Bei 1000 Atm und 12000° K ist das Verhältnis $\frac{I_\nu(\mathrm{H}^-)}{I_\nu(\mathrm{H})} \approx 30$, der zugehörige Meßpunkt liegt, wie man in Fig. 31 erkennt, bereits im Gebiet der schwarzen Strahlung.

Eine quantitative Darstellung der Frequenzverteilung der Strahlung einer 2 mm dicken Plasmaschicht im wasserstabilisierten Hochdruckbogen bei 12000° K und verschiedenen Gasdrucken gibt Fig. 32. Bei 1 Atm ist der Beitrag des H⁻-Kontinuums noch gering. Auf die frequenzunabhängige Emission im Ultraroten folgt nach Gl. (15.3) die Überlagerung der einzelnen Seriengrenzkontinua, insbesondere ein rascher Abfall nach Überschreiten der BALMER-Grenze und ein starker Anstieg im LYMAN-Kontinuum, welches bei Drucken von über 3 atm bereits an die PLANCK-Kurve anstößt. Da im vorliegenden Bogen ein sehr kleiner Ionisationsgrad herrscht, sollte die Strahlung allgemein proportional zum Gasdruck anwachsen. Nun wird aber bei genügend hohen Drucken das H⁻-Kontinuum maßgebend, für das nach Gl. (15.5)

$$I_\nu(\mathrm{H}^-) \sim N_\mathrm{H}\, p_e \qquad\qquad (15.6)$$

gilt. Da bei kleinem Ionisationsgrad $N_H \sim p$ und $p_e \sim p^{\frac{1}{2}}$ ist, wächst $I_\nu(\mathrm{H}^-)$ mit dem Druck wie $p^{\frac{3}{2}}$. Weiterhin liegt nach der Rechnung das Maximum der H⁻-Strahlung, wie man aus Fig. 32 erkennt, bei etwa $1\,\mu$ ($\nu = 3 \cdot 10^{14}$ Hz), während das Maximum der auf die Wellenlängeneinheit bezogene Intensität I_λ bei etwa 8000 Å liegt. Folglich wölbt sich die Intensität im Gebiet zwischen 10^{14} und 10^{15} Hz allmählich auf, bis bei 1000 Atm die der PLANCK-Funktion entsprechende Intensität erreicht wird. Entsprechende Rechnungen für die viel komplizierteren Verhältnisse in den Sternatmosphären sind von VITENSE [455] durchgeführt worden. Die experimentelle Bestätigung der in Fig. 32 graphisch dargestellten Rechnungen findet man in den Spektren der Fig. 33. Das 1000 Atm-Spektrum zeigt eine kontinuierliche Energieverteilung, die praktisch der eines schwarzen Körpers von 12000° K Temperatur entspricht. Neben der H⁻-Strahlung liefert natürlich auch die der stark verbreiterten Linien der BALMER-Serie einen Beitrag zu der hohen Intensität. Die Existenz des H⁻-Kontinuums ist auch für den geringen Intensitätsunterschied zwischen dem BALMER-Grenzkontinuum und dem PASCHEN-Grenzkontinuum zwischen den BALMER-Linien bei den Spektren aus optisch dünner Schicht verantwortlich. Dieser „BALMER-Sprung" (s. Ziff. 24) müßte erheblich größer sein, wenn das H⁻-Kontinuum nicht an der Strahlung beteiligt wäre.

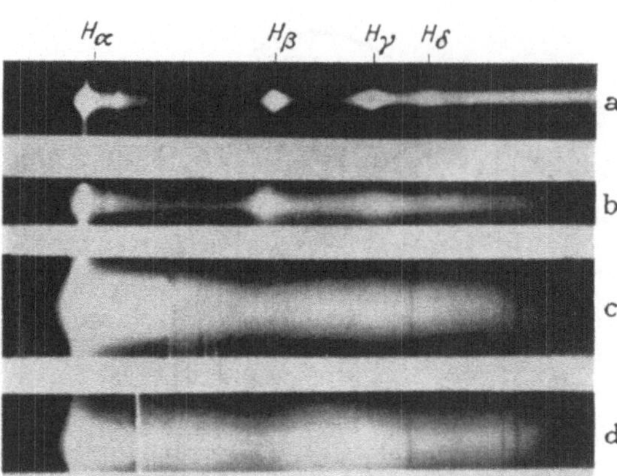

Fig. 33a—d. Spektren des wasserstabilisierten Hochdruckbogens nach PETERS [355]. a 1 Atm, 50 Amp. Die Linie Hδ und der große Intensitätsunterschied zwischen BALMER- und PASCHEN-Kontinuum sind noch deutlich erkennbar. b 100 Atm, 100 Amp. Nur noch Hγ ist vom Kontinuum zu unterscheiden. c 700 Atm, 200 Amp. Die Linie Hα ist gerade noch vom Kontinuum getrennt zu erkennen. Schwache Kupferlinien in Emission und Calciumlinien in Absorption. d 1000 Atm, 200 Amp. Das Spektrum entspricht praktisch demjenigen eines schwarzen Strahlers von 12000° K. Bei den letzten beiden Aufnahmen wurden zwei verschiedene Plattensorten benutzt, wie am Einschnitt hinter Hα zu erkennen ist.

Im Wasserstoffspektrum tritt weiterhin besonders augenfällig ein für alle Spektren von Hochdruckplasmen gleich bedeutender Effekt auf, den man gewöhnlich als das „Übergreifen" des Kontinuums nach größeren Wellenlängen bezeichnet und der im Grunde zur Theorie der Linienverbreiterung gehört. Die höheren Terme werden von den im Plasma vorhandenen Mikrofeldern so stark verbreitert, daß sie praktisch mit zum kontinuierlichen Bereich zu zählen sind (Erniedrigung der Ionisierungsspannung). Die höheren Glieder einer Linienserie verschmelzen dabei nach und nach mit wachsender Störung vollständig zu einem Kontinuum, welches sich stetig an das wahre Seriengrenzkontinuum anschließt. Besonders eindrucksvoll zeigen dies die von FINKELNBURG [141] an einer kondensierten Funkenentladung in Wasserstoff gewonnenen Spektren (Fig. 34). Quantitative Messungen im Übergangsgebiet zwischen reinem Linienspektrum und Kontinuum des BALMER-Spektrums wurden von JÜRGENS [246] am wasserstabilisierten Lichtbogen von Atmosphärendruck durchgeführt. Die Achsentemperatur von 12600° K wurde sowohl aus den Absolutintensitäten und Profilen der BALMER-Linien (STARK-Effekt-Verbreiterung) als auch aus der absoluten Intensität und der Energieverteilung des BALMER-Kontinuums entnommen. In Fig. 35 ist der von JÜRGENS ausgemessene Bereich in der Nähe der BALMER-Grenze dargestellt. Während die wahre Seriengrenze bei $\lambda = 3647$ Å liegt, beginnt das Kontinuum bereits bei etwa 3900 Å (s. auch Fig. 32). H_ε ist die letzte vom Kontinuum zu unterscheidende BALMER-Linie.

Diese Erscheinung kann nach INGLIS und TELLER [231] zur Bestimmung der Konzentration der Ladungsträger, welche die Mikrofelder aufbauen und somit die Störung verursachen, ausgenutzt werden. Bezeichnet man die Hauptquantenzahl der letzten getrennt sichtbaren Linie mit n^*, so gilt

$$\left.\begin{aligned}\mathrm{Log}_{10}\,N = 23{,}26 - \\ -\,7{,}5\,\mathrm{Log}_{10}\,n^*.\end{aligned}\right\} (15.7)$$

Dabei ist für Temperaturen

$$T < 10^5/n^*;\quad N = N_e + N_i = 2\,N_i$$

bei Quasineutralität und für

$$T > 10^5/n^* \quad N = N_i,$$

weil im letzteren Temperaturbereich die Elektronen wegen zu großer thermischer Geschwindigkeit keinen merklichen Beitrag zur Verbreiterung der Terme leisten. Bei der vorliegenden Temperatur von 12600° K ist nicht klar zu unterscheiden, welche der beiden Gleichungen anzuwenden ist. Es ergibt sich somit eine Unsicherheit in der Bestimmung der Ladungsträgerdichte um einen Faktor 2. JÜRGENS erhält mit $n^* = 7$ (H_ε als letzte Linie) eine Elektronendichte von $N_e = 4$ bis $8 \cdot 10^{16}$ cm^{-3}. Andererseits haben wir eingangs im theoretischen Teil festgestellt, daß die Erniedrigung der Ionisierungsspannung das Übergreifen des Kontinuums nach größeren Wellenlängen verursacht. UNSÖLD [447] berechnet diese Erniedrigung für wasserstoffähnliche Elemente zu

$$\left.\begin{aligned}\Delta E_i = 7 \cdot 10^{-7}\,N_e^{\frac{1}{3}} \\ \text{gemessen in eV.}\end{aligned}\right\} (15.8)$$

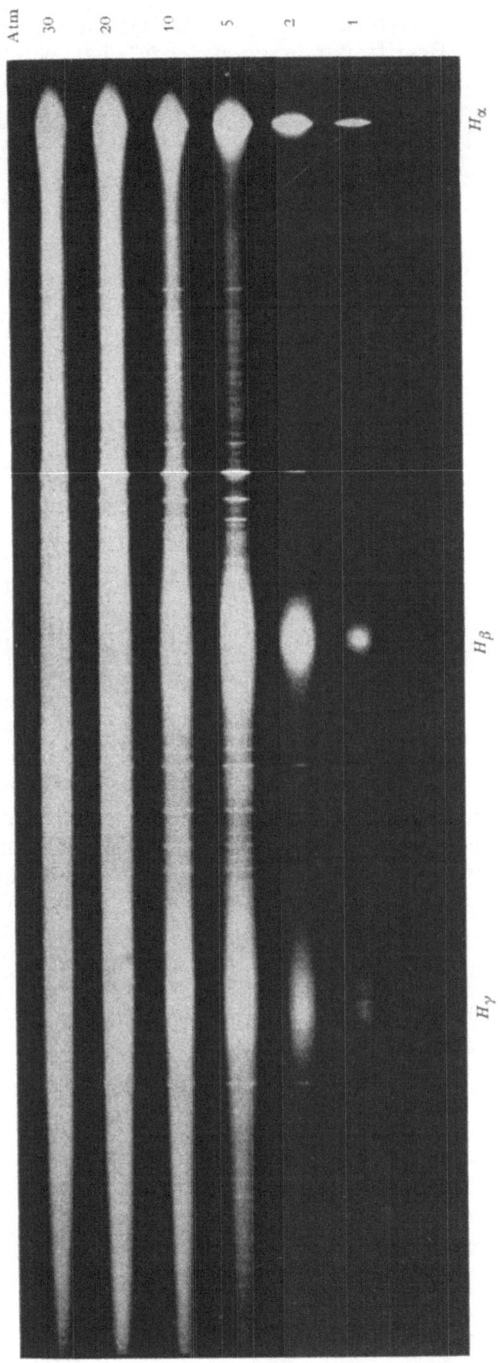

Fig. 34. Das BALMER-Spektrum des Wasserstoffs in kondensierten Funkenentladungen in Abhängigkeit vom Gasdruck. Übergang vom Linienspektrum zum Druckkontinuum. (Nach FINKELNBURG [141].)

Da die höheren Terme der übrigen Elemente praktisch alle wasserstoffähnlich sind, kann dieser Ausdruck in erster Näherung auch auf andere Elemente angewandt werden. Setzt man nun die obigen Elektronenkonzentrationen ein, so folgt daraus ein Vorziehen der Seriengrenze um etwa 300 Å, was man in der Tat im Spektrum (Fig. 35) bestätigt findet.

Der Begriff der „Seriengrenze" ist, wie man ebenfalls dieser Darstellung entnimmt, mit Vorsicht aufzufassen. Da die an das Kontinuum anschließenden Linien schon erheblich verbreitert sind, überlappen sich deren Linienflügel weitgehend und bilden somit ein Scheinkontinuum, welches sich an das eigentliche Kontinuum anschließt. Man kann daher in keinem Fall einen regelrechten Sprung beim Übergang vom Linienspektrum zum zugehörigen Grenzkontinuum erwarten. Für das vorgetäuschte Kontinuum unter den Linien ist die Unsöldsche Rechnung nicht mehr zuständig, wohl aber die spezielle Theorie der Linienverbreiterung und Autoionisierung im Feld, auf die hier nicht näher eingegangen werden kann.

Bislang haben wir so gerechnet, als ob die Strahlungsquellen eine einheitliche Temperatur über eine gewisse Schichtdicke besäßen, und haben im wesentlichen die spektrale Energieverteilung bei dieser Temperatur betrachtet. In Wirklichkeit liegt in jedem Lichtbogen oder Funken eine räumliche Temperaturverteilung vor, so daß im allgemeinen bei optisch dünnen Schichten über die aus verschiedenen Temperaturbereichen emittierte Strahlung integriert wird. Die Messung der radialen Verteilung der kontinuierlichen Strahlung kann mithin auch zur Bestimmung der radialen Temperaturverteilung dienen. Allgemein ist

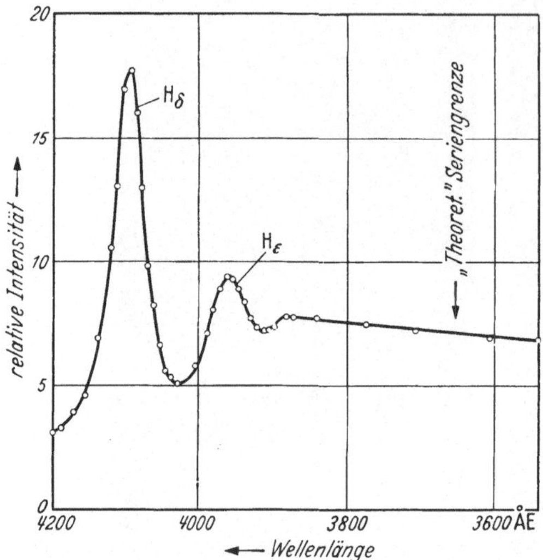

Fig. 35. Intensitätsverteilung an der Grenze der Balmer-Linien, gemessen von Jürgens [246].

$$\varepsilon_\nu \sim \frac{N_e N_i}{(kT)^{\frac{1}{2}}}. \tag{15.9}$$

Bei kleinem Ionisationsgrad berechnet sich das Produkt $N_e N_i$ aus der Saha-Gleichung zu

$$N_e N_i \approx 2 \frac{Z_+}{Z_0} \frac{(2\pi m)^{\frac{3}{2}}}{h^3} p \, (kT)^{\frac{1}{2}} e^{-E_i/kT}, \tag{15.10}$$

so daß gilt

$$\varepsilon_\nu \sim p \, e^{-E_i/kT}. \tag{15.11}$$

Die Emission steigt bei konstantem Druck rasch mit der Temperatur an. Bei Lichtbögen mit geringem Ionisationsgrad beobachtet man demnach eine mehr oder minder steile Strahlungsdichte-Verteilung über den Querschnitt. Ganz anders liegen die Verhältnisse indessen bei einem Plasma, das praktisch vollständig ionisiert ist. Hier ist

$$N_e = N_i \approx \frac{p}{2kT} \tag{15.12}$$

und damit

$$\varepsilon_\nu \sim \frac{p^2}{(kT)^{\frac{5}{2}}}. \tag{15.13}$$

Abgesehen davon, daß die Emission jetzt quadratisch mit dem Gasdruck anwächst, fällt sie bei $p = \text{const}$ mit steigender Temperatur ab, weil bei vollständiger Ionisation die maximale Elektronenkonzentration erreicht ist und N_e mit wachsendem T nach dem Gasgesetz mit $1/T$ abnimmt (vgl. Ziff. 6). Zwischen dem raschen Anstieg der Strahlung mit der Temperatur bei geringer Ionisation und ihrem Abfall mit $T^{-\frac{3}{2}}$ bei vollständiger Ionisation muß demnach im Gebiet mittleren Ionisationsgrades ein Strahlungsmaximum liegen. Das Ergebnis einer genauen Rechnung für das BALMER-Kontinuum ist in Fig. 36 wiedergegeben. Das Maximum liegt, Atmosphärendruck vorausgesetzt, bei etwa 16000° K. Eine ähnliche Temperaturabhängigkeit zeigen übrigens auch die entsprechenden Spektrallinien. LARENZ [274] hat aus

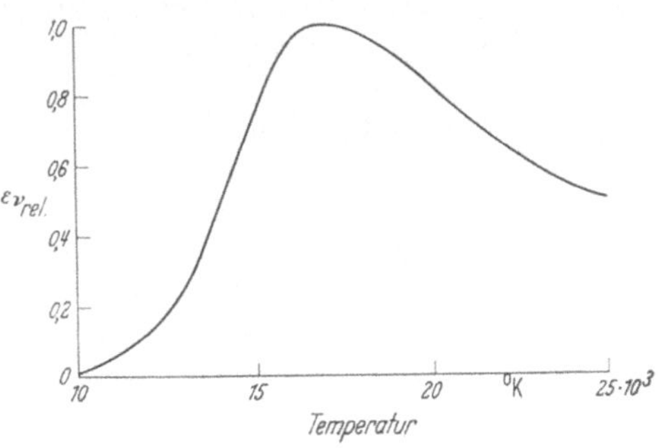

Fig. 36. Relativer Intensitätsverlauf des BALMER-Kontinuums als Funktion der Temperatur bei $p = 1$ Atm. Maximum bei 16000° K.

diesem Verhalten eine Temperatur-Meßmethode für Hochtemperaturbögen entwickelt, und diese zuerst auf den wasserstabilisierten Lichtbogen von GERDIEN

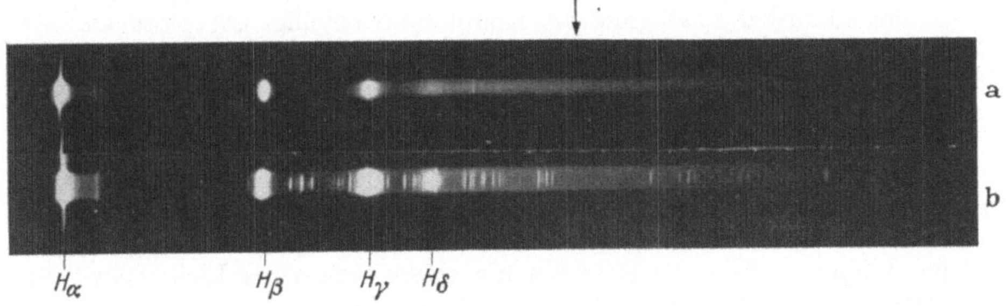

Fig. 37a u. b. Spektren des wasserstabilisierten Hochleistungsbogens nach MAECKER [294]. a 70 Amp, 15000° K Achsentemperatur. b 200 Amp, 30000° K Achsentemperatur. Wird in Richtung des Pfeiles photometriert, so erhält man die Photometerkurven in Fig. 38a, b.

und LOTZ [175] angewandt. Die BALMER-Spektren der Fig. 37 sind an dem von MAECKER [294] aus dem GERDIEN-LOTZ-Bogen entwickelten wasserstabilisierten Hochleistungsbogen aufgenommen. Die zugehörigen Photometerkurven in Fig. 38 stellen die radiale Schwärzungsverteilungen senkrecht zur λ-Richtung dar. Bei 70 Amp Bogenstrom und 15000° K Achsentemperatur tritt der theoretischen Kurve entsprechend noch keine Einsenkung der kontinuierlichen Strahlung in der Achse auf. Hingegen

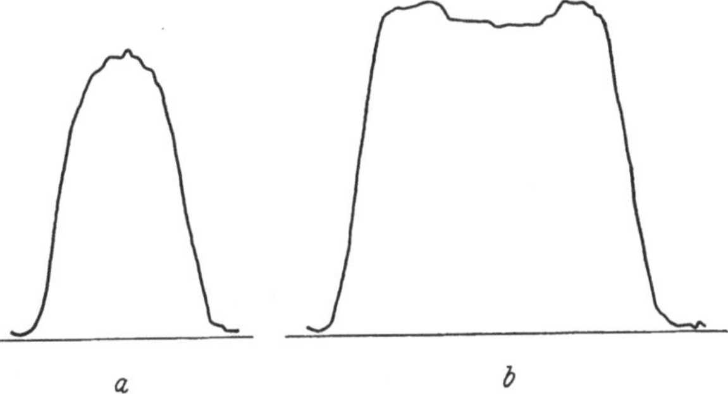

Fig. 38 a u. b. Photometerkurven zu den Spektren in Fig. 37a, b.

ist die Strahlungsverteilung bei 200 Amp und 30000° K Achsentemperatur im mittleren Bogenteil ($T > 16000$° K) deutlich eingesenkt. Die hohe Temperatur wird dort durch das Erscheinen der O II-Linien veranschaulicht. Die Einsenkung des Kontinuums und ebenso die der Wasserstofflinien ist auch visuell ohne Schwierigkeiten im Spektrum zu erkennen.

Eine bei *konstantem Druck* betriebene Strahlungsquelle zeigt also keineswegs stets einen ständigen Anstieg der Strahlungsleistung mit wachsender Temperatur. Nach Überschreiten eines bestimmten Maximalwertes kann die Lichtquelle wieder dunkler werden, weil die Anzahl der strahlenden Atome bzw. Ionen nach dem Gasgesetz mit steigender Temperatur abnimmt. Erst eine genaue Untersuchung kann zeigen, ob in einem bestimmten Fall die durch Überlagerung der Beiträge der verschiedenen Ionisationsstufen entstehende Gesamtstrahlung mit der Temperatur monoton zunimmt oder nicht.

Eine große Anzahl von Arbeiten befaßt sich mit der kontinuierlichen Strahlung des Quecksilber-Hochdruckplasmas, wobei als Strahlungsquellen meistens die bekannten Arten der Quecksilber-Hochdrucklampen verwendet werden.

Von Rompe, Schulz und Thouret [382] wurde die Abhängigkeit der Kontinuumintensität von Gasdichte und Stromstärke gemessen. Die Zunahme der Intensität mit der Stromstärke wie $I^{\frac{3}{2}}$ und eine fast lineare Zunahme mit der Dichte konnte nach der Unsöldschen Theorie gedeutet werden. Unter Benutzung der Temperaturmessungen von Schulz [427] aus den Elektronenstoßbreiten der Spektrallinien ergab sich auch eine befriedigende Übereinstimmung im Absolutwert. Andererseits konnte aus einigen Beobachtungen über die Druckabhängigkeit (s. Busz und Schulz [79], Elenbaas [132]) vermutet werden, daß noch ein anderer Prozeß an der Erzeugung des Kontinuums beteiligt ist. Rössler [388] hat deshalb in einer Reihe von Arbeiten nachzuweisen versucht, daß das Quecksilber-Bogenkontinuum im Druckbereich von 1,5 bis 85 Atm als Molekülkontinuum in dem Sinne aufzufassen sei, daß es im Stoß angeregter mit unangeregten Hg-Atomen, d. h. also von quasimolekülartigen Stoßpaaren emittiert wird, wie das für gewisse kontinuierliche Spektren von Hg-Niederdruckbögen (vgl. Ziff. 45) bekannt ist. Rösslers Schluß stützt sich besonders auf die von ihm aus der radialen Leuchtdichteverteilung des Kontinuums bestimmte „Anregungsenergie" des Kontinuums von 8,6 eV, während man nach der gemessenen Elektronendichte eine „effektive" Ionisierungsenergie von 9,5 bis 10 eV erwarten müßte. Ein vielfach gegen die Rekombinationstheorie und für die Molekülstrahlungstheorie angeführtes Argument, nämlich die ungenügende Erfüllung der Forderung nach frequenzunabhängiger Kontinuumsemission beruht u. E. auf einem Mißverständnis. Das Hg-Atom gehört nämlich zu dem Typ von Elementen mit sehr vielen isoliert liegenden Termen, und für solche Atome ist die Voraussetzung für die Frequenzunabhängigkeit besonders im sichtbaren und ultravioletten Spektralbereich wenig gut erfüllt (Typ I). Die Unsöldsche Theorie kann hier nur einen, wenn auch recht brauchbaren, Mittelwert für die Absolutintensität liefern. Daß die Schwankungen der Kontinuumsintensität über die Frequenzskala nach Fig. 39 dennoch gering sind, spricht um so mehr für die Gültigkeit der Elektronenstrahlungstheorie. Bei Lichtbögen mit kleinem Ionisationsgrad kommt hinzu, daß die Strahlungsintensität rasch mit der Temperatur zunimmt; geringe Änderungen in den für die Strahlung nach Gl. (5.48) maßgebenden Größen T und ΔE_i ergeben nach Fig. 40 beträchtliche Änderungen in der Strahlungsintensität. Es kann daher allein aus einer von der Theorie abweichenden Druckabhängigkeit der Strahlungsdichte, ohne genaue Angabe der gleichzeitigen Änderung der Bogentemperatur und der Erniedrigung der Ionisierungsspannung, nicht eindeutig auf das Vorhandensein eines zusätzlichen Mechanismus der Kontinuumserzeugung geschlossen werden.

Zur genauen Prüfung der Theorie haben daher Göing, Meier und Meinen [183] die Energieverteilung des Kontinuums, die Temperatur und die „effektive"

Ionisierungsenergie an mehreren Hg-Höchstdrucklampen vom Typ HBO 200 (200 W, 60 Atm) vermessen. Die Temperaturbestimmung erfolgte nach dem von BARTELS [31] angegebenen Verfahren an Linien mit Selbstumkehr und ergab im Mittel den Wert $(7800\pm200)°$ K. Aus dem Übergreifen des Seriengrenz-

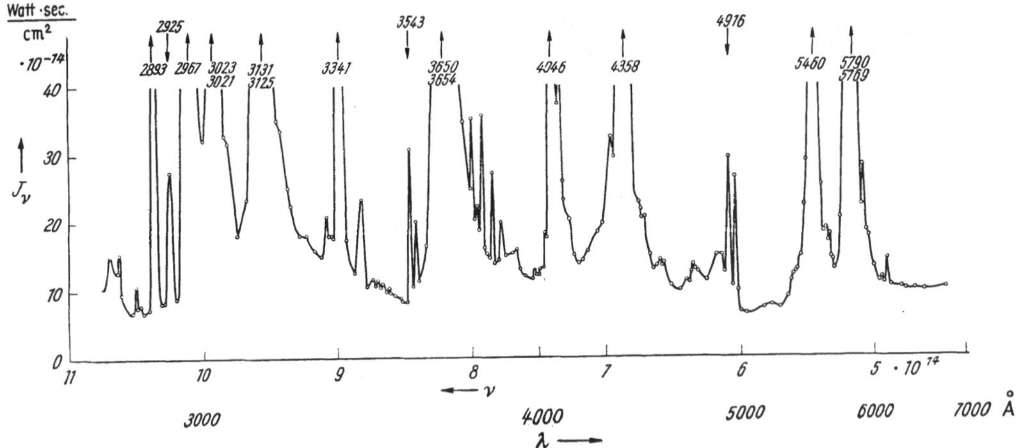

Fig. 39. Energieverteilung im Spektrum einer von GÖING, MEIER und MEINEN [183] untersuchten Hg-Höchstdrucklampe vom Typ HBO 200.

kontinuums folgte nach dem oben besprochenen Verfahren eine effektive Ionisierungsenergie von 10,1 eV. Die UNSÖLDsche Formel (15.8) ergibt in bester Übereinstimmung mit diesem Ergebnis für $N_e = 3 \cdot 10^{17}$ und $\Delta E_i = 0,45$ eV den

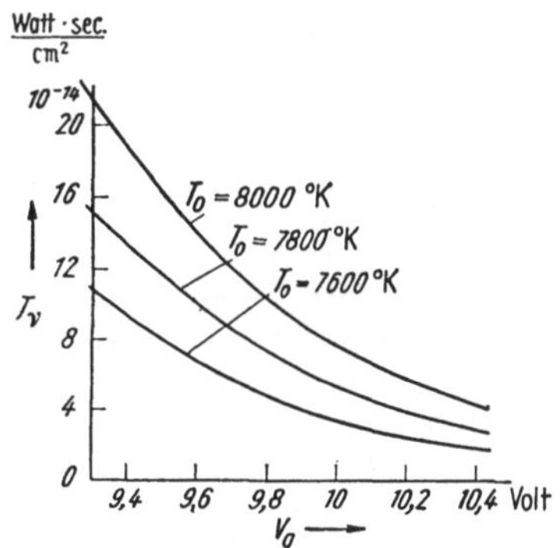

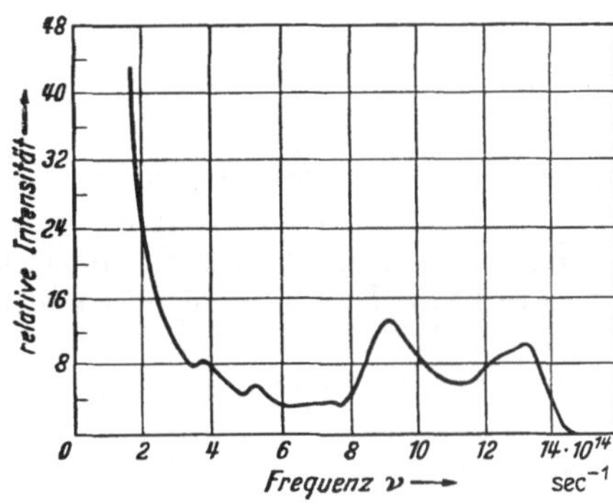

Fig. 40. Strahlungsdichte in Abhängigkeit von der Ionisierungsspannung bei verschiedenen Temperaturen nach GÖING, MEIER und MEINEN [183].

Fig. 41. Kontinuierliche Strahlungsverteilung im UV-Standard. (Nach RÖSSLER [388].)

Wert $E_i - \Delta E_i = 10$ eV. Die gemessene Energieverteilung ergibt nach Fig. 39 für den kontinuierlichen Untergrund eine Strahlungsdichte zwischen $6 \cdot 10^{-14}$ und $10 \cdot 10^{-14}$ Wsec cm⁻². Nach UNSÖLD errechnet sich, wie aus Fig. 40 zu entnehmen ist, bei 7800° K und 10,1 V Ionisierungsspannung eine Strahlungsdichte von $4,7 \cdot 10^{-14}$ Wsec cm⁻². Dieses ist innerhalb der Meßgenauigkeit von $\pm 200°$ K als gute Übereinstimmung zu bezeichnen, zumal in der Rechnung $Z_{\text{eff}}^2 = \overline{(Z+s)^2} = 1$ gesetzt worden ist. Eine ähnliche gute Übereinstimmung fand RÖSSLER [388] bereits bei dem Osram-UV-Normal (1,6 Atm). Die einzelnen Maxima in Fig. 41 wurden von ihm als Beiträge der bereits erwähnten Molekülstrahlung gedeutet, wofür außerdem z. B. auch eine gewisse „Struktur" im Spektrum der Hochdrucklampe HgQ 500 sprechen soll. Der rasche Abfall der Strahlung für $\nu \geq 13 \times 10^{14}\ sec^{-1}$ kann nach Maecker und Peters [295] aber ebensogut auf Grund der Rekombinationstheorie erklärt werden, weil nach jeder Seriengrenze ein Abfall mit $\exp(-h\nu/kT)$ eintritt und daher zwischen der Seriengrenze des ersten Anregungsniveaus ($6\,^3P_0$) und der Hauptseriengrenze nicht einmal mehr angenäherte Frequenzunabhängigkeit zu erwarten ist. Dieses erste Anregungsniveau liegt 5,72 eV unterhalb der Ionisierungsgrenze. Rechnet man bei einer Temperatur von 7500° K mit einer Erniedrigung von 0,3 eV, so liegt bei $\nu = 13 \cdot 10^{14}\ sec^{-1}$ gerade die Seriengrenze des ersten Anregungsniveaus, von der aus der exponentielle Abfall des Kontinuums nach höheren Frequenzen hin erfolgt. Es soll nicht von der Hand gewiesen werden, daß die Stoßmoleküle auch bei Hochdruckentladungen einen gewissen Beitrag zur Emission liefern können. Aber dieser Beitrag kann gegenüber der Rekombinations- und Bremsstrahlung nach den neueren Messungen nur gering sein.

Zu dieser Auffassung kommt auch Hettner [212], der an Hand verschiedener Untersuchungen über die Emission und Absorption von Hg-Hochdruckbögen im Gebiet der Zehntelmillimeterwellen (vgl. Ziff. 19) nachweist, daß es sich hierbei um Elektronenbremsstrahlung im Sinne der Unsöldschen Theorie handeln muß, während Rössler diese längstwellige Ultrarotstrahlung wiederum als Quasimolekülstrahlung gemäß Ziff. 26 deutet. In einer von Hettner angeregten theoretischen Arbeit über den Wirkungsquerschnitt und das „Sammlungsvermögen" einer Potentialschale in einem Teilchenstrom kommt Westpfahl [470] in Übereinstimmung mit früheren Überlegungen von Rompe und Steenbeck [383] zu dem Ergebnis, daß die Gesamtzahl der gebildeten Stoßmoleküle in dem Hochdruckplasma viel zu gering ist, um einen merklichen Beitrag zur Intensität des Kontinuums zu leisten. Das Problem scheint zwar noch nicht vollständig geklärt, aber die Untersuchungen zeigen, daß der von den Stoßmolekülen herrührende Beitrag zum Kontinuum der Hochdruckbögen nur gering sein kann.

β) Emissionskontinua von Elementen des Typs II. Hier sind vor allem die Arbeiten über die zuerst von Schulz [425] untersuchten Edelgasbogenlampen (Ne, A, Kr, X) zu nennen. Die intensive kontinuierliche Strahlung dieser Bögen, die von Schulz als Rekombinations- und Bremsstrahlung gedeutet wurde, nimmt für die verschiedenen Edelgase in der Reihenfolge Neon, Argon, Krypton, Xenon, also mit abnehmender Ionisierungsenergie zu, wie es nach der Beziehung $\varepsilon_\nu \sim p\, e^{-E_i/kT}$ auch zu erwarten ist. Die Xenonentladung ist daher die intensivste überhaupt bekannte Lichtquelle; nach Arbeiten von Larché [273] und Schirmer [417] werden vor der Kathode auf allerdings sehr kleinem Raum Leuchtdichten von 10^6 candela/cm² [1] erreicht (s. Ziff. 6). In den Termanordnungen der genannten Edelgase existieren im Anschluß an eine relativ dichte Termfolge — bis etwa 4 eV unterhalb der Ionisierungsgrenze — überhaupt keine weiteren Anregungszustände mehr. Berücksichtigen wir noch die Erniedrigung der Ionisierungsenergie, die 0,5 bis 1 eV betragen kann, so folgt, daß die Anwendung der Gl. (5.43) nur bis zu (negativen) Energiewerten von 3 bis 3,5 eV erlaubt ist. Entsprechend wird eine nahezu konstante Intensitätsverteilung im Spektrum nur vom Ultrarot bis zu Wellenlängen von 3500 bis 4000 Å zu erwarten sein; anschließend folgt der Abfall mit $\exp(-h\nu/kT)$. Die Energieverteilung in der X-Entladung ist von Schulz [426], Kienle, Baum und Dunkelmann [41], Anderson [14] und

---

[1] Candela (auch als Neue Kerze NK bezeichnet) entspricht etwa der älteren Einheit der Hefnerkerze (HK): 1 cd ≈ 1,16 HK.

HELLER [203] gemessen worden (Fig. 42). Der Verlauf ist bei Krypton sehr ähnlich. Wie alle Verfasser feststellten, kann die relative Verteilung recht gut durch
eine PLANCK-Funktion für $T = 5500°$ K (Farbtemperatur) angenähert werden,
was als Zufall anzusehen ist und zeigt, daß die Frequenzunabhängigkeit für größere Wellenlängen als 4000 Å nicht besonders gut erfüllt ist. Der Abfall für
$\lambda < 4000$ Å kann aber nach MAECKER und PETERS [295] zur Temperaturbestimmung ausgenutzt werden. Nach Gl. (5.45) ist

$$\frac{d \ln \varepsilon_\nu}{d\nu} = -\frac{h}{kT}. \qquad (15.14)$$

Bei der 160 W-Xenonlampe berechnet sich aus (15.4) eine Temperatur von
$7400°$ K, während der Absolutwert der Strahlung $7800°$ K liefert. Die Differenz
ist nicht überraschend, weil die Messung
aus dem Abfall der Intensität nach der
Grenzfrequenz ν_g nicht sehr genau ausgeführt werden kann. Immerhin sind die
beiden Temperaturwerte, die sich aus der
Anwendung der UNSÖLDschen Theorie
ergeben, recht plausibel.

Nach neueren Messungen am Argon-
Hochstrombogen von BUSZ und FIN
KELNBURG [80] bei 200 Amp Bogenstrom ist die Kontinuumsintensität in
der Frequenzskala im Sichtbaren etwa
konstant und fällt im UV ab. Der Argon-
Hochstrombogen scheint also die Forderung der Theorie wesentlich besser zu
erfüllen als der Krypton- und Xenonbogen.

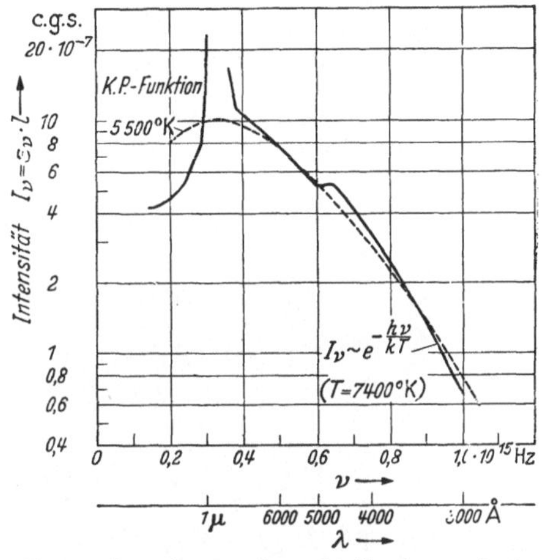

Fig. 42. Intensitätsverteilung des Kontinuums in der
160 Watt-Xenon-Lampe. (Nach SCHULZ und KIENLE.)

Eingehende Untersuchungen über das
intensive Kontinuum des Hochstromkohlebogens (Fig. 43) sind von MAECKER [292], [293] und MAECKER und PE
TERS [295] durchgeführt worden. Im Hochstromkohlebogen herrscht bei 200 Amp
eine Achsentemperatur von $11\,000°$ K mit einer Elektronendichte von $7 \cdot 10^{16}$ cm⁻³.
Zur Bestimmung der Grenzfrequenz ν_g, bis zu der eine konstante Emission in
der Frequenzskala zu erwarten ist, muß die Gaszusammensetzung bekannt sein.
Aus der spektroskopischen Beobachtung ergab sich, daß 30% C, 55% N und
15% O am Plasma beteiligt sind. Diese Elemente gehören, wie erwähnt, zum
Typ II. Die Grenzfrequenz ν_g liegt bei Mittelung über alle drei Atomarten und
Berücksichtigung der Erniedrigung der Ionisierungsenergie um 0,33 eV bei etwa
$9 \cdot 10^{14}$ cm⁻¹ (etwa 3500 Å), von wo ab die Kontinuumsintensität exponentiell
nach hohen Frequenzen hin abfallen sollte. Der Vergleich zwischen den gemessenen Werten und dem theoretischen Verlauf ist in Fig. 44 dargestellt. Die
Abweichung im Absolutbetrag liegt bei 20%, was leicht auf einen Fehler von
10% bei der Bestimmung der Elektronenkonzentration zurückgeführt werden
kann. Ähnliche Messungen sind von TER HORST und RUTGERS [222] am parallel
zur Achse angeblasenen Hochstromkohlebogen mit Wechselstrombetrieb gemacht
worden. In ihrem relativen Verlauf decken sich beide Messungen von 6000 bis
3000 Å recht gut.

γ) *Kontinua der Elemente des Typs III.* Die Ausmessung der Kontinua von
Elementen mit einer hohen Termmultiplizität ist außerordentlich schwierig, weil
in den linienreichen Spektren nur vereinzelt Lücken zu finden sind, in denen die

9*

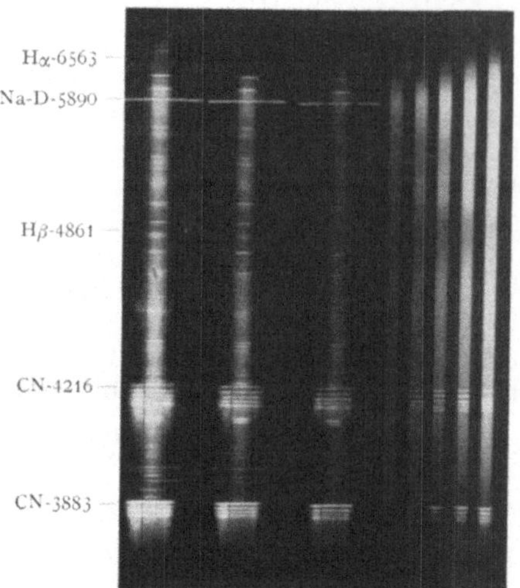

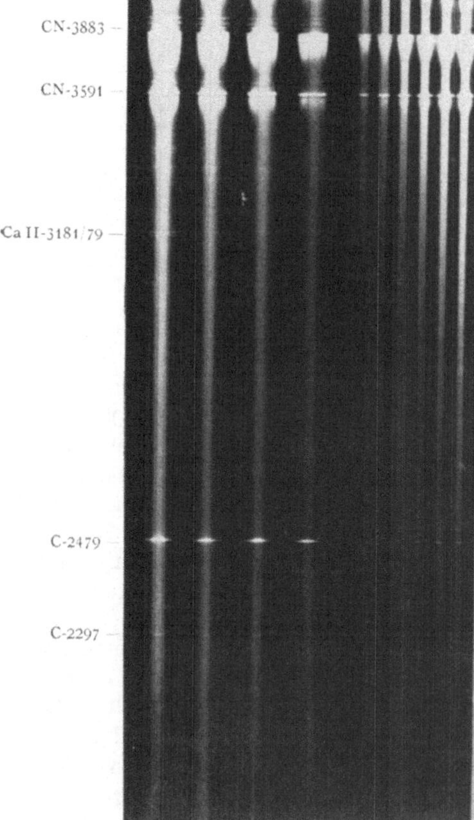

Fig. 43. Querspektren von der Säule des 200 Amp-Kohle-bogens 1,4 cm über der Kathode (linke Spektren). Strah-lungsnormal (rechte Spektren). (Nach Maecker und Peters [295].)

wahre Kontinuumsintensität meßbar ist. Zu erwähnen ist hier eine Arbeit von Burhorn [74] über das thermische Gleichgewicht im Eisenbogen. Die Gastemperatur wurde dabei aus der Doppler-Breite einiger günstiger Eisenlinien bestimmt und mit der aus der Absolutintensität des Kontinuums nach Gl. (5.43) berechneten Elektronentemperatur verglichen. Innerhalb der Meßgenauigkeit waren beide Temperaturen gleich. Die Frequenzabhängigkeit des Fe-Kontinuums konnte mit einer hochauflösenden Gitteranordnung vermessen werden. Dabei ergab sich der in Fig. 45 dargestellte Verlauf, der für $\lambda < 3000$ Å im Widerspruch steht mit der Theorie, nach der man auf Grund der dichten Termfolge eine Frequenzunabhängigkeit bis etwa 2200 Å erwarten sollte. Ähnliche Abweichungen findet man übrigens auch bei den Metallkontinua in Sternspektren.

δ) *Kontinua von Funkenentladungen und Verdichtungsstößen.* Von den zahlreichen Untersuchungen über Kondensatorentladungen können an dieser Stelle nur diejenigen berücksichtigt werden, die in gewissem Umfange einen quantitativen Vergleich mit der Theorie der kontinuierlichen Strahlung zulassen.

Zu dem Ergebnis einer sicheren Deutung als Elektronenbrems- und Rekombinationsstrahlung kamen zuerst Hahn und Finkelnburg [195] für den Fall der kontinuierlichen Emission kondensierter Funkenentladungen durch Capillaren mit etwas verdünnten verschiedenen Gasen bei Stromdichten zwischen 40000 und 170000 Amp/cm². Oberhalb 70000 Amp/cm² ist das Kontinuum vorherrschend gegenüber der Linienstrahlung. Sorgfältige Messung der relativen spektralen Intensitätsverteilung ergab unabhängig von der Stromdichte, der Gasart und dem Gasdruck die der Unsöldschen Theorie entsprechende frequenzunabhängige Emission mit einer mittleren Abweichung von nur $\pm 8\%$. Die Abhängigkeit der gesamten Strahlungsintensität von der magnetisch gemessenen maximalen Funkenstromstärke wurde in Übereinstimmung mit der Erwartung für Stöße zwischen Ionen und Elektronen als quadratisch (mit nur geringen, erklärbaren Abweichungen) gefunden.

Recht eingehende und umfangreiche Strahlungsmessungen hat dann GLASER [181] an Hochleistungsfunken (Kondensatorentladungen in einer Hochdruckatmosphäre bei hoher Spannung und kurzer Entladungsdauer) durchgeführt. Im ersten Maximum der periodischen Entladung zeigte sich ein nahezu reines Kontinuum, dessen Verteilung sich eher einer PLANCKschen Spektralverteilung als der UNSÖLDschen Verteilung angleichen ließ. Unter der Annahme quasischwarzer Strahlung erhielt GLASER Temperaturwerte bis etwa 40000° K zu Beginn der Entladung. Da das Funkenplasma bei diesen Temperaturen aus-

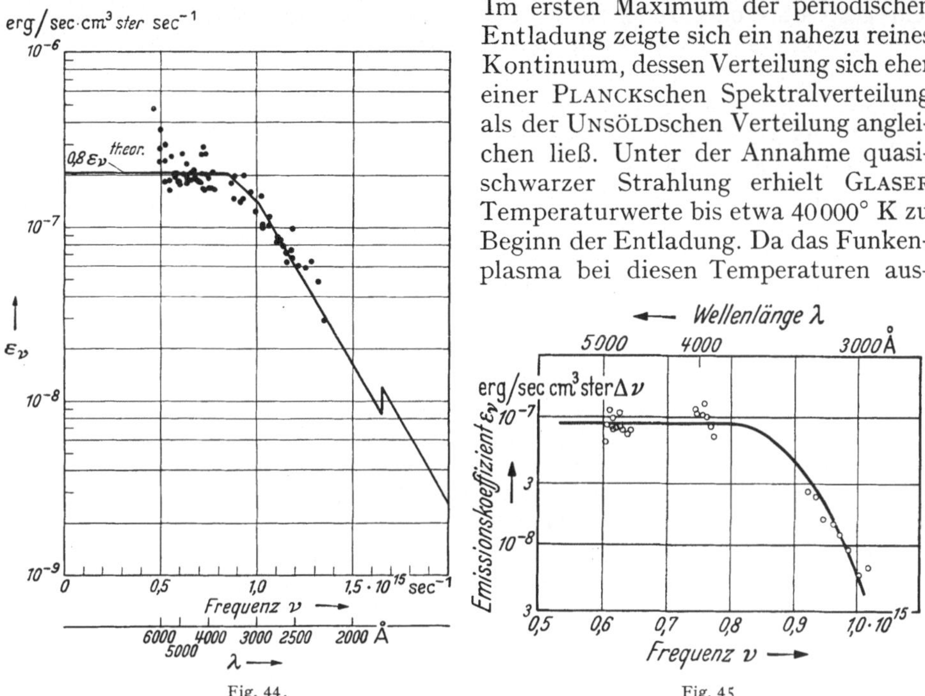

Fig. 44.

Fig. 45

Fig. 44. Emissionskoeffizient in der Achse des 200 Amp-Kohlebogens (Punkte). Theoretische Kurve um 20% herabgesetzt (ausgezogen). Beides als Funktion der Frequenz. (Nach MAECKER und PETERS [295].)

Fig. 45. Kontinuierlicher Emissionskoeffizient in der Achse des Eisenbogens nach BURHORN [74].

schließlich aus Elektronen und mehrfach ionisierten Atomen besteht, ist die Verwendung der von GLASER zur Temperaturbestimmung herangezogenen Gl. (15.11), die mit $N_0 kT \approx p$ nur für schwach ionisierte Gase gilt, allerdings unzulässig, weil die Temperaturabhängigkeit bei hoher Ionisierung eine völlig andere ist [s. Gl. (15.13)].

Daß die Funkenstrahlung zu Beginn der Entladung als schwarze Strahlung hoher Temperatur aufzufassen ist, wurde schon von ANDERSON [13] auf Grund seiner Arbeiten über Metalldrahtentladungen (explodierende Drähte) und stromstarke Kondensatorentladungen durch Capillaren bei sehr geringen Gasdrucken erkannt. Nach ANDERSON absorbierte bei einer Stromdichte von 30000 Amp/cm² eine Gasschicht von 1 cm Dicke vollständig, bei 100000 Amp/cm² bereits eine solche von 1 mm Dicke. Die Intensität des Kontinuums entsprach in Übereinstimmung mit GLASER den Intensitäten schwarzer Strahler von Temperaturen bis zu 50000° K.

Möglicherweise sind durch hohe Absorption auch die Messungen von WYNEKEN [490] und SCHUBERT [421] am Unterwasserfunken zu verstehen, die im Gegensatz zur UNSÖLDschen Theorie keine frequenzunabhängige Energieverteilung des Kontinuums, sondern eher die eines schwarzen Strahlers ergab. Die angegebenen Werte für den Gasdruck von 12 Atm und für die Temperatur von

5000° K lassen zwar auf einen sehr geringen Absorptionskoeffizienten schließen, jedoch ist der kleine Unterschied in der Absoluthöhe zwischen H_α und dem Kontinuum im UV ohne Annahme starker Absorption unverständlich.

Die Polarisation des von Metalldrahtentladungen emittierten Lichtes hat Cohn [95] untersucht. Da er keine meßbare Polarisation fand und daher auch kein gestreutes oder reflektiertes Licht am Spektrum beteiligt sein konnte, schließt Cohn auf Elektronenbremsstrahlung oder „Temperaturstrahlung" als Ursache der kontinuierlichen Emission.

Durch Überlagerung einer Kondensatorentladung über einen Kupferbogen erhielt Levintov [282] ein Kontinuum, welches ebenfalls in Übereinstimmung mit der Unsöldschen Theorie stand.

Zu interessanten Ergebnissen gelangten ferner Fowler, Goldstein und Clotfelter [158] sowie Fowler, Atkinson und Marks [159] bei der Erzeugung von Stoßwellen durch Funkenentladungen. In ein seitlich an das eigentliche Entladungsgefäß angebrachtes Rohr läuft nach dem Funkendurchschlag eine Stoßwelle mit Geschwindigkeiten bis zu 18 km/sec (in Wasserstoff) hinein. Dieser Vorgang wurde mit einer rotierenden Spiegelanordnung zeitlich aufgelöst. Bei Wasserstoff-Füllung des Entladungsgefäßes zeigten sich im Spektrum die Balmer-Linien bis H_ε mit anschließendem Rekombinationskontinuum. Aus den Halbwertsbreiten der Linien H_α bis H_δ wurde eine Ionenkonzentration von 2 bis $5 \cdot 10^{16}$ cm^{-3} gefunden. Ein Hinzufügen von Wasserdampf zur Wasserstoff-Füllung ergab keine Änderung des kontinuierlichen Spektrums, so daß offensichtlich kein H_2-Molekülkontinuum beteiligt sein konnte.

Die Aufklärung dieser Strahlungserscheinungen als Stoßfrontleuchten steht im Widerspruch mit den früheren Annahmen von Rayleigh [370] und Zanstra [495], daß es sich hierbei um einen Nachleuchteffekt (afterglow) handeln müßte.

Sehr genaue Messungen der kontinuierlichen Strahlung von Verdichtungsstößen haben schließlich Petschek, Rose, Glick, Kane und Kantrowitz [356] durchgeführt. Die Verdichtungsstöße wurden dabei nicht elektrisch durch Funken, sondern in einem Stoßrohr (shocktube) erzeugt. Die kontinuierliche Strahlung des hinter der Stoßfront gebildeten hochverdichteten Argonplasmas zeigte eine frequenzunabhängige Verteilung in dem ausgemessenen Spektralbereich von 5800 bis 4200 Å. Die Absolutwerte des Kontinuums stimmten im angegebenen Temperaturbereich von 10000 bis 13000° K mit den aus Gl. (5.43) berechneten Werten gut überein, wenn $Z_{eff} = 1$ gesetzt wurde.

16. Spezielle Serien-Grenzkontinua in Niederdruck-Entladungen. Im Gegensatz zu den im vorigen Abschnitt behandelten Entladungstypen bei hohen Drucken ist bei Niederdruckentladungen nicht mehr mit thermischem Gleichgewicht des Entladungsplasmas zu rechnen. Im nichtthermischen Plasma muß jeder einzelnen Energieform eine eigene „Temperatur" zugeordnet werden. So unterscheidet man zwischen den „kinetischen Elektronen- und Gastemperaturen", der „Ionisationstemperatur" und den verschiedenen „Anregungstemperaturen". Zur Berechnung dieser verschiedenen „Temperaturen" ist die Behandlung der einzelnen Elementarprozesse wie Anregung, Ionisation, Rekombination usw. erforderlich, wozu die betreffenden verschiedenen Wahrscheinlichkeitskoeffizienten bekannt sein müssen. Da deren Theorie aber noch in den Anfängen steckt, ist eine geschlossene Darstellung der Theorie dieser Prozesse in einem nichtthermischen Plasma kaum möglich. Wir weisen in diesem Zusammenhang auf die neueren Arbeiten von Miyamoto [316], Biermann [63], Elwert [136], [137], Wolley und Allen [482], [483] hin. Die Wiedervereinigung von Elektronen und Ionen geht ja nicht nur im Zweierstoß unter Emission eines Lichtquants im kontinuierlichen Spektrum vor sich, sondern vielfach auch strahlungslos im Dreierstoß, wobei die Überschußenergie an ein drittes Teilchen abgegeben wird. Bei den Gasentladungen spielt außerdem die Abdiffusion von Ladungsträgern an die Wände eine erhebliche Rolle, so daß ohne Kenntnis der ganzen Entladungsparameter kaum zu übersehen ist, welche Prozesse bei der Trägererzeugung und -vernichtung die Oberhand gewinnen. In unserem Zusammenhang interessiert aber nur die zur Emission eines Kontinuums führende Zweierstoß-Rekombination.

Qualitative Beobachtungen von Emissions-Grenzkontinua sind recht zahlreich. Den ersten gesicherten Nachweis des Na-Nebenseriengrenzkontinuums im Vakuumbogen erbrachte BARTELS [29]; gleichzeitig fand LYMAN [289], [290] das He-Hauptserienkontinuum in einer intermittierenden Entladung (s. Fig. 46). Es folgten Beobachtungen des Nebenserienkontinuums des Hg im Nachleuchten eines Hg-Bogens von Lord RAYLEIGH [369], der an die Grenzen $2S$ und $2P$ sich anschließenden Kontinua des He in der Hohlkathode durch PASCHEN [353], und endlich die Beobachtung des BALMER-Grenzkontinuums in einer elektrodenlosen Ringentladung durch HERZBERG [208].

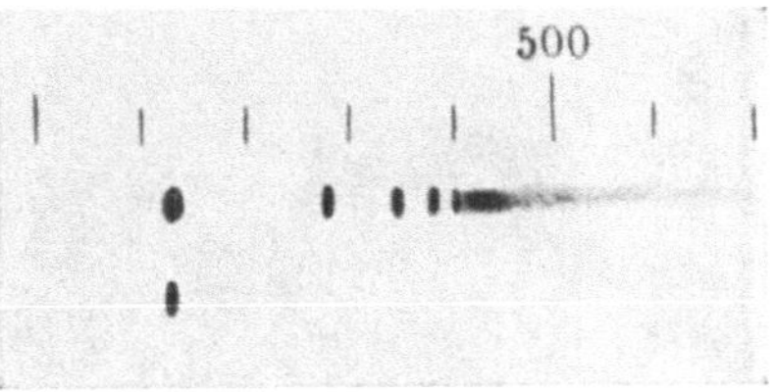

Fig. 46. Die höheren Glieder der He-Hauptserie mit dem anschließenden Grenzkontinuum in Emission im äußersten Ultraviolett. Aufgenommen mit intermittierender Entladung von LYMAN [290].

Die Grenzkontinua der Haupt-, Neben- und BERGMANN-Serie der Alkalien sowie mehrerer Elemente der zweiten und dritten Spalte des periodischen Systems sind namentlich von KREFFT [258] bis [260] in der positiven Säule von Glühkathodenentladungen bei einigen Ampere Stromstärke und Gasdrucken zwischen 1 und 100 mm Hg beobachtet worden (s. Fig. 47 und 48). Bei dem in Fig. 47 wiedergegebenen Nebenserien-Grenzkontinuum des Thalliums ist wiederum das Übergreifen des Kontinuums nach längeren Wellenlängen mit zunehmender Stromdichte, also auch zunehmender Elektronendichte, deutlich zu erkennen.

Mit Sicherheit nachgewiesen sind Emissionskontinua in Niederdruckentladungen bisher bei den Elementen H, He, Na, K, Rb, Cs, Mg, Ca, Sr, Ba, Hg, Cd, Zn, Tl und In.

Wie bereits betont, hängen Intensität und Ausdehnung der Kontinua weitgehend von den Versuchsbedingungen ab, d.h. in erster Linie von der Dichte und Geschwindigkeitsverteilung der Elektronen und Ionen. Bei der Ableitung empirischer Gesetzmäßigkeiten ist daher Vorsicht geboten. Deutlich feststellbar ist allerdings, daß Grenzkontinua intensiver bei Elementen mit Dublettspektren als bei solchen mit Singulett- und Triplettspektren auftreten, und daß Triplettkontinua intensiver auftreten als Singulettkontinua. Allgemein sind auch die Grenzkontinua der Nebenserien und der BERGMANN-Serie intensiver als die der Hauptserie.

Quantitative Untersuchungen über Emissions-Grenzkontinua in Niederdruckentladungen liegen vor von MOHLER [317], [323] bis [326] und MOHLER und BOECKNER [318] an Cs- und He-Entladungen, sowie von JANCKE [240] ebenfalls an He-Entladungen. Die Untersuchungen der erstgenannten Autoren über die Grenzkontinua der Haupt-, Neben- und BERGMANN-Serie des Caesiums sind an Gleichstromentladungen bei Drucken zwischen 10^{-3} und 1 mm Hg und bei Stromdichten zwischen 0,03 und 2,5 Amp/cm² ausgeführt worden. Fig. 49 zeigt die Intensitätsverteilung der Emissions-Grenzkontinua in Abhängigkeit von der Wellenlänge und läßt die schon erwähnte geringere Intensität des Hauptserien-Grenzkontinuums gegenüber denen der Nebenserien und BERGMANN-Serie

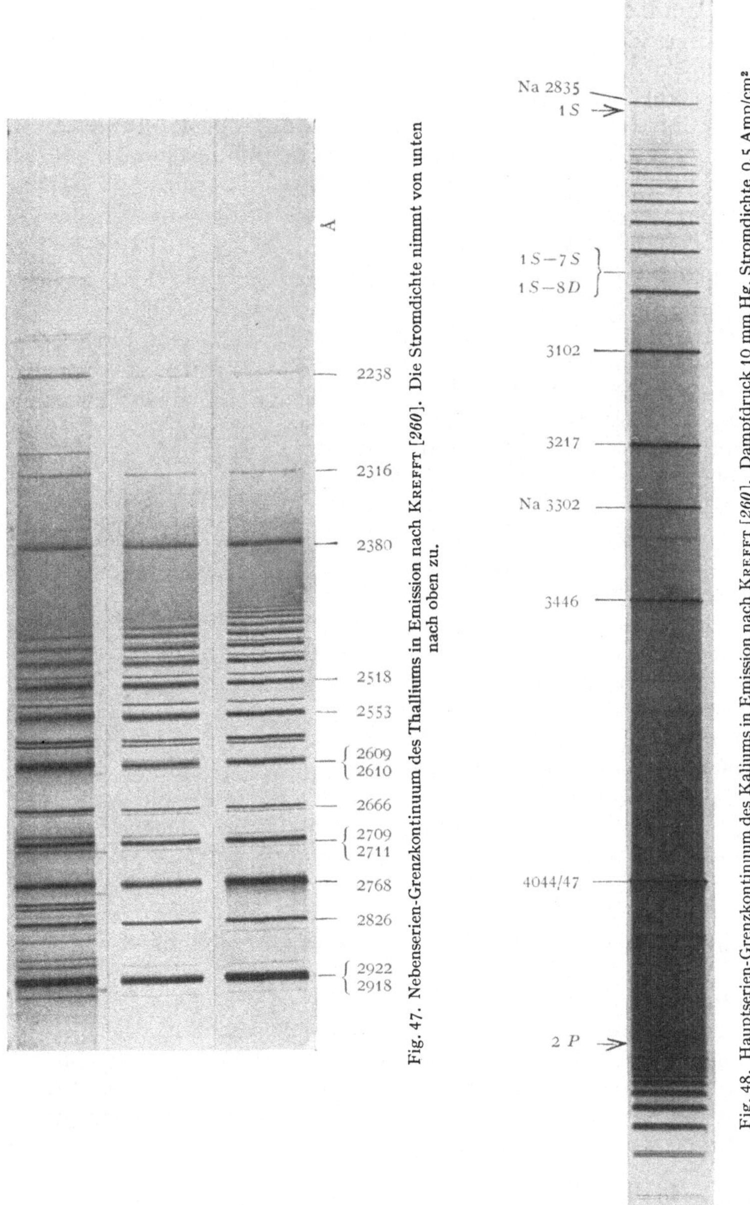

Fig. 47. Nebenserien-Grenzkontinuum des Thalliums in Emission nach Krefft [260]. Die Stromdichte nimmt von unten nach oben zu.

Fig. 48. Hauptserien-Grenzkontinuum des Kaliums in Emission nach Krefft [260]. Dampfdruck 10 mm Hg, Stromdichte 0,5 Amp/cm².

erkennen. Die Gesamtintensität der Kontinua steigt bei niedrigen Drucken und Stromdichten, wie zu erwarten, mit dem Quadrat der durch Sondenmessungen ermittelten Elektronendichte. Die Messung der Elektronendichte ermöglicht nach

$$\left.\begin{aligned} \frac{dN_e}{dt} &= -\,\alpha_n\,N_e\,N_i = -\,v\,q_n\,(v)\,N_e\,N_i \\ \text{mit } v &= \text{Elektronengeschwindigkeit} \end{aligned}\right\} \tag{16.1}$$

die Berechnung absoluter Werte des Rekombinationskoeffizienten α_n bzw. des Einfang-Querschnittes q_n, der gegeben ist durch die Beziehung

$$q_n = \frac{c_n}{v^2\,(v - v_n)}\,, \qquad (16.2)$$

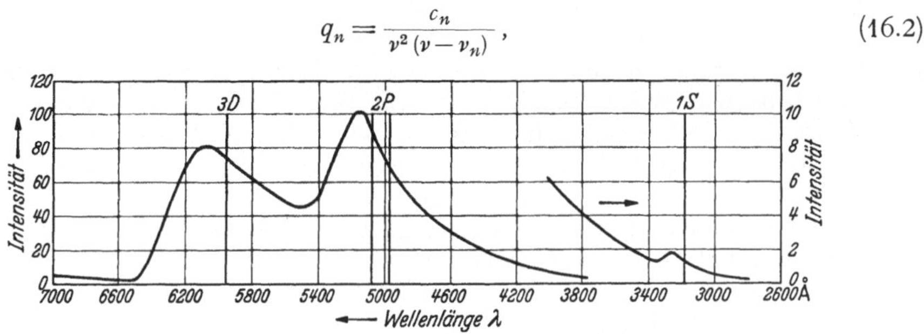

Fig. 49. Intensitätsverlauf der verschiedenen Seriengrenzkontinua des Cs-Atoms in Emission. (Nach MOHLER [317a].)

wobei v_n die Seriengrenze angibt und c_n eine für das betreffende Grenzkontinuum charakteristische Konstante ist. Den gemessenen Verlauf für das Nebenserien-Grenzkontinuum des Cs zeigt Fig. 50. Als Absolutwert von q ergibt sich für die Rekombination von 0,3 eV-Elektronen in den P-Zustand $q_{2P} = 1,7 \cdot 10^{-21}$ cm².

Die Grenzkontinua des Heliums hat JANCKE [240] in Gleich- und Wechselstromentladungen bei Drukken bis zu 12 mm Hg in Abhängigkeit von den elektrischen Bedingungen untersucht. Auch er fand eine quadratische Abhängigkeit der Intensität von der Stromstärke. Von besonderem Interesse ist ferner, daß die Intensität des Triplettkontinuums stets wesentlich über der des Sigulett-Grenzkontinuums liegt und auch mit steigendem Gasdruck viel stärker anwächst als diese. JANCKE schließt daher auf eine deutliche Bevorzugung der Rekombination in Triplettzustände, namentlich bei höheren Drucken. Den Intensitätsverlauf der Serien-

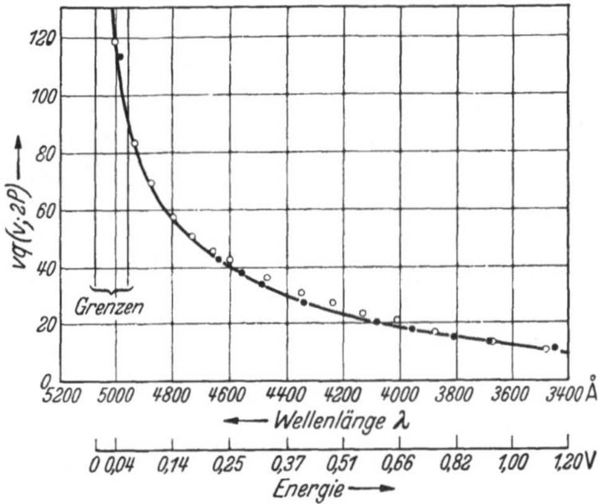

Fig. 50. Gemessener Verlauf des Rekombinationskoeffizienten für Rekombination in den 2 P-Zustand des Cs-Atoms, aufgetragen gegen die Wellenlängen. Unterer Maßstab: Voltenergie der rekombinierenden Elektronen. (Nach MOHLER [317a].)

grenzkontinua im nahen UV nach den Messungen von MOHLER und BOECKNER zeigt Fig. 51.

In engem Zusammenhang mit der Emission der Grenzkontinua und der Bestimmung der Rekombinationskoeffizienten bzw. -querschnitte stehen die Untersuchungen über Nachleuchterscheinungen in elektrisch

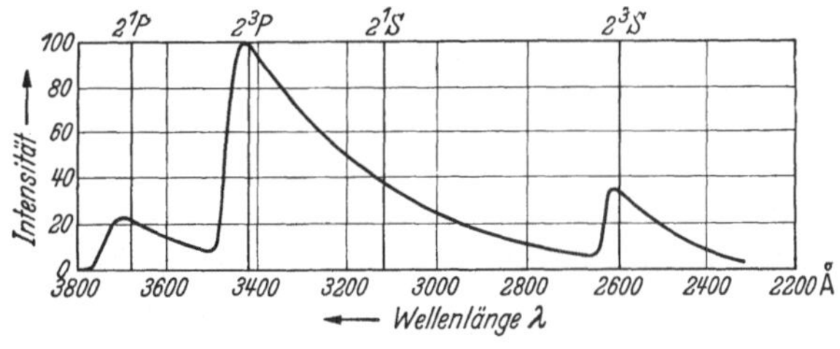

Fig. 51. Intensitätsverlauf der Emissions-Seriengrenzkontinua des Heliums im nahen Ultraviolett nach Messungen von MOHLER und BOECKNER [318]. (Aus MOHLER [317a].)

angeregten Gasen und Dämpfen. Die älteren Versuche darüber sind von SEELIGER [401] eingehend beschrieben und diskutiert worden. Daß an dem Nachleuchten der meisten Gase Rekombinationsvorgänge maßgebend beteiligt sind, steht außer Zweifel. Bei der Berechnung von Rekombinationskoeffizienten ist aber Vorsicht am Platze, weil, wie erwähnt, noch andere Prozesse an der Ladungsträgervernichtung beteiligt sind, deren Einfluß nur in den wenigsten Fällen abschätzbar ist. Wir möchten an dieser Stelle nur auf die neue, insbesondere von Biondi [66], [67] und Mitarbeitern angewendete Mikrowellentechnik und die Arbeiten von Holt und Mitarbeitern [215], [110] hinweisen. Für genauere Einzelheiten über Rekombinationsvorgänge sei aber auf die Gasentladungsbände XXI, XXII dieses Handbuches verwiesen. Eine Diskussion einiger noch ungeklärter Versuchsergebnisse bei Niederdruckentladungen und ihrer Deutungsversuche, z.B. durch die Annahme, daß Rekombinationen vorzugsweise in den höheren Quantenzuständen erfolgen sollen, findet man in Finkelnburgs „Kontinuierliche Spektren" [149]. Die Diskussion zeigt, daß bezüglich der Seriengrenzkontinua in Emission Theorie und Beobachtung noch keineswegs übereinstimmen.

c) Kontinuierliche Emission und Absorption freier Elektronen.

17. Allgemeines. Im Anschluß an die Ionisations- und Rekombinationskontinua haben wir als dritte, zur Emission und Absorption kontinuierlicher Strahlung führende Gruppe von Elektronenübergängen diejenige zu behandeln, bei denen das Elektron im Anfangs- wie im Endzustand frei ist und bei der Emission bzw. Absorption lediglich kinetische Energie verliert bzw. gewinnt. Bei der Übersicht über die grundsätzlich möglichen Elektronenübergänge in Ziff. 4 wurden diese sog. frei-frei-Übergänge bereits kurz besprochen und in das Übergangsschema der Fig. 1 aufgenommen. Es handelt sich also um Übergänge zwischen zwei verschiedenen Energiezuständen im kontinuierlichen Energiebereich. Im Bilde der klassischen Elektrodynamik entspricht dem Emissionsvorgang die Bremsung eines Elektrons im Feld eines positiven Ions unter Aussendung von Dipolstrahlung; in der Ausdrucksweise der Bohrschen Theorie handelt es sich um Übergänge zwischen Hyperbelbahnen verschiedener Energie. Die Frequenzbedingung für die kontinuierliche Emission und Absorption freier Elektronen lautet also

$$h\nu = E_1 - E_2 = \frac{m}{2}\left(v_1^2 - v_2^2\right), \tag{17.1}$$

wobei E_1 und E_2 die Energie und v_1, v_2 die Geschwindigkeiten des Elektrons im Anfangs- und Endzustand bedeuten. Für die kurzwellige Grenze des aus den Übergängen entstehenden Kontinuums gilt

$$\nu_0 = \frac{m\,v_1^2}{2h} \tag{17.2}$$

das sog. Duane-Huntsche Gesetz.

Die Theorie der Bremsstrahlung wurde zur Erklärung der empirischen Gesetzmäßigkeiten des *Röntgen-Bremsspektrums* zuerst auf klassisch-korrespondenzmäßiger Grundlage entwickelt. Wie wir in Ziff. 5 erwähnt haben, ergeben die quantenmechanischen Rechnungen grundsätzlich das gleiche Ergebnis.

Das Röntgen-Bremsspektrum wird in Ziff. 18 behandelt. Auf die bei der *Bremsung langsamer Elektronen* emittierte kontinuierliche Strahlung im sichtbaren und ultraroten Spektralbereich gehen wir in Ziff. 19 ein. Eine besondere Rolle spielt die kontinuierliche Emission und Absorption freier Elektronen in der Astrophysik. Hier bildet die Theorie der Bremsstrahlung die Grundlage für die Untersuchung und Deutung der *kosmischen Radiofrequenzstrahlung* (vgl. Abschnitt IV). In Ziff. 20 wird schließlich noch kurz auf den Zusammenhang zwischen den *Plasmaschwingungen* und der frei-frei-Strahlung hingewiesen.

18. Röntgenbremsstrahlung. Die erste Vergleichsmöglichkeit zwischen Theorie und Beobachtung bot das bei der Bremsung schneller Elektronen an der Antikathode einer Röntgenröhre entstehende Röntgenbremskontinuum. Dieses Kontinuum ist seit seiner ersten spektroskopischen Untersuchung von MOSELEY und DARWIN [331] Gegenstand zahlreicher Arbeiten geworden. Da uns an dieser Stelle die kontinuierlichen Röntgenspektren nur soweit interessieren, als der physikalische Zusammenhang mit den „optischen" Kontinua in Erscheinung treten soll, müssen wir uns auf die wesentlichsten Gesetzmäßigkeiten der Röntgenbremsstrahlung beschränken und wegen näherer Einzelheiten auf Bd. XXX dieses Handbuches verweisen. Die zahlreichen Arbeiten auf diesem Gebiet behandeln außer der Feststellung des Zusammenhanges der kurzwelligen Grenze mit der maximalen Elektronengeschwindigkeit [Gl. (17.2)] besonders die Energieverteilung bei massiver wie bei äußerst dünner Antikathode zum Vergleich mit der Theorie. Untersucht wurden ferner die Abhängigkeit der Strahlungsintensität vom Emissionswinkel, die Polarisation, sowie schließlich die Gesamtintensität und der Wirkungsgrad der Umsetzung von kinetischer Elektronenenergie in Bremsstrahlung. DUANE und HUNT [122] zeigten zuerst, daß die theoretische Beziehung zwischen der maximalen Voltgeschwindigkeit und der kurzwelligen Grenze in der Form $h\nu = eV_{\max}$ erfüllt ist. Diese ist später von zahlreichen Beobachtern bei großen wie kleinen Geschwindigkeiten als unabhängig vom Material der Anti-

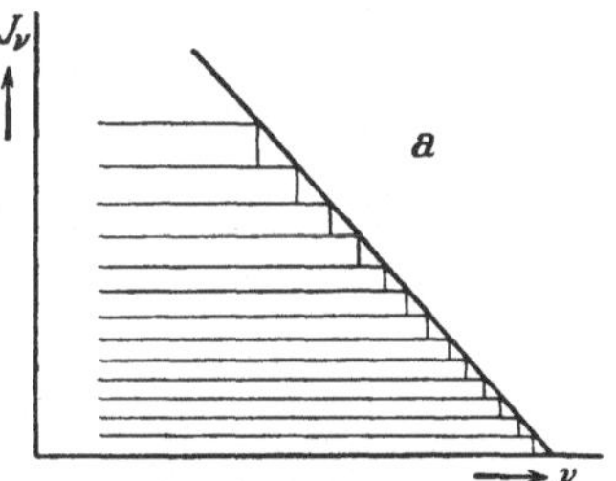
Fig. 52. Schematischer Aufbau der Bremsstrahlung einer dicken Antikathode nach KULENKAMPFF.

kathode und von der Emissionsrichtung bestätigt worden. Bei genauer Kenntnis von $V_{\max}$ läßt sich aus der Messung von ν_0 mit großer Genauigkeit der Wert des Wirkungsquantums h bestimmen (vgl. DuMOND [125]).

Die *Energieverteilung* der kontinuierlichen Röntgenstrahlen ist nach KRAMERS leicht zu berechnen. Aus der Integration der Gl. (5.30) folgt, wenn S Elektronen der Geschwindigkeit v pro sec eine sehr dünne Schicht, die A Atome der Ordnungszahl Z pro Flächeneinheit enthält, durchsetzen:

$$i_\nu = \frac{32\,\pi^2}{3\,\sqrt{3}}\,\frac{Z^2 e^6}{c^3\,m^2\,v^2}\cdot S\,A\,g. \qquad (18.1)$$

Dieser Ausdruck enthält nicht mehr die Frequenz ν, so daß die Intensität konstant ist für $0 \leqq \nu \leqq \nu_0$. Führen wir mit $eV = mv^2/2$ in (18.1) die Spannung ein, so ergibt sich ein hyperbolischer Verlauf der Isochromaten. Diese Gesetzmäßigkeiten können experimentell geprüft werden, wenn als Antikathodenmaterial sehr dünne Folien (z.B. KULENKAMPFF [270]) oder auch Hg-Dampfstrahlen geringer Dichte (DUANE [123]) verwendet werden.

In einer massiven Antikathode verlieren die eindringenden Kathodenstrahl-Elektronen nach und nach ihre kinetische Energie. Dieser Verlust ist aber nur in gewissem Umfang der Strahlungserzeugung zuzuschreiben. Hauptsächlich werden Elektronen aus dem Inneren des Atoms herausgeschlagen, was Anlaß gibt zur Emission der sog. *charakteristischen Röntgenstrahlung*. Vernachlässigt man weiterhin die Absorption der Strahlung und die Reflexion und Diffusion der Elektronen in der endlich dicken Antikathode, so kann die Strahlung schematisch als Summe der Strahlungen einer großen Anzahl dünner Antikathoden aufgefaßt werden, deren kurzwellige Grenzen sich gemäß Fig. 52 wegen der mit wachsender Eindringtiefe abnehmenden Elektronengeschwindigkeit immer mehr nach langen Wellen zu verschieben. An Stelle der konstanten Intensitätsverteilung bei sehr dünner Antikathode tritt daher bei der massiven Antikathode eine gleichmäßige Intensitätsabnahme nach kurzen Wellen zu.

Zur Berechnung der Intensitätsverteilung integrieren wir über die verschiedenen Geschwindigkeiten der Elektronen in der Antikathode:

$$I_\nu = \int\limits_{v_0}^{v} i_\nu \frac{d s}{d v}\, d v . \tag{18.2}$$

Unter Anwendung des Thomson-Whiddingtonschen Gesetzes über die Geschwindigkeitsänderung mit der Eindringtiefe

$$\frac{d v^4}{d s} = -16\pi \left(\frac{e}{m}\right)^2 Z A\, l \tag{18.3}$$

mit $A =$ Anzahl der Atome pro cm³ und der Konstanten $l \sim 6$ (für das Röntgengebiet) erhält man

$$I_\nu = S\, \frac{4\pi e^2}{3\sqrt{3}\, l\, c^3}\, Z\, (v_0^2 - v^2) , \tag{18.4}$$

oder mit Einführung der Frequenz nach $h\nu = m v^2/2$ den Ausdruck

$$I_\nu = S\, \frac{8\pi e^2 h}{3\sqrt{3}\, l\, c^3\, m}\, Z\, (\nu_0 - \nu) . \tag{18.5}$$

Auch dieses Gesetz steht in guter Übereinstimmung mit den Beobachtungen, obwohl beträchtliche Vernachlässigungen vorliegen.

Die gesamte von einem einzigen Elektron emittierte Strahlung ist

$$I = \frac{8\pi e^2 h}{3\sqrt{3}\, l\, c^3\, m}\, Z \int\limits_{0}^{\nu_0} (\nu_0 - \nu)\, d\nu = \frac{4\pi e^2 h}{3\sqrt{3}\, l\, c^3\, m}\, Z\, \nu_0^2 = \frac{\pi e^2 m}{3\sqrt{3}\, l\, c^3\, h}\, Z\, v^4 . \tag{18.6}$$

Diese Bremsstrahlungsenergie, ins Verhältnis gesetzt zur kinetischen Energie des Elektrons, ergibt den *Wirkungsgrad* η der Röntgenstrahlerzeugung

$$\eta = \frac{I}{E_{\mathrm{kin}}} = \frac{2\pi}{3\sqrt{3}\, l}\, \frac{e^2}{c\, h}\, Z \left(\frac{v}{c}\right)^2 . \tag{18.7}$$

Der Nutzeffekt beträgt maximal einige Promille.

Sehr ausführliche Untersuchungen über die *Richtungsverteilung* der Bremsstrahlung sind insbesondere von Kulenkampff durchgeführt worden. Die *Polarisation* hat erstmalig Barkla [27] nachgewiesen.

19. Bremsstrahlung im langwelligen Spektralgebiet. α) *Kontinuierliche Bremsstrahlung in Gasentladungen.* Am übersichtlichsten liegen die Verhältnisse bei ionisierten Gasen. Nach Fig. 6 in Ziff. 5, welche die kontinuierliche Strahlungsverteilung eines „Modellgases" angibt, ist frei-frei-Strahlung zwar im gesamten Frequenzbereich zu erwarten; wesentlichen Anteil am gesamten Kontinuum hat sie allerdings nur im fernen Ultrarot. Im Radiofrequenzgebiet besteht die kontinuierliche Emission praktisch nur noch aus Bremsstrahlung freier Elektronen.

Eine Emission im Gebiet der Zehntel-Millimeterwellen fanden zuerst Rubens und Baeyer [390] (1911) bei der Untersuchung eines Hg-Hochdruckbogens. Da zu jener Zeit die Grundlagen der kontinuierlichen Emission noch nicht klar waren und zudem die spektrale Intensitätsverteilung diejenige einer Doppelbande zu sein schien, wurden Franck und Grotrian [160] zu der Vermutung geführt, daß es sich um eine Rotationsschwingungsbande des Hg_2-Moleküls handeln könnte. Weitere experimentelle Untersuchungen über die Emission und Absorption von Hochdruckbögen in Hg, Cd, Zn und Tl von Koch [*250*] und Dahlke [*108*], [*109*] (s. Fig. 53) im Gebiet der Zehntelmillimeterwellen (100 bis 400 μ) ergaben dann aber, daß die beobachteten Selektivitäten der Strahlung auf dem Wege von der Strahlungsquelle zum Empfänger, in der Hauptsache durch Wasserdampfabsorption der Luft, zustande kommen.

Während Rössler [*389*] noch eine Deutung im Sinne der Hg-Quasimolekülstrahlung versuchte, hat Hettner [*212*] durch eingehende Rechnungen auf der Grundlage der Kramers-Unsöldschen Theorie gezeigt, daß der experimentelle Befund von Dahlke eindeutig auf Bremsstrahlung freier Elektronen hinweist. Hettners Überlegungen führten zu dem Ergebnis, daß in dem betrachteten Spektralbereich eine erhebliche, mit der Wellenlänge zunehmende Absorption vorhanden ist (Tabelle 3). Unter Berücksichtigung dieser Absorption in Verbindung mit einem plausiblen Ansatz für

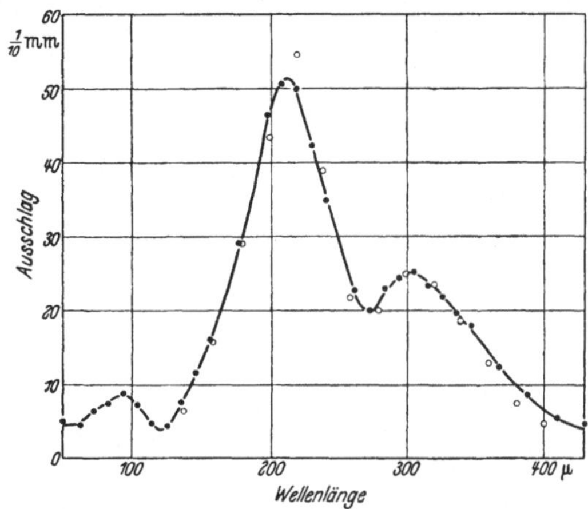

Fig. 53.   Ultrarot-Spektrum der Osram-Hg-Normallampe nach Dahlke [*108*].

die Temperaturverteilung über den Bogenradius wurde das Emissionsvermögen E_i des Bogens bestimmt, wobei zu beachten ist, daß E_i sich aus der eigentlichen Bogenemission und der des Bogenhintergrundes, also der Intensität $E_0^{(s)}$ der

Tabelle 3. *Osram-UV-Normal: Berechnetes Absorptionsvermögen des Bogens.*

λ	100 μ	200 μ	300 μ	400 μ	500 μ
$A(\lambda)$ . . .	12%	39%	67%	86%	95%

Hohlraumstrahlung von Zimmertemperatur (300° K) zusammensetzt. Die Strahlung der den Bogen umgebenden Quarzwandung konnte gesondert berücksichtigt werden. Die Ergebnisse der Rechnungen für das Osram-UV-Normal sind in Tabelle 4 aufgetragen. Zeile (2) gibt das berechnete Emissionsvermögen des eigentlichen Bogen E_B an. $E_0^{(s)}$ ist die Hohlraumstrahlung des Zimmers. Aus

$$E_i = E_B + D\,E_0^{(s)}$$

Tabelle 4. *Osram-UV-Normal: Emissionsvermögen E_B des Bogens. Emittierte Strahlungsintensität E_i und E_a und ihre schwarzen Temperaturen $T_i^{(s)}$ und $T_a^{(s)}$ innerhalb und außerhalb des Rohres.*

(1)	λ in μ	100	200	300	400	500
(2)	$E_B(\lambda) \cdot 10^{-3}$ in erg/cm² sec .	558	124	37,3	14,0	5,98
(3)	$E_0^{(s)}(\lambda) \cdot 10^{-3}$ in erg/cm² sec .	194	14	2,9	1,0	0,40
(4)	$E_i(\lambda) \cdot 10^{-3}$ in erg/cm² sec .	728	132	38,3	14,1	6,00
(5)	$T_i^{(s)}(\lambda)$ in °K	950	2550	3730	4360	4530
(6)	D_{Qu}	0,14	0,45	0,69	0,74	0,75
(7)	$E_a(\lambda) \cdot 10^{-3}$ in erg/cm² sec .	268	67	27,2	10,7	4,60
(8)	$T_a^{(s)}(\lambda)$ in °K	392	1300	2660	3310	3480

ergibt sich dann das gesamte Emissionsvermögen in Zeile (4) (ohne Quarzstrahlung), wobei

$$D = e^{-\int_{-R}^{+R} \varkappa_\nu \, dl} \qquad (R = \text{Bogenradius})$$

die Gesamtdurchlässigkeit senkrecht zur Bogenachse bedeutet. Nach der Vorschrift von Ziff. 3 kann dieser Intensität eine schwarze Temperatur $T_i^{(s)}$ zugeordnet werden. Berücksichtigt man nun noch die Durchlässigkeit D_{Qu} der 1 mm starken Quarzwand nach den Messungen von Cartwright [83] und Koch [250], so läßt sich das Emissionsvermögen E_a der Lampe im Außenraum [Zeile (7)] und schließlich die zugehörige schwarze Temperatur $T_a^{(s)}$ [Zeile (8)] berechnen.

Wie man sieht, übertrifft das Emissionsvermögen E_a bei 100 μ die Zimmertemperaturstrahlung $E_0^{(s)}$ nur um etwa 30%. Das Ergebnis von Dahlke, daß die Bogenstrahlung in der Gegend von 100 μ neben der Strahlung des heißen Quarzrohres überhaupt nicht mehr nachweisbar ist, wird somit richtig wiedergegeben. Bei den längeren Wellen erreicht dagegen die berechnete Strahlungsintensität das 4- bis 12fache der Zimmertemperaturstrahlung. Nach Angaben von Koch und Dahlke liegen die Werte der gemessenen schwarzen Temperatur $T_i^{(s)}$ im Wellenlängenbereich von etwa 200 bis 300 μ zwischen 2500 und 4000° K, was als befriedigende Übereinstimmung mit den in Zeile (5) errechneten Werten angesehen werden kann.

Hettner hat fernerhin auch das Emissionsvermögen und die schwarzen Temperaturen im gleichen Wellenlängenbereich für den Hg-Bogen des Lampentyps HQ 500 berechnet. Die schwarzen Temperaturen erreichen schon bei kürzeren Wellenlängen infolge des höheren Druckes und der höheren Achsentemperatur (10 Atm und 6500 bis 7000° K gegenüber etwa 1 Atm und 6000° K des UV-Normals) Werte von der Größenordnung der Bogentemperatur. Auch hier spielt also die Selbstabsorption des Bogenplasmas eine erhebliche Rolle.

In Ziff. 6 haben wir bereits die Bedingungen für die Emission von Hohlraumstrahlung durch ionisierte Gase behandelt, und zwar unter der Voraussetzung einer homogenen Gasschicht einheitlicher Temperatur. Bei den wirklichen Gasentladungslichtquellen liegt jedoch eine mehr oder minder steile Temperaturverteilung quer zur Achse des Entladungskanals vor. Bei genügend hohen Drucken kann nun das Absorptionsvermögen des Gases in der Umgebung der Entladungsachse (Gebiet höchster Temperatur innerhalb der Entladung) so groß werden, daß praktisch keine Strahlung aus diesem Bereich mehr austreten kann und somit die gesamte beobachtete Emission von äußeren, kälteren Schichten der Lampe herrührt. Zur Erzeugung von Hohlraumstrahlung möglichst hoher Temperatur ist daher gar nicht erwünscht, das Absorptionsvermögen der achsennahen Gebiete bzw. den Gasdruck zu stark hochzutreiben. Diese Aussage entspricht genau dem experimentellen Befund von Dahlke über das Emissionsvermögen bei einer mittleren Wellenlänge von 300 μ als Funktion des Gasdruckes (Fig. 54). Die Tatsache, daß Bögen von 1 bis 10 Atm am günstigsten sind und Höchstdrucklampen hinsichtlich des vorliegenden Wellenlängenbereiches mindestens keinen Vorteil bringen, wird durch die starke Selbstabsorption erklärt.

Eine weitere Prüfung dieser auf der Kramers-Unsöldschen Theorie basierenden Ergebnisse wurde von Säufferer [395] durchgeführt, der die Welligkeit der Strahlung an einer mit Wechselstrom betriebenen Lampe vom Typ HQ 500 untersuchte. Da nach Messungen von Kern [248] die Achsentemperatur des Bogens während jeder Halbperiode um etwa 1000° K schwankt, kann aus der beobachteten Strahlungsschwankung Aufschluß erzielt werden über die Abhängigkeit des Emissionsvermögens von der Temperatur. Auch diese Ergebnisse lassen sich gut mit der vorstehenden Theorie erklären.

β) Kontinuierliche Bremsstrahlung an Metall- und Kristalloberflächen. Hierher gehören die Leuchterscheinungen an *Metaloberflächen* in Niederdruck-Gasentladungen, die zum Teil auf Bremsstrahlung der aus dem Entladungsraum auf die Oberfläche auftreffenden langsamen Elektronen zurückzuführen sind (MOHLER und BOECKNER [320] bis [322]).

Weiter treten Strahlungserscheinungen auf an *Halbleitergrenzschichten*, über die eine umfangreiche Literatur vorliegt. Wieweit es sich dabei um echte Bremsstrahlung, und wieweit um Elektronenstoßanregung von Energiebändern mit darauffolgender Emission, oder um Rekombination von Elektronen mit „Löchern" (Defektelektronen) handelt, ist nur von Fall zu Fall zu entscheiden. Eine Diskussion dieser Phänomene geht indessen weit über den Rahmen des vorliegenden Beitrages hinaus. Eine ausführliche Darstellung dieser Erscheinungen findet man unter anderen bei LEVERENZ [281a].

γ) Optische kontinuierliche Strahlung schneller Elektronen. Auch auf die zahlreichen Untersuchungen der Strahlung *schneller* Elektronen im Sichtbaren und UV können wir nur ganz kurz

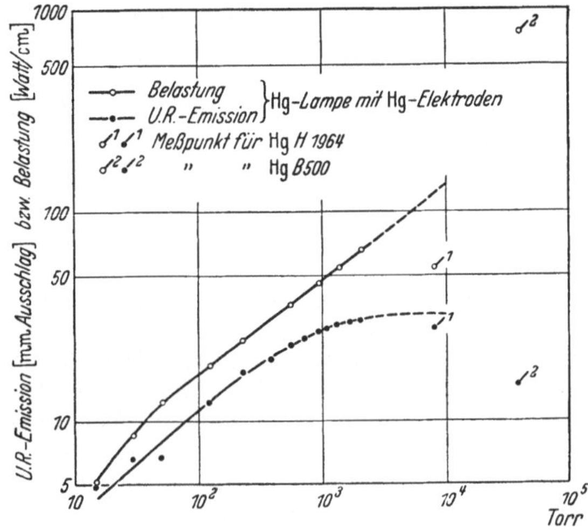

Fig. 54. Emissionsvermögen im Ultrarot in Abhängigkeit vom Gasdruck nach DAHLKE [108].

hinweisen. Schon in den älteren Arbeiten von LILIENFELD [284], FOOTE, MEGGERS und CHENAULT [155] wurde die an den beschossenen Metallen (Antikathoden) beobachtete sichtbare kontinuierliche Strahlung als Bremsstrahlung gedeutet. COHN [94] hat besonders eingehend den *Äonaeffekt*, d.h. das Ausströmen von Elektronen aus einer feinen Spitze auf einen gegenüberstehenden Zylinder im Hochvakuum und das dabei auftretende Leuchten, ferner das *Antikathodenleuchten* im Hochvakuum und das kontinuierliche Leuchten von *Metallen, Gasen und Dämpfen* bei Beschießung mit schnellen Elektronen untersucht. (Eine Darstellung der älteren Arbeiten gibt FINKELNBURG [149].) Neuerdings wird die Deutung der Leuchterscheinungen an der Antikathode als Bremsstrahlung allerdings von SOMMERMEYER [411] abgelehnt und die Effekte als Plasmaschwingungen von Gasresten an der Metalloberfläche gedeutet.

20. Plasmaschwingungen und frei-frei-Strahlung. Die bisherigen Betrachtungen über die kontinuierliche Emission oder Absorption ionisierter Gase zeigten, daß bei kleinen Frequenzen praktisch nur die frei-frei-Strahlung ausschlaggebend ist. Wir dürfen aber den Gültigkeitsbereich der für das Radiofrequenzgebiet abgeleiteten Gl. (5.42) nicht bis zu beliebig kleinen Frequenzen ausdehnen. Jedes ionisierte Gas besitzt nämlich wegen der Wechselwirkung der Elektronen untereinander eine bestimmte *Elektroneneigenfrequenz* (LANGMUIR).

Werden durch irgendeine Kraft (z.B. durch das elektrische Feld einer Lichtwelle), welche wegen des großen Massenunterschiedes auf die Elektronen und Ionen verschieden wirkt, die Ladungsträgerarten gegeneinander um einen Vektor s

verschoben, so entstehen Raumladungen der Dichte

$$\varrho = \mathrm{div}\,(e\,N_e s),\qquad(20.1)$$

die ihrerseits wiederum ein elektrisches Feld nach

$$\mathrm{div}\,E = 4\pi\varrho\qquad(20.2)$$

erzeugen. Aus (20.1) und (20.2) erhält man für dieses Feld

$$E = 4\pi\,e\,N_e s.\qquad(20.3)$$

Durch die an jedes Elektron angreifende rücktreibende Kraft $-eE$ entstehen nun nach der Schwingungsgleichung

$$m\,\frac{d^2 s}{d t^2} = -4\pi\,e^2\,N_e\,s\qquad(20.4)$$

Plasmaschwingungen der Frequenz

$$\nu_0 = \sqrt{\frac{e^2}{\pi\,m}\,N_e}.\qquad(20.5)$$

Diese Plasmaschwingungen spielen in der *Gasentladungsphysik* und besonders in der *Radioastronomie* (z.B. gestörte Sonnenstrahlung) eine wichtige Rolle.

Im Prinzip haben wir hier eine Schwingung fester Frequenz („Spektrallinie") vor uns. In Wirklichkeit werden die Elektronen beim Durchschwingen durch das Ionengas von den einzelnen Ionen abgelenkt, und da infolge des großen Massenunterschiedes praktisch keine kinetische Energie übertragen wird, muß die Energie in Form von Bremsstrahlung emittiert werden. Durch diese Energieabgabe wird die Eigenschwingung gedämpft, die „*Plasmaschwingungslinie*" erhält folglich „*Dämpfungsflügel*". Der kurzwellige Teil ist, wie sich zeigen läßt, für $\nu \gg \nu_0$ identisch mit der *frei-frei-Strahlung*.

Nach Burkhardt [78] ergibt sich nämlich für die allgemeine Dispersionsformel des Plasmas der Ausdruck

$$\mathrm{n}^2 - 1 = -\omega_0^2\,\frac{1 + i\,\gamma/\omega}{\omega^2 + \gamma^2}\qquad(20.6)$$

mit

$$\omega_0^2 = (2\pi\,\nu_0)^2 = \frac{4\pi\,e^2}{m}\,N_e$$

und γ als Dämpfungsglied. Durch Trennung von Real- und Imaginärteil erhält man den *Brechungsindex n* und den „*Absorptionskoeffizienten*" $n\,\varkappa$, der mit dem bisher definierten Absorptionskoeffizienten pro cm verknüpft ist durch die Beziehung

$$\varkappa_\nu = 2\,\frac{\omega}{c}\,n\,\varkappa.\qquad(20.7)$$

Vernachlässigt man die Dämpfung ($\gamma = 0$), so geht (20.6) in die bekannte Form

$$n^2 - 1 = -\frac{e^2}{\pi\,m}\,\frac{N_e}{\nu^2}\qquad(20.8)$$

über. Berechnet man γ unter der Voraussetzung, daß die Energie in Form von Bremsstrahlung emittiert wird, so erhält man [78]

$$\gamma = \frac{16\pi^2\,N_i\,e^4}{3\,m^{\frac12}\,(2\pi\,kT)^{\frac32}}\,\ln\frac{kT}{1{,}44\,Z\,e^2\,N_i^{\frac13}}.\qquad(20.9)$$

Aus (20.6) folgt andererseits für $v \gg v_0$ der Ausdruck

$$n \varkappa \approx \frac{\gamma}{4\pi} \frac{v_0^2}{v^3}$$ (20.10)

und zusammen mit Gl. (20.7) für den Absorptionskoeffizienten

$$\varkappa_\nu = \frac{\gamma}{c} \left(\frac{v_0}{v}\right)^2 = \frac{16\pi\, e^6 Z^2 N_e N_i}{3c\,(2\pi\, m\, kT)^{\frac{3}{2}}} \cdot \frac{1}{v^2} \ln \frac{kT}{1{,}44\, Z\, e^2\, N_i^{\frac{1}{3}}},$$ (20.11)

woraus durch Einsetzen von γ aus Gl. (20.9) wieder die Formel (5.42b) aus Tabelle 1 für den Absorptionskoeffizienten im Radiofrequenzgebiet folgt.

Es ist also zu beachten, daß diese Gleichung nur für $v \gg v_0$ gültig ist, d.h. für Frequenzen, die erheblich größer sind als die betreffende Plasmaeigenfrequenz.

III. Sonderfälle kontinuierlicher Strahlung freier Ladungsträger.

21. Allgemeines. Die bisher behandelten Elektronenkontinua in Emission oder Absorption entstanden durch Primärprozesse in *atomaren* Dimensionen. Demgegenüber wollen wir in diesem Kapitel auf zwei Erscheinungen eingehen, die schon vom *makroskopischen* Standpunkt aus zu verstehen sind, also keiner mikroskopischen Betrachtungen bedürfen. Kontinuierliche Strahlung beliebiger Frequenz entsteht nach der klassischen Elektrodynamik z.B. schon bei *Beschleunigung von Ladungsträgern* in makroskopischen elektromagnetischen Feldern. In den modernen Zirkulationsbeschleunigern werden Ladungsträger auf sehr hohe Energien gebracht, wobei eine intensive kontinuierliche Strahlung entsteht, deren Frequenzverteilung im wesentlichen von der Energie der Teilchen abhängt. In Ziff. 22 werden wir über einige Untersuchungen über die kontinuierliche Strahlung beschleunigter *Elektronen* berichten. Sodann wird in Ziff. 23 die sog. Čerenkov-Strahlung behandelt, die zwar auf der Wechselwirkung schnell bewegter Ladungsträger (Elektronen, Positronen, Protonen) mit der Materie beruht, aber ebenfalls schon in der phänomenologischen Maxwellschen Theorie enthalten ist, weil nur der Brechungsindex n der von den Ladungsträgern mit „Überlichtgeschwindigkeit" durcheilten Materie in die Betrachtung eingeht.

22. Kontinuierliche Strahlung beschleunigter Elektronen im Betatron und Synchrotron. Nach der klassischen Elektrodynamik senden auf Kreisbahnen bewegte Elektronen (bzw. Ionen) elektromagnetische Strahlung aus. Diese Strahlung kann in den modernen Beschleunigungsmaschinen beobachtet werden. Theoretische Betrachtungen wurden bereits 1898 von Lienard [283] und 1907 von Schott [419] in Verbindung mit den ersten Atommodellen durchgeführt. Da der Strahlungsverlust erheblichen Einfluß auf die Wirkungsweise der Elektronenbeschleuniger hoher Energie hat, sind auch in neuerer Zeit eingehende Rechnungen insbesondere von Schiff [416], Schwinger [428], Arzimovich und Pomeranchuk [16] und von Abragam [3] veröffentlicht worden.

Nach Schwinger ist die Energieverteilung der Strahlung eines Elektrons konstanter Energie E gegeben durch

$$P(\omega)\, d\omega = \frac{3\sqrt{3}}{4\pi}\, \omega_0\, \frac{e^2}{R} \left(\frac{E}{m\, c^2}\right)^4 \left[\int\limits_{\omega/\omega_c}^{\infty} K_{\frac{5}{3}}(x)\, dx\right] \frac{\omega}{\omega_c}\, d\omega.$$ (22.1)

$P(\omega)\, d\omega$ ist die Strahlungsleistung eines mit der Kreisfrequenz ω umlaufenden Elektrons im Intervall $d\omega$. R ist der Radius in cm, $\omega_0 = v/R$ die Winkelgeschwindigkeit des Elektrons. $\omega_c = \dfrac{3}{2}\,\omega_0\left(\dfrac{E}{m\,c^2}\right)^3$ ist eine kritische Frequenz, die näherungsweise die obere Grenze des Spektrums angibt. $K_{\frac{5}{3}}$ schließlich ist eine Zylinderfunktion, die folgenden Zusammenhang mit den Bessel-Funktionen $J_n(x)$ besitzt:

$$
\left.
\begin{aligned}
K_{\frac{5}{3}}(x) &= \frac{4}{3\,x}\,K_{\frac{2}{3}}(x) + K_{\frac{1}{3}}(x)\;, \\[4pt]
K_{\frac{1}{3}}(x) &= \frac{\pi}{\sqrt{3}}\left[i^{\frac{1}{3}}\,J_{-\frac{1}{3}}(i\,x) - i^{-\frac{1}{3}}\,J_{\frac{1}{3}}(i\,x)\right], \\[4pt]
K_{\frac{2}{3}}(x) &= \frac{\pi}{\sqrt{3}}\left[i^{\frac{2}{3}}\,J_{-\frac{2}{3}}(i\,x) - i^{-\frac{2}{3}}\,J_{\frac{2}{3}}(i\,x)\right],
\end{aligned}
\right\}
\tag{22.2}
$$

Für die gesamte Strahlungsleistung gilt

$$
P = \int_0^\infty P(\omega)\,d\omega = \frac{2}{3}\,\omega_0\,\frac{e^2}{R}\left(\frac{E}{m\,c^2}\right)^4.
\tag{22.3}
$$

Sie wächst also mit der vierten Potenz der Elektronenenergie an. Der größte Teil der Strahlungsenergie liegt im Bereich der kritischen Frequenz und wird in einem schmalen Raumwinkel in Richtung der Elektronenbewegung ausgestrahlt. Das Licht ist polarisiert mit dem elektrischen Vektor in der Ebene der Elektronenbahn.

Bei der Beobachtung am Synchrotron wird nun allerdings nicht Strahlung von Elektronen *konstanter* Energie gemessen. Wächst das Magnetfeld sinusförmig an, so daß der Maximalwert in der Zeit T erreicht ist, so gilt

$$
E = E_{\max}\sin\left(\frac{\pi}{2\,T}\right)t
\tag{22.4}
$$

und ferner

$$
\omega_c = \frac{3}{2}\,\omega_0\left(\frac{E_{\max}}{m\,c^2}\right)^3\sin^3\left(\frac{\pi}{2\,T}\right)t = \omega_m\sin^3\left(\frac{\pi}{2\,T}\right)t.
\tag{22.5}
$$

Gewöhnlich wird die Größe

$$
\tau = \frac{\omega_m}{\omega_c} = \sin^{-3}\left(\frac{\pi}{2\,T}\right)t
\tag{22.6}
$$

in die Rechnung eingeführt. Die über T gemittelte Strahlungsleistung ergibt sich dann zu

$$
\bar{P}(\omega)\,d\omega = \frac{3\sqrt{3}}{2\pi^2}\,\omega_0\,\frac{e^2}{R}\left(\frac{E_{\max}}{m\,c^2}\right)^4\left(\frac{\omega}{\omega_m}\right)^2\left[\int_1^\infty (\tau^{\frac{2}{3}} - 1)^{\frac{1}{2}}\cdot K_{\frac{5}{3}}\left(\frac{\omega}{\omega_m}\,\tau\right)d\tau\right]\frac{d\omega}{\omega_m}.
\tag{22.7}
$$

Als Beispiel für den Vergleich zwischen Theorie und Beobachtung wollen wir die Untersuchungen von Elder, Langmuir und Pollock [*131*] am Schenectady-Synchrotron anführen, in dem Elektronen auf maximal 80 MeV beschleunigt werden. Von etwa 30 MeV ab tritt sichtbare Strahlung mit zunächst dunkelroter Farbe auf. Bei 80 MeV wird intensives bläulichweißes Licht emittiert. Mit dem vorgegebenen Radius $R = 29{,}2$ cm haben die Verfasser nach Gl. (22.1) zunächst die Strahlung für Elektronen *konstanter* Energie E berechnet (Fig. 55). Die Strahlungsverluste für Elektronen mit *sinusförmiger* Beschleunigung über eine Viertelperiode nach Gl. (22.7) sind mit verschiedenen Maximalenergien in Fig. 56 wiedergegeben. Die Strahlung wurde durch ein Quarzfenster beobachtet und schließlich das Spektrum mit Gitter und Photomultiplier registriert. Als Strahlungsnormal diente eine Wolframbandlampe. In Fig. 57 sind die nach Gl. (22.7) berechneten Kurven für verschiedene Maximalenergien zusammen mit

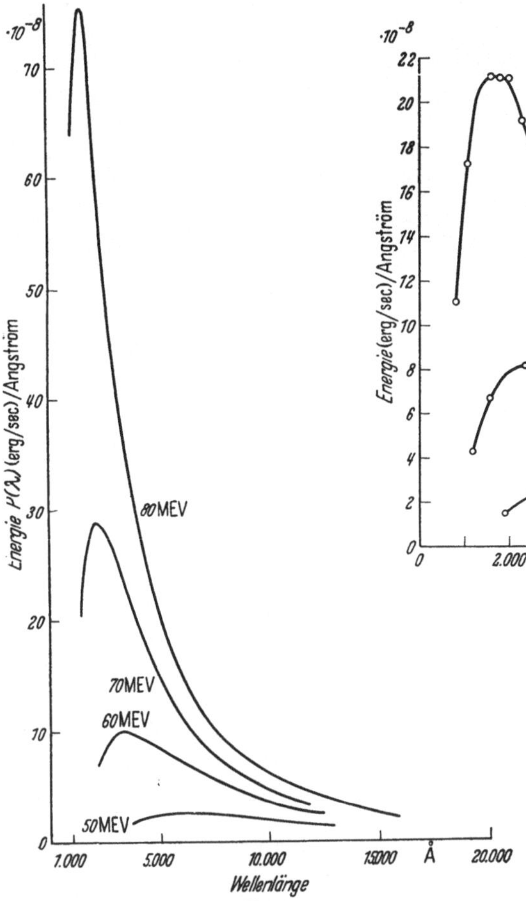

Fig. 55.

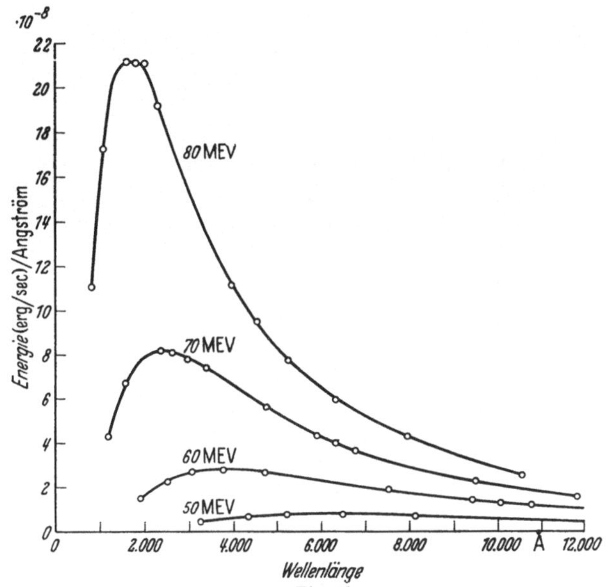

Fig. 56.

Fig. 55. Energieverlust pro Ångström-Einheit eines Elektrons konstanter Energie nach ELDER, LANGMUIR und POLLOCK [131].

Fig. 56. Energieverlust pro Ångström-Einheit eines sinusförmig beschleunigten Elektrons ($\frac{1}{4}$ Periode) nach ELDER, LANGMUIR und POLLOCK ([131].

Fig. 57 a—d. Vergleich der gemessenen Energieverteilung der Elektronenstrahlung mit den theoretischen Kurven. Elektronen beschleunigt auf maximale Energien von a 80 MeV, b 70 MeV, c 60 MeV, d 42,5 MeV. (Nach ELDER, LANGMUIR und POLLOCK [131].)

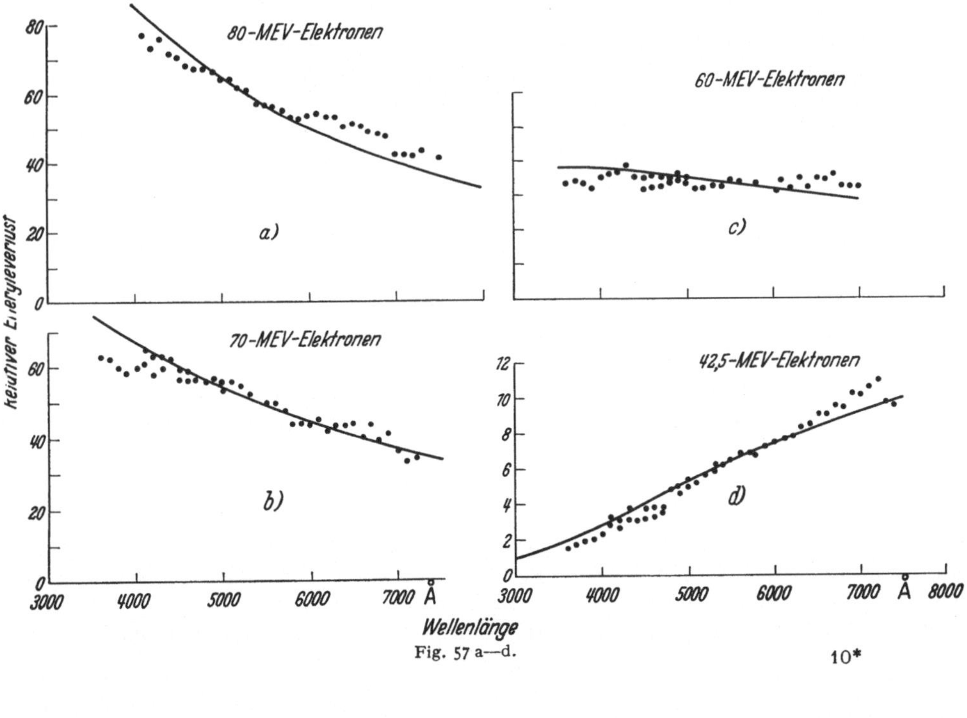

Fig. 57 a—d.

den Meßpunkten aufgetragen. Nach Ansicht der Verfasser rührt die Streuung in den Meßpunkten möglicherweise von Schwingungen des Elektronenstrahls her.

Hartmann und Tomboulian [200] führten ähnliche Untersuchungen an dem Cornell-Synchrotron durch. Die Maximalenergie betrug hierbei 310 MeV, so daß der Hauptteil der Strahlung in das Schumann-Gebiet (50 bis 500 Å) fiel. Bei 310 MeV war im Bereich zwischen 50 und 400 Å keine merkliche Intensitätsschwankung festzustellen. Bei 220 MeV hingegen ergab sich ein Anwachsen der Intensität von etwa 60 Å an bis zu einem Maximum bei 170 Å mit anschließendem Abfall zu längeren Wellenlängen.

23. Čerenkov-Strahlung. Im Jahre 1934 entdeckte Čerenkov [85] eine bei Beschuß von durchsichtigen Flüssigkeiten (Wasser, Benzol usw.) mit γ-Strahlen

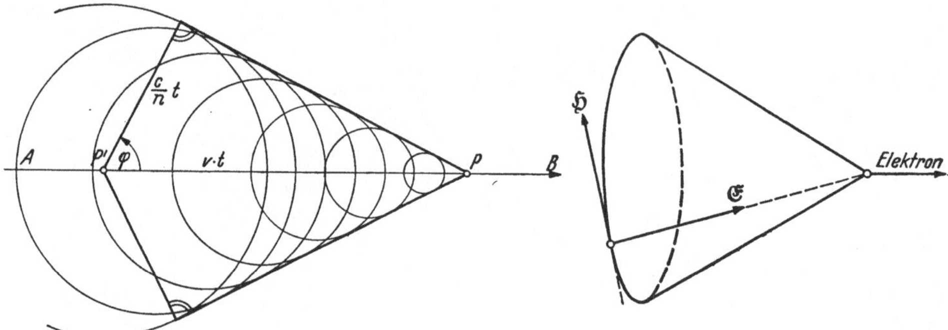

Fig. 58. Entstehung der Čerenkov-„Kopfwelle". (Aus Maurer und Kolz [305].)   Fig. 59. Lage des elektrischen und magnetischen Vektors in der Kegelfläche. (Aus Maurer und Kolz [305].)

auftretende eigentümliche *kontinuierliche* Strahlung im blau-violetten Spektralbereich. Wie bereits sein Mitarbeiter Wawilow [462] vermutete, war die Ursache dieser Strahlung primär nicht in den γ-Strahlen, sondern in den von ihnen ausgelösten Compton-*Elektronen* zu suchen. Diese Vermutung wurde durch die 1937 von Čerenkov durchgeführten Versuche [86] mit harten β-Strahlen (Radiumpräparat) bestätigt. Frank und Tamm [163] gelang es im gleichen Jahre, diese Strahlung auf der Grundlage der Maxwellschen Theorie zu erklären.

Bewegen sich Elektronen mit gleichförmiger Geschwindigkeit durch ein Dielektrikum, und ist diese Geschwindigkeit v größer als die Phasengeschwindigkeit c/n des Lichtes in der betreffenden Materie mit dem Brechungsindex n, so emittiert das Elektron eine kohärente elektromagnetische Strahlung in Analogie zur akustischen *Kopfwelle* an Geschoßen mit Überschallgeschwindigkeit. Die Verhältnisse lassen sich an Hand von Fig. 58 leicht übersehen. Alle von dem mit der Geschwindigkeit v bewegten Elektron zwischen P' und P erzeugten Kugelwellen werden von einer Kegelfläche eingehüllt, der „Kopfwelle" des Elektrons, die identisch ist mit der Čerenkov-Strahlung.

Diese Strahlung wird also nur unter einem bestimmten Winkel ausgesandt, der gegeben ist durch die Beziehung

$$\cos \varphi = \frac{c}{n\,v} = \frac{1}{\beta\,n}, \quad \left(\beta = \frac{v}{c}\right). \tag{23.1}$$

Die Lage des elektrischen und magnetischen Vektors in der Kegelfläche zeigt Fig. 59. Der Poyntingsche Strahlungsvektor steht also senkrecht auf dem Kegelmantel. Die Beziehung (23.1) ist besonders von Wykoff und Henderson [489] genau geprüft worden, und zwar mit Elektronenenergien von 240 bis 815 kV

und Glimmerplättchen von 1 bis 25 μ Dicke. Die ausgezogene Kurve in Fig. 60
wurde nach Gl. (23.1) berechnet.

Aus der Theorie ergibt sich fernerhin für die pro Längeneinheit ausgestrahlte
Energie im Frequenzintervall $d\nu$ der Ausdruck

$$d E = \left(\frac{2\pi e}{c}\right)^2 \left(1 - \frac{1}{\beta^2 n^2}\right) \nu\, d\nu. \tag{23.2}$$

Durch Integration über alle Frequenzen erhält man den gesamten Energieverlust
des Elektrons beim Durchgang durch Materie. Der Verlust ist bei endlichen

Materieschichten allerdings so klein
gegenüber der kinetischen Energie, daß
er in sehr guter Näherung vernachlässigt
werden kann und die Annahme konstanter Elektronengeschwindigkeit v bei
den Rechnungen berechtigt ist.

Aus (23.2) ist sofort abzulesen, daß
die Strahlung wegen der *Dispersions-
eigenschaften* der Materie vorwiegend
im optischen Spektralbereich liegen
muß. (Im Röntgengebiet müßte wegen
$n<1$ die Elektronengeschwindigkeit
$v>c$ werden.)

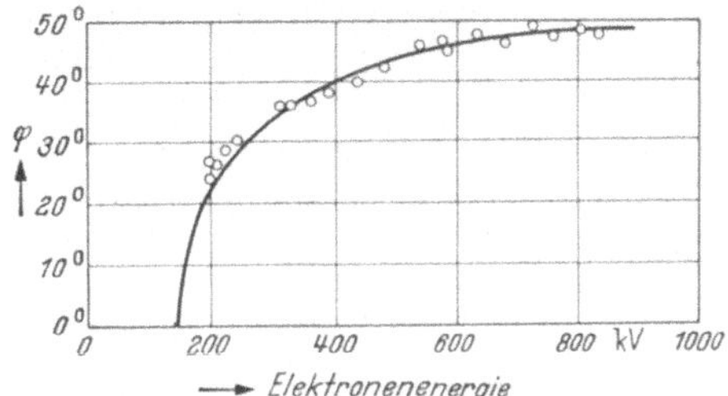

Fig. 60. Ausstrahlungswinkel als Funktion der Elektronenenergie. Theoretische Kurve nach Gl. (23.1). Die Kreise
geben die Meßwerte von Wykoff und Henderson wieder.
(Aus Maurer und Kolz [305].)

Weitere theoretische Arbeiten über Čerenkov-Strahlung liegen vor von
Beck [48], Jauch und Watson [242], Dedrick [112], Ginsburg [178], Yin-
Yuan Li [492], [493] und Cox [105]. Letzterer behandelt die Impuls- und

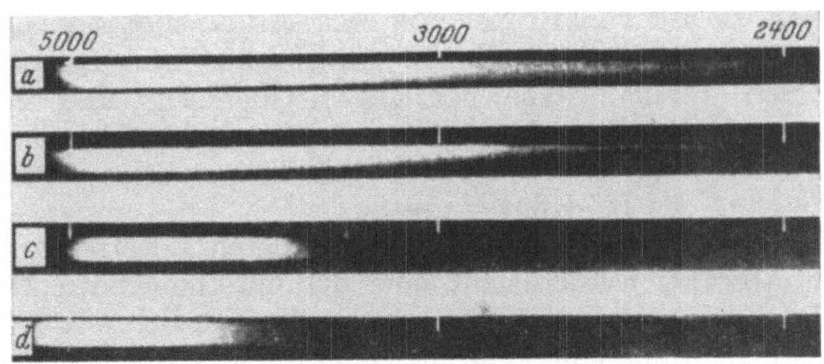

Fig. 61 a—d. Kontinuierliche Spektren der Čerenkov-Strahlung, aufgenommen von Collins und Reiling. a Wasser
b Alkohol, c Benzol, d Vergleichsspektrum (Wolframbandlampe). (Aus Maurer und Kolz [305].)

Energiebilanz zwischen Elektron und emittiertem Lichtquant, was auf ein allerdings vernachlässigbares Zusatzglied in Gl. (23.1) führt. Ginsburg [179] hat
die Frage diskutiert, ob Mikrowellen durch den Čerenkov-Effekt erzeugt werden
können.

Von den experimentellen Arbeiten ist insbesondere diejenige von Collins
und Reiling [96] aus dem Jahre 1938 zu nennen. Die Verfasser verwandten
zur Erzeugung eines homogenen Elektronenstrahls hoher Energie einen Hoch-
spannungsgenerator von 2000 kV ($v/c = 0{,}98$) mit einer Stromstärke von 10 μAmp.
Als Dielektrika dienten dünne Filme von 29 μ bis 1 mm Dicke. Auch Flüssigkeiten wurden als kapillare Schichten untersucht. Fig. 61 zeigt einige von Collins und Reiling an Wasser, Alkohol und Benzol gewonnene kontinuierliche
Spektren der Čerenkov-Strahlung.

Der Čerenkov-Effekt wird neuerdings in verschiedenen Anwendungsgebieten
ausgenutzt:

1. *Zählung* von geladenen Teilchen ausreichend hoher Energie.

2. *Bestimmung der Geschwindigkeit* der Teilchen mit $v > c/n$ und zugleich Bestimmung der Energie bei bekannter Masse.

3. *Unterscheidung* von Teilchen verschiedener Masse bei gleicher Energie usw.

1947 hat Getting [176] einen Zähler für schnelle Elektronen und geladene Mesonen konstruiert. Neuere Arbeiten über Čerenkov-Zähler sind von Marshall [297] publiziert. Als Dielektrikum dient dabei vielfach der optisch gut geeignete Kunststoff „Lucit" $(n = 1,5)$.

Ein Instrument mit direkter optischer Registrierung der Čerenkov-Strahlung von *Protonen* des 460 cm Synchrocyclotrons in Berkeley wurde von Mather [302] angegeben. Der Fehler der Geschwindigkeitsmessung der Protonen betrug bei 340 MeV etwa 1%. Čerenkov-Strahlung von *Positronen* ist von Wilfong und Henderson [477] beobachtet worden. Schließlich sei erwähnt, daß die genannten Zähler und Meßmethoden auch bei der Untersuchung der kosmischen Strahlung Verwendung finden.

Kurze zusammenfassende Darstellungen über das gesamte Erscheinungsgebiet findet man bei Maurer und Kolz [305] (Literatur bis 1949) und Jelley [243] (Literatur bis in die neueste Zeit), ferner in dem Artikel von R. P. Sutton in Bd. XLV dieses Handbuches.

IV. Emissions- und Absorptionskontinua in der Astrophysik[1].

24. Kontinuierliche Emission und Absorption im Sterninneren und in Sternatmosphären. Für die Astrophysik spielt die Emission und Absorption kontinuierlicher Strahlung eine besonders wichtige Rolle. Wie bereits in Ziff. 15 erwähnt, wurde die von Kramers entwickelte korrespondenzmäßige Theorie kontinuierlicher Absorption zuerst von Eddington und Milne auf das Innere und die Atmosphären der Sterne angewandt und erst später von Finkelnburg und Unsöld auf Gasentladungssysteme übertragen. Überhaupt sind ähnlich wie bei der Untersuchung der Linienverbreiterungseffekte viele experimentelle Arbeiten über Kontinua den Anregungen von astrophysikalischer Seite her zu verdanken, z.B. die Entdeckung des H⁻-Kontinuums.

Für die *Theorie des Sterninneren* ist neben anderen Materialeigenschaften der kontinuierliche Absorptionskoeffizient eine der entscheidenden Größen, deren Kenntnis die Berechnung von Temperatur, Druck, Gasdichte usw. als Funktion des Sternradius erst ermöglicht. Die im Inneren eines Sternes bei etwa 10^7 °K durch Kernreaktionen (Umwandlung von Wasserstoff in He) frei werdende Energie wird ja fast ausschließlich durch Strahlung an die Oberfläche transportiert. Unter der gut begründeten Annahme, daß im gesamten Stern — mit Ausnahme der äußeren Schichten der Atmosphäre — lokales thermisches Gleichgewicht herrscht, und somit Emission und Absorption durch das Kirchhoffsche Gesetz verknüpft sind, läßt sich bei gegebenem kontinuierlichen Absorptionskoeffizienten der Strahlungstransport bzw. das Strahlungsgleichgewicht berechnen. Bei den im Inneren herrschenden Temperaturen von 10^6 bis 10^7 °K liegt das Strahlungsmaximum im Röntgengebiet, so daß die in Ziff. 11 behandelte Theorie der Röntgenabsorptionskanten anzuwenden ist. Daneben spielt aber auch die Thomson-*Streuung* an freien Elektronen, die als *Grenzfall des* Compton-*Effektes* für große Wellenlängen zu betrachten ist, eine erhebliche Rolle. Die Strahlung wird dabei nun auf dem Weg zur Sternoberfläche Schritt für Schritt von immer kühleren Schichten absorbiert und in diesen in Form schwarzer Strahlung der betreffenden Schichttemperatur weiter nach außen abgestrahlt,

---

[1] Ausführlichere Darstellung in den Bänden L bis LIII dieses Handbuches.

wobei sich das Strahlungsmaximum zu immer größeren Wellenlängen verschiebt. Von der Sternoberfläche schließlich wird vorwiegend Strahlung der Umgebung des sichtbaren Bereiches emittiert, deren Energieverteilung in erster Näherung der PLANCK-Kurve für die Oberflächentemperatur (Temperatur der Photosphäre) des Sternes entspricht und an Hand der Sternspektrogramme ausmeßbar ist.

Diese Sternspektren bestehen alle aus einem von Absorptions- bzw. Emissionslinien unterbrochenem Kontinuum. Sie werden in zehn Hauptklassen

$$\begin{matrix} & & & & & & S \\ O-B-&A-&F-&G-&K-&M \\ & & & & R-&N \end{matrix}$$

eingeteilt, deren jede wieder zehn Unterklassen enthält.

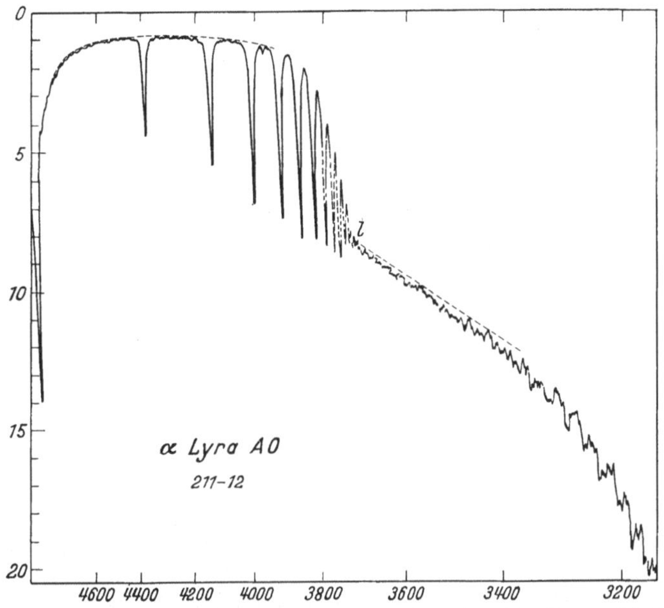

Fig. 62. Photometerkurve des Spektrums von α Lyra A 0. (Aus UNSÖLD [449].)

Die außerhalb der Sequenz stehenden planetarischen Nebel (P) und die ersten Unterklassen der O-Sterne zeigen im Spektrum Emissionslinien von H, He, O, N, C, usf. während bei allen folgenden Klassen praktisch nur Absorptionslinien zu finden sind. Von $K\,5$ ab treten auch Molekülbanden auf. Am auffälligsten aber ist die von O bis $A\,2$ zunehmende und dann bis zur Klasse G wieder abnehmende Intensität der Wasserstoff-Absorptionslinien und des die gleiche Intensitätsabhängigkeit zeigenden schon von HUGGINS [228] entdeckten BALMER-Kontinuums. Diese Intensitätsabhängigkeit spiegelt sich in dem sog. „BALMER-Sprung" wieder, dem Intensitätssprung an der BALMER-Grenze, welcher ebenfalls ein Maximum bei den $A\,2$-Sternen aufweist (BARBIER und CHALONGE [25]). Ein Beispiel dafür gibt die Registrierkurve des $A\,0$-Sterns α-Lyrae in Fig. 62 mit der BALMER-Serie und dem anschließenden Kontinuum in Absorption. Einen ähnlichen, aber kleineren Sprung im Kontinuum findet man erwartungsgemäß an der PASCHEN-Grenze bei $\lambda \sim 8200$ Å. Die Erklärung für diesen Intensitätsverlauf längs der Hauptreihe liegt in der SAHAschen Theorie der thermischen Ionisation. Es handelt sich um die Erscheinung, die bereits am wasserstabilisierten Lichtbogen besprochen und dort von LARENZ [274] zur Temperaturbestimmung benutzt wurde, nämlich das Absinken des BALMER-Kontinuums und der entsprechenden Linien von einem Maximalwert am Bogenrand zu der Bogenmitte auf Grund des in Fig. 36 dargestellten Verlaufs der Strahlungsintensität als Funktion der Temperatur. Die Spektralsequenz ist nämlich im wesentlichen nach abnehmender Temperatur geordnet (Fowler und Milne [156]). Entsprechend ändern sich auch die Farben der Sterne von bläulich-weiß (Sirius A 0) über gelb (Capella und Sonne G 2) nach rot (Beteigeuze M 0). Die zugehörigen Farbtemperaturen, die man durch Angleichung der Energieverteilung der Sternspektren an eine Planck-Funktion in einem gewissen Spektralbereich gewinnt, zeigen daher den gleichen Temperaturgang, wobei die Temperaturwerte allerdings besonders bei den „frühen" Spektraltypen höher liegen als die eigentlichen mittleren Gastemperaturen in den Atmosphären. Die Farbtemperaturen sind in Tabelle 5 eingetragen; von G 0 ab muß man aber zwischen den Hauptsequenzsternen und den Riesen unterscheiden.

Tabelle 5. *Farbtemperaturen der Sterne bezogen auf A 0 = 15 000° K. (Aus Unsöld [449].)*

Spektren	Hauptsequenz	Riesen
O	(71 000)	—
B 0	55 000	—
B 5	28 000	—
A 0	15 000	—
A 5	10 600	—
F 0	8 700	—
F 5	7 300	—
G 0	6 700	5 600
G 5	6 300	4 700
K 0	5 900	4 100
K 5	5 000	3 700
M 0	3 850	3 400
M 5		2 850

Bei der Sonne (Spektraltyp G 2) kann außer der Energieverteilung, die besonders präzise von Abbot [1] und Wilsing [478] im Spektralbereich von 3000 bis 25 000 Å und neuerdings von Pettit [357] im Bereich von 2920 bis 7000 Å vermessen wurde, auch die Gesamtintensität (Solarkonstante) mit Hilfe der verschiedenen Typen von Pyrheliometern bestimmt werden, woraus unter Verwendung des Stefan-Boltzmannschen Gesetzes eine effektive Temperatur $T_e = 5780°$ K folgt. Wie Fig. 63 zeigt, stimmt die Energieverteilung des Sonnenspektrums

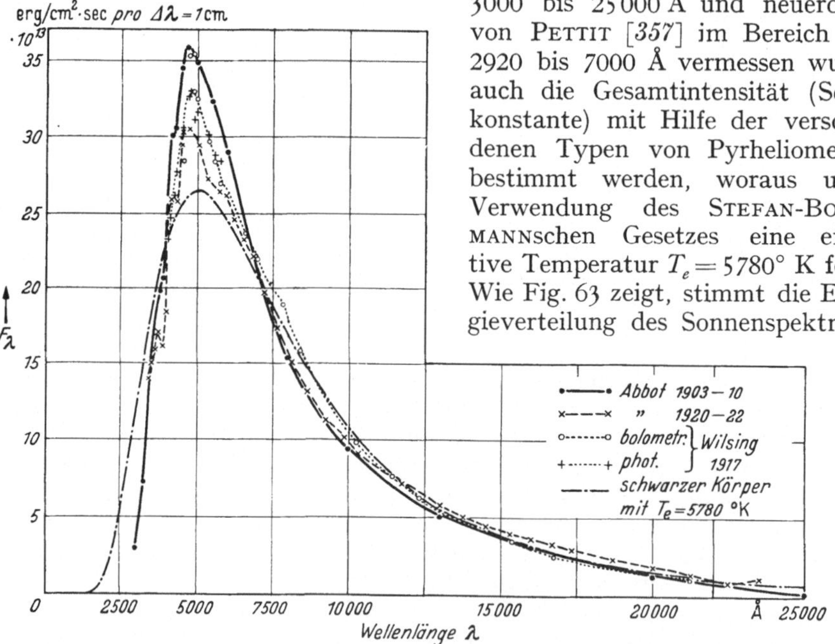

Fig. 63. Energieverteilung im Spektrum der integrierten Sonnenstrahlung F_λ erg/cm² sec pro $\Delta\lambda = 1$ cm. Messungen von Abbot (1903 bis 1910 und 1920 bis 1922); reduziert nach Minnaert, dessen Werte mit einer Skalenkorrektion von −2,4% versehen wurden. Messungen von Wilsing, multipliziert mit passender Skalenkonstante. Schwarzer Körper von $T_e = 5780°$ K. (Aus Unsöld [449].)

auch einigermaßen mit der eines schwarzen Körpers der Photosphärentemperatur überein, so daß die Annahme thermischen Gleichgewichtes näherungsweise berechtigt erscheint. Trotzdem weist schon die Abnahme der Intensität von der Sonnenmitte zum Sonnenrand (Verhältnis 1:0,4), die sog. „*Mitte-Rand-Variation*", darauf hin, daß die Sonne nicht wie ein schwarzer Körper strahlt, denn ein solcher müßte wegen der Isotropie der Hohlraumstrahlung als gleichmäßige helle Scheibe erscheinen. Meßtechnisch besteht bei der Sonne wie bei allen auf den F-Typ folgenden Spektralklassen die Schwierigkeit, daß die große Anzahl der dicht aufeinanderfolgenden FRAUNHOFER-Linien, die „wahre" kontinuierliche Verteilung erheblich verzerrt. Die ersten eingehenden Untersuchungen über diesen Gegenstand sind von MULDERS [333] durchgeführt worden. Der Bruchteil der in den Linien absorbierten Gesamtstrahlung wird

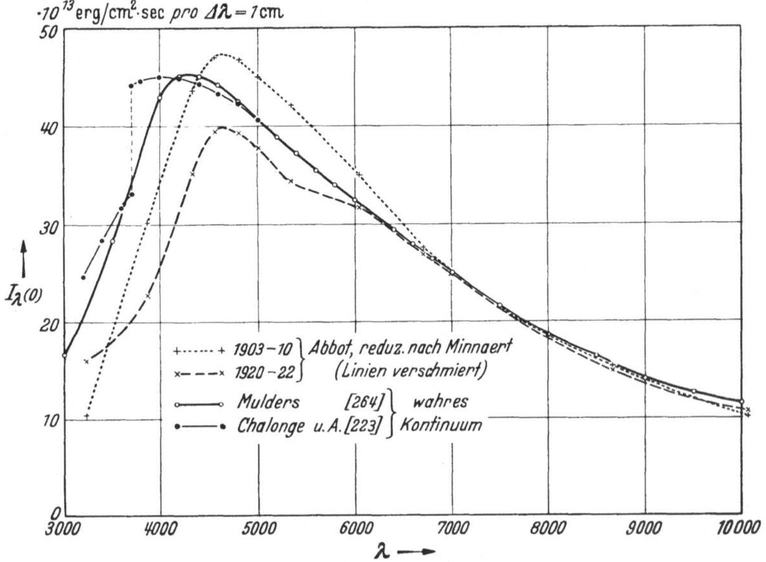

Fig. 64. $I_\lambda(0)$ = Intensität des kontinuierlichen Spektrums der Sonnenmitte. (Aus UNSÖLD [449].)

neuerdings von MICHARD [313] mit 12,4% angegeben. CHALONGE und DÉJARDIN [88] erkannten 1940, daß zwischen den dicht liegenden Linien im Sichtbaren und Ultravioletten noch einige teilweise weniger als 1 Å breite „Fenster" übrig bleiben, in denen das wahre Kontinuum sichtbar ist. Berücksichtigt man die Linienabsorption, so ergibt sich eine Energieverteilung des kontinuierlichen Spektrums der Sonne, wie sie in Fig. 64 nach den Arbeiten verschiedener Autoren angegeben ist. Hier tritt auch der „BALMER-Sprung" bei $\lambda \sim 3700$ Å deutlich zutage.

Intensität und Intensitätsverteilung der kontinuierlichen Sternspektren einschließlich des Intensitätssprungs an der BALMER-Grenze sind wie das Linienspektrum Funktionen der chemischen Zusammensetzung und den Zustandsgrößen der betreffenden Sternatmosphäre. Vor 1938 bestanden erhebliche Meinungsverschiedenheiten bezüglich des Häufigkeitsverhältnisses der Atomzahlen von Wasserstoff (H) und den Metallen (M). RUSSELL [393] hatte aus der Lage der Maxima einiger Metall-Funkenlinien in der Spektralsequenz und aus der Stärke verschiedener Hydridbanden im Sonnenspektrum ein Verhältnis von H:$M \sim 1000$ bis 2000 berechnet. Die von UNSÖLD [442] untersuchte Temperaturabhängigkeit des BALMER-Sprunges im Bereich der Spektralklassen $A - G$, insbesondere die Abnahme des Intensitätssprunges nach der Klasse $A\,2$, dagegen konnte nur durch die Annahme erklärt werden, daß bei Temperaturen $< 9000°$ K die kontinuierliche Absorption der Metalle gegenüber derjenigen des Wasserstoffes mehr und mehr an Bedeutung gewinne, woraus ein Verhältnis von maximal 50 für H:M folgen würde. Bei einer solchen Metallhäufigkeit hätten andererseits wenigstens einige Metall-Absorptionskanten, z.B. die des Mg, im Sonnenspektrum beobachtet werden müssen, was nicht der Fall ist.

Diese Schwierigkeiten wurden 1938 behoben, als WILDT [475], [476] den Hauptanteil an der kontinuierlichen Emission in den Atmosphären der kühleren Sterne dem negativen Wasserstoffion H⁻ zuschrieb (vgl. Ziff. 9 und 15). Außerdem wurde das Verhältnis von Wasserstoff zu Metallen noch größer angenommen, als von RUSSELL. Die auf dieser Grundlage durchgeführten quantitativen Rechnungen insbesondere von CHANDRASEKHAR und BREEN [91] ergaben eine Strahlungsverteilung, die das Sonnenspektrum auf Grund dieser Vorstellung vollständig zu erklären vermochte. Auch die Ergebnisse von CHALONGE und KOURGANOFF [89], die die Wellenlängenabhängigkeit des Absorptionskoeffizienten bzw. der optischen Tiefe als Funktion der Temperatur in der Sonnenatmosphäre aus der Theorie des Strahlungsgleichgewichtes in Verbindung mit den Randverdunklungsmessungen von ABBOT [2] ermittelten, wiesen eindeutig auf das negative Wasserstoffion als Hauptquelle der kontinuierlichen Emission bzw. Absorption hin. Der hohe Wasserstoffgehalt der Sterne ist also reell. Weiterhin erwies sich die chemische Zusammensetzung nach den von UNSÖLD und Mitarbeitern [448], durchgeführten Analysen der drei Hauptsequenzsterne: Sonne (G 2) [446], τ Scorpii (B 0) [444], 10 Lacertae (0 8,5) [449] und des Übergiganten 55 Cygni (B 2) [457] innerhalb Beobachtungsgenauigkeit konstant. Das Ergebnis ist in Tabelle 6 in der Reihenfolge der Häufigkeit der wichtigsten Elemente zusammengestellt.

In den Atmosphären der kühleren Sterne ist die Bildung des negativen Wasserstoffions offensichtlich begünstigt durch die große Anzahl freier Elektronen, die den weitgehend ionisierten Metallatomen entstammen und durch die hohe Konzentration des neutralen Wasserstoffes, da nach Gl. (15.5) der Absorptionskoeffizient der H⁻-Strahlung natürlich proportional zu $N_H \cdot p_e$ ist. Sobald aber der Wasserstoff merklich ionisiert ist, wie bei den heißeren Sternen, tritt auch die Bedeutung der H⁻-Absorption zurück zugunsten der Absorptionsgrenzkontinua und der frei-frei-Absorption des Wasserstoffes. Bei den O- und B-Sternen müssen schließlich auch die Helium-Absorptionskontinua und die THOMSON-Streuung berücksichtigt werden. Im anderen Extremfall der kühleren Sterne, deren Spektren weitgehend durch das Auftreten von Molekülbanden charakterisiert sind, werden hingegen die Molekülkontinua und die RAYLEIGH-Streuung an Molekülen (wie in der Erdatmosphäre) wesentlich.

Alle im vorangehenden entwickelten Vorstellungen und Rechnungen über kontinuierliche Spektren sind 1950 von VITENSE [455] zusammengefaßt worden, um die gesamte kontinuierliche Absorption und Streuung stellarer Materie der in Tabelle 6 angegebenen chemischen Zusammensetzung als Funktion von Temperatur und Druck zu berechnen. Als Parameter wurden die für die Rechnung übersichtlicheren Größen $\Theta = 5040/T$ und der dekadische Logarithmus des Elektronendruckes $\log P_e$ benutzt.

(Die Funktion $\Theta = 5040/T$ tritt in der logarithmischen Schreibweise des SAHAschen Ionisationsgleichgewichtes auf, und die Angabe des Elektronenpartialdruckes P_e hat gegenüber der des Gesamtdruckes den Vorteil, daß bei ionisierten Gasgemischen P_e in allen SAHA-Gleichungen der einzelnen Gemischkomponenten vorkommt.)

Gerechnet wurde für Temperaturen zwischen $T = 3877$ und $100000°$ K, d.h. Θ-Werte zwischen 1,3 und 0,05. Für die Absorption des Wasserstoffes gilt die in Ziff. 15 angegebene Formel von UNSÖLD (vgl. die quantitative Bestätigung

von JÜRGENS bezüglich des BALMER-Grenzkontinuums an dem von MAECKER
für spektroskopische Zwecke entwickelten wasserstabilisierten Rohrbogen). He II
kann völlig wasserstoffähnlich mit $Z=2$ behandelt werden. Dem Absorptionskoeffizienten des He I liegen die bereits in Ziff. 9 besprochenen Untersuchungen
von WHEELER, VINTI, KÖRVIEN, SU-SHU HUANG und GOLDBERG zugrunde.
Kohlenstoff, Sauerstoff und Stickstoff und die wichtigsten Metalle Fe, Mg, Si,
wurden — in Ermangelung besserer quantenmechanischer Rechnungen — nach
den in Ziff. 5 angegebenen allgemeinen Grundlagen behandelt. Wie wir in Ziff. 15
aus den Messungen von MAECKER und PETERS am Hochstromkohlebogen gesehen haben, erfüllen C, O, N diese Bedingungen zumindest im Sichtbaren und
Ultravioletten recht gut, während die Formeln
für die eigentlichen Metallkontinua im UV
nur noch als grobe Näherungen zu betrachten
sind. Die Grundlagen für die Berechnung
der H⁻-Absorption und deren experimentelle
Bestätigung im Laboratorium wurden in
Ziff. 9 und 15 ebenfalls schon besprochen.

Tabelle 6. *Elementhäufigkeiten, bezogen
auf je 100 000 Wasserstoffatome.*
(Nach UNSÖLD.)

Z	Element	Atomzahl
1	Wasserstoff	100 000
2	Helium	18 000
6	Kohlenstoff	25
7	Stickstoff	54
8	Sauerstoff	98
10	Neon	115
11	Natrium	0,23
12	Magnesium	6,2
13	Aluminium	0,38
14	Silicium	5,6
16	Schwefel	1,5
20	Calcium	0,30
26	Eisen	9,3

Von den zahlreichen graphischen Darstellungen der Rechnungsergebnisse von VITENSE
sei als Beispiel nur diejenige für $\Theta = 0{,}9$
($T = 5600°$ K) wiedergegeben, weil darin die
Wellenlängenabhängigkeit des Absorptionskoeffizienten in der Sonnenatmosphäre
($\log P_e \approx 1{,}5$) enthalten ist. Aufgetragen ist
(mit — und $\log P_e$ als Parameter) $\log \frac{\varkappa_\nu}{P_e}$, wobei $\varkappa_\nu = \sum \varkappa_\nu (1 - e^{-h\nu/kT}) + \sigma_\nu$ den gesamten
effektiven Absorptions- und Streukoeffizien-
ten darstellt. Skala I gibt $\log \frac{\varkappa_\nu}{P_e}$ je schweres Teilchen, Skala II die Werte je
Gramm Sternmaterie. Der Absorptionskoeffizient der Wasserstoffatome allein ist
ohne den Faktor $(1 - e^{-h\nu/kT})$ als punktierter Kurvenzug in Fig. 65 eingezeichnet
und der Übersichtlichkeit wegen nach unten verschoben. Die gestrichelten
Kurven (---) geben den Einfluß des RAYLEIGH-Streukoeffizienten mit $\log P_e$ als
Parameter wieder.

Wie man sieht, ist für $\log P_e = 0 \rightarrow 3$ im optisch erfaßbaren Spektralbereich
($\log \lambda \sim 3{,}5$ bis 4) der diskret $\rightarrow$ kontinuierliche Anteil des H⁻-Kontinuums (d.h.
das H⁻-Grenzkontinuum) ausschlaggebend; lediglich der Intensitätssprung an
der BALMER-Grenze tritt daneben noch etwas in Erscheinung. Der Anstieg des
Absorptionskoeffizienten für $\log \lambda > 4{,}3$ (Ultrarot) wird im wesentlichen bedingt
durch den frei-frei-Anteil des H⁻-Kontinuums. Im fernen UV tritt schließlich
die RAYLEIGH-Streuung, überlagert von den Hauptseriengrenzkontinua der
Metalle Mg, Fe, Si und des Kohlenstoffs, stark hervor.

Alle hier nur kurz darstellbaren Ergebnisse über die kontinuierliche Absorption der stellaren Materie bilden den Ausgangspunkt für die an anderer
Stelle (Bd. L) eingehend behandelte Berechnung der Temperatur- und Druck-
schichtung in Sternatmosphären, wobei neben der Absorption stets auch die
Reemission zu berücksichtigen ist (Theorie des Strahlungsgleichgewichtes). Gelegentlich arbeitet man bei der Berechnung von Atmosphärenmodellen genau
wie bei der Theorie des Sterninneren mit einem mittleren frequenzunabhängigen
Absorptionskoeffizienten (Opazitätskoeffizient). Dieser wird nach ROSSELAND[385]
so definiert, daß der über alle Frequenzen integrierte Gesamtstrahlungsstrom

bei der Mittelung richtig wiedergegeben wird, was unter der Voraussetzung thermischen Gleichgewichtes und Berücksichtigung der „erzwungenen Emission"

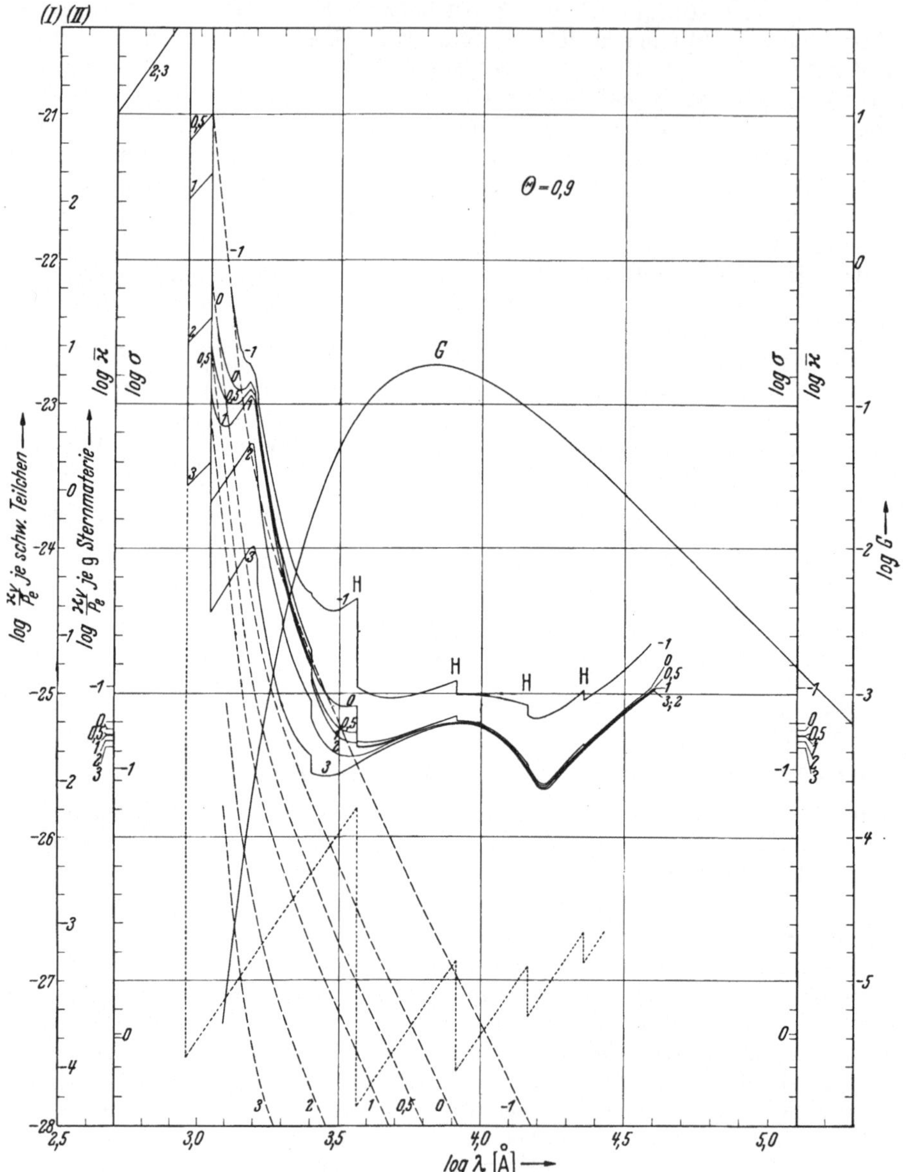

Fig. 65. $\log \varkappa_\nu/P_e$ mit $\log P_e$ als Parameter nach Vitense [455]. (Aus Unsöld [449].

zur Mittelungsvorschrift

$$\frac{1}{\overline{\varkappa}} = \int\limits_0^\infty \frac{1}{\varkappa_\nu} G(u)\, du \tag{24.1}$$

mit $u = \dfrac{h\nu}{kT}$ und $\varkappa_\nu = \sum \varkappa_\nu (1 - e^{-u}) + \sigma_\nu$ führt. $G(u)$ ist die Rosselandsche Gewichtsfunktion

$$G(u) = \frac{15}{4\pi^4}\, \frac{u^4 e^u}{(e^u - 1)^2}, \tag{24.2}$$

die ebenfalls in dem Diagramm (Fig. 65) enthalten ist (mit G bezeichnete Kurve) und ein anschauliches Bild über die Wellenlängenbereiche vermittelt, welche für die Mittelbildung des Absorptionskoeffizienten wesentlich sind.

Bei der Behandlung der äußeren Schichten der Sternatmosphären, der Sonnenchromosphäre, der Korona oder der Gasnebel treten, genau wie in der Gasentladungsphysik beim Übergang von der thermischen Bogenentladung zur Niederdruckentladung, die äußerst schwierigen Probleme der Nichtgleichgewichtszustände auf. In den tieferen Schichten der Chromosphäre ist das kontinuierliche Spektrum noch durch das H^--Ion bestimmt, während in größeren Höhen die THOMSON-Streuung an freien Elektronen überwiegt. Das BALMER-Grenzkontinuum trägt demgegenüber nach den Rechnungen am Chromosphärenmodell von BÖHM-VITENSE [68] zur Strahlung aller Schichten der Chromosphäre wesentlich bei. An die in heftiger turbulenter Bewegung befindlichen Chromosphäre von $T \sim 4000°$ K schließt sich nach außen die Korona an, in der die „Temperatur" sehr rasch mit der Höhe auf $10^{6°}$ K ansteigt (EDLÉN [130]). Ihr von Korona-Emissionslinien überlagertes Kontinuum entsteht nach GROTRIAN [194] und ALLEN [7], wie schon SCHWARZSCHILD 1905 vermutete, durch Streuung des Sonnenlichtes an freien Elektronen, deren hohe thermische Geschwindigkeiten bei 10^6 °K (~ 5500 km/sec) zu völliger Verwaschung der FRAUNHOFER-Linien führt. Auf Grund dieses Streueffektes ist das Koronakontinuum teilweise polarisiert, und zwar bevorzugt der magnetische Vektor die radiale und der elektrische Vektor die tangentiale Schwingungsrichtung relativ zur Sonnenoberfläche. Vom thermischen Gleichgewicht ist das stark verdünnte Gas ziemlich weit entfernt, denn während, wie erwähnt, die „Translations- wie die Ionisationstemperatur" $10^{6°}$ K beträgt, liegen die „Anregungstemperaturen" nur zwischen 13000 und 29000° K. Die für die Ionisation und Rekombination verantwortlichen Elementarprozesse sind z.B. in den Arbeiten von BIERMANN [63] und ELWERT [136], [137] im einzelnen behandelt und dürften auch für die Theorie der Gasnebel (vgl. ZANSTRA [494]) von Bedeutung sein.

Schließlich sei erwähnt, daß kontinuierliche Spektren auch in den *Sonnen-Protuberanzen* zu beobachten sind. Außer dem BALMER- und PASCHEN-Kontinuum fand LYOT [291] bei ihnen in der Umgebung von $\lambda = 6200$ Å ein Kontinuum, dessen Strahlung senkrecht zum Sonnenrand polarisiert war, was wiederum auf THOMSON-Streuung der kontinuierlichen Photosphärenstrahlung an freien Elektronen schließen läßt. Ein Anteil des H^--Kontinuums ist nach UNSÖLD [449] nicht zu erwarten, weil die Konzentration der neutralen Wasserstoffatome viel zu gering ist.

25. Kontinuierliche Radiofrequenzstrahlung. Einen ganz erstaunlichen Aufschwung hat die Erforschung der äußersten Schichten der Sternatmosphären durch die *Radioastronomie* erhalten. Daß die Milchstraße Radiowellen aussendet, hatte schon JANSKY [241] 1932 festgestellt. Die Radiofrequenzstrahlung der Sonne wurde zuerst 1942 von HEY ($\lambda = 4$ bis 6 m) und SOUTHWORTH ($\lambda = 3,2$ cm) entdeckt. Es wird dabei zwischen der „ruhigen" und der „gestörten" solaren Radiofrequenzstrahlung unterschieden. Erstere ist, wie MARTYN [298] 1946 schon richtig erkannte, eine kontinuierliche thermische Strahlung, welche im Gebiet der Zentimeterwellen vorwiegend von der Chromosphäre und im Gebiet der Meterwellen von der Korona emittiert wird (Fig. 66). Die gestörte Strahlung wird im allgemeinen auf Anfachung von Plasmaschwingungen (s. Ziff. 20) durch turbulente Plasmaströmungen bzw. Stoßwellen zurückgeführt. Denselben Ursprung hat offenbar auch die galaktische Radiofrequenzstrahlung.

Die *thermische Emission* („ruhige" Strahlung) vollständig ionisierter Gase im Radiofrequenzgebiet ist, wie wir in Ziff. 5 gesehen haben, praktisch kontinuierliche frei-frei-Strahlung. Für den Absorptionskoeffizienten gilt nach Gl. (5.42b) bzw. (20.11)

$$\varkappa_\nu = \frac{\varepsilon_\nu}{B_\nu} = \frac{16\pi\,e^6\,Z^2\,N_e\,N_i}{3c\,(2\pi\,m\,kT)^{\frac{3}{2}}}\,\frac{1}{\nu^2}\,\ln\left(\frac{kT}{1{,}44\,Z\,e^2\,N_i^{\frac{1}{3}}}\right). \tag{25.1}$$

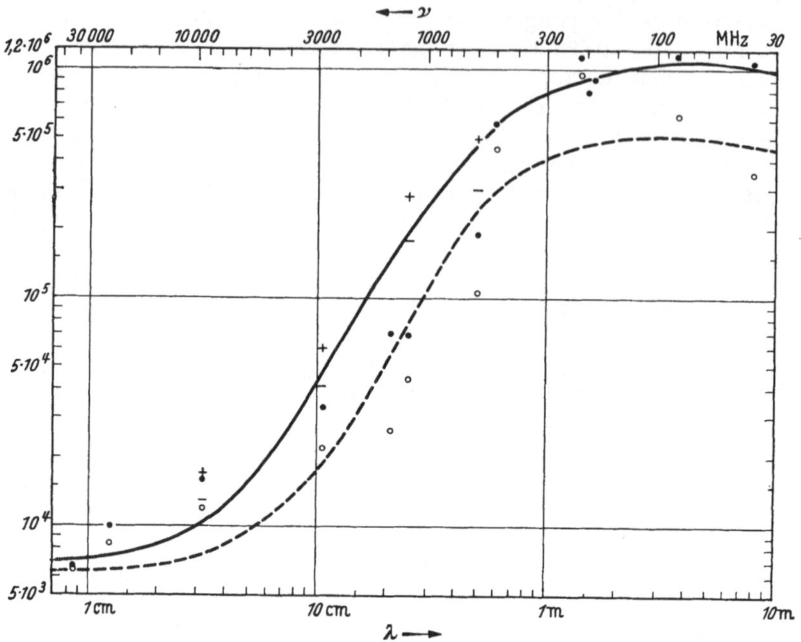

Fig. 66. Radiofrequenzstrahlung der ruhigen Sonne (ohne Aktivitätszentren) als Funktion der Wellenlänge λ bzw. Frequenz ν: Effektive Temperatur T_e ● gemessen, — berechnet. Strahlungstemperatur Sonnenmitte $T_s(0)$ ○ gemessen - - - berechnet. + + T_e, — — $T_s(0)$ fleckenfreie Sonne ∼1948. Theorie für N_e nach Baumbach-Allen. (Aus Unsöld [449].)

Rechnet man ferner mit vollständig ionisierter Stellarmaterie (80% H, 20% He), so ist $N_i\,N_e\,Z^2 = 1{,}35\,N_e^2$ und man erhält

$$\varkappa_\nu = 1{,}318 \cdot 10^{-2}\,\frac{N_e^2}{T^{\frac{3}{2}}\nu^2}\,\ln(367\,T\cdot N_e^{-\frac{1}{3}}). \tag{25.2}$$

Unter Verwendung der Formel von Baumbach-Allen über die Verteilung der Elektronendichte N_e in der Korona

$$N_e(r) = 10^8\,(2{,}99\,r^{-16} + 1{,}55\,r^{-6}) \tag{25.3}$$

und der Annahme konstanter Korona- und Chromosphärentemperatur erhält man nach Rechnungen von Martyn [298], Unsöld [445], Waldmeier und Müller [459] u.a. zunächst Aussagen über die optischen Tiefen der Korona und Chromosphäre bei verschiedenen Wellenlängen und schließlich die Intensitätsverteilung $I_\nu(\varrho)$ als Funktion des Radius ϱ (in Einheiten des Sonnenradius) Die Theorie ergibt (s. Fig. 67), daß die Korona für Wellenlängen > 50 cm auf der ganzen Sonnenscheibe und darüber hinaus optisch dick ist, so daß eine gleichmäßige „helle Scheibe" entsteht, die aber erheblich größer ist als die „optische" Sonne. Wegen $\varkappa_\nu$ bzw. $\tau_\nu \sim \nu^{-2}$ nimmt für kürzere Wellen die Ausdehnung dieser „Scheibe" erheblich ab. Im Zentimetergebiet überwiegt immer mehr die Strahlung der Chromosphäre ($T < 6500°$ K), und da diese in der Mitte der Sonnenscheibe bei abnehmender Wellenlänge optisch dünn wird, am Sonnenrand wegen der

großen Ausdehnung — der Beobachter sieht ja in einen strahlenden Ring hinein —
aber noch optisch dick bleibt, ergibt sich bei kürzeren Wellenlängen ein „heller"
Ring am Sonnenrand. Diese Aussagen der Theorie wurden durch die Interfero-
meter-Messungen von CHRISTIANSEN und WARBURTON [93] bei $\lambda = 21$ cm und
von ALON, ARSAC und STEINBERG [8] bei $\lambda = 3,2$ cm auf das beste bestätigt.

Im Gebiet der Meterwellen erweist es sich als notwendig, die Brechung der
Strahlen zu berücksichtigen. Nach Gl. (20.8) ist der Brechungsindex eines Plas-
mas <1 und geht mit $\nu \to \nu_0 = \sqrt{\dfrac{e^2}{\pi\,m}\,N_e}$ (LANGMUIR-Plasmafrequenz) gegen Null.

Ein Strahl, der in ein Gebiet ansteigender Elektronendichte hineinläuft, wird
daher zurückgebogen (ent-
sprechend den Reflexionen
der Radiowellen an der
HEAVISIDE-Schicht). Die Berechnung solcher gekrümm-
ter Bahnen in der Korona
haben unter anderen BURK-
HARDT und SCHLÜTER [77]
durchgeführt. Auch diese
Rechnungen stehen in guter
Übereinstimmung mit den
Beobachtungen, obwohl diese
darauf hindeuten, daß auch
eine unregelmäßige Brechung
der Radiofrequenzstrahlung
im Gebiet der Meterwellen
infolge Dichteschwankungen
in der Korona eine wichtige
Rolle spielt.

Bei der „gestörten" Strahlung der Sonne handelt es
sich, ähnlich wie bei der

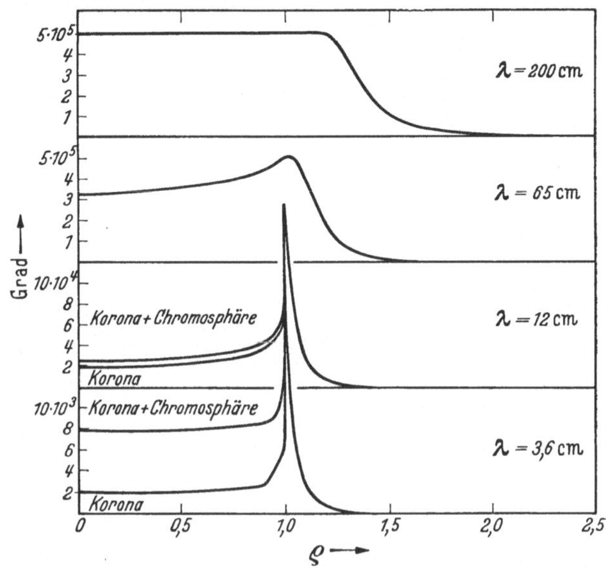

Fig. 67. Verteilung der Intensität $I_\nu(\varrho)$ bzw. Strahlungstemperatur $T_\nu(\varrho)$
über die Sonnenscheibe. (Aus UNSÖLD [449].)

galaktischen Radiofrequenzstrahlung nicht eigentlich um kontinuierliche Strahlung. Das Spektrum setzt sich vielmehr aus einer großen Anzahl mehr oder
weniger dicht liegender „Plasmaschwingungslinien" zusammen, die dadurch
entstehen, daß sich turbulente Gasmassen mit örtlich und zeitlich sehr verschiedener Elektronendichte durcheinander bewegen, wobei das Elektronengas zu
kohärenten Schwingungen der sog. Plasmafrequenz angeregt wird. Auf welche
Weise dabei überhaupt elektromagnetische Strahlung emittiert werden kann, ist
zur Zeit noch weitgehend ungeklärt. Neuere Arbeiten von LARENZ [275] und
LÜST [288] deuten darauf hin, daß neben der hohen Materiegeschwindigkeit
auch die in turbulenten Plasmen sicher vorhandenen Magnetfelder für diesen
Prozeß von Bedeutung sind.

B. Molekülkontinua.

I. Systematik der Molekülkontinua.

26. Allgemeine Übersicht über kontinuierliche und diskrete Molekülspektren.
Wie die Atomkontinua, so sind auch die echten Molekülkontinua das Ergebnis
von Energiezustandsänderungen, hier von Molekülen, bei denen der eine der
beiden kombinierenden Zustände nicht stationär, gebunden, sondern ein freier
dissoziierter Zustand des Moleküls ist. Ist der Anfangszustand diskret, während

der Endzustand im Kontinuum liegt, so geht ein anfänglich stabiles Molekül unter Absorption oder Emission eines kontinuierlichen Spektrums in einen instabilen Zustand über, dissoziiert also in Atome oder Atomgruppen. Liegt umgekehrt der Anfangszustand im Kontinuum, während der Endzustand diskret ist, so bildet sich aus zusammenstoßenden Atomen oder Atomgruppen unter Emission oder Absorption kontinuierlicher Strahlung ein stabiles Molekül (Molekülrekombination). Der dritte noch mögliche Fall, die Kombination zweier nichtstationärer, freier, im Kontinuum liegender Energiezustände, würde bedeuten, daß ein freies System (Atome im Stoß) unter Emission oder Absorption kontinuierlicher Strahlung in einen andersartig freien Zustand mit anderer Elektronenanordnung der Stoßpartner übergeht. Auch dieser (praktisch nicht sehr wichtige) Fall ordnet sich zwanglos in das angedeutete Ordnungsschema ein, wenn wir *als „Molekül" jedes in Wechselwirkung befindliche System von zwei oder mehr Atomen bzw. Atomgruppen definieren*, ohne Rücksicht darauf, ob ein solches System stabil ist oder nicht. Im letzteren Fall spricht man bei zwei Atomen gern von einem *Stoßpaar*.

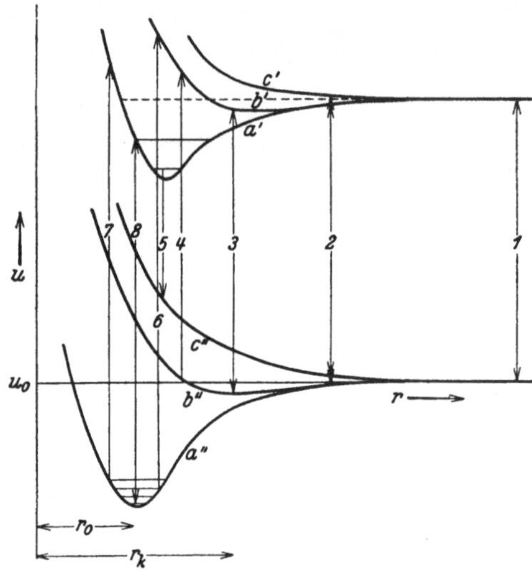

Fig. 68. Potentialkurvenschema zum Verständnis der wichtigsten Typen von Molekülkontinua.

Jedes Molekül in diesem weiteren Sinne ist charakterisiert durch ein Potentialkurvenschema, das für jeden möglichen Elektronenzustand des der Einfachheit halber als zweiatomig angenommenen Moleküls die Abhängigkeit seiner potentiellen Energie vom Abstand der beiden Atomkerne zeigt. Fig. 68 zeigt für den Grundzustand und einen angeregten Elektronenzustand die wichtigsten Typen der Wechselwirkung zweier Atome, die für das Verständnis der kontinuierlichen Molekülspektren von Bedeutung sind. Besitzt das Molekül bei einer bestimmten Elektronenanordnung einen stabilen, stationären Zustand, so besitzt die diese Elektronenanordnung beschreibende Potentialkurve bei einem bestimmten Kernabstand ein Minimum (Kurve a' bzw. a''). Überwiegen bei jedem Kernabstand die abstoßenden über die anziehenden Kräfte, so gibt es kein stabiles Molekül dieser Elektronenanordnung, und ein solcher Zustand wird durch „Abstoßungskurven" wie c' bzw. c'' beschrieben. Es gibt schließlich Elektronenanordnungen, bei denen eine bei relativ großen Kernabständen (3 bis 5 Å) wirksame Anziehung durch van der Waals-Kräfte zu einem hier gelegenen flachen Potentialminimum führt (Kurven b' bzw. b''); in diesem Elektronenzustand gibt es dann ein zwar stabiles, aber meist nur sehr schwach gebundenes „van der Waals-Molekül".

Moleküle in einem Zustand unterhalb der jeweiligen Dissoziationsgrenze (rechte Asymptote der Kurven) sind stabil; ihre Kerne führen stationäre Schwingungen gegeneinander aus, während Moleküle oberhalb der Dissoziationsgrenze ihres jeweiligen Elektronenzustandes nichtstationär, dissoziierend oder dissoziiert sind, ihre Zustände also dem kontinuierlichen Energiebereich angehören, der sich gemäß Fig. 69 an die Konvergenzstelle der stationären Schwingungszustände anschließt.

Die diskreten wie die kontinuierlichen Molekülspektren im sichtbaren oder ultravioletten Spektralbereich kommen durch Übergänge zwischen zwei der angedeuteten Potentialkurven zustande; bei solchen Übergängen ändert sich im allgemeinen also sowohl die potentielle Energie der Elektronenanordnung wie die der Kerne gegeneinander, d. h. die potentielle Energie der gegeneinander schwingenden bzw. zusammenstoßenden Atome des Moleküls. Die bei Übergängen zwischen zwei Potentialkurven absorbierten oder emittierten Spektren folgen aus dem sog. FRANCK-CONDON-Prinzip. Es besagt in seiner von FRANCK [161] stammenden klassischen Fassung, daß die dem Übergang entsprechende Elektronenumordnung im Molekül so schnell erfolgt, daß Lage und Geschwindigkeit

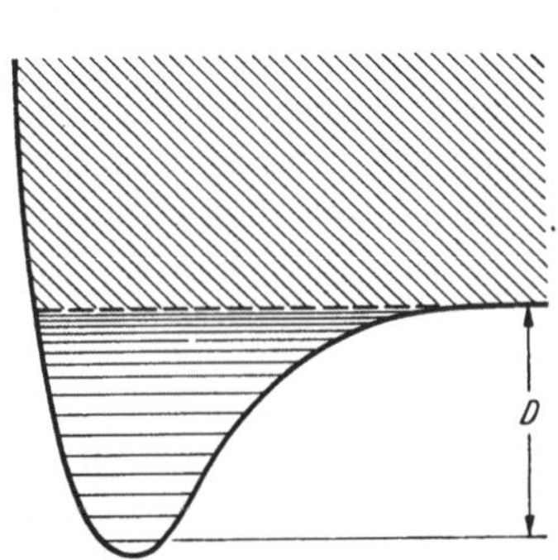

Fig. 69. Potentialkurve eines Molekül-Elektronenzustandes mit diskreten Schwingungszuständen und kontinuierlichem, zur Moleküldissoziation gehörenden Energiebereich.

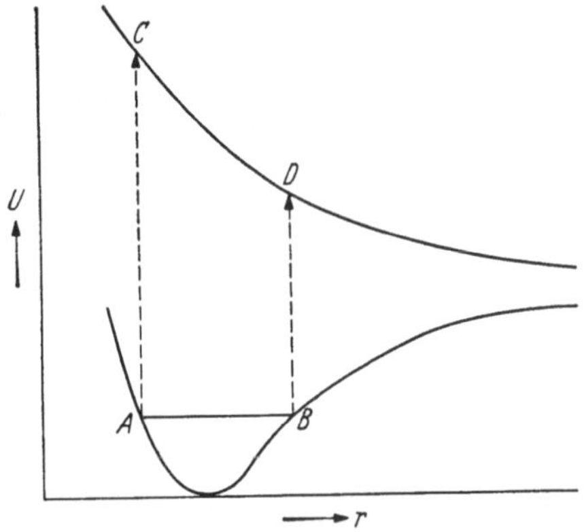

Fig. 70. Potentialkurven zur Darstellung der Wellenlängenausdehnung eines kontinuierlichen Molekülspektrums.

der schweren Atomkerne sich während des Übergangs nicht merklich ändern. In der Potentialkurvendarstellung (Fig. 68) heißt das, daß Übergänge ohne Änderung des Kernabstandes, d. h. senkrecht erfolgen, und zwar bevorzugt von den Schwingungsumkehrpunkten, d. h. den Schnittpunkten zwischen Potentialkurve und Energiegerade zu den entsprechenden Schnittpunkten der anderen Potentialkurve. Wenn nun der Kernabstand des Moleküls im Anfangszustand bekannt ist, so ergibt sich aus dem senkrechten Abstand der beiden Potentialkurven bei diesem Kernabstand r der Betrag der absorbierten oder ausgestrahlten Energie und daraus mit $E = hc/\lambda$ die entsprechende Wellenlänge des Spektrums. Dabei ist natürlich zu berücksichtigen, daß die Kernabstände der absorbierenden oder emittierenden Moleküle innerhalb der durch die Potentialkurve bestimmten Grenzen variieren. So entsteht beispielsweise durch Absorptionsübergänge von stabilen mit der Schwingungsenergie E_v schwingenden Molekülen (Fig. 70) zu einer höheren Abstoßungskurve ein kontinuierliches Spektrum, dessen Wellenlängengrenzen durch die Energiewerte der Übergänge $A \rightarrow C$ und $B \rightarrow D$ gegeben sind. Da das schwingende Molekül sich im Zeitmittel am längsten in den Schwingungsumkehrpunkten A und B aufhält, sollen nach dem FRANCKschen Prinzip auch die von hier ausgehenden Übergänge intensiver sein als die von mittleren Kernabständen, die schneller durchlaufen werden.

In Fig. 68 sind nun die verschiedenen möglichen Übergänge zwischen den beiden angenommenen Potentialkurven-Scharen angegeben, aus denen unter Berücksichtigung der Kernabstandsvariation alle grundsätzlich möglichen diskreten wie kontinuierlichen Molekülspektren folgen.

Während Übergang 1 zwischen den ungestörten Atomzuständen rechts eine scharfe Atomlinie ergibt, entstehen durch Übergänge 2 verbreiterte Linien, wobei ersichtlich Größe und Charakter der Verbreiterung vom mittleren Abstand der Teilchen (Funktion von Gasdichte und Temperatur) sowie vom Verlauf der beiden kombinierenden Potentialkurven abhängen. In dieser Darstellung erscheint somit die in Bd. XXVII dieses Handbuches gesondert behandelte Linienverbreiterung zwanglos als Grenzfall der Molekülkontinua.

Die Übergänge 3 und 4 gehören zu schwach gebundenen van der Waals-Molekülen bzw. -Stoßpaaren; die ihnen entsprechenden Spektren sind schmale kontinuierliche Bänder, die verbreiterten Linien nicht unähnlich sehen. Übergang 3 ergibt dabei eines der erwähnten unechten Kontinua, weil die Energiezustände in den beiden flachen Potentialminima zwar diskret sind, aber so dicht liegen, daß das bei Übergängen zwischen ihnen entstehende Bandenspektrum bei ungenügender Auflösung leicht kontinuierlich erscheinen kann. Bei Übergang 4 dagegen liegen beide Zustände im Energiekontinuum; es entsteht ein echt kontinuierliches Band, dessen Ausdehnung und Intensitätsverteilung durch die relative Neigung der beiden Potentialkurven und die Gastemperatur bestimmt ist, weil von letzterer die Besetzung der Molekül- und Stoßpaarzustände abhängt.

Die Übergänge 5 bis 7 finden in dem Kernabstandsgebiet statt, in dem die Elektronenwolken der beiden Atome sich schon stark durchdringen ($r < r_k$) und dadurch große Energieänderungen gegenüber den ungestörten Atomzuständen bewirken; sie stellen also Molekülkontinua im engeren Sinne dar. Bei Übergang 5 ist der angeregte obere Zustand stabil und diskret, der untere unstabil mit kontinuierlichem Energiebereich. Berücksichtigt man wieder, daß Übergänge 5 von allen Energieniveaus der Potentialkurve a' zur Kurve c'' erfolgen können, so erkennt man, daß aus der verschiedenen Neigung der beiden Kurven ein weit ausgedehntes kontinuierliches Spektrum in Emission resultiert.

Die Umkehrung des Überganges 5 ist Übergang 6. Hier ist der untere Zustand a'' stabil mit diskreten Energiezuständen, der obere b' (analog ist es bei Übergängen zu c') eine Abstoßungskurve mit kontinuierlichem Energiebereich. Es resultiert (wieder unter Berücksichtigung aller übrigen nicht eingezeichneten Übergänge $a'' \rightarrow b'$) ein ausgedehntes kontinuierliches Spektrum in Absorption. Ebenfalls ein ausgedehntes Absorptionskontinuum erhält man durch Übergänge 7 von den diskreten Zuständen der Kurve a'' zum kontinuierlichen Energiebereich der Kurve a'. Daß es sich hier, im Gegensatz zu Kurve b' und c', um eine Kurve mit ausgebildetem Potentialminimum handelt, also um den kontinuierlichen Bereich eines Zustandes, der auch stabil existieren kann, ersieht man aus der Beobachtung, daß sich an das kontinuierliche Spektrum Übergang 7 ein diskretes Bandenspektrum infolge der Übergänge 8 zwischen diskreten Zuständen derselben beiden Kurven anschließt.

Das Franck-Condon-Prinzip in seiner einfachsten klassischen Form zusammen mit der anschaulichen Potentialkurvendarstellung der Elektronenzustände der Moleküle gestattet also, bereits einen guten Überblick über die bei den verschiedenen Molekülübergängen zu erwartenden kontinuierlichen wie diskreten Spektren zu gewinnen.

27. Theorie der Potentialkurven und Molekülkontinua. Aus Ziff. 26 folgt, daß Ausdehnung und Intensitätsverteilung der kontinuierlichen Molekülspektren vom relativen Verlauf der beiden unter Absorption und Emission kombinierenden Molekülzustände, von der Besetzung der entsprechenden diskreten oder kontinuierlichen Zustände der Ausgangspotentialkurve und von den Übergangswahrscheinlichkeiten zwischen den verschiedenen Potentialkurventeilen abhängen. Hieraus folgt weiter, daß es *eine quantitative Theorie der Molekülkontinua bisher nicht geben kann, weil wir weder eine quantitative Theorie der Potentialkurven* *noch eine solche der Übergangswahrscheinlichkeit als Funktion des Kernabstandes besitzen.*

Eine Potentialkurve $U(r)$ gibt ja das Potential *aller* zwischen den Kernen eines zweiatomigen Moleküls wirkenden Kräfte als Funktion von deren Abstand an; ihre Berechnung würde also eine quantitative Theorie der chemischen Bindung einschließlich aller Kräfte höherer Ordnung voraussetzen. Nun läßt sich zwar die Bindung in den einfachsten Molekülen wie H_2^+ und H_2 mit entsprechendem Rechenaufwand wellenmechanisch berechnen (z.B. [101], [214], [232], [235] bis [238], [343], [360]), doch versagen diese Methoden schon bei den zweiatomigen Mehrelektronenmolekülen (vgl. [87], [229], [249]), von den vielatomigen Molekülen ganz zu schweigen, und auch die Verwendung der elektronischen Rechenmaschinen dürfte an diesem Zustand nur langsam etwas ändern können. Aus der Analyse der entsprechenden Bandenspektren lassen sich zwar (vgl. z.B. Band XXXVII dieses Handbuches) ziemlich sichere Aussagen über den Verlauf der Potentialkurven stabiler Molekülzustände in der Gegend der Potentialminima ableiten; für den für die Kontinuaforschung interessanten Verlauf in der Nähe der Dissoziationsgrenze und oberhalb dieser aber versagen diese Methoden vollständig. Hier ist man also auf Näherungsrechnungen angewiesen, deren Genauigkeit meist schwer abzuschätzen ist [101]. Die Spektroskopie der Molekülkontinua hat daher zunächst die umgekehrte Aufgabe, nämlich aus den beobachteten kontinuierlichen Spektren auf den Verlauf der Potentialkurve des nichtstationären Molekülzustandes zu schließen. Voraussetzung hierfür ist natürlich die Kenntnis des Potentialkurvenverlaufes des meist stabilen Ausgangszustandes. Auf Einzelheiten wird in Ziff. 30 und 31 eingegangen.

28. Das Franck-Condon-Prinzip und die Struktur von Molekülkontinua. Bei Kenntnis des Verlaufes der kombinierenden Potentialkurven und ihrer (jedenfalls für den Fall thermischen Gleichgewichts berechenbaren) Besetzung erfordert die Berechnung des mit dem Übergang verknüpften Molekülspektrums die Kenntnis der Übergangswahrscheinlichkeit als Funktion des Kernabstandes. Nun erfolgt bei einem bestimmten Übergang im allgemeinen sowohl eine Änderung der Elektronenanordnung wie der potentiellen Energie der zwischen den Kernen wirkenden Kräfte. Das in Ziff. 26 in seiner klassischen Form schon angedeutete Franck-Condon-Prinzip geht nun von der Annahme aus, daß diese beiden Anteile der Übergangswahrscheinlichkeit voneinander unabhängig sind, man das elektrische Moment eines Überganges zwischen zwei Molekülzuständen also schreiben kann als

$$\mathfrak{M}_{e'k' \leftrightarrow e''k''}(r) = \int \psi_{e'}\, \mathfrak{M}_{e' \leftrightarrow e''}\, \psi_{e''}^*\, d\tau \cdot \int \psi_{k'}\, \mathfrak{M}_{k' \leftrightarrow k''}(r)\, \psi_{k''}^*\, d\tau, \qquad (28.1)$$

wo der erste Ausdruck nur von den Koordinaten bzw. Eigenfunktionen ψ_e (oder Quantenzahlen) der Elektronen, der zweite nur vom Kernabstand r und den Eigenfunktionen ψ_k der Kernbewegung abhängt (Condon [97], [98]). Dabei kann [143] entgegen der früheren üblichen Annahme der Elektronenanteil der Übergangswahrscheinlichkeit vom Kernabstand nicht unabhängig sein, weil es Übergänge gibt, die für getrennte Atome (r groß) verboten sind, für aus solchen Atomen zusammengesetzte Moleküle (r klein) aber erlaubt sind.

In der klassischen Fassung sagt das Franck-Condon-Prinzip über die relative Intensität von Übergängen zwischen zwei verschiedenen Potentialkurven aus, daß unter der Annahme eines konstanten Elektronenanteils der Übergangswahrscheinlichkeit die Intensität der Übergänge der Aufenthaltswahrscheinlichkeit der Kerne im Anfangszustand proportional sei, und daß daher mit größter Intensität die Übergänge von den Schwingungsumkehrpunkten der

Ausgangspotentialkurve zu den senkrecht über bzw. unter ihnen gelegenen Punkten der Potentialkurve des Endzustandes erfolgen sollten. Nach der Wellenmechanik ist die Häufigkeitsverteilung der verschiedenen Kernabstände durch das Quadrat der betreffenden Schwingungseigenfunktionen ψ_k gegeben, während die Wahrscheinlichkeit von Übergängen durch das Übergangsintegral

$$\int \psi_{k'}\, \mathfrak{r}\, \psi_{k''}^{*}\, d\tau \tag{28.2}$$

bestimmt ist, also für einen bestimmten Kernabstand r vom Betrag der Schwingungseigenfunktionen $\psi_{k'}$ und $\psi_{k''}$ der kombinierenden Zustände bei diesem r-Wert abhängt. Da nach Fig. 71 die ψ_k-Werte der höheren Schwingungszustände

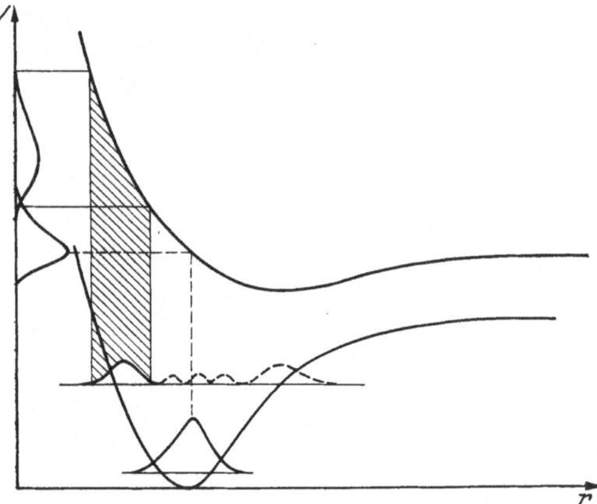

Fig. 71. Potentialkurven und Schwingungseigenfunktionen zur Verdeutlichung des Franck-Condon-Prinzips.

Maxima nahe den klassischen Schwingungsumkehrpunkten besitzen, ist das Franck-Condon-Prinzip auch wellenmechanisch richtig, jedoch mit zwei wesentlichen Modifikationen. Einmal nämlich, gehört zum Schwingungsgrundzustand nur *ein* ψ_k-Maximum statt der zwei klassischen Schwingungsumkehrpunkte der Nullpunktsschwingung. Zum anderen ist wellenmechanisch klar, daß die Struktur des durch Übergang von *einem* Schwingungszustand aus entstehenden Spektrums von der Breite der ψ_k^2-Maxima der kombinierenden Zustände abhängt. Condon hat nun gezeigt [97], [98] wie man unter gewissen vereinfachten Bedingungen aus den ψ_k^2-Maxima des Elektronengrundzustandes die Intensitätsverteilung der durch Übergang zu einer höheren und steilen Abstoßungskurve entstehenden Absorptionskontinua ermitteln kann. Zum Verständnis seiner sog. Reflexionsmethode dient Fig. 71. In ihr sind die ψ_k^2-Maxima für den Schwingungsgrundzustand und einen höheren Schwingungszustand des Elektronengrundzustandes des Moleküls gezeichnet. Betrachtet werden Übergänge von dem einen Maximum des Schwingungsgrundzustandes zu dem darüberliegenden Teil der Abstoßungskurve, sowie zweitens die Übergänge vom kernnahen Maximum des höheren Schwingungszustandes zu dem entsprechenden, über ihm liegenden Teil der Abstoßungskurve. Jeder der unendlich dicht liegenden Energiezustände der oberen Abstoßungskurve entspricht ja einem aperiodischen Zusammenstoß der beiden Atome des dissoziierten Moleküls. Jede entsprechende ψ_k-Funktion besitzt, wie die wellenmechanische Behandlung zeigt, im wesentlichen ein einziges scharfes Maximum am Schnittpunkt der Abstoßungskurve mit dem betreffenden Energiezustand. Deshalb wird die Intensitätsverteilung des durch Übergänge von unten nach oben in Fig. 71 entstehenden Spektrums nur von der Form der ψ_k^2-Kurve des unteren Zustandes bestimmt und kann durch „Reflexion" der ψ_k^2-Maxima der Grundzustandskurve an der oberen Abstoßungskurve erhalten werden. Die so sich ergebenden Intensitätsverteilungen sind in Fig. 71 links über der Ordinate aufgetragen. Bei dieser Konstruktion sind die Übergänge von dem kleinen mittleren und dem größeren rechten ψ_k^2-Maximum des höheren der beiden betrachteten Schwingungszustände unberücksichtigt geblieben, weil über ihnen gebundene Schwingungszustände der oberen Kurve liegen, diese Übergänge also nicht zu kontinuierlichen Spektren führen. Für alle Einzelheiten sei auf Finkelnburgs Monographie [149] und die neuere Literatur (z.B. [211],

[*234*], [*338*], [*339*]) verwiesen. Schon aus unserer kurzen Diskussion aber dürfte klar sein, daß weder die Theorie der Potentialkurven noch die der Übergangswahrscheinlichkeiten für gleichzeitige Änderung von Elektronenbewegung und Schwingung genügend quantitativ durchgeführt ist, um einen genauen Vergleich von Theorie und Beobachtung für die Molekülkontinua zu erlauben. Umgekehrt sind auch aus der spektroskopischen Beobachtung nur relativ grobe Schlüsse auf den Verlauf der Potentialkurve des unstabilen Zustandes möglich.

29. Dissoziation und Prädissoziation. Ein wesentlicher Teil des Interesses an den Absorptionskontinua der Moleküle beruht auf der Tatsache, daß sie direkte Hinweise auf den wichtigen Prozeß der Photodissoziation erlauben. Es soll deshalb auf die Bestimmung der Dissoziationsenergie der Moleküle aus der langwelligen Grenze ihrer Photodissoziationskontinua etwas näher eingegangen werden. Bei der in Fig. 72 angegebenen gegenseitigen Lage der Potentialkurven finden vom Schwingungsgrundzustand des Elektronengrundzustandes Übergänge besonders zu den verschiedenen höheren Schwingungszuständen des angeregten Moleküls (Absorption eines Bandenzuges), und in den daran anschließenden, der Dissoziation des angeregten Moleküls in ein normales und ein angeregtes Atom

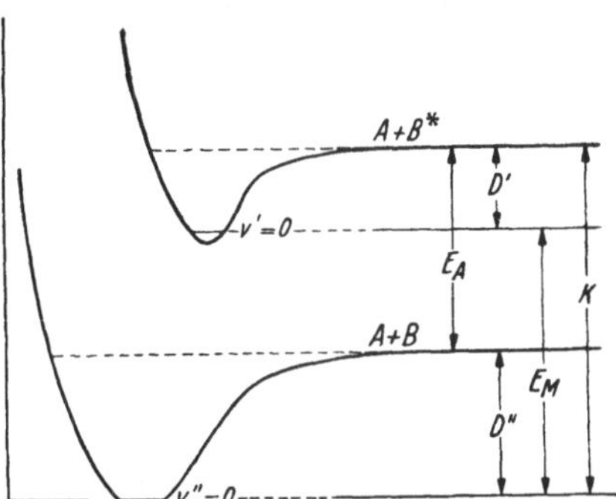

Fig. 72. Potentialkurvenschema zur Erklärung des Zusammenhangs von Dissoziation und kontinuierlicher Absorption von Molekülen.

$(A + B^+)$ entsprechenden kontinuierlichen Energiebereich statt. Der langwelligen Grenze des Kontinuums, aus der TRAUTZ [*435*] zuerst auf die Dissoziationsenergie des Cl_2 geschlossen hat, entspricht die Energie K, die man, wenn das Dissoziationskontinuum wegen ungünstiger Lage der oberen Potentialkurve nicht beobachtet wird, auch durch Extrapolation der Banden auf deren Konvergenzstelle ermitteln kann. Beobachtet man im Spektrum auch die dem Übergang zum Schwingungsgrundzustand $(v' = 0)$ des angeregten Moleküls entsprechende $0 \leftrightarrow 0$-Bande mit der Energie E_M, so ist ersichtlich

$$D' = K - E_M \tag{29.1}$$

die Dissoziationsenergie des *angeregten* Moleküls. Die besonders interessierende Dissoziationsenergie D'' des *normalen* Moleküls für Dissoziation in normale Atome $A + B$ ist nach Fig. 72

$$D'' = K - E_A, \tag{29.2}$$

wo E_A die Anregungsenergie des bei der Photodissoziation entstehenden angeregten Atoms B ist, deren Kenntnis für eine sichere Bestimmung der Dissoziationsenergie also Voraussetzung ist. Die Aufklärung dieser Zusammenhänge zwischen Moleküldissoziation und kontinuierlicher Absorption gelang FRANCK [*161*] 1926.

Mit der Dissoziation verwandt ist die bei manchen zweiatomigen, besonders aber bei zahlreichen mehratomigen Molekülen zu beobachtende, von HENRI [*205*] entdeckte Erscheinung der *Prädissoziation*. Sie äußert sich in einem Diffuswerden der normalerweise scharfen Absorptionsbanden *vor* Erreichen der Konvergenzstelle, und photochemisch in einer Dissoziation durch diese Strahlung, die langwelliger ist als die der Grenze des Dissoziationskontinuums. Ursache der Prädissoziation ist eine solche Überlagerung der Potentialkurven zweier angeregter Elektronenzustände eines Moleküls (vgl. Fig. 73), daß strahlungslose

Übergänge aus diskreten Schwingungszuständen der einen Elektronenanordnung (a') zu dissoziierten, dem Kontinuum angehörenden Zuständen der anderen Elektronenanordnung (Potentialkurve b') möglich sind. Bei Übergängen von dem in Fig. 73 nicht gezeichneten Molekülgrundzustand nach oben beobachtet man dann einen Absorptionsbandenzug, der bei Abwesenheit des durch die Potentialkurve b' beschriebenen Zustandes erst oberhalb der Konvergenzstelle zu Photodissoziation Anlaß geben würde. Bei Überlagerung der beiden Kurven aber besteht, wenn gewisse Auswahlregeln der Elektronenquantenzahlen beider Zustände erfüllt sind, schon oberhalb der Dissoziationsgrenze von b' die Möglichkeit der Dissoziation des Moleküls durch strahlungslose Übergänge von der linken

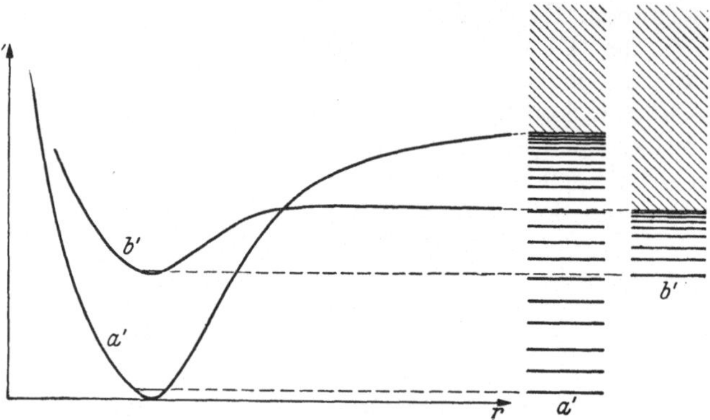

Fig. 73. Potentialkurven und Schwingungsniveaus zweier angeregter Molekülzustände, deren Überlagerung zur Prädissoziation führt.

zur rechten der beiden gezeichneten Energiezustandsfolgen. Da diese strahlungslosen Übergänge die Lebensdauer der oberhalb der Dissoziationsenergie von b' liegenden stationären Schwingungszustände von a' stark verkleinern, verbreitern nach der Ungenauigkeitsbeziehung diese Zustände und verursachen damit das Diffuswerden der durch Übergänge zu ihnen entstehenden Absorptionsbanden. Für alle Einzelheiten sei auf [149] und die dort aufgeführte weitere Literatur verwiesen. Für den Photochemiker folgt aus diesen Überlegungen das wichtige Ergebnis: *Lichteinstrahlung solcher Wellenlängen, denen scharfe Absorptionsbanden der bestrahlten Moleküle entsprechen, führt zur Bildung angeregter (und damit unter Umständen schon reaktionsfähiger) Moleküle; Lichteinstrahlung solcher Wellenlängen, denen kontinuierliche Absorptionsspektren oder diffuse Absorptionsbanden (die nicht nur infolge zu geringer Dispersion des Spektralapparats diffus erscheinen!) entsprechen, ergeben Moleküldissoziation.*

30. Die typischen Fälle von Molekül-Absorptionskontinua. Die Übersicht über die Vielfalt der tatsächlich beobachteten Molekülkontinua und ihrer Eigenschaften wird durch die Besprechung typischer Fälle [149] erleichtert. Die Potentialkurven der verschiedenen Typen von *Absorptions*kontinua sind in Fig. 74 zusammengestellt.

Besitzt das absorbierende Molekül einen *stabilen* Grundzustand (Potentialkurve U'' in Fig. 74 I, II, III), so erhält man die Fälle I, II bzw. III der Absorptionskontinua, je nachdem, welchen Verlauf die Potentialkurve des angeregten Zustandes zeigt. Bei reiner Abstoßungskurve des oberen Zustandes findet man das rein kontinuierliche Absorptionsspektrum des Falles I, wobei die Ausdehnung des vom Schwingungsgrundzustand von U'' aus absorbierten Kontinuums von der Neigung von U' abhängt und mittels der Reflexionsmethode (Ziff. 28) ermittelt werden kann. Mit zunehmender Temperatur des absorbierenden

Gases verbreitert sich das Kontinuum nach beiden Seiten, weil nach Fig. 74 I
Übergänge auch von höheren Schwingungsniveaus von U'' aus erfolgen. Die

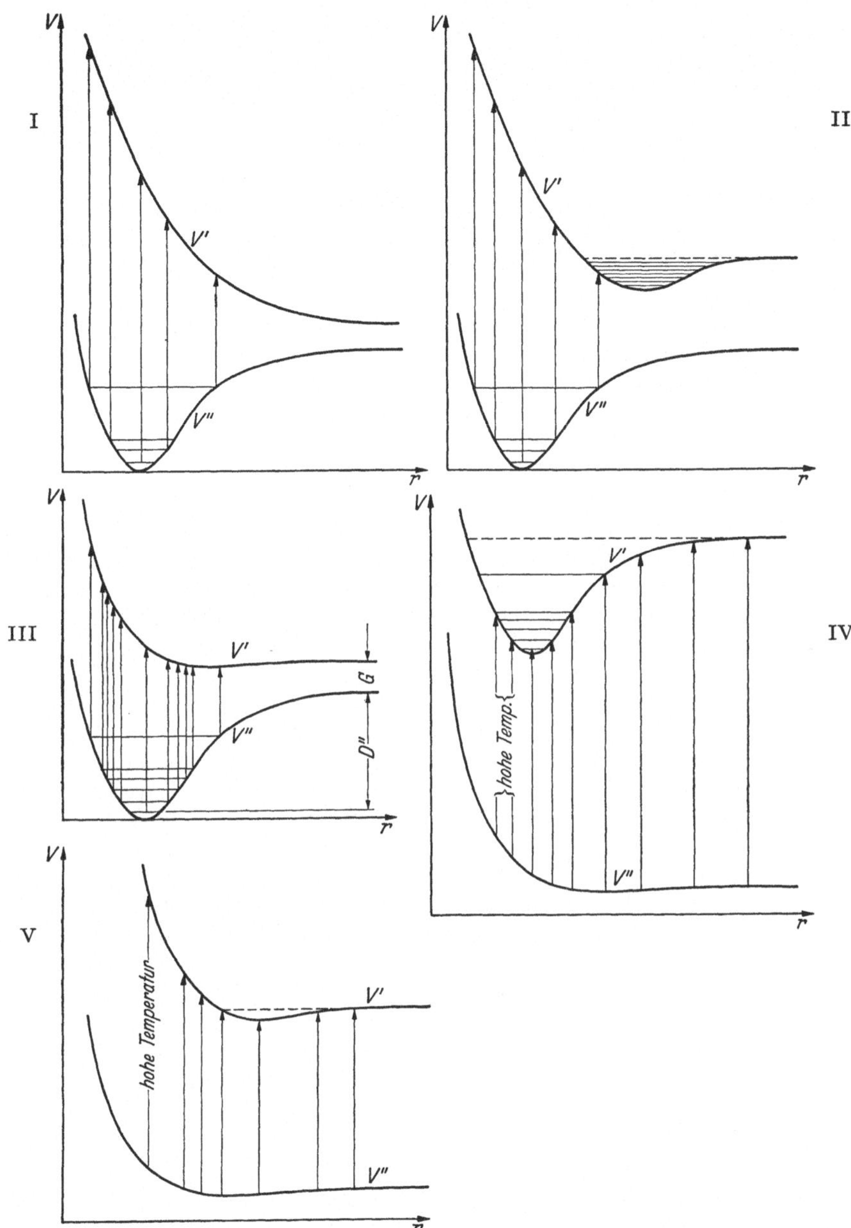

Fig. 74. Potentialkurvenschemata zur Darstellung der verschiedenen Typen von Absorptionskontinua.

Dissoziationsenergie D des Grundzustandes kann aus dem Kontinuum Fall I
ersichtlich *nicht* bestimmt werden. Die bekanntesten Beispiele für Fall I sind
die im langwelligen Ultraviolett gelegenen Absorptionskontinua der Halogenwasserstoffe (Ziff. 34); auch zahlreiche Absorptionskontinua im äußersten Ultraviolett dürften diesem Typ angehören, da es theoretisch unter den hoch

angeregten Molekülzuständen zahlreiche mit Abstoßungspotentialkurven geben sollte.

Beim Fall II besitzt die Potentialkurve des angeregten Zustandes U' zwar ein Minimum, doch liegt dieses infolge Bindungslockerung bei der Molekülanregung bei größerem Kernabstand. Bei tiefer Temperatur (Absorption vom Schwingungsgrundzustand von U'' aus) ergibt sich dann wieder ein reines Absorptionskontinuum. Bei hoher Temperatur dagegen erscheint an der langwelligen Seite des Kontinuums Bandenstruktur infolge von Übergängen zu den diskreten Schwingungsniveaus des oberen Zustandes. Der Fall II eignet sich zur Bestimmung der Dissoziationsenergie des normalen Moleküls, wenn erstens die Dissoziationsprodukte des angeregten Zustandes und zweitens die Anregungsenergie des dabei entstehenden angeregten Atoms bekannt sind (Ziff. 29), und wenn man drittens durch die Bandenanalyse sicherstellen kann, daß die langwellige Grenze des Kontinuums wirklich dem Übergang vom *untersten* Schwingungsniveau des Grundzustandes zur Dissoziationsgrenze des angeregten Zustandes entspricht. Gut untersuchte Beispiele für Fall II sind die Absorptionskontinua der Halogenmoleküle (Ziff. 35) sowie das im fernen Ultraviolett unterhalb 1750 Å gelegene Absorptionskontinuum des O_2-Moleküls (Ziff. 36), an das sich nach langen Wellen die sog. Schumann-Runge-Füchtbauer-Banden anschließen. Auch die im äußersten Ultraviolett jenseits 1000 Å liegenden Absorptionskontinua des H_2 und N_2 gehören diesem Typ an.

Bei Fall III der Absorptionskontinua ist nach Fig. 74 III die Potentialkurve des angeregten Zustandes eine Abstoßungskurve oder eine van der Waals-Kurve mit flachem Minimum. Der Unterschied gegenüber Fall I und II, der den Kontinua ein ganz anderes Aussehen gibt, liegt darin, daß die obere Kurve in dem über dem Minimum der unteren liegenden Teil nicht mehr durch eine Gerade großer Neigung angenähert werden kann, sondern entweder schon ganz flach verläuft oder in den flachen Verlauf umbiegt. Bei Übergängen vom Schwingungsgrundzustand aus erhält man ein Kontinuum, das wesentlich schmaler ist als in Fall I und II, und dessen Breite in der üblichen Weise nach Ziff. 28 zu ermitteln ist. Bei Übergängen auch von den höheren Schwingungsniveaus aus erhält man infolge des annähernden Zusammenfallens der kernnahen Übergänge ein homogenes kurzwelliges Kontinuum, während die kernfernen Übergänge wegen des flachen Verlaufes von U' einzelne schmale Kontinua ergeben, sog. *Fluktuationen*. Auch ihre Breite ergibt sich direkt aus der Breite der Maxima der Eigenfunktionen ψ_k des Grundzustandes. Ihre Wellenlängenabstände sind annähernd durch die Abstände der Schwingungsterme v'' bestimmt, die nach der Dissoziationsgrenze konvergieren. Die Absorptionskontinua nach Fall III bestehen also aus kontinuierlichen Fluktuationen, die nach langen Wellen zu konvergieren, während sie nach kurzen Wellen hin bei genügend hoher Temperatur in ein homogenes Kontinuum auslaufen.

Bei völlig horizontalem Verlauf der oberen Kurve konvergiert die Folge der Fluktuationen nach Fig. 74 III gegen eine Grenze G, die der Anregungsenergie des bei der Dissoziation des oberen Zustandes entstehenden angeregten Atoms entspricht. Die Energiedifferenz dieser Konvergenzgrenze und des von $v'' = 0$ herrührenden Intensitätsmaximums ist nach Fig. 74 III gleich der Dissoziationsenergie D'' des normalen Moleküls, falls die obere Kurve U' auch über dem Minimum der unteren noch horizontal verläuft. Steigt die obere Kurve in diesem Gebiet bereits, so ergibt sich nach dieser Methode die Dissoziationsenergie um den Betrag zu hoch, um den U' bereits von der Horizontalen abweicht (Pseudokonvergenz nach Kuhn [*266*]). Bei der Bestimmung von Dissoziationsenergien aus Fluktuationskontinua Fall III ist dies zu beachten.

Besitzt die Potentialkurve U' infolge bindender VAN DER WAALS-Kräfte ein flaches Minimum mit eng liegenden Schwingungsniveaus, so treten die gleichen Fluktuationen auf, wie sie eben abgeleitet wurden; doch sind sie nun nicht echt kontinuierlich, sondern bestehen aus den Übergängen zu den dicht liegenden Schwingungsniveaus des oberen Zustandes, stellen also ein sehr enges Bandensystem dar. Praktisch zeigt sich dieser Fall durch eine den Fluktuationen überlagerte ganz feine Struktur an (unechte Kontinua).

Beispiele für Fall III der Absorptionskontinua bieten die Absorptionsspektren der Alkalihalogenide (Ziff. 38) sowie der Thalliumhalogenide (Ziff. 37).

Der horizontale Verlauf der oberen Kurve ist bei ersteren in guter Annäherung erfüllt; die Abstände der kontinuierlichen Fluktuationen stimmen erstaunlich genau mit den theoretisch berechneten Abständen der Schwingungsniveaus des Grundzustandes überein.

Im Gegensatz zu den bisher besprochenen Fällen gehören die Fälle IV und V der Absorptionskontinua nicht zu stabilen Molekülen, sondern die Absorption erfolgt durch zwei Atome, die entweder im Stoß absorbieren oder während der Absorption infolge VAN DER WAALSscher Wechselwirkungen ein schwach gebundenes Molekül bilden. Ihr Grundzustand wird also durch eine flache Potentialkurve charakterisiert. Je nachdem nun der angeregte Zustand U', zu dem die Absorption erfolgt, ein ausgeprägtes Minimum besitzt (Fig. 74 IV) oder auch ein VAN DER WAALS-Zustand ist (Fig. 74 V), erhalten wir die Spektren der Fälle IV oder V.

Fall IV stellt gewissermaßen eine Umkehrung von Fall III dar. Auch hier fallen die den kernnahen Übergängen entsprechenden Einzelkontinua zusammen, während im kernfernen Gebiet wegen der entgegengesetzten Neigung der Kurven das Kontinuum sich in Fluktuationen auflösen kann, wenn die Schwingungsniveaus von U' genügend weit getrennt sind. Aus Fig. 74 IV folgt aber, daß die Fluktuationen jetzt auf der *kurzwelligen* Seite auftreten und nach kurzen Wellen konvergieren, während sich nach langen Wellen hin ein homogenes Kontinuum anschließt.

Bei dem Versuch der Ermittlung der Intensitätsverteilung eines Kontinuums Fall IV ist nicht nur die Besetzung der im wesentlichen kontinuierlichen Zustände der unteren Kurve mit Molekülen zu berücksichtigen, sondern auch die Tatsache, daß die Eigenfunktionen im flachen und im steilen Bereich der Potentialkurve einen ganz verschiedenen Verlauf zeigen. Nur eine verfeinerte Anwendung des FRANCK-CONDON-Prinzips führt also hier zum Ziel. Weiter ist zu beachten, daß stets mehr Atompaare in großem Abstand voneinander vorhanden sind als solche in kleinem Abstand. Auch Absorption wird daher, falls nicht Auswahlregeln dies verhindern, bevorzugt bei großem Kernabstand stattfinden, was zur Folge hat, daß die kernfernen, kurzwelligsten Übergänge zunächst am intensivsten auftreten und bei Steigerung von Druck und Temperatur das Spektrum sich von der kurzwelligen Seite her nach langen Wellen zu entwickelt. Der langwellige, homogene Teil des Kontinuums kommt dabei oft gar nicht zur Ausbildung, weil der kernnahe Teil von U'' nicht genügend besetzt wird. Beispiele für Absorptionskontinua Fall IV finden sich in den Spektren der VAN DER WAALS-Moleküle Hg_2 und Cd_2 (Ziff. 41), die im Grundzustand Dissoziationsenergien von nur einigen Hundertstel eV haben, dagegen angeregte Elektronenzustände mit ausgeprägten Minima besitzen. Das gleiche gilt für das He_2-Molekül, das bei genügend hohem Gasdruck im äußersten Ultraviolett ebenfalls Absorptionskontinua dieses Typs besitzen sollte.

Sind die Potentialkurven *beider* kombinierender Molekülzustände flache Kurven vom VAN DER WAALSschen Typ, so erhalten wir den Fall V der

Molekülkontinua in Absorption, relativ schmale kontinuierliche Bänder, die den Linienverbreiterungen verwandt sind. Die geringe Ausdehnung rührt von dem annähernd parallelen Verlauf der beiden Potentialkurven her. Je nachdem, ob die Potentialkurve des angeregten Zustands ein tieferes oder ein flacheres Minimum besitzt als der Grundzustand (bzw. überhaupt kein Minimum), liegt das bei dem Übergang entstehende Spektrum beiderseits der Atomlinie oder nur auf der kurzwelligen Seite. Dabei sind nach Fig. 74 V die auf der kurzwelligen Seite auftretenden Spektren stets echte Kontinua, die auf der langwelligen Seite der Atomlinie liegenden dagegen als Übergänge zwischen den dicht liegenden Schwingungsniveaus der flachen Potentialmulden in Wirklichkeit enge Bandensysteme. Beispiele für kontinuierliche Absorptionsbänder Fall V finden sich namentlich wieder in den Spektren der van der Waals-Moleküle Hg_2, Cd_2, Zn_2, Ziff. 41.

31. Die typischen Fälle von Molekül-Emissionskontinua. Zu den grundsätzlich möglichen Typen von Molekülkontinua in *Emission* gelangt man durch Vertauschung der beiden Potentialkurven in Fig. 74 und Betrachtung der Übergänge von oben nach unten (vgl. Fig. 75). Der Allgemeinheit wegen sei bemerkt, daß die unteren Zustände *nicht* die Grundzustände der Moleküle zu sein brauchen, sondern ihrerseits angeregt sein können.

Wird der untere Zustand (Fall I) durch eine reine Abstoßungskurve beschrieben, so entspricht dem Übergang ein gleichförmiges Emissionskontinuum, dessen Ausdehnung in bekannter Weise mit der Neigung der Abstoßungskurve verknüpft ist und das am kurzwelligen Ende Intensitätsfluktuationen zeigen sollte, falls sehr hohe Schwingungszustände von U' an der Emission beteiligt sind. Wieweit letzteres der Fall ist, läßt sich bei *Emission*kontinua im allgemeinen allerdings schwer feststellen, da die Besetzung der Schwingungszustände des angeregten Moleküls im allgemeinen sehr empfindlich von den speziellen Anregungsbedingungen abhängt. Das klassische Beispiel für ein Emissionskontinuum Fall I ist das große H_2-Kontinuum, das sich von etwa 5000 Å im Sichtbaren bis ins ferne Ultraviolett erstreckt und durch Übergänge aus dem stabilen angeregten $1\,s\sigma\,2s\sigma\,{}^3\Sigma_g$-Zustand des H_2 in den instabilen, in normale Atome dissoziierenden $1\,s\sigma\,2p\sigma\,{}^3\Sigma_u^+$-Zustand entsteht (vgl. Ziff. 40).

Emissionskontinua nach Fig. 75 II mit diskreter Bandenstruktur am kurzwelligen Ende scheinen noch nicht beobachtet worden zu sein. Theoretisch sind sie zwar möglich, aber darum unwahrscheinlich, weil eine Molekülanregung beim zweiatomigen Molekül fast stets eine Bindungslockerung ergibt und das Potentialminimum der oberen Kurve dann nicht über dem kontinuierlichen Bereich der unteren liegen kann.

Der in Fig. 75 III dargestellte Potentialkurvenverlauf dagegen kommt bei van der Waals-Molekülen nicht selten vor. Er gibt Anlaß zu einem homogenen Emissionskontinuum, dessen kurzwelliges Ende sich in Fluktuationen auflösen sollte, falls Übergänge von genügend hohen Schwingungszuständen des angeregten Moleküls aus vorkommen. Beispiele solcher Kontinua finden sich in den Spektren des Hg_2, Cd_2 und Zn_2 (vgl. Ziff. 41).

Bei einem Potentialkurvenverlauf der beiden kombinierenden Molekülzustände nach Fig. 75 IV erwarten wir kontinuierliche, nach langen Wellen konvergierende Fluktuationen, die bei genügend dicht liegenden Schwingungsniveaus von U'' zu einem homogenen Kontinuum zusammenfließen können. Besitzt die obere Potentialkurve U' ein flaches Minimum mit diskreten Schwingungszuständen, so sind die bei Emissionsübergängen entstehenden Fluktuationen nur scheinbar kontinuierlich, in Wirklichkeit aber enge diskrete Bandenspektren. Wir kommen auf Fall IV der Emissionskontinua wegen deren spezieller Bedeutung in Ziff. 32 noch einmal zurück.

Bei Übergängen zwischen zwei VAN DER WAALS-Potentialkurven gemäß Fig. 75 V entstehen, wie bei dem entsprechenden Absorptionsfall V, echte oder

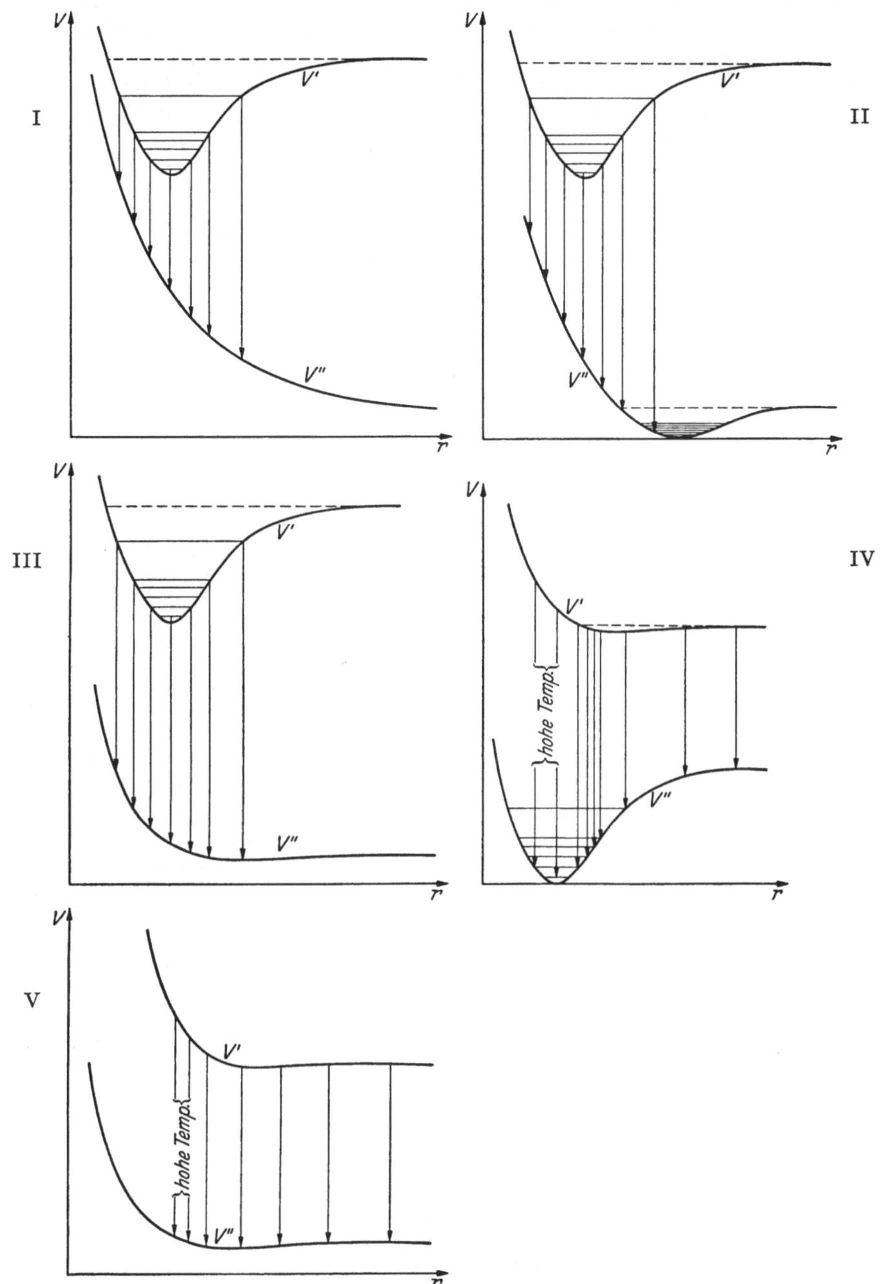

Fig. 75. Potentialkurvenschemata zur Darstellung der verschiedenen Typen von Molekül-Emissionskontinua.

scheinbar kontinuierliche Bänder, wie sie als Begleiter von Atomlinien bei genügend dichten Plasmen (große Stoßwahrscheinlichkeit) häufig beobachtet werden (z. B. Ziff. 44 und 45).

32. Molekülrekombinationsspektren. Kehren wir noch einmal zu Fig. 75 IV mit einer oberen Potentialkurve ohne ausgeprägtem Minimum zurück, so erkennen wir, daß es sich hier um ein Molekülrekombinationskontinuum handelt. Ein normales und ein angeregtes Atom stoßen zusammen und bilden unter Emission der Bindungs- und Anregungsenergie ein normales, fest gebundenes Molekül. Da eine genügende Dichte angeregter Atome und ein spezieller Potentialkurvenverlauf Voraussetzungen für diesen Prozeß sind, werden solche Rekombinationskontinua relativ selten beobachtet. Ein in einer Tellurentladung von Rompe [380], [381] gefundenes Emissionskontinuum dürfte wohl das bisher sicherste Beispiel für diesen Fall IV sein.

Es sei in diesem Zusammenhang erwähnt, daß es eine direkte Molekülrekombination zweier zusammenstoßender *normaler* Atome ebensowenig gibt, wie den Umkehrfall der Photodissoziation ohne Molekülanregung, weil beide Prozesse, wie man sich an Hand der Potentialkurvendarstellung leicht überzeugt, dem Franck-Condon-Prinzip widersprechen würden.

Dagegen ist eine Strahlungsrekombination zweier normaler Atome möglich, wenn der Grundzustand des aus ihnen entstehenden Moleküls in guter Näherung als aus Ionen aufgebaut angesehen werden kann. Die Verhältnisse werden dann etwa durch die Potentialkurven Fig. 76 beschrieben und ergeben eine wichtige

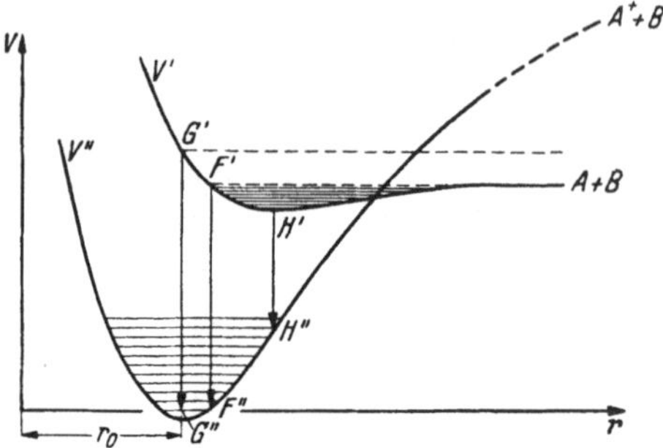

Fig. 76. Potentialkurvenschema zur Strahlungsrekombination eines Ionenmoleküls.

Abart des Falles IV der Emissionskontinua: Zwei normale Atome können im Stoß den zwischen A' und B' gelegenen Teil der Kurve U' erreichen und unter kontinuierlicher Emission einen Übergang zur Ionenmolekül-Potentialkurve U'' (Elektronensprung vom elektronegativen zum elektropositiven Atom) ausführen. Erleidet aber das Stoßpaar, bevor es zur Ausstrahlung kommt, einen Zusammenstoß mit einem dritten Partner, so kann dieser einen Teil der Energie des Stoßpaares übernehmen und dieses in einem der diskreten Schwingungszustände der Potentialmulde von U' zurücklassen. Dieses im Dreierstoß gebildete angeregte Molekül geht dann in den Grundzustand über, und zwar unter Ausstrahlung des den Übergängen von den diskreten Niveaus von U' zu denen von U'' entsprechenden Bandenspektrums, das sich dann an die langwellige Grenze des Rekombinationskontinuums anschließen muß. Es scheint ziemlich sicher, daß bei den Alkalihalogeniden der Potentialkurvenverlauf recht gut durch Fig. 76 dargestellt wird und nach Ziff. 38 die bei der Rekombination von Alkali- und Halogenatomen emittierten Kontinua und Banden als Rekombinationsspektren dieser Abart von Fall IV zu deuten sind. Für Einzelheiten sei auf die eingehende Diskussion in [149] verwiesen.

33. Kontinuierliche Spektren mehratomiger Moleküle. Bei den kontinuierlichen Spektren der mehr- und vielatomigen Moleküle handelt es sich grundsätzlich wie bei denen der zweiatomigen um das Ergebnis von Übergängen zwischen stabilen Schwingungszuständen einerseits und freien, in Atome oder Atomgruppen dissoziierten Molekülzuständen mit kontinuierlichen Energiebereichen andererseits. Die wesentlichsten Ergebnisse der Systematik zweiatomiger Moleküle können also zur Deutung der Kontinua mehratomiger Moleküle mit gebührender Vorsicht herangezogen werden. Die entscheidende Schwierigkeit aber liegt in der großen Zahl der miteinander gekoppelten Schwingungs- und Zerfallsmöglichkeiten vielatomiger Moleküle und in der komplizierten Wirkung der zahlreichen Bindungselektronen auf diese. Die Möglichkeit der Anregung zahlreicher verschiedener Bindungselektronen mit ihrem verschiedenen Einfluß auf die Bindungsverhältnisse bedingt äußerst komplizierte Energieniveauschemata, in denen häufig diskrete Schwingungszustände und kontinuierliche, zu dissoziierten Zuständen gehörende Energiebereiche sich überlagern und damit nach Ziff. 29 zur Prädissoziation Anlaß geben können.

Als Leitprinzip für die Deutung der kontinuierlichen Spektren mehratomiger Moleküle in Absorption wie Emission dient wiederum das FRANCK-CONDON-Prinzip (vgl. Ziff. 28), dessen Anwendbarkeit besonders von HERZBERG und TELLER [210] untersucht wurde. FRANCK, SPONER und TELLER [162] behandelten theoretisch die Prädissoziation mehratomiger Moleküle. Zur Sicherung der Zuordnung der Absorptionskontinua zu bestimmten Dissoziationsvorgängen müssen häufig Ergebnisse der Photochemie sowie Fluoreszenzbeobachtungen der Dissoziationsprodukte und natürlich, wie bei den zweiatomigen Molekülen, die Temperaturabhängigkeit der Kontinua herangezogen werden (vgl. [81], [253], [332]), [349], [434], [463], [467], [472]. Wie ungeklärt die Lage noch ist, zeigen theoretische Meinungsverschiedenheiten (vgl. [50] und [351]). Zusammenfassende Berichte über die Kontinua mehratomiger Moleküle liegen vor von SPONER und TELLER [412], ROLLEFSON und BURTON [376] sowie GROTH [193]. Für die sehr ausgedehnte Spezialliteratur muß auf diese Berichte verwiesen werden.

II. Die kontinuierlichen Spektren zweiatomiger Moleküle.

Nach der Behandlung der Systematik der Molekülkontinua bringen wir im folgenden eine knappe Besprechung des gegenwärtigen Standes der Kenntnis von den wichtigsten Kontinua zweiatomiger Moleküle. Die Anordnung entspricht einigermaßen der der typischen Fälle (Ziff. 30 und 31) und schreitet von den einfacheren zu den komplizierteren Spektren vor. Die Darstellung soll zeigen, daß zwar ein gewisser Grad des allgemeinen Verständnisses der Molekülkontinua erreicht ist, daß aber bei kaum einem die Sicherheit und Klarheit der Deutung erzielt ist, die unsere Kenntnis der diskontinuierlichen Molekülspektren auszeichnet.

34. Die Spektren der Halogenwasserstoffe. Die Halogenwasserstoffe HF, HCl, HBr und HJ sind wohl die einzigen zweiatomigen Moleküle, von denen in Absorption wie in Emission lediglich kontinuierliche Spektren bekannt sind. Die ältere Literatur ist bei FINKELNBURG [149] eingehend diskutiert; aus neuerer Zeit gibt es nur einige Arbeiten von ROMAND und VODAR [377] bis [379] sowie DATTA und CHAKRAVARTY [111] mit Messungen über die Absorptionsspektren der schwereren Halogenwasserstoffe sowie eine entsprechende Arbeit von SAFARY, ROMAND und VODAR [394] über das HF-Spektrum.

Lediglich die Absorptionsspektren des HJ und HBr sind so genau bekannt, daß ein Vergleich mit der Theorie möglich ist. So konnten GOODEVE und TAYLOR [186] durch Messung der Temperaturabhängigkeit der Absorption sicherstellen, daß die von etwa 4000 bzw. 3200 Å nach kurzen Wellen sich erstreckende kontinuierliche Absorption jedenfalls in ihrem langwelligen Bereich durch Übergänge vom untersten Schwingungszustand der normalen Atome zu einer reinen Abstoßungskurve des Molekülzustandes entsteht, der in normale, d.h. nicht angeregte Atome dissoziiert. Es handelt sich also um Absorptionskontinua des Falles I. Nach der theoretischen Diskussion durch MULLIKEN [337] sollten allerdings gemäß Fig. 77 vom Molekülgrundzustand auch Übergänge zu höheren

Abstoßungskurven möglich sein, deren Zustände in ein normales und ein angeregtes Atom dissoziieren. In ihrem kurzwelligeren Bereich dürften die Absorptionskontinua der Halogenwasserstoffe also wahrscheinlich Überlagerungen mehrerer Kontinua des Falles I darstellen.

Emissionskontinua, die den Halogenwasserstoffen zugeschrieben werden, sind beim HCl, HBr und HJ verschiedentlich gefunden und als Beispiele für die Emissionstypen III und IV gedeutet worden. Weder diese Deutung noch die Zuordnung zu den Halogenwasserstoffen ist aber wirklich gesichert. Für Einzelheiten vgl. [149]; neuere Veröffentlichungen liegen nicht vor.

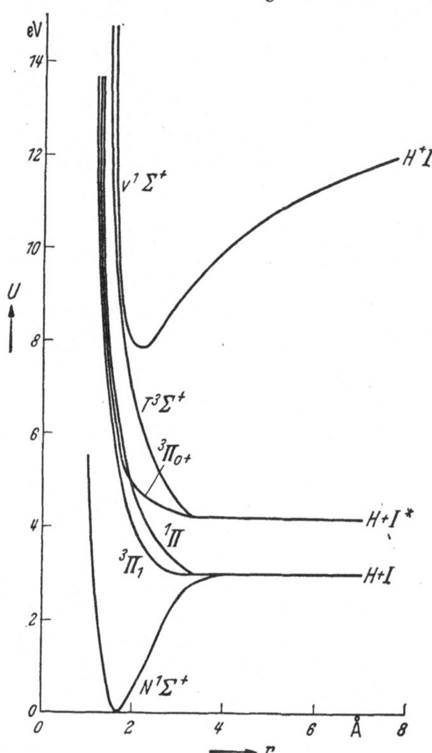

Fig. 77. Potentialkurvenschema des Jodwasserstoff-Moleküls nach Mulliken [337].

35. Die Spektren der Halogenmoleküle. Von den Halogen-Doppelmolekülen F_2, Cl_2, Br_2 und J_2 sowie den Mischmolekülen BrCl, JCl und JBr sind neben hier weniger interessierenden Banden zahlreiche kontinuierliche Spektren in Absorption wie in Emission bekannt. Die ungefähren Wellenlängen und Maxima der Absorptionskontinua sind aus Tabelle 7 zu ersehen. Wie allgemein sind die kurzwelligen Absorptionskontinua von geringerem theoretischen Interesse, weil es unter den hoch angeregten Elektronenzuständen der Moleküle fast stets solche mit Abstoßungspotentialkurven oder Potentialkurven mit Minima bei sehr großem Kernabstand gibt, die die Möglichkeit der Absorption von Kontinua der Fälle I und II bieten. Die großen langwelligen Absorptionskontinua der Halogene aber sind von besonderem Interesse, weil an ihnen zuerst der Zusammenhang zwischen kontinuierlicher Absorption und Moleküldissoziation entdeckt und aufgeklärt wurde (Trautz [435], Mecke [308], Dymond [127], Franck [161] und Kuhn [265]), und weil am großen langwelligen Cl_2-Kontinuum von Gibson, Rice und Bayliss [177] der erste und bisher fast einzige quantitative Vergleich zwischen Theorie und Beobachtung durchgeführt wurde.

Tabelle 7. *Die Absorptionskontinua der Halogenmoleküle*

| | Kontinuum I | | | | Kontinuum II | | Kontinuum III |
| | Konverg.-Stelle | Grenzen | Maxima | | Grenzen | Maximum | |
			A	B			
J_2 · · ·	4995	4995—4500	5200	7320	—	—	< 1515
Br_2 · ·	5107	5107—3400	4150	4950	—	—	< 2700
Cl_2 · · ·	4785	4200—3000	3300	4250	—	—	< 1900
F_2 · · ·	?	3700—2000	2900	?	—	—	?
JBr · ·	?	5700—3800	4950	?	?	4050	< 1950
JCl · ·	5744	5744—3500	4700	?	2700—2200	2400	< 2200
BrCl · ·	?	?	3750	?	2400— ?	2150	< 1685

Bei diesen langwelligen Absorptionskontinua der Halogene handelt es sich um charakteristische Vertreter des Absorptionsfalles II, da die Potentialkurve des ersten angeregten Molekülzustandes zwar ein Minimum besitzt, dieses aber gegenüber dem des Grundzustandes zu größeren Kernabständen hin verschoben ist. Diese Vergrößerung des Gleichgewichtskernabstandes durch die Molekülanregung ist relativ gering bei dem schwersten Molekül J_2, wo das durch Übergang vom Molekülgrundzustand entstehende Absorptionsmaximum noch im Bandengebiet liegt. Bei den leichten Halogenen Cl_2 und F_2 umgekehrt ist die Bindungslockerung bei der Molekülanregung so groß, daß über dem Minimum der Grundzustandskurve gemäß Fig. 74 II der steile über der Dissoziationsenergie liegende Abstoßungsast der oberen Potentialkurve liegt und die Absorptionskontinua daher rein kontinuierlich sind.

GIBSON, RICE und BAYLISS haben nun zuerst beim Cl_2 die Intensitätsverteilung im großen langwelligen Absorptionskontinuum in Abhängigkeit von der Temperatur bis über 1000° K gemessen und daraus die Beiträge der Übergänge vom Schwingungsgrundzustand und von den ersten angeregten Schwingungszuständen zum gesamten Absorptionskontinuum isoliert. Die Ergebnisse stehen in bester Übereinstimmung mit der theoretischen Erwartung nach dem FRANCK-CONDON-Prinzip, in dessen einfachster Form, nach der der Elektronenanteil der Übergangswahrscheinlichkeit vom Kernabstand unabhängig angenommen wird. Ähnliche Untersuchungen sind am Br_2 [4] und dem kurzwelligen Kontinuum des JCl [65] ausgeführt worden. Mit der Elektronenkonfiguration der verschiedenen an den Kontinua beteiligten angeregten Zustände und den entsprechenden Photodissoziationsprodukten befaßt sich eine Reihe von Arbeiten von MULLIKEN [335], [336], [340] bis [342], [344] u. a. Die Frage ist theoretisch besonders interessant und schwierig, weil die Kopplung zwischen Elektronenbewegung und Molekülrotation bei den leichten und den schweren Halogenmolekülen sehr verschieden ist und zu erheblichen Unterschieden in der relativen Intensität verschiedener Elektronenübergänge führt. Bezüglich der Einzelheiten wie der gesamten älteren Literatur muß auf [149] verwiesen werden. In neuerer Zeit ist im Zusammenhang mit dem zunehmenden Interesse an der Physik der Flüssigkeiten und Festkörper auch der Frage der Veränderung der Intensitätsverteilung der Halogenabsorptionskontinua durch Zusatz von Fremdgas sowie durch Einbettung der Moleküle in verschiedenartige Flüssigkeiten besondere Aufmerksamkeit geschenkt worden. Auf die diesbezüglichen Arbeiten [5], [42], [43], [84], [371] und vier neue Arbeiten von BAYLISS [44], [45], [46], [372] kann hier nur hingewiesen werden.

Unsere Kenntnis der Emissionskontinua der Halogene ist trotz einer sehr umfangreichen Literatur über diese [82], [134], [450], [473] noch sehr unvollständig und widerspruchsvoll. Einigermaßen sicher scheint nach den Untersuchungen von KONDRATJEW und LEIPUNSKY [251], [252] sowie UCHIDA [441], daß die von den hoch erhitzten Halogendämpfen emittierten Kontinua und Banden als echte Molekülrekombinationsspektren (Emissionsfall IV) anzusehen sind, die im Stoß zwischen normalen und im untersten Zustand angeregten metastabilen Halogenatomen emittiert werden. Daß nur 10^{-9} der Stöße zur Molekülbildung unter Emission der Bindungsenergie führen, entspricht der theoretischen Erwartung.

In stromschwachen Entladungen aller Art, bei Fluoreszenzanregung und Anregung durch aktiven Stickstoff emittieren die Halogene ferner kontinuierliche oder oft kontinuierlich erscheinende Spektren, die sich beim J_2, Br_2 und Cl_2 von einer ziemlich scharfen Kante bei 4800 bzw. 4200 bzw. 3200 Å aus nach kurzen Wellen erstrecken, und denen sich im Ultraviolett noch je ein schmales, meist kontinuierlich erscheinendes Band von etwa 100 Å Breite mit einer langwelligen Kante bei 3460 bzw. 2930 bzw. 2610 Å überlagert. Während beim J_2 diese „Kontinua" anscheinend weitgehend (besonders deutlich bei Stickstoffzusatz) in Bandengruppen aufgelöst werden konnten, gilt das bei Cl_2 wohl nur für den langwelligeren Teil, während unterhalb 2600 Å echte Kontinua vorliegen sollen. Für Einzelheiten und ältere Literatur sei wieder auf [149] verwiesen.

36. Die Photodissoziationskontinua des O_2-Moleküls. Das Sauerstoffmolekül O_2 besitzt im Ultraviolett ein äußerst intensives Absorptionskontinuum zwischen 1750 und 1300 Å mit ausgeprägtem Maximum bei 1450 Å, an das sich nach langen Wellen zu die sog. Schumann-Runge-Füchtbauer-Banden anschließen. Ferner gibt es zwischen 2800 und 2400 Å noch zwei Bandensysteme mit nach kurzen Wellen hin anschließender kontinuierlicher Absorption, die als verbotene Elektronenübergänge im O_2 nur bei sehr großer Schichtdicke oder hohem Druck beobachtbar sind. Alle diese Kontinua sind schöne Beispiele für den Absorptionsfall II (Fig. 74 II). Der obere Zustand des Hauptkontinuums besitzt einen Gleichgewichtskernabstand (Potentialminimum) bei 1,60 Å gegenüber 1,20 Å für den O_2-Grundzustand; er dissoziiert in ein normales und ein ^1D-Atom. Die oberen Zustände

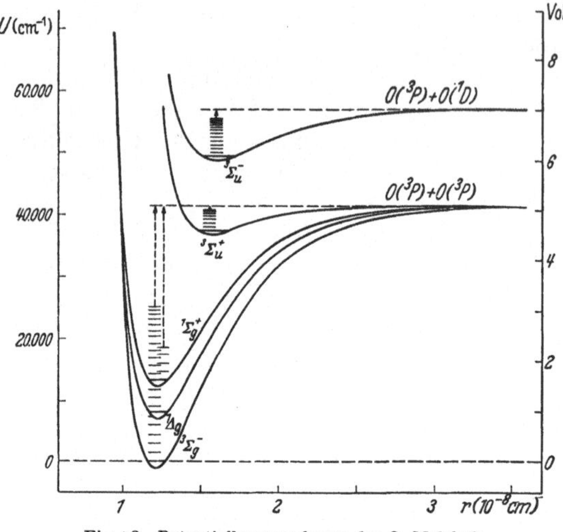

Fig. 78. Potentialkurvenschema des O_2-Moleküls nach Herzberg [211].

der beiden verbotenen langwelligeren Kontinua dagegen dissoziieren wie das normale O_2-Molekül in nichtangeregte Atome Fig. 78. Für das Hauptkontinuum hat Stueckelberg [432] unter Benutzung der Absorptionsmessungen von Ladenburg, Boyce und van Voorhis [271], [272] gezeigt, daß die spektrale Intensitätsverteilung im Kontinuum dem Franck-Condon-Prinzip entspricht. Wegen des metereologischen Interesses am Sauerstoff sind die Absorptionsmessungen im Kontinuum neuerdings mit verbesserten Mitteln wiederholt worden [119], [460], [464]; neuere Arbeiten liegen ferner über die Temperatur- und Druckabhängigkeit der nach langen Wellen sich anschließenden Absorptionsbanden [206] sowie über die Emission des bei der Photodissoziation entstehenden ^1D-Sauerstoffatoms vor [207].

37. Die Absorptionskontinua der Silber- und Thalliumhalogenide. Bei Temperaturen zwischen 700 und 900° C besitzen die Silberhalogenide AgF, AgCl, AgBr und AgJ einen zur Aufnahme der Absorptionsspektren ausreichenden Dampfdruck und absorbieren alle im nahen wie weiteren Ultraviolett (das AgJ auch noch im sichtbaren Gebiet) eine Reihe von Bandensystemen und Kontinua. Die ältere Literatur findet sich bei Finkelnburg [149]; aus neuerer Zeit liegen nur zwei Arbeiten über die sichtbare und die im kurzwelligen UV liegende Absorption des AgBr und AgJ vor [310], [311]. Die Deutung aller dieser Spektren geht auf eine grundlegende Arbeit von Mulliken [337] zurück und ist von besonderem theoretischen Interesse, weil sie nach Fig. 79 die Annahme von Potentialkurven angeregter Molekülzustände mit einem Potential*maximum* und einem *über* der Dissoziationsenergie liegenden Potentialminimum erfordert. Dieser bisher in der Molekültheorie einmalige Potentialkurvenverlauf (Kurve B in Fig. 79) ist nach MULLIKEN so zu verstehen, daß der kernnahe Teil und das Minimum der Potentialkurve zu einer Elektronenanordnung des Moleküls gehört, die „eigentlich" in ein normales Halogen- und ein mit 3,7 eV angeregtes Ag-Atom dissoziieren sollte. Wegen der in diesem Fall erforderlichen Überschneidung der Potentialkurve mit den beiden höher gelegenen ordnen sich die Kurven aber so um, daß die Dissoziation aus dem B-Zustand tatsächlich in ein normales Silber- und ein $^2P_{\frac{1}{2}}$-Jod-Atom erfolgt. Alle wesentlichen Beobachtungen über die Silberhalogenide lassen sich aus diesen Annahmen über den Potentialkurvenverlauf verstehen, obwohl weitere experimentelle wie theoretische Untersuchungen zur Stützung dieser interessanten Deutung notwendig erscheinen.

Die Absorptionsspektren der Thalliumhalogenide TlF, TlCl, TlBr und TlJ sind insofern von grundsätzlichem Interesse, als in ihnen neben mehreren Gebieten kontinuierlichen Absorption im nahen, mittleren und fernen Ultraviolett noch kontinuierlicher Fluktuationen beobachtet werden, die sich teilweise scharfen Bandensystemen überlagern. Bei den im Gebiet

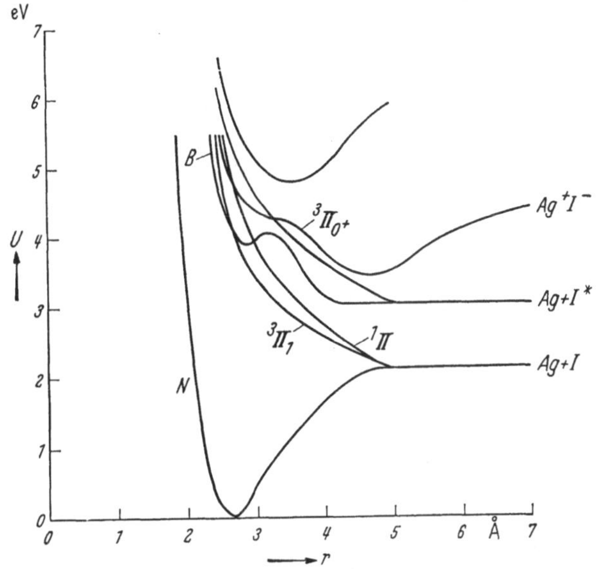

Fig. 79. Potentialkurvenschema des Silberjodid-Moleküls nach MULLIKEN [*337*].

4200 und 3800 Å auftretenden kontinuierlichen Fluktuationen des TlJ handelt es sich nach KUHN [*266*] um den Absorptionsfall II (vgl. Fig. 74 III), d.h. Übergänge vom Minimum der Grundzustandskurve zu einer über diesem annähernd horizontal verlaufenden höheren Potentialkurve. Das Thallium-Jodidspektrum ist in seinem langwelligen Bereich also der charakteristische Vertreter des Absorptionsfalles III. Die drei kurzwelligeren Absorptionskontinua wurden früher Übergängen zu angeregten Molekülzuständen zugeschrieben, deren unterster in ein normales Tl-Atom und ein $^2P_{\frac{1}{2}}$-Halogenatom dissoziieren sollte, während die den kurzwelligeren Kontinua entsprechenden Molekülzustände in normale Halogenatome und Tl-Atome im $^2P_{\frac{3}{2}}$- bzw. ^2S-Zustand dissoziieren sollten. Diese letzte Zuordnung wird in interessanter Weise dadurch gestützt, daß bei Absorption dieses Kontinuums die Emission der Tl-Linien beobachtet wird, die beim Übergang des ^2S-Tl-Atoms in die beiden Komponenten des Dublett-Grundzustandes emittiert werden. Bei dieser älteren Deutung (für Einzelheiten und Literatur vgl. [*149*]) werden also nur solche Molekülzustände der Tl-Halogenide berücksichtigt, die in normale oder angeregte *Atome* zerfallen. Demgegenüber erwähnt HERZBERG [*211*], daß die Tl-Halogenidmoleküle über dem in Atome dissoziierenden Grundzustand auch einen in *Ionen* dissoziierenden angeregten Zustand besitzen sollten, dessen Minimum bei so großem Kernabstand läge, daß vom Grundzustand Absorptionsübergänge nur in den kontinuierlichen Bereich dieses Ionenzustandes erfolgen könnten. Welches der drei kurzwelligen Kontinua dieser Photodissoziation in Ionen entsprechen soll, wird jedoch nicht angegeben. Neuere Literatur seit dem Erscheinen von Finkelnburgs Buch [211], [223] bis [225] u. a.

Je zwei kontinuierliche Absorptionsspektren, die dem Absorptionsfall I zugeordnet wurden, finden sich nach [245] auch beim SnF, beim SnCl [157] und PbF [375], während die Kontinua der Halogenide des Hg, Cd und Zn von Wieland [474] untersucht wurden.

38. Die kontinuierlichen Spektren der zweiatomigen Alkalihalogenidkontinuamoleküle. Die Spektren der elf untersuchten zweiatomigen Alkalihalogenidmoleküle bestehen im Dampfzustand in Absorption wie Emission aus zahlreichen Kontinua, Fluktuationen und mehr oder weniger diffusen Bandenzügen. Sie bilden daher ein gutes Beispiel für die Deutung eines komplexen kontinuierlichen Molekülspektrums. Jedes der Alkalihalogenidmoleküle absorbiert eine Reihe von Kontinua im Ultraviolett, die sich sämtlich bei Temperaturerhöhung nach langen Wellen zu ausdehnen. Tabellen der Wellenlängen und sonstigen Daten finden sich bei Finkelnburg [149]. Am langwelligen Ende des jeweils langwelligsten Absorptionskontinuums erscheinen bei genügend hoher Temperatur Fluktuationen, die nach langen Wellen zu konvergieren, und denen sich eine feine Struktur überlagert. Schon aus diesen wenigen Angaben folgt, daß es sich um Fall III der Absorptionskontinua (Fig. 74 III) handelt, d. h. um Übergänge von einem festgebundenen Molekülgrundzustand zu einem höheren van der Waals-Zustand, der wegen der feinen Struktur ein

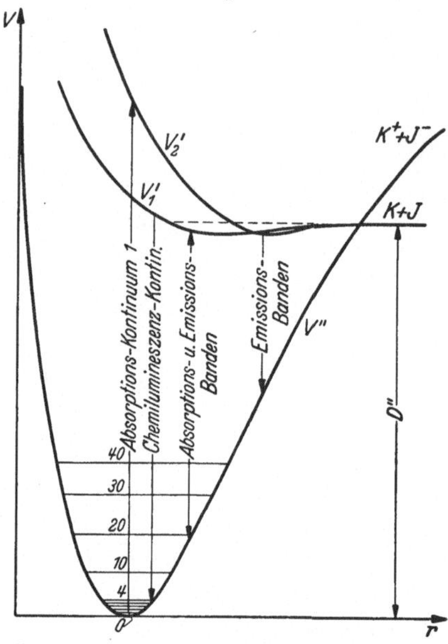

Fig. 80. Potentialkurvenschema und Übergang im Kaliumjodid-Molekül

flaches Minimum besitzen muß. Dieser Schluß wird gestützt durch den Emissionsbefund. In Entladungen beobachtet man bei mehreren Alkalihalogeniden diffuse Banden, die Übergängen von den flachen Potentialminima der angeregten Kurven zum Grundzustand entsprechen, während in der Chemilumineszenz (Rekombination im Stoß von Alkali- mit Halogenatomen) an diese diffusen Banden sich ein kurzwelliges Kontinuum anschließt, so daß der Fall IV der Emissionskontinua (Fig. 75 IV) erfüllt ist.

Aus den Abständen der Fluktuationen, die identisch in Absorption (Fall III) und Emission (Fall IV) auftreten, folgen unter der Voraussetzung angenähert horizontalen Verlaufes der oberen Kurve die Abstände der Schwingungsniveaus des Grundzustandes, und diese stimmen bis auf wenige Prozent mit den Werten überein, die sich unter der Annahme des Aufbaues der Moleküle aus negativen Halogen- und positiven Alkali-Ionen berechnen lassen. Diese Übereinstimmung spricht für die Auffassung der Alkalihalogenide als Ionenmoleküle, die bei reiner Schwingungsdissoziation auch in Ionen zerfallen würden. Der erste angeregte Molekülzustand dagegen dissoziiert in normale Atome, wie namentlich die Chemilumineszenz-Versuche eindeutig zeigen. Die Verhältnisse werden also durch ein Potentialkurvenschema gemäß Fig. 80 dargestellt. Daß die flachen Potentialmulden der ersten angeregten Zustände, wie es die Theorie verlangt, bei wesentlich größerem Kernabstand liegen als das Minimum des Grundzustandes, folgt zwingend aus der Deutung der Spektren. Bei mäßiger Temperatur erfolgt die Absorption zu einem weit oberhalb der Dissoziationsgrenze gelegenen Punkt von U_2' und ergibt ein Kontinuum, dessen Breite die Konstruktion des über dem Minimum von U'' gelegenen Teiles von U_2' erlaubt. Bei Temperatursteigerung dehnt sich dieses Absorptionskontinuum infolge der kernfernen Übergänge nach langen Wellen zu aus, wobei es sich wegen der entgegengesetzten Neigung von U' und U'' in Fluktuationen aufzulösen beginnt. Die kernnahen Äste von U_2' und U'' müssen im Übergangsgebiet annähernd parallel laufen, da nur dann die einzelnen Übergänge gleiche Wellenlängen besitzen, und damit ein schmales, aber intensives Absorptionskontinuum ergeben. Für alle Einzelheiten, die die Deutung und das Potentialkurvenschema Fig. 80 sehr gut stützen, vgl. [*149*], wo sich auch die gesamte Literatur findet. Dort wird auch gezeigt, welche Elektronenanordnung zu den angeregten Elektronenzuständen gehört, und daß auch in dieser Beziehung Übereinstimmung zwischen Theorie und Experiment besteht. Während die Übergänge zu den ersten angeregten Molekülzuständen nach Fig. 80 eine Dissoziation in normale Atome ergeben, führen die bis zu vier kurzwelligeren Absorptionskontinua, die bei einzelnen Alkalihalogeniden beobachtet worden sind, zur Dissoziation in je ein normales und ein angeregtes Atom. Der Anregungszustand der den verschiedenen Kontinua entsprechenden angeregten Atome folgt eindeutig aus der Beobachtung der von ihnen emittierten Atomlinien und steht in bester Übereinstimmung mit der theoretischen Erwartung.

Auch die Mehrzahl der beobachteten Emissionskontinua fügt sich der Deutung befriedigend ein, doch gibt es auch hier Unstimmigkeiten. Eine molekültheoretische Behandlung der Alkalihalogenidmoleküle und ihrer Spektren findet sich bei MULLIKEN [*337*], Einzelheiten über den Potentialkurvenverlauf bei VERWEY und DE BOER [*451*].

39. Die kontinuierlichen Spektren der zweiatomigen Metalloxyde und -sulfide. Eine ganze Anzahl zweiatomiger Metalloxyde und -sulfide zeigt ausschließlich kontinuierliche Absorptionsspektren. Die bestuntersuchten Beispiele dürften die Sulfide des Zn, Cd und Hg sein, die als Ionenmoleküle aufzufassen sind, bei denen im Grundzustand ein doppelt positives Metallion gegen das doppelt negative Schwefelion schwingt. Für alle Einzelheiten sei auf [*149*] verwiesen. Die jeweils langwelligsten der bis zu drei beobachteten Kontinua entsprechen der Dissoziation in normale Atome, während die beiden kurzwelligeren der Dissoziation in ein normales Metallatom und ein im ^1D- bzw. ^1S-Zustand angeregtes Schwefelatom entsprechen.

Bei den Oxyden des Zn, Cd und Hg ist nur je ein Absorptionskontinuum bekannt, dessen Absorption zum Zerfall in normale Atome führt. Bei den Seleniden der drei genannten Metalle ebenso wie bei den Oxyden und Sulfiden der Eisenmetalle, des Kupfers, der Erdalkalien und des Bleies sind meist zwei Kontinua bekannt, deren langwelligeres dem Zerfall in normale Atome entspricht, während bei Absorption des kurzwelligeren das Sauerstoff- bzw. Schwefelatom im ^1D-Zustand angeregt ist. Literatur: [*28*], [*303*], [*402*] bis [*404*], [*436*].

40. Das große Emissionskontinuum des H_2-Moleküls. Von den verschiedenen, von Wasserstoffentladungen emittierten Emissionskontinua ist das vom Grünen bis ins Ultraviolett sich erstreckende große H_2-Kontinuum von besonderem Interesse, weil es als bisher einziges Emissionskontinuum einen quantitativen Vergleich von Theorie und Beobachtung erlaubt. Beim Experimentieren muß man aber sicherstellen, daß man es wirklich mit dem H_2-Kontinuum zu tun hat und nicht mit den Kontinua des H-Atoms, des H^--Atoms oder des H_2^+, die andere Anregungsbedingungen haben. Für eine ausführliche Darstellung der weit über 100 Arbeiten unfassenden Literatur vgl. Finkelnburgs Buch [149].

Die Intensitätsverteilung des H_2-Kontinuums, das nach Fig. 81 bei Übergängen von einem angeregten Molekülzustand mit ausgeprägtem Potentialminimum ($1s\sigma\, 2s\sigma^3\Sigma_g$) zu dem instabilen Triplett-Grundzustand des Moleküls ($1s\sigma\, 2p\sigma^3\Sigma_u$) mit Abstoßungskurve emittiert wird, hängt von den Anregungsbedingungen ab, weil diese bestimmen, welche Schwingungszustände des angeregten Elektronenzustandes sich an der Emission beteiligen. Die in [149] besprochenen Anregungsuntersuchungen bestätigen aber die obige von Winans und Stueckelberg [479] stammende und von Finkelnburg und Weizel [140] weitergeführte Deutung in schöner Weise. Der quantitative Vergleich von Theorie und Messung wurde von Coolidge, James und Present [99], [100], [101], [237] durchgeführt, die dazu zunächst nach einem Variationsverfahren mit Hilfe einer Integrieranlage die beiden kombinierenden Potentialkurven, und anschließend die Beiträge des 1., 2., 3. und 4. Schwingungszustandes des angeregten Moleküls zu dem beobachteten Emissionskontinuum mit großer Genauigkeit berechneten. Dabei mußte die Annahme gemacht werden, daß für die Elektronenstoßanregung der verschiedenen Schwingungszustände des $^3\Sigma_g$-Zustandes vom Molekülgrundzustand aus auch das Franck-Condon-Prinzip gilt. Der Vergleich des auf diese Weise berechneten Emissionskontinuums mit dem gemessenen ergab aber nur Übereinstimmung, wenn angenommen wurde, daß (vgl. Ziff. 28) der Elektronenanteil der Anregungswahrscheinlichkeit vom Kernabstand abhängt, und zwar in der Form

$$M(r) \sim 1 + 0{,}43\, r. \tag{40.1}$$

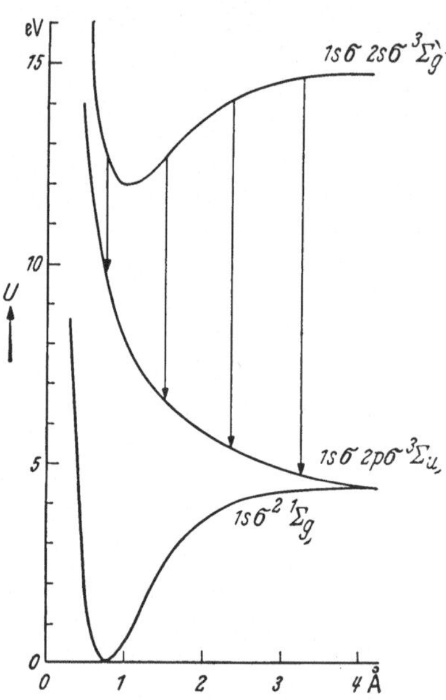

Fig. 81. Potentialkurvenschema des H_2-Moleküls mit den zum großen Emissionskontinuum führenden Übergängen.

Von Coolidge und James [102], [239] wurde die Berechnung der Intensitätsverteilung dann auch auf das D_2-Emissionskontinuum ausgedehnt und ein Vergleich der berechneten mit der bei genau kontrollierter Anregung mit einem Elektronenstrahl gemessenen Intensitätsverteilung des D_2- und des H_2-Kontinuums vorgenommen. Wie wichtig die Kontrolle der Anregungsbedingungen für einen Vergleich von Theorie und Beobachtung ist, zeigt eine neue Mitteilung von Dieke und Rainwater [114], die in ihrer Entladung das früher beobachtete Intensitätsmaximum bei 2500 Å nicht finden konnten und zur Erklärung der weiteren Intensitätssteigerung zum UV hin die Annahme machten, daß höhere Elektronenzustände des H_2 an der Emission des Kontinuums beteiligt seien. Die Anregung dieser höheren Elektronenzustände unterdrückt man z.B. durch Heliumzusatz und erhält dann das Kontinuum mit Maximum bei 2500 Å. Über den Einfluß der Anregungsbedingungen einschließlich des Zusatzes von Edelgas auf die Intensitätsverteilung des im UV auch als Intensitätsstandard verwendeten H_2-Kontinuums berichtet Grandsire [187]. Auf die zahlreichen veröffentlichten H_2-Lampenkonstruktionen kann hier nicht eingegangen werden.

Wie das große H_2-Kontinuum ist wahrscheinlich auch ein bei geringem Gasdruck im Blau-Violett beobachtetes Emissionskontinuum als Emissionsfall I zu deuten, das dem H_2^+ zugeschrieben wird (vgl. [149]).

Auch das Helium, das bekanntlich angeregte He_2-Moleküle mit ausgeprägten Potentialminima zu bilden vermag, emittiert nach HOPFIELD [204], [218], [219], [433] zwei verschiedene als Fall I zu deutende Emissionskontinua. Das kurzwelligste zwischen 1100 und 600 Å beobachtete soll Übergängen von einem angeregten He_2 zur flachen Abstoßungskurve des He_2-Grundzustandes entsprechen, während ein den gesamten sichtbaren Spektralbereich überdeckendes Emissionskontinuum, dessen genaue Untersuchung allerdings noch aussteht, Übergängen zwischen angeregten He_2-Zuständen zugeordnet werden muß [465].

41. Die kontinuierlichen Spektren des Hg_2, Cd_2 und Zn_2 sowie anderer VAN DER WAALS-Moleküle.

Die Absorptions- und Emissionsspektren der im Grundzustand nur schwach durch VAN DER WAALS-Kräfte gebundenen Moleküle Hg_2, Cd_2 und Zn_2 gehören mit ihrer Vielzahl von echten wie unechten Kontinua, kontinuierlichen und diffusen Bändern, Bandenzügen und Fluktuationen zu den kompliziertesten und meist untersuchten aller Molekülspektren. Es finden sich in ihnen Beispiele aller in Ziff. 30 und 31 behandelten Typen von Molekülkontinua, und darin liegt die grundsätzliche Bedeutung dieser Spektren. Die über 200 Nummern umfassende ältere Literatur ist bei FINKELNBURG [149] behandelt, der in einer Tabelle für jedes der drei Moleküle 13 verschiedene Spektren aufführt, deren Deutung mittels des Potentialkurvenschemas (Fig. 82) dort eingehend diskutiert ist und sich aus den eingezeichneten Übergängen ergibt.

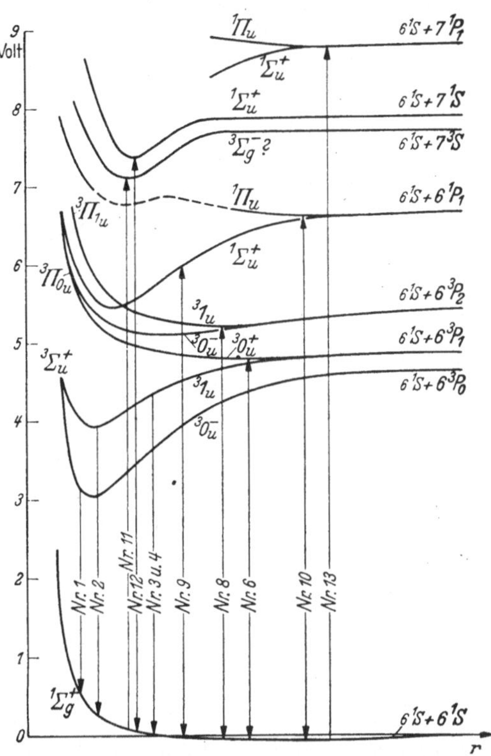

Fig. 82. Potentialkurvenschema des Hg_2-Moleküls nach MROZOWSKI (mit geringen Änderungen [149]).

Fig. 82 dürfte im wesentlichen noch dem heutigen Kenntnisstand entsprechen, nur scheint nach McCOUBREY [306] die unterste angeregte Potentialkurve (des $^30_u^-$-Zustandes) ganz dicht unter der nächst höheren ^31_u-Kurve zu verlaufen. Das früher dem Übergang von der $^30_u^-$-Kurve zum Grundzustand zugeordnete langwellige Emissionskontinuum Nr. 1 mit Maximum bei 4850 Å soll nicht direkt vom Hg_2 emittiert werden, sondern durch einen im einzelnen noch nicht geklärten Dreierstoßprozeß zustande kommen.

Daß bei neueren Untersuchungen immer häufiger eine feine Bandenstruktur in einzelnen der Kontinua gefunden wird, scheint verständlich. In allen Elektronenzuständen mit Potentialminima muß es ja diskrete, stationäre Schwingungsniveaus geben, die sich bei Übergängen zu den flachen Potentialkurventeilen anderer Elektronenzustände als diskrete Struktur bemerkbar machen sollten. Für alle Einzelheiten der Deutung und der sie stützenden vielfältigen

Beobachtungen sei auf [*149*] und die neuere Literatur [*15*], [*61*], [*152*], [*168*], [*281*], [*287*], [*330*] verwiesen.

Die behandelten Moleküle Hg_2, Cd_2 und Zn_2 sind nur die bestuntersuchten Beispiele für van der Waals-Moleküle. Kontinuierliche Bänder als Begleiter von Atomlinien (gelegentlich auch verbotenen) sind aber auch in den Spektren der zweiatomigen Alkalimetall-Moleküle [*149*], [*92*], der Alkali-Edelgasmoleküle [*496*], der Hg-Edelgasmoleküle [*267*], [*269*] sowie der van der Waals-Moleküle Hg-O_2, Hg-N_2, Hg-CO_2 (Literatur: [*329*], [*392*]), O_2-O_2, O_2-Edelgas, O_2-N_2 und O_2-CO_2 (Literatur: [*149*]) gefunden worden. Diese kontinuierlichen Bänder dürften, obwohl eine detaillierte Diskussion noch aussteht, wieder als Typ V (vgl. Ziff. 30) Übergängen zwischen den für van der Waals-Moleküle charakteristischen flachen Potentialkurven zuzuordnen sein.

III. Übersicht über die kontinuierlichen Gasspektren, nach Elementen geordnet.

Zur Erleichterung der Übersicht über die vorliegenden Beobachtungen kontinuierlicher Gasspektren, sowie zum praktischen Gebrauch wird in diesem Kapitel eine nach Elementen geordnete Übersicht über die bisher bekannten und die zu erwartenden Absorptions- und Emissionskontinua mit kurzen Hinweisen auf Anregungsbedingungen und Deutung gebracht. Für die Literatur sei, soweit nicht angegeben, auf [*149*] verwiesen.

42. Die Kontinua des Wasserstoffs. In Absorption zeigt normales, aus H_2-Molekülen bestehendes Wasserstoffgas lediglich ein Kontinuum, das sich im äußersten Ultraviolett $\lambda < 850$ Å an ein ausgedehntes Bandensystem anschließt, und dessen Intensität nach dem fernen UV hin abfällt. Es ist das Photodissoziationskontinuum (Absorptionsfall II) des H_2. Literatur: [*57*], [*113*], [*217*], [*278*]. Atomarer Wasserstoff absorbiert das astrophysikalisch wichtige Lyman-Grenzkontinuum (Ziff. 9) $\lambda < 911$ Å, bei genügend hoher Temperatur (Atmosphäre der Fixsterne) auch die Grenzkontinua der Balmer- und Paschen-Serie $\lambda < 3650$ und $\lambda < 8200$ Å. Die ebenfalls astrophysikalisch interessante kontinuierliche Absorption des H^--Ions ist in Ziff. 24 schon im einzelnen behandelt worden. Schließlich soll [*73*] auch das H_2^+ ein dem Absorptionsfall I entsprechendes Photodissoziationskontinuum ($\lambda < 4500$ Å) besitzen.

In Emission finden sich bei genügend intensiver, zur Dissoziation des H_2 führender Anregung die Seriengrenzkontinua der Lyman-, Balmer- und Paschen-Serie, bei geeigneten Anregungsbedingungen (Entladungen in Wasserstoff von hohem Druck) auch das H^--Emissionskontinuum (für alle diese Kontinua vgl. Ziff. 15) bei schwächeren Anregungen (Glimmentladungen) ferner das ausgedehnte intensive Emissionskontinuum des H_2 im Sichtbaren und UV (Ziff. 40) und bei sehr geringem Druck ein schwer zu beobachtendes schwaches Emissionskontinuum des H_2^+ im blau-grünen Spektralgebiet.

43. Die Kontinua der Alkalien. Die Dämpfe der Alkalimetalle absorbieren je ein an die Grenze der Hauptserie nach kurzen Wellen zu sich anschließendes Photoionisationskontinuum, dessen Intensität vom Lithium zum Caesium hin stark abfällt. Die langwelligen Grenzen dieser Absorptionskontinua sind in Spalte 1 von Tabelle 8 angegeben. Ein bei geringer Dispersion zu beobachtender kontinuierlicher Untergrund zwischen den Atomlinien besteht in Wirklichkeit aus eng liegenden Absorptionsbanden der zweiatomigen Alkalimetall-Moleküle. Schmale kontinuierliche, die Atomlinien in Absorption begleitende Bänder werden nach Ziff. 30 von den nach der Theorie ebenfalls zu erwartenden zweiatomigen

VAN DER WAALS-Molekülen der Alkalimetalle absorbiert (Fall V der Absorptions-kontinua).

In Emission beobachtet man in allen genügend stromstarken Alkalientladungen die Grenzkontinua der Hauptserie, der Nebenserien und der BERGMANN-Serie, die sich von den in Tabelle 8 angegebenen Grenzen nach kurzen Wellen zu erstrecken. Nebenserien- und BERGMANN-Serienkontinua sind im allgemeinen wesentlich intensiver als die Hauptserien-Kontinua; mit zunehmendem Dampfdruck wird dieser Unterschied immer ausgeprägter. Die Intensität aller Seriengrenzkontinua in Emission nimmt ferner mit zunehmender Ordnungszahl vom Lithium zum Caesium hin stark zu. Bei hohen Dampf- und Stromdichten verbreitern die Resonanzlinien der Alkalien gelegentlich bis zur Größenordnung 10^3 Å und erwecken damit oft den Eindruck selbständiger Emissionskontinua. Wie in Absorption treten auch in Emission die schmalen kontinuierlichen Bänder der Alkali-VAN DER WAALS-Moleküle (Ziff. 31) auf. Noch ungedeutet sind drei schmale, bei hohem Dampfdruck beobachtete Caesium-Emissionskontinua bei 7060, 7126 und 7187 Å [147], ferner zwei von BARTELS [30] gefundene Natrium-Emissions-kontinua mit Maxima bei 4364 und 4526 Å, die KUHN [268] als Molekül-Emissionskontinua Fall I deuten möchte.

Tabelle 8. *Langwellige Grenzen der Alkaligrenzkontinua in Å.*

	H. S.	N. S.	B. S.
Li . .	2299	3497	8192
Na . .	2412	4085	8144
K . .	2856	4552	7423
Rb . .	2967	4790	6974
Cs . .	3183	5082	5948

44. Die Kontinua der Erdalkalien. Tabelle 9 gibt eine Zusammenstellung der Seriengrenzen der Erdalkalien, die als langwellige Grenzen von Grenzkontinua in Frage kommen. In Absorption können unter normalen Bedingungen nur die Hauptseriengrenzkontinua (Grenze ^1S), in Emission alle angeführten Grenzkontinua auftreten. Wie allgemein ist die Intensität der Triplettkontinua größer als die der Singulettkontinua, und unter ersteren wieder die der Nebenseriengrenzkontinua am größten.

Tabelle 9. *Langwellige Grenzen der wichtigsten möglichen Seriengrenzkontinua der Erdalkalien in* Å

	^1S	^1P	^1D	^3S	$^3P_2$	$^3D_2$
Be	1330	3064	9286	4326	1880	7610
Mg	1630	3756	6574	4883	2514	7290
Ca	2027	3897	3641	5627	2941	3455
Sr	2177	4126	3878	5920	3222	3621
Ba	2378	4171	3263	6300	3506	3082

In der Hohlkathodenentladung in Erdalkalidämpfen beobachtet man ferner nach HAMADA [196] eine Reihe kontinuierlicher Spektren, die anscheinend VAN DER WAALS-Molekülen zuzuordnen sind. In Calciumdampf soll um die Resonanzlinie 4227 Å ein von 5100 bis 3980 Å reichendes Kontinuum auftreten. Auch in der Gegend der Mg-Resonanzlinie 2852 Å wird ein Kontinuum gefunden, das sich von 2670 Å mit mehreren Maxima bis 4412 Å erstrecken und im Bereich 3681 bis 3265 Å Fluktuationen besitzen soll, die zur Linie hin konvergieren. In der gleichen Entladung wurden schmale kontinuierliche Bänder bei 4662, 4608 und 2713 Å beobachtet, die wohl als Emissionsfall V zu deuten sind.

45. Die Kontinua des Quecksilbers, Cadmiums und Zinks. Tabelle 10 gibt die langwelligen Grenzen der möglichen Seriengrenzkontinua der Atome Zn, Cd und Hg, darunter in der ersten Spalte die in Absorption möglichen Hauptseriengrenzkontinua. In Emission sind bisher nur die Triplett-Nebenserienkontinua des Zn und Cd sowie alle Triplettgrenzkontinua des Hg beobachtet worden.

Außerdem zeigt Zinkdampf in Absorption ein bei hohem Druck kontinuierlich erscheinendes, bei niedrigem Druck in Fluktuationen sich auflösendes Spektrum im Gebiet 2700 bis 2400 Å, ferner ein kontinuierliches Band bei der Resonanzlinie 2139 Å, das von einer scharfen Kante bei 2064 Å aus sich je nach Druck und Temperatur verschieden weit nach langen Wellen hin erstreckt, endlich noch ein schmales Kontinuum bei 2002 Å. In verschiedensten Entladungen wie in Fluoreszenz treten in Emission auf: ein intensives Kontinuum zwischen 4900

Tabelle 10. *Seriengrenzen von* Zn, Cd, Hg *in* Å.

	^1S	^1P	^1D	^3S	$^3P_2$	$^3D_2$
Zn	1320	3445	7512	4525	2331	7697
Cd	1379	3465	7506	4748	2456	7677
Hg	1188	3320	7781	4579	2491	7841

und 4000 Å mit Maximum bei 4450 Å, ein Zug Fluktuationen im Gebiet 3980 bis 2400 Å, ein von 2140 Å nach langen Wellen sich erstreckendes Band, sowie das schmale Kontinuum bei 2002 Å. Alle diese Spektren gehören zum Zn_2-VAN DER WAALS-Molekül (vgl. Ziff. 41).

Cadmiumdampf besitzt ein Absorptionskontinuum mit Fluktuationen, das sich von 2212 Å aus bis 3050 Å ausdehnen kann. Schmale Absorptionsbänder, die gelegentlich auch Struktur zeigen, liegen bei 3261 Å, 3180 bis 3140 Å und bei 2140 bis 1990 Å, ein Zug diffuser Banden im Gebiet 2800 bis 2650 Å. In Entladungen und in Fluoreszenz beobachtet man als Emissionsspektren ein intensives Kontinuum im Gebiet 5100 bis 3800 Å mit Maximum bei 4650 Å und Fluktuationen zwischen 4500 und 4200 Å, ein die Resonanzlinie 2288 einschließendes kontinuierliches Band von 2212 bis 3000 Å, ferner schmale Bänder, z. T. mit Struktur, bei 3261, 3180 bis 3140, 2900 bis 2650 sowie 2136 bis 2112 Å. Alle diese Spektren gehören dem Cd_2-VAN DER WAALS-Molekül an.

Quecksilberdampf absorbiert ein Kontinuum mit Fluktuationen im Gebiet 2610 bis 2950 Å, ein die Resonanzlinie 2537 Å begleitendes Absorptionskontinuum, das sich von einer scharfen Grenze bei 2540 Å nach langen Wellen zu erstreckt, einen Zug intensiver Fluktuationen im Gebiet 2300 bis 2000 Å, ein die Resonanzlinie 1849 Å begleitendes, von 1807 Å nach langen Wellen sich erstreckendes kontinuierliches Band, sowie schmale kontinuierliche Absorptionsbänder bei 1755, 1690 und 1403 Å. Bei hohem Druck und hoher Temperatur können die langwelligeren Spektren zu einem einzigen Absorptionskontinuum zwischen 3000 und 1800 Å zusammenfließen.

In Emission (und Fluoreszenz) beobachtet man bei Quecksilber zwei intensive Kontinua in den Gebieten 5300 bis 4000 und 3700 bis 3000 Å mit Maxima bei 4850 und 3350 Å, zwei verschiedene Züge von Fluktuationen zwischen 3000 und 2600 Å, ein schwaches Kontinuum bei 2650, das die Resonanzlinie 2537 Å begleitende Kontinuum mit scharfer kurzwelliger Grenze bei 2540, einen Zug Fluktuationen im Gebiet 2300 bis 2000 Å, und schließlich schmale kontinuierliche Bänder bei 1850 und 1690 Å. Auch diese Spektren, die in [*149*] sehr eingehend beschrieben sind, können bei hohem Dampfdruck zu einem einzigen, von 5500 bis 1600 Å reichenden Kontinuum zusammenfließen; sie gehören nach Ziff. 41 wie die entsprechenden Absorptionsspektren sämtlich dem Hg_2-VAN DER WAALS Molekül an.

Im Spektrum stromstarker Hg-Bogenlampen beobachtet man bei Dampfdrucken zwischen 25 und 200 atm außer starken Linienverbreiterungen das in
Ziff. 15 eingehend beschriebene intensive Elektronenkontinuum, das im Extremfall mit den verbreiterten Linien ein einheitliches Kontinuum bildet.

Im Spektrum aller Quecksilber-Bogenentladungen (sowie in zahlreichen anderen Metalldampfbögen) findet man schließlich ein im fernsten Ultrarot (100
bis 400 μ) gelegenes Kontinuum, das für Absorptionsmessungen in diesem Gebiet
viel Verwendung findet und nach HETTNER [212] als Elektronenbremskontinuum
zu deuten ist (vgl. Ziff. 19).

46. Die Kontinua der Elemente der dritten Gruppe des Periodischen Systems.
Kontinua des Elementes Bor scheinen weder in Absorption noch in Emission
beobachtet worden zu sein. Dagegen emittieren die Aureolen aller Bogenentladungen und aller Flammen, in denen Spu-
ren von Bor und Sauerstoff zusammen vorkommen, das sog. Borsäurekontinuum. Es
handelt sich um ein vom Rot bis ins nahe
Ultraviolett sich erstreckendes rein kontinuierliches Spektrum mit einer Reihe ausgeprägter Maxima verschiedenster Intensität, deren
wichtigste bei den Wellenlängen 6400, 6000,
5800, 5480, 5180, 4930, 4720, 4535, 4370,

Tabelle 11. *Seriengrenze der Erdmetalle in Å.*

	$^2P_{\frac{3}{2}}$	^2S	$^2D_{\frac{3}{2}}$
Al . .	2075	4359	6310
Ga . .	2102	4237	7355
In . .	2249	4484	7270
Tl . .	2410	3724	6580

4195, 4078, 3894 und 3768 Å liegen. Das früher kurzzeitig dem Ozon zugeschrie-
bene Kontinuum hat nach [148] mit Sicherheit eine Borsauerstoff-Verbindung
als Träger, und zwar wahrscheinlich nicht das BO, sondern ein mehratomiges
Bor-Sauerstoffmolekül. Zur Deutung kommt wohl wegen der welligen Struktur
in erster Linie Fall III der Emissionskontinua bei sinngemäßer Übertragung auf
mehratomige Moleküle in Frage.

Die als langwellige Kanten der Seriengrenzkontinua der Erdmetalle Al, Ga,
In und Tl in Frage kommenden Wellenlängen gibt Tabelle 11, wobei die in der
ersten Spalte aufgeführten Nebenseriengrenzkontinua in Absorption auftreten
können, da der $^2P_{\frac{1}{2}}$-Zustand der Grundzustand der Atome ist. Wegen der geringen
Dampfdrucke scheinen bisher nur die Emissionskontinua des Tl beobachtet zu
sein [260]. In der Hohlkathodenentladung in Tl-Dampf wurde ferner von HA-
MADA [196] eine ganze Anzahl ausgedehnter Kontinua, teilweise mit Fluktuationen,
im Gebiet 6500 bis 2740 Å beobachtet, die bei höherem Dampfdruck zu einem
einzigen welligen Kontinuum zusammenfließen können. Da die einzelnen Kontinua in der für VAN DER WAALS-Moleküle üblichen Weise mit den Atomlinien
in Zusammenhang stehen, hat man an locker gebundene Tl_2-Moleküle als Träger
dieser Spektren zu denken.

47. Die Kontinua der Elemente der vierten und fünften Gruppe. Von den
Elementen der vierten Gruppe des Periodischen Systems sind Kontinua bisher
nicht bekannt geworden, von denen der fünften Gruppe nur das Photodissoziationskontinuum des N_2, das sich als Absorptionsfall II von einer bei 990 Å liegenden
kurzwelligen Grenze eines Bandensystems aus nach kurzen Wellen erstreckt.

48. Die Kontinua der Sauerstoffgruppe. Sauerstoffgas absorbiert im fernen
Ultraviolett ein äußerst intensives, von 1750 bis 1300 Å reichendes Dissoziationskontinuum mit Maximum bei 1450, das Ziff. 36 bereits als Photodissoziationskontinuum des O_2 nach Fall II gedeutet wurde. Außerdem absorbiert Sauerstoffgas zwei äußerst schwache Kontinua mit langwelliger Grenze bei 2400 Å, die
nur bei großer Schichtdicke oder hohem Druck beobachtbar sind und offenbar
verbotenen Übergängen zugehören, bei denen das Molekül in zwei normale Atome

dissoziiert [142], [209]. Price und Collins [362] haben ferner zwei schwache Absorptionskontinua des O_2 mit langwelligen Grenzen bei 1105 und 740 Å beobachtet, die Dissoziations- oder Ionisationskontinua sein können. Eine Anzahl weiterer Absorptionskontinua tritt nur im komprimierten, flüssigen und festen Sauerstoff auf und hat ein O_2-van der Waals-Molekül als Träger (Ziff. 41). Das Absorptionskontinuum mit langwelliger Grenze bei 2400 Å wurde eben bereits erwähnt; absorbiert wird ferner eine Reihe breiter kontinuierlicher Bänder, deren intensivste bei 6900, 6300, 5780, 5335, 4775, 3815, 3617 und 3453 Å liegen, und die die blaue Farbe des komprimierten, flüssigen und festen Sauerstoffs bedingen. Bezüglich der Deutung sei auf die Darstellung bei Finkelnburg [149] verwiesen.

Zu den Sauerstoffspektren gehört auch das des *Ozons*. Außer einer Gruppe diffuser Banden zwischen 6100 und 4300 Å absorbiert das Ozon sehr intensiv ein Kontinuum im Gebiet 2900 bis 2300 Å mit Maximum bei 2500 Å. Absorption dieses Dissoziationskontinuums führt zum Zerfall des O_3 in $O_2 + O$. Dieses Kontinuum ist es, das die Beobachtung der ultravioletten Spektren der Sonne und der Fixsterne unmöglich macht, weil infolge seines großen Absorptionskoeffizienten der geringe Ozongehalt unserer Atmosphäre zur völligen Absorption der Ultraviolettstrahlung in diesem Gebiet ausreicht.

Emissionskontinua des Sauerstoffs selbst scheinen nicht bekannt zu sein. Dagegen wird gelegentlich als „kontinuierliches Sauerstoff-Nachleuchten" das kontinuierliche Spektrum einer Nachleuchterscheinung bezeichnet, die man in schwachen Entladungen in Luft von 1 bis 3 Torr und besonders gut in der elektrodenlosen Ringentladung in stickstoffverunreinigtem Sauerstoff beobachtet. Das gleiche Leuchten tritt bei der direkten Einwirkung von Ozon auf NO, beim thermischen Ozonzerfall in Luft, in thermisch angeregtem NO_2 sowie in NO und O enthaltenden Flammen (Ziff. 56) auf, dagegen nicht in Entladungen in ganz reinem Sauerstoff. Das Spektrum überdeckt das gesamte sichtbare Spektralgebiet von 6700 bis jenseits 3800 Å mit einem Maximum im Gelbrot und nach kurzen Wellen zu abnehmender Intensität. Es soll teilweise in diffuse Banden auflösbar sein. Nach Gaydon [173] soll dieses „grüne Kontinuum" bei der Reaktion von O-Atomen mit NO-Molekülen emittiert werden. Die ältere Literatur findet sich in [149]; neuere Literatur in [26], [173].

Die Absorptionsspektren des *Schwefeldampfes* zeigen eine starke Abhängigkeit von Temperatur und Druck, was auf die Bildung höherer Komplexe S_4, S_6, S_8 zurückzuführen ist, die im wesentlichen kontinuierlich absorbieren, während S_2 teils scharfe, teils infolge von Prädissoziation diffuse Banden absorbiert. Bei hohem Dampfdruck absorbiert Schwefeldampf im ganzen Ultraviolett von 3500 Å an praktisch kontinuierlich, während bei hoher Temperatur im gleichen Gebiet eine Reihe teils scharfer, teils diffuser Bandensysteme auftritt, die sich bis ins sichtbare Gebiet erstrecken. In Emission scheint nur ein von Schröer [420] beschriebenes S_2-Rekombinationskontinuum bekannt zu sein, dessen Maximum bei 3500 Å liegt.

Selendampf zeigt bei geringem Druck Banden bis 3238 Å, an die sich nach kürzeren Wellen hin kontinuierliche Absorption anschließt, bei hohem Dampfdruck aber kontinuierliche Absorption von λ 4900 Å bis ins Ultraviolett; über Emissionskontinua des Selens scheint nichts bekannt zu sein.

Die Absorptionsspektren des *Tellurdampfes* sind denen des Schwefels und Selens sehr ähnlich, nur daß das große Trägheitsmoment des Te_2 eine Struktur mit kleinen Spektralapparaten noch weniger aufzulösen gestattet. So absorbiert Tellurdampf bei hohem Druck im ganzen Gebiet von λ 6000 Å abwärts fast kontinuierlich; bei niedrigem Druck zeigen sich im sichtbaren Gebiet Banden, z.T.

mit Prädissoziation, an die sich von 3720 Å an nach dem Ultraviolett zu ein Kontinuum anschließt, das wohl als Dissoziationskontinuum Fall II (Ziff. 30) zu deuten ist. Emissionsspektren des Te_2 sind in einer Bogenentladung großer Stromdichte von ROMPE [381] untersucht worden. Vom Rot bis 5200 Å finden sich diffuse Banden, anschließend bis 4360 Å ziemlich scharfe Banden, aber auf einem kontinuierlichen Untergrund, anschließend bis 3890 Å wieder diffuse Banden und von dieser Wellenlänge an ein echtes Emissionskontinuum, das bei der Wiedervereinigung von normalen und angeregten Te-Atomen zu Molekülen emittiert wird.

49. Die Kontinua der Halogene. Die Absorptions- und Emissionskontinua der Halogene wurden in Ziff. 35 behandelt; Tabelle 12 gibt eine Übersicht über alle beobachteten Spektren. Wir erwähnen deshalb hier nur die für den Experimentator und den Photochemiker wichtigsten, auffallendsten Kontinua.

Alle Halogene zeigen je ein besonders intensives Absorptionsgebiet, das den Molekülen zuzuordnen ist, und zwar Fluor bei 3800 bis 2000 Å mit Maximum bei 2900 Å, Chlor bei 4000 bis 2800 Å mit Maximum bei 3300 Å, Brom bei 5000 bis 3500 Å mit Maximum bei 4200 Å und Jod vom Rot bis 4500 Å mit Maximum bei 5200 Å.

Tabelle 12. *Emissionskontinua der Halogene.*

	λ Å	λ in Å
Cl_2 . .	3180	2610—2500
Br_2 . .	4200	2930—2800
J_2 . .	4800	3460—3350

Diese Absorptionsgebiete sind bei F_2 und Cl_2 rein kontinuierlich und zeigen den Zerfall in Atome an; bei Br_2 und J_2 besteht der langwellige Teil aus Banden, an die sich nach kurzen Wellen zu das Dissoziationskontinuum anschließt. Außer diesen Hauptabsorptionsgebieten besitzen alle Halogene mehrere Absorptionsgebiete im Ultraviolett. Das gleiche gilt für die aus zwei verschiedenen Halogenatomen zusammengesetzten Mischmoleküle.

In Emission beobachtet man bei allen Halogenentladungen je zwei kontinuierlich erscheinende breite Bänder mit ziemlich scharfen langwelligen Grenzen und langsamem Intensitätsabfall nach kurzen Wellen zu. Tabelle 12 gibt in Spalte 1 die Grenzen des breiteren Bandes, das mit abnehmender Intensität noch über das schmale, kurzwelligere Band hinausreicht, dessen ungefähre Grenzen Spalte 2 angibt (vgl. [133]).

Beim Erhitzen von Halogendämpfen auf über 600° C treten ferner schwache Emissionskontinua, zum Teil in Begleitung von Banden auf, deren Wellenlängengebiete ungefähr denen der großen Absorptionskontinua entsprechen, und die als Molekülrekombinationskontinua nach Ziff. 32 gedeutet werden müssen.

50. Die Kontinua der Edelgase. Helium absorbiert im äußersten Ultraviolett $\lambda < 504$ Å das der Photoionisation des normalen Atoms entsprechende Hauptseriengrenzkontinuum, das in Ziff. 9 behandelt worden ist. In Emission können außer diesem Hauptserienkontinuum, das von LYMAN [290] in einer Ausdehnung von 504 bis 400 Å beobachtet worden ist, noch die anderen in Tabelle 13 angeführten Grenzkontinua auftreten. Wieder sind die Triplettkontinua intensiver als die Singulettkontinua, die Nebenserienkontinua intensiver als die der Hauptserie.

Tabelle 13. *Die langwelligen Grenzen der Heliumgrenzkontinua in Å.*

^1S	^1P	^3S	^3P
507	3678	2600	3421

Außer diesen Atomgrenzkontinua emittiert Helium bei Drucken von einigen cm Hg, namentlich bei leicht kondensierten Entladungen in nicht zu weiten Rohren, ein sich über das ganze sichtbare Gebiet erstreckendes Kontinuum, sowie im äußersten Ultraviolett ein sehr intensives Kontinuum im Bereich 1100 bis 600 Å. Beide Kontinua werden beim Zerfall angeregter He_2-Moleküle emittiert (Ziff. 40).

Das kurzwellige Kontinuum hat eine besondere Bedeutung als kontinuierliche Lichtquelle für Absorptionsversuche im äußersten Ultraviolett.

Absorptionskontinua der schwereren Edelgase Neon bis Xenon scheinen nicht beobachtet zu sein, desgleichen keine Atomkontinua in Emission. Dagegen emittieren alle Edelgase in elektrischen Entladungen geringer Stromdichte bei Drucken oberhalb 10 Torr ziemlich intensive kontinuierliche Spektren, die sich von einer für die verschiedenen Edelgase zwischen 7500 und 9000 Å liegenden langwelligen Grenze über das gesamte sichtbare und ultraviolette Spektralgebiet bis zu mindestens 1800 Å (unterhalb nicht untersucht) erstrecken und im Sichtbaren jeweils 4 bis 5 ausgeprägte Maxima aufweisen. Die Abhängigkeit der Kontinua und ihrer Intensitätsverteilung von den Versuchsbedingungen, besonders Stromstärke und Druck, wurde von Vogel [456] eingehend studiert. Es handelt sich zweifellos um Molekülkontinua, die dem großen sichtbaren He_2-Kontinuum (Ziff. 40) entsprechen dürften.

Hochdruckbögen und -funken genügender Stromdichte in Edelgasen emittieren ferner nach Schulz [424], [426], Glaser [180] u.a. die in Ziff. 15 behandelten kontinuierlichen Spektren hoher Intensität, die als Elektronenbrems- und Rekombinationskontinua zu deuten sind. Namentlich die Xenon-Hochdrucklampe mit ihrem weit ins Ultraviolett reichenden intensiven Emissionskontinuum hat wissenschaftlich wie technisch große Bedeutung erlangt.

51. Die Kontinua der wichtigsten zweiatomigen Mischmoleküle. Die wichtigsten kontinuierlichen Spektren der aus verschiedenen Elementen bestehenden zweiatomigen Mischmoleküle sind in Ziff. 34 bis 39 besprochen. Intensive Absorptionskontinua besitzen die Halogenwasserstoffe (Ziff. 34) im gesamten Ultraviolett, ferner die meisten Halogenide (Ziff. 37 und 38), von denen die Alkalihalogenide am besten untersucht und theoretisch von größtem Interesse sind. Absorptionskontinua im Ultraviolett besitzen ferner eine große Anzahl zweiatomiger Metalloxyde, Metallsulfide und Metallselenide. Emissionskontinua sind bei den Halogenwasserstoffen im Ultraviolett, beim HBr allerdings bis ins Grün reichend, gefunden (Ziff. 34), ferner bei den Alkalihalogeniden besonders in der Chemilumineszenz.

IV. Die kontinuierlichen Emissionsspektren
nach Erzeugungsmethoden geordnet.

In Ergänzung zu Ziff. 42 bis 51 bringen wir jetzt eine Übersicht über die wichtigsten Emissionskontinua, nach ihren Erzeugungsmethoden geordnet. Dabei werden besonders diejenigen Kontinua ausführlich behandelt, die in den vorhergehenden Kapiteln im Zusammenhang mit der Theorie nicht ausführlich besprochen worden sind.

52. Emissionskontinua in Glimmentladungen und verwandte Niederdruckentladungen. Daß die Glimmentladung eine Quelle zahlreicher Emissionskontinua darstellt, ist seit Doves ersten Untersuchungen [121] im Jahre 1858 bekannt. Das Gleiche gilt vom Niederdruckglimmbogen, der elektrodenlosen Ringentladung und ähnlichen Tesla-Entladungen. Äußerst intensiv tritt in der Glimmentladung in Wasserstoff das H_2-Kontinuum auf (Ziff. 40); auch die Emissionskontinua der Halogenwasserstoffe (Ziff. 34) und der Halogene (Ziff. 35), sowie die zahlreichen kontinuierlichen Spektren des Quecksilbers, des Cadmiums und des Zinks (Ziff. 41) werden am einfachsten in der Glimmentladung erzeugt. Bei den Alkalien sowie bei manchen anderen Metallen bildet die Glimmentladung (dann meist mit Glühkathode) auch eine intensive Quelle der Atomgrenzkontinua in Emission (Ziff. 16). Diese sind ferner besonders gut in der elektrodenlosen Ringentladung [*208*], [*19*] bis [*24*] zu erzeugen, die des Heliums auch in der Hohlkathode einer Glimmentladung [*353*]. Die Hohlkathode ist ferner eine sehr brauchbare Lichtquelle für die kontinuierlichen Spektren des Calciums, des Magnesiums und des Thalliums (Ziff. 44 und 46). In Luft, verunreinigtem Sauerstoff und besonders gut in Sauerstoff-Stickstoffgemischen beobachtet man in der Niederdruckentladung das in Ziff. 48 besprochene sogenannte Nachleuchtkontinuum des Sauerstoffs. Niederdruckentladungen nicht zu geringen Druckes sind nach Ziff. 50 eine Quelle für alle Edelgas-Molekülkontinua. *Tesla*-Entladungen geringen Druckes sowie andere stromschwache Glimmentladungen bilden eine Quelle zahlreicher Emissionskontinua mehratomiger Moleküle, besonders, wenn man nach SCHÜLER [*422*] durch entsprechende Ausgestaltung des Entladungsrohres den Zerfall der oft wenig stabilen Moleküle bei Berührung mit den Elektroden verhindert. Allgemein beobachtet man bei Glimmentladungen in Dämpfen mehratomiger Moleküle nicht selten unechte Kontinua, die sich bei größerer Dispersion als diffuse Banden erweisen.

53. Emissionskontinua in Bogenentladungen[1]. Fast alle elektrischen Lichtbögen emittieren in einzelnen Entladungsgebieten mehr oder weniger intensive kontinuierliche Spektren, die grundsätzlich vier verschiedene Ursachen haben können:

Die Elektroden sowie feste glühende Teilchen, falls solche im Bogen vorhanden sind, emittieren kontinuierliche Spektren, deren Intensität und spektrale Intensitätsverteilung durch Temperatur und Emissionsvermögen dieser festen Strahler bestimmt ist. Die wichtigste dieser kontinuierlichen Lichtquellen ist der positive Krater des Homogenkohlebogens, der nach den neuesten Messungen von EULER [*139*] in einem weiten Spektralbereich die gleiche kontinuierliche Strahlung emittiert wie ein schwarzer Strahler der Temperatur von 3800° K.

Bei nicht zu hoher Stromdichte können von Lichtbögen, besonders von deren Außenzonen, auch Molekülkontinua (Ziff. 26f.) emittiert werden. Zu diesen gehört das Ziff. 46 behandelte grüne sog. Borsäurekontinuum, das z.B. von allen Kohlelichtbögen mit borhaltigen Elektroden emittiert wird und sich mit zahlreichen Maxima über das gesamte sichtbare Spektralgebiet erstreckt, nach BURHORN [*74*] ferner ein in den Außenzonen des PFUND-Eisenbogens auftretendes, wahrscheinlich vom FeO emittiertes Molekülkontinuum. Bei den von MOORE [*328*] in vielen Metallbögen geringer Stromstärke gefundenen Emissionskontinua dürfte es sich ebenfalls um Molekülkontinua handeln, da diese bei höherer Stromstärke zu verschwinden scheinen. Auch die von WENIGER und HERMAN [*466*] bei sehr geringen Stromstärken in Argon von 3 bis 120 Atm Druck gefundenen Emissionskontinua müssen den Bedingungen nach Molekülkontinua wie die in Ziff. 50 besprochenen sein. Die Frage, wie weit an der kontinuierlichen Strahlung zahlreicher Quecksilberbögen Hg_2-VAN DER WAALS-Moleküle beteiligt sind, wurde Ziff. 15 schon im einzelnen diskutiert.

Bei Lichtbögen in Gasen und Dämpfen mit zahlreichen dicht liegenden oder wenigen besonders stark verbreiternden Emissionslinien können kontinuierliche Spektren auch durch echtes Zusammenfließen stark verbreiterter Linien entstehen oder durch sehr viele sehr dicht liegende Linien nur vorgetäuscht werden. Zum ersten Typ gehören die Quecksilber-Höchstdruckbögen (Fig. 83), sowie die Bögen in Wasserstoff, Wasserdampf oder Alkalimetalldämpfen, nach HUMPHREYS [*230*] auch ein unter 100 Atm Druck brennender Eisenbogen. Ein unechtes,

---

[1] Vgl. auch den Artikel über elektrische Bögen und thermisches Plasma in Bd. XXII dieses Handbuches.

also nur scheinbar kontinuierliches Spektrum dagegen emittiert beispielsweise das aus der Positivkohle des Beck-Bogens [150] verdampfende Cer-Plasma; sein Spektrum besteht in Wirklichkeit aus tausenden scharfer Linien, und selbst der zwischen diesen sichtbare Untergrund beruht wahrscheinlich auf äußerst zahlreichen diskreten Atom- und Moleküllinien.

Schließlich werden von Lichtbögen intensive Elektronenkontinua (Rekombinations- und frei-frei-Strahlung) nach Ziff. 15 immer dann emittiert, wenn eine genügend hohe Ladungsträgerdichte vorhanden ist. Da diese Kontinua Ziff. 15 ausführlich diskutiert wurden, beschränken wir uns hier auf die Erwähnung der wichtigsten Bogenarten, in denen solche Kontinua beobachtet werden.

Als intensivste Lichtquellen mit in weiten Spektralbereichen besonders des Ultraviolett rein kontinuierlicher Elektronenstrahlung sind zu nennen die Bogenentladungen zwischen Wolframelektroden in Edelgasen hohen Druckes, besonders

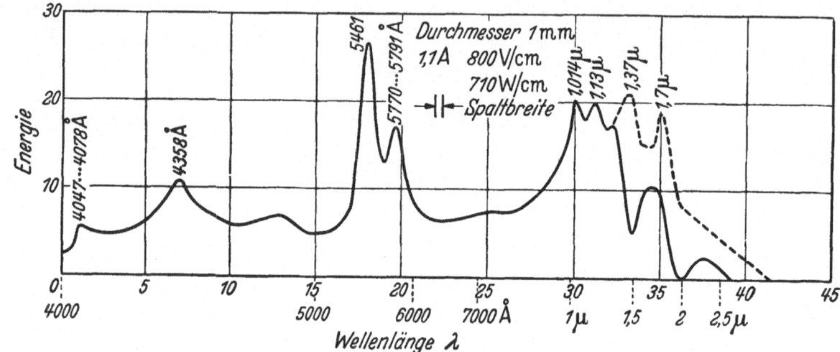

Fig. 83. Emissionskontinuum eines 200 Atm-Quecksilber-Höchstdruckbogens nach Elenbaas [132].

die im Xenon. Sie gehen auf Arbeiten von Schulz [424] bis [426] und Larché [273] zurück. Ein ähnlich intensives Elektronenkontinuum wird neben den verbreiterten Linien von den schon länger bekannten Quecksilber-Höchstdruckbögen emittiert, wobei die Intensität des Kontinuums nach [79], [132], [382], [423] mit der Stromstärke und dem Hg-Dampfdruck wächst. Auf die sehr langwellige, ebenfalls von den freien Elektronen herrührende kontinuierliche Emission des Hg-Bogens und anderer Metallbögen wurde in Ziff. 19 bereits ausführlich eingegangen. Besonders sorgfältig untersucht ist (vgl. Ziff. 15) das Elektronen-Emissionskontinuum der Säule des 200 Amp-Hochstromkohlebogens. Überhaupt unterliegt es heute keinem Zweifel mehr, daß der kontinuierliche Untergrund *aller* Lichtbögen nicht zu geringer Stromdichte Rekombinations- und frei-frei-Strahlung der Entladungselektronen als Ursache hat. Die zahlreichen Sonderformen von stabilisierten Lichtbögen mit kontinuierlicher Säulenstrahlung sind ebenfalls in Ziff. 15 bereits behandelt worden.

54. Emissionskontinua in Funkenentladungen. Unter Funkenentladungen sollen hier im engeren Sinne stromstarke Kondensatorentladungen verstanden werden, da die spektralen Erscheinungen unkondensierter, d.h. leistungsschwacher Funken je nach dem Gasdruck, denen der Glimm- oder der Bogenentladung zuzurechnen sind. Alle Funken zeigen bei höheren Drucken kontinuierliche Spektren mit nur geringen individuellen Unterschieden. Bei allen ist ferner das UV gegenüber den entsprechenden Bogenspektren wesentlich verstärkt, was auf eine gegenüber dem Bogenplasma erhöhte Temperatur als Folge der höheren Stromdichte hindeutet. Am deutlichsten prägen sich alle diese Erscheinungen in einer Reihe extremer Fälle von Kondensatorentladungen aus, wie dem Unterwasserfunken, dem stromdichten Vakuumfunken in Kapillaren und den elektrischen Metalldrahtexplosionen, auf die wir im folgenden näher eingehen werden.

Schon seit den ersten Untersuchungen von Wüllner [487], [488] im Jahre 1869 ist bekannt, daß in kondensierten Entladungen bei allen Gasen und Elektrodenmetallen kontinuierliche Spektren und Linienverbreiterungen ungewöhnlichen Ausmaßes vorkommen. Seine Beobachtung, daß namentlich die Balmer-Linien des Wasserstoffes enorm verbreitern und bei höheren Drukken zu einem Kontinuum zusammenfließen können, hat erst in viel späterer Zeit zu exakteren Untersuchungen angeregt (s. Finkelnburg [141], ferner Fischer [153]). Allgemein nimmt mit wachsendem Druck und wachsender Stromdichte die Intensität des kontinuierlichen Unter-

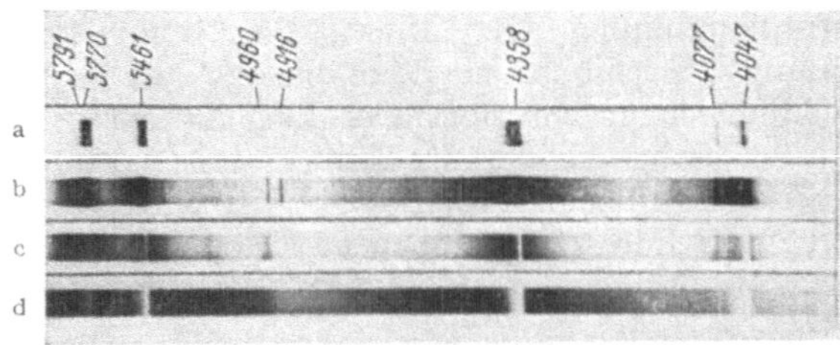

Fig. 85a—d. Stoßspektren der Bogenmitte von Quecksilberhöchstdrucklampen, im Maximum des Stoßleuchtens aufgenommen. $U =$ Stoßspannung (Kondensator 1 µF), $T =$ Achsentemperatur in Bogenmitte (über 12000° nur Minimalwert). a Vergleichsspektrum ohne Stoß $T = 7700°$; b wie a, nur längere Belichtungszeit; c $U = 400$ V bzw. $T = 11700°$; d $U = 1200$ V bzw. $T_{min} = 16000°$. (Nach Göing [182]).

grundes und fast stets auch die Größe der Linienverbreiterung stark zu. Bei linienreichen Spektren, wie etwa dem des Eisens, wachsen die einzelnen verbreiterten Linien dann zu einem intensiven Emissionskontinuum zusammen. Da der Funkenkanal dabei stets von einem Mantel aus unangeregtem Gas umgeben ist, erscheint das bei geringem Druck normale Emissionslinienspektrum nun umgekehrt als Absorptionslinienspektrum auf hellem kontinuierlichen Grund [s. Fig. 84 und 85].

In einer Anzahl von Arbeiten (Lawrence und Dunnington [277], Hamos [197], Craggs und Meek [106], Fünfer [167], Glaser [181], Steinhaus, Crosswhite und Dieke [430] u. a.) wird die Abhängigkeit der spektralen Emission vom *zeitlichen Ablauf* des Funkens behandelt. Über die sehr ausführlichen Untersuchungen insbesondere von Glaser wurde bereits in Ziff. 15 berichtet. Danach wird das intensivste Kontinuum zusammen mit den stark verbreiterten Emissionslinien im Zeitraum höchster Stromdichte ausgestrahlt, während sich das Spektrum mit abnehmender Stromdichte jedes einzelnen Entladungsstoßes allmählich zum normalen Linienspektrum ohne kontinuierlichen Untergrund weiterentwickelt. Das gewöhnlich beobachtete Funkenspektrum, wie es etwa auch der Blitz zeigt (s. Israel und Wurm [233]), stellt also eine Überlagerung der in den verschiedenen einander folgenden Funkenstadien emittierten Spektren dar.

Eine besonders ausgeprägte Form des Hochdruckfunkens bilden die bereits erwähnten, zuerst von Anderson [9] bis [12] untersuchten Entladungen

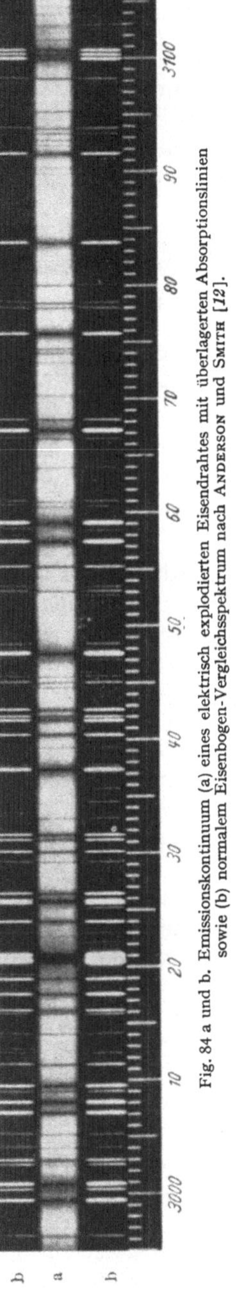

Fig. 84 a und b. Emissionskontinuum (a) eines elektrisch explodierten Eisendrahtes mit überlagerten Absorptionslinien sowie (b) normalem Eisenbogen-Vergleichsspektrum nach Anderson und Smith [12].

eines auf 70 kV aufgeladenen 0,4 μF-Kondensators durch sehr dünne Metall-drähte *(Metalldrahtentladungen)*. Durch den sehr großen Momentanstrom ver-dampfen die dünnen Drähte in äußerst kurzer Zeit („exploded wires"), und die eigentliche Funkenentladung findet in dem so erzeugten Metalldampf statt. Geht die Entladung in freier Luft vor sich, so zeigt ihr Spektrum ein nur schwaches Kontinuum, „explodiert" der Draht dagegen in dem schmalen Schlitz eines Holzblockes oder in einem engen Holzrohr, welches eine Ausdehnung der hoch erhitzten Metalldampfwolke verhindert, so beobachtet man ein äußerst intensives Kontinuum, von dem sich in Absorption die Linien des Drahtmetalls abheben. Ähnliche Erscheinungen treten auch bei Entladungen durch mit Salzlösungen getränkte Asbestfasern auf [128], [396].

Das Kontinuum ist um so intensiver, je leichter das Metall verdampft und je geringer seine Ionisierungsenergie ist. Die Intensität wächst ebenfalls stark mit Erhöhung der Kondensatorspannung. Bei Einschalten immer größerer Selbstinduktion in den Entladungskreis nimmt sie allmählich so weit ab, bis das typische Funkenspektrum schließlich in ein normales Bogenspektrum über-gegangen ist. Wegen der starken Abhängigkeit des Kontinuums vom äußeren Druck wird es in einer sehr ausführlichen Arbeit von Vaudet [449a] als Verdichtungsstoßleuchten gedeutet. Für eine ganz neue Übersicht vgl. Conn [103].

Mit den Metalldrahtentladungen nahe verwandt sind die *Unterflüssigkeits-funken*, bei denen die den Funkenkanal umgebende Flüssigkeit einer Ausdehnung des gebildeten Metalldampfes während der Funkenzeit von unter 10^{-6} sec einen sehr großen Widerstand entgegensetzt, so daß wir es ebenfalls mit einem typischen Hochdruckfunken zu tun haben. Konen [255] hat den Unterwasserfunken zuerst genauer untersucht und auch auf die Brauchbarkeit dieser Lichtquelle für Absorptionsversuche im UV hingewiesen. Die quantitativen Messungen über die Intensitätsverteilung im Unterwasserfunken von Wyneken [490] wurden bereits erwähnt. Ähnliche Untersuchungen stammen von Wrede [484], vgl. auch Assmus [17] und Schubert [421].

Zu den Funkenentladungen gehören auch die stromstarken Kondensator-entladungen durch Kapillaren bei sehr *geringem Gasdruck*, die Lyman [290] als kontinuierliche Absorptionslichtquellen für das äußerste Ultraviolett entwickelt hat und die physikalisch besonders von Anderson [13] untersucht worden sind. Anderson arbeitet mit einem Kondensator von 1 bis 2 μF, einer Spannung von 35 kV und bei Gasdrucken zwischen 0,1 und 10 mm Hg. Dabei werden Strom-dichten bis zu 10^5 Amp/cm² erreicht. Nach seinen Beobachtungen hängt die spektrale Emission nur von der Stromdichte ab; sie ist unabhängig von der Natur des Füllgases und in den angegebenen Grenzen auch vom Druck. Auf die sich zu etwas höheren Drucken erstreckende, die Deutung des Kapillarfunken-kontinuums als Elektronenstrahlung belegende Untersuchung von Hahn und Finkelnburg [195] wurde in Ziff. 15 schon hingewiesen. Zu ähnlichen Ergebnissen gelangt neuerdings Greiner [192], der die kontinuierliche Emission im Druck-bereich zwischen $5 \cdot 10^{-3}$ und 100 mm Hg bei Stromdichten von $14,7 \cdot 10^3$ bis $36,8 \cdot 10^3$ Amp/cm² untersucht hat.

Schließlich sind an dieser Stelle auch die sog. *Kondensatorstoßentladungen* zu erwähnen, die einer normalen Bogenentladung überlagert werden können (Rompe und Schulz [384]). Ein interessantes Zeitlupenverfahren zur zeitlichen Auf-lösung des Stoßleuchtens in einer Quecksilber-Höchstdrucklampe (HBO 200-Osram) hat Göing [182] entwickelt (s. Fig. 85). Über ähnliche Untersuchungen berichtet Euler [138].

55. Emissionskontinua in Stoßwellen. Die bei der stoßartigen Verdichtung von Gasen auftretenden Leuchterscheinungen gewinnen neuerdings immer mehr an Interesse. Quantitative Untersuchungen von FOWLER und Mitarbeitern [158] sowie von KANTROWITZ und Mitarbeitern [356] wurden in Ziff. 15 bereits behandelt. Eine relativ schwache Ausstrahlung hatte schon RAMSAUER [368] beobachtet, der durch Hineinschießen eines Geschosses in einen Gewehrlauf eine stoßartige Kompression des im Lauf eingeschlossenen Gases bewirkte.

Zahlreiche eingehende Untersuchungen über das Verdichtungsstoßleuchten bei der Detonation von Sprengstoffen stammen von MURAOUR und MICHEL-LÉVY [345] bis [347].Werden beispielsweise zwei Mengen eines äußerst brisanten Sprengstoffes in geringem Abstand voneinander zur Detonation gebracht, so wird in dem umgebenden Gas je eine kugelförmig sich ausbreitende Stoßwelle erzeugt, und beim Zusammenprall beider beobachtet man eine 4 μsec dauernde, sehr intensive Leuchterscheinung mit kontinuierlichem Spektrum, das vom roten Spektralgebiet unter günstigen Bedingungen bis zur Grenze der Beobachtung unterhalb 2000 Å reicht. Das Leuchten ist unter sonst gleichen Bedingungen um so intensiver, je höher die Detonationsgeschwindigkeit des Sprengstoffes ist, d.h. je brisanter dieser ist. Bei gegebenem Sprengstoff und gegebenem Zentrenabstand ist das Leuchten nur von der Natur der Gasatmosphäre abhängig. Seine Intensität ist am größten in einatomigen Gasen höchster Dichte und nimmt in der Reihenfolge Kr, Ar, Cl_2, O_2, N_2, CO_2, H_2, C_4H_{10} ab. MICHEL-LÉVY, MURAOUR und VASSY [314] bestimmten weiterhin die Energieverteilung im kontinuierlichen Spektrum der Stoßwellen in A-, He-, O_2-Gas, wobei festgestellt wurde, daß die Verteilung nicht derjenigen eines schwarzen Strahlers entsprach. Einen Vergleich zwischen den in kondensierten Funkenentladungen auftretenden Spektren und den bei Stoßwellen beobachteten führen MURAOUR, ROMAND und VODAR [348] durch.

56. Emissionskontinua in Flammen. Bei den Flammen [173], hat man zwischen zwei Arten echter Emissionskontinua und drittens den unechten, in Wirklichkeit aus diffusen oder sehr dicht liegenden Banden bestehenden scheinbaren Kontinua zu unterscheiden. Letztere werden besonders bei Flammen in Dämpfen mehratomiger Moleküle beobachtet. Die zwei verschiedenen Gruppen echter Flammenkontinua sind:

1. ein von Druck, Temperatur und Ionisierungsenergie der Flammenatome abhängiger kontinuierlicher Grund, dessen Intensität weitgehend vom Ionisierungsgrad in der Flamme abzuhängen scheint und als Elektronenrekombinations-Kontinuum (Ziff. 14) zu deuten ist.

2. die sog. Chemilumineszenz-Kontinua, die wohl sicher als Rekombinationskontinua (Ziff. 32) zwei- oder mehratomiger, in der Flamme sich bildender Moleküle anzusehen sind.

Die Elektronenkontinua in Flammen sind (Literatur vgl. [149]) besonders an der mit Metalldämpfen versetzten, ohne solche nur ein äußerst schwaches Kontinuum zeigenden Bunsenflamme untersucht worden. Die Intensität dieses kontinuierlichen Grundes steigt ganz beträchtlich bei Zusatz eines Alkalisalzes; gleichzeitig steigt infolge der niedrigen Ionisierungsspannung der Alkalien die Elektronen- und Ionendichte gegenüber der nichtleuchtenden Flamme um einige Größenordnungen. Wegen der Abhängigkeit von der Ionisierungsenergie steigt die Kontinuumsintensität ferner vom Lithium ($E_i = 5,37$ eV) bis zum Caesium ($E_i = 3,88$ eV) sehr stark an. Da das Kontinuum besonders intensiv in den Flammenteilen emittiert wird, in denen auch die Nebenserien der Alkalien beobachtet werden, dürfte es sich im wesentlichen um die nach Ziff. 16 ja stets besonders intensiv auftretenden Nebenserienkontinua handeln. Mit diesem Schluß stimmt neben anderen Befunden auch die Beobachtung überein, daß die flachen Intensitätsmaxima der Kontinua sich mit abnehmender Ionisierungsenergie vom Blau-Violett ins Grün verschieben. Schließlich schwächt Anlegen eines starken, die Rekombination von Elektronen und Ionen behindernden elektrischen Feldes ganz allgemein die Intensität dieser Art von Flammenkontinua.

Wie in der Bunsenflamme tritt auch im Spektrum der Knallgasflamme ein kontinuierlicher Grund auf, jedoch infolge der höheren Temperatur auch mit größerer Intensität. Auch hier erhält man bei Alkalizusatz eine starke Intensitätszunahme. Im übrigen verhalten sich die Spektren genau wie die in der Bunsenflamme, woraus zu schließen ist, daß der Chemismus der Flamme keinen direkten Einfluß auf diese Kontinua hat. Aber auch den Chemilumineszenz-Kontinua ist, sofern die Flammentemperatur genügend hoch ist, stets ein kontinuierlicher Grund unterlagert, dessen Intensität mit zunehmendem Ionisierungsgrad wächst.

Das beste Beispiel für die zweite Gruppe von Flammenkontinua, nämlich die von der speziellen in der Flamme ablaufenden Reaktion abhängenden Molekül-Emissionskontinua, ist das viel untersuchte „grüne Kontinuum" einer in Sauerstoff brennenden CO-Flamme, das von 5800 bis etwa 3000 Å reicht. Es wird offenbar bei der Bildung von CO_2 aus CO und O_2 emittiert und dürfte damit als CO_2-Molekül-Rekombinationskontinuum nach Ziff. 32 anzusehen sein. Für diese Deutung spricht z.B., daß nach Hornbeck und Hopfield [221] bei der dem CO_2 entsprechenden stöchiometrischen Zusammensetzung des CO-O_2-Gemisches ein reines und sehr intensives Kontinuum beobachtet wird, während ein Überschuß von O_2 zur gleichzeitigen Emission der O_2-Bandenspektren führt. Zu den Flammen sind natürlich auch die entsprechenden Explosionen (aber nicht Detonationen, auf die in Ziff. 55 eingegangen wurde) zu rechnen, und so ist die Emission des grünen CO_2-Kontinuums bei Explosionen ebenfalls genau studiert worden [220].

Dem besprochenen Kontinuum verwandt ist wahrscheinlich das in Ziff. 48 schon erwähnte gelb-grüne Kontinuum, das nach Gaydon [172] überall dort emittiert wird, wo NO mit Sauerstoffatomen reagiert, besonders also auch von allen NO- und O-Atome enthaltenden Flammen. Gaydon benutzt dieses gelb-grüne Kontinuum deshalb als empfindlichen Nachweis für die Existenz von O-Atomen in Flammen, denen er geringe Mengen von NO beimischt. Ob das in einer C_6H_6-NO-Flamme gefundene intensive Kontinuum [49] mit dem NO-O-Kontinuum identisch ist, muß wohl noch untersucht werden.

Emissionskontinua findet man außer bei der CO-Flamme auch bei vielen anderen Gasflammen. So brennt Wasserstoff in Sauerstoff, NO_2 und Chlor stets unter Emission eines lichtstarken Kontinuums mit Maximum im grünen Spektralgebiet. Die Schwefelflamme wie die des Schwefelwasserstoffes zeigt ebenfalls einen kontinuierlichen Grund mit intensivem Maximum im Grün-Blau, und das Gleiche gilt für die Flamme des Schwefel-Kohlenstoffes in den verschiedensten Gasen. Dagegen wurde das früher für kontinuierlich gehaltene Spektrum der „kalten" Phosphor-Luft-Flamme inzwischen in Banden aufgelöst und diese dem PO-Molekül zugeordnet [386]. Bezüglich der in der älteren Literatur besprochenen Flammenkontinua ist also anscheinend eine gewisse Vorsicht am Platze. Emissionskontinua, die aller Wahrscheinlichkeit nach Rekombinations- oder Dissoziationskontinua mehratomiger Moleküle sind, werden auch in Kohlenwasserstoff-Sauerstoff-Flammen sowie -Explosionen, z.B. in allen Verbrennungsmotoren, beobachtet. Veröffentlichungen über ihre speziellen Eigenschaften und ihre detaillierte Deutung scheinen noch zu fehlen, obwohl Beziehungen zwischen der kontinuierlichen Intensität, dem Reaktionsmechanismus und der Reaktionsgeschwindigkeit gefunden wurden (z. B. [47]). Über die Druckabhängigkeit der Intensität der Flammenkontinua liegen nur wenige Veröffentlichungen vor.

Bekannt ist, daß Wasserstoff in Sauerstoff von höherem Druck mit äußerst heller, im gesamten Sichtbaren kontinuierlich strahlender Flamme brennt, wobei die Intensität dem Druck proportional sein soll.

Einwandfreie Chemilumineszenzkontinua mit Maxima im Blau bzw. Grün ohne unterlagerten, auf der Wechselwirkung von Elektronen und Iònen beruhenden kontinuierlichen Grund beobachtet man nach POLANYI [359] bei der Reaktion hochverdünnter Alkalidämpfe (Na oder K) mit den Zinnhalogeniden $SnCl_4$ oder $SnBr_4$. Ganz ähnliche Kontinua erhält man übrigens bei der Einwirkung von aktivem Stickstoff auf $SnCl_4$ [431] und bei Entladungen in $SnCl_4$-Dampf [374]. Auch bei der Reaktion von Kaliumdampf mit Cl_2 und Br_2 erhält man nach [262] Emissionskontinua, die als Emissionsfall I (Ziff. 31) dem Zerfall angeregter K_2-Moleküle in normale K-Atome zugeschrieben worden sind, ohne daß diese Deutung völlig gesichert scheint.

Allgemein steht die Untersuchung der kontinuierlichen Flammenspektren noch sehr am Anfang, dürfte aber für die Aufklärung der Reaktionsmechanismen in Flammen noch erhebliche Bedeutung gewinnen.

57. Nachleuchtkontinua. Als Nachleuchterscheinung bezeichnet man eine Strahlung, die nach vorhergegangener Anregung emittiert wird, falls die zwischen Anregung und Emission liegende Zeit erheblich größer ist als die mittlere Lebensdauer normaler angeregter Zustände. Am einfachsten sind Nachleuchtkontinua beobachtbar in den zwischen den einzelnen Stromstößen liegenden Pausen intermettierender Entladungen, ferner räumlich getrennt von der eigentlichen Entladung durch Absaugen der angeregten Gase oder Dämpfe aus der Entladung und Beobachtung in einem seitlichen Rohr. Drei verschiedene Gruppen kontinuierlicher Emissionsspektren können im Nachleuchten beobachtet werden.

In stromstarken Funkenentladungen bieten die stromlosen Pausen zwischen den einzelnen Durchschlägen den Ionen und Elektronen Gelegenheit zur Wiedervereinigung unter Emission von Seriengrenzkontinua. Diese werden daher, auch besonders intensiv in intermettierenden Entladungen, d.h. eigentlich im Nachleuchten, beobachtet.

Zweitens werden im Nachleuchten solche Strahlungsvorgänge beobachtet, die durch Stöße zweiter Art und namentlich solchen mit metastabilen Teilchen überhaupt erst ermöglicht werden. Kontinuierliche Spektren dieser Art werden z.B. besonders von den VAN DER WAALS-Molekülen wie Hg_2, Cd_2 und Zn_2 emittiert; ihr bekanntester Vertreter ist das große grüne Quecksilberkontinuum mit Maximum bei 4850 Å (Ziff. 41).

Als dritte Gruppe von Nachleuchterscheinungen sind endlich die Strahlungsvorgänge zu nennen, deren Träger durch chemische Reaktionen in der Entladung erst gebildet werden. Die entsprechenden Kontinua stellen also einen Sonderfall der Chemilumineszenzspektren dar, die oben bereits erwähnt wurden. Dieser Gruppe gehört z.B. das Ziff. 48 erwähnte Nachleuchtkontinuum an, das nach Entladungen in Sauerstoff-Stickstoffgemischen beobachtet wird und einer Reaktion von O-Atomen und NO-Molekülen zuzuschreiben ist, die beide in der Entladung erst gebildet werden.

Literaturverzeichnis.

[1] ABBOT, C. G., and F. E. FOWLE: Ann. Smithson Astrophys. Obs. 2 (1908).

[2] ABBOT, C. G., F. E. FOWLE and L. B. ALDRICH: Ann. Smithson Astrophys. Obs. 4 (1922).

[3] ABRAGAM, A.: J. Phys. Radium 9, 143 (1948).

[4] ACTON, A. P., R. G. AICKIN and N. S. BAYLISS: J. Chem. Phys. 4, 474 (1936).

[5] AICKIN, R. G., and N. S. BAYLISS: Trans. Faraday Soc. 34, 1371 (1938).

[6] ALLEN, C. W.: Phys. Rev. 39, 42 (1932).

196   W. FINKELNBURG und TH. PETERS: Kontinuierliche Spektren.

[7] ALLEN, C. W.: Monthly Not. **106**, 137 (1946).
[8] ALON, I., J. ARSAC et J. L. STEINBERG: C. R. Acad. Sci., Paris **237**, 300 (1953).
[9] ANDERSON, J. A.: Astrophys. J. **51**, 37 (1920).
[10] ANDERSON, J. A.: Proc. Nat. Acad. Amer. **6**, 42 (1920).
[11] ANDERSON, J. A.: Proc. Nat. Acad. Amer. **8**, 231 (1922).
[12] ANDERSON, J. A., and S. SMITH: Astrophys. J. **64**, 295 (1926).
[13] ANDERSON, J. A.: Astrophys. J. **75**, 394 (1932).
[14] ANDERSON, W. T.: J. Opt. Soc. Amer. **41**, 385 (1951).
[15] ARNOT, F. L., and M. B. M'EWEN: Proc. Roy. Soc. Lond. **165**, 133 (1938).
[16] ARZIMOVICH, L., i I. POMERANCHUK: J. Phys. USSR. **9**, 267 (1945).
[17] ASSMUS, F.: Ann. Phys. **36**, 737 (1940).
[18] AUGER, P.: Thèses, Paris 1926.
[19] BALASSE, G., et O. GOCHE: Bull. de Belge **12**, 385 (1926).
[20] BALASSE, G.: C. R. Acad. Sci., Paris **184**, 1002, 1320 (1927).
[21] BALASSE, G.: Bull. de Belge **13**, 543 (1927).
[22] BALASSE, G.: C. R. Acad. Sci., Paris **186**, 310 (1928).
[23] BALASSE, G.: J. Phys. Radium **5**, 304 (1934).
[24] BALASSE, G.: Bull. de Belge **20**, 563 (1934).
[25] BARBIER, D., et D. CHALONGE: Ann. d'Astrophys. **2**, 254 (1939).
[26] BARBIER, D., D. CHALONGE, et M. MASRIERA: C. R. Acad. Sci., Paris **212**, 984 (1941).
[27] BARKLA, C. G.: Phil. Trans. Roy. Soc. Lond. **204**, 467 (1905).
[28] BARROW, R. F., G. DRUMMOND and H. C. ROWLINSON: Proc. Phys. Soc. Lond. **66**, 885 (1953).
[29] BARTELS, H.: Z. Physik **25**, 378 (1924).
[30] BARTELS, H.: Z. Physik **73**, 203 (1931).
[31] BARTELS, H.: Z. Physik **128**, 546 (1950).
[32] BATES, D. R.: Monthly Not. **100**, 25 (1939).
[33] BATES, D. R., R. A. BUCKINGHAM, H. S. W. MASSEY and J. J. UNWIN: Proc. Roy. Soc. Lond. A **170**, 322 (1939).
[34] BATES, D. R., and H. S. W. MASSEY: Proc. Roy. Soc. Lond. A **177**, 281 (1941).
[35] BATES, D. R., and H. S. W. MASSEY: Proc. Roy. Soc. Lond. A **177**, 329 (1941).
[36] BATES, D. R., and H. S. W. MASSEY: Trans. Roy. Soc. Lond. **239**, 269 (1943).
[37] BATES, D. R.: Monthly Not. **106**, 128 (1946).
[38] BATES, D. R.: Monthly Not. **106**, 423, 432 (1946).
[39] BATES, D. R.: Proc. Roy. Soc. Lond. A **188**, 350 (1947).
[40] BATES, D. R., and M. J. SEATON: Monthly Not. Roy. Astr. Soc. **109**, 698 (1949).
[41] BAUM, W. A., and L. DUNKELMAN: J. Opt. Soc. Amer. **40**, 782 (1950).
[42] BAYLISS, N. S., and A. L. G. REES: Nature, Lond. **143**, 560 (1939).
[43] BAYLISS, N. S., and A. L. G. REES: Trans. Faraday Soc. **35**, 792 (1939).
[44] BAYLISS, N. S., and A. L. G. REES: J. Chem. Phys. **8**, 377 (1940).
[45] BAYLISS, N. S.: Letter in Nature, Lond. **163**, 764 (1949).
[46] BAYLISS, N. S.: J. Chem. Phys. **18**, 292 (1950).
[47] BECK, G., u. C. ERICHSEN: Forsch. Ing.-Wes. **7**, 1 (1936).
[48] BECK, G.: Phys. Rev. **74**, 795 (1948).
[49] BEHRENS, H., u. F. RÖSSLER: Naturwiss. **36**, 218 (1949).
[50] BERGMANN, E., u. R. SAMUEL: Nature, Lond. **141**, 832 (1938).
[51] BEUTLER, H.: Z. Physik **86**, 495, 710 (1933).
[52] BEUTLER, H.: Z. Physik **87**, 19 (1934).
[53] BEUTLER, H., u. K. GUGGENHEIMER: Z. Physik **87**, 176, 188 (1934).
[54] BEUTLER, H., u. K. GUGGENHEIMER: Z. Physik **88**, 25 (1934).
[55] BEUTLER, H.: Z. Physik **91**, 131 (1934).
[56] BEUTLER, H., u. W. DEMETER: Z. Physik **91**, 143, 202, 218 (1934).
[57] BEUTLER, H.: Z. phys. Chem. **29**, 315 (1935).
[58] BEUTLER, H.: Z. Physik **93**, 177 (1935).
[59] BEUTLER, H., u. H. O. JÜNGER: Z. Physik **100**, 80 (1936).
[60] BEUTLER, H., u. H. O. JÜNGER: Z. Physik **101**, 285 (1936).
[61] BHATTACHARGGA u. J. MURARI: Indian J. Phys. **19**, 20 (1945).
[62] BIERMANN, L.: Nachr. Ges. Wiss. Göttingen Nr. 1, 12 (1947).

[63] BIERMANN, L.: Naturwiss. **34**, 87 (1947).

[64] BIERMANN, L., u. K. LÜBECK: Z. Astrophys. **26**, 43 (1949).

[65] BINDER, J. L.: Phys. Rev. **54**, 114 (1938).

[66] BIONDI, M. A., and S. C. BROWN: Phys. Rev. **75**, 1700 (1949).

[67] BIONDI, M. A., and S. C. BROWN: Phys. Rev. **76**, 1697 (1949).

[68] BÖHM-VITENSE, E.: Z. Astrophys. **36**, 145 (1955).

[69] BOTT, J.: Ann. Phys. **35**, 314 (1939).

[70] BRADDICK, H. J. J., and R. W. DITCHBURN: Proc. Roy. Soc. Lond., Ser. A **150**, 478 (1935).

[71] BRANSDEN, B. H., and A. DALGARUO: Proc. Phys. Soc. Lond. A **66**, 904, 911 (1953).

[72] BROGLIE, M. DE: C. R. Acad. Sci., Paris **163**, 87 (1916).

[73] BUCKINGHAM, R. A., S. REID and R. SPENCE: Monthly Not. **112**, 382 (1952).

[74] BURHORN, F.: Z. Physik **140**, 440 (1955).

[75] BURKHARDT, G.: Ann. Phys. **26**, 567 (1936).

[76] BURKHARDT, G., G. ELWERT u. A. UNSÖLD: Z. Astrophys. **25**, 310 (1948).

[77] BURKHARDT, G., u. A. SCHLÜTER: Z. Astrophys. **26**, 295 (1949).

[78] BURKHARDT, G.: Ann. Phys. **5**, 373 (1950).

[79] BUSZ, G., u. P. SCHULZ: Ann. Phys. **6**, 232 (1949).

[80] BUSZ, G., u. W. FINKELNBURG: Z. Physik **139**, 212 (1954).

[81] BUTKOW, K.: J. exp. theor. Phys. USSR. **9**, 561 (1939).

[82] CAMERON, W. H. B., and A. ELLIOT: Proc. Roy. Soc. Lond. **169**, 463 (1939).

[83] CARTWRIGHT, C. H.: Z. Physik **90**, 480 (1934).

[84] CENNAMO, F.: Nuovo Cim. (N.S.) **16**, 355 (1939).

[85] CERENKOV, P. A.: C. R. Acad. Sci. URSS. **2**, 451, 455 (1934).

[86] CERENKOV, P. A.: C. R. Acad. Sci. URSS **3**, 414 (1936); **14**, 101, 105 (1937). — Phys. Rev. **52**, 378 (1937).

[87] CHAKRAVORTI, S. K.: Z. Physik **109**, 25 (1938).

[88] CHALONGE, D., et G. DÉJARDIN: Bull. Soc. franç. Phys. **441** (1940).

[89] CHALONGE, D., et V. KOURGANOFF: Ann. d'Astrophys. **9**, 69 (1946).

[90] CHANDRASEKHAR, S.: Astrophys. J. **102**, 223, 395 (1945).

[91] CHANDRASEKHAR, S., and F. H. BREEN: Astrophys. J. **104**, 430 (1946).

[92] CH'EN SHANG-YI: Phys. Rev. **58**, 884 (1940).

[93] CHRISTIANSEN, W. N., and J. A. WARBURTON: Austral. J. Phys. **6**, 190, 262 (1953).

[94] COHN, W. M.: Z. Physik **75**, 544 (1932).

[95] COHN, W. M.: Phys. Rev. **58**, 50 (1940).

[96] COLLINS, G., and G. H. REILING: Phys. Rev. **54**, 499 (1938).

[97] CONDON, E. U.: Phys. Rev. **28**, 1182 (1926).

[98] CONDON, E. U.: Phys. Rev. **32**, 858 (1928).

[99] COOLIDGE, A. S., H. M. JAMES and R. D. PRESENT: J. Chem. Phys. **4**, 193 (1936).

[100] COOLIDGE, A. S., and H. M. JAMES: J. Chem. Phys. **6**, 730 (1938).

[101] COOLIDGE, A. S., H. M. JAMES and E. L. VERNON: Phys. Rev. **54**, 726 (1938).

[102] COOLIDGE, A. S.: Phys. Rev. **65**, 236 (1944).

[103] CONN, W. M.: Z. angew. Phys. **7**, 539 (1955).

[104] COOPER, J. N.: Phys. Rev. **65**, 155 (1944).

[105] COX, R. T.: Phys. Rev. **66**, 106 (1944).

[106] CRAGGS, J. D., and J. M. MEEK: Proc. Roy. Soc. Lond. **186**, 241 (1946).

[107] CRINO, B. R.: Ann. Astrophys. **14**, 105, 221 (1951).

[108] DAHLKE, W.: Z. Physik **114**, 205, 672 (1939).

[109] DAHLKE, W.: Z. Physik **115**, 1 (1940).

[110] DANDURAND, P., and R. B. HOLT: Phys. Rev. **82**, 868 (1951).

[111] DATTA, S., and B. CHAKRAVARTY: Proc. Nat. Inst. Sci., Indian **7**, 297 (1941).

[112] DEDRICK, K. G.: Phys. Rev. **85**, 765 (1952).

[113] DIEKE, G. H., u. J. J. HOPFIELD: Z. Physik **40**, 299 (1927).

[114] DIEKE, G. H., and C. S. RAINWATER: Department of Physics. Baltimore: The John Hopkins University 1950.

[115] DITCHBURN, R. W.: Proc. Roy. Soc. Lond. A **117**, 486 (1928).

[116] DITCHBURN, R. W., J. TUNSTEAD and J. G. YATES: Proc. Roy. Soc. Lond. A **181**, 386 (1943).

[*117*] Ditchburn, R. W., and P. J. Jutsum: Nature, Lond. **165**, 723 (1950).

[*118*] Ditchburn, R. W., and G. V. Marr: Proc. Phys. Soc. Lond. A **66**, 655 (1953).

[*119*] Ditchburn, R. W., and D. W. O. Heddle: Proc. Roy. Soc. Lond. A **220**, 61 (1953).

[*120*] Ditchburn, R. W., P. J. Jutsum and G. V. Marr: Proc. Roy. Soc. Lond. A **219**, 89 (1953).

[*121*] Dove, H. W.: Poggendorfs Ann. **104**, 186 (1858).

[*122*] Duane, W., and F. L. Hunt: Phys. Rev. **6**, 166 (1915).

[*123*] Duane, W.: Proc. Nat. Acad. Sci. USA. **13**, 662 (1927).

[*124*] DuMond, J. W.: Rev. Mod. Phys. **5**, 1 (1933).

[*125*] DuMond, J. W., and V. Bollman: Phys. Rev. **51**, 400 (1937).

[*126*] DuMond, J. W., and H. A. Kirkpatrick: Phys. Rev. **52**, 419 (1937).

[*127*] Dymond, E. G.: Z. Physik **34**, 553 (1925).

[*128*] Eckstein, L., u. I. M. Freeman: Z. Physik **64**, 547 (1930).

[*129*] Eddington, A. S.: Monthly Not. **84**, 104 (1924).

[*130*] Edlén, B.: Ark. Mat. Astr. Och. Fys. B **28**, Nr. 1 (1941).

[*131*] Elder, F. R., R. V. Langmuir and H. C. Pollock: Phys. Rev. **74**, 52 (1948).

[*132*] Elenbaas, W.: The High Pressure Mercury Vapour Discharge. Amsterdam: North-Holland Publishing Co. 1951.

[*133*] Elliot, A., and W. H. B. Cameron: Proc. Roy. Soc. Lond. **158**, 681 (1931).

[*134*] Elliot, A.: Proc. Roy. Soc. Lond. **174**, 273 (1940).

[*135*] Elwert, G.: Z. Naturforsch. **3**a, 477 (1948).

[*136*] Elwert, G.: Z. Naturforsch. **7**a, 432 (1952).

[*137*] Elwert, G.: Z. Naturforsch. **7**a, 703 (1952).

[*138*] Euler, J.: Z. angew. Phys. **4**, 82 (1952).

[*139*] Euler, J.: Ann. Phys. **11**, 203 (1953).

[*140*] Finkelnburg, W., u. W. Weizel: Z. Physik **68**, 577 (1931).

[*141*] Finkelnburg, W.: Z. Physik **70**, 375 (1931).

[*142*] Finkelnburg, W., u. W. Steiner: Z. Physik **79**, 69 (1932).

[*143*] Finkelnburg, W.: Z. Physik **81**, 781 (1933).

[*144*] Finkelnburg, W.: Phys. Rev. **45**, 341 (1934).

[*145*] Finkelnburg, W.: Z. Physik **88**, 297, 768 (1934).

[*146*] Finkelnburg, W.: Z. Physik **93**, 201 (1935).

[*147*] Finkelnburg, W., u. O. Th. Hahn: Phys. Z. **39**, 98 (1938).

[*148*] Finkelnburg, W., u. H. Hess: Phys. Z. **39**, 666 (1938).

[*149*] Finkelnburg, W.: Kontinuierliche Spektren. Berlin: Springer 1938.

[*150*] Finkelnburg, W.: Hochstromkohlebogen. Berlin-Göttingen-Heidelberg: Springer 1948.

[*151*] Finkelnburg, W.: J. Opt. Soc. Amer. **39**, 185 (1949).

[*152*] Fischer, E., u. H. König: Phys. Z. **39**, 313 (1938).

[*153*] Fischer, H.: Phys. Rev. **89**, 336 (1953).

[*154*] Fock, V.: Z. Physik **61**, 126 (1930).

[*155*] Foote, P. D., W. F. Meggers and R. L. Chenault: J. Opt. Soc. Amer. **9**, 541 (1924).

[*156*] Fowler, R. H., and E. A. Milne: Monthly Not. **83**, 403 (1923).

[*157*] Fowler, C. A.: Phys. Rev. **62**, 141 (1942).

[*158*] Fowler, R. G., J. S. Goldstein and B. E. Clotfelter: Phys. Rev. **82**, 879 (1951).

[*159*] Fowler, R. G., W. R. Atkinson and L. W. Marks: Phys. Rev. **87**, 966 (1952).

[*160*] Franck, J., u. W. Grotrian: Z. Physik **4**, 89 (1921).

[*161*] Franck, J.: Z. phys. Chem. **120**, 144 (1926).

[*162*] Franck, J., H. Sponer u. E. Teller: Z. phys. Chem. **18**, 88 (1932).

[*163*] Frank, I., i. I. Tamm: C. R. Acad. Sci. URSS. **14**, 109 (1937).

[*164*] Franz, W.: Ann. Phys. **29**, 721 (1937).

[*165*] Fuchs, R.: Z. Physik **130**, 69 (1951).

[*166*] Fues, E.: Z. Physik **43**, 726 (1927).

[*167*] Fünfer, E.: Z. angew. Phys. **1**, 295 (1949).

[*168*] Garth, R. C., and G. E. Moore: Phys. Rev. **60**, 208 (1941).

[*169*] Garton, W. R. S.: Proc. Phys. Soc. Lond. A **65**, 268 (1952).

[*170*] Gaunt, J. A.: Phil. Trans. Roy. Soc. Lond. **229**, 163 (1930).

[*171*] Gaunt, J. A.: Proc. Roy. Soc. Lond. A **126**, 654 (1930).

[172] GAYDON, A. G.: Proc. Roy. Soc. Lond. A **183**, 111 (1944).

[173] GAYDON, A. G.: Spectroscopy and Combustion Theory, 2. Aufl. London: Chapman & Hall 1948.

[174] GEHLHOFF, G.: Lehrbuch der Technischen Physik, Bd. III. Leipzig 1929.

[175] GERDIEN, H., u. A. LOTZ: Wiss. Veröff. Siemenskonz. **2**, 489 (1922).

[176] GETTING, J. A.: Phys. Rev. **71**, 123 (1947).

[177] GIBSON, G. E., O. K. RICE and N. S. BAYLISS: Phys. Rev. **44**, 193 (1933).

[178] GINSBURG, V. L.: J. Phys. USSR. **2**, 441 (1940).

[179] GINSBURG, V. L.: Dokl. Acad. Nauk. USSR. **56**, 253 (1947).

[180] GLASER, G.: Optik **7**, 33, 61, 78 (1950).

[181] GLASER, G.: Z. Naturforsch. **6**a, 706 (1951).

[182] GÖING, W.: Naturwiss. **24**, 558 (1950).

[183] GÖING, W., H. MEIER u. H. MEINEN: Z. Physik **140**, 376 (1955).

[184] GOLDBERG, L.: Astrophys. Z. **90**, 414 (1939).

[185] GOLDBERG, L.: Astrophys. Z. **93**, 244 (1941).

[186] GOODEVE, C. F., and A. W. C. TAYLOR: Proc. Roy. Soc. Lond. **152**, 221 (1935).

[187] GRANDSIRE, G.: Ann. Astrophys. **17**, 287 (1954).

[188] GREAVES, W. M. H.: Monthly Not. **108**, 131 (1948).

[189] GREEN, L. C.: Astrophys. J. **109**, 289 (1949).

[190] GREEN, L. C., and N. E. WEBER: Astrophys. J. **111**, 587 (1950).

[191] GREEN, L. C., N. E. WEBER and E. KRAWITZ: Astrophys. J. **113**, 690 (1951).

[192] GREINER, H.: Note in Naturwiss. **40**, 238 (1953).

[193] GROTH, W.: Z. Elektrochem. **45**, 262 (1939).

[194] GROTRIAN, W.: Z. Astrophys. **8**, 124 (1934).

[195] HAHN, O. TH., u. W. FINKELNBURG: Z. Physik **122**, 36 (1944).

[196] HAMADA, H.: Phil. Mag. **12**, 50 (1931).

[197] HAMOS, V. v.: Ann. Phys. **7**, 857 (1930).

[198] HARGREAVES, J.: Proc. Cambridge Phil. Soc. **25**, 75 (1929).

[199] HARRISON, G. R.: Phys. Rev. **24**, 466 (1924).

[200] HARTMANN, P. L., and D. H. TOMBOULIAN: Phys. Rev. **91**, 1577 (1953).

[201] HARTREE, D. R.: Proc. Cambridge Phil. Soc. **24**, 89 (1928).

[202] HARTREE, D. R.: Phys. Z. **30**, 517 (1929).

[203] HELLER, TH.: Z. Astrophys. **38**, 55 (1955).

[204] HENNING, H. J.: Ann. Phys. **13**, 599 (1932).

[205] HENRI, V.: Structure des Molecules. Paris 1925.

[206] HERCZOG, A., u. K. WIELAND: Helv. phys. Acta **23**, 432 (1950).

[207] HERMAN, R., C. WENIGER and L. HERMAN: Phys. Rev. **82**, 751 (1951).

[208] HERZBERG, G.: Ann. Phys. **84**, 565 (1927).

[209] HERZBERG, G.: Naturwiss. **20**, 577 (1932).

[210] HERZBERG, G., u. E. TELLER: Z. phys. Chem. **21**, 409 (1933).

[211] HERZBERG, G.: Molecular Spectra and Molecular Structure, Bd. 1, 2. Aufl. New York: Van Nostrand 1953.

[212] HETTNER, G.: Z. Physik **131**, 385 (1952).

[213] HICKS, B.: Phys. Rev. **52**, 436 (1937).

[214] HIRSCHFELDER, J. O., and J. W. LINNETT: J. Chem. Phys. **18**, 130 (1950).

[215] HOLT, R. B., J. M. RICHARDSON, B. HOWLAND and B. T. McCLURE: Phys. Rev. **77**, 239 (1950).

[216] HOLTSMARK, J.: Phys. Z. **20**, 88 (1919).

[217] HOPFIELD, J. J.: Phys. Rev. **20**, 573 (1922).

[218] HOPFIELD, J. J.: Phys. Rev. **35**, 1133 (1930).

[219] HOPFIELD, J. J.: Phys. Rev. **36**, 784 (1930).

[220] HORNBECK, G. A.: J. Chem. Phys. **16**, 845 (1948).

[221] HORNBECK, G. A., and H. S. HOPFIELD: J. Chem. Phys. **17**, 982 (1949).

[222] HORST, D. TH. J. TER, u. G. A. W. RUTGERS: Physica, Haag **19**, 565 (1953).

[223] HOWELL, H. G.: Proc. Roy. Soc. Lond. **163**, 242 (1937).

[224] HOWELL, H. G., and N. COULSON: Proc. Roy. Soc. Lond. **166**, 238 (1938).

[225] HOWELL, H. G., and N. COULSON: Proc. Phys. Soc. Lond. **53**, 706 (1941).

[226] HUANG, K.: Astrophys. J. **101**, 196 (1945).

[227] Huang, S. S.: Astrophys. J. **108**, 354 (1948).
[228] Huggins, W., and L.: Atlas of Representative Stellar Spectra. London 1899.
[229] Hulburt, H. M., and J. O. Hirschfelder: J. Chem. Phys. **9**, 61 (1941).
[230] Humphreys, W. J.: Astrophys. J. **26**, 18 (1907).
[231] Inglis, D. R., and E. Teller: Astrophys. J. **90**, 439 (1939).
[232] Ishèguro, E.: J. Phys. Soc. Japan **3**, 129 (1948).
[233] Israel, H., u. K. Wurm: Naturwiss. **29**, 778 (1941).
[234] Jablonski, A.: Acta phys. Pol. **6**, 350 (1938).
[235] James, H. M., and A. S. Coolidge: J. Chem. Phys. **1**, 825 (1933).
[236] James, H. M.: J. Chem. Phys. **2**, 794 (1934).
[237] James, H. M., A. S. Coolidge and R. D. Present: J. Chem. Phys. **4**, 187 (1936).
[238] James, H. M., and A. S. Coolidge: Phys. Rev. **51**, 1001 (1937).
[239] James, H. M., and A. S. Coolidge: Phys. Rev. **55**, 184 (1939).
[240] Jancke, H. O.: Z. Physik **99**, 169 (1936).
[241] Jansky, K. G.: Proc. Inst. Radio Engrs. **20**, 1920 (1932).
[242] Jauch, J. M., and K. M. Watson: Phys. Rev. **74**, 1485 (1948).
[243] Jelley, J. V.: Brit. J. Appl. Phys. **6**, 227 (1955).
[244] Jen, C. K.: Phys. Rev. **43**, 540 (1933).
[245] Jenkins, F. A., and G. D. Rochester: Phys. Rev. **52**, 1135 (1937).
[246] Jürgens, G.: Z. Physik **134**, 21 (1952).
[247] Kappeler, H.: Ann. Phys. **27**, 129 (1936).
[248] Kern, J.: Z. Physik **114**, 552 (1939).
[249] King, G. W., and J. H. van Vleck: Phys. Rev. **55**, 1165 (1939).
[250] Koch, B.: Ann. Phys. **33**, 335 (1938).
[251] Kondratjew, V., u. A. Leipunsky: Z. Physik **50**, 366 (1928).
[252] Kondratjew, V., u. A. Leipunsky: Z. Physik **56**, 353 (1929).
[253] Kondratjew, V., and A. Jakowlewa: J. Phys. Chem. **14**, 853 (1940).
[254] Kondratjew, V., and A. Jakowlewa: J. Phys. Chem. **14**, 859 (1940).
[255] Konen, H.: Ann. Phys. **9**, 742 (1902).
[256] Körwien, H.: Z. Physik **91**, 1 (1934).
[257] Kramers, H. A.: Phil. Mag. **46**, 836 (1923).
[258] Krefft, H.: Naturwiss. **19**, 269 (1931).
[259] Krefft, H.: Phys. Z. **32**, 948 (1931).
[260] Krefft, H.: Z. Physik **77**, 752 (1932).
[261] Krefft, H., F. Rössler u. A. Rüttenauer: Z. techn. Phys. **18**, 20 (1937).
[262] Krocsak, M., u. G. Schay: Z. phys. Chem. **19**, 344 (1932).
[263] Kruger, P. G., u. F. Paschen: Ann. Phys. **8**, 1005 (1931).
[264] Kuhn, W.: Z. Physik **33**, 408 (1925).
[265] Kuhn, H.: Z. Physik **39**, 77 (1926).
[266] Kuhn, H.: Z. Physik **63**, 458 (1930).
[267] Kuhn, H., and O. Oldenberg: Phys. Rev. **41**, 72 (1932).
[268] Kuhn, H.: Z. Physik **76**, 782 (1932).
[269] Kuhn, H.: Proc. Roy. Soc. Lond. **158**, 230 (1937).
[270] Kulenkampff, H.: Ann. Phys. **69**, 548 (1922).
[271] Ladenburg, R., C. C. van Voorhis and J. C. Boyce: Phys. Rev. **40**, 1018 (1932).
[272] Ladenburg, R., and C. C. van Voorhis: Phys. Rev. **43**, 315 (1933).
[273] Larché, K.: Z. Physik **136**, 74 (1953).
[274] Larenz, R. W.: Z. Physik **129**, 327, 343 (1951).
[275] Larenz, R. W.: Z. Naturforsch. **10**a, 761, 766 (1955).
[276] Lawrence, E. O., and N. E. Edlefsen: Phys. Rev. **34**, 1056 (1929).
[277] Lawrence, E. O., and F. G. Dunnington: Phys. Rev. **35**, 396 (1930).
[278] Lee, P., u. G. L. Weissler: Astrophys. J. **115**, 570 (1952).
[279] Lee, P., and G. L. Weissler: J. Opt. Soc. Amer. **42**, 214 (1952).
[280] Lee, P., and G. L. Weissler: Phys. Rev. **92**, 533 (1953).
[281] Lennuier, R.: C. R. Acad. Sci., Paris **213**, 169 (1941).
[281a] Leverenz, H. W.: Luminescence of Solids. London: Chapman & Hall 1950.
[282] Levintov, I. I.: Bull. Acad. Sci. USSR. **11**, 229 (1947).

[*283*] LIENARD, A.: L'Eclairage Elec. **16**, 5 (1898).

[*284*] LILIENFELD, J. E.: Phys. Z. **20**, 280 (1919).

[*285*] LOCHTE-HOLTGREVEN, W., u. W. NISSEN: Z. Physik **133**, 124 (1952).

[*286*] LORD, M. P.: Proc. Phys. Soc. Lond. **58**, 477 (1946).

[*287*] LUYCKX, A.: Proc. Roy. Soc. Lond. **172**, 492 (1939).

[*288*] LÜST, R.: Z. Astrophys. **37**, 67 (1955).

[*289*] LYMAN, TH.: Nature, Lond. **113**, 785 (1924).

[*290*] LYMAN, TH.: Astrophys. J. **60**, 1 (1924).

[*291*] LYOT, B.: C. R. Acad. Sci., Paris **202**, 392 (1936).

[*292*] MAECKER, H.: Z. Physik **114**, 500 (1939).

[*293*] MAECKER, H.: Z. Physik **116**, 257 (1940).

[*294*] MAECKER, H.: Z. Physik **129**, 108 (1951).

[*295*] MAECKER, H., u. TH. PETERS: Z. Physik **139**, 448 (1954).

[*296*] MARR, G. V.: Proc. Roy. Soc. Lond. A **224**, 83 (1954).

[*297*] MARSHALL, J.: Phys. Rev. **86**, 583, 685 (1952).

[*298*] MARTYN, D. F.: Nature, Lond. **158**, 632 (1946).

[*299*] MASSEY, H. S. W., and R. A. SMITH: Proc. Roy. Soc. Lond. A **155**, 472 (1936).

[*300*] MASSEY, H. S. W.: Negative Ions. Cambridge U. P. 1938.

[*301*] MASSEY, H. S. W., and D. R. BATES: Astrophys. J. **91**, 202 (1940).

[*302*] MATHER, R. L.: Phys. Rev. **84**, 181 (1951).

[*303*] MATHUR, L. S.: Indian J. Phys. **11**, 177 (1937).

[*304*] MAUE, A. W.: Ann. Phys. **13**, 161 (1932).

[*305*] MAURER, E., u. H. KOLZ: Z. angew. Phys. **2**, 223 (1950).

[*306*] McCOUBREY, A. O.: Phys. Rev. **93**, 1249 (1954).

[*307*] McPHERSON, H. G.: J. Opt. Soc. Amer. **30**, 189 (1940).

[*308*] MECKE, R.: Ann. Phys. **71**, 104 (1924).

[*309*] MENZEL, D. H., and C. L. PEKERIS: Monthly Not. **96**, 77 (1936).

[*310*] METROPOLIS, N.: Phys. Rev. **55**, 636 (1939).

[*311*] METROPOLIS, N., and H. BEUTLER: Phys. Rev. **55**, 1113 (1939).

[*312*] MICHARD, R.: Rec. Opt. **28**, 479 (1949).

[*313*] MICHARD, R.: Bull. Astr. Netherl. **11**, 227 (1950).

[*314*] MICHEL-LÉVY, A., H. MURAOUR et E. VASSY: L'Optique **20**, 149 (1941).

[*315*] MILNE, E. A.: Monthly Not. **85**, 750 (1925).

[*316*] MIYAMOTO, S.: Publ. Astr. Soc. Japan **1**, 10 (1949).

[*317*] MOHLER, F. L.: Phys. Rev. **31**, 187 (1928).

[*317a*] MOHLER, F. L.: Rev. Mod. Phys. **1**, 216 (1929).

[*318*] MOHLER, F. L., and C. BOECKNER: Bur. Stand. J. Res. Wash. **2**, 489 (1929).

[*319*] MOHLER, F. L., and C. BOECKNER: Bur. Stand. J. Res. Wash. **3**, 303 (1929).

[*320*] MOHLER, F. L., and C. BOECKNER: Bur. Stand. J. Res. Wash. **6**, 673 (1931).

[*321*] MOHLER, F. L., and C. BOECKNER: Bur. Stand. J. Res. Wash. **7**, 751 (1931).

[*322*] MOHLER, F. L.: Bur. Stand. J. Res. Wash. **8**, 357 (1932).

[*323*] MOHLER, F. L.: Phys. Rev. **43**, 374 (1933).

[*324*] MOHLER, F. L.: Bur. Stand. J. Res. Wash. **10**, 771 (1933).

[*325*] MOHLER, F. L.: Bur. Stand. J. Res. Wash. **17**, 45, 849 (1936).

[*326*] MOHLER, F. L.: Bur. Stand. J. Res. Wash. **19**, 447, 559 (1937).

[*327*] MOHLER, F. L.: Nat. Bur. Stand. **21**, 697, 873 (1938).

[*328*] MOORE, B. E.: Astrophys. J. **54**, 191, 246 (1921).

[*329*] MOORE, H. R.: Science, Lancaster, Pa. **66**, 543 (1927).

[*330*] MOORE, G. E., and R. C. GARTH: Phys. Rev. **60**, 216 (1941).

[*331*] MOSELEY, J. H., and C. G. DARWIN: Phil. Mag. **26**, 210 (1913).

[*332*] MÜLLER, P., u. M. WEHRLI: Helv. phys. Acta **15**, 397 (1942).

[*333*] MULDERS, G. F. W.: Z. Astrophys. **11**, 132 (1935).

[*334*] MULLIKEN, R. S.: J. Chem. Phys. **1**, 492 (1933).

[*335*] MULLIKEN, R. S.: Phys. Rev. **46**, 549 (1934).

[*336*] MULLIKEN, R. S.: J. Chem. Phys. **4**, 620 (1936).

[*337*] MULLIKEN, R. S.: Phys. Rev. **51**, 310 (1937).

[*338*] MULLIKEN, R. S.: J. Chem. Phys. **7**, 14 (1939).

[*339*] Mulliken, R. S.: J. Chem. Phys. **7**, 20, 121, 339 (1939).
[*340*] Mulliken, R. S.: Phys. Rev. **57**, 500 (1940).
[*341*] Mulliken, R. S.: J. Chem. Phys. **8**, 234 (1940).
[*342*] Mulliken, R. S.: J. Chem. Phys. **8**, 382 (1940).
[*343*] Mulliken, R. S.: Proc. Nat. Acad. Sci. U.S.A. **38**, 160 (1952).
[*344*] Mulliken, R. S.: J. Amer. Chem. Soc. **77**, 884 (1955).
[*345*] Muraour, H., et A. Michel-Lévy: C. R. Acad. Sci. Paris **200**, 924 (1935).
[*346*] Muraour, H., et A. Michel-Lévy: C. R. Acad. Sci. Paris **201**, 828 (1935).
[*347*] Muraour, H.: J. Phys. Radium **7**, 411 (1936).
[*348*] Muraour, H., J. Romand et B. Vodar: C. R. Acad Sci. Paris **223**, 620 (1946).
[*349*] Nicolle, J., et B. Vodar: C. R. Acad. Sci. Paris **210**, 142 (1940).
[*350*] Nissen, W.: Z. Physik **139**, 638 (1954).
[*351*] Norrish, R. G. W.: Nature, Lond. **101**, 1138 (1938).
[*352*] Oppenheimer, J. R.: Z. Physik **55**, 725 (1929).
[*353*] Paschen, F.: Berl. Ber. **135** (1926).
[*354*] Patzelt, F., u. K. Baldewein: Wiss. Veröff. Siemenskonz. **21**, 2, 213 (1942).
[*355*] Peters, Th.: Z. Physik **135**, 573 (1953).
[*356*] Petschek, H. E., P. H. Rose, H. S. Glick, A. Kane and A. Kantrowitz: J. Appl. Phys. **26**, 83 (1955).
[*357*] Pettit, E.: Astrophys. J. **91**, 159 (1940).
[*358*] Phillips, M.: Phys, Rev. **39**, 905 (1932).
[*359*] Polanyi, M., u. G. Schay: Z. Physik **47**, 814 (1928).
[*360*] Present, R. D.: J. Chem. Phys. **3**, 122 (1935).
[*361*] Price, W. C.: Phys. Rev. **47**, 419, 444, 510 (1935).
[*362*] Price, W. C.. and G. Collins: Phys. Rev. **48**, 714 (1935).
[*363*] Price, W. C., J. Chem. Phys. **3**, 256 (1935).
[*364*] Price, W. C., and R. W. Wood: J. Chem. Phys. **3**, 439 (1935).
[*365*] Price, W. C.: J. Chem. Phys. **4**, 147, 539, 547 (1936).
[*366*] Price, W. C., and W. M. Evans: Proc. Roy. Soc. Lond. A **162**, 110 (1937).
[*367*] Price, W. C., and W. M. Evans: Nature, Lond. **139**, 630 (1937).
[*368*] Ramsauer, C.: Phys. Z. **34**, 890 (1933).
[*369*] Rayleigh, Lord: Proc. Roy. Soc. Lond. A **112**, 14 (1926).
[*370*] Rayleigh, Lord: Proc. Roy. Soc. Lond. A **183**, 26 (1944).
[*371*] Rees, A. L. G.: J. Chem. Phys. **8**, 429 (1940).
[*372*] Rees, A. L. G.: Proc. Phys. Soc. Lond. **59**, 1008 (1947).
[*373*] Reiche, F., u. W. Thomas: Z. Physik **34**, 510 (1925).
[*374*] Reismann: Z. wiss. Photograph. **13**, 282 (1914).
[*375*] Rochester, G. D.: Proc. Roy. Soc. Lond. **167**, 567 (1938).
[*376*] Rollefson, G. H., and M. Burton: J. Chem. Phys. **6**, 674 (1938).
[*377*] Romand, J., et B. Vodar: C. R. Acad. Sci. Paris **226**, 238 (1948).
[*378*] Romand, J., et B. Vodar: C. R. Acad. Sci. Paris **226**, 890 (1948).
[*379*] Romand, J.: Ann. Phys., Paris **4**, 527 (1949).
[*380*] Rompe, R: Z. techn. Phys. **17**, 381 (1936).
[*381*] Rompe, R.: Z. Physik **101**, 214 (1936).
[*382*] Rompe, R., P. Schulz u. W. Thouret: Z. Physik **112**, 369 (1939).
[*383*] Rompe, R. u. M. Steenbeck: Ergebn. exakt. Naturw. **18**, 350 (1939).
[*384*] Rompe, R., u. P. Schulz: Phys. Z. **42**, 105 (1941).
[*385*] Rosseland, S.: Monthly Not. **84**, 525 (1924).
[*386*] Rössler, F.: Z. Physik **110**, 495 (1938).
[*387*] Rössler, F.: Z. Physik **125**, 427 (1949).
[*388*] Rössler, F.: Z. Physik **133**, 80 (1952).
[*389*] Rössler, F.: Z. Physik **139**, 56 (1954).
[*390*] Rubens, H., u. O. v. Baeyer: Berl. Ber. **339**, 666 (1911).
[*391*] Rudkjöbing, M.: Kgl. danske Vid. Selsk., mat.-fys. Medd. **18**, 125 (1940).
[*392*] Rühmkorf, H. A.: Ann. Phys. **33**, 21 (1938).
[*393*] Russell, H. N.: Astrophys. J. **78**, 239 (1933).
[*394*] Safary, E., J. Romand and B. Vodar: J. Chem. Phys. **19**, 379 (1951).
[*395*] Säufferer, H.: Z. Physik **131**, 376 (1952).

[396] SAWYER, R. A., and L. BECKER: Astrophys. J. **57**, 98 (1923).

[397] SEATON, M. J.: Proc. Roy. Soc. Lond. A **208**, 408 (1951).

[398] SEATON, M. J.: Proc. Roy. Soc. Lond. A **208**, 418 (1951).

[399] SEATON, M. J.: Phil. Trans. Roy. Soc. Lond., Ser. A **245**, 469 (1953).

[400] SEATON, M. J.: Proc. Phys. Soc. Lond. A **67**, 927 (1954).

[401] SEELIGER, R.: Phys. Z. **30**, 329 (1929).

[402] SEN-GUPTA, P. K.: Bull. Acad. Sci. U. P. Allahabad **2**, 245 (1933).

[403] SEN-GUPTA, P. K.: Proc. Roy. Soc. Lond. **143**, 438 (1934).

[404] SHARMA, R. S.: Bull. Acad. Sci. U. P. Allahabad **3**, 17 (1933).

[405] SHENSTONE, A. G.: Phys. Rev. **38**, 873 (1931).

[406] SHENSTONE, A. G.: Phys. Rev. **57**, 894 (1940).

[407] SHENSTONE, A. G.: Phil. Trans. Roy. Soc. Lond., Ser. A **241**, 297 (1948).

[408] SOMMERFELD, A., u. G. SCHUR: Ann. Phys. **4**, 409, 433 (1930).

[409] SOMMERFELD, A.: Ann. Phys. **11**, 257 (1931).

[410] SOMMERFELD, A.: Ann. Phys. **29**, 715 (1937).

[411] SOMMERMEYER, K.: Z. Naturforsch. **4**a, 440 (1949).

[412] SPONER, H., and E. TELLER: Rev. Mod. Phys. **13**, 75 (1941).

[413] SUGIURA, Y.: Sci. Pap. Inst. Phys. Chem. Res., Tokio **11**, 1 (1929).

[414] SUGIURA, Y.: Sci. Pap. Inst. Phys. Chem. Res., Tokio **13**, 23 (1930).

[415] SCHERZER, O.: Ann. Phys. **13**, 137 (1932).

[416] SCHIFF, L. I.: Rev. Sci. Instrum. **17**, 6 (1946).

[417] SCHIRMER, H.: Z. Physik **136**, 87 (1935).

[418] SCHNAIDT, F.: Ann. Phys. **21**, 89 (1934).

[419] SCHOTT, G. A.: Ann. Phys. **24**, 641 (1907).

[420] SCHRÖER, E.: Z. phys. Chem. **42**, 117 (1939).

[421] SCHUBERT, H.: Ann. Phys. **39**, 295 (1941).

[422] SCHÜLER, H., u. A. WOELDIKE: Phys. Z. **42**, 390 (1941).

[423] SCHULZ, P.: Z. Physik **119**, 167 (1942).

[424] SCHULZ, P.: Reichsber. Physik **1**, 147 (1944).

[425] SCHULZ, P.: Ann. Phys. **1**, 95, 107 (1947).

[426] SCHULZ, P.: Z. Naturforsch. **2**a, 583 (1947).

[427] SCHULZ, P.: Ann. Phys. **3**, 280 (1948).

[428] SCHWINGER, J. S.: Phys. Rev. **70**, 798 (1946).

[429] STAVER, T. B.: Ark. Math. og Naturvidenskab, Oslo **51**, 29 (1949).

[430] STEINHAUS, D. W., H. M. CROSSWHITE and G. H. DIEKE: J. Opt. Soc. Amer. **43**, 257 (1953).

[431] STRUTT, R. J.: Proc. Roy. Soc. Lond. **85**, 219 (1911).

[432] STUECKELBERG, E. C. G.: Phys. Rev. **42**, 518 (1932).

[433] TAKAMINE, T., u. T. TANAKA: Astrophys. J. **93**, 386 (1941).

[434] TERRIEN, J.: Ann. de Phys. **9**, 477 (1938).

[435] TRAUTZ, M.: Z. anorg. allg. Chem. **102**, 119 (1918).

[436] TRIVEDI, H.: Proc. Acad. Sci. U. P. Indian **5**, 27 (1935).

[437] TRUMPY, B.: Z. Physik **47**, 804 (1928).

[438] TRUMPY, B.: Z. Physik **54**, 372 (1929).

[439] TRUMPY, B.: Z. Physik **71**, 720 (1931).

[440] TUNSTEAD, J.: Proc. Phys. Soc. Lond. A **66**, 304 (1953).

[441] UCHIDA, Y.: Sci. Pap. Inst. Phys. Chem. Res., Tokio **30**, 71 (1936).

[442] UNSÖLD, A.: Z. Astrophys. **8**, 225 (1934).

[443] UNSÖLD, A.: Ann. Phys. **33**, 607 (1938).

[444] UNSÖLD, A.: Z. Astrophys. **21**, 22 (1941).

[445] UNSÖLD, A.: Naturwiss. **34**, 194 (1947).

[446] UNSÖLD, A.: Z. Astrophys. **24**, 306 (1948).

[447] UNSÖLD, A.: Z. Astrophys. **24**, 355 (1948).

[448] UNSÖLD, A.: Phys. Bl. **7**, 453 (1951).

[449] UNSÖLD, A.: Physik der Sternatmosphären, 2. Aufl. Berlin: Springer 1955.

[449a] VAUDET, M.: Ann. de Phys. **9**, 645 (1938).

[450] VENKATESWARLU, P.: Proc. Ind. Acad. Sci. A **26**, 22 (1947).

[451] VERWEY, E. J. W., u. J. H. DE BOER: Rec. Trav. chim. Pays-Bas **59**, 633 (1940).
[452] VINTI, J. P.: Phys. Rev. **42**, 632 (1932).
[453] VINTI, J. P.: Phys. Rev. **44**, 524 (1933).
[454] VITENSE, E.: Himmelswelt **56**, 33 (1949).
[455] VITENSE, E.: Z. Astrophys. **28**, 81 (1951).
[456] VOGEL, B.: Ann. Phys. **41**, 196 (1942).
[457] VOIGT, H. H.: Z. Astrophys. **31**, 48 (1952).
[458] WAGNER, A.: Phys. Z. **18**, 433 (1917).
[459] WALDMEIER, M., u. H. MÜLLER: Astr. Mitt., Zürich **1948**, Nr. 155.
[460] WATANABE, K., E. C. Y. INN and M. ZELIKOFF: Phys. Rev. **90**, 359 (1953).
[461] WATANABE, K., F. MARMO and E. C. Y. INN: Phys. Rev. **91**, 436, 1155 (1953).
[462] WAWILOW, S. J..: C. R. Acad. Sci. URSS. **2**, 457 (1934).
[463] WEHRLI, M., u. W. WENK: Helv. phys. Acta **12**, 559 (1939).
[464] WEISSLER, G. L.: Mem. Soc. Roy. Sci. Liège **12**, 281 (1952).
[465] WEIZEL, W.: Handbuch der Experimentalphysik, Erg.-Bd. 1. 1931.
[466] WENIGER, S., et R. HERMAN: C. R. Acad. Sci. Paris **232**, 951 (1951).
[467] WENK, W.: Helv. phys. Acta **14**, 355 (1941).
[468] WENTZEL, G.: Z. Physik **43**, 524 (1927).
[469] WENTZEL, G.: Z. Physik **43**, 1, 779 (1927).
[470] WESTPFAL, K.: Z. Physik **140**, 414 (1955).
[471] WHEELER, J. A.: Phys. Rev. **43**, 258 (1933).
[472] WIELAND, K.: Habil.-Schr. Zürich 1941.
[473] WIELAND, K.: Vol. Commé m. V. Henri 1947/48.
[474] WIELAND, K.: Helv. phys. Acta **21**, 434 (1949).
[475] WILDT, R.: Astrophys. J. **89**, 295 (1939).
[476] WILDT, R.: Astrophys. J. **90**, 611 (1939).
[477] WILFONG jr. T. C., and J. E. HENDERSON: Phys. Rev. **85**, 765 (1952).
[478] WILSING, I.: Potsd. Publ. **23**, Nr. 72 (1917).
[479] WINANS, J. G., and E. C. G. STUECKELBERG: Proc. Nat. Acad. Amer. **14**, 867 (1928).
[480] WOOD, R. W.: Astrophys. J. **29**, 100 (1909).
[481] WOOD, R. W.: Phil. Mag. **18**, 530 (1909).
[482] WOOLLEY, R. v. D. R., and C. W. ALLEN: Monthly Not. **108**, 292 (1948).
[483] WOOLLEY, R. v. D. R., and C. W. ALLEN: Monthly Not. **110**, 358 (1950).
[484] WREDE, B.: Ann. Phys. **3**, 823 (1929).
[485] WU, T. Y.: Phys. Rev. **66**, 291 (1944).
[486] WU, T. Y., and L. OUROM: Canad. J. Res. **28**A, 542 (1950).
[487] WÜLLNER, A.: Poggendorfs Ann. **137**, 337 (1869).
[488] WÜLLNER, A.: Poggendorfs Ann. **147**, 321 (1872).
[489] WYCKOFF, H. O., and J. E. HENDERSON: Phys. Rev. **64**, 1 (1943).
[490] WYNEKEN, J.: Ann. Phys. **86**, 1071 (1928).
[491] YAMANOUCHI, and KOTANI: Proc. Math. Phys. Soc. Japan **24**, 351 (1942).
[492] YIN-YUAN, L.: Phys. Rev. **80**, 104 (1950).
[493] YIN-YUAN, L.: Phys. Rev. **82**, 281 (1951).
[494] ZANSTRA, H.: Astrophys. J. **65**, 50 (1927).
[495] ZANSTRA, H.: Proc. Roy. Soc. Lond. **186**, 236 (1946).
[496] ZÉ, N. T., et CH'EN SHANG-YI: J. Phys. Radium **9**, 169 (1938).

Kristallspektren.

Von

E. FICK und G. JOOS.

Mit 30 Figuren.

A. Allgemeiner Überblick.

I. Einleitung und Abgrenzung des Stoffs.

1. Bindungstypen bei Kristallen. An die Spektroskopie der freien Atome und Molekeln schließt sich logischerweise die Spektroskopie der im Kristallverband befindlichen Atome bzw. Molekeln an. Beim Einzelatom gehört jedes Elektron dem einen Kern an, bei Molekeln sind die für die sichtbaren und ultravioletten Spektren maßgebenden Elektronen mehr oder weniger zwischen den einzelnen Kernen aufgeteilt, es gibt aber auch Fälle, bei denen die Bindung durch VAN DER WAALSsche Kräfte so lose ist, daß man von den Quantenzuständen des freien Atoms ausgehen und die Beeinflussung durch die Nachbarn als Störung einführen kann, z.B. bei den „Quasimolekeln" Hg_2 u. dgl. Wenn wir diese Einteilung auf die Kristalle übertragen, so entspricht der durch gemeinsame Elektronen zusammengehaltenen Molekel der Metallkristall, in welchem einige der Atomelektronen allen Atomen des Kristalls zugehören. Der Zustand eines Ionengitters entspricht dagegen viel mehr der VAN DER WAALSschen Molekel. Die Bindung kann zwar sehr fest sein. Sie wird durch die elektrostatische Anziehung gewährleistet, die Elektronen gehören aber jeweils einem bestimmten Ion zu. Die FOURIER-Synthese der Röntgenbeugungsaufnahmen, welche die Ladungsdichte ergibt, zeigt, daß in einfachen polaren Kristallen die Elektronendichte zwischen dem positiven und dem benachbarten negativen Ion nahezu auf null heruntergeht. Eine Mittelstellung nehmen nichtpolar gebundene Isolatorkristalle wie Diamant ein, bei denen die Atome durch gemeinsame Elektronenpaare — „Bindungsstriche" — zunächst mit ihren nächsten Nachbarn verkettet sind. Wegen der Ununterscheidbarkeit der Elektronen gehören diese Bindeelektronen damit wieder zum ganzen Gitter, da sie sozusagen von Atom zu Atom weitergegeben werden können. Auf Grund dieser Einteilung ist zu erwarten, daß man ausgehend von den Quantenzuständen der freien atomaren Systeme nur bei den Ionenkristallen zu einer Deutung ihrer Spektren und damit ihrer Quantenzustände kommen kann, während die Quantenzustände der nichtpolaren Kristalle sehr viel schwieriger zu behandeln sind. Die Spektren der Metalle werden im Zusammenhang mit der metallischen Leitung behandelt. Die Spektren der nichtpolar gebundenen Isolatorkristalle, welche breite Absorptionsgebiete aufweisen, werden im Zusammenhang mit der lichtelektrischen Leitfähigkeit dieser Kristalle erörtert. Dagegen sind noch Kristalle denkbar, bei denen die Bausteine geladene oder ungeladene Molekeln sind, welche ihre Elektronenhülle unter geringer Veränderung beibehalten. Dazu gehören z.B. Molekülionen wie MnO_4^- und UO_2^{++}, dann die große Zahl fest gebundener organischer Molekeln wie Benzol, die im Kristall durch van der Waalssche Kräfte zusammengehalten werden. Bei diesen Kristallen kann man von den freien Molekeln ausgehen und die Einwirkung der Nachbarn als Störung behandeln. Praktisch wird nun die Möglichkeit, einigermaßen definierte Spektren in Form von Linien oder schmalen Absorptions- bzw. Emissionsgebieten zu erhalten, noch weiter eingeengt: Wohl sind im Grundzustand z.B. in einem NaCl-Kristall die Elektronenhüllen von Na^+ und Cl^- sauber getrennt. Jeder Anregungszustand aber, bei dem die Hauptquantenzahl n geändert wird, bewirkt eine größere Raumerfüllung der Elektronenwolke, die damit in die des Nachbarions übergreift, womit die einfachen Verhältnisse, die es ermöglichen, vom Zustand des freien Ions auszugehen, zerstört sind. Angeregte Zustände, bei denen die Hauptquantenzahl unverändert bleibt, sind aber nur bei unvollständigen Elektronenschalen möglich. So sind eigentliche Linienspektren im oben umrissenen Sinn nur bei den Ionen der Übergangselemente zu erwarten. Diese Übergänge sind aber beim freien Ion fast immer „verboten", d.h. sie geben keine elektrische Dipolstrahlung. Die Übergangswahrscheinlichkeiten von magnetischer Dipol-, elektrischer Quadrupol- und erzwungener elektrischer Dipolstrahlung sind aber viele Zehnerpotenzen kleiner als die der „erlaubten" Übergänge (einfache Dipolstrahlung). Dies hat auch sein Gutes: Da die Untersuchungen fast immer in Absorption ausgeführt werden, müßte man, um überhaupt Licht durchzubringen, bei erlaubten Übergängen ungewöhnlich dünne Schichten herstellen. Bei $KMnO_4$, wo ein erlaubter Dipolübergang vorliegt, ist ein Kristall von 0,1 mm Dicke schon praktisch undurchsichtig. Hier kann man sich oft nur durch isomorphen Einbau des absorbierenden Ions in einen nicht absorbierenden Kristall helfen.

Neben den bisher besprochenen Kristallen einheitlicher Zusammensetzung, haben die Spektren *solcher* Kristalle große Bedeutung, bei denen der Träger der Spektren eine in kleinen Konzentrationen vorhandenen Verunreinigung ist. Hierher gehören die Spektren der Alkalihalogenide, welche Alkali oder Halogen im Überschuß besitzen, aber auch die Spektren der mit einem Aktivator versehenen anorganischen Phosphore. Dies letzte Gebiet ist so umfangreich, daß es an anderer Stelle bearbeitet wird[1]. Lediglich die Alkalihalogenide, die wieder als Modellsubstanzen z.B. für die Silberhalogenide dienen, sind auch spektroskopisch so wichtig, daß sie hier kurz behandelt werden sollen.

2. Zusammenhang der Kristallspektren mit andern Gebieten der Festkörperphysik. Wie wir sehen werden, spalten im Feld der Nachbarionen die Terme der meist paramagnetischen, als Träger der Spektren wirkenden Ionen in einzelne oft eng benachbarte Teilniveaus auf. Die Trennung der durch diese Aufspaltung bedingten Liniengebilde ist sehr oft erst bei tiefsten Temperaturen möglich, da bei höheren die Wärmebewegung die Linien unscharf macht. So ist die Kristallspektroskopie im wesentlichen an tiefe Temperaturen gebunden und gerade in diesem Gebiet zeigen auch andere Eigenschaften der Kristalle sehr auffällige Änderungen. Kennt man aus der Spektroskopie die Teilniveaus und ihre magnetischen Momente, so ergibt sich daraus der oft sehr merkwürdige Gang der paramagnetischen Suzeptibilität dieser Stoffe, da bei tiefen Temperaturen die Besetzung der höheren Niveaus zurückgeht[2]. Im Zusammenhang damit steht wieder der paramagnetische Anteil der Drehung der Polarisationsebene im Magnetfeld. Ebenso ist bei tiefer Temperatur eine Besonderheit im Gang der spezifischen Wärmen zu erwarten, da neben der Energiezufuhr zur Anregung der Gitterschwingungen noch eine solche zur Besetzung der höheren Teilniveaus erforder-

---

[1] Vgl. den Beitrag von G. F. J. Garlick in Bd. XXVI dieses Handbuches.
[2] Vgl. den Beitrag von J. van den Handel: Low Temperature Magnetism in Bd. XV.

lich ist[1]. Beide Einflüsse zusammen haben große Bedeutung für die durch adiabatische Entmagnetisierung bewirkte Erzeugung allertiefster Temperaturen. Wenn wir von einer quantitativen Beherrschung dieser Zusammenhänge auch noch weit entfernt sind, so sind sie doch in vielen Fällen schon qualitativ gut geklärt. Diese Zusammenhänge werden beim Magnetismus und der Theorie der spezifischen Wärmen behandelt werden.

Eine noch weiter reichende Anwendung ist die Erforschung der Struktur fester Körper mittels Sonden aus Ionen, deren Kristallspektren geklärt sind. Die dabei zu gewinnenden Erkenntnisse gehen über die Röntgenanalyse weit hinaus, da sie die Felder an der Oberfläche der Ionen unmittelbar anzeigen. Sie stehen in einem gewissen Zusammenhang mit den Untersuchungen der Feinstruktur der Röntgenabsorptionskanten, die ja auch zu Übergängen in „optische" Niveaus der ursprünglich in Kernnähe befindlichen Elektronen gehören.

II. Experimentelle Erzeugung von Kristallspektren.

3. Absorptionsspektren. Da die Erzeugung von Emissionskristallspektren wesentlich schwieriger ist und an Einkristallen nur in wenigen Fällen gelingt, sind die meisten Untersuchungen in Absorption vorgenommen. Hier ist die Technik verhältnismäßig einfach. Lediglich die Notwendigkeit, die Kristalle auf möglichst tiefer Temperatur zu halten, verursacht einige technische Schwierigkeiten. Bei Zimmertemperatur sind die Linien so breit, daß viele Einzelheiten der Feinstruktur verschmiert werden. Abkühlung auf die Temperatur der flüssigen Luft ist mindestens zu fordern. Viel mehr Einzelheiten erkennt man aber bei der Temperatur des flüssigen Wasserstoffs (20° K), die durch Abpumpen auf 14° K gesenkt werden kann. Selbst diese Temperatur ist nicht immer tief genug, in letzter Zeit sind daher viele Untersuchungen bei der Temperatur des flüssigen Heliums (4,2° K) ausgeführt worden. Beim Arbeiten mit flüssiger Luft oder flüssigem Wasserstoff genügt es, den zu durchstrahlenden Kristall direkt in die Flüssigkeit zu tauchen. Dem dabei verwendeten DEWAR-Gefäß gibt man zweckmäßigerweise die Form der Fig. 1. Dadurch daß der Kristall im verengten Teil sitzt, vermeidet man die Durchstrahlung größerer Schichten flüssiger Luft bzw. Wasserstoffs und hat bei ZEEMAN-Untersuchungen die Möglichkeit, zur Erzeugung eines hohen Feldes die Polschuhe eng zusammenzuschieben. Die größere Weite im oberen Teil ist zur Aufnahme einer größeren Menge Kühlflüssigkeit erforderlich. Beim Arbeiten mit flüssigem Helium muß dieses durch einen Mantel mit flüssiger Luft vor allzuschneller Erwärmung geschützt werden. Bei den meisten Anordnungen werden einfach zwei DEWAR-Gefäße ineinandergestellt, so daß ein genügender Zwischenraum zur Füllung mit flüssiger Luft, besser flüssigem Stickstoff, bleibt. Eine kompliziertere Konstruktion von H. MEISSNER[2] verwendet metallene DEWAR-Gefäße, das innere trägt ein durch Covarverbindung angeschlossenes, unten verschlossenes Quarzrohr. Am Mantel des äußeren sind wie in Fig. 2 Rohrstücke mit Quarzfenstern angesetzt, die Schutzkühlung wird lediglich durch ein Kupferrohr bewirkt, das an das mit flüssiger Luft gefüllte DEWAR-Gefäß angelötet ist. Für solche Fälle, in denen die Kühlflüssigkeit optisch stört, sind Vorrichtungen

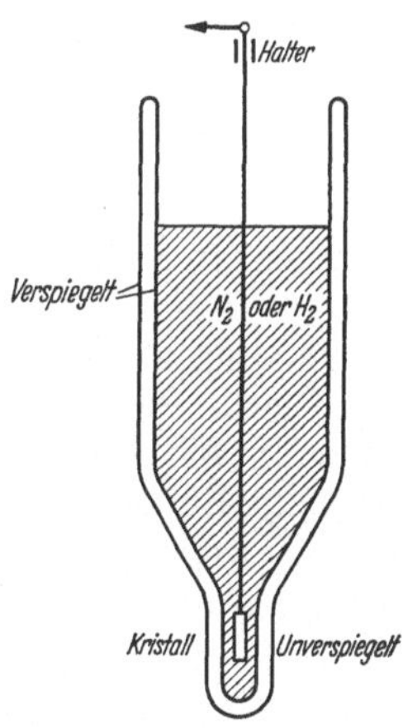

Fig. 1. Einfache Anordnung zur Kühlung der Kristalle.

---

[1] Vgl. z.B. L. D. ROBERTS, C. C. SARTAIN u. B. BORIE: Rev. Mod. Phys. **25**, 170 (1953).
[2] Beschrieben bei G. H. DIEKE u. L. HEROUX: Phys. Rev. **103**, 1227 (1956).

gebaut worden, bei denen der Kristall sich im Vakuum befindet (Fig. 2). Zur Kühlung ist er mit möglichst gutem Wärmekontakt in einen Kupferklotz eingesetzt, der seinerseits an den Boden eines mit flüssiger Luft bzw. flüssigem Wasserstoff gefüllten metallenen Dewar-Gefäß angelötet ist. Die wahre Temperatur muß durch ein am Kristall anliegendes Thermoelement ermittelt werden. Bei gutem Wärmekontakt liegt sie nur wenige zehntel Grad höher als die der Kühlflüssigkeit.

Die Kristalle werden im allgemeinen, da es sich um Salze handelt, aus der Lösung gezüchtet. Ihre Dimensionen können sehr klein sein, einige mm und weniger. Sie sind in einen lichtundurchlässigen Schirm mit entsprechender Öffnung einzusetzen. Bildet man eine konzentrierte Lichtquelle (Kohlebogen oder Autoscheinwerferlampe) auf den Kristall und von dort auf den Spektrographenspalt ab, so geht noch genügend Licht zur Erzeugung eines Spektrums durch. Will man die Beobachtungen auf das kürzerwellige Ultraviolett ausdehnen, so müssen natürlich alle Teile, die vom Strahlengang durchsetzt werden, aus Quarz bzw. Quarzglas gemacht werden, als Lichtquelle dient dann am besten eine Xenon-Hochdrucklampe.

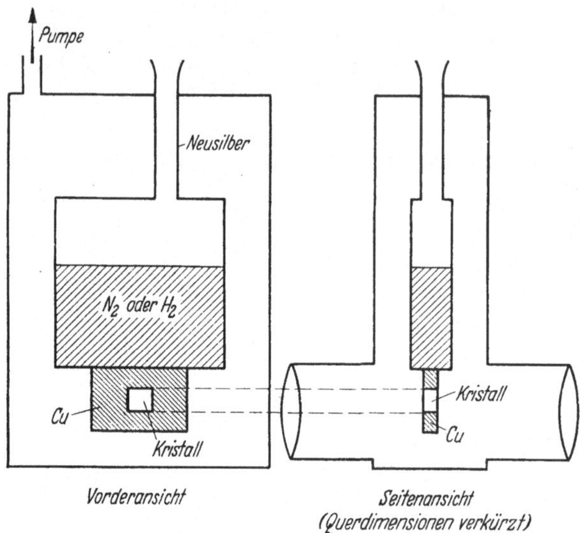

Fig. 2. Indirekte Kristallkühlung ohne Durchstrahlung des Kühlmittels.

Aus den Kristallen werden nach den in der Kristallographie üblichen Methoden orientierte Platten ausgeschnitten bzw. geschliffen. Wegen der sehr verschiedenen Stärke der Linien macht man für ein Spektrum Platten, deren Dicken etwa im Bereich 1 bis 10 variieren.

Die Platten werden dann mittels Reflexion an der Oberfläche zum Lichtstrahl senkrecht gestellt und alle gewünschten Stellungen durch Drehung des Kristallhalters, der mit einem Goniometertisch verbunden ist, erzielt. Einkristalle werden in polarisiertem Licht untersucht.

Für Stoffe, von denen keine größeren Kristalle gewonnen werden können, kann man eine Pulvermethode verwenden, bei der man aber auf alle mit der Orientierung des elektrischen Vektors zur Kristallachse zusammenhängenden Feinheiten verzichten muß. Man bringt zu diesem Zweck das einigermaßen durchsichtige Pulver zwischen zwei Glas- bzw. Quarzplatten und durchstrahlt die Schicht. Dabei geht naturgemäß viel Licht durch Streuung aus dem Strahlengang verloren. Selbst Remissionsspektren im Auflicht sind mitunter wertvoll. Wegen der geringen Absorption kommt das Licht auf vielen Umwegen wieder an die Oberfläche und wenigstens teilweise in den Spektrographen. Man erhält so im allgemeinen wieder die Absorptionslinien (auf diese Weise suchten früher Prospektoren mit dem Taschenspektroskop auf sonnenbeschienenen Flächen den Seltene Erden enthaltenden Monazitsand). Bei hinreichend starker Absorption tritt in einigen wenigen Fällen die nach den Gesetzen der klassischen Optik zu erwartende verstärkte Reflexion auf und die Linien erscheinen als „Emissionslinien" entsprechend den Reststrahlen[1].

Bei Messungen im Infrarot stört die starke Absorption des Kristallwassers. GOBRECHT[2] verwandte daher in diesem Spektralgebiet eine Boraxperle, in welche die Salze eingeschmolzen wurden.

4. Emissionsspektren. Einige der in betracht kommenden Kristalle lassen sich durch Ultraviolettlicht zur Fluorescenz anregen. Wie oben betont, müssen in diesem Fall die Hauptteile der Kühlgefäße aus Quarz hergestellt werden. Das anregende Licht muß senkrecht zur Beobachtungsrichtung einfallen. Das Streulicht wird durch geeignete Filter vom Fluorescenzlicht getrennt. Die Lichtanregung muß sehr kräftig sein, man bildet am besten die Flamme eines BECK-Bogens oder einen Xenon-Hochdruckbogen mit Quarzlinsen auf den Kristall ab. Merkwürdigerweise lassen sich einige Ionen der Seltenen Erden, die im reinen Kristall nicht leuchten, in der Boraxperle zum Fluorescenzleuchten bringen.

Eine Anregung, die bei Versagen der Ultraviolettanregung oft noch gelingt, ist die Anregung durch Elektronenstoß im Kathodenstrahl. Eine einfache Anordnung dazu beschreibt TOMASCHEK[3]: Die meist nur in Pulverform vorliegende Substanz wird in die Rillen eines Kupferrohrs eingerieben, das innen mit dem Kühlmittel gefüllt ist (Fig. 3). Seitlich ist eine Oxydkathode angebracht, zwischen Kupferrohr und Kathode werden einige tausend Volt Spannung angelegt. Ebensogut kann man einen größeren Kristall in das Kupferrohr einsetzen und sein Licht ungefähr senkrecht zur Kathodenstrahlrichtung beobachten.

5. Spektralapparate. Als Spektralapparate kommen ungefähr alle Typen vom kleinen Taschenspektroskop bis zum 6 m-Gitter in betracht. Das hohe Auflösungsvermögen eines großen Gitters lohnt sich bei den tiefsten Temperaturen, wo es sich um Linientrennungen von einigen hundertstel Å handelt. In anderen Fällen wird man lichtstarken Prismenspektrographen den Vorzug geben, insbesondere bei der Aufnahme der stets lichtschwachen Luminescenzspektren. Die Absorptionsmessungen im Infrarot wurden mit Infrarotspektrograph und Thermosäule durchgeführt.

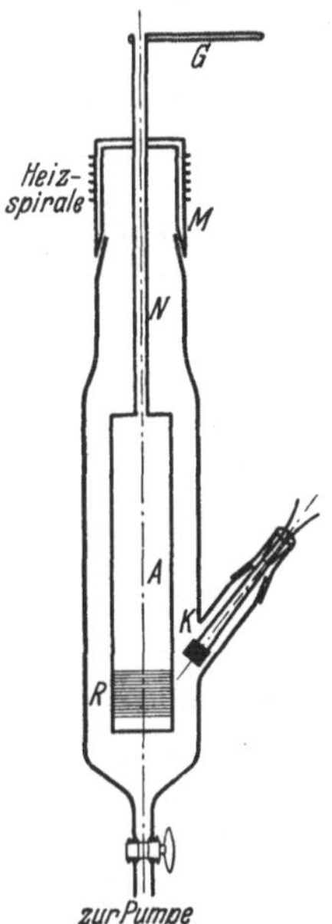

Fig. 3. Anregung der Luminescenz von Kristallpulvern durch Kathodenstrahlen nach TOMASCHEK.

6. Hilfsmittel zu ZEEMAN-Effekt-Messungen. Bei der Untersuchung des ZEEMAN-Effekts braucht man möglichst große Magnete, da die DEWAR-Gefäße einen verhältnismäßig großen Polabstand benötigen. So sind die schönen Aufnahmen von BROCHARD und HELLWEGE (Ziff. 47) mit dem großen Bellevue-Magneten gemacht worden.

ZEEMAN-Effektmessungen an Pulvern haben keinen Sinn, da die Aufspaltung von der Orientierung des Magnetfelds zur Kristallachse abhängt, man also bei Pulvern eine verwaschene Mittelbildung erhält.

Die ZEEMAN-Aufspaltung des Grundterms und damit sein g-Faktor läßt sich noch auf ganz andere Weise mit hoher Genauigkeit ermitteln: Während die gewöhnliche Spektroskopie die Übergänge zwischen den ZEEMAN-Niveaus zweier weit auseinanderliegender Terme gibt, erhält man aus der paramagnetischen

[1] F. H. SPEDDING u. R. S. BEAR: Phys. Rev. **39**, 948 (1932).
[2] H. GOBRECHT: Ann. d. Phys. (5) **31**, 755 (1938).
[3] R. TOMASCHEK: Ergebn. exakt. Naturw. **20**, 268 (1942).

Resonanz die Übergänge zwischen den Zeeman-Niveaus des Grundterms selbst. Bei dem geringen energetischen Abstand der Zeeman-Niveaus fällt die dem Übergang zwischen ihnen entsprechende Frequenz bzw. Wellenlänge in den Bereich der cm-Wellen. Bringt man daher den Kristall in einen Hohlraumleiter der mit einem Klystron angeregt wird, so tritt für $\nu = \Delta E/h$ eine erhöhte Dämpfung des Hohlraums auf[1].

7. Hilfsmittel zum Studium des Druckeinflusses. Wenn man den Einfluß allseitigen Drucks auf die Kristallspektren studieren will, braucht man Drucke von der Größenordnung 100 Atm. Diese lassen sich nach Paetzold[2] folgendermaßen erzeugen: Das Absorptionsrohr ist durch Stahlkapillaren mit einem größeren Stahlbehälter und beide mit einer N_2-Druckflasche verbunden. Zuerst bringt man beide Gefäße in flüssigen Stickstoff und stellt die Verbindung mit der Vorratsbombe her. Dabei kondensiert sich in beiden Gefäßen flüssiger Stickstoff. Nach Schließen des zur Bombe führenden Ventils hält man das Absorptionsgefäß weiter auf tiefer Temperatur, während man den Stahlbehälter auf Zimmertemperatur bringt. Je nach der Menge des eingebrachten Stickstoffs erhält man so Drucke von einigen Hundert Atm, die an einem angelöteten Manometer abgelesen werden. Das Absorptionsgefäß hat ein konisches Glasfenster. Hinter dem zu durchstrahlenden Kristall befindet sich ein Spiegel, so daß man ein Fenster spart und das vorhandene im Hin- und Rückgang benützt.

B. Theorie der Kristallspektren.

8. Die Behandlungsweise der Termbeeinflussung durch ein Kristallfeld. Auf ein Ion oder Atom im Kristall wirkt ein von allen übrigen Ionen herrührendes Kristallfeld bestimmter Symmetrie, das eine Aufspaltung eines im freien Ion entarteten Termes hervorruft. Wenn man diese Vorstellung der Beschreibung von Kristallspektren zugrunde legt, so wird dabei ein Elektronenaustausch mit den benachbarten Ionen nicht berücksichtigt. Da wir aber gesehen haben, daß *scharfe* Kristallspektren nur bei den Salzen der Übergangselemente auftreten, bei welchen in den beiden zu einer Spektrallinie gehörenden Zuständen kein wesentliches Übergreifen von Elektronen benachbarter Ionen stattfindet, so genügt es, ein Ion unter dem pauschalen Einfluß eines Kristallfeldes zu betrachten.

Den Ausgangspunkt dieser von H. Bethe[3] entwickelten Theorie bildet also das freie Ion oder Atom, dessen Hamilton-Funktion sich bei einer beliebigen Drehung des Koordinatensystems sowie bei einer Inversion am Kern nicht ändert. Diese Invarianz bedingt eine $(2J+1)$-fache Entartung des Energieeigenwertes. Die beim Einbringen des Ions in den Kristall hinzukommende Störung durch das Kristallfeld ruft nun je nach dessen Symmetrie eine teilweise oder ganze Aufhebung der Entartung hervor: Die Energieterme spalten auf. Mit Hilfe der Gruppentheorie gelingt es, allein aus der Kenntnis der Symmetrie des Kristallfeldes, die durch die Symmetrie der Lage des betrachteten Ions, d. h. durch die zugehörige Kristallklasse, gegeben ist, die *Anzahl und Vielfachheit* der auftretenden Terme zu bestimmen. Dies ist der Inhalt von Abschnitt B I dieses Artikels. Die Frage der Termaufspaltung wurde neuerdings auch von K. H. Hellwege unter Vermeidung gruppentheoretischer Methoden aufgegriffen und eingehend behandelt. Diese auf die Elektroneneigenfunktionen zurückgreifende Theorie, in

---

[1] Vgl. z. B. den Bericht von G. Joos: Z. angew. Phys. **6**, 43 (1954).
[2] H. K. Paetzold: Ann. d. Phys. (5) **37**, 470 (1940).
[3] H. Bethe: Ann. Phys. (5) **3**, 133 (1929)

der die einzelnen Terme durch *Kristallquantenzahlen* beschrieben werden, soll in B II besprochen werden. Im Abschnitt B III werden wir die *Größe* der Termaufspaltung explizit berechnen und in B IV den Einfluß eines äußeren homogenen Magnetfeldes auf die Kristallterme (ZEEMAN-*Effekt der Kristallspektren*) untersuchen.

I. Gruppentheoretische Behandlung.

Wir wollen die gruppentheoretische Lösung des Aufspaltungsproblemes an Hand einer speziellen Symmetrie, nämlich der tetragonalen durchführen. Obwohl Kristallfelder dieser Symmetrie verhältnismäßig selten vorkommen, wählen wir gerade dieses Beispiel, da man an ihm alle wesentlichen Züge der Methode in besonders übersichtlicher Weise erkennen kann. Für die Beweise der verwendeten Sätze aus der Theorie der Gruppen und ihrer Darstellungen sei auf den Bd. II (G. FALK) des Handbuches und die einschlägigen Lehrbücher[1] verwiesen. Die benötigten kristallographischen Begriffe finden sich im Bd. VII/1 (H. JAGODZINSKI).

9. Die Symmetriegruppe der Kristallklasse D_4. Wir wollen in einem tetragonalen System alle Drehachsen betrachten, die einen Quader mit quadratischer Basis in sich überführen. Es entspricht dies der tetragonal-trapezoedrischen Kristallklasse D_4. Wie aus Fig. 4 ersichtlich, erlaubt die 4zählige tetragonale Drehachse (■) die Drehungen a um 180°, b um +90° und c um −90°. Die vier zur tetragonalen Achse senkrechten 2zähligen Drehachsen (⬤) d, f und g, h stellen je Drehungen um 180° dar. Mit der Einheit e (Drehung 0°) ergeben sich somit 8 Drehungen, die eine *Gruppe* $\mathfrak{G}$ bilden

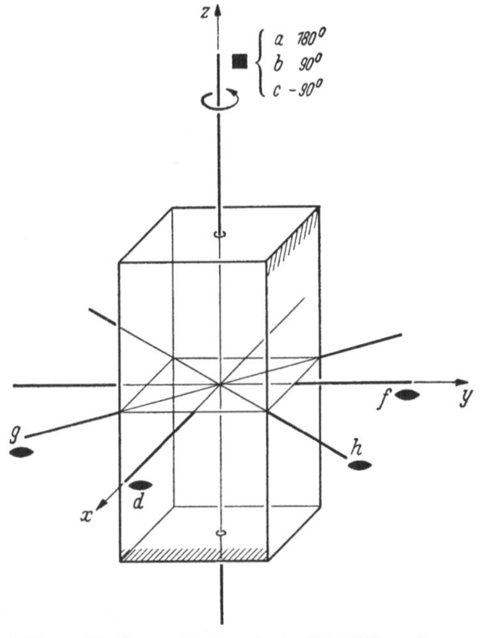

Fig. 4. Die Symmetrieelemente der Kristallklasse D_4: Drehungen a bis h (z.B. gehen bei der Operation $b \cdot d = h$ die schraffierten Ebenen ineinander über).

($n=$ Ordnung der Gruppe = Anzahl ihrer Elemente = 8). Die Gruppe ist endlich, d.h. sie besitzt endlich viele Elemente. Man überzeugt sich leicht, daß die bekannten Gruppenaxiome erfüllt sind, z.B. liefert zuerst d und dann b ausgeführt dieselbe Stellung des Quaders wie die Drehung h, d.h. $b \cdot d = h$[2]. Die umgekehrte Reihenfolge ergibt indes $d \cdot b = g$, d.h. die Gruppe ist nicht kommutativ (nicht abelsch). Die gesamten Eigenschaften dieser Gruppe können wir am besten aus der Gruppentafel (9.1) ablesen, in der alle Multiplikationsergebnisse

---

[1] H. BOERNER: Darstellungen von Gruppen. Berlin-Göttingen-Heidelberg 1955. — B. L. v. D. WAERDEN: Algebra I und II. Berlin-Göttingen-Heidelberg 1955. — A. SPEISER: Theorie der Gruppen von endlicher Ordnung. Berlin 1937. — E. WIGNER: Gruppentheorie und ihre Anwendung auf die Quantenmechanik der Atomspektren. Braunschweig 1931. — B. L. v. D. WAERDEN: Die gruppentheoretische Methode in der Quantenmechanik. Berlin 1932. — G. LUDWIG: Die Grundlagen der Quantenmechanik. Berlin-Göttingen-Heidelberg 1954.

[2] Die Produkte sind also von rechts nach links zu lesen. — Wir treffen für die Ausführung zweier aufeinanderfolgender Drehungen die Konvention, daß die Drehachsen *raumfest* gehalten, d.h. nicht mitgedreht werden.

aufgeführt sind.

Rechter Faktor

(1. Operation)

		e	a	b	c	d	f	g	h
	e	e	a	b	c	d	f	g	h
	a	a	e	c	b	f	d	h	g
	b	b	c	a	e	h	g	d	f
linker Faktor	c	c	b	e	a	g	h	f	d
(2. Operation)	d	d	f	g	h	e	a	b	c
	f	f	d	h	g	a	e	c	b
	g	g	h	f	d	c	b	e	a
	h	h	g	d	f	b	c	a	e

$$(9.1)$$

Daraus ist ersichtlich, daß z.B. $a = a^{-1}$ ist, denn es gilt $a \cdot a = e$.

Greift man irgendein Element r der Gruppe $\mathfrak{G}$ heraus und bildet damit nach der Multiplikationsvorschrift (9.1) $x^{-1} r x$, wobei x alle Elemente der Gruppe durchläuft, so stellt die Gesamtheit dieser zu r „konjugierten" Elemente $x^{-1} r x = a_{1r}, a_{2r}, \ldots$ die *Klasse*[1] $\mathfrak{C}_r$ dar, wofür man symbolisch schreibt

$$\mathfrak{C}_r = \sum_{\nu=1}^{h_r} a_{\nu r}, \qquad (9.2)$$

h_r ist die Anzahl der in $\mathfrak{C}_r$ vereinigten Elemente; sie ist ein Teiler der Gruppenordnung $n \left(= \sum_r h_r \right)$. In unserem Beispiel erhalten wir dadurch 5 verschiedene Klassen

$$\left. \begin{aligned}
\mathfrak{C}_e &= & e & = \mathfrak{C}_1, & (h_1 = 1); \\
\mathfrak{C}_a &= & a & = \mathfrak{C}_2, & (h_2 = 1); \\
\mathfrak{C}_b = \mathfrak{C}_c &= & b + c & = \mathfrak{C}_3, & (h_3 = 2); \\
\mathfrak{C}_d = \mathfrak{C}_f &= & d + f & = \mathfrak{C}_4, & (h_4 = 2); \\
\mathfrak{C}_g = \mathfrak{C}_h &= & g + h & = \mathfrak{C}_5, & (h_5 = 2).
\end{aligned} \right\} \qquad (9.3)$$

Wie man aus Fig. 4 ersieht, werden gerade die geometrisch gleichwertigen Drehachsen zu einer Klasse vereinigt. Jedes Element von $\mathfrak{G}$ gehört genau einer Klasse an

$$\mathfrak{G} = \sum_{r=1}^{c} \mathfrak{C}_r. \qquad (9.4)$$

Dabei numeriert jetzt r die verschiedenen Klassen, deren Anzahl c sei [für das Beispiel (9.3) ist also $c = 5$]. Bei abelschen Gruppen ist jedes Element eine Klasse für sich (alle $h_r = 1$, $c = n$).

10. Zur Darstellungstheorie endlicher Gruppen. Wenn man den Elementen $e, a, b, \ldots$ einer (endlichen) Gruppe $\mathfrak{G}$ solche endliche, quadratische Matrizen $E, A, B, \ldots$ (mit von Null verschiedener Determinante) zuordnet, daß das Produkt zweier Gruppenelemente dem Produkt der zugeordneten Matrizen entspricht

$$a \cdot b = c \;\to\; A \cdot B = C \quad \text{(Homomorphie } \mathfrak{G} \to \mathfrak{M}), \qquad (10.1)$$

---

[1] Man verwechsle die *Klassen* $\mathfrak{C}_r$ einer Gruppe nicht mit den 32 möglichen *Kristallklassen* (z.B. D_4), welche die Symmetriegruppe bestimmen. Wir sagen deshalb statt Kristallklasse auch manchmal *Kristallgruppe*.

so erhält man eine „*Darstellung*" Γ *der Gruppe* $\mathfrak{G}$ *durch die Matrizen* $\mathfrak{M}$. Diese Zuordnung ist *homomorph*, d.h. sie braucht nicht umkehrbar eindeutig zu sein, es können vielmehr verschiedene Gruppenelemente derselben Matrix zugeordnet sein, wenn nur dabei die Produkteigenschaften von $\mathfrak{G}$ auch in $\mathfrak{M}$ gewahrt bleiben. Der Grad t der verwendeten Matrizen (d.h. die Anzahl ihrer Zeilen) heißt die „*Dimension*" der Darstellung $[A = A_{\lambda\mu}(\lambda, \mu = 1, \ldots, t)]$; man schreibt $^t\Gamma$ für eine Darstellung der Dimension t.

Jede Matrix A der Darstellung läßt sich als eine Abbildung eines Vektors x auf einen Vektor y in einem t-dimensionalen Vektorraum $\mathfrak{R}_t$ ansehen: $y = A\,x$. Führt man eine Koordinatentransformation S im $\mathfrak{R}_t$ aus, also $x' = S\,x$, $y' = S\,y$, so lautet die Abbildung in den neuen Koordinaten $y' = A'\,x'$ und ist gegeben durch

$$A' = S \cdot A \cdot S^{-1}. \tag{10.2}$$

A und A' heißen zueinander äquivalent. Übt man auf *jede* Matrix der Darstellung Γ *dieselbe* Transformation S aus, so erhält man eine *äquivalente Darstellung* Γ', die zur ursprünglichen isomorph[1] ist

$$A \cdot B = C \xleftrightarrow{S} A' \cdot B' = C'. \tag{10.3}$$

Äquivalente Darstellungen können wir daher als nicht wesentlich verschieden ansehen.

Eine Darstellung ist *(vollständig) reduzibel*[2], wenn eine Transformation S möglich ist, so daß *alle* Matrizen der Darstellung die Gestalt

$$A = \begin{pmatrix} A_1 & & & & 0 \\ & A_2 & & & \\ & & A_3 & & \\ 0 & & & A_4 & \\ & & & & \ddots \end{pmatrix}, \quad B = \begin{pmatrix} B_1 & & & & 0 \\ & B_2 & & & \\ & & B_3 & & \\ 0 & & & B_4 & \\ & & & & \ddots \end{pmatrix}, \ldots \tag{10.4}$$

annehmen. Die quadratischen Matrizen $A_1, B_1, \ldots$ bzw. $A_2, B_2, \ldots, \ldots$ (mit dem Grad $t_1, t_2, \ldots$) bilden dann Gruppen, die zu $\mathfrak{M}$ homomorph sind; denn es ist

$$A \cdot B = \begin{pmatrix} A_1 B_1 & & & & 0 \\ & A_2 B_2 & & & \\ & & A_3 B_3 & & \\ 0 & & & A_4 B_4 & \\ & & & & \ddots \end{pmatrix}. \tag{10.5}$$

---

[1] Eine *umkehrbar eindeutige* Zuordnung, bei der die Rechenregeln erhalten bleiben, nennt man einen *Isomorphismus*.

[2] Eine Darstellung heißt *reduzibel* im weiteren Sinn, wenn in den Matrizen an Stelle von (10.4) rechts oben statt der Nullen von Null verschiedene Elemente stehen, links unten aber die Elemente wieder Null sind. Bei endlichen Gruppen ist eine reduzible Darstellung stets vollständig reduzibel; sie „zerfällt" durch eine geeignete Transformation S in die Form (10.4) (MASCHKE).

Wenn eine solche Transformation nicht möglich ist, so heißt die Darstellung *irreduzibel*. Die Zerlegung einer reduziblen Darstellung in ihre irreduziblen Bestandteile $\Gamma_1, \Gamma_2, \ldots$, d.h. in die irreduziblen Kästchenmatrizen von (10.4), läßt sich auf genau eine Weise durchführen. Man schreibt symbolisch

$$\Gamma = \sum_k q_k \Gamma_k, \tag{10.6}$$

wodurch ausgedrückt wird, daß die irreduzible Darstellung Γ_k gerade q_k mal in Γ vorkommt.

Beispiel. Man erhält eine dreidimensionale Darstellung von D_4, wenn man berücksichtigt, daß sich jede Drehung mit Hilfe der EULERschen Winkel Φ, Θ und Ψ durch die Transformationsmatrix

$$\begin{pmatrix} \cos\Phi & -\sin\Phi & 0 \\ \sin\Phi & \cos\Phi & 0 \\ 0 & 0 & 1 \end{pmatrix} \cdot \begin{pmatrix} 1 & 0 & 0 \\ 0 & \cos\Theta & -\sin\Theta \\ 0 & \sin\Theta & \cos\Theta \end{pmatrix} \cdot \begin{pmatrix} \cos\Psi & -\sin\Psi & 0 \\ \sin\Psi & \cos\Psi & 0 \\ 0 & 0 & 1 \end{pmatrix} \tag{10.7}$$

beschreiben läßt [vgl. Bd. II, S. 135 (Tietz)], wobei die Drehung zusammengesetzt wird aus den Drehungen mit Ψ um die z-Achse, dann mit Θ um die x-Achse und schließlich mit Φ wieder um die z-Achse:

$$^3\Gamma \begin{cases} E = \begin{pmatrix} 1 & 0 & 0 \\ 0 & 1 & 0 \\ 0 & 0 & 1 \end{pmatrix}, & A = \begin{pmatrix} -1 & 0 & 0 \\ 0 & -1 & 0 \\ 0 & 0 & 1 \end{pmatrix}, & B = \begin{pmatrix} 0 & -1 & 0 \\ 1 & 0 & 0 \\ 0 & 0 & 1 \end{pmatrix}, & C = \begin{pmatrix} 0 & 1 & 0 \\ -1 & 0 & 0 \\ 0 & 0 & 1 \end{pmatrix} \\[12pt] D = \begin{pmatrix} 1 & 0 & 0 \\ 0 & -1 & 0 \\ 0 & 0 & -1 \end{pmatrix}, & F = \begin{pmatrix} -1 & 0 & 0 \\ 0 & 1 & 0 \\ 0 & 0 & -1 \end{pmatrix}, & G = \begin{pmatrix} 0 & -1 & 0 \\ -1 & 0 & 0 \\ 0 & 0 & -1 \end{pmatrix}, & H = \begin{pmatrix} 0 & 1 & 0 \\ 1 & 0 & 0 \\ 0 & 0 & -1 \end{pmatrix}. \end{cases} \tag{10.8}$$

Diese Darstellung ist bereits von der Form (10.4); sie zerfällt in eine zweidimensionale Darstellung (= die zweidimensionalen Matrizen links oben) und in eine eindimensionale (± 1 rechts unten):

$$^3\Gamma = {}^2\Gamma_5 + {}^1\Gamma_2.$$

Diese beiden Darstellungen sind irreduzibel [vgl. (12.2)].

Als *Charakter* $\chi(A)$ einer Matrix A bezeichnet man die Summe ihrer Diagonalelemente (= Summe der charakteristischen Wurzeln)

$$\chi(A) = \sum_{\lambda=1}^{t} A_{\lambda\lambda} \tag{10.9}$$

Für äquivalente Matrizen ist der Charakter gleich: $\chi(A) = \chi(A')$. Die Zahlen $\chi(E) = t, \chi(A), \chi(B), \ldots$ nennt man das *Charakterensystem* $\chi(\Gamma)$ einer Darstellung Γ. Zwei Darstellungen sind dann und nur dann äquivalent, wenn die Charakterensysteme gleich sind. *Eine Darstellung ist durch ihr Charakterensystem bis auf Äquivalenzen bestimmt.* — Für konjugierte Gruppenelemente ist χ gleich:

$$\chi(A) = \chi(X^{-1}AX) \qquad (X \text{ Matrix aus } \Gamma) \tag{10.10}$$

d.h. der Charakter hat für alle Elemente einer Klasse $\mathfrak{C}$ denselben Wert, χ ist eine *Klassenfunktion*.

11. Sätze über die Charaktere von Darstellungen. Es gibt genau c nichtäquivalente, *irreduzible* Darstellungen $\Gamma_1, \Gamma_2, \ldots, \Gamma_c$ einer (endlichen) Gruppe, d.h. gerade soviel wie die Gruppe verschiedene Klassen konjugierter Elemente enthält. Die Dimension t_k (= positiv ganz) der irreduziblen Darstellung $^{t_k}\Gamma_k$

$(k = 1, \ldots, c)$ ist ein Teiler der Gruppenordnung n. Es gilt ferner

$$\sum_{k=1}^{c} l_k^2 = n. \tag{11.1}$$

Die Charaktere χ_{kr} der irreduziblen Darstellungen Γ_k können also in einem quadratischen Schema nach den Klassen $\mathfrak{C}_r$ geordnet werden

$$
\begin{array}{c|cccc|l}
 & \mathfrak{C}_1 & \mathfrak{C}_2 & \cdots & \mathfrak{C}_c & = \text{Klassen} \\
\hline
 & h_1 & h_2 & \cdots & h_c & = \text{Anzahl der Klassenelemente} \\
\hline
{}^{l_1}\Gamma_1 & \chi_{11} & \chi_{12} & \cdots & \chi_{1c} & \\
{}^{l_2}\Gamma_2 & \chi_{21} & \chi_{22} & \cdots & \chi_{2c} & \\
\vdots & \vdots & \vdots & & \vdots & \\
{}^{l_c}\Gamma_c & \chi_{c1} & \chi_{c2} & \cdots & \chi_{cc} & \\
\end{array}
\tag{11.2}
$$

Zwischen den Charakteren zweier irreduzibler Darstellungen[1] Γ_k und $\Gamma_{k'}$ gilt

$$\sum_{r=1}^{c} h_r\, \chi_{kr}\, \chi_{k'r}^{*} = n\, \delta_{kk'}, \tag{11.3}$$

während zwischen den Charakteren zweier Klassen $\mathfrak{C}_r$ und $\mathfrak{C}_{r'}$ die Gleichung

$$h_r \sum_{k=1}^{c} \chi_{kr}\, \chi_{kr'}^{*} = n\, \delta_{rr'} \tag{11.4}$$

besteht.

Zur Berechnung der Charaktere kann man etwa folgenden Weg einschlagen: Multipliziert man jedes Element einer Klasse $\mathfrak{C}_r = \sum\limits_{v=1}^{h_r} a_{vr}$ mit jedem Element einer anderen Klasse $\mathfrak{C}_{r'} = \sum\limits_{v'=1}^{h_{r'}} a_{v'r'}$ nach den Gesetzen der Gruppentafel, so läßt sich die Gesamtheit der entstehenden Elemente $\mathfrak{C}_r \mathfrak{C}_{r'} = \sum\limits_{v=1}^{h_r} \sum\limits_{v'=1}^{h_{r'}} a_{vr}\, a_{v'r'}$ in der Form

$$\mathfrak{C}_r \mathfrak{C}_{r'} = \sum_{s=1}^{c} c_{rr's}\, \mathfrak{C}_s = \mathfrak{C}_{r'} \mathfrak{C}_r \tag{11.5}$$

schreiben. Die Koeffizienten $c_{rr's}$ geben an, wie oft die Klasse $\mathfrak{C}_s$ in dem Klassenprodukt $\mathfrak{C}_r \mathfrak{C}_{r'}$ enthalten ist. Die $c_{rr's}$ sind damit also bekannte, nichtnegative, ganze Zahlen. Ersetzt man in der Gl. (11.5) die Symbole $\mathfrak{C}_r$ formal durch Zahlen η_r, so kann man aus den (nicht unabhängigen) Gleichungen

$$\eta_r \eta_{r'} = \sum_{s=1}^{c} c_{rr's}\, \eta_s \tag{11.6}$$

die Unbekannten η_r bestimmen. Die Gruppencharaktere ergeben sich dann gemäß

$$\chi_{kr} = l_k\, \frac{\eta_r}{h_r}. \tag{11.7}$$

---

[1] Im folgenden werden die Darstellungen als unitär vorausgesetzt, was durch eine Transformation S stets erreicht werden kann. Es gilt dann also für die Darstellungsmatrizen $A \cdot A^{\dagger} = E$, wobei $A^{\dagger}$ aus A durch Vertauschung von Zeilen mit Spalten und Übergang zum Konjugiert-komplexen entsteht ($E = $ Einheitsmatrix).

Für den Charakter χ_r einer *reduziblen* Darstellung Γ gilt gemäß Gl. (10.6)

$$\chi_r = \sum_{k=1}^{c} q_k \chi_{kr}; \tag{11.8}$$

d.h. die Darstellung Γ zerfällt bei der Ausreduktion in q_1 irreduzible Darstellungen Γ_1 mit dem Charakter χ_{1r}, etc. Die nichtnegativen, ganzen Zahlen q_k sind eindeutig bestimmt; es ist nach Gl. (11.3)

$$q_k = \frac{1}{n} \sum_{r=1}^{c} h_r \chi_r \chi_{kr}^{*}. \tag{11.9}$$

Außerdem erhält man, wenn man über alle Elemente g der Gruppe summiert

$$\sum_{g} |\chi_{\text{red}}(g)|^2 = n \sum_{k=1}^{c} q_k^2 > n, \tag{11.10}$$

im Gegensatz zu den Charakteren einer irreduziblen Darstellung [Gl. (11.3)], für die das Gleichheitszeichen gilt,

$$\sum_{g} |\chi_{\text{irr}}(g)|^2 = n. \tag{11.11}$$

12. Die Charaktere der irreduziblen Darstellungen von D_4. Für das in Ziff. 9 behandelte Beispiel war $n = 8$ und $c = 5$. Wir erhalten also genau fünf irreduzible Darstellungen dieser Gruppe. Ihre Dimensionen t_k findet man nach Gl. (11.1) durch die Zerlegung der Zahl 8 in die fünf Quadrate

$$8 = 1^2 + 1^2 + 1^2 + 1^2 + 2^2. \tag{12.1}$$

Vier der Darstellungen sind also eindimensional $(t_1 = t_2 = t_3 = t_4 = 1)$, d.h. die Matrizen werden gewöhnliche Zahlen, während eine Darstellung $(t_5 = 2)$ durch zweidimensionale Matrizen erfolgt.

Die zugehörigen Charaktere ergeben sich, indem man nach Gl. (11.5) je zwei Klassen aus (9.3) miteinander multipliziert, z.B.

$$\mathfrak{C}_3 \mathfrak{C}_4 = (b + c)(d + f) = 2h + 2g = 2\mathfrak{C}_5.$$

In dem Ergebnis ersetzen wir die Klassensymbole $\mathfrak{C}_r$ durch die Zahlen η_r und erhalten das Gleichungssystem

$$\eta_1 \eta_r = \eta_r \quad \text{für} \quad r = 1, \ldots, 5,$$
$$\eta_2^2 = \eta_1, \qquad \eta_3^2 = \eta_4^2 = \eta_5^2 = 2\eta_1 + 2\eta_2,$$
$$\eta_2 \eta_3 = \eta_3, \qquad \eta_3 \eta_4 = 2\eta_5,$$
$$\eta_2 \eta_4 = \eta_4, \qquad \eta_3 \eta_5 = 2\eta_4,$$
$$\eta_2 \eta_5 = \eta_5, \qquad \eta_4 \eta_5 = 2\eta_3$$

mit den Lösungen

η_1	η_2	η_3	η_4	η_5
1	1	2	2	2
1	1	2	−2	−2
1	1	−2	2	−2
1	1	−2	−2	2
1	−1	0	0	0

Das gesuchte Charakterensystem $\chi_{k\,r}$ ergibt sich hieraus gemäß Gl. (11.7) und (11.3) zu

	$\mathfrak{C}_1$	$\mathfrak{C}_2$	$\mathfrak{C}_3$	$\mathfrak{C}_4$	$\mathfrak{C}_5$
$h_r =$	1	1	2	2	2
eindimensionale Darstellungen $\quad {}^1\Gamma_1 = A_1{}^*$	1	1	1	1	1
${}^1\Gamma_2 = A_2$	1	1	1	-1	-1
${}^1\Gamma_3 = B_1$	1	1	-1	1	-1
${}^1\Gamma_4 = B_2$	1	1	-1	-1	1
zweidimensionale Darstellung $\quad {}^2\Gamma_5 = E$	2	-2	0	0	0

$$(12.2)$$

Damit sind die irreduziblen Darstellungen der Symmetriegruppe der Kristallklasse D_4 charakterisiert.

13. Darstellungstheorie und Eigenfunktionen der Wellenmechanik. Wenn der HAMILTON-Operator $\mathscr{H}$ eines physikalischen Systems invariant gegen bestimmte Transformationen R des Konfigurationsraumes

$$x' = R \cdot x, \quad \text{d.h.} \quad x_i' = \sum_k R_{ik} \cdot x_k \tag{13.1}$$

ist, so ist für einen vorgegebenen Energieeigenwert E mit $\psi(x)$ auch $\psi(Rx)$ eine Eigenfunktion. Die Transformationen R, die gleichwertige Stellen ineinander überführen, bilden die *Symmetriegruppe* des Operators $\mathscr{H}$.

Man bezeichnet mit $\mathscr{P}_R \psi$ jene Funktion, für die

oder
$$\left. \begin{aligned} \mathscr{P}_R \psi(R x) &\equiv \mathscr{P}_R \psi(x') \equiv \psi(x) \\ \mathscr{P}_R \psi(x) &\equiv \psi(R^{-1} x) \end{aligned} \right\} \tag{13.2}$$

identisch vermöge (13.1) gilt. $\mathscr{P}_R \psi$ ist dabei als ein Funktionszeichen anzusehen und $\mathscr{P}_R \psi(x)$ ist der Wert dieser Funktion an der Stelle x. Die erste Gleichung von (13.2) sagt aus, daß die Funktion $\mathscr{P}_R \psi$ an der Stelle $x' = R \cdot x$ denselben Wert annehmen soll, den die Funktion ψ an der Stelle x besitzt. Der Operator $\mathscr{P}$ ist linear, d.h. bei Anwendung auf eine Linearkombination zweier Funktionen ψ_1 und ψ_2 gilt

$$\mathscr{P}_R (a \psi_1 + b \psi_2) = a \mathscr{P}\psi_1 + b \mathscr{P}\psi_2 \quad (a, b = \text{const}). \tag{13.3}$$

Sind die Punkte x und $x' = R \cdot x$ des Konfigurationsraumes physikalisch gleichwertig, so sind auch die Eigenfunktionen ψ und $\mathscr{P}_R \psi$ des Systems gleichwertig, da beide denselben Energieeigenwert besitzen

$$\mathscr{H} \psi = E \psi, \quad \mathscr{H} \mathscr{P}_R \psi = E \mathscr{P}_R \psi. \tag{13.4}$$

Hieraus folgt
$$\mathscr{H} \mathscr{P}_R = \mathscr{P}_R \mathscr{H}, \tag{13.5}$$

d.h. die Operatoren $\mathscr{H}$ und $\mathscr{P}_R$ sind vertauschbar. Diese Symmetrieoperatoren $\mathscr{P}_R$ sind zur Symmetriegruppe R isomorph; es gilt für zwei Transformationen R und S der Symmetriegruppe

$$\mathscr{P}_{R \cdot S} = \mathscr{P}_R \cdot \mathscr{P}_S. \tag{13.6}$$

Ist der Eigenwert E t-fach [d.h. $(t-1)$-fach entartet], gehören zu ihm also t linear unabhängige Eigenfunktionen $\psi^1, \psi^2, \ldots, \psi^t$, so müssen sich die ebenfalls zu E gehörigen Eigenfunktionen $\mathscr{P}_R \psi^1, \mathscr{P}_R \psi^2, \ldots, \mathscr{P}_R \psi^t$ als Linearkombinationen

---

* In der Molekülphysik und in der Theorie der chemischen Bindungen sind die Bezeichnungen A, B für eindimensionale, E für zweidimensionale und T für dreidimensionale Darstellungen üblich (vgl. auch Ziff. 17).

der $\psi^1, \psi^2, \ldots, \psi^t$ schreiben lassen:

$$\mathscr{P}_R \psi^\mu(x) = \sum_{\lambda=1}^{t} D^{\lambda\mu}(R)\, \psi^\lambda(x) \qquad (\mu = 1, \ldots, t)\,, \tag{13.7}$$

oder

$$\psi^\mu(x') = \sum_{\lambda=1}^{t} D^{\lambda\mu}(R^{-1}) \cdot \psi^\lambda(x)\,. \tag{13.8}$$

Die t-dimensionalen Matrizen $D(R)$ haben die Produkteigenschaft

$$D(R \cdot S) = D(R) \cdot D(S)\,, \tag{13.9}$$

d.h. sie bilden eine Darstellung der Symmetriegruppe der SCHRÖDINGER-Gleichung. Die Dimension t der Darstellung ist gleich der Anzahl der linear unabhängigen Eigenfunktionen $\psi^1, \psi^2, \ldots, \psi^t$, die zum t-fachen Eigenwert E gehören. Geht man von einem anderen Satz von linear unabhängigen Eigenfunktionen, die ebenfalls zu E gehören, aus, so unterscheiden sich die Darstellungen nur durch eine Ähnlichkeitstransformation, sie sind äquivalent. Zu jedem Eigenwert E gehört in diesem Sinne eindeutig eine t-dimensionale Darstellung der Symmetriegruppe. Sind die zu E gehörigen Eigenfunktionen orthonormiert, so ist die Darstellung unitär.

Mit Hilfe der Gruppentheorie lassen sich über die *Eigenwerte eines gestörten Problems*, für das sich der HAMILTON-Operator $\mathscr{H} = \mathscr{H}^0 + \mathscr{H}'$ aus dem ungestörten $\mathscr{H}^0$ und dem Störoperator $\mathscr{H}'$ zusammensetzt, folgende Aussagen machen:

a) Hat die Störung $\mathscr{H}'$ *dieselbe Symmetrie* wie $\mathscr{H}^0$, so kann die Störung, wie groß sie auch sein mag, niemals eine Termaufspaltung hervorrufen, wenn für das ungestörte Problem eine irreduzible Darstellung der Symmetriegruppe vorliegt. Ist dagegen für das Problem ohne Störung die Darstellung der Dimension t vollständig reduzibel, d.h. zerfällt sie in s irreduzible Darstellungen

$$\Gamma^0 = \sum_{k=1}^{s} q_k^0\, \Gamma_k^0\,,$$

so kann der t-fache Term durch die Störung aufspalten und zwar in höchstens $\sum_{k=1}^{s} q_k^0$ Terme.

b) Ist die Störung $\mathscr{H}'$ *nicht invariant* gegenüber der ganzen Symmetriegruppe von $\mathscr{H}^0$, sondern nur gegenüber einer Untergruppe, so werden gegenüber dieser kleineren Gruppe die Darstellungen i.a. reduzibel sein. Die Ausreduzierung der ursprünglichen Darstellung bezüglich dieser Untergruppe liefert die Aufspaltung der Energieterme durch die Störung. Man hat also aus der irreduziblen Darstellung der Symmetriegruppe von $\mathscr{H}^0$ jene Gruppenelemente auszuwählen, welche die Symmetriegruppe von $\mathscr{H}'$ bilden, und diese reduzible Darstellung Γ der Symmetriegruppe von $\mathscr{H}'$ entsprechend Gl. (10.6) auszureduzieren:

$$\Gamma = \sum q_k\, \Gamma_k\,,$$

d.h. also nachzusehen, wie oft (q_k mal) die irreduzible Darstellung Γ_k der Symmetriegruppe von $\mathscr{H}'$ in Γ enthalten ist. Der ungestörte Term spaltet dann bei der vorgenommenen Verminderung der Symmetrie auf in q_1 Terme, die zur Darstellung Γ_1, in q_2 Terme, die zu Γ_2, usw. gehören. (Die q_k sind natürlich für verschiedene ungestörte Terme verschieden; sie hängen von der Quantenzahl des ungestörten Termes ab.) Die Gesamtzahl der durch die Störung $\mathscr{H}'$ aus einem ungestörten Energieterm entstehenden Terme ist $\sum q_k$. Die Vielfachheit der einzelnen Terme wird durch die Dimension der irreduziblen Darstellungen Γ_k^i gegeben. Für die Bestimmung von Anzahl und Vielfachheit der Terme ist es belanglos, wie groß die Störung ist.

Der eben besprochene Fall b) ist nun für die Theorie der Kristallspektren von ausschlaggebender Bedeutung. Während die HAMILTON-Funktion des freien Atoms oder Ions gegen beliebige Drehungen des Koordinatensystems (Nullpunkt = Kern) und Inversion am Kern invariant ist, reduziert sich die Symmetriegruppe durch das Kristallfeld auf jene Symmetrieoperationen, welche die Lage des Kernes ungeändert lassen und gleichzeitig den Kristall in sich überführen. Die maximale[1] Anzahl der Terme im Kristallfeld ist uns bekannt, wenn wir die Größen q_k kennen. Diese lassen sich in eindeutiger Weise nach Gl. (11.9) aus den Charakteren bestimmen. Während wir bereits am Beispiel der Kristallklasse D_4 gesehen haben, wie man die Charaktere der irreduziblen Darstellungen der Symmetriegruppe von Kristallklassen bestimmt, müssen wir — um Gl. (11.9) anwenden zu können — die Darstellungen der dreidimensionalen Drehgruppe, die die Symmetriegruppe des freien Atoms darstellt, angeben. Von Drehspiegelungen und der Inversion am Kern sehen wir zunächst ab (vgl. Ziff. 18).

14. Die dreidimensionale Drehgruppe. Im Gegensatz zu den bisher besprochenen endlichen Gruppen, d.h. Gruppen mit endlich vielen Elementen, stellen sämtliche Drehungen im dreidimensionalen Raum eine unendliche Gruppe dar. Sie ist eine dreiparametrige kontinuierliche Gruppe, d.h. sie hängt von drei sich stetig verändernden Parametern, z.B. den drei EULERschen Winkeln ab. Es lassen sich eine große Zahl der Sätze, die bei endlichen Gruppen gelten, einfach auf kontinuierliche Gruppen erweitern, wenn man an Stelle der Summen über die Gruppenelemente entsprechende Integrale einführt. Es ist hier nicht der Platz, darauf näher einzugehen.

Alle räumlichen Drehungen um einen Winkel α sind geometrisch gleichwertig und bilden eine Klasse $\mathfrak{C}(\alpha)$ der dreidimensionalen Drehgruppe. Da der Charakter χ, der eine Darstellung eindeutig definiert, eine Klassenfunktion ist, $\chi = \chi(\alpha)$, genügt es, zu seiner Bestimmung den Winkel α speziell als Drehung um die z-Achse eines Kugelkoordinatensystemes (r, ϑ, φ) anzunehmen. Mit Hilfe der auf die Drehachse bezogenen Kugelfunktionen $P_J^M(\cos\vartheta)\, e^{iM\varphi}$ ergibt sich bei der Drehung $\varphi' = \varphi + \alpha$ nach Gl. (13.2) und (13.7)

$$\mathscr{P}_\alpha P_J^M(\cos\vartheta)\, e^{iM\varphi} = P_J^M(\cos\vartheta)\, e^{iM(\varphi-\alpha)} = \sum_{M'=-J}^{+J} \mathfrak{D}_J^{M'M}(\alpha)\, P_J^{M'}(\cos\vartheta)\, e^{iM'\varphi},$$

durch Koeffizientenvergleich

$$\mathfrak{D}_J^{M'M}(\alpha) = e^{-iM\alpha}\, \delta_{M'M}.$$

Zur Drehung α gehört also die $(2J+1)$-dimensionale Darstellung

$$\mathfrak{D}_J(\alpha) = \begin{pmatrix} e^{-iJ\alpha} & & & 0 \\ & e^{-i(J-1)\alpha} & & \\ & & \ddots & \\ 0 & & & e^{iJ\alpha} \end{pmatrix}, \tag{14.1}$$

da M die $2J+1$ Werte $J, J-1, J-2, \ldots, -J$ annehmen kann. J durchläuft dabei die ganzen Zahlen $0, 1, 2, 3, \ldots$ und stellt physikalisch die Drehimpulsquantenzahl des freien Atoms dar. Die Anzahl der Darstellungen der Dreh-

---

[1] Wenn man über die Natur des Kristallfeldes keine einschränkenden Annahmen macht, ergibt sich diese Anzahl. Wird jedoch das Kristallfeld als rein elektrisch angesehen, so kann eine sog. KRAMERS-Entartung auftreten (vgl. Ziff. 20), so daß sich die Termzahl verringert.

gruppe ist unendlich, weil es unendlich viele Klassen gibt. Der zum Winkel α gehörige Charakter $\chi(\alpha)$ berechnet sich als Summe der Diagonalglieder von Gl. (14.1) zu

$$\chi_J(\alpha) = \sum_{M=J}^{-J} e^{-iM\alpha} = \frac{\sin(2J+1)\frac{\alpha}{2}}{\sin\frac{\alpha}{2}}. \tag{14.2}$$

Für eine beliebige Drehung nimmt natürlich die Darstellung $\mathfrak{D}_J$ eine kompliziertere Form als Gl. (14.1) an: Die Matrixelemente, die dann auch in den Nichtdiagonalgliedern von Null verschieden sind, hängen von den die Drehung beschreibenden drei EULERschen Winkeln ab. Diese $(2J+1)$-dimensionalen Darstellungen $\mathfrak{D}_0, \mathfrak{D}_1, \ldots, \mathfrak{D}_J, \ldots$ stellen die sämtlichen, miteinander nicht äquivalenten, irreduziblen Darstellungen der dreidimensionalen (reinen) Drehgruppe dar. Da aber die durch die drei EULERschen Winkel beschriebene Drehung stets *einer* um eine geeignet gewählte Achse ausgeführte Drehung α entspricht, ist der Charakter der irreduziblen Darstellung stets durch (14.2) gegeben.

Außer den ganzen Zahlen kann J auch *halbzahlige* Werte annehmen: $J = \frac{1}{2}, \frac{3}{2}, \frac{5}{2}, \ldots.$ Gruppentheoretisch entspricht diese dem Elektronenspin gerecht werdende Tatsache der Homomorphie zwischen der Gruppe der zweidimensionalen unitären Matrizen mit der Determinante 1 ($=$ *unitäre* Gruppe)

$$\mathfrak{u} = \begin{pmatrix} a & b \\ -b^* & a^* \end{pmatrix}, \quad |a^2| + |b^2| = 1 \tag{14.3}$$

und der dreidimensionalen Drehgruppe, bei der jedem unitären Matrizenpaar $\mathfrak{u}, -\mathfrak{u}$ umkehrbar eindeutig eine Drehung zugeordnet ist

$$\mathfrak{u}(\Phi, \Theta, \Psi) = \pm \begin{pmatrix} e^{-\frac{i}{2}\Phi} & 0 \\ 0 & e^{+\frac{i}{2}\Phi} \end{pmatrix} \cdot \begin{pmatrix} \cos\frac{\Theta}{2} & -i\sin\frac{\Theta}{2} \\ -i\sin\frac{\Theta}{2} & \cos\frac{\Theta}{2} \end{pmatrix} \cdot \begin{pmatrix} e^{-\frac{i}{2}\Psi} & 0 \\ 0 & e^{\frac{i}{2}\Psi} \end{pmatrix} = \mathfrak{D}_{\frac{1}{2}}(\Phi, \Theta, \Psi) \tag{14.4}$$

$[\Phi, \Theta, \Psi =$ EULERsche Winkel, vgl. Gl. (10.7)]. Die irreduziblen Darstellungen der unitären Gruppe umfassen neben den irreduziblen Darstellungen der Drehgruppe mit ganzzahligem J die *zweideutigen*, irreduziblen Darstellungen der Drehgruppe mit halbzahligen J-Werten, d.h. Darstellungen gerader Dimension $(=2J+1)$. Insbesondere bilden die Matrizen (14.4) selbst die Darstellung $\mathfrak{D}_{\frac{1}{2}}$. Die zweideutigen Darstellungen sind keine eigentlichen Darstellungen, da für sie gilt

$$\mathfrak{D}_J(R)\,\mathfrak{D}_J(S) = \pm\,\mathfrak{D}_J(R\cdot S) \quad (J \text{ halbzahlig}). \tag{14.5}$$

Für die Charaktere $\chi(\alpha)$ der zweideutigen Darstellungen gilt wieder

$$\chi_J(\alpha) = \frac{\sin(J+\frac{1}{2})\alpha}{\sin\frac{\alpha}{2}},$$

wobei also $J + \frac{1}{2}$ eine ganze Zahl ist. Bei Vermehrung des Winkels α um 2π erhält man im Gegensatz zu den eindeutigen Darstellungen nicht wieder $\chi(\alpha)$, sondern $-\chi(\alpha)$:

$$\chi_J(\alpha + 2\pi) = -\chi_J(\alpha) \quad (J \text{ halbzahlig}). \tag{14.6}$$

Der Charakter der Darstellung wird zweideutig. [Lediglich $\chi(\pi) = \chi(3\pi) = 0$ ist eindeutig.] Hierin liegt die Ursache dafür, daß die Aufspaltung der Terme mit halbzahligen Quantenzahlen eine etwas andere Behandlung als die der Terme mit ganzen Quantenzahlen nötig macht. Es sei darauf hingewiesen, daß wegen des Auftretens der Spinvariablen bei halbzahligem J in der zu Gl. (13.7) analogen

Gleichung
$$\mathscr{O}_R \psi_J^M = \sum_{M'} \mathfrak{D}_J^{M'M}(R) \cdot \psi_J^{M'} \tag{14.7}$$

der lineare Operator $\mathscr{O}_R$ keine Punkttransformation mehr darstellt[1].

15. Die Anzahl der Terme im Kristallfeld $\boldsymbol{D_4}$ bei gerader Elektronenzahl. Es möge die Termaufspaltung in einem Kristallfeld nach der in Ziff. 13 allgemein besprochenen Methode an Hand des Beispieles der Kristallklasse D_4 besprochen werden, wobei zunächst der einfachere Fall einer geraden Elektronenzahl, d.h. $J =$ ganzzahlig, behandelt werde, in welchem also die eindeutige Darstellung der Drehgruppe benützt wird. Für die acht Drehelemente der Kristallklasse D_4, bzw. ihre fünf Klassen $\mathfrak{C}_1$ bis $\mathfrak{C}_5$, erhält man in der Darstellung der dreidimensionalen Drehgruppe die Charaktere $\chi_J(\alpha_r)$, wenn man in Gl. (14.2) den zur r-ten Klasse gehörigen Drehwinkel α_r einsetzt

$$\xrightarrow{\quad} r$$

	$\mathfrak{C}_1$	$\mathfrak{C}_2$	$\mathfrak{C}_3$	$\mathfrak{C}_4$	$\mathfrak{C}_5$
$\alpha_r =$	0	π	$\pm\dfrac{\pi}{2}$	π	π
$\chi_J(\alpha_r) = \dfrac{\sin(2J+1)\frac{\alpha_r}{2}}{\sin\frac{\alpha_r}{2}}$	$2J+1$	$(-1)^J$	$(-1)^{\left[\frac{J}{2}\right]*}$	$(-1)^J$	$(-1)^J$
z.B. $J=0$	1	1	1	1	1
1	3	-1	1	-1	-1
2	5	1	-1	1	1
3	7	-1	-1	-1	-1
4	9	1	1	1	1
$\vdots$					

$$\tag{15.1}$$

Die Anzahl der Terme im Kristallfeld findet man, indem man nach Gl. (11.9) die Zahlen q_{kJ} bestimmt:

$$q_{kJ} = \frac{1}{n}\sum_{r=1}^{c} h_r \chi_J(\alpha_r)\, \chi_{kr}^*. \tag{15.2}$$

Die Charaktere $\chi_J(\alpha_r)$ sind dabei aus der Tabelle (15.1), die Charaktere χ_{kr} aus Tabelle (12.2) zu entnehmen. Es ergibt sich damit ($n=8, c=5$)

		$J =$	0	1	2	3	4	5 …
einfach	$^1\Gamma_1 = A_1$	$q_{1,J} =$	1	0	1	0	2	1
	$^1\Gamma_2 = A_2$	$q_{2,J} =$	0	1	0	1	1	2
	$^1\Gamma_3 = B_1$	$q_{3,J} =$	0	0	1	1	1	1
	$^1\Gamma_4 = B_2$	$q_{4,J} =$	0	0	1	1	1	1
zweifach	$^2\Gamma_5 = E$	$q_{5,J} =$	0	1	1	2	2	3
Gesamtzahl der einfachen Terme			1	1	3	3	5	5
Anzahl der zweifachen Terme			0	1	1	2	2	3
Gesamttermzahl			1	2	4	5	7	8
$(2J+1)$ Zeeman-Terme			1	3	5	7	9 11

$$\tag{15.3}$$

---

[1] Vgl. z.B. E. Wigner: Gruppentheorie und ihre Anwendung auf die Quantenmechanik der Atomspektren. Braunschweig 1931, S. 240.

* Unter $[a]$ verstehen wir jene ganze Zahl, die man durch Abrunden der im allgemeinen nicht ganzen Zahl a erhält, z.B. $[\frac{3}{2}] = 1$. Ist a selbst ganzzahlig, so ist $[a] = a$.

Hieraus können wir z.B. entnehmen, daß ein ungestörter Term mit $J = 5$ in einem Kristallfeld der Symmetrie D_4 in fünf einfache und drei zweifache Terme aufspaltet. Insgesamt ergeben sich also acht verschiedene Terme. Bei großen J-Werten gehören zu einer Darstellung Γ_k im allgemeinen mehrere Terme; z.B. sind bei $J = 5$ unter $^1\Gamma_2$ zwei verschiedene einfache Terme und unter $^2\Gamma_5$ drei verschiedene zweifache Terme vereinigt.

Im Kristallfeld D_4 sind die Terme höchstens zweifach, da die irreduziblen Darstellungen dieser Kristallklasse maximal zweidimensional sind.

16. Die Anzahl der Terme im Kristallfeld D_4 bei ungerader Elektronenzahl. Bei ungerader Anzahl von Elektronen ist J halbzahlig. Hierzu gehört nach Ziff. 14 die *zweideutige* Darstellung der Drehgruppe. Diese kann natürlich auch nur zweideutige Darstellungen der Kristallgruppe enthalten. Um diese Darstellungen zu erhalten, stellen wir uns mit H. Bethe und W. Opechowski[1] vor, daß der Kristall bei einer Drehung um irgendeine Achse um 2π nicht in sich übergeht, sondern erst bei Drehungen um 4π. (In Analogie zur Riemannschen Fläche mehrdeutiger Funktionen.) Der Gruppeneinheit entspricht also die Drehung um 4π, während der Drehung um 2π ein neues Gruppenelement e' zugeordnet wird ($e'^2 = e$). Diese „*Kristalldoppelgruppe*" $\mathfrak{G}^\dagger$ enthält damit doppelt soviel Elemente (x und x') als die entsprechende gewöhnliche Kristallgruppe $\mathfrak{G}$. $\mathfrak{G}^\dagger$ ist die Gruppe jener $2n$ Matrizen aus der unitären Gruppe $\mathfrak{u}$ [Gl. (14.4)], die den n Elementen von $\mathfrak{G}$ entsprechen. Es gilt

$$x \cdot e = x = e \cdot x, \quad x \cdot e' = x' = e' \cdot x, \quad x' \cdot e = x' = e \cdot x', \quad x' \cdot e' = x = e' \cdot x'. \tag{16.1}$$

An Hand des Beispieles D_4 werde das Verfahren erläutert. Die Doppelgruppe enthält $n = 2 \cdot 8 = 16$ Elemente, zu denen nach Gl. (14.4) die Matrizen

$$e = \begin{pmatrix} 1 & 0 \\ 0 & 1 \end{pmatrix}, \quad a = \begin{pmatrix} -i & 0 \\ 0 & i \end{pmatrix}, \quad b = \begin{pmatrix} \varepsilon^* & 0 \\ 0 & \varepsilon \end{pmatrix}, \quad c = \begin{pmatrix} \varepsilon & 0 \\ 0 & \varepsilon^* \end{pmatrix}, \quad d = \begin{pmatrix} 0 & -i \\ -i & 0 \end{pmatrix},$$
$$f = \begin{pmatrix} 0 & -1 \\ 1 & 0 \end{pmatrix}, \quad g = \begin{pmatrix} 0 & \varepsilon^* \\ -\varepsilon & 0 \end{pmatrix}, \quad h = \begin{pmatrix} 0 & -\varepsilon \\ \varepsilon^* & 0 \end{pmatrix}, \quad e' = \begin{pmatrix} -1 & 0 \\ 0 & -1 \end{pmatrix}, \quad x' = e'x, \left(\varepsilon = e^{i\frac{\pi}{4}}\right) \tag{16.2}$$

gehören. Als Gruppentafel erhält man damit

	$e\ e'$	$a\ a'$	$b\ b'$	$c\ c'$	$d\ d'$	$f\ f'$	$g\ g'$	$h\ h'$
$\begin{matrix}e\\e'\end{matrix}$	E	A	B	C	D	F	G	H
$\begin{matrix}a\\a'\end{matrix}$	A	$\tilde{E}$	$\tilde{C}$	B	F	$\tilde{D}$	H	$\tilde{G}$
$\begin{matrix}b\\b'\end{matrix}$	B	$\tilde{C}$	A	E	H	$\tilde{G}$	D	F
$\begin{matrix}c\\c'\end{matrix}$	C	B	E	$\tilde{A}$	G	H	$\tilde{F}$	D
$\begin{matrix}d\\d'\end{matrix}$	D	$\tilde{F}$	G	H	$\tilde{E}$	A	$\tilde{B}$	$\tilde{C}$
$\begin{matrix}f\\f'\end{matrix}$	F	D	H	$\tilde{G}$	$\tilde{A}$	$\tilde{E}$	C	$\tilde{B}$
$\begin{matrix}g\\g'\end{matrix}$	G	$\tilde{H}$	$\tilde{F}$	D	$\tilde{C}$	B	$\tilde{E}$	A
$\begin{matrix}h\\h'\end{matrix}$	H	G	D	F	$\tilde{B}$	$\tilde{C}$	$\tilde{A}$	$\tilde{E}$

$$\tag{16.3}$$

[1] H. Bethe: Ann. Phys. (5) **3**, 133 (1929). — W. Opechowski: Physica, Haag **7**, 552 (1940).

dabei bedeutet z.B. $F=\begin{pmatrix} f & f' \\ f' & f \end{pmatrix}$ und $\widetilde{F}=\begin{pmatrix} f' & f \\ f & f' \end{pmatrix}$. Aus der Tabelle liest man z.B. ab, daß $b\cdot d=h$ und $b\cdot d'=h'$ ist.

Wie in Ziff. 9 können wir die folgenden sieben Klassen konjugierter Elemente bestimmen

$$\left.\begin{aligned}
\mathfrak{C}_{\bar 1}&=e & (h_{\bar 1}&=1), & \mathfrak{C}_{\bar 3}&=b+c & (h_{\bar 3}&=2), \\
\mathfrak{C}_{\bar{\bar 1}}&=e' & (h_{\bar{\bar 1}}&=1), & \mathfrak{C}_{\bar{\bar 3}}&=b'+c' & (h_{\bar{\bar 3}}&=2), \\
\mathfrak{C}_2&=a+a' & (h_2&=2), & \mathfrak{C}_4&=d+d'+f+f' & (h_4&=4), \\
& & & & \mathfrak{C}_5&=g+g'+h+h' & (h_5&=4),
\end{aligned}\right\} \qquad (16.4)$$

wobei die Numerierung im Anschluß an Gl. (9.3) gewählt wurde, damit der Zusammenhang mit den eindeutigen Darstellungen ersichtlich werde [vgl. Gl. (16.6)]. (Der Index r in den obigen Formeln durchläuft also jetzt die sieben Zahlen $\bar 1$, $\bar{\bar 1}$, 2, $\bar 3$, $\bar{\bar 3}$, 4 und 5.) Man sieht, daß bei Doppelgruppen die Anzahl der Klassen und damit die Anzahl der irreduziblen Darstellungen größer ist als bei gewöhnlichen Gruppen, aber nicht doppelt so groß. Die Zerlegung von $n=16$ in die Summe von sieben Quadraten

$$16 = 1^2 + 1^2 + 1^2 + 1^2 + 2^2 + 2^2 + 2^2 \qquad (16.5)$$

ergibt, daß von den sieben irreduziblen Darstellungen vier eindimensional und drei zweidimensional sind $(t_1=t_2=t_3=t_4=1\,;\ t_5=t_6=t_7=2)$. Durch Multiplikation von zwei Klassen, z.B.

$$\mathfrak{C}_2\,\mathfrak{C}_{\bar 3} = (a+a')\,(b+c) = c' + b + c + b' = \mathfrak{C}_{\bar 3} + \mathfrak{C}_{\bar{\bar 3}}$$

erhält man die folgenden Gleichungen für η

$$\eta_{\bar 1}\cdot\eta_r = \eta_r$$

$$
\begin{array}{l|l|l|l}
\cdots = \eta_{\bar 1} & \eta_2^2 = 2\,(\eta_{\bar 1} + \eta_{\bar{\bar 1}}) & \eta_{\bar 3}^2 = \eta_{\bar{\bar 3}}^2 = 2\eta_{\bar 1} + \eta_2 & \eta_4^2 = \eta_5^2 = 4\,(\eta_{\bar 1} + \eta_{\bar{\bar 1}} + \eta_2) \\
\cdots = \eta_2 & \eta_2\cdot\eta_{\bar 3} = \eta_2\cdot\eta_{\bar{\bar 3}} = \eta_{\bar 3} + \eta_{\bar{\bar 3}} & \eta_{\bar 3}\cdot\eta_{\bar{\bar 3}} = 2\eta_{\bar{\bar 1}} + \eta_2 & \eta_4\cdot\eta_5 = 4\,(\eta_{\bar 3} + \eta_{\bar{\bar 3}}) \\
\cdots = \eta_{\bar 3} & \eta_2\cdot\eta_4 = 2\eta_4 & \eta_{\bar 3}\cdot\eta_4 = \eta_{\bar{\bar 3}}\eta_4 = 2\eta_5 & \\
\cdots = \eta_{\bar{\bar 3}} & \eta_2\cdot\eta_5 = 2\eta_5 & \eta_{\bar 3}\cdot\eta_5 = \eta_{\bar{\bar 3}}\eta_5 = 2\eta_4 & \\
\cdots = \eta_4 & & & \\
\cdots = \eta_5 & & &
\end{array}
$$

deren Lösungen mit Gl. (11.7) die Charaktere $\chi_{k\,r}$ der tetragonalen Doppelgruppe ergeben

$$\xrightarrow{\quad r\quad}$$

	$\mathfrak{C}_{\bar 1}$	$\mathfrak{C}_{\bar{\bar 1}}$	$\mathfrak{C}_2$	$\mathfrak{C}_{\bar 3}$	$\mathfrak{C}_{\bar{\bar 3}}$	$\mathfrak{C}_4$	$\mathfrak{C}_5$
$h_r =$	1	1	2	2	2	4	4
$^1\Gamma_1$	1	1	1	1	1	1	1
$^1\Gamma_2$	1	1	1	1	1	-1	-1
$^1\Gamma_3$	1	1	1	-1	-1	1	-1
$^1\Gamma_4$	1	1	1	-1	-1	-1	1
$^2\Gamma_5$	2	2	-2	0	0	0	0
$^2\Gamma_6$	2	-2	0	$\sqrt{2}$	$-\sqrt{2}$	0	0
$^2\Gamma_7$	2	-2	0	$-\sqrt{2}$	$\sqrt{2}$	0	0

eindimensionale Darstellungen: $^1\Gamma_1$, $^1\Gamma_2$, $^1\Gamma_3$, $^1\Gamma_4$; zweidimensionale Darstellungen: $^2\Gamma_5$, $^2\Gamma_6$, $^2\Gamma_7$ $\qquad (16.6)$

Man beachte, daß sich für die ersten fünf Darstellungen dieselben Charaktere wie in Gl. (12.2) ergeben.

Von OPECHOWSKI wurde allgemein gezeigt, daß jede irreduzible Darstellung einer Kristallgruppe $\mathfrak{G}$ eine irreduzible Darstellung von $\mathfrak{G}^\dagger$ ist. Man nennt die Darstellungen von $\mathfrak{G}^\dagger$, die keine (eindeutigen) Darstellungen von $\mathfrak{G}$ sind, *spezifische* Darstellungen von $\mathfrak{G}^\dagger$ [in (16.6) also $^2\Gamma_6$ und $^2\Gamma_7$]. Eine spezifische Darstellung kann nur in spezifische, irreduzible Darstellungen zerfallen. Jeder Klasse $\mathfrak{C}$ aus $\mathfrak{G}$, die Drehungen um π enthält, welche auf wenigstens einer anderen Drehung um π von $\mathfrak{G}$ senkrecht stehen, entspricht nach OPECHOWSKI *eine* Klasse von $\mathfrak{G}^\dagger$, deren irreduzible spezifische Darstellungen stets den Charakter Null besitzen [z.B. $\mathfrak{C}_2$ in (16.6)]. Alle anderen Klassen $\mathfrak{C}$ von $\mathfrak{G}$ entsprechen *zwei* Klassen $\overline{\mathfrak{C}}$ und $\overline{\overline{\mathfrak{C}}}$ von $\mathfrak{G}^\dagger$, deren irreduzible, spezifische Darstellungen Charaktere mit entgegengesetztem Vorzeichen haben: $\overline{\chi} = -\overline{\overline{\chi}}$ (wobei auch Null möglich ist). Ist $\mathfrak{G}^\dagger$ nicht abelsch, so bilden die die Gruppe definierenden Matrizen aus $\mathfrak{u}$ selbst eine spezifische, irreduzible Darstellung [(16.2) ergibt z.B. $^2\Gamma_6$].

Nun sollen die zweideutigen Darstellungen der dreidimensionalen Drehgruppe als Darstellungen der tetragonalen Doppelgruppe ausreduziert werden. Die Charaktere $\chi_J(\alpha_r)$ der Klassen der tetragonalen Doppelgruppe in der $(2J+1)$-dimensionalen Darstellung der dreidimensionalen Drehgruppe sind

	$\mathfrak{C}_{\overline{1}}$	$\mathfrak{C}_{\overline{\overline{1}}}$	$\mathfrak{C}_2$	$\mathfrak{C}_{\overline{3}}$	$\mathfrak{C}_{\overline{\overline{3}}}$	$\mathfrak{C}_4$	$\mathfrak{C}_5$
$\alpha_r =$	0	2π	$\pi, 3\pi$	$\dfrac{\pi}{2}, \dfrac{7\pi}{2}$	$\dfrac{5\pi}{2}, \dfrac{3\pi}{2}$	$\pi, 3\pi$	$\pi, 3\pi$
$\chi_J(\alpha_r) = \dfrac{\sin\left(J+\frac{1}{2}\right)\alpha_r}{\sin\dfrac{\alpha_r}{2}}$	$2J+1$	$-(2J+1)$	0	$\begin{aligned}&\sqrt{2}, \text{ für}^{1}\ J \equiv \tfrac{1}{2} \pmod 4\\ &0, \text{ für }\ J \equiv \tfrac{3}{2}, \tfrac{7}{2} \pmod 4\\ &-\sqrt{2}, \text{ für }\ J \equiv \tfrac{5}{2} \pmod 4\end{aligned}$	$-\chi_{\overline{3}}$	0	0

$$(16.$$

Nach Gl. (15.2) findet man damit

$$q_{1,J} = q_{2,J} = q_{3,J} = q_{4,J} = q_{5,J} = 0, \tag{16.8}$$

d.h. es gibt für halbzahliges J sicher *keine einfachen* Kristallterme, was man auch aus der Zerlegung $16 = 8 + 2^2 + 2^2$ ersieht. ($8 =$ Quadratsumme der Dimensionszahlen der eindeutigen Darstellungen.) Bei halbzahligem J bestimmen allein die *spezifischen*, irreduziblen Darstellungen von $\mathfrak{G}^\dagger$ die Anzahl der Kristallterme. Für $^2\Gamma_6$ und $^2\Gamma_7$ erhält man das folgende Ergebnis für die Aufspaltung in zweifache Terme

	$J =$	$\frac{1}{2}$	$\frac{3}{2}$	$\frac{5}{2}$	$\frac{7}{2}$	$\frac{9}{2}$ ...
$^2\Gamma_6$	$q_{6,J} =$	1	1	1	2	3
$^2\Gamma_7$	$q_{7,J} =$	0	1	2	2	2
Gesamttermzahl		1	2	3	4	5
$(2J+1) =$ Anzahl der ZEEMAN-Terme		2	4	6	8	10

$$(16.9)$$

---

[1] $a \equiv b \pmod c$ heißt $a = b + fc$ mit $f = 0, \pm 1, \pm 2 \ldots$

17. Die Kristallklassen ohne Inversionsachsen. α) *Gewöhnliche Kristallgruppen.* In der folgenden Zusammenstellung der Charaktere der irreduziblen Darstellungen der Kristallgruppen ohne Inversionsachsen bezeichnen wir in der ersten Zeile — abweichend von der bisherigen Symbolik — die Klasse, die h p-zählige Drehachsen umfaßt, mit hC_p^λ, wobei C_p^λ die λ-te Potenz der Drehung C_p bedeutet ($C_p^p = E = \text{Einheit}$). Die Darstellungen der einzelnen Gruppen sind in der ersten Spalte durch die gruppentheoretischen Symbole, in der zweiten durch die Symbole der Molekülphysik und in der dritten durch Kristallquantenzahlen (Abschnitt B II) bezeichnet. Die vorletzte und letzte Spalte beziehen sich auf die KRAMERSsche Entartung (Ziff. 20).

ische Gruppen C_p (alle $h=1$; ABELsche Gruppen).

C_1	$\mu=$	E	β	v
A	0	1	1 1	

C_2		$\mu=$	E	C_2	β	v
$^1\Gamma_1$	A	0	1	1	1	1
$^1\Gamma_2$	B	1	1	-1	1	1

C_3		$\mu=$	E	C_3	C_3^2	β	v
$^1\Gamma_1$	A	0	1	1	1	1	1
$^1\Gamma_2$	$E\Big\{$	-1	1	ε	ε^*	$0\Big\}$	2
$^1\Gamma_3$		1	1	ε^*	ε	0	

$$\varepsilon = e^{\frac{2\pi i}{3}}$$

C_4	$\mu=$	E	C_4	$C_4^2=C_2$	C_4^3	β	v
A	0	1	1	1	1	1	1
B	2	1	-1	1	-1	1	1
$E\Big\{$	-1	1	i	-1	$-i$	$0\Big\}$	2
	1	1	$-i$	-1	i	0	

C_6		$\mu=$	E	C_6	C_3	C_2	C_3^2	C_6^5	β	v
$^1\Gamma_1$	A	0	1	1	1	1	1	1	1	1
$^1\Gamma_2$	B	3	1	-1	1	-1	1	-1	1	1
$^1\Gamma_3$	$E_1\Big\{$	-1	1	ε	$-\varepsilon^*$	-1	$-\varepsilon$	ε^*	$0\Big\}$	2
$^1\Gamma_4$		1	1	ε^*	$-\varepsilon$	-1	$-\varepsilon^*$	ε	0	
$^1\Gamma_5$	$E_2\Big\{$	-2	1	$-\varepsilon^*$	$-\varepsilon$	1	$-\varepsilon^*$	$-\varepsilon$	$0\Big\}$	2
$^1\Gamma_6$		2	1	$-\varepsilon$	$-\varepsilon^*$	1	$-\varepsilon$	$-\varepsilon^*$	0	

$$\varepsilon = e^{\frac{2\pi i}{6}}$$

$$\left.\begin{array}{}\end{array}\right\}\;(17.1)$$

Diedergruppen D_p.

$D_2 = V$		v	μ	E	$C_2(z)$	$C_2(y)$	$C_2(x)$	β	v
$^1\Gamma_1$	A	0	0	1	1	1	1	1	1
$^1\Gamma_2$	B_1	1	0	1	1	-1	-1	1	1
$^1\Gamma_3$	B_2	0	1	1	-1	1	-1	1	1
$^1\Gamma_4$	B_3	1	1	1	-1	-1	1	1	1

D_3		v	μ	E	$2C_3$	$3C_2$	β	v
$^1\Gamma_1$	A_1	0	0	1	1	1	1	1
$^1\Gamma_2$	A_2	1	0	1	1	-1	1	1
$^2\Gamma_3$	E	$-$	$\{\pm 1\}$	2	-1	0	1	2

D_4		v	μ	E	$2C_4$	C_2	$2C_2'$	$2C_2''$	β	v
$^1\Gamma_1$	A_1	0	0	1	1	1	1	1	1	1
$^1\Gamma_2$	A_2	1	0	1	1	1	-1	-1	1	1
$^1\Gamma_3$	B_1	0	2	1	-1	1	1	-1	1	1
$^1\Gamma_4$	B_2	1	2	1	-1	1	-1	1	1	1
$^2\Gamma_5$	E	$-$	$\{\pm 1\}$	2	0	-2	0	0	1	2

D_6		v	μ	E	$2C_6$	$2C_3$	C_2	$3C_2'$	$3C_2''$	β	v
$^1\Gamma_1$	A_1	0	0	1	1	1	1	1	1	1	1
$^1\Gamma_2$	A_2	1	0	1	1	1	1	-1	-1	1	1
$^1\Gamma_3$	B_1	0	3	1	-1	1	-1	1	-1	1	1
$^1\Gamma_4$	B_2	1	3	1	-1	1	-1	-1	1	1	1
$^2\Gamma_5$	E_1	$-$	$\{\pm 1\}$	2	1	-1	-2	0	0	1	2
$^2\Gamma_6$	E_2	$-$	$\{\pm 2\}$	2	-1	-1	2	0	0	1	2

$$\left.\begin{array}{}\end{array}\right\}\;(17.2)$$

Kubische Gruppen.

T		$\varkappa$	μ	E	$4C_3$	$4C_3^2$	$3C_2$	β	v
$^1\Gamma_1$	A	0	0	1	1	1	1	1	1
$^1\Gamma_2$	$E\;\{$	0	-1	1	ε	ε^*	1	$0\}$	2
$^1\Gamma_3$		0	1	1	ε^*	ε	1	$0\}$	
$^3\Gamma_4$	T	$-$	$\{0,\pm1\}$	3	0	0	-1	1	3

O		$\varkappa$	μ	E	$8C_3$	$3C_2$	$6C_4$	$6C_2'$	β	v
$^1\Gamma_1$	A_1	0	0	1	1	1	1	1	1	1
$^1\Gamma_2$	A_2	0	2	1	1	1	-1	-1	1	1
$^2\Gamma_3$	E	$-$	$\{0,\;2\}$	2	-1	2	0	0	1	2
$^3\Gamma_4$	T_1	$-$	$\{0,\pm1\}$	3	0	-1	1	-1	1	3
$^3\Gamma_5$	T_2	$-$	$\{2,\pm1\}$	3	0	-1	-1	1	1	3

$$\varepsilon = e^{\frac{2\pi i}{3}} \qquad\qquad (17.3)$$

(Die Darstellungen T werden manchmal auch mit F bezeichnet.)

β) *Kristalldoppelgruppen.* Die *spezifischen*, irreduziblen Darstellungen der Kristalldoppelgruppen $\mathfrak{G}^\dagger$ ohne Inversionsachsen sind in den nachstehenden Tabellen angegeben. Wir bezeichnen mit $h^\dagger C_{p(p^\dagger)}$ eine Klasse aus $\mathfrak{G}^\dagger$, die $h^\dagger$ Elemente der Ordnung $p^\dagger$ (d.h. $C_{p(p^\dagger)}^{p^\dagger}=E$) aus $\mathfrak{G}^\dagger$ umfaßt, welchen in $\mathfrak{G}$ Drehungen C_p um den Winkel $2\pi/p$ entsprechen. Zwei Klassen aus $\mathfrak{G}^\dagger$, die einer Klasse in $\mathfrak{G}$ entsprechen, werden mit einem Querstrich bzw. Doppelquerstrich versehen.

Zyklische Doppelgruppen $C_p^\dagger$ (alle $h^\dagger = 1$; Abelsche Gruppen). $C_p^\dagger$ isomorph C_{2p}.

$C_1^\dagger$	$\mu=$	$\bar E$	$\bar{\bar E}$	β	v
$^1\Gamma_2$	$\tfrac{1}{2}$	1	-1	1	2

$C_2^\dagger$	$\mu=$	$\bar E$	$\bar{\bar E}$	$\overline{C}_{2(4)}$	$\overline{\overline{C}}_{2(4)}$	β	v
$^1\Gamma_3$	$-\tfrac{1}{2}$	1	-1	i	$-i$	$0\}$	
$^1\Gamma_4$	$\tfrac{1}{2}$	1	-1	$-i$	i	$0\}$	2

$C_3^\dagger$	$\mu=$	$\bar E$	$\bar{\bar E}$	$\overline{C}_{3(6)}$	$\overline{\overline{C}}_{3(3)}$	$\overline{C}_{3(6)}^2$	$\overline{\overline{C}}_{3(3)}^2$	β	v
$^1\Gamma_4$	$\tfrac{3}{2}$	1	-1	-1	1	-1	1	1	2
$^1\Gamma_5$	$-\tfrac{1}{2}$	1	-1	ε	$-\varepsilon$	ε^*	$-\varepsilon^*$	$0\}$	
$^1\Gamma_6$	$\tfrac{1}{2}$	1	-1	ε^*	$-\varepsilon^*$	ε	$-\varepsilon$	$0\}$	2

$$\varepsilon = e^{i\frac{\pi}{3}}$$

$C_4^\dagger$	$\mu=$	$\bar E$	$\bar{\bar E}$	$\overline{C}_{4(8)}$	$\overline{\overline{C}}_{4(8)}$	$\overline{C}_{2(4)}$	$\overline{\overline{C}}_{2(4)}$	$\overline{C}_{4(8)}^3$	$\overline{\overline{C}}_{4(8)}^3$	β	v
$^1\Gamma_5$	$-\tfrac{1}{2}$	1	-1	ε	$-\varepsilon$	i	$-i$	ε^*	$-\varepsilon^*$	$0\}$	2
$^1\Gamma_6$	$\tfrac{1}{2}$	1	-1	ε^*	$-\varepsilon^*$	$-i$	i	ε	$-\varepsilon$	$0\}$	
$^1\Gamma_7$	$\tfrac{3}{2}$	1	-1	$-\varepsilon$	ε	i	$-i$	$-\varepsilon^*$	ε^*	$0\}$	2
$^1\Gamma_8$	$-\tfrac{3}{2}$	1	-1	$-\varepsilon^*$	ε^*	$-i$	i	$-\varepsilon.$	ε	$0\}$	

$$\varepsilon = e^{i\frac{\pi}{4}} \qquad\qquad (17.4)$$

$C_6^\dagger$	$\mu=$	$\bar E$	$\bar{\bar E}$	$\overline{C}_{6(12)}$	$\overline{\overline{C}}_{6(12)}$	$\overline{C}_{3(6)}$	$\overline{\overline{C}}_{3(3)}$	$\overline{C}_{2(4)}$	$\overline{\overline{C}}_{2(4)}$	$\overline{C}_{3(6)}^2$	$\overline{\overline{C}}_{3(3)}^2$	$\overline{C}_{6(12)}^5$	$\overline{\overline{C}}_{6(12)}^5$	β	v
$^1\Gamma_7$	$-\tfrac{1}{2}$	1	-1	ε	$-\varepsilon$	ε^2	$-\varepsilon^2$	i	$-i$	$-\varepsilon^4$	ε^4	$-\varepsilon^5$	ε^5	$0\}$	2
$^1\Gamma_8$	$\tfrac{1}{2}$	1	-1	$-\varepsilon^5$	ε^5	$-\varepsilon^4$	ε^4	$-i$	i	ε^2	$-\varepsilon^2$	ε	$-\varepsilon$	$0\}$	
$^1\Gamma_9$	$-\tfrac{3}{2}$	1	-1	i	$-i$	-1	1	$-i$	i	-1	1	$-i$	i	$0\}$	2
$^1\Gamma_{10}$	$\tfrac{3}{2}$	1	-1	$-i$	i	-1	1	i	$-i$	-1	1	i	$-i$	$0\}$	
$^1\Gamma_{11}$	$-\tfrac{5}{2}$	1	-1	ε^5	$-\varepsilon^5$	$-\varepsilon^4$	ε^4	i	$-i$	ε^2	$-\varepsilon^2$	$-\varepsilon$	ε	$0\}$	2
$^1\Gamma_{12}$	$\tfrac{5}{2}$	1	-1	$-\varepsilon$	ε	ε^2	$-\varepsilon^2$	$-i$	i	$-\varepsilon^4$	ε^4	ε^5	$-\varepsilon^5$	$0\}$	

$$\varepsilon = e^{i\frac{\pi}{6}}$$

Diederdoppelgruppen $D_p^\dagger$.

$D_2^\dagger$	$\mu=$	$\bar E$	$\bar{\bar E}$	$2C_{2(4)}(z)$	$2C_{2(4)}(y)$	$2C_{2(4)}(x)$	β	v
$^2\Gamma_5$	$\{\pm\tfrac12\}$	2	-2	0	0	0	-1	2

$D_3^\dagger$	$v=$	$\mu=$	$\bar E$	$\bar{\bar E}$	$2\bar C_{3(6)}$	$2\bar{\bar C}_{3(3)}$	$3\bar C_{2(4)}$	$3\bar{\bar C}_{2(4)}$	β	v
$^1\Gamma_4$	$-\tfrac12$	$\tfrac32$	1	-1	-1	1	i	$-i$	0}	2
$^1\Gamma_5$	$\tfrac12$	$\tfrac32$	1	-1	-1	1	$-i$	i	0}	
$^2\Gamma_6$		$\{\pm\tfrac12\}$	2	-2	1	-1	0	0	-1	2

$D_4^\dagger$	$\mu=$	$\bar E$	$\bar{\bar E}$	$2\bar C_{4(8)}$	$2\bar{\bar C}_{4(8)}$	$2C_{2(4)}$	$4C'_{2(4)}$	$4C''_{2(4)}$	β	v
$^2\Gamma_6$	$\{\pm\tfrac12\}$	2	-2	$\sqrt2$	$-\sqrt2$	0	0	0	-1	2
$^2\Gamma_7$	$\{\pm\tfrac32\}$	2	-2	$-\sqrt2$	$\sqrt2$	0	0	0	-1	2

$D_6^\dagger$	$\mu=$	$\bar E$	$\bar{\bar E}$	$2\bar C_{6(12)}$	$2\bar{\bar C}_{6(12)}$	$2\bar C_{3(6)}$	$2\bar{\bar C}_{3(3)}$	$2C_{2(4)}$	$6C'_{2(4)}$	$6C''_{2(4)}$	β	v
$^2\Gamma_7$	$\{\pm\tfrac12\}$	2	-2	$\sqrt3$	$-\sqrt3$	1	-1	0	0	0	-1	2
$^2\Gamma_8$	$\{\pm\tfrac52\}$	2	-2	$-\sqrt3$	$\sqrt3$	1	-1	0	0	0	-1	2
$^2\Gamma_9$	$\{\pm\tfrac32\}$	2	-2	0	0	-2	2	0	0	0	-1	2

$$(17.5)$$

Kubische Doppelgruppen.

$T^\dagger$	$\mu=$	$\bar E$	$\bar{\bar E}$	$4\bar C_{3(6)}$	$4\bar{\bar C}_{3(3)}$	$4\bar C^2_{3(6)}$	$4\bar{\bar C}^2_{3(3)}$	$6C_{2(4)}$	β	v
$^2\Gamma_5$	$\{\pm\tfrac12\}$	2	-2	1	-1	1	-1	0	-1	2
$^2\Gamma_6$	$\{-\tfrac12,\ \tfrac32\}$	2	-2	ε	$-\varepsilon$	ε^*	$-\varepsilon^*$	0	0}	4
$^2\Gamma_7$	$\{\ \tfrac12,\ -\tfrac32\}$	2	-2	ε^*	$-\varepsilon^*$	ε	$-\varepsilon$	0	0}	

$$\varepsilon = e^{\frac{2\pi i}{3}}$$

$O^\dagger$	$\mu=$	$\bar E$	$\bar{\bar E}$	$8\bar C_{3(6)}$	$8\bar{\bar C}_{3(3)}$	$6C_{2(4)}$	$6\bar C_{4(8)}$	$6\bar{\bar C}_{4(8)}$	$12C'_{2(4)}$	β	v
$^2\Gamma_6$	$\{\pm\tfrac12\}$	2	-2	1	-1	0	$\sqrt2$	$-\sqrt2$	0	-1	2
$^2\Gamma_7$	$\{\pm\tfrac32\}$	2	-2	1	-1	0	$-\sqrt2$	$\sqrt2$	0	-1	2
$^4\Gamma_8$	$\{\pm\tfrac12,\ \pm\tfrac32\}$	4	-4	-1	1	0	0	0	0	-1	4

$$(17.6)$$

18. Die Kristallklassen mit Inversionsachsen. *α) Gewöhnliche Kristallgruppen*: Enthält eine Kristallklasse neben den Drehungen auch p-zählige Drehinversionsachsen I_p ($I_1 = I =$ Inversionszentrum, $I_2 =$ Spiegelebene), so ist die Gruppe entweder *isomorph* zu einer der in Ziff. 17α aufgeführten Kristallklassen und besitzt daher dieselben irreduziblen Charaktere oder die Gruppe ist das *direkte Produkt* einer dieser Gruppen mit der Gruppe C_i (bestehend aus E und I) oder C_s (bestehend aus E und I_2):

	p gerade	p ungerade	
$C_{pv}=$	D_p	D_p	$T_d=O$
$C_{ph}=$	$C_p\times C_i$	$C_p\times C_s$	$T_h=T\times C_i$
$S_{2p}=$	C_{2p}	$C_p\times C_i$	$O_h=O\times C_i$
$D_{pd}=$	D_{2p}	$D_p\times C_i$	
$D_{ph}=$	$D_p\times C_i$	D_{2p}	

$$(18.1)$$

Man nennt eine Gruppe $\mathfrak{G}$ das *direkte Produkt* zweier Untergruppen $\mathfrak{G}_1$ und $\mathfrak{G}_2$: $\mathfrak{G}=\mathfrak{G}_1\times\mathfrak{G}_2$, wenn 1. $\mathfrak{G}_1$ und $\mathfrak{G}_2$ nur die Einheit gemeinsam haben, 2. jedes Element g_1 aus $\mathfrak{G}_1$ mit jedem Element g_2 aus $\mathfrak{G}_2$ vertauschbar ist, 3. jedes Element g

aus $\mathfrak{G}$ als Produkt $g = g_1 g_2 = g_2 g_1$ geschrieben werden kann. Sind g_1 und g_1' in $\mathfrak{G}_1$, g_2 und g_2' in $\mathfrak{G}_2$ konjugiert, so ist $g = g_1 g_2$ und $g' = g_1' g_2'$ in $\mathfrak{G}$ konjugiert. Die Ordnung n, die Klassenzahl c (= Anzahl der irreduziblen Darstellungen) und die Charaktere χ ergeben sich als die Produkte

$$n = n_1 n_2, \qquad c = c_1 c_2, \qquad \chi = \chi_1 \chi_2 \tag{18.2}$$

der entsprechenden Größen von $\mathfrak{G}_1$ und $\mathfrak{G}_2$.

Für die Charaktere der irreduziblen Darstellungen der Gruppen C_i und C_s, die zueinander und zu C_2 isomorph sind, findet man

C_i						E	I			
				C_s		E	I_2	β	v	
		μ_I			μ_I					
$^1\Gamma_1$	A_g	0	$^1\Gamma_1$	A'	0	1	1	1	1	
$^1\Gamma_2$	A_u	$\tfrac{1}{2}$	$^1\Gamma_2$	A''	1	1	-1	1	1	

$$\tag{18.3}$$

Die Gruppen $\mathfrak{G} = \mathfrak{G}_1 \times C_i$ bzw. $\mathfrak{G} = \mathfrak{G}_1 \times C_s$ in (18.1) enthalten somit doppelt so viel Elemente in doppelt so vielen Klassen ($\mathfrak{C}$ und $I\mathfrak{C}$ bzw. $I_2\mathfrak{C}$) wie $\mathfrak{G}_1$ und besitzen daher doppelt so viele irreduzible Darstellungen wie $\mathfrak{G}_1$: 1. Die *geraden* Darstellungen (Index g), bei welchen die Klassen $I\mathfrak{C}$ bzw. $I_2\mathfrak{C}$ denselben Charakter haben wie $\mathfrak{C}$. 2. Die *ungeraden* Darstellungen (Index u), bei welchen die Charaktere von $I\mathfrak{C}$ bzw. $I_2\mathfrak{C}$ die negativen Charaktere von $\mathfrak{C}$ sind.

Für das *freie* Atom ergibt sich in analoger Weise aus der Drehspiegelungsgruppe $\mathfrak{D}' = \mathfrak{D} \times C_i$ eine Unterscheidung von geraden und ungeraden Thermen [vgl. Gl. (21.7)]. Ein gerader Term des freien Atoms zerfällt in einem Kristallfeld der Symmetrie $\mathfrak{G} = \mathfrak{G}_1 \times C_i$ bzw. $\mathfrak{G} = \mathfrak{G}_1 \times C_s$ in lauter gerade, ein ungerader Term in lauter ungerade Kristallterme. Im übrigen braucht man sich um die Inversion nicht weiter zu kümmern: z.B. spaltet bei tetragonaler Holoedrie D_{4h} ein Term in ebensoviele Kristallterme auf wie bei der Kristallklasse D_4.

β) Kristalldoppelgruppen. $\mathfrak{G}^\dagger$ ist eine abstrakte Gruppe, deren Elemente von der Form $J\mathfrak{u}$ sind, wobei J entweder die Einheit E oder die Inversion $I(I^2 = E)$ und $\mathfrak{u}$ ein Element der unitären Gruppe ist, das durch die Drehung bestimmt ist. Die Multiplikation zweier Elemente ist gegeben durch $J_1 \mathfrak{u}_1 \cdot J_2 \mathfrak{u}_2 = J_1 J_2 \mathfrak{u}_1 \mathfrak{u}_2$. Die Kristalldoppelgruppen mit Drehinversionsachsen sind ebenfalls entweder isomorph zu den in Ziff. 17 β besprochenen Gruppen oder als direktes Produkt darstellbar

	p gerade	p ungerade	
$C_{pv}^\dagger =$	$D_p^\dagger$	$D_p^\dagger$	$T_d^\dagger = O^\dagger$
$C_{ph}^\dagger =$	$C_p^\dagger \times C_i$	$C_p \times C_s^\dagger$	$T_h^\dagger = T^\dagger \times C_i$
$S_{2p}^\dagger =$	$C_{2p}^\dagger$	$C_p^\dagger \times C_i$	
$D_{pd}^\dagger =$	$D_{2p}^\dagger$	$D_p^\dagger \times C_i$	$O_h^\dagger = O^\dagger \times C_i$
$D_{ph}^\dagger =$	$D_p^\dagger \times C_i$	$D_{2p}^\dagger$	

$$\tag{18.4}$$

Die spezifischen, irreduziblen Darstellungen von $C_i^\dagger$ und $C_s^\dagger$, die nicht zueinander isomorph sind, lauten

$C_i^\dagger$		$\bar{E}$	$\bar{\bar{E}}$	$\bar{I}$	$\bar{\bar{I}}$	β	v
	$\mu_I =$						
$^1\Gamma_3$	0	1	-1	1	-1	1	2
$^1\Gamma_4$	$\tfrac{1}{2}$	1	-1	-1	1	1	2

$$C_i^\dagger = C_1^\dagger \times C_i$$

$C_s^\dagger$		$\bar{E}$	$\bar{\bar{E}}$	$\bar{I}_{2(4)}$	$\bar{\bar{I}}_{2(4)}$	β	v
	$\mu_I =$						
$^1\Gamma_3$	$-\tfrac{1}{2}$	1	-1	i	$-i$	0	2
$^1\Gamma_4$	$\tfrac{1}{2}$	1	-1	$-i$	i	0	2

$$C_s^\dagger \text{ isomorph } C_2^\dagger$$

$$\tag{18.5}$$

19. Überblick über die Anzahl der Kristallterme. In der folgenden Tabelle sind die Anzahl und Vielfachheit der Kristallterme der verschiedenen Kristallklassen für niedrige Quantenzahlen J *ohne* Berücksichtigung der Kramers-Entartung (Ziff. 20) wiedergegeben

Kristallsystem	Kristallklasse	$\longrightarrow J$ 0	$\frac{1}{2}$	1	$\frac{3}{2}$	2	$\frac{5}{2}$	3	$\frac{7}{2}$	4	$\frac{9}{2}$	5	$\frac{11}{2}$	Vielfachheit t
Triklin	C_1, C_i						$2J+1$							einfach
Monoklin	C_s, C_2, C_{2h}						$2J+1$							einfach
Rhombisch	C_{2v}, D_2, D_{2h}	1	0	3	0	5	0	7	0	9	0	11	0	einfach
		0	1	0	2	0	3	0	4	0	5	0	6	zweifach
Trigonal	C_3, C_{3i}						$2J+1$							einfach
	C_{3v}, D_3, D_{3d}	1	0	1	2	1	2	3	2	3	4	3	4	einfach
		0	1	1	1	2	2	2	3	3	3	4	4	zweifach
Tetragonal	S_4, C_4, C_{4h}						$2J+1$							einfach
	$D_{2d}, C_{4v}, D_4, D_{4h}$	1	0	1	0	3	0	3	0	5	0	5	0	einfach
		0	1	1	2	1	3	2	4	2	5	3	6	zweifach
Hexagonal	C_{3h}, C_6, C_{6h}						$2J+1$							einfach
	$D_{3h}, C_{6v}, D_6, D_{6h}$	1	0	1	0	1	0	3	0	3	0	3	0	einfach
		0	1	1	2	2	3	2	4	3	5	4	6	zweifach
Kubisch	T, T_h	1	0	0	0	2	0	1	0	3	0	2	0	einfach
		0	1	0	2	0	3	0	4	0	5	0	6	zweifach
		0	0	1	0	1	0	2	0	2	0	3	0	dreifach
	T_d, O, O_h	1	0	0	0	0	0	1	0	1	0	0	0	einfach
		0	1	0	0	1	1	0	2	1	1	1	2	zweifach
		0	0	1	0	1	0	2	0	2	0	3	0	dreifach
		0	0	0	1	0	1	0	1	0	2	0	2	vierfach

$$(19.1)$$

20. Kramers-Entartung. Für die bisher besprochene Termaufspaltung durch ein Kristallfeld war nur die Kenntnis der Symmetrie des Kristalls nötig. Über die Natur dieses Feldes wurden keine Aussagen gemacht. Da jedoch das Kristallfeld vorwiegend *elektrostatisch* ist, kann eine an sich mögliche magnetische Termaufspaltung gegenüber der elektrischen vernachlässigt werden. Die sich so ergebende Entartung wurde zuerst von H. A. Kramers[1] untersucht.

Mit Hilfe der Paulischen Spinfunktionen

$$\xi^{(\frac{1}{2})}(\sigma) = \delta_{\sigma,\frac{1}{2}} = \begin{pmatrix} 1 \\ 0 \end{pmatrix} \quad \text{und} \quad \xi^{(-\frac{1}{2})}(\sigma) = \delta_{\sigma,-\frac{1}{2}} = \begin{pmatrix} 0 \\ 1 \end{pmatrix} \tag{20.1}$$

läßt sich der Zustand eines Atoms mit N Spinelektronen durch die Wellenfunktion

$$\psi = \psi(\mathfrak{r}_1, \ldots, \mathfrak{r}_N, \sigma_1, \ldots, \sigma_N) = \sum_{s_1=\frac{1}{2}}^{-\frac{1}{2}} \cdots \sum_{s_N=\frac{1}{2}}^{-\frac{1}{2}} \psi^{s_1 \ldots s_N}(\mathfrak{r}_1, \ldots, \mathfrak{r}_N)\, \xi_1^{(s_1)}(\sigma_1) \ldots \xi_N^{(s_N)}(\sigma_N) \tag{20.2}$$

---

[1] H. A. Kramers: Proc. Kon. Akad. Amsterd. **33**, 959 (1930).

beschreiben:

$$H(\mathfrak{r}_1, \ldots, \mathfrak{r}_N, \mathfrak{p}_1, \ldots, \mathfrak{p}_N, \mathfrak{S}_1, \ldots, \mathfrak{S}_N)\,\psi = E\psi, \tag{20.3}$$

wobei die Komponenten von $\mathfrak{S}_k$ $(k = 1, \ldots, N)$ die Paulischen Spinmatrizen[1]

$$S_{xk} = \frac{1}{2}\begin{pmatrix} 0 & 1 \\ 1 & 0 \end{pmatrix}, \quad S_{yk} = \frac{1}{2}\begin{pmatrix} 0 & -i \\ i & 0 \end{pmatrix}, \quad S_{zk} = \frac{1}{2}\begin{pmatrix} 1 & 0 \\ 0 & -1 \end{pmatrix} \tag{20.4}$$

bedeuten, die auf die Spinfunktionen des k-ten Elektrons wirken:

$$\left.\begin{aligned}
S_{xk}\,\xi_k^{(s_k)} &= \frac{1}{2}\,\xi_k^{(-s_k)}, \quad S_{yk}\,\xi_k^{(s_k)} = \frac{i}{2}\,(-1)^{s_k-\frac{1}{2}}\,\xi_k^{(-s_k)} = i\,s_k\,\xi_k^{(-s_k)}, \\
S_{zk}\,\xi_k^{(s_k)} &= \frac{1}{2}\,(-1)^{s_k-\frac{1}{2}}\,\xi_k^{(s_k)} = s_k\,\xi_k^{(s_k)}.
\end{aligned}\right\} \tag{20.5}$$

Der *Satz von* Kramers besagt nun: Wenn (20.2) eine Eigenfunktion von (20.3) ist, so ist der sog. Kramerssche Zustand

$$\left.\begin{aligned}
\widetilde{\psi} &= \widetilde{\psi}\,(\mathfrak{r}_1, \ldots, \mathfrak{r}_N, \sigma_1, \ldots, \sigma_N) \\
&= \sum_{s_1=\frac{1}{2}}^{-\frac{1}{2}} \cdots \sum_{s_N=\frac{1}{2}}^{-\frac{1}{2}} (-1)^{\frac{N}{2}-\sum\limits_{l=1}^{N} s_l}\,\psi^{s_1\ldots s_N *}(\mathfrak{r}_1, \ldots, \mathfrak{r}_N)\,\xi_1^{(-s_1)}(\sigma_1) \ldots \xi_N^{(-s_N)}(\sigma_N)
\end{aligned}\right\} \tag{20.6}$$

eine Eigenfunktion von

$$H(\mathfrak{r}_1, \ldots, \mathfrak{r}_N, -\mathfrak{p}_1, \ldots, -\mathfrak{p}_N, -\mathfrak{S}_1, \ldots, -\mathfrak{S}_N)\,\widetilde{\psi} = E\widetilde{\psi} \tag{20.7}$$

und zwar für denselben Energieeigenwert E.

Beweis: Bilden wir die zu (20.3) konjugiert komplexe Gleichung

$$H(\mathfrak{r}_1, \ldots, \mathfrak{r}_N, \mathfrak{p}_1^*, \ldots, \mathfrak{p}_N^*, \mathfrak{S}_1^*, \ldots, \mathfrak{S}_N^*)\,\psi^* = E\psi^*,$$

so können wir nach Gl. (20.4) und $\mathfrak{p} = -i\hbar\,\mathrm{grad}$ dafür schreiben

$$H(\mathfrak{r}_1, \ldots, \mathfrak{r}_N, -\mathfrak{p}_1, \ldots, -\mathfrak{p}_N, +S_{x1}, -S_{y1}, +S_{z1}, \ldots, +S_{xN}, -S_{yN}, +S_{zN})\,\psi^* = E\psi^*. \tag{20.8}$$

Führen wir vorübergehend die Funktion

$$\varphi(\mathfrak{r}_1, \ldots, \mathfrak{r}_N, \sigma_1, \ldots, \sigma_N) = \sum_{s_1=\frac{1}{2}}^{-\frac{1}{2}} \cdots \sum_{s_N=\frac{1}{2}}^{-\frac{1}{2}} \psi^{s_1\ldots s_N *}(\mathfrak{r}_1, \ldots, \mathfrak{r}_N)\,\xi_1^{(-s_1)}(\sigma_1) \ldots \xi_N^{(-s_N)}(\sigma_N) \tag{20.9}$$

ein, so folgt dafür aus Gl. (20.8), weil nach Gl. (20.5) beim Übergang $\xi_k^{(s_k)} \to \xi_k^{(-s_k)}$ für die Spinmatrizen $S_{xk} \to S_{xk}$, $S_{yk} \to -S_{yk}$, $S_{zk} \to -S_{zk}$ gilt,

$$H(\mathfrak{r}_1, \ldots, \mathfrak{r}_N, -\mathfrak{p}_1, \ldots, -\mathfrak{p}_N, +S_{x1}, +S_{y1}, -S_{z1}, \ldots, +S_{xN}, +S_{yN}, -S_{zN})\,\varphi = E\varphi. \tag{20.10}$$

Schließlich bilden wir einerseits

$$S_{xk}\,\varphi = \frac{1}{2}\sum_{s_1} \cdots \sum_{s_k} \cdots \sum_{s_N} \psi^{s_1\ldots s_k\ldots s_N *}(\mathfrak{r}_1, \ldots, \mathfrak{r}_N)\,\xi_1^{(-s_1)}(\sigma_1) \ldots \xi_k^{(+s_k)}(\sigma_k) \ldots \xi_N^{(-s_N)}(\sigma_N)$$

$$= \frac{1}{2}\sum_{s_1} \cdots \sum_{s_k} \cdots \sum_{s_N} \psi^{s_1\ldots -s_k\ldots s_N *}(\mathfrak{r}_1, \ldots, \mathfrak{r}_N)\,\xi_1^{(-s_1)}(\sigma_1) \ldots \xi_k^{(-s_k)}(\sigma_N) \ldots \xi_N^{(-s_k)}(\sigma_N),$$

andererseits liefert die Anwendung von $-S_{xk}$ auf (20.6)

$$-S_{xk}\,\widetilde{\psi} = -\frac{1}{2}\sum_{s_1} \cdots \sum_{s_k} \cdots \sum_{s_N} (-1)^{\frac{N}{2}-\sum\limits_{l} s_l}\,\psi^{s_1\ldots s_k\ldots s_N *}(\mathfrak{r}_1, \ldots, \mathfrak{r}_N)\,\xi_1^{(-s_1)}(\sigma_1) \ldots \xi_k^{(+s_k)}(\sigma_k) \ldots \xi_N^{(-s_N)}(\sigma_N),$$

$$= -\frac{1}{2}\sum_{s_1} \cdots \sum_{s_k} \cdots \sum_{s_N} (-1)^{\frac{N}{2}-\sum\limits_{l} s_l + 2s_k}\,\psi^{s_1\ldots -s_k\ldots s_N *}(\mathfrak{r}_1, \ldots, \mathfrak{r}_N)\,\xi_1^{(-s_1)}(\sigma_1) \ldots \xi_k^{(-s_k)}(\sigma_k) \ldots \xi_N^{(-s_N)}(\sigma_N),$$

$$= +\frac{1}{2}\sum_{s_1} \cdots \sum_{s_k} \cdots \sum_{s_N} (-1)^{\frac{N}{2}-\sum\limits_{l} s_l}\,\psi^{s_1\ldots -s_k\ldots s_N *}(\mathfrak{r}_1, \ldots, \mathfrak{r}_N)\,\xi_1^{(-s_1)}(\sigma_1) \ldots \xi_k^{(-s_k)}(\sigma_k) \ldots \xi_N^{(-s_N)}(\sigma_N).$$

---

[1] $\mathfrak{S}$ ist der in Einheiten von $\hbar$ gemessene Spindrehimpuls.

Führt man analoge Rechnungen für S_{yk} und S_{zk} durch, so ergibt sich, daß die Anwendung von S_{xk}, S_{yk} bzw. S_{zk} auf φ dieselbe Wirkung hat, wie die Anwendung von $-S_{xk}$, $-S_{yk}$, $+S_{zk}$ auf $\tilde{\psi}$. Aus Gl. (20.10) folgt damit die zu beweisende Gl. (20.7).

Beschreibt man mit E. Wigner[1,2] die Umkehr aller Impuls- und Spinrichtungen (= Zeitumkehr) durch den nichtlinearen Operator $\mathscr{K}$

$$\mathscr{K}\psi = \tilde{\psi}$$

$$\mathscr{K}^2\psi = (-1)^N\psi, \quad \mathscr{K}(a\,\psi_1 + b\,\psi_2) = a^*\,\mathscr{K}\psi_1 + b^*\,\mathscr{K}\psi_2 \quad (a,b = \text{const}), \quad (20.11)^3$$

so gehören zum Eigenwert E des Hamilton-Operators $\mathscr{H}$ eines Atoms in einem rein elektrostatischen Kristallfeld die beiden Eigenfunktionen ψ und $\mathscr{K}\psi = \tilde{\psi}$, da in diesem Fall $\mathscr{H}$ invariant ist gegenüber einer Umkehr aller Impuls- und Spinrichtungen. Die Anzahl und Vielfachheit der Terme in einem solchen Feld ergibt sich, indem man die Wirkung des Operators $\mathscr{K}$ auf die Darstellungen $D^{\lambda\mu}(R)$ der räumlichen Symmetriegruppe untersucht. Es zeigt sich, daß das Auftreten einer zusätzlichen Entartung der Betheschen Terme davon abhängt, ob I. die Darstellung D einer reellen Darstellung äquivalent ist, und wenn nicht ob II. die Darstellung D mit D^* nicht äquivalent oder III. die Darstellung D mit D^* äquivalent ist. Der letzte Fall kann nur eintreten, wenn die Dimension der Darstellung geradzahlig ist. Welcher dieser drei *Arten* eine gegebene Darstellung angehört, ist durch die Charaktere $\chi(R)$ eindeutig bestimmt. Es gilt[4] das Kriterium

$$\sum_R \chi(R^2) = \beta\,n, \tag{20.12}$$

wonach die Summe über die Charaktere der Quadrate aller Gruppenelemente R einer Darstellung gleich βmal der Gruppenordnung n ist, wobei $\beta = 1, 0$ oder -1 ist, je nachdem ob die Darstellung von der I., II. oder III. Art ist. Der Zusammenhang zwischen der Art der Darstellung und der Vielfachheit v der Terme in einem rein elektrostatischen Kristallfeld ist durch folgende Tabelle gegeben ($t = $ Vielfachheit der Betheschen Terme)

Darstellung	$\chi(R)$	β	N gerade $v=$	N ungerade $v=$	
I. Art	alle reell	$+1$	t	$2t$	(20.13)
II. Art	nicht alle reell	0	$2t$	$2t$	
III. Art	alle reell	-1	$[2t]$	t	

Für $\beta = 1$ und ungerader Elektronenzahl N fallen zwei Terme derselben Darstellung zusammen, während sich für $\beta = 0$ immer zwei Terme aus D und D^* vereinen. Der Fall $\beta = -1$ ist bei geradem N nicht realisiert.

In den Tabellen der Ziff. 17 sind die Art β und die Vielfachheit v für jede Darstellung angegeben. Man sieht, daß z.B. für die Gruppen C_p und $C_p^\dagger$ die Terme mit den Kristallquantenzahlen $+\mu$ und $-\mu$ zu zweifachen Termen zusammenfallen[5]. Alle anderen Terme bleiben einfach. Bei den Diedergruppen D_p und $D_p^\dagger$ liefert die Symmetrie $\mathscr{K}$ keine weitere Entartung. *Bei ungerader Elektronenzahl ist die Vielfachheit v stets geradzahlig.* Dieser Satz gilt nach Kramers unabhängig von jeder Symmetrie des elektrostatischen Feldes. — Die

---

[1] E. Wigner: Nachr. Ges. Wiss. Göttingen, Math.-phys. Kl. **1932**, 546.

[2] E. Fick: Z. Physik **147**, 307 (1957).

[3] Bei gerader Elektronenzahl N werden die Relationen (20.11) zu denselben, die für den Übergang zum Konjugiert-komplexen ($\mathscr{K}^0\psi = \psi^*$) gelten. Es ist $\mathscr{K} = \mathscr{U}\mathscr{K}^0$, wobei $\mathscr{U}$ eine unitäre Transformation ist, die auf die Spinfunktionen wirkt. Vgl. Fußnote 1.

[4] G. Frobenius u. I. Schur: Sitzgsber. Akad. Wiss. Berlin **1906**, 186.

[5] K. H. Hellwege: Ann. Phys. **4**, 143 (1949).

Art der Darstellung für die Kristallklassen mit Inversionsachsen, welche in (18.1) und (18.4) als direktes Produkt geschrieben sind, ist durch die Art des ersten Faktors bestimmt: Die ungerade Darstellung ist von derselben Art wie die entsprechende gerade Darstellung.

In der folgenden Tabelle (20.14) ist die Anzahl und Vielfachheit v der Terme in *rein elektrostatischen* Kristallfeldern angegeben, die zu einem J-Wert des freien Atoms gehören. Diese Anzahl ist — im Gegensatz zu den Betheschen Termen (Ziff. 19) — für alle Kristallklassen eines Kristallsystems gleich. Bei halbzahligem J liefert jedes nichtkubische Kristallfeld $J+\frac{1}{2}$ zweifache Werte.

Kristall-system	$J=0\ 1\ 2\ 3\ 4\ \ 5\ldots$ ganzzahlig	v	$J=\frac{1}{2}\ \frac{3}{2}\ \frac{5}{2}\ \frac{7}{2}\ \frac{9}{2}\ \frac{11}{2}\ldots$ halbzahlig	v
triklin monoklin rhombisch	$\left.\begin{array}{}\\ \\ \\\end{array}\right\}\ 1\ \ 3\ \ 5\ \ 7\ \ 9\ \ 11\ldots 2J+1$	1	$\left.\begin{array}{}\\ \\ \\ \\\end{array}\right\}$	
trigonal hexagonal	$\left.\begin{array}{}\\ \\\end{array}\right\}\begin{array}{l}1\ \ 1\ \ 1\ \ 3\ \ 3\ \ \ 3\ldots 2\left[\dfrac{J}{3}\right]+1\\[2ex] 0\ \ 1\ \ 2\ \ 2\ \ 3\ \ \ 4\ldots\left[\dfrac{2J-1}{3}\right]+1\end{array}$	1 2	$\left.\begin{array}{}\\ \\\end{array}\right\}\ 1\ 2\ 3\ 4\ 5\ 6\ldots J+\frac{1}{2}$	2
tetragonal	$\begin{array}{l}1\ \ 1\ \ 3\ \ 3\ \ 5\ \ \ 5\ldots 2\left[\dfrac{J}{2}\right]+1\\[2ex] 0\ \ 1\ \ 1\ \ 2\ \ 2\ \ \ 3\ldots\left[\dfrac{J+1}{2}\right]\end{array}$	1 2		
kubisch	$\begin{array}{l}1\ 0\ 0\ 1\ 1\ \ 0\ldots\varphi_1(x)+\left[\dfrac{J}{6}\right]\\[2ex] 0\ 0\ 1\ 0\ 1\ \ 1\ldots\varphi_2(x)+\left[\dfrac{J}{6}\right]\\[2ex] 0\ 1\ 1\ 2\ 2\ \ 3\ldots\left[\dfrac{J+1}{2}\right]\end{array}$	1 2 3	$\begin{array}{l}1\ 0\ 1\ 2\ 1\ 2\ldots\varphi(y)+\left[\dfrac{J}{3}\right]\\[2ex] 0\ 1\ 1\ 1\ 2\ 2\ldots\left[\dfrac{J+\frac{3}{2}}{3}\right]\end{array}$	2 4

$$(20.14)\,[1]$$

Die im kubischen System auftretenden Funktionen $\varphi_1(x)$, $\varphi_2(x)$ und $\varphi(y)$ sind definiert durch

x	0 1 2 3 4 5
$\varphi_1(x)$	1 0 0 1 1 0
$\varphi_2(x)$	0 0 1 0 1 1

y	$\frac{1}{2}\ \frac{3}{2}\ \frac{5}{2}$
$\varphi(y)$	1 0 1

wobei sich x bzw. y bei gegebenem J aus $J=x+6\lambda$ bzw. $J=y+3\lambda\,(\lambda=0,1,2,\ldots)$ bestimmen.

II. Analytische Behandlung.

Zur Bestimmung der Termaufspaltung im Kristallfeld gibt es neben der eben besprochenen gruppentheoretischen Methode eine von K. H. Hellwege[2] angegebene Behandlungsweise, die auf den Elektroneneigenfunktionen basiert und explizite gruppentheoretische Sätze nur an einigen Stellen benötigt. In diesem Verfahren werden die Elektronenzustände im Kristall in nullter Näherung bestimmt und entsprechend ihrem Transformationsverhalten gegenüber Deckoperationen, welche das betrachtete Kristallfeld erlaubt, durch *Kristall-Quantenzahlen* klassifiziert. Diese mehr anschauliche Bedeutung besitzenden Größen treten an die Stelle der formalen Darstellungssymbole $^t\Gamma_k$ in der Betheschen Theorie, wodurch gleichzeitig auch eine einfache Behandlung der Auswahlregeln

---

[1] Das Symbol $[a]$ ist auf S. 221 (Fußnote) erklärt.
[2] K. H. Hellwege: Ann. Phys. **4**, 95, 127, 136, 143, 150, 357 (1949).

für Übergänge zwischen Kristalltermen möglich wird. Systeme mit halb- bzw. ganzzahligem Drehimpuls können — im Gegensatz zur gruppentheoretischen Lösungsmethode — nach denselben Ansätzen behandelt werden.

21. Die Zustände eines freien Atoms. In der SCHRÖDINGER-Gleichung für ein *freies* Atom (Ion)

$$\mathscr{H}^0 \psi_{\gamma J}^M = E_{\gamma J} \psi_{\gamma J}^M \tag{21.1}$$

ist der HAMILTON-Operator $\mathscr{H}^0$ invariant gegen alle Drehungen um den Kern, gegen Inversion am Kern und Spiegelungen an allen Ebenen durch den Kern. Zum Energieeigenwert $E_{\gamma J}$ gehören die $2J+1$ Eigenfunktionen $\psi_{\gamma J}^M$, die sich in der magnetischen Quantenzahl

$$M = J, \quad J-1, \quad \ldots, \quad -J \tag{21.2}$$

unterscheiden. Nach CONDON und SHORTLEY[1] bezeichnet γ die Eigenwerte aller mit den Drehimpulsoperatoren $\mathfrak{J}^2$ und J_z vertauschbaren Operatoren; also z.B. die l_k der Einzelelektronen.

Bei einer *Drehung* transformieren sich die Eigenfunktionen $\psi_{\gamma J}^M$ des freien Atoms wie die Monome

$$e^{iM\pi} \left(\frac{2J}{J+M} \right)^{\frac{1}{2}} \xi^{J+M} \eta^{J-M}. \tag{21.3}$$

Die Größen ξ und η stellen dabei die PAULIschen Spinfunktionen dar (vgl. Ziff. 20), für die bei einer Drehung (EULERsche Winkel Φ, Θ und Ψ [vgl. Gl. (10.7)]) gilt

$$\left. \begin{aligned} \xi' &= e^{\frac{i}{2}(\Phi+\Psi)} \cos \frac{\Theta}{2} \cdot \xi + i\, e^{-\frac{i}{2}(\Phi-\Psi)} \sin \frac{\Theta}{2} \cdot \eta, \\ \eta' &= i\, e^{\frac{i}{2}(\Phi-\Psi)} \sin \frac{\Theta}{2} \cdot \xi + e^{-\frac{i}{2}(\Phi+\Psi)} \cos \frac{\Theta}{2} \cdot \eta. \end{aligned} \right\} \tag{21.4}$$

Das Transformationsverhalten (21.3) entspringt aus der in Ziff. 14 erwähnten Homomorphie zwischen der Gruppe der zweidimensionalen unitären Matrizen [vgl. Gl. (14.3)] im Raum der Vektoren ξ, η und der dreidimensionalen Drehgruppe[2]. Die Transformation (21.4) entspricht der zweideutigen Darstellung $\mathfrak{D}_{\frac{1}{2}}(\Phi, \Theta, \Psi)$.

Für die Inversion am Nullpunkt sind die Formeln (21.3) nicht gültig. Das Verhalten der Eigenfunktionen eines Atoms bei der *Inversion* erhält man jedoch leicht, wenn man bedenkt, daß sich in den Eigenfunktionen des *Ein*elektronenproblems

$$R_{nl}(r)\, Y^{m_j \mp \frac{1}{2}}(\vartheta, \varphi) \cdot \begin{cases} \xi \\ \eta \end{cases} \tag{21.5}$$

die Kugelfunktionen

$$Y_l^m(\vartheta, \varphi) = \frac{(-1)^{m+l}}{2^l\, l!} \sqrt{\frac{2l+1}{4\pi} \frac{(l+m)!}{(l-m)!}} \frac{1}{\sin^m \vartheta} \frac{d^{l-m} \sin^{2l} \vartheta}{d(\cos\vartheta)^{l-m}} \cdot e^{im\varphi} \tag{21.6}$$

bei der Inversion ($\vartheta \to \pi - \vartheta$, $\varphi \to \pi + \varphi$) mit $(-1)^l$ multiplizieren. Bei einem Atom mit N *Elektronen* hat man, wenn man zunächst die gegenseitige Störung nicht berücksichtigt, als Eigenfunktionen Produkte aus N Funktionen der Form

---

[1] E. CONDON u. G. SHORTLEY: Theory of Atomic Spectra. Cambridge 1953.
[2] Vgl. dazu die auf S. 211 zitierte Literatur.

(21.5), d.h. den *Spiegelungscharakter*

$$I = (-1)^{\sum\limits_{k=1}^{N} l_k}, \tag{21.7}$$

der auch nach Einschaltung der wechselseitigen Störung erhalten bleibt. Die Energieterme heißen *gerade* oder *ungerade*, je nachdem $I = +1$ oder -1 ist.

Man erhält damit das Transformationsverhalten der $\psi_{\gamma J}^{M}$ für die im folgenden wichtigen Fälle:

a) Drehung $\dfrac{2\pi}{p}$ um die z-Achse, d.h. Drehung $\left(\Phi = \dfrac{2\pi}{p}, \Theta = 0, \Psi = 0\right)$

$$\psi_{\gamma J}^{M}\left(r, \vartheta, \varphi' = \varphi + \frac{2\pi}{p}\right) = e^{iM\frac{2\pi}{p}} \psi_{\gamma J}^{M}(r, \vartheta, \varphi). \tag{21.8}$$

b) Drehung π um die y-Achse, d.h. Drehung $(0, \pi, \pi)$

$$\psi_{\gamma J}^{M}(r, \vartheta' = \pi - \vartheta, \varphi' = \pi - \varphi) = (-1)^{M-J} \psi_{\gamma J}^{-M}(r, \vartheta, \varphi). \tag{21.9}$$

c) Inversion

$$\psi_{\gamma J}^{M}(x' = -x, y' = -y, z' = -z) = (-1)^{\Sigma l_k} \psi_{\gamma J}^{M}(x, y, z). \tag{21.10}$$

d) p-zählige Drehinversion, d.h. Drehung $\left(\dfrac{2\pi}{p}, 0, 0\right)$ und Inversion

$$\psi_{\gamma J}^{M}\left(r, \vartheta' = \pi - \vartheta, \varphi' = \varphi + \frac{2\pi}{p} - \pi\right) = e^{i\frac{2\pi}{p}\left(M + \frac{p}{2}\Sigma l_k\right)} \psi_{\gamma J}^{M}(r, \vartheta, \varphi). \tag{21.11}$$

e) Spiegelung an der xz-Ebene, d.h. Drehung $(0, \pi, \pi)$ und Inversion

$$\psi_{\gamma J}^{M}(x' = x, y' = -y, z' = z) = (-1)^{M-J+\Sigma l_k} \psi_{\gamma J}^{-M}(x, y, z). \tag{21.12}$$

22. Das Atom im Kristallfeld. Die Lösungen der Schrödinger-Gleichung für das Atom im Kristallfeld

$$\mathscr{H} U \equiv (\mathscr{H}^0 + \mathscr{H}') U = W U \tag{22.1}$$

können als Linearkombinationen

$$U = \sum_{\gamma, J, M} a_{\gamma J}^{M} \psi_{\gamma J}^{M} \tag{22.2}$$

der Zustände $\psi_{\gamma J}^{M}$ des freien Atoms angenommen werden. Der Hamilton-Operator $\mathscr{H}'$ des Kristallfeldes und damit auch die gesamte Hamilton-Funktion $\mathscr{H} = \mathscr{H}^0 + \mathscr{H}'$ sind gegen die endliche Anzahl der Drehungen und Spiegelungen invariant, welche in der Punktsymmetrie des betrachteten Atoms im Kristallfeld enthalten sind. Nach den gruppentheoretischen Überlegungen (Ziff. 13) transformieren sich bei einer Koordinatentransformation, gegen die $\mathscr{H}$ invariant ist, die zu einem Energieeigenwert W gehörenden entarteten Zustände linear untereinander. Ein Kristallterm ist einfach, wenn der Zustand U bei allen erlaubten Symmetrieoperationen bis auf einen Faktor vom Betrage 1 in sich übergeht. Der Eigenwert ist zweifach, wenn es zwei linear unabhängige Eigenfunktionen U_1 und U_2 gibt, die sich bei den Symmetrieoperationen linear untereinander transformieren; usw. bei mehrfachen Termen. In der Sprache der Gruppentheorie heißt dies, daß die Darstellungen der Terme ein-, zwei- oder mehrdimensional sind.

Zur Bestimmung der *Anzahl der Terme* in einem (beliebig großen) Kristallfeld genügt es nach Ziff. 13 die Kristallzustände in nullter Näherung zu betrachten, d.h. das Kristallfeld bei festgehaltener Symmetrie verschwinden zu lassen. Dann reduziert sich U auf eine Linearkombination der $2J+1$ Eigenfunktionen, die zum Eigenwert $E_{\gamma J}$ des freien Atoms gehören, d. h. in Gl. (22.2) bleibt γ und J *fest*:

$$u_{\gamma J} = \sum_M a_{\gamma J}^M \psi_{\gamma J}^M, \tag{22.3}$$

wobei wir der Einfachheit halber im folgenden die festen Indices γ und J am u weglassen. Man bestimmt zunächst alle einfachen Kristallterme, indem man die u untersucht, die sich bei den erlaubten Symmetrieoperationen nur durch eine Größe vom Betrag 1 unterscheiden. Hierauf sucht man alle zweifachen Terme auf, usw. Es zeigt sich, daß bei nichtkubischen Kristallen höchstens zweifache Terme, bei kubischen Kristallen maximal vierfache Terme auftreten können.

Zyklische Kristallklassen C_p ($p = 1, 2, 3, 4, 6$). Im einfachsten Fall der zyklischen Kristallklassen C_p ist als einziges Symmetrieelement eine p-zählige Drehachse vorhanden, die in Richtung der z-Achse gelegt werde. Für die einfachen Zustände muß sich bei der Drehung um $2\pi/p$ die Kristalleigenfunktion u mit einem Faktor D vom Betrag 1 multiplizieren

$$u\left(r, \vartheta, \varphi' = \varphi + \frac{2\pi}{p}\right) = D \cdot u(r, \vartheta, \varphi). \tag{22.4}$$

Andererseits ist das Transformationsverhalten der Eigenfunktionen des freien Atoms, aus denen sich u in nullter Näherung nach Gl. (22.3) linear zusammensetzt, bei der Drehung um $2\pi/p$ durch Gl. (21.8) gegeben. Die Gln. (21.8), (22.3) und (22.4) sind nur dann miteinander verträglich, wenn nur solche ψ^M zur Funktion u gehören, deren M-Werte sich um Vielfache von p unterscheiden:

$$M = \mu + f p, \quad (\text{Laufzahl } f = 0, \pm 1, \pm 2, \ldots), \tag{22.5}$$

wofür man auch schreibt

$$M \equiv \mu \;(\mathrm{mod}\, p). \tag{22.5}$$

Damit wird in Gl. (22.3) die Summation über M durch die Summation über die Laufzahlen f ersetzt

$$u^\mu = \sum_f a^{M=\mu+fp}\, \psi^{M=\mu+fp}. \tag{22.6}$$

Da nach (22.5) μ insgesamt p Werte annehmen kann, erhält man p Funktionen u^μ. Für D ergibt sich damit

$$D = \mathrm{e}^{\,i\mu\frac{2\pi}{p}}, \tag{22.7}$$

d. h. die u^μ transformieren sich bei der Drehung $2\pi/p$ gemäß

$$u^\mu(r, \vartheta, \varphi') = \mathrm{e}^{\,i\mu\frac{2\pi}{p}}\, u^\mu(r, \vartheta, \varphi). \tag{22.8}$$

Die ordnenden Zahlen μ können wir als *Kristallquantenzahlen* ansehen. Sie entsprechen den Darstellungssymbolen Γ in der gruppentheoretischen Behandlung. Für die μ benützt man in den verschiedenen Kristallklassen C_p die folgenden Werte[1]

	N gerade (M, μ ganz)	N ungerade (M, μ halbzahlig)
$p = 1$	$\mu = \quad 0$	$\mu = \quad \frac{1}{2}$
2	$0, 1$	$-\frac{1}{2}, \frac{1}{2}$
3	$-1, 0, 1$	$-\frac{1}{2}, \frac{1}{2}, \frac{3}{2}$
4	$-1, 0, 1, 2$	$-\frac{3}{2}, -\frac{1}{2}, \frac{1}{2}, \frac{3}{2}$
6	$-2, -1, 0, 1, 2, 3$	$-\frac{5}{2}, -\frac{3}{2}, -\frac{1}{2}, \frac{1}{2}, \frac{3}{2}, \frac{5}{2}$

$$\tag{22.9}$$

---

[1] Man könnte z. B. aber auch setzen

$$\mu = 0, 1, 2, \ldots, p-1, \;(N \text{ gerade}); \quad \mu = \frac{1}{2}, \frac{3}{2}, \ldots, \frac{2p-1}{2}, \;(N \text{ ungerade}).$$

wobei bei gerader Elektronenanzahl N die μ ganz, bei ungeradem N halbzahlig werden.

Da es immer möglich ist, die ψ_M nach den μ zu ordnen, sind *alle Terme einfach*; man erhält bei einem Störpotential zyklischer Symmetrie $2J+1$ Kristallterme.

Zu einem μ-Wert gehören im allgemeinen mehrere einfache Terme. Ihre Anzahl z_μ ist die Anzahl der M-Werte, die bei festem J und p zu einem μ gehören; sie ist also gleich der Anzahl der f-Werte. Da $f_{\max} = \left[\dfrac{J-\mu}{p}\right]^\dagger$ und $f_{\min} = -\left[\dfrac{J+\mu}{p}\right] < 0$ ist, so erhält man, wenn man noch $f=0$ berücksichtigt, für die Anzahl der Terme, die zur Quantenzahl μ gehören

$$z_\mu = 1 + \left[\frac{J+\mu}{p}\right] + \left[\frac{J-\mu}{p}\right] = z_{-\mu}. \qquad (22.10)$$

Ein homogenes Magnetfeld parallel zur z-Achse gehört mit $p=\infty$ als Grenzfall zur zyklischen Klasse. Es wird $\mu=M$, $f=0$ und $z_M=1$. Zu jedem M gehört jetzt genau ein Term.

Zyklisch-inverse Kristallklassen: C_i, C_s, C_{3i}, S_4 und C_{3h}. Bei den zyklisch-inversen Kristallklassen C_i, C_s, C_{3i}, S_4 und C_{3h}, die eine p-zählige Inversionsachse ($p=1, 2, 3, 4, 6$) besitzen, welche wieder zur z-Achse gemacht werde, gilt für einfache Kristallzustände

$$u\left(r, \vartheta' = \pi - \vartheta, \varphi' = \varphi + \frac{2\pi}{p} - \pi\right) = D_I \cdot u(r, \vartheta, \varphi). \qquad (22.11)$$

Aus Gl. (21.11) und (22.3) folgt damit, daß jene ψ^M zu u^{μ_I} gehören, für die gilt

$$M + \frac{p}{2} \sum l_k \equiv \mu_I \pmod{p}. \qquad (22.12)$$

Bei der p-zähligen Drehinversion multiplizieren sich also die Kristallzustände u^{μ_I} mit

$$D_I = e^{i\mu_I \frac{2\pi}{p}}. \qquad (22.13)$$

Die hier auftretenden Kristallquantenzahlen μ_I lassen sich formal auch durch die μ darstellen:

$$\mu_I \equiv \mu + \frac{p}{2} \sum l_k \pmod{p}. \qquad (22.14)$$

Entsprechend Gl. (22.12) ist μ_I der folgenden Werte fähig:

		N gerade		N ungerade	
		$\sum l_k =$ gerade	$\sum l_k =$ ungerade	$\sum l_k =$ gerade	$\sum l_k =$ ungerade
C_i $p=1$		$\mu_I = 0$	$\frac{1}{2}$	$\frac{1}{2}$	0
C_{3i} 3		$\mu_I = -1, 0, 1$	$-\frac{1}{2}, \frac{1}{2}, \frac{3}{2}$	$-\frac{1}{2}, \frac{1}{2}, \frac{3}{2}$	$-1, 0, 1$
C_s 2		$\mu_I = 0, 1$		$-\frac{1}{2}, \frac{1}{2}$	
S_4 4		$\mu_I = -1, 0, 1, 2$		$-\frac{3}{2}, -\frac{1}{2}, \frac{1}{2}, \frac{3}{2}$	
C_{3h} 6		$\mu_I = -2, -1, 0, 1, 2, 3$		$-\frac{5}{2}, -\frac{3}{2}, -\frac{1}{2}, \frac{1}{2}, \frac{3}{2}, \frac{5}{2}$	

$$(22.15)$$

Sämtliche Terme sind einfach. Die Anzahl z der Terme, die zu einer Kristallquantenzahl gehören, finden sich bei K. H. Hellwege[1].

---

† Vgl. Fußnote *, S. 221.

[1] K. H. Hellwege: Ann. Phys. (6), **4**, 95 (1949).

Zyklische Kristallklassen mit vertikalen Spiegelebenen C_{pv} ($p=2,3,4,6$). Neben der p-zähligen Drehachse in der z-Richtung sind durch diese hindurchgehende vertikale Spiegelebenen vorhanden. Da nur die voneinander unabhängigen Drehoperationen behandelt zu werden brauchen, genügt es *eine* Spiegelebene zu betrachten, die in die xz-Ebene gelegt werde. Für die *einfachen* Kristallzustände muß damit gleichzeitig neben der Bedingung (22.4) noch die Forderung

$$u(x' = x, y' = -y, z' = z) = S \cdot u(x, y, z) \tag{22.16}$$

erfüllt sein. Da in Gl. (21.12) bei der Spiegelung an der xz-Ebene ψ^M bis auf einen Faktor in $\psi^{(-M)}$ übergeht, müssen die einfachen Kristallzustände neben ψ^M auch $\psi^{(-M)}$ enthalten, d.h. es muß entsprechend Gl. (22.5) gelten

$$M \equiv \mu \;(\mathrm{mod}\,p) \quad und \quad -M \equiv \mu \;(\mathrm{mod}\,p), \tag{22.17}$$

woraus folgt

$$\mu = 0 \quad \text{oder} \quad \frac{p}{2}. \tag{22.18}$$

Da die einfachen Kristallfunktionen $u = \sum (a^M \psi^M + a^{(-M)} \psi^{(-M)})$ die Transformationseigenschaft (22.16) besitzen, erhält man aus Gl. (21.12)

$$S \cdot a^M = (-1)^{\Sigma l_k - J - M} a^{(-M)} \quad und \quad S \cdot a^{(-M)} = (-1)^{\Sigma l_k - J + M} a^M,$$

woraus folgt

$$S = \pm (-1)^{\Sigma l_k - J} \tag{22.19}$$

(für $M = 0$ gilt nur das positive Zeichen) und außerdem

$$a^{(-M)} = \pm (-1)^M a^M.$$

Die einfachen Kristallterme sind also durch die *beiden* Quantenzahlen μ und S charakterisiert.

Für *zweifache* Kristallterme (höhere Entartung tritt nicht auf) ist

$$\mu \neq 0, \quad \frac{p}{2}. \tag{22.20}$$

Ein S kann nicht mehr definiert werden. Bei der Spiegelung transformieren sich die beiden Zustände

$$u_+ = \sum a^M \psi^M, \qquad u_- = \sum a^{(-M)} \psi^{(-M)}$$

linear untereinander:

$$\left. \begin{aligned} u_+(x' = x, y' = -y, z' = z) &= A_1 u_+(x, y, z) + A_2 u_-(x, y, z), \\ u_-(x' = x, y' = -y, z' = z) &= B_1 u_+(x, y, z) + B_2 u_-(x, y, z). \end{aligned} \right\} \tag{22.21}$$

Nach Gl. (21.12) ergibt sich, daß $A_1 = B_2 = 0$ und $|A_2| = |B_1| = 1$ ist. Da sich entsprechend Gl. (22.8) der Zustand u_+ bei der Drehung um die p-zählige Achse mit $D_+ = e^{i\mu \frac{2\pi}{p}}$, hingegen der Zustand u_- mit $D_- = e^{-i\mu \frac{2\pi}{p}}$ multipliziert, bezeichnet man nach K. H. Hellwege die zweifachen Terme mit $\{\pm\mu\}$. Es fallen also Zustände mit entgegengesetztem Drehsinn um die p-zählige Achse energetisch zusammen.

Für die verschiedenen Kristallklassen C_{pv} sind damit die folgenden Quantenzahlen möglich:

	N gerade		N ungerade	
	einfache Terme	zweifache Terme	einfache Terme	zweifache Terme
$p=2$ C_{2v}	$\mu=0\;\;\;0\;\;\;1\;\;\;1$ $S=1\;-1\;\;\;1\;-1$			$\{\pm\tfrac{1}{2}\}$ $-$
$p=3$ C_{3v}	$\mu=0\;\;\;0$ $S=1\;-1$	$\{\pm 1\}$ $-$	$\tfrac{3}{2}\;\;\;\tfrac{3}{2}$ $i\;\;\;-i$	$\{\pm\tfrac{1}{2}\}$ $-$
$p=4$ C_{4v}	$\mu=0\;\;\;0\;\;\;2\;\;\;2$ $S=1\;-1\;\;\;1\;-1$	$\{\pm 1\}$ $-$		$\{\pm\tfrac{1}{2}\}\{\pm\tfrac{3}{2}\}$ $-\;\;\;\;-$
$p=6$ C_{6v}	$\mu=0\;\;\;0\;\;\;3\;\;\;3$ $S=1\;-1\;\;\;1\;-1$	$\{\pm 1\}\{\pm 2\}$ $-\;\;\;\;-$		$\{\pm\tfrac{1}{2}\}\{\pm\tfrac{3}{2}\}\{\pm\tfrac{5}{2}\}$ $-\;\;\;\;-\;\;\;\;-$

$$(22.\;)$$

Bei geradzahligem p und ungerader Elektronenanzahl N (M halbzahlig) gibt es nach Gl. (22.17) und (22.18) keine einfachen Terme.

Ein homogenes elektrisches Feld ist mit $p=\infty$ als Grenzfall enthalten: $\mu=M$. Nur die Zustände mit $M=0$ sind einfach.

Diederklassen D_p ($p=2, 3, 4, 6$). Bei den Diederklassen treten neben der p-zähligen Drehachse in der z-Richtung noch zweizählige Drehachsen in der x, y-Ebene auf (vgl. z.B. D_4 in Fig. 4). Es genügt wiederum nur *eine* zweizählige Drehachse zu betrachten, die zur y-Achse gemacht werde. Neben Gl. (22.4) muß damit für die einfachen Zustände die Bedingung

$$u(x'=-x, y'=y, z'=-z)=D_y\cdot u(x,y,z) \qquad (22.23)$$

erfüllt sein. Da bei der Drehung π um die y-Achse [Gl. (21.9)] ψ^M in ψ^{-M} übergeht, besteht völlige Analogie zum soeben behandelten Fall C_{pv}, wenn man bedenkt, daß jetzt $\sum l_k$ im Exponenten [vgl. Gl. (22.19)] nicht auftritt. Der zweizähligen Nebenachse wird eine Kristallquantenzahl ν durch

$$D_y=e^{i\nu\pi} \qquad (22.24)$$

zugeordnet $\big($analog zu $D=e^{i\mu\frac{2\pi}{p}}$ für die p-zählige Hauptachse$\big)$, so daß

$$\left.\begin{aligned} \nu&=0,1 \quad \text{für gerades } N, \\ \nu&=\pm\tfrac{1}{2} \quad \text{für ungerades } N \end{aligned}\right\} \qquad (22.25)$$

entsprechend Gl. (22.19) gilt. Für die einfachen Zustände muß wieder $\mu=0$ oder $p/2$ sein, während die zweifachen Zustände ($\mu\neq 0, p/2$) durch $\{\pm\mu\}$ gekennzeichnet werden.

Die Kristallquantenzahlen μ und ν, die für die verschiedenen Diederklassen D_p möglich sind, erhält man, indem man einfach in (22.22) S durch $e^{i\nu\pi}$ ersetzt. So findet man z.B. für D_4

	N gerade					N ungerade	
$\mu=0$	0	2	2	$\{\pm 1\}$		$\{\pm\tfrac{1}{2}\}$	$\{\pm\tfrac{3}{2}\}$
$\nu=0$	1	0	1	$-$		$-$	$-$
gruppentheoretisch: ${}^1\Gamma_1$	${}^1\Gamma_2$	${}^1\Gamma_3$	${}^1\Gamma_4$	${}^2\Gamma_5$		${}^2\Gamma_6$	${}^2\Gamma_7$

$$(22.26)$$

Wir können an diesem Beispiel den *Zusammenhang mit der gruppentheoretischen Behandlungsweise* erkennen. Bei den einfachen Zuständen ergeben sich für

$D = e^{i\mu\frac{\pi}{2}}$ der Reihe nach die Werte $1, 1, -1, -1$, d.h. mit diesen Zahlen multiplizieren sich die Zustände bei der Drehung $2\pi/p = \pi/2$ um die tetragonale z-Achse (Drehung b in Fig. 4). Da es sich um einfache Zustände handelt, sind diese Werte von D die zur Drehung $b^{-1} = c$ gehörigen Charaktere[1]. (Die Darstellungsmatrizen sind eindimensional und daher gleich ihrem Charakter.) Vergleichen wir dieses Ergebnis mit der Tabelle (12.2), so sehen wir, daß gerade die zur Drehung c gehörige Klasse $\mathfrak{C}_3$ diese Charaktere besitzt. Analog stellen die Werte von $D_\nu = e^{i\nu\pi} = 1, -1, 1, -1$ die Charaktere für die zweizählige Drehung um die y-Achse (Drehung f in Fig. 4) dar, wie sie in (12.2) unter der Klasse $\mathfrak{C}_4$ zu finden sind. Wir haben gesehen, daß bei den zweifachen Termen in der Matrix (22.21) für die Drehung um die y-Achse die Glieder $A_1 = B_2 = 0$ sind, so daß ihr Charakter zu Null wird. Genau dasselbe ersieht man aus (12.2): der Charakter von $\mathfrak{C}_4$ in $^2\Gamma_5$ ist Null; ebenso verschwindet in (16.6) der entsprechende Charakter von $\mathfrak{C}_4$ in $^2\Gamma_6$ und $^2\Gamma_7$. Damit ist der Zusammenhang zwischen der gruppentheoretischen und der analytischen Methode hergestellt. Wir sehen, daß sie prinzipiell *gleichwertig* sind. Die Bestimmung der $z_{\mu,\nu}$ [analog Gl. (22.10)] entspricht dem Aufsuchen der q_k in der gruppentheoretischen Behandlungsweise.

Zyklisch-inverse Klassen mit Spiegelebenen D_{2d} und D_{3h}. Es liegt eine 4- bzw. 6zählige Inversionsachse mit durch sie hindurchgehenden Spiegelebenen vor. Die möglichen Quantenzahlen μ_I und S für D_{2d} bzw. D_{3h} ergeben sich aus der Tabelle (22.22) für C_{4v} bzw. C_{6v}, wenn man dort μ durch μ_I ersetzt.

Kristallklassen mit Inversionszentrum C_{ph}, D_{ph} ($p = 2, 4, 6$) und D_{3d}. Bei der Inversion am Kern multiplizieren sich nach Gl. (21.10) alle Kristalleigenfunktionen mit dem Symmetriecharakter $I = (-1)^{\Sigma l_k}$

$$u(x' = -x, y' = -y, z' = -z) = I \cdot u(x, y, z). \tag{22.27}$$

Das Symmetriezentrum in C_{ph}, D_{ph} ($p = 2, 4, 6$) und D_{3d} bewirkt also gegenüber C_p, D_p ($p = 2, 4, 6$) und D_3 nur eine Unterscheidung zwischen geraden ($I = 1$) und ungeraden ($I = -1$) Zuständen.

Kubisches Kristallsystem T, O, T_d, T_h und O_h. Macht man in der Tetraederklasse T eine dreizählige Drehachse zur z-Achse, so gilt für einfache Zustände

$$M \equiv \mu \pmod 3,$$

die sich bei der dreizähligen Drehung mit $D = e^{i\mu\frac{2\pi}{3}}$ multiplizieren. Der Drehung um eine der zweizähligen Drehachsen ist die Quantenzahl $\varkappa$ durch $K = e^{i\pi\varkappa}$ zugeordnet. Man kann sich jedoch auf $\varkappa = 0$ beschränken, so daß man die folgenden möglichen Quantenzahlen erhält

		N gerade				N ungerade		
T	$\mu = 0$	1	-1	$\{0, \pm 1\}$	$\{\pm\frac{1}{2}\}$	$\{\frac{1}{2}, -\frac{3}{2}\}$	$\{-\frac{1}{2}, \frac{3}{2}\}$
	$\varkappa = 0$	0	0	$-$	$-$	$-$	$-$

$$\tag{22.28}$$

wobei für die entarteten Terme $\varkappa$ wiederum nicht definiert ist, und die zwei- bzw. dreifachen Terme durch zwei bzw. drei μ-Werte bestimmt werden.

Für die Oktaederklasse O, deren vier- und dreizählige Drehachse durch μ bzw. $\varkappa$ charakterisiert seien, erhält man entsprechend

		N gerade				N ungerade		
O	$\mu = 0$	2	$\{0, 2\}$	$\{0, \pm 1\}$	$\{2, \pm 1\}$	$\{\pm\frac{1}{2}\}$	$\{\pm\frac{3}{2}\}$	$\{\pm\frac{1}{2}, \pm\frac{3}{2}\}$
	$\varkappa = 0$	0	$-$	$-$	$-$	$-$	$-$	$-$

$$\tag{22.29}$$

---

[1] Nach (13.8) ist $\chi(C_p^{-1}) = \sum\limits_{\tau=1}^{t} \exp\left[\frac{2\pi i}{p}\mu_\tau(C_p)\right]$ für eine Drehung C_p [Ziff. 17].

Die Kristallklasse T_d (vierzählige Inversionsachse) geht aus O hervor, wenn man μ durch μ_I substituiert. Die Klassen T_h und O_h unterscheiden sich von T und O nur durch den zusätzlich hinzukommenden Symmetriecharakter $I = (-1)^{\Sigma l_k}$.

Für *rein elektrostatische* Kristallfelder hat K. H. Hellwege[1] durch Anwendung des Kramersschen Satzes (Ziff. 20) auf die Kristallzustände gezeigt, daß zu dem Zustand

$$U = \sum_{\gamma, J, M} a^M_{\gamma J} \psi^M_{\gamma J} \qquad (22.30)$$

der Kramerssche Zustand

$$\tilde{U} = \sum_{\gamma, J, M} (-1)^{J+M-\Sigma l_k} a^{M*}_{\gamma J} \psi^{-M}_{\gamma J} \qquad (22.31)$$

gehört.

23. Auswahlregeln. Die Übergangswahrscheinlichkeit zwischen den durch die Kristallquantenzahlen gekennzeichneten Termen ist bestimmt durch die Operatoren des elektrischen Dipolmoments, des magnetischen Dipolmoments und des elektrischen Quadrupolmoments, usw. Aus der Forderung, daß die Matrixelemente dieser Operatoren für einen Übergang von einem Term i zu einem Term l invariant sein müssen gegen die möglichen Symmetrieoperationen, ergeben sich nach K. H. Hellwege[2] *Auswahlregeln* als notwendige Bedingungen für nichtverschwindende Matrixelemente.

Die Matrixelemente z.B. der x-Komponente des elektrischen Dipolmoments lauten (bis auf einen unwesentlichen Faktor)

$$\langle i\,|\,x\,|\,l\rangle \equiv (u_i,\, x\,u_l) \equiv \sum \int u_i^*\, x\, u_l\, d\tau, \qquad (23.1)$$

wobei über alle Ortskoordinaten der Elektronen integriert und über ihre Spinkoordinaten $(+\tfrac{1}{2}$ und $-\tfrac{1}{2})$ summiert wird. x steht dabei für $\sum\limits_{k=1}^{N} x^{(k)}$.

Ist das Matrixelement $\langle i\,|\,z\,|\,l\rangle \neq 0$, so ist bei Beobachtung senkrecht zur kristallographischen Achse (z-Achse) eine π-*Komponente*, d.h. eine Schwingung mit $\mathfrak{E}\,\|\,z$ vorhanden. Wenn dagegen $\langle i\,|\,x\pm iy\,|\,l\rangle \neq 0$ ist, so ergibt sich eine σ-*Komponente*, d.h. $\mathfrak{E}\perp z$. Beobachtet man parallel zur z-Achse, so ist im letzteren Fall die Polarisation elliptisch, da das $+$-Zeichen einem in der x, y-Ebene negativ drehenden Dipol und das $-$-Zeichen einem positiv drehenden Dipol entspricht.

Betrachten wir z.B. den Fall zyklischer Kristallklassen C_p, so erhalten wir die Auswahlregeln für die Kristallquantenzahl μ bei elektrischer Dipolstrahlung. Aus der Transformationsformel (22.8) für die Eigenfunktion u^μ folgt

$$\langle i\,|\,z\,|\,l\rangle = \left(u^{\mu_i}(r, \vartheta, \varphi'),\, r\cos\vartheta\, u^{\mu_l}(r, \vartheta, \varphi')\right)$$
$$= e^{-i\Delta\mu\frac{2\pi}{p}} \left(u^{\mu_i}(r, \vartheta, \varphi),\, r\cos\vartheta\, u^{\mu_l}(r, \vartheta, \varphi)\right) = e^{-i\Delta\mu\frac{2\pi}{p}} \langle i\,|\,z\,|\,l\rangle$$

und entsprechend

$$\langle i\,|\,x\pm iy\,|\,l\rangle = \left(u^{\mu_i}(r, \vartheta, \varphi'),\, r\sin\vartheta\, e^{\pm i\varphi'} u^{\mu_l}(r, \vartheta, \varphi')\right)$$
$$= e^{-i\Delta\mu\frac{2\pi}{p}\pm i\frac{2\pi}{p}} \left(u^{\mu_i}(r, \vartheta, \varphi),\, r\sin\vartheta\, e^{\pm i\varphi} u^{\mu_l}(r, \vartheta, \varphi)\right)$$
$$= e^{-i(\Delta\mu\mp 1)\frac{2\pi}{p}} \langle i\,|\,x\pm iy\,|\,l\rangle,$$

wobei $\Delta\mu = \mu_i - \mu_l$ die Differenz zwischen den beiden Kristallquantenzahlen μ_i und μ_l bedeutet. Hieraus folgt, daß die Matrixelemente nur dann von Null

---

[1] K. H. Hellwege: Ann. Phys. **4**, 143 (1949).
[2] K. H. Hellwege: Ann. Phys. **4**, 95, 127 (1949).

verschieden sein können, wenn die Auswahlregeln

$$\left.\begin{aligned}\Delta\mu &\equiv 0 \,(\mathrm{mod}\,p) &&\text{für}\quad \langle i|z|l\rangle \neq 0 &&(\pi\text{-Komponenten}),\\ \Delta\mu &\equiv \pm 1 \,(\mathrm{mod}\,p) &&\text{für}\quad \langle i|x\pm iy|l\rangle \neq 0 &&(\sigma\text{-Komponenten})\end{aligned}\right\} \quad (23.2)$$

erfüllt sind (vgl. Fig. 15 und 16).

Für die verschiedenen im Kristall möglichen Quantenzahlen ergeben sich analog die folgenden Auswahlregeln:

Elektrische Dipolstrahlung: Kristallklassen ohne Inversionsachsen.

		$\Delta\mu$	$\Delta\nu$	I	S
π	$\langle i|z|l\rangle$	$\equiv 0 \;(\mathrm{mod}\,p)$	± 1		$S_i = S_l$
	$\langle i|x\pm iy|l\rangle$		$-\,^*$		$-\,^*$
σ	$\langle i|x|l\rangle$	$\equiv \pm 1 \;(\mathrm{mod}\,p)$	± 1	$I_i = -I_l$	$S_i = S_l$
	$\langle i|y|l\rangle$		0		$S_i = -S_l$

$$(23.3)$$

Die Forderung $I_i = -I_l$, die bei Kristallklassen mit Inversionszentrum wirksam wird, entspricht der LAPORTEschen Regel, wonach nur gerade Terme mit ungeraden und umgekehrt kombinieren ($\sum l_k$ ändert sich um eine ungerade Zahl). Da die Spektren der Seltenen Erden innerhalb der $4f$-Schale entstehen ($\sum l_k = $ fest), kann bei den Kristallklassen mit Inversionszentrum keine elektrische Dipolstrahlung auftreten.

Elektrische Dipolstrahlung: Kristallklassen mit Inversionsachsen.

		$\Delta\mu_I$	S
π	$\langle i|z|l\rangle$	$\equiv \dfrac{p}{2} \;(\mathrm{mod}\,p)$	$S_i = S_l$
	$\langle i|x\pm iy|l\rangle$		$-$
σ	$\langle i|x|l\rangle$	$\equiv \pm\left(1+\dfrac{p}{2}\right)\;(\mathrm{mod}\,p)$	$S_i = S_l$
	$\langle i|y|l\rangle$		$S_i = -S_l$

$$(23.4)$$

Für die *magnetische* Dipolstrahlung, die von DEUTSCHBEIN und ROSA[1] in den Spektren der Seltenen Erden experimentell nachgewiesen wurde, sind die Matrixelemente $(u_i, \mathfrak{M}u_k)$ des magnetischen Dipolmoments

$$\mathfrak{M} \sim \mathfrak{L} + 2\mathfrak{S}$$

verantwortlich. Nach K. H. HELLWEGE findet man für die Auswahlregeln dieser Strahlung:

Magnetische Dipolstrahlung: Kristallklassen ohne Inversionsachsen.

		$\Delta\mu$	$\Delta\nu$	I	S
π	$\langle i|z|l\rangle$	$\equiv 0 \;(\mathrm{mod}\,p)$	± 1		$S_i = -S_l$
	$\langle i|x\pm iy|l\rangle$		$-$		$-$
σ	$\langle i|x|l\rangle$	$\equiv \pm 1 \;(\mathrm{mod}\,p)$	± 1	$I_i = I_l$	$S_i = -S_l$
	$\langle i|y|l\rangle$		0		$S_i = S_l$

$$(23.5)$$

---

* Das heißt es gibt keine in der x-y-Ebene drehenden Dipole.

[1] O. DEUTSCHBEIN: Ann. Phys (5). **36**, 183 (1939). — A. ROSA: Ann. Phys. (5) **43**, 161 (1943).

Magnetische Dipolstrahlung: Kristallklassen mit Inversionsachsen.

	$\Delta\mu_I$	S
π $\langle i\,|z|\,l\rangle$	$\equiv 0 \pmod{p}$	$S_i = -\,S_l$
$\sigma\begin{cases}\langle i\,|x\pm iy|\,l\rangle\\ \langle i\,|x|\,l\rangle\\ \langle i\,|y|\,l\rangle\end{cases}$	$\equiv \pm 1 \pmod{p}$	$\begin{array}{c}-\\ S_i=-\,S_l\\ S_i=S_l\end{array}$

$$(23.6)$$

Man sieht, daß alle Kristalloperationen, die aus reinen Drehungen bestehen, für elektrische und magnetische Dipolstrahlung dieselben Auswahlregeln liefern. Die Inversion, die Drehinversionen und Spiegelebenen hingegen bedingen für elektrische und magnetische Dipolstrahlung gerade entgegengesetztes Verhalten.

Die Auswahlregeln für die bisher experimentell nicht beobachtete *elektrische Quadrupolstrahlung* sind ebenfalls von K. H. Hellwege[1] abgeleitet worden.

Liegen *entartete* Zustände vor, so sind auf die zusammenfallenden Werte von μ bzw. μ_I die Auswahlregeln für $\Delta\mu$ bzw. $\Delta\mu_I$ getrennt anzuwenden (vgl. Fig. 15 und 16).

Bei der magnetischen Dipolstrahlung müssen für die Gesamtdrehimpulsquantenzahl J die Auswahlregeln

$$\Delta J = 0, \pm 1 \qquad (0 \to 0 \text{ verboten}) \qquad (23.7)$$

wie beim freien Ion erfüllt sein. Für die elektrische Dipolstrahlung ist eine solche Auswahlregel nach van Vleck[2] *nicht* nötig, da diese Strahlung erst vom Kristallfeld *erzwungen* wird. Im freien Ion ist ja diese Strahlung wegen der Laporteschen Regel für Übergänge zwischen Termen mit gleichen $\sum l_k$ verboten.

Den von den Einkristallen absorbierten oder emittierten Spektrallinien sind als Strahlungsquellen korrespondenzmäßig elektrische oder magnetische Dipole (im allgemeinen Multipole) zugeordnet. Aus der Polarisation der Spektrallinien kann mit Hilfe der Auswahlregeln die Orientierung dieser Elementarstrahler relativ zum Kristall und ihre Natur (elektrischer oder magnetischer Dipol) bestimmt werden[3].

III. Die Größe der Kristallfeldaufspaltung.

24. Der Einfluß von Kristallfeldern verschiedener Größenordnung. Bisher wurde das Verhalten eines Atoms oder Ions mit einem bestimmten Drehimpuls in einem Kristallfeld vorgegebener Symmetrie untersucht. Je nach der Größenordnung des Kristallfeldes gegenüber den Termen der Hamilton-Funktion des freien Atoms müssen wir verschiedene Fälle unterscheiden:

$\alpha)$ „*Kleines*" *Kristallfeld.* Die Beeinflussung des Atoms durch das Kristallfeld ist klein gegenüber der Spin-Bahn-Wechselwirkung (LS-Koppelung), aber groß gegenüber der Wechselwirkung der Elektronen mit dem magnetischen Moment und dem elektrischen Quadrupolmoment des Kernes. Die Kristallfeldaufspaltung ist also klein gegen den Termabstand innerhalb eines Multipletts. Der Gesamtdrehimpuls $\mathfrak{J}$ des Atoms stellt sich dabei relativ zu den Kristallachsen ein[4]. Für die Gesamtdrehimpulsquantenzahl J, die auch noch für das Atom im Kristallfeld in guter Näherung definiert ist, gelten alle bisher abgeleiteten Beziehungen. Die Größe der Kristallfeldaufspaltung erhält man

---

[1] K. H. Hellwege: Ann. Phys. **4**, 136, 150 (1949). — Z. Physik **129**, 626 (1951); **131**, 98 (1951).

[2] J. van Vleck: Z. phys. Chem. **41**, 67 (1937).

[3] K. H. Hellwege u. H. G. Kahle: Z. Physik **129**, 62 (1951). — K. H. Hellwege: Z. Physik **143**, 340 (1955).

[4] Natürlich braucht $\mathfrak{J}$ im Kristall nicht konstant zu sein.

durch eine Störungsrechnung, bei der in den Eigenfunktionen nullter Näherung die Spin-Bahn-Wechselwirkung bereits berücksichtigt ist. Dieser Fall ist bei den Salzen der *Seltenen Erden* realisiert, bei denen die $5s$- und $5p$-Schalen die Wirkung des Kristallfeldes auf die $4f$-Elektronen stark abschirmen.

β) *„Mittleres" Kristallfeld.* Dieser Fall tritt ein, wenn das Kristallfeld klein ist gegenüber der COULOMB-Wechselwirkung zwischen den Elektronen, aber groß gegen die Spin-Bahn-Wechselwirkung, so daß die Quantenzahl J ihren Sinn verliert, und der Gesamtbahndrehimpuls $\mathfrak{L}$ und der Gesamtspin $\mathfrak{S}$ im Kristallfeld getrennt zu behandeln sind. Die Aufspaltung der $(2L+1)\cdot(2S+1)$-fachen Ausgangszustände wird erhalten, indem man *in der bisher entwickelten Theorie überall J durch L ersetzt* [also z.B. auch in der Übersicht (20.14)], da das elektrische Kristallfeld nur auf die Bahnbewegung einwirkt, so daß man $v\cdot(2S+1)$-fache Kristallterme erhält. [Vielfachheit v nach (20.14): $1 \leq v \leq 4$.] Erst in höherer Näherung, bei der die Spin-Bahn-Wechselwirkung Berücksichtigung findet, spalten auch die Spin-Zustände schwach auf. Der Fall eines „mittleren" Kristallfeldes, bei dem also die Kristallfeldaufspaltung klein gegenüber dem Abstand verschiedener Multipletts, aber groß gegen die Termdifferenzen ein und desselben Multipletts des freien Atoms ist, wird bei den Salzen der *Eisenreihe* realisiert. Die bekannte Auslöschung der magnetischen Wirkung des Bahnmoments ("quenching of the orbitals") der $3d$-Eektronen, die sich beim g-Faktor dieser Verbindungen zeigt, läßt sich durch diese Einwirkung des Kristallfeldes auf die Bahnbewegung verstehen.

γ) *„Großes" Kristallfeld.* Die Wechselwirkung der Elektronen untereinander wird durch das Kristallfeld aufgehoben. Auf die Bahnimpulse $\mathfrak{l}_k$ der einzelnen Elektronen wirkt das Kristallfeld, so daß die Kristallfeldaufspaltung groß wird gegenüber dem Abstand verschiedener Multipletts des freien Atoms. In diesem von H. BETHE (vgl. Fußnote 3 auf S. 210) eingehend untersuchten Fall zerfällt eine aus $2(2l+1)$-Quantenzellen bestehende Schale des freien Atoms entsprechend der vorliegenden Kristallsymmetrie in mehrere Unterschalen, von denen jede einem anderen Elektronenterm entspricht.

Bevor wir zur Beschreibung des *Überganges* etwa von einem „schwachen" zu einem „mittleren" Kristallfeld kommen, ist es vorteilhaft den Begriff der Produktdarstellung einzuführen.

25. Produktdarstellung, Vektoradditionskoeffizienten. Betrachtet man zwei Vektoren $u = u^\mu$ ($\mu = 1, \ldots, p$) und $v = v^\nu$ ($\nu = 1, \ldots, q$) aus zwei Vektorräumen $\mathfrak{R}_p$ und $\mathfrak{R}_q$, so kann man die $p \cdot q$ Produkte $u^\mu v^\nu$ als Komponenten eines Vektors in einem $\mathfrak{R}_{pq}$ auffassen. Transformiert man den Vektor u mit der Matrix $A = A^{\mu'\mu}$ und den Vektor v mit $B = B^{\nu'\nu}$,

$$u^{\mu'} = \sum_\mu A^{\mu'\mu} u^\mu \quad \text{und} \quad v^{\nu'} = \sum_\nu B^{\nu'\nu} v^\nu,$$

so transformiert sich der Vektor $u^\mu v^\nu$ nach der Produkttransformation

$$u^{\mu'} v^{\nu'} = \sum_\mu \sum_\nu A^{\mu'\mu} B^{\nu'\nu} u^\mu v^\nu.$$

Diese sog. KRONECKERsche Produktmatrix $A \times B$ mit den Elementen $(A \times B)^{\mu'\nu',\mu\nu} = A^{\mu'\mu} B^{\nu'\nu}$ hat $p \cdot q$ Zeilen und Spalten. Für ihre Spur gilt

$$\chi(A \times B) = \chi(A)\,\chi(B). \tag{25.1}$$

Bilden die A und B zwei Darstellungen Γ_1 und Γ_2 einer Gruppe $\mathfrak{G}$, so bildet $A \times B$ wieder eine Darstellung, die *Produktdarstellung* $\Gamma_1 \times \Gamma_2$. Der Charakter einer Produktdarstellung ist gleich dem Produkt der Charaktere der Faktordarstellungen.

Wenden wir uns der wellenmechanischen Anwendung zu: Wenn die Eigenfunktionen zweier quantenmechanischer Systeme (1) und (2) eine Darstellung $\mathfrak{D}_{j_1}$ bzw. $\mathfrak{D}_{j_2}$ der Drehgruppe erlauben, so daß also die Eigenfunktionen $\psi_{j_1}^{m_1}(|m_1| \leq j_1)$ den Raum der Darstellung $\mathfrak{D}_{j_1}$ und $\psi_{j_2}^{m_2}$ den der Darstellung $\mathfrak{D}_{j_2}$ aufspannen, so bilden die $(2j_1+1)\cdot(2j_2+1)$ Funktionen $\psi_{j_1}^{m_1}\psi_{j_2}^{m_2}$ des Gesamtsystemes eine Basis für den Raum der Produktdarstellung $\mathfrak{D}_{j_1}\times\mathfrak{D}_{j_2}$. Die in $\mathfrak{D}_{j_1}\times\mathfrak{D}_{j_2}$ enthaltenen irreduziblen Darstellungen der Drehgruppe ergeben sich aus

$$\chi(\mathfrak{D}_{j_1}\times\mathfrak{D}_{j_2}) = \sum_{m_1=j_1}^{-j_1} e^{im_1\alpha} \sum_{m_2=j_2}^{-j_2} e^{im_2\alpha} = \chi(\mathfrak{D}_{j_1+j_2}) + \chi(\mathfrak{D}_{j_1+j_2-1}) + \cdots + \chi(\mathfrak{D}_{|j_1-j_2|})$$

zu

$$\mathfrak{D}_{j_1}\times\mathfrak{D}_{j_2} = \mathfrak{D}_{j_1+j_2} + \mathfrak{D}_{j_1+j_2-1} + \cdots + \mathfrak{D}_{|j_1-j_2|} = \sum_{J=j_1+j_2}^{|j_1-j_2|} \mathfrak{D}_J. \tag{25.2}$$

Dies ist die gruppentheoretische Formulierung der Vektoraddition der Drehimpulsquantenzahlen.

Die Eigenfunktionen $\psi_{Jj_1j_2}^{M}(|M| \leq J;\ M=m_1+m_2)$, die sich nach der Darstellung $\mathfrak{D}_J (J=j_1+j_2, j_1+j_2-1, \ldots |j_1-j_2|)$ transformieren, sind mit den $\psi_{j_1}^{m_1}\psi_{j_2}^{m_2}$ durch eine unitäre Transformation

$$\psi_{Jj_1j_2}^{M} = \varrho_{Jj_1j_2} \sum_{m_1=j_1}^{-j_1} \sum_{m_2=j_2}^{-j_2} c_J^{M\,m_1m_2}{}_{j_1\,j_2} \psi_{j_1}^{m_1}\psi_{j_2}^{m_2} \tag{25.3}$$

miteinander verknüpft. Die $\psi_{j_1}^{m_1}\psi_{j_2}^{m_2}$ sind Eigenfunktionen der Darstellung, in welcher j_1^2, j_2^2, j_{1z} und j_{2z} diagonal sind. Die $\psi_{Jj_1j_2}^{M}$ hingegen sind Eigenfunktionen der Darstellung, in welcher j_1^2, j_2^2, $\mathfrak{J}^2$ und $J_z=j_{z1}+j_{z2}$ diagonal sind. Die auftretenden Koeffizienten $c_J^{M\,m_1m_2}{}_{j_1\,j_2}$ werden *Vektoradditionskoeffizienten* oder WIGNER-*Koeffizienten* genannt; sie berechnen sich zu[1]

$$c_J^{M\,m_1m_2}{}_{j_1\,j_2} = \sum_\nu (-1)^\nu \frac{\sqrt{(j_1+m_1)!\,(j_1-m_1)!\,(j_2+m_2)!\,(j_2-m_2)!\,(J+M)!\,(J-M)!}}{(j_1-m_1-\nu)!\,(j_2+m_2-\nu)!\,(j_1+j_2-J-\nu)!\,(J-j_1-m_2+\nu)!\,(J-j_2+m_1+\nu)!\,\nu!}\,\delta_{M,\,m_1+m_2} \tag{25.4}$$

Der Faktor $\varrho_{Jj_1j_2}$ hängt nur von J, j_1 und j_2 ab, und wird meist so gewählt, daß $\psi_{Jj_1j_2}^{M}$ ein *normiertes* Orthogonalsystem darstellt[2]. Da nur solche M-Werte zugelassen sind, für die $M=m_1+m_2$ gilt, tritt in (25.4) $\delta_{M,\,m_1+m_2}$ auf. In Gl. (25.3) kann daher etwa die Summe über m_2 sofort ausgeführt werden. Die WIGNER-Koeffizienten sind für spezielle Werte von j_2 in Tabellen explizit zu finden[1,3]; außerdem gibt es zahlreiche Rekursionsformeln[4].

Auf Grund der Unitarität lautet die Auflösung der Gl. (25.3)

$$\psi_{j_1}^{m_1}\psi_{j_2}^{m_2} = \sum_J \varrho_{Jj_1j_2} c_J^{M\,m_1m_2}{}_{j_1\,j_2} \psi_{Jj_1j_2}^{M} \quad (M=m_1+m_2). \tag{25.5}$$

---

[1] B. L. VAN DER WAERDEN: Die gruppentheoretische Methode in der Quantenmechanik. Berlin 1932.

[2] Die Bezeichnung der WIGNER-Koeffizienten ist nicht einheitlich. VAN DER WAERDEN (l. c.) schreibt dafür $c_{m_1m_2}^J$, während CONDON und SHORTLEY dafür $(j_1j_2m_1m_2|j_1j_2JM)$ setzen. Häufig wird auch der Faktor $\varrho_{Jj_1j_2}$ in die Formel für die WIGNER-Koeffizienten mit hineingenommen.

[3] E. CONDON u. G. SHORTLEY: The Theory of Atomic Spectra, S. 76. Cambridge 1953.

[4] G. RACAH: Phys. Rev. **62**, 438 (1942). — M. ROSE: Multipole Fields. New York 1955.

26. Der Übergang vom „schwachen" zum „mittleren" Kristallfeld. Bei einem
„mittleren" Kristallfeld stellt sich nach Ziff. 24 der Bahndrehimpuls $\mathfrak{L}$ zum Kristall ein. Die Anzahl der Kristallterme ist gegeben durch die Zahlen q_{kL} in

$$\mathfrak{D}_L = \sum q_{kL}\, \Gamma_k, \tag{26.1}$$

die sich wie in Gl. (15.2) bestimmen. Berücksichtigt man noch die Wechselwirkung zwischen Spin und orientiertem Bahnimpuls, so tritt eine weitere Aufspaltung ein

$$\Gamma_k \times \mathfrak{D}_S = \sum p_{ki}\, \Gamma_i. \tag{26.2}$$

Insgesamt müssen damit natürlich ebensoviel Terme entstehen, wie bei der Aufspaltung im „schwachen" Kristallfeld, bei der sich der Gesamtdrehimpuls $\mathfrak{J}$
zum Feld einstellt.

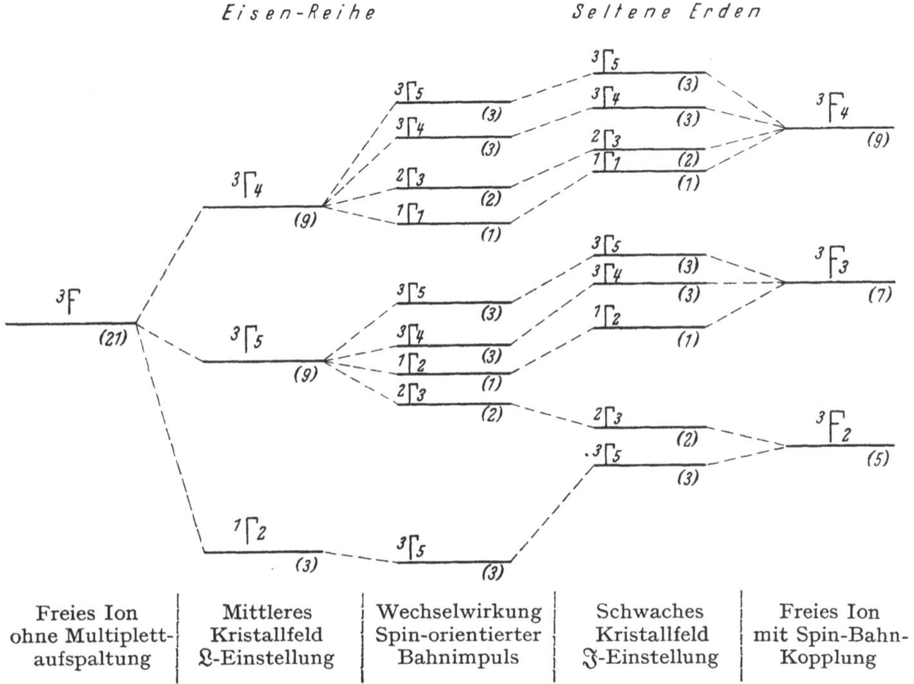

Fig. 5. Der Übergang vom mittleren zum schwachen kubischen Kristallfeld für das Beispiel $L = 3$, $S = 1$ (^3F) (Vielfachheit der Terme in Klammern).

In der Fig. 5 ist für den Fall $L = 3$, $S = 1$ der Zusammenhang zwischen einem
„schwachen" und einem „mittleren" Kristallfeld kubischer Symmetrie O gezeigt.
Nach (17.3) spaltet der ^3F-Zustand im mittleren Kristallfeld in den $1 \cdot (2S+1) = 3$-
fachen Term $^1\Gamma_2$ und in die beiden $3 \cdot (2S+1) = 9$-fachen Terme $^3\Gamma_4$ und $^3\Gamma_5$
auf. Tritt nun noch eine Spin-Bahn-Wechselwirkung auf, so sind entsprechend
(26.2) die irreduziblen Bestandteile der Produktdarstellungen $^1\Gamma_2 \times \mathfrak{D}_1$, $^3\Gamma_4 \times \mathfrak{D}_1$
und $^3\Gamma_5 \times \mathfrak{D}_1$ aufzusuchen. Die Multiplikation der Charaktere (17.3), (14.2) liefert

$$\chi(^1\Gamma_2 \times \mathfrak{D}_1) = 1 \cdot 3 = 3,\ 1 \cdot (-1) = -1,\ (-1) \cdot 1 = -1,\ (-1) \cdot (-1) = 1,\ 1 \cdot 0 = 0$$

für die Klassen E, $3C_2$, $6C_4$, $6C_2'$, $8C_3$ bei kubischer Symmetrie; d.h. $^1\Gamma_2 \times \mathfrak{D}_1 = {}^3\Gamma_5$.
Entsprechend erhält man $\chi(^3\Gamma_4 \times \mathfrak{D}_1) = 9, 1, 1, 1, 0$, also $^3\Gamma_4 \times \mathfrak{D}_1 = {}^1\Gamma_1 + {}^2\Gamma_3 +$
$^3\Gamma_4 + {}^3\Gamma_5$ und entsprechend $\chi(^3\Gamma_5 \times \mathfrak{D}_1) = 9, 1, -1, -1, 0$; d.h. $^3\Gamma_5 \times \mathfrak{D}_1 = {}^1\Gamma_2 +$
$^2\Gamma_3 + {}^3\Gamma_4 + {}^3\Gamma_5$. Man findet also — wie es sein muß — dieselben Terme, die sich

bei der Einstellung des Gesamtdrehimpulses mit $J = 2, 3, 4$ zum Kristallfeld ergeben. Die Zuordnung der so entstehenden Terme bei mittlerem Kristallfeld zu jenen des schwachen Kristallfeldes ergibt sich aus der Forderung, daß Terme, die zur *gleichen* Darstellung der Symmetriegruppe des Kristalls gehören, sich nicht überschneiden dürfen. Der Reihenfolge der gruppentheoretischen Reduktionen entspricht die Reihenfolge der Störungsrechnungen: bei „mittlerem" Kristallfeld tritt die Aufspaltung durch die Spin-Bahn-Wechselwirkung erst in höherer Näherung auf.

Der Übergang vom „mittleren" zum „starken" Kristallfeld wurde von H. Bethe (l. c.) in analoger Weise behandelt.

27. Das Kristallfeld und seine Matrixelemente. Bei der Berechnung des Kristallfeldes am Ort des herausgegriffenen Ions hat man sich dieses aus dem Kristall entfernt zu denken. Wenn man annimmt, daß die Ladungen der Nachbarionen nicht bis an den Ort des herausgegriffenen Ions reichen, muß das Potential V des Kristallfeldes Lösung der Laplaceschen Differentialgleichung $\Delta V = 0$ sein, das sich nach (normierten) Kugelfunktionen $Y_\lambda^\varkappa(\vartheta, \varphi)$ [vgl. (21. 6)] entwickeln läßt:

$$V = \sum_\lambda \sum_{\varkappa = +\lambda}^{-\lambda} \sum_{k=1}^{N} A_\lambda^\varkappa \cdot (r_k)^\lambda \cdot Y_\lambda^\varkappa(\vartheta_k, \varphi_k) = \sum_\lambda \sum_\varkappa V_\lambda^\varkappa \tag{27.1}$$

(N = Anzahl der Elektronen in der nichtabgeschlossenen Schale). Mit Hilfe der Störungsrechnung für entartete Zustände $\psi_{\gamma J}^M$ (= Eigenfunktionen des freien Ions) ergibt sich die *Größe* der Kristallfeldaufspaltung in erster Näherung, wenn man die Säkulargleichung der Matrixelemente des Potentials, die im folgenden durch das Symbol

$$\int \psi_{\gamma J}^{M*} V \psi_{\gamma J}^{M'} d\tau \equiv \langle \gamma\, J\, M \,|\, V \,|\, \gamma\, J\, M' \rangle \tag{27.2}$$

bezeichnet werden, löst. Man kann sich daher im Potential (27.1) auf jene Indices λ und $\varkappa$ beschränken, die nichtverschwindende Matrixelemente liefern.

Das Potential $V_\lambda^\varkappa$ transformiert sich bei einer Drehung nach einer irreduziblen Darstellung der Drehgruppe, d.h. wie ein Vektor im Raum $\mathfrak{D}_\lambda$. Daher lassen sich (K. W. H. Stevens[1]) die Matrixelemente von $V_\lambda^\varkappa$ durch die Wigner-Koeffizienten ausdrücken

$$\langle \gamma\, J\, M \,|\, V_\lambda^\varkappa \,|\, \gamma'\, J'\, M' \rangle = (\alpha_\lambda)_{\gamma J, \gamma' J'} \cdot c_{J \lambda J'}^{M \varkappa M'}, \tag{27.3}$$

wobei der Faktor $(\alpha_\lambda)_{\gamma J, \gamma' J'}$ von M, $\varkappa$ und M' unabhängig ist. Man erhält also [Gl. (25.4)] nur dann von Null verschiedene Matrixelemente, wenn

$$M = \varkappa + M' \tag{27.4}$$

erfüllt ist. Außerdem ergibt sich aus der „Vektoraddition" $J = \lambda + J', \dots |\lambda - J'|$ für die Elemente (27.2) die Bedingung

$$\lambda \leq 2J. \tag{27.5}$$

Hiermit tritt eine weitere Bedingung in Konkurrenz, wenn man bedenkt, daß sich die $\psi_{\gamma J}^M$ auf die Eigenfunktionen der Einzelelektronen $R(r_k)\, Y_l^m(\vartheta_k, \varphi_k)$ zurückführen lassen ($l = 2$ in der Eisenreihe, $l = 3$ bei den Seltenen Erden). Die in den Matrixelementen auftretenden Produkte zweier dieser Eigenfunktionen lassen sich nach ersten $2l$ Kugelfunktionen entwickeln, so daß die Matrixelemente von $V_\lambda^\varkappa$ wegen der Orthogonalität nur für

$$\lambda \leq 2l \tag{27.6}$$

---

[1] K. W. H. Stevens: Proc. Phys. Soc. Lond. A **65**, 209 (1952).

von Null verschieden sind. Da sich bei der Spiegelung am Nullpunkt der Faktor $(-1)^{2l+\lambda}=(-1)^{\lambda}$ ergibt, sind nur *gerade* λ sinnvoll

$$\lambda = 2, 4 \qquad \text{(Eisenreihe)}$$

oder

$$\lambda = 2, 4, 6 \qquad \text{(Seltene Erden)}, \qquad (27.7)$$

wobei der Term mit $\lambda = 0$ unterdrückt werden kann, da er nur eine additive Konstante liefert.

Die Symmetrie des betrachteten Kristalls ergibt eine Einschränkung für den oberen Index $\varkappa\,(|\varkappa| \leq \lambda)$. Fordert man nämlich z.B. tetragonale Symmetrie um die z-Achse, d.h. $V(r, \vartheta, \varphi) = V\left(r, \vartheta, \varphi + \dfrac{\pi}{2}\right)$ so kann, da $Y_\lambda^\varkappa \sim e^{i\varkappa\varphi}$ ist, $\varkappa$ nur die Werte 0 und ± 4 annehmen. Allgemein gilt

$$\varkappa \equiv 0 \ (\mathrm{mod}\, p) \qquad (p = \text{Zähligkeit der Kristallachse}). \qquad (27.8)$$

Damit genügt es z.B. für das Potential eines *tetragonalen* Kristallfeldes in der *Eisenreihe* die Glieder

$$V_{\text{tetrag}} = \sum_k [A_2^0\, r^2\, Y_2^0 + A_4^0\, r^4\, Y_4^0 + A_4^4\, r^4\, Y_4^4 + A_4^{-4}\, r^4\, Y_4^{-4}] \qquad (27.9)$$

mit der Realitätsbedingung $A_4^{-4*} = A_4^4$ zu berücksichtigen. $\sum\limits_k$ bedeutet die Summation über die Koordinaten der N Elektronen [Gl. (27.1)]; der Einfachheit halber lassen wir den Index k an den Koordinaten r, ϑ, φ bzw. x, y, z weg. — Die Umrechnung von Gl. (27.9) auf kartesische Koordinaten liefert[1]

$$V_{\text{tetrag.}} = \sum_k [B_2^0(3z^2 - r^2) + B_4^0(35z^4 - 30r^2 z^2 + 3r^4) + B_4^4(x^4 - 6x^2 y^2 + y^4)], \qquad (27.10)$$

wobei für die Koeffizienten der Zusammenhang gilt

$$B_2^0 = \frac{1}{4}\sqrt{\frac{5}{\pi}}\,A_2^0; \qquad B_4^0 = \frac{3}{16\sqrt{\pi}}\,A_4^0; \qquad B_4^4 = \frac{3}{8}\sqrt{\frac{35}{2\pi}}\,A_4^4.$$

Im Fall der *kubischen* Symmetrie wird $A_2^0 = 0$ und $A_4^4 = \sqrt{\tfrac{5}{14}}\,A_4^0$, so daß das Potential in kartesischen Koordinaten nur vierte Potenzen enthält:

$$\begin{aligned}
V_{\text{kub}} &= \sum_k A_4^0\, r^4 \left\{ Y_4^0 + \sqrt{\frac{5}{14}}\,(Y_4^4 + Y_4^{-4}) \right\} \\
&= \sum_k D\left\{ \frac{1}{20}(35z^4 - 30r^2 z^2 + 3r^4) + \frac{1}{8}[(x + iy)^4 + (x - iy)^4] \right\} \\
&= \sum_k D\left(x^4 + y^4 + z^4 - \frac{3}{5}r^4 \right), \qquad \text{mit} \quad D = \frac{15}{4\sqrt{\pi}}\,A_4^0.
\end{aligned} \qquad (27.11)$$

Bezeichnungen. a) Wir verwenden die DIRACsche Schreibweise für eine Eigenfunktion $\psi_n = |n\rangle$ („ket"-Vektor), für $\psi_n^* = \langle n|$ („bra"-Vektor) und für die damit gebildeten Matrixelemente eines Operators $\int \psi_m^*\, A\, \psi_n\, d\tau = \langle m|A|n\rangle$ („bra(c)ket"). Damit nehmen die Matrixelemente von V, gebildet mit den Eigenfunktionen $\psi_{\gamma J}^M = |\gamma J M\rangle$, die Form (27.2) an.

b) Im folgenden unterdrücken wir der Einfachheit halber den [in Gl. (27.2) festen] Index γ. Die Abhängigkeit der Größe $\langle \gamma J M| V |\gamma J M'\rangle$ von γ kommt in den folgenden Ziff. 29 und 30 dadurch zum Ausdruck, daß die Eigenfunktionen $|\gamma J M\rangle$ auf die der Einzelelektronen zurückgeführt werden.

c) Werden in die Matrixelemente spezielle Zahlenwerte etwa für J und M eingesetzt, so schreiben wir z.B. $\langle J = 3,\ M = 2| V | J = 3,\ M' = 1 \rangle$ oder kürzer $\langle J = 3,\ M = 2| V | 3,\ 1 \rangle$. Besteht kein Zweifel darüber, welche Quantenzahlen (z.B. J, M oder L, M_L usw.) gemeint sind, so setzen wir einfach $\langle 3, 2| V | 3, 1 \rangle$.

---

[1] Die 2zähligen Nebenachsen von D_4 (d, f in Fig. 4) ergeben $\mathfrak{Im}(A_4^4) = 0$.

d) Die antimetrischen Elektroneneigenfunktionen eines Atoms werden in geschwungenen Klammer $\{...\}$ geschrieben; z.B. bedeutet $\{3^+, 2^-\}$ den Zustand zweier Elektronen mit $m_l = 3$ und $m_s = +\tfrac{1}{2}$ (oberer Index $+$) und $m_l = 2$, $m_s = -\tfrac{1}{2}$ (obererIndex $-$). $\{3^+, 2^-\}$ ist die SLATER-Determinante

$$\{3^+, 2^-\} = \frac{1}{\sqrt{2!}} \begin{vmatrix} |m_l = 3, m_s = \tfrac{1}{2}\rangle_1 & |m_l = 2, m_s = -\tfrac{1}{2}\rangle_1 \\ |m_l = 3, m_s = \tfrac{1}{2}\rangle_2 & |m_l = 2, m_s = -\tfrac{1}{2}\rangle_2 \end{vmatrix},$$

wobei die Indizes 1 und 2 an $|\rangle$ die Koordinaten der beiden Elektronen bezeichnen.

e) Die in den folgenden Ziffern eingeführten Faktoren α_J, β_J und γ_J sind proportional zu $(\alpha_\lambda)_{\gamma J, \gamma' J'}$

$$(\alpha_\lambda)_{\gamma J, \gamma J} \sim \begin{cases} \overline{r^2} \cdot \alpha_J & \text{für} \quad \lambda = 2 \\ \overline{r^4} \cdot \beta_J & \text{für} \quad \lambda = 4 \\ \overline{r^6} \cdot \gamma_J & \text{für} \quad \lambda = 6, \end{cases}$$

wobei der Index γ wieder unterdrückt wird und ein von den radialen Eigenfunktionen abhängiger Faktor herausgezogen wird [Gl. (29.10)].

28. Operatoräquivalenz. Die Berechnung der Matrixelemente (27.2) gelingt in besonders einfacher Weise, wenn man sich der von K. W. H. STEVENS (l. c.) eingeführten Methode der Operatoräquivalenz bedient. Danach bestehen zwischen den Matrixelementen des Potentials und denen von geeigneten Kombinationen der Drehimpulskomponenten einfache Beziehungen, die man erhält, wenn man z.B. beachtet, daß sich die beiden Sätze von Funktionen $x^2 - y^2$, $3z^2 - r^2$, xy, yz, zx und $J_x^2 - J_y^2$, $3J_z^2 - J(J+1)$, $\tfrac{1}{2}(J_x J_y + J_y J_x)$, $\tfrac{1}{2}(J_y J_z + J_z J_y)$, $\tfrac{1}{2}(J_{zz} + J_{xz})$, welche jeweils die unabhängigen Komponenten eines symmetrischen Tensors bilden, nach der Darstellung $\mathfrak{D}_2$ transformieren. Hieraus folgt, daß innerhalb einer Mannigfaltigkeit, in der J konstant ist, z.B. gilt

$$\left. \begin{aligned} \langle J, J_z | \textstyle\sum (x^2 - y^2) | J, J_z' \rangle &= \overline{r^2}\, \alpha_J \langle J, J_z | J_x^2 - J_y^2 | J, J_z' \rangle, \\ \langle J, J_z | \textstyle\sum (3z^2 - r^2) | J, J_z \rangle &= \overline{r^2}\, \alpha_J (3J_z^2 - J(J+1)), \\ \langle J, J_z | \textstyle\sum xy | J, J_z' \rangle &= \overline{r^2}\, \alpha_J \tfrac{1}{2} \langle J, J_z | J_x J_y + J_y J_x | J, J_z' \rangle, \\ \cdots\cdots\cdots\cdots\cdots\cdots\cdots\cdots\cdots\cdots\cdots\cdots \end{aligned} \right\} \quad (28.1)$$

wobei der Nichtkommutativität der Drehimpulskomponenten Rechnung getragen ist. Es tritt hierbei $\overline{r^2}$ auf, eine Größe, die sich bei der Integration über die Radialabhängigkeit ergibt [vgl. Gl. (29.10)]. Für die übrigen Glieder in Gl. (27.10) erhält man entsprechend

$$\left. \begin{aligned} &\langle J, J_z | \textstyle\sum (x^4 - 6x^2 y^2 + y^4) | J, J_z' \rangle \\ &= \tfrac{1}{2}\overline{r^4}\, \beta_J \langle J, J_z | (J_x + i J_y)^2 + (J_x - i J_y)^2 | J, J_z' \rangle, \quad [J_z = \pm 4 + J_z'; (27.4)] \\ &\text{wegen} \\ &\qquad x^4 - 6x^2 y^2 + y^4 = \tfrac{1}{2}\{(x + i y)^4 + (x - i y)^4\} \end{aligned} \right\} \quad (28.2)$$

und

$$\left. \begin{aligned} &\langle J, J_z | \textstyle\sum (35z^4 - 30 r^2 z^2 + 3 r^4 | J, J_z \rangle \\ &= \overline{r^4}\, \beta_J \{35 J_z^4 - 30 J(J+1) J_z^2 + 25 J_z^2 - 6 J(J+1) + 3 J^2 (J+1)^2\}. \end{aligned} \right\} \quad (28.3)$$

Die praktische Verwendung dieser Formeln werde an zwei Beispielen gezeigt.

29. Größe der Kristallfeldaufspaltung in der Eisenreihe. Betrachten wir die Aufspaltung des Termes $3d^2\,{}^3F$, des Grundzustandes von V^{+++}, in einem Kristallfeld kubischer Symmetrie (vgl. Fig. 5). Wenn wir in den obigen Formeln J durch $L = 3$ und $J_z = M$ durch $M_L = 3, 2, 1, 0, -1, -2, -3$ ersetzen, so lautet nach Gln. (27.2), (27.4) die Säkulardeterminante dieses bezüglich L siebenfachen Terms für die Matrixelemente $\langle LM\,|\,V\,|\,LM'\rangle$ des Potentials (27.11)

$$\begin{vmatrix} |3,3\rangle-W_1 & 0 & 0 & 0 & \langle 3,3\,|V_4^4|\,3,-1\rangle & 0 & 0 \\ 0 & \langle 3,2\,|V_4^0|\,3,2\rangle-W_1 & 0 & 0 & 0 & \langle 3,2\,|V_4^4|\,3,-2\rangle & 0 \\ 0 & 0 & \langle 3,1\,|V_4^0|\,3,1\rangle-W_1 & 0 & 0 & 0 & \langle 3,1\,|V_4^4|\,3,-3\rangle \\ 0 & 0 & 0 & \langle 3,0\,|V_4^0|\,3,0\rangle-W_1 & 0 & 0 & 0 \\ |V_4^{-4}|\,3,3\rangle & 0 & 0 & 0 & \langle 3,-1\,|V_4^0|\,3,-1\rangle-W_1 & 0 & 0 \\ 0 & \langle 3,-2\,|V_4^{-4}|3,2\rangle & 0 & 0 & 0 & \langle 3,-2\,|V_4^0|3,-2\rangle-W_1 & 0 \\ 0 & 0 & \langle 3,-3\,|V_4^{-4}|3,1\rangle & 0 & 0 & 0 & \langle 3,-3\,|V_4^0|3,-3\rangle-W_1 \end{vmatrix} \tag{29.1}$$

Für die Diagonalglieder erhält man aus der Äquivalenz (28.3)

$$\begin{aligned} \langle 3,3\,|V_4^0|3,3\rangle &= \frac{D\,\overline{r^4}}{20}\,\beta_{L=3}\left\{35\cdot 3^4-30\cdot 3\cdot 4\cdot 3^2+25\cdot 3^2-6\cdot 3\cdot 4+3\cdot 3^2\cdot 4^2\right\} \\ &= 9C = \langle 3,-3\,|V_4^0|3,-3\rangle, \\ \langle 3,2\,|V_4^0|3,2\rangle &= -21\,C = \langle 3,-2\,|V_4^0|3,-2\rangle, \\ \langle 3,1\,|V_4^0|3,1\rangle &= 3\,C = \langle 3,-1\,|V_4^0|3,-1\rangle, \\ \langle 3,0\,|V_4^0|3,0\rangle &= 18\,C \quad \text{mit} \quad C = D\,\overline{r^4}\,\beta_{L=3}. \end{aligned} \tag{29.2}$$

Die Berechnung der Nichtdiagonalglieder erfolgt unter Ausnützung der Beziehung

$$(L_x\pm i\,L_y)\,\psi_L^{M_L} = \sqrt{(L\mp M_L)\,(L\pm M_L+1)}\;\psi_L^{M_L\pm 1}, \tag{29.3}$$

also [vgl. Gl. (28.2)]

$$\begin{aligned} \langle 3,3\,|V_4^4|3,-1\rangle &= \langle 3,3\,|\sum\frac{D}{8}(x+i\,y)^4|3,-1\rangle = \frac{D\,\overline{r^4}}{8}\,\beta_{L=3}\langle 3,3\,|(L_x+i\,L_y)^4|3,-1\rangle \\ &= \frac{C}{8}\sqrt{4\cdot 3}\;\sqrt{3\cdot 4}\;\sqrt{2\cdot 5}\;\sqrt{1\cdot 6} = 3\sqrt{15}\;C = \langle 3,1\,|V_4^4|3,-3\rangle \\ &= \langle 3,-1\,|V_4^{-4}|3,3\rangle = \langle 3,-3\,|V_4^{-4}|3,1\rangle, \\ \langle 3,2\,|V_4^4|3,-2\rangle &= 15\,C = \langle 3,-2\,|V_4^{-4}|3,2\rangle. \end{aligned} \tag{29.4}$$

Das Verschwinden der Säkulardeterminante (29.1) ergibt eine Gleichung siebten Grades für die Änderung W_1 des Energieeigenwertes in erster Näherung durch das Kristallfeld

$$[18\,C-W_1]\,[(3\,C-W_1)\,(9\,C-W_1)-135\,C^2]^2\,[(21\,C+W_1)^2-225\,C^2] = 0$$

mit den Lösungen

$$\left.\begin{aligned} W_1 &= 18\,C \quad \text{(dreifach)}, \\ W_1 &= -6\,C \quad \text{(dreifach)}, \\ W_1 &= -36\,C \quad \text{(einfach)}. \end{aligned}\right\} \tag{29.5}$$

Die Anzahl und Vielfachheit der Kristallterme stimmt natürlich mit den gruppentheoretischen Ergebnissen (20.14) überein. Solange der Wert von C nicht bekannt ist, ist also nur die *relative* Größe der Kristallfeldaufspaltungen bestimmt (vgl. Fig. 6).

Mit Hilfe der Operatoräquivalenz läßt sich über C eine weitere Aussage machen. Innerhalb der Mannigfaltigkeit $L=3$ gilt nach Gl. (29.2) für $M_L=3$

$$\langle L=3,\,M_L=3\,|\,V_4^0\,|\,L=3,\,M_L=3\rangle = 9\,C = 9\,D\,\overline{r^4}\,\beta_{L=3}. \tag{29.6}$$

Da zwei $3d$-Elektronen $(l=2)$ vorliegen, läßt sich der Zustand $M_L=3$ (und $M_S=1$) auf die antimetrisierten Elektroneneigenfunktionen $\{2^+, 1^+\}$, d.h. $m_l=2$, $m_s=+\tfrac{1}{2}$ und $m_l=1$, $m_s=+\tfrac{1}{2}$, zurückführen (für den Fall, daß *mehrere* solche Determinantenzustände $\{\ldots\}$ zu einem M_L, M_S-Zustand gehören, vgl. Ziff. 30):

$$\begin{aligned}
\langle L=3, M_L=3 | V_4^0 | L=3, M_L=3 \rangle &= \int \{2^+ 1^+\}^* V_4^0 \{2^+ 1^+\} \, d\tau \\
&= \langle l=2, m_l=2 | V_4^0 | l=2, m_l=2 \rangle + \langle l=2, m_l=1 | V_4^0 | l=2, m_l=1 \rangle,
\end{aligned} \qquad (29.7)$$

wobei die Spinabhängigkeit nicht interessiert, da V nur die Bahnbewegung beeinflußt. Wir wenden auf die beiden letzten Glieder die Operatoräquivalenz (28.3) mit $J \to l=2$ und $J_z \to m_l=2$ bzw. $=1$ an

$$\begin{aligned}
\langle l=2, m_l=2 | V_4^0 | 2, 2 \rangle &= \frac{D \overline{r^4}}{20} \beta_{l=2} \cdot 12, \\
\langle l=2, m_l=1 | V_4^0 | 2, 1 \rangle &= \frac{D \overline{r^4}}{20} \beta_{l=2} \cdot (-48)
\end{aligned} \qquad (29.8)$$

und berechnen etwa das erste explizit zu $[\psi_{l=2}^{m_l=2} = R(r)\, Y_2^2(\vartheta, \varphi)]$

$$\begin{aligned}
\langle l=2, m_l=2 | V_4^0 | 2, 2 \rangle &= A_4^0 \iiint R\, Y_2^{2*}\, r^4\, Y_4^0\, R\, Y_2^2\, r^2\, dr\, \sin\vartheta\, d\vartheta\, d\varphi \\
&= \frac{1}{14\sqrt{\pi}}\, \overline{r^4}\, A_4^0 = \frac{2}{105}\, \overline{r^4}\, D.
\end{aligned} \qquad (29.9)$$

Integrale über drei Kugelfunktionen finden sich bei Bethe (l.c. S. 210); allgemein lassen sie sich durch die Wigner-Koeffizienten[1] darstellen. Während die Größe A_4^0 bzw. D durch die *Stärke* des Kristallfeldes bestimmt ist, geht in den Wert von

$$\overline{r^\lambda} = \int R^2(r)\, r^\lambda r^2\, dr \qquad (29.10)$$

wesentlich die Abschirmung der Kernladung durch die übrigen Elektronen ein, so daß das Integral (29.10) nur näherungsweise mittels Hartree-Funktionen berechnet werden kann. — Aus Gl. (29.6) bis (29.9) folgt

$$\beta_{l=2} = \frac{2}{63}; \qquad \beta_{L=3} = -\frac{2}{315} \qquad (29.11)$$

also

$$C = -\frac{2}{315}\, D\,\overline{r^4}. \qquad (29.12)$$

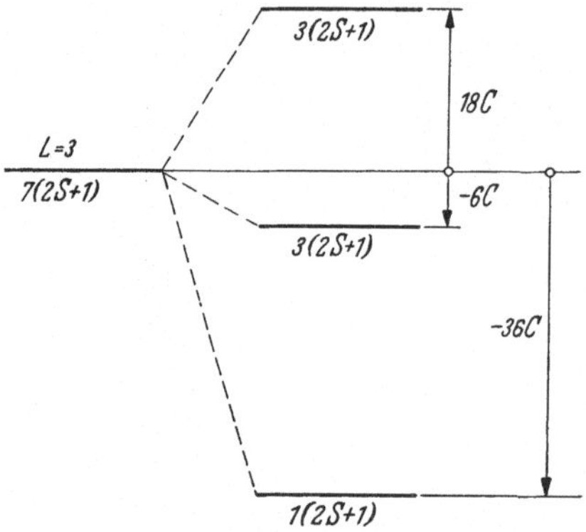

Fig. 6. Die Größe der Aufspaltung des Termes $L=3$ im kubischen Kristallfeld.

In der Eisenreihe tritt zur Wirkung des Kristallfeldes auf das Bahnmoment erst in höherer Ordnung die Störung durch Spin-Bahn- bzw. Spin-Spin-Kopplung (vgl. Ziff. 26). Diese die Spinentartung aufhebende Wechselwirkung wurde von B. Bleaney und K. W. H. Stevens[2] durch Einführung einer Spin-Hamilton-Funktion in ähnlicher Weise mittels Operatoräquivalenzen behandelt. Diese Art von Störung erhält insbesondere bei den *S-Termen* eine besondere Bedeutung (z.B. $^3d^5\,^6S_{\frac{5}{2}}$, der Grundzustand von Mn^{++} und Fe^{+++}). In diesem Fall eines verschwindenden Bahndrehimpulses liefert nämlich das Kristallfeld keine Beeinflussung des Bahnzustandes. Eine Aufspaltung kommt überhaupt erst durch die genannten Spin-Wechselwirkungen zustande.

---

[1] M. Rose: Multipole Fields. New York 1955.
[2] B. Bleaney u. K. W. H. Stevens: Rep. Progr. Phys. **16**, 108 (1953).

30. Die Größe der Kristallfeldaufspaltung bei den Seltenen Erden. Die Berechnung der Kristallfeldaufspaltung bei den Seltenen Erden geht analog zu der bei der Eisenreihe, wenn man bedenkt, daß sich jetzt der Gesamtdrehimpuls $\mathfrak{J}$ zum Kristallfeld einstellt. Wir betrachten mit K. W. H. STEVENS[1] die Aufspaltung des Grundterms von $4f^5\,{}^6H_{\frac{5}{2}}$ ($J=\frac{5}{2}$, $L=5$, $S=\frac{5}{2}$) von Sm^{+++}. Alles wesentliche ist dabei etwa aus der Berechnung der Matrixelemente von

$$V_4^0 = \frac{D}{20} \sum (35 z^4 - 30 r^2 z + 3 r^4)$$

bei Zugrundelegen von L-S-Koppelung ersichtlich. Für $J=\frac{5}{2}$ und $J_z=M_J$ folgt aus der Äquivalenz (28.3)

$$\left.\begin{aligned}
\left\langle J=\frac{5}{2}, M_J\left|\frac{20}{D} V_0^4\right| J=\frac{5}{2}, M_J\right\rangle &= \overline{r^4}\,\beta_{J=\frac{5}{2}} \times \\
\times \left\{35 M_J^4 - 30\cdot\frac{5}{2}\cdot\frac{7}{2} M_J^2 + 25 M_J^2 - 6\cdot\frac{5}{2}\cdot\frac{7}{2} + 3\left(\frac{5}{2}\right)^2\left(\frac{7}{2}\right)^2\right\}&,
\end{aligned}\right\} \quad (30.1)$$

also z.B.

$$\left\langle J=\frac{5}{2}, M_J=\frac{5}{2}\left|\frac{20}{D} V_4^0\right|\frac{5}{2},\frac{5}{2}\right\rangle = \overline{r^4}\,\beta_{J=\frac{5}{2}}\cdot 60. \qquad (30.2)$$

Zur Bestimmung von $\beta_{J=\frac{5}{2}}$ aus dieser Gleichung beachten wir, daß innerhalb $J=\frac{5}{2}$, $L=5$, $S=\frac{5}{2}$ zu $M_J=\frac{5}{2}$ die Zustände

M_L	5	4	3	2	1	0
M_S	$-\frac{5}{2}$	$-\frac{3}{2}$	$-\frac{1}{2}$	$\frac{1}{2}$	$\frac{3}{2}$	$\frac{5}{2}$

gehören, so daß für die Eigenfunktion $\psi_{J=\frac{5}{2}}^{M_J=\frac{5}{2}}\equiv|J=\frac{5}{2}, M_J=\frac{5}{2}\rangle$ geschrieben werden kann

$$\left.\begin{aligned}
|J=\tfrac{5}{2}, M_J=\tfrac{5}{2}\rangle = a\,|M_L=5, M_S=-\tfrac{5}{2}\rangle + b\,|4,-\tfrac{3}{2}\rangle + c\,|3,-\tfrac{1}{2}\rangle + d\,|2,\tfrac{1}{2}\rangle + \\
+ e\,|1,\tfrac{3}{2}\rangle + f\,|0,\tfrac{5}{2}\rangle \qquad\qquad (\text{stets } L=5, S=\tfrac{5}{2})
\end{aligned}\right\} \quad (30.3)$$

mit $a^2 + b^2 + c^2 + d^2 + e^2 + f^2 = 1$. Da das Kristallfeld nicht auf den Spin einwirkt, werde im folgenden die Spinquantenzahl unterdrückt und man erhält für die linke Seite von (30.2) wegen (27.4)

$$\left.\begin{aligned}
&\langle J=\tfrac{5}{2}, M_J=\tfrac{5}{2}|V_4^0|\tfrac{5}{2},\tfrac{5}{2}\rangle \\
&= a^2\langle L=5, M_L=5|V_4^0|5,5\rangle + b^2\langle 5,4|V_4^0|5,4\rangle + \cdots + f^2\langle 5,0|V_4^0|5,0\rangle.
\end{aligned}\right\} \quad (30.4)$$

Auf jedes Glied läßt sich wieder (28.3) anwenden ($J\to L=3$, $J_z\to M_L$),

$$\left\langle L=5, M_L=5\left|\frac{20}{D} V_4^0\right|5,5\right\rangle = \overline{r^4}\,420\cdot 6\cdot\beta_{L=5}\,,\,\cdots \qquad (30.5)$$

so daß man erhält

$$\left\langle J=\frac{5}{2}, M_J=\frac{5}{2}\left|V_4^0\right|\frac{5}{2},\frac{5}{2}\right\rangle = 420\,\overline{r^4}\,\beta_{L=5}\,\frac{D}{20}\,(6a^2 - 6b^2 - 6c^2 - d^2 + 4e^2 + 6f^2). \quad (30.6)$$

Die linke Seite von Gl. (30.5) führt man auf die antimetrisierten Elektroneneigenfunktionen $\{3^-, 2^-, 1^-, 0^-, -1^-\}$ zurück

$$\left.\begin{aligned}
\langle L=5, M_L=5|V_4^0|5,5\rangle &= \int \{3^-,2^-,1^-,0^-,-1^-\}^* V_4^0 \{3^-,2^-,1^-,0^-,-1^-\}\,d\tau \\
&= \langle l=3, m_l=3|V_4^0|3,3\rangle + \langle l=3, m_l=2|V_4^0|3,2\rangle + \cdots + \langle l=3, m_l=-1|V_4^0|3,-1\rangle \\
&= \frac{D}{20}\,\overline{r^4}\,\beta_{l=3}\,(180 - 420 + 60 + 360 + 60) = \frac{D}{20}\,r^4\,\beta_{l=3}\cdot 240,
\end{aligned}\right\} \quad (30.7)$$

---

[1] K. W. H. STEVENS: Proc. Phys. Soc. Lond. A **65**, 209 (1952).

wobei die Äquivalenz (28.3) mit $J \to l = 3, J_z \to m_l$ wieder Verwendung findet. Die Ausrechnung des Integrals

$$\langle l = 3, m_l = 3 \mid V_4^0 \mid 3, 3 \rangle$$
$$= \iiint R^2 Y_3^{3*} V_4^0 Y_3^3 \, r^2 \, dr \, \sin \vartheta \, d\vartheta \, d\varphi = A_4^0 \overline{r^4} \frac{3}{22 \sqrt{\pi}} = D \overline{r^4} \frac{2}{55} \Bigg\} \qquad (30.8)$$

liefert

$$\beta_{l=3} = \frac{2}{5 \cdot 9 \cdot 11} \qquad (30.9)$$

und nach Gl. (30.5) und (30. 7)

$$\beta_{L=5} = \frac{4}{11 \cdot 21 \cdot 45} \qquad (30.10)$$

Um die gesuchte Größe $\beta_{J=\frac{5}{2}}$ angeben zu können, müssen zunächst die noch unbekannten Koeffizienten $a, b, \ldots, f$ bestimmt werden. Dazu bedient man sich nach K. W. H. Stevens des folgenden Kunstgriffes. Betrachtet man die Gleichung $\mathfrak{J}^2 = \mathfrak{L}^2 + \mathfrak{S}^2 + 2\,\mathfrak{L}\,\mathfrak{S}$, so sieht man, daß der Operator $2\,\mathfrak{L}\,\mathfrak{S}$ innerhalb einer Mannigfaltigkeit mit festem $\mathfrak{L}$, $\mathfrak{S}$ und $\mathfrak{J}$ äquivalent zu $J(J+1) - S(S+1) - L(L+1)$ ist, d.h. in unserem Fall $2\,\mathfrak{L}\,\mathfrak{S} = -30$. Bei Anwendung von $2\,\mathfrak{L}\,\mathfrak{S}$ auf Gl. (30.3) erhält man damit

$$-30\,a = \langle M_L = 5, M_S = -\tfrac{5}{2} \mid -30\,(a \mid 5, -\tfrac{5}{2}\rangle + b \mid 4, -\tfrac{3}{2}\rangle + \cdots),$$
$$= \langle M_L = 5, M_S = -\tfrac{5}{2} \mid 2\,\mathfrak{L}\,\mathfrak{S}\,(a \mid 5, -\tfrac{5}{2}\rangle + b \mid 4, -\tfrac{3}{2}\rangle + \cdots), \Bigg\} \qquad (30.11)$$
$$= -25\,a + \sqrt{50}\,b.$$

Die dabei auftretenden Matrixelemente $\langle M_L, M_S \mid 2\,\mathfrak{L}\,\mathfrak{S} \mid M_L', M_S' \rangle$ finden sich z. B. bei Condon und Shortley[1]. Auf diese Weise erhält man genügend Beziehungen für die Konstanten $a, b, \ldots, f$. Es ergibt sich in unserem Fall

$$a = \left(\frac{6}{11}\right)^{\frac{1}{2}}, \; b = -\left(\frac{3}{11}\right)^{\frac{1}{2}}, \; c = \left(\frac{4}{33}\right)^{\frac{1}{2}}, \; d = -\left(\frac{1}{22}\right)^{\frac{1}{2}}, \; e = \left(\frac{1}{77}\right)^{\frac{1}{2}}, \; f = \left(\frac{1}{6 \cdot 77}\right)^{\frac{1}{2}} \quad (30.12)$$

und damit nach Gl. (30.2) und (30.6) für die gesuchte Größe

$$\beta_{J=\frac{5}{2}} = \frac{26}{7 \cdot 33 \cdot 45}. \qquad (30.13)$$

Wenn zu einem M_L, M_S-Zustand *mehrere* Determinantenzustände $\{\cdots\}$ der Einzelelektronen (m_l, m_s) gehören, so hat man ähnlich zu verfahren, wie beim hier durchgeführten Übergang von M_J nach M_L, M_S. Die auftretenden Koeffizienten bestimmt man dabei statt mit $2\,\mathfrak{L}\,\mathfrak{S}$ mittels des Operators $\Sigma(\mathfrak{l}_i \mathfrak{l}_j)$.

Bei den Seltenen Erden muß nach Gl. (27.7) die Entwicklung des Potentials bis $\lambda = 6$ durchgeführt werden; die dabei nötigen Operatoräquivalenzen $(\sim \gamma_J)$ sind bei K. W. H. Stevens[2] angegeben. In unserem speziellen Beispiel $^6H_{\frac{5}{2}}$ verschwindet jedoch entsprechend Gl. (27.5) der dabei auftretende Faktor $\gamma_{J=\frac{5}{2}}$. Für Terme mit $J = 0$ ist natürlich $\alpha_0 = \beta_0 = \gamma_0 = 0$, da diese Terme nicht entartet sind. Dasselbe ist auch bei *S-Zuständen* (z.B. $^8S_{\frac{7}{2}}$, dem Grundzustand von Gd^{+++})

---

[1] E. Condon u. G. Shortley: The Theory of Atomic Spectra, S. 222. Cambridge 1953.
[2] K. W. H. Stevens: Proc. Phys. Soc. Lond. A **65**, 209 (1952). — R. Elliott u. K. W. H. Stevens: Proc. Roy. Soc. Lond., Ser. A **219**, 387 (1953).

der Fall, da das Kristallfeld nur die Bahnbewegung beeinflußt. Für diesen Fall gilt das am Ende von Ziff. 29 Gesagte: Erst durch die Spin-Bahn- bzw. Spin-Spin-Wechselwirkung kommt eine Aufspaltung im Kristallfeld zustande (bei $^8S_{\frac{7}{2}}$ des Gd^{+++} von der Größenordnung 1 cm^{-1}).

Von R. J. ELLIOTT und K. W. H. STEVENS[1] wird der Einfluß des Kernspins auf die Terme und außerdem Abweichungen von der RUSSELL-SAUNDERS-Koppelung diskutiert. In dieser Arbeit finden sich auch Operatoräquivalenzen für Matrixelemente zwischen J und $J+1$, die für eine Störungsrechnung zweiter Ordnung von Bedeutung sind.

IV. Der ZEEMAN-Effekt in Kristallen.

31. Störung durch ein äußeres, homogenes Magnetfeld. Die erste quantenmechanische Behandlung des ZEEMAN-Effektes an Absorptionslinien von Kristallen wurde von H. BETHE[2] gegeben, die eine Deutung des von J. BECQUEREL[3] gefundenen Unterschiedes zwischen dem Effekt von freien Atomen und Ionen und dem an Ionen im Kristallgitter ergab. Diese Abweichungen, die darauf beruhen, daß der Einfluß des elektrischen Kristallfeldes den des äußeren Magnetfeldes im allgemeinen überwiegt, so daß dieses als eine kleine Störung angesehen werden kann, wurden von H. BETHE an Hand des speziellen Beispieles tetragonaler Symmetrie (einschließlich des Grenzfalles kubischer Symmetrie) beschrieben. Von K. H. HELLWEGE[4] wurde die Behandlung auf sämtliche 32 Kristallklassen ausgedehnt.

Die Ausgangsfunktionen für eine Störungsrechnung, bei der die ZEEMAN-Aufspaltung klein gegen die Kristallfeldaufspaltung bleibt, bilden die in (22.2) angegebenen Kristallzustände

$$U = \sum_{\gamma} \sum_{J} \sum_{M} a_{\gamma J}^{M} \psi_{\gamma J}^{M}, \tag{31.1}$$

wobei die Entwicklungskoeffizienten $a_{\gamma J}^{M}$ den durch die Kristallsymmetrie gegebenen Bedingungen genügen. Bei Einschalten des äußeren homogenen Magnetfeldes gehen diese Symmetrieeigenschaften bis auf jene, die dem Kristallfeld und dem Magnetfeld gemeinsam sind, — d.h. höchstens eine Drehachse, die parallel zu $\mathfrak{H}$ liegt — verloren. *Damit spalten sämtliche entarteten Kristallterme im äußeren Magnetfeld in einfache Terme auf.* Neben der sog. Symmetrieentartung wird auch die KRAMERSsche Entartung (Ziff. 20), bedingt durch den rein elektrischen Charakter des Kristallfeldes, aufgehoben. Das Ziel der Untersuchungen von H. BETHE und K. H. HELLWEGE ist die Klärung der Frage, ob diese Aufspaltung in erster Näherung, d.h. proportional zu H (linearer ZEEMAN-Effekt) oder ob sie erst in höherer Näherung, das bedeutet im allgemeinen proportional zu H^2 (quadratischer ZEEMAN-Effekt), stattfindet.

Betrachtet man die Bewegung von Elektronen in einem äußeren Magnetfeld $\mathfrak{H}$, so ergibt sich unter Berücksichtigung des Elektronenspins für die HAMILTON-Funktion das zusätzliche Glied

$$\mu_B \, (\mathfrak{L} + 2\,\mathfrak{S}, \, \mathfrak{H}) \tag{31.2}$$

in dem μ_B das BOHRsche Magneton $= e\hbar/2mc$ und $\mathfrak{L}$ bzw. $\mathfrak{S}$ den in Einheiten von $\hbar$ gemessenen gesamten Bahn- bzw. Spindrehimpuls bedeuten. (Vgl. den Artikel über den ZEEMAN-Effekt in diesem Bande.) Diese Störung (31.2) liefert

---

[1] R. J. ELLIOTT u. K. W. H. STEVENS: Proc. Roy. Soc. Lond., Ser. A **218**, 553 (1953).
[2] H. BETHE: Z. Physik. **60**, 218 (1930).
[3] J. BECQUEREL: Z. Physik **58**, 205 (1929).
[4] K. HELLWEGE: Z. Physik **127**. 513, (1950); **128**, 172, (1950).

die Energieänderung der Kristallterme im Magnetfeld. Wir beschränken uns im folgenden auf den Fall der Seltenen Erden, in welchem sich für $\mathfrak{H}=0$ der Gesamtdrehimpuls $\mathfrak{J}=\mathfrak{L}+\mathfrak{S}$ relativ zum Kristallfeld einstellt.

32. Magnetfeld parallel zur p-zähligen Drehachse. Zunächst besprechen wir den Fall, daß das Magnetfeld parallel zur p-zähligen Drehachse (Kristallklasse C_p) angelegt wird, die zur z-Achse gemacht werde ($H_x = H_y = 0$, $H_z = H$). Da dann diese Hauptachse als Symmetrieelement ebenfalls im Magnetfeld erhalten bleibt, erfolgt auch im Magnetfeld die Summation über M gerade über jene Werte, für die gilt [vgl. Gl. (22.5)]

$$M = \mu + f\,p \qquad (f = 0,\, \pm 1,\, \pm 2,\, \ldots)\,. \tag{32.1}$$

Die Energiestörung erster Näherung beträgt, da vom Operator

$$\mathfrak{T} = \mathfrak{L} + 2\,\mathfrak{S} \tag{32.2}$$

nur die z-Komponente eingeht,

$$W_{\parallel}^{\mu} = \mu_B H(U^{\mu}, T_z U^{\mu}) = \mu_B H \sum_{\gamma, J, f} \sum_{\gamma', J', f'} a_{\gamma J}^{M*}\, a_{\gamma' J'}^{M'} \langle \gamma\, J\, M \,|\, T_z \,|\, \gamma'\, J'\, M' \rangle\,. \tag{32.3}$$

Zur Ermittlung der Matrixelemente $\langle \gamma\, J\, M \,|\, T_z \,|\, \gamma'\, J'\, M' \rangle = \int \psi_{\gamma J}^{M*}\, T_z\, \psi_{\gamma' J'}^{M'}\, d\tau$ benützen wir das folgende *Theorem*[1]: Wenn ein Vektor-Operator $\mathfrak{T}$ bezüglich des Drehimpulses $\mathfrak{J}$ (in der Einheit $\hbar$) die Vertauschungsrelationen

$$J_x T_x - T_x J_x = 0, \qquad J_x T_y - T_y J_x = T_x J_y - J_y T_x = i T_z \tag{32.4}$$

$$\cdots \cdots \cdots \cdots \cdots \cdots \cdots \cdots \cdots$$

erfüllt, so lauten die nichtverschwindenden Matrixelemente von T_x

$$\left.\begin{aligned}
\langle \gamma\, J\, M \,|\, T_x \,|\, \gamma',\, J+1,\, M\pm 1 \rangle &= \mp \tfrac{1}{2}\, T_{\gamma J, \gamma' J+1} \sqrt{(J\pm M+1)\,(J\pm M+2)}\,,\\
\langle \gamma\, J\, M \,|\, T_x \,|\, \gamma',\, J,\, M\pm 1 \rangle &= \ \tfrac{1}{2}\, T_{\gamma J, \gamma' J} \sqrt{(J\mp M)\,(J\pm M+1)}\,,\\
\langle \gamma\, J\, M \,|\, T_x \,|\, \gamma',\, J-1,\, M\pm 1 \rangle &= \pm\tfrac{1}{2}\, T_{\gamma J, \gamma' J-1} \sqrt{(J\mp M)\,(J\mp M-1)}\,,
\end{aligned}\right\} \tag{32.5}$$

und jene von T_z

$$\left.\begin{aligned}
\langle \gamma\, J\, M \,|\, T_z \,|\, \gamma',\, J+1,\, M \rangle &= \ T_{\gamma J, \gamma' J+1} \sqrt{(J+1)^2 - M^2}\,,\\
\langle \gamma\, J\, M \,|\, T_z \,|\, \gamma',\, J,\, M \rangle &= \ T_{\gamma J, \gamma' J}\, M\,,\\
\langle \gamma\, J\, M \,|\, T_z \,|\, \gamma',\, J-1,\, M \rangle &= \ T_{\gamma J, \gamma' J-1} \sqrt{J^2 - M^2}\,,
\end{aligned}\right\} \tag{32.6}$$

wobei die Größen $T_{\gamma J, \gamma' J'}$ nicht mehr von M, wohl aber vom Kopplungstypus abhängen. Da die Voraussetzungen für dieses Theorem durch $\mathfrak{T} = \mathfrak{L} + 2\,\mathfrak{S}$ erfüllt sind und in diesem Fall außerdem noch

$$T_{\gamma J, \gamma' J'} = \delta_{\gamma \gamma'}\, T_{J J'} = \delta_{\gamma \gamma'}\, T_{J' J} \tag{32.7}$$

gilt[1], erhält man für die Energiestörung durch das zur Kristallachse parallele Magnetfeld

$$\left.\begin{aligned}
W_{\parallel}^{\mu} = \mu_B H \sum_{\gamma, J, f} \Big\{ a_{\gamma J}^{M*}\, a_{\gamma, J-1}^{M} \sqrt{J^2 - M^2}\, T_{J, J-1} + |a_{\gamma J}^{M}|^2\, M\, T_{J J} + \\
+\, a_{\gamma J}^{M*}\, a_{\gamma J+1}^{M} \sqrt{(J+1)^2 - M^2}\, T_{J, J+1} \Big\}\,.
\end{aligned}\right\} \tag{32.8}$$

---

[1] E. Condon u. G. Shortley: The Theory of Atomic Spectra, S. 59ff. Cambridge 1953.

Bei rein elektrischem Kristallfeld gehört zum Kristallzustand U ein KRAMERSscher Zustand [vgl. Gl. (22.31)]

$$\widetilde{U} = \sum_{\gamma, J, f} (-1)^{J+M-\Sigma l_k} a_{\gamma J}^{M*} \psi_{\gamma J}^{-M}, \tag{32.9}$$

für dessen Aufspaltung im Magnetfeld sich ergibt

$$\widetilde{W}_\| = \mu_B H (\widetilde{U}, T_z \widetilde{U}) = - W_\|. \tag{32.10}$$

Besteht für den Kristallterm eine KRAMERSsche Entartung, so spalten also die beiden Zustände im Magnetfeld symmetrisch und linear mit H auf, falls $W \neq 0$ ist. Bei einem *einfachen* Kristallterm, bei dem $\widetilde{U}$ mit U bis auf eine Phase übereinstimmt, wird $\widetilde{W} = W = - W = 0$, so daß kein linearer ZEEMAN-Effekt auftreten kann; eine Termverschiebung ist erst in höherer Näherung möglich. Dieser Fall tritt nur bei gerader Elektronenzahl bei den Termen $\mu = 0$ oder $p/2$ auf.

Ist der Einfluß des Kristallfeldes als klein gegenüber den Multiplettabständen anzusehen, so wird es näherungsweise erlaubt sein, an Stelle der wahren Kristallzustände [Gl. (31.1)] die Kristallzustände nullter Näherung [Gl. (22.3) und (22.6)]

$$w^\mu = \sum_f a^M \psi^M, \tag{32.11}$$

zu benützen (wobei wir die *festen* Indices γ und J wieder weglassen) und mit diesen die Termänderung im Magnetfeld

$$W_\|^\mu = \mu_B H(w^\mu, T_z w^\mu) \tag{32.12}$$

zu bestimmen. An Stelle von Gl. (32.8) erhält man damit

$$W_\|^\mu = \mu_B H \, T_{JJ} \sum_f M \, |a^M|^2 = \mu_B H \, g \sum_f M \, |a^M|^2. \tag{32.13}$$

Beschreibt man diesen linearen ZEEMAN-Effekt eines Magnetfeldes parallel zur kristallographischen Achse mit einem *Aufspaltungsfaktor* $\gamma_\|$[1]

$$W_\|^\mu = \mu_B \gamma_\| H, \tag{32.14}$$

so folgt

$$\gamma_\| = g \sum_f M \, |a^M|^2. \tag{32.15}$$

Im Fall reiner RUSSELL-SAUNDERS-Koppelung ergibt sich der LANDÉsche g-Faktor in der üblichen Weise zu

$$T_{JJ} \equiv g = 1 + \frac{J(J+1) - L(L+1) + S(S+1)}{2J(J+1)}, \tag{32.16}$$

während für $T_{J, J+1}$ gilt[2]

$$T_{J,J+1} = \sqrt{\frac{(J+L+S+2)(-J+S+L)(J+S-L+1)(J+L-S+1)}{4(J+1)^2(2J+1)(2J+3)}}. \tag{32.17}$$

33. Magnetfeld senkrecht zur p-zähligen Drehachse. Legt man das äußere Magnetfeld senkrecht zur p-zähligen Drehachse ($p > 1$, da für $p = 1$ jede Richtung

---

[1] Die anisotropen Aufspaltungsfaktoren $\gamma_\|, \gamma_\perp$ und γ_{ik} (vgl. Ziff. 36) sind nicht mit der Größe γ zu verwechseln, die die Eigenwerte aller mit $\mathfrak{J}^2$ und J_z vertauschbaren Operatoren symbolisiert.

[2] J. van VLECK: Electric and Magnetic Susceptibilities, S. 167. Oxford 1932.

zur Drehachse gemacht werden kann), so zerstört das Magnetfeld die p-zählige Symmetrie, so daß nur Linearkombinationen der Zustände (31.1) und (32.9)

$$\left.\begin{aligned} V &= \alpha\,U + \beta\,\tilde{U}, \\ \tilde{V} &= -\beta^*\,U + \alpha^*\,\tilde{U} \quad \text{mit} \quad \alpha\,\alpha^* + \beta\,\beta^* = 1, \end{aligned}\right\} \tag{33.1}$$

als Ausgangsfunktionen zur Bestimmung der Energieänderungen der Terme im Magnetfeld zu verwenden sind. Legt man das $\mathfrak{H}$-Feld etwa in die x-Richtung $(H_x = H,\ H_y = H_z = 0)$, so ist

$$\left.\begin{aligned} W_\perp^\mu &= \mu_B\,H\,(V^\mu,\,T_x\,V^\mu) \\ &= \mu_B\,H\,\{|\alpha|^2(U^\mu,\,T_x\,U^\mu) + |\beta|^2(\tilde{U}^\mu,\,T_x\,\tilde{U}^\mu) + \alpha^*\beta\,(U^\mu,\,T_x\,\tilde{U}^\mu) + \alpha\,\beta^*\,(\tilde{U}^\mu,\,T_x\,U^\mu)\}. \end{aligned}\right\} \tag{33.2}$$

Durch Einsetzen von Gln. (31.1) und (32.9) erhält man für die einzelnen Glieder wie in (32.3) Summen über die Matrixelemente

$$\langle\gamma J M\,|\,T_x\,|\,\gamma' J' M'\rangle, \quad \langle\gamma,\,J,\,-M\,|\,T_x\,|\,\gamma',\,J',\,-M'\rangle, \quad \langle\gamma J M\,|\,T_x\,|\,\gamma',\,J',\,-M'\rangle$$

und

$$\langle\gamma,\,J,\,-M\,|\,T_x\,|\,\gamma' J' M'\rangle.$$

Diese Matrixelemente sind nach Gl. (32.5) nur dann von Null verschieden, wenn sich die vorderen und hinteren magnetischen Quantenzahlen um ± 1 unterscheiden. Für $\langle\gamma J M\,|\,T_x\,|\,\gamma' J' M'\rangle$ ergibt sich hieraus die Forderung $M' = \mu + f'p = M \pm 1 = \mu + fp \pm 1$ oder $(f' - f)\,p = \pm 1$, was für $p > 1$ mit ganzzahligen $f,\,f'$ nicht erfüllbar ist. Da dieser Schluß genauso für $\langle\gamma,\,J,\,-M\,|\,T_x\,|\,\gamma',\,J',\,-M'\rangle$ durchführbar ist, verschwinden also in Gl. (33.2) die Terme bei $|\alpha|^2$ und $|\beta|^2$ vollständig. Die verbleibenden Glieder sind konjugiert komplex

$$(U^\mu,\,T_x\,\tilde{U}^\mu) = (\tilde{U}^\mu,\,T_x\,U^\mu)^*, \tag{33.3}$$

so daß es genügt, allein etwa $\langle\gamma,\,J,\,M\,|\,T_x\,|\,\gamma,\,J',\,-M'\rangle$ zu betrachten. Damit in erster Näherung überhaupt ein Effekt eintreten kann $(W_\perp^\mu \neq 0)$, muß die Bedingung $-M' = +M \pm 1$, d.h.

$$(f' + f)\,p = -2\mu \mp 1 \tag{33.4}$$

wenigstens für ein Wertetripel $\mu,\,f,\,f'$ erfüllt sein.

Die Diskussion der Gl. (33.4) für ganz- bzw. halbzahliges μ (gerade oder ungerade Elektronenzahl N) ergibt, daß nur in den folgenden Fällen ein linearer Zeeman-Effekt bei einem Magnetfeld senkrecht zur p-zähligen Achse möglich ist:

$$\begin{array}{c|c|c} N & p & \mu \\ \hline \text{gerade} & 3 & \pm 1 \\ \hline & 2,\,3 & \pm \dfrac{1}{2} \\ \text{ungerade} & & \\ & 4,\,6 & \pm \dfrac{1}{2},\ \pm \dfrac{p-1}{2} \end{array} \tag{33.5}$$

Die sich dabei ergebenden Matrixelemente sind nach Gl. (32.5) nur für solche $J,\,J'$-Werte von Null verschieden, für die $J' = J,\,J \pm 1$ erfüllt ist. Geht man näherungsweise wieder, wie am Ende von Ziff. 32, von den Kristalleigenfunktionen nullter Näherung [Gl. (32.11)] aus, so treten in $W_\perp^\mu$ nur solche Matrixelemente auf, für die J fest bleibt $(J' = J)$, also nur jene, die in Gl. (32.5) in der zweiten Zeile stehen.

Für die Energieänderung $\tilde{W}$ ergibt sich

$$\tilde{W}_\perp = \mu_B\,H\,(\tilde{V},\,T_x\,\tilde{V}) = -W_\perp, \tag{33.6}$$

d.h. die lineare Aufspaltung der Terme (33.5) erfolgt symmetrisch zur Nullage. Alle übrigen Terme spalten in erster Näherung nicht auf und bleiben konstant $(\widetilde{W} = W = -W = 0)$.

34. Der ZEEMAN-Effekt in Kristallklassen mit verschiedenen Symmetrieelementen. Bei den nicht-kubischen Kristallklassen, die neben der p-zähligen senkrechten Hauptachse noch zweizählige horizontale *Nebenachsen* (D_p, D_{3d}, D_{ph}) oder *vertikale Spiegelebenen* (C_{pv}) enthalten, besitzen einige Terme aus Symmetriegründen eine zweifache Entartung (vgl. Tabelle 19.1). Da dieselben Terme auch schon auf Grund der reinen KRAMERS-Entartung zweifach sind, ergibt sich für den ZEEMAN-Effekt von Kristallen dieser Symmetrie gegenüber dem bisherigen nichts Neues.

Wird in einem Kristall, der als Symmetrieelement eine *Drehinversionsachse* besitzt, das Magnetfeld *parallel* zur Achse gelegt, so geht die Inversion als Symmetrieelement verloren, und es resultiert im Magnetfeld als Symmetrieelement eine reine Drehachse, deren Zähligkeit unter Umständen geringer ist als die der Drehinversionsachse:

$$\mathfrak{H} = 0 \quad \begin{array}{c|ccccccc} & C_i & C_s & C_{3i} & S_4 & D_{2d} & C_{3h} & D_{3h} \\ p = & 1 & 2 & 3 & 4 & 4 & 6 & 6 \\ \hline \mathfrak{H} \neq 0 & C_1 & C_1 & C_3 & C_2 & C_2 & C_3 & C_3 \end{array} \tag{34.1}$$

Mit zunehmendem Magnetfeld tauchen also z.B. bei S_4 solche $\psi_{\gamma J}^{M}$ auf, für die $M = \mu + 2f$ gilt. Für die Energiestörung ergibt sich dasselbe Ergebnis wie in Ziff. 32.

In ähnlicher Weise wie in Ziff. 33 kann für ein Magnetfeld *senkrecht* zur p-zähligen Drehinversionsachse ein linearer ZEEMAN-Effekt nur in den folgenden Fällen auftreten

N	$\sum l_k$	p	μ_I
gerade	gerade	3	± 1
ungerade	ungerade		
gerade	ungerade	3	$\pm \dfrac{1}{2}$
ungerade	gerade		
ungerade	beliebig	2, 4, 6	$\pm \dfrac{1}{2},\ \pm \dfrac{p-1}{2}$

$$\tag{34.2}$$

In *kubischen* Kristallen ist es für die Entscheidung, wann ein linearer ZEEMAN-Effekt auftreten kann, gleichgültig in welcher Richtung das Magnetfeld angelegt wird, da es stets Parallelkomponenten zu den drei- bzw. vierzähligen Achsen besitzt. Man findet

Kristall-klasse	T, T_h		O, O_h, T_d [1]	
$N =$	gerade	ungerade	gerade	ungerade
$\mu =$	$0\ \{\pm 1\}\ \{0, \pm 1\}$	$\{\pm \tfrac{1}{2}\}\ \{\pm \tfrac{1}{2}, \pm \tfrac{3}{2}\}$	$0\ \ 2\ \ \{0, 2\}\ \{0, \pm 1\}\ \{2, \pm 1\}$	$\{\pm \tfrac{1}{2}\}\ \{\pm \tfrac{3}{2}\}\ \{\pm \tfrac{1}{2}, \pm \tfrac{3}{2}\}$
Linearer ZEEMAN-Effekt für $\mu =$	$-\quad \pm 1 \qquad \pm 1$	$\pm \tfrac{1}{2}\quad \pm \tfrac{1}{2}, \pm \tfrac{3}{2}$	$-\ -\ -\ \quad \pm 1 \qquad \pm 1$	$\pm \tfrac{1}{2}\quad \pm \tfrac{3}{2}\quad \pm \tfrac{1}{2}, \pm \tfrac{3}{2}$

$$\tag{34.3}$$

[1] Da T_d vierzählige *Inversions*achsen besitzt, ist μ durch μ_I zu ersetzen.

35. Auswertung von Zeeman-Effekt-Messungen. Die Frage, ob die durch das Magnetfeld bedingte Änderung der Energie des Kristallterms positiv oder negativ ist, d.h. welcher der Terme $+\mu$ oder $-\mu$ bei der Aufspaltung nach oben und welcher nach unten geht, hängt nach Gl. (32.13) ganz vom Gewicht a^M der ψ^M ab, die diese in u^μ besitzen. Aus dem *experimentell* bestimmten Verhalten der Kristallspektren im Magnetfeld lassen sich diese und zahlreiche andere Fragen, die durch die Theorie bis jetzt noch gar nicht oder nur sehr unsicher beantwortet werden können, empirisch erledigen, wie von J. Brochard und K. H. Hellwege[1] an Hand des PrMg-Nitrat gezeigt wurde.

Bei Anwesenheit eines Magnetfeldes *parallel* zur p-zähligen kristallographischen Hauptdrehachse bleiben die Auswahlregeln für die elektrische Dipolstrahlung (23.3)

$$\Delta\mu \equiv 0 \quad (\mathrm{mod}\, p) \quad \text{für } \pi\text{-Komponenten,} \left.\begin{array}{}\\\\\end{array}\right\}$$
$$\Delta\mu \equiv \pm 1 \quad (\mathrm{mod}\, p) \quad \text{für } \sigma\text{-Komponenten} \qquad (35.1)$$

erhalten, da die p-zählige Symmetrie dabei nicht zerstört wird. Die Auswahlregeln, die sich jedoch z.B. aus der Existenz von Spiegelebenen für S ableiten, werden beim Anwachsen dieses Magnetfeldes immer mehr aufgehoben. Aus der experimentell beobachteten, verschieden starken Aufspaltung der π- und σ-Komponenten zwischen den einzelnen Termen sind Schlüsse auf die Lage der Terme $+\mu$ und $-\mu$ möglich.

Die *Größe* der Zeeman-Aufspaltung eines zweifachen Kristallterms läßt sich in der Form

$$\tilde{\nu}_+ = \frac{W_+}{h\,c} = \gamma' H + \delta H^2,$$
$$\tilde{\nu}_- = \frac{W_-}{h\,c} = -\gamma' H + \delta H^2 \qquad \left(\gamma' = \frac{\mu_B}{h\,c}\gamma_{\parallel}\right) \qquad (35.2)$$

als Summe eines (symmetrischen) linearen Effektes und eines quadratischen Effektes schreiben. Für den Übergang zwischen zwei Termen (0) und (1) erhält man hieraus vier Gleichungen für die Wellenzahldifferenzen

$$\tilde{\nu}_{+1} - \tilde{\nu}_{+0} = (\gamma_1' - \gamma_0') H + (\delta_1 - \delta_0) H^2,$$
$$\tilde{\nu}_{+1} - \tilde{\nu}_{-0} = (\gamma_1' + \gamma_0') H + (\delta_1 - \delta_0) H^2,$$
$$\tilde{\nu}_{-1} - \tilde{\nu}_{+0} = -(\gamma_1' + \gamma_0') H + (\delta_1 - \delta_0) H^2,$$
$$\tilde{\nu}_{-1} - \tilde{\nu}_{-0} = -(\gamma_1' - \gamma_0') H + (\delta_1 - \delta_0) H^2, \qquad (35.3)$$

aus denen sich die unbekannten Größen γ_0', γ_1', $\delta_1 - \delta_0$ berechnen lassen, wenn man die experimentell gemessene Aufspaltung der Spektrallinie im Magnetfeld einsetzt. Für ein Zeeman-Triplett wird $\gamma_1' = \gamma_0'$, für ein Dublett $\gamma_0' = 0$, und schließlich für ein Singulett $\gamma_1' = \gamma_0' = 0$.

Aus dem so bestimmten Wert von γ' eines Kristallterms läßt sich nach Gl. (32.15) sein g-Faktor bestimmen

$$\frac{\gamma' h\,c}{\mu_B} = \gamma_{\parallel} = g\sum_f |a^M|^2 \cdot M \qquad (M = \mu + f p), \qquad (35.4)$$

wenn die rechts stehende Summe bekannt ist. J. Brochard und K. H. Hellwege finden z.B. auf diese Weise bei Pr-Mg-Nitrat für den Term $^3P_1$, der im Kristallfeld der Symmetrie C_{3v} in den einfachen Term $\mu = 0$ (mit quadratischem

---

[1] J. Brochard u. K. H. Hellwege: Z. Physik **135**, 620 (1953).

ZEEMAN-Effekt) und den zweifachen Term $\mu = \{\pm 1\}$ (mit symmetrischem, linearen ZEEMAN-Effekt) aufspaltet (Fig. 19), den Wert $g = 1,17$. (In diesem Fall ist die rechts stehende Summe gleich Eins.) Dieser Wert spricht für eine Abweichung von der reinen RUSSELL-SAUNDERS-Koppelung [Gl. (32.16)], nach der sich $g = 1,5$ ergeben müßte.

Mit Hilfe der experimentellen Werte von γ' sind die genannten Autoren auch in der Lage gewisse *Matrixelemente* des Kristallfeldes empirisch zu bestimmen.

36. Magnetfeld in beliebiger Richtung. Es werde nach der Aufspaltung etwa eines nach KRAMERS zweifachen Termes gefragt, wenn ein äußeres Magnetfeld $\mathfrak{H}$ in *irgendeiner* Richtung zum Kristall angelegt wird. Für die paramagnetische Resonanz ist dieses Problem besonders für den Grundzustand von Bedeutung. Nach R. J. ELLIOTT und K. W. H. STEVENS[1] kann man die Störung $\mu_B (\mathfrak{L} + 2\,\mathfrak{S}, \mathfrak{H})$ durch den Ausdruck

$$2\mu_B \sum_{i,\,k = x,\,y,\,z} \gamma_{ik}\, S_i'\, H_k \tag{36.1}$$

ersetzen, in welchem γ_{ik} ein effektiver, anisotroper Aufspaltungsfaktor (Tensor) ist und $\mathfrak{S}' = S_x',\, S_y',\, S_z'$ aus den 2×2-reihigen PAULISchen Spinmatrizen (20.4) für Spin $\frac{1}{2}$ besteht, wodurch man die Aufspaltung des *zwei*fachen Termes erfaßt. Die Wirkung dieses Spins kann nach (36.1) als eine Präzession um ein effektives Magnetfeld $\mathfrak{H}'$ mit den Komponenten $H_i' = \sum_k \gamma_{ik} H_k$ angesehen werden. Die Eigenwerte der Matrix

$$2\mu_B\, \mathfrak{S}'\, \mathfrak{H}' = \mu_B \begin{pmatrix} H_z' & H_x' - i H_y' \\ H_x' + i H_y' & -H_z' \end{pmatrix} \tag{36.2}$$

sind $W = \pm \mu_B H'$. Ist die Symmetrie des Kristallfeldes höher als rhombisch, so liegt die ausgezeichnete Hauptachse des rotationssymmetrischen Tensorellipsoids parallel zur Kristallachse (z-Achse)

$$H_x' = \gamma_\perp H_x, \qquad H_y' = \gamma_\perp H_y, \qquad H_z' = \gamma_\parallel H_z. \tag{36.3}$$

Für die Energieänderung eines Termes durch ein Magnetfeld $\mathfrak{H}$, das mit der z-Achse den Winkel ϑ einschließt, gilt daher

$$W = \pm \mu_B\, \gamma(\vartheta) \cdot H \tag{36.4}$$

mit

$$\gamma^2(\vartheta) = \gamma_\parallel^2 \cos^2 \vartheta + \gamma_\perp^2 \sin^2 \vartheta. \tag{36.5}$$

Die Größen $\gamma_\parallel$ bzw. $\gamma_\perp$ messen also die Aufspaltung durch ein Magnetfeld parallel bzw. senkrecht zur Kristallachse (Ziff. 32 und 33).

37. Der Übergang zu einem starken Magnetfeld. Die bisher gemachten Aussagen über die Abhängigkeit des ZEEMAN-Effektes von der Feldstärke sind nur für solche Magnetfelder gültig, die eine gegen die Kristallfeldaufspaltung kleine ZEEMAN-Aufspaltung liefern. Ist jedoch umgekehrt das Magnetfeld groß gegenüber dem Kristallfeld, so erhält man den ZEEMAN-Effekt des freien Ions. Nach Rechnungen von KITTEL und LUTTINGER[2] erhält man z.B. für den Grundzustand $^8S_{\frac{7}{2}}$ von Gd^{+++}, der im kubischen Kristallfeld in die beiden zweifachen Terme $^2\Gamma_6$ ($\mu = \pm\frac{1}{2}$), $^2\Gamma_7$ ($\mu = \pm\frac{3}{2}$) und den vierfachen Term $^4\Gamma_8$ ($\mu = \pm\frac{1}{2},\, \pm\frac{3}{2}$) (Termabstände $^4\Gamma_8 - {}^2\Gamma_7 = 5\,C$ und $^2\Gamma_6 - {}^2\Gamma_7 = 8\,C$) zerfällt, in einem Magnetfeld $\mathfrak{H}$ der

---

[1] R. J. ELLIOTT u. K. W. H. STEVENS: Proc. Roy. Soc. Lond., Ser. A **218**, 533 (1953).
[2] C. KITTEL u. J. LUTTINGER: Phys. Rev. **73**, 162 (1948).

[001]-Richtung die Aufspaltung in acht Zeeman-Terme

$$
\begin{aligned}
\eta_{1,2} &= \pm\tfrac{3}{2}\xi + \tfrac{13}{2} \pm \tfrac{1}{2}\sqrt{\left(3\mp\tfrac{2}{3}\xi\right)^2 + \tfrac{140}{9}\xi^2} \rightarrow \begin{cases} 8+\tfrac{25}{8}\xi \text{ aus } {}^2\Gamma_6' \\ 5-\tfrac{29}{8}\xi \text{ aus } {}^4\Gamma_8' \end{cases} \rightarrow \pm\tfrac{7}{2}\xi \\[2mm]
\eta_{3,4} &= \pm\tfrac{1}{2}\xi + \tfrac{5}{2} \pm \tfrac{1}{2}\sqrt{(5\mp 2\xi)^2 + 12\xi^2} \rightarrow \begin{cases} 5+\tfrac{3}{10}\xi \text{ aus } {}^4\Gamma_8 \\ 0-\tfrac{7}{2}\xi \text{ aus } {}^2\Gamma_7 \end{cases} \rightarrow \pm\tfrac{5}{2}\xi \\[2mm]
\eta_{5,6} &= \mp\tfrac{1}{2}\xi + \tfrac{5}{2} \pm \tfrac{1}{2}\sqrt{(5\pm 2\xi)^2 + 12\xi^2} \rightarrow \begin{cases} 5-\tfrac{3}{10}\xi \text{ aus } {}^4\Gamma_8 \\ 0+\tfrac{7}{2}\xi \text{ aus } {}^2\Gamma_7 \end{cases} \rightarrow \pm\tfrac{3}{2}\xi \\[2mm]
\eta_{7,8} &= \mp\tfrac{3}{2}\xi + \tfrac{13}{2} \pm \tfrac{1}{2}\sqrt{\left(3\pm\tfrac{2}{3}\xi\right)^2 + \tfrac{140}{9}\xi^2} \rightarrow \begin{cases} 8-\tfrac{25}{8}\xi \text{ aus } {}^2\Gamma_6 \\ 5+\tfrac{29}{8}\xi \text{ aus } {}^4\Gamma_8 \end{cases} \rightarrow \pm\tfrac{1}{2}\xi,
\end{aligned} \tag{37.1}
$$

wobei $\eta = \dfrac{W}{C}$ und $\xi = \dfrac{g\,\mu_B\,H}{C}$ (wegen $L=0$ ist $g=2$) bedeutet. (Der Kristallterm $^2\Gamma_7$ wurde zum Nullpunkt gemacht.) Die angegebene Entwicklung für kleine ξ zeigt den nach (34.3) zu erwartenden linearen Zeeman-Effekt symmetrisch zu den Kristalltermen. Bei einem Magnetfeld hingegen, das groß gegen die Kristallfeldaufspaltung ist ($\xi \gg 1$), erhält man den Zeeman-Effekt des freien Ions ($M = \pm\tfrac{7}{2}, \pm\tfrac{5}{2}, \pm\tfrac{3}{2}, \pm\tfrac{1}{2}$), der bei Gd^{+++} wegen der Kleinheit der Kristallfeldaufspaltung ($\approx 1\ \mathrm{cm}^{-1}$) beobachtet wird (Ziff. 47 ε).

C. Weitere Erscheinungen an Kristallspektren.

I. Überlagerung von Gitterschwingungsfrequenzen.

38. Deutung der überzähligen Linien. Beim ersten Versuch einer Analyse der Absorptionslinien von Salzen der Seltenen Erden schreckt die große Zahl von Linien, die bei dicken durchstrahlten Schichten immer zahlreicher werden (vgl. Fig. 8, S. 261). Diese Zahl ist viel größer als man irgend durch Kristallfeldaufspaltung der freien Ionenterme erwarten kann. Auch die Linienzahl im Spektrum der viel untersuchten Chromalaune ist für die einfache Erklärung durch Kristallfeldaufspaltung viel zu groß. Hier treten aber Folgen schwacher Linien auf, die mit ihren oft gleichen Abständen von vornherein an Banden zweiatomiger Molekeln erinnern. So fand Sauer[1] im Chromselenat-Alaun eine äquidistante Linienfolge mit 42 cm⁻¹, die als die Überlagerung einer Gitterschwingung angesprochen wurde. Die dazugehörige Infrarotwellenlänge ist 232 μ. Das Auftreten von Elektronensprung-Schwingungskombinationen erscheint in Analogie zu den Molekeln bei den Chromsalzen, bei denen die verantwortlichen Elektronen in unmittelbarem Kontakt mit den andern Gitterbausteinen stehen, nicht besonders verwunderlich. Überraschender ist das Bild bei den Salzen der Seltenen Erden: Joos und Ewald[2] fanden zuerst beim $Nd(NO_3)_3$ Wiederholungen charakteristischer Liniengruppen mit Schwingungszahlendifferenzen, die den bekannten Eigenschwingungen des NO_3^--Ions entsprechen, dann auch solche, die den Wasserschwingungen zugehören und außerdem noch kleinere $\Delta\nu$, die als eigentliche („äußere", im Gegensatz zu den obigen Molekülionen-Eigenschwingungen, die als „innere" bezeichnet werden) Gitterschwingungen gedeutet werden; kurz es

[1] H. Sauer: Ann. d. Phys. (4) **87**, 197 (1928). Eine weitgehende Schwingungsanalyse führte E. Golling, Ann. d. Phys. (6) **9**, 181 (1951) durch.

[2] G. Joos u. H. Ewald: Göttinger Nachr. (II) **3**, 71 (1938).

scheint so, als ob alle im Kristall vorkommenden Schwingungsfrequenzen zu Kombinationslinien, die wesentlich schwächer als die Elektronensprunglinien sind, Anlaß geben (vgl. Fig. 7 und 8).

Da diese Deutung der vielen überzähligen Linien Zweifeln ausgesetzt ist, seien die Gründe dafür aufgezählt. Als Alternativerklärung wird vielfach vorgeschlagen, daß es sich bei den Trägern dieser überzähligen Linien um Ionen Seltener Erden handelt, die an verschiedenen Gitterplätzen sitzen. Diese Erklärung ist nicht immer von der Hand zu weisen, die Wiederholung charakteristischer Liniengruppen mit genau gleicher Aufspaltung wie sie beispielsweise in Fig. 7 wiedergegeben ist, bliebe dann jedoch unverständlich; denn es müßte sich gerade die *Aufspaltung* verändern, wenn das Ion an einem Platz säße, wo ein anderes Kri-

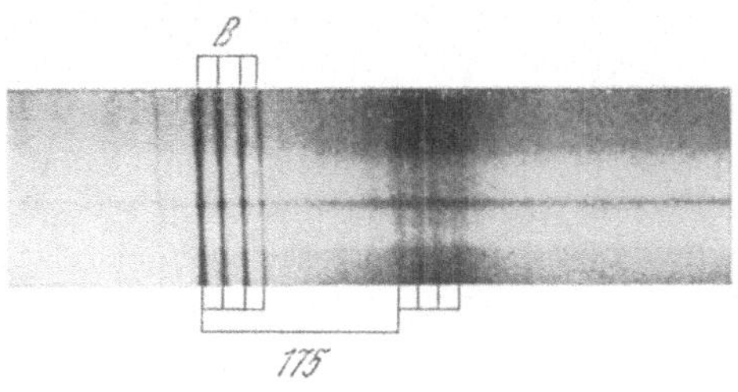

Fig. 7. Wiederholung einer Liniengruppe durch Überlagerung einer äußeren Gitterschwingung mit $\tilde{\nu} = 175$ cm^{-1}.

stallfeld herrscht. Soweit die Struktur der betreffenden Salze geklärt ist, befindet sich meist in der Elementarzelle nur ein Seltenes-Erd-Ion, so daß mindestens bei diesen Salzen sich alle Ionen in gleichen Feldern befinden. Als Beweise *für* die

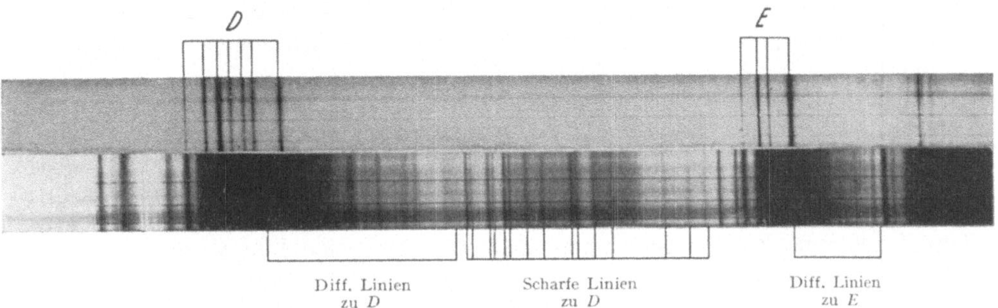

Fig. 8. Wiederholung der Liniengruppen D und E durch Überlagerung von mehreren Gitterschwingungen. Oben dünne Pulverschicht, unten dicke Schicht.

Richtigkeit der Deutung sei angeführt: 1. In Kristallen, die wie NdF$_3$ keine Anionmolekülionen und kein Kristallwasser enthalten, fehlen die inneren Schwingungen von Molekeln zugeordneten $\Delta\nu$ [1]. 2. Die Abstände der Wiederholungslinien sind anders bei den Nitraten wie bei den Sulfaten, sie entsprechen den Eigenschwingungen dieser Gruppen, die im Kristallfeld wieder aufgespalten sind. Tabelle 1 gibt ein Beispiel [2]. 3. Die den Wasserschwingungen zugeordneten $\Delta\nu$ werden beim Einbau von schwerem Wasser stark verschoben [3].

Bei den ersten Arbeiten dieser Art wurden Pulver aus kleinen Kristallen verwendet; die Untersuchungen an Einkristallen im polarisierten Licht lassen gerade bei den Schwingungsüberlagerungen viele mit den theoretischen Erwartungen vergleichbare Einzelheiten erkennen.

So wie man klassisch-qualitativ den RAMAN-Effekt durch die im Takt der Molekülschwingung stattfindende Amplitudenmodulation des Streulichts verstehen

---

[1] Y. K. CHOW: Z. Physik **124**, 52 (1945).

[2] K. H. HELLWEGE u. A. M. ROEWER: Z. Physik **114**, 564 (1939). — A. M. HELLWEGE: Ann. d. Phys. (5) **37**, 226 (1940).

[3] H. EWALD: Ann. d. Phys. (5) **34**, 209 (1939).

kann, ist die Überlagerung von Gitterschwingungen qualitativ wie bei den Molekeln als eine Frequenzmodulation im Takt der Schwingungen der Nachbarionen zu erklären. Die spektrale Zerlegung gibt dann die Nebenfrequenzen.

39. Quantenmechanische Ableitung der Schwingungsüberlagerung. A. M. und K. H. Hellwege[1] gaben eine quantenmechanische Ableitung des Zustandekommens der Schwingungskombinationslinien in Analogie zur Theorie der Bandenspektren. Sie gehen von der Separierbarkeit der Schrödinger-Gleichung des ganzen Kristalls in eine Elektronen- und eine Gitterschwingungsfunktion aus, so daß man die gesamte Eigenfunktion als Produkt einer Elektronenfunktion $\psi_{\mathrm{el}}^{(s)}$ und einer Schwingungsfunktion $\psi_k^{(t)}$ schreiben kann (s und t Laufzahlen). Die Schwingungsfunktion wird wieder als Produkt der zu einer Zelle gehörenden Normalschwingungen geschrieben

$$\psi_k^{(t)} = \prod_n \varphi_n(v_n^{(t)}) \tag{39.1}$$

(n Laufzahl, $v_n^{(t)}$ Schwingungsquantenzahl). Diese Funktionen sind im Kristallgitter der vorgegebenen Symmetrie (bei den Pr- und Nd-Zn-Nitraten C_{3v}) durch Kristallquantenzahlen, in unserem Beispiel C_{3v} [vgl. (22.22), wo $S = e^{i\bar{v}\pi}$ ist]

$$\mu \equiv (\mu_{\mathrm{el}}^{(s)} + \mu_k^{(t)}) \bmod 3 \quad \text{und} \quad \bar{v} \equiv (\bar{v}_{\mathrm{el}}^{(s)} + \bar{v}_k^{(t)}) \bmod 2, \tag{39.2}$$

gekennzeichnet. Für diese werden die Übergänge, welche zu einem nichtverschwindenden Dipolmoment gehören, aus Symmetriebetrachtungen abgeleitet und tabuliert.

Für die inneren Schwingungen des NO_3^--Ions erhält man 6 Normalschwingungen. Da aber in einer Zelle 12 NO_3-Ionen sitzen, spalten durch Koppelung diese Schwingungen noch in je 12 auf, sofern keinerlei Symmetriebedingungen des Gitters diese Zahl vermindern. Bei der Symmetrie C_{3v} werden sie auf 8 reduziert, von denen je 4 zum totalsymmetrischen Typ A und 4 zum unsymmetrischen Typ E gehören. Zu der Schwingung 730 cm^{-1} des freien Ions wurden 7 dieser 8 Schwingungen, nämlich

710,5;   715,3;   715,8;   718,2;   718,4;   723,8;   724,4 cm^{-1}

in Kombinationslinien identifiziert. Die fehlende dürfte unaufgelöst geblieben sein. Fig. 9 gibt diese Linien mit den aus der Polarisation folgenden Symmetriecharakteren dieser Schwingungen. Tabelle 1 enthält alle in Pr- und Nd-Salzen beobachteten Schwingungen.

Auch die äußeren Gitterschwingungen zerfallen, wie schon bei der Theorie des Raman-Effekts von Kristallen[2] erkannt wurde, in verschiedene Symmetrietypen. Bei strenger Separierbarkeit der Schrödinger-Gleichung können bei C_{3v} mit einem Elektronenübergang aus dem schwingungslosen Zustand nur Schwingungen des Typs A_1 (totalsymmetrische Schwingung aller Gitterpunkte $\mu_k = \bar{v}_k = 0$) kombinieren. Wenn aber die Separierbarkeit nur näherungsweise gilt, kommen auch Kombinationen mit Schwingungen vom Typ A_2 (antisymmetrische Schwingung $\mu_k = 0$, $\bar{v}_k = 1$) und Typ E (entartete unsymmetrische Schwingung $\mu_k = \pm 1$) vor. Diese sind aber schwächer als die Kombinationen mit A_1. Die mit A_1 kombinierenden Linien haben dieselbe Polarisation wie die schwingungsfreien Linien, bei

---

[1] A. M. u. K. H. Hellwege: Z. Physik **133**, 174 (1952).
[2] G. Placzek: Rayleigh-Streuung und Raman-Effekt. Handbuch der Radiologie, Bd. VI, 2. Leipzig 1934. — Eine Zusammenstellung der Symmetrieeigenschaften der Schwingungen findet sich auch bei K. W. F. Kohlrausch: Der Smekal-Raman-Effekt, S. 35. Berlin 1938.

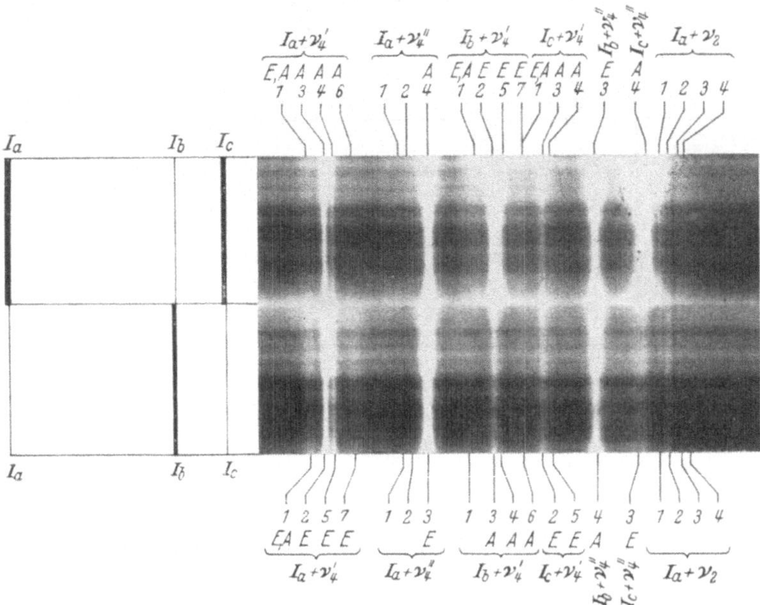

Fig. 9. Überlagerung einer entarteten Schwingung ($\tilde{\nu}_4$) des NO$_3$-Ions über den Elektronenübergang D (Linien I_a, I_b, I_c) im NdZn-Nitrat. Die NO$_3$-Frequenz $\tilde{\nu}_4$ spaltet infolge Symmetrieverringerung auf in $\tilde{\nu}_4'$ und $\tilde{\nu}_4''$. Jede dieser beiden Frequenzen spaltet weiter auf in mindestens 7 Koppelfrequenzen. Am rechten Bildrand schwach die Überlagerung der einfachen NO$_3$-Frequenz $\tilde{\nu}_2$. Links schematisch die reinen Elektronensprunglinien. Polarisation: Oben $\mathfrak{E} \parallel X$, unten $\mathfrak{E} \perp X$ (X = optische Achse). (Negative.)

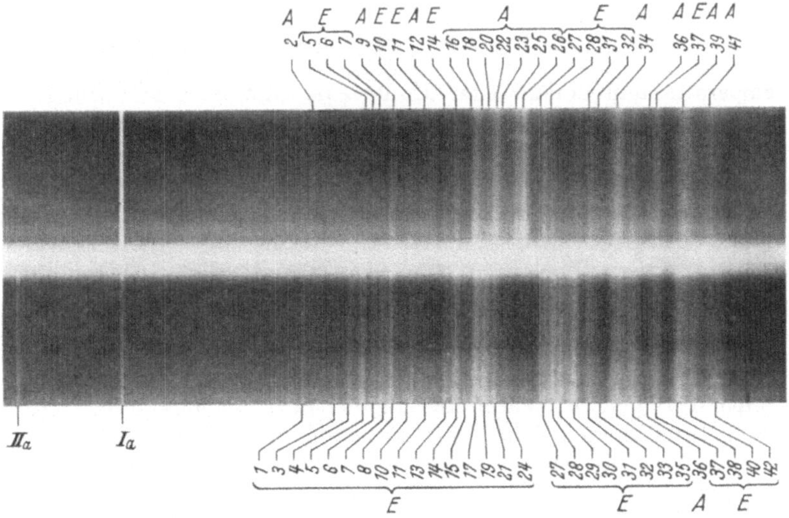

Fig. 10. Überlagerung äußerer Gitterschwingungen über den Übergang $^3H_4 \rightarrow {}^3P_0$ im PrZn-Nitrat. Strahlvektor $\perp X$, oben $\mathfrak{E} \perp X$, unten $\mathfrak{E} \parallel X$ (X = optische Achse) Frequenz nach rechts wachsend. (Negative.)

den andern kommen beide Polarisationsrichtungen vor. Fig. 10 zeigt, daß entsprechend den Auswahlregeln der Kristallquantenzahlen die schwingungsfreie Linie nur für $\mathfrak{E} \perp X$ eine wesentliche Intensität aufweist (ihr allerdings schwaches Auftreten für $\mathfrak{E} \parallel X$ — unteres Spektrum — ist nicht ohne weiteres zu erklären).

Tab. 1. *Eigenfrequenzen v des* NO_3^- *und* SO_4^{---} *Ions aus* RAMAN- *und Ultrarotmessungen (vorwiegend an Lösungen und in den Kristallspektren beobachtete* Δv *(in* cm^{-1}*).*

Bezeichnung	v	Δv in cm^{-1} aus			
	cm^{-1}	PrZn-Nitrat	NdZn-Nitrat	Pr-Nitrat	Nd-Nitrat
v_1	725	715, 727, 744	716, 725, 746	702, 725, 750	709, 723, 745
v_2	835	805, 813	815, 826	—	—
v_3	1055	1040, 1045	1046, 1051	1028, 1046	1034, 1049
v_4	1370	—	1310, 1356	1294, 1377	1294
$v_1 + v_3$	1780		1764, 1794		
$2v_1$	1450		1430, 1460		
$v_1 + v_2$	1560		1535, 1555		

Bezeichnung	v	Δv in cm^{-1} aus	
	cm^{-1}	Pr-Sulfat	Nd-Sulfat
v_2	455	—	450
v_3	615	—	—
v_1	990	994, 1006	995, 1011
v_4	1125	1125, 1136, 1156	1120, 1160
$2v_2$	910	898	892

Die starken Linien im oberen Spektrum entsprechen Kombinationen mit A_1-Schwingungen; sie fehlen fast ganz im unteren. Hier dürfen nur Kombinationen mit E bzw. auch mit A_2-Schwingungen auftreten. Der Symmetriecharakter der einzelnen Gitterschwingungen ist zum Teil aus RAMAN-Effektmessungen bekannt und deckt sich mit den aus den Kombinationsschwingungen erschlossenen.

II. Temperatur- und Druckbeeinflussung der Kristallspektren.

40. Temperatur und Linienbreite. Es ist eine alltägliche Erfahrung, daß die Linien von Kristallen erst bei sehr tiefen Temperaturen einigermaßen scharf werden. Schon 1907 hat BECQUEREL[1] für die Halbwertsbreite das Gesetz angegeben, daß sie proportional mit $\sqrt{T}$ wächst. Das gilt einigermaßen für höhere Temperaturen. Die Verbreiterung ist natürlich durch die infolge der Wärmeschwingungen wechselnde Größe der Kristallfelder bedingt. In Wirklichkeit hat aber jede Kristallinie auch bei $T = 0$ eine gewisse Breite, die vom Grad der Wechselwirkung des absorbierenden Ions mit der Umgebung abhängt. Das $\sqrt{T}$-Gesetz gilt daher erst von solchen Temperaturen an, bei denen diese Grenzbreite klein gegen die Temperaturbreite ist.

41. Temperatur- und Druckbeeinflussung der Lage des Maximums. Neben der Verbreiterung bewirkt steigende Temperatur auch eine Verschiebung der Lage des Maximums, wobei die Richtung für die verschiedenen Linien verschieden ist. Wenn man davon ausgeht, daß nur die Mittellage der Nachbaratome maßgebend ist, muß eine Temperatur*erniedrigung* in gleicher Richtung wirken wie eine Kompression durch Druckerhöhung. PAETZOLD[1] hat dies bei Seltenen Erden (Pr-Salzen) auch gefunden, dagegen sind bei Rubin die Vorzeichen entgegengesetzt, obwohl gute Proportionalität zwischen Temperatur- und Druckverschiebung besteht. Bei einem so stark anisotropen Kristall wie dem Rubin kann aber die Wärmekontraktion ganz anders verlaufen als die Kompression bei allseitigem

---

[1] H. K. PAETZOLD: Ann. d. Phys. (5) **37**, 470 (1940). — Z. Physik **129**, 123 (1951).

Druck. HELLWEGE und SCHRÖCK-VIETOR[1] untersuchten die kristallspektroskopisch in allen Einzelheiten eingeordneten Linien des Europium-Zinknitrats. Wenn es nur auf die Mittellage und nicht auf die Schwingungsamplitude ankommt, muß für alle Linien einer zur gleichen Linie des freien Ions gehörenden Gruppe dasselbe Verhältnis zwischen Temperatur- und Druckverschiebung, das sich leicht durch Kompressibilität und Ausdehnungskoeffizient ausdrücken läßt, bestehen. Dies ist aber nicht der Fall. Es ist also offenbar die Termverschiebung eine Funktion von Mittellage *und* Schwingungsamplitude. Dabei ist die Druckverschiebung streng druckproportional, während die Temperaturverschiebung sich besser durch eine parabolische Funktion von T darstellen läßt.

III. Die Kristallhyperfeinstruktur (KHFS).

42. Bei tiefer Temperatur und hoher spektraler Auflösung findet man, daß die einzelnen Linien noch eine weitere Aufspaltung besitzen, die von der Größenordnung 0,1 bis 1 cm^{-1} ist. Diese weitere Differenzierung wurde von HELLWEGE[2] entdeckt und als Kristallhyperfeinstruktur bezeichnet, womit nicht gesagt sein soll, daß es sich wie bei der Hyperfeinstruktur der freien Atome um Kernmomentwirkungen handeln soll, die nach dem ganzen Verhalten wohl als Ursache ausscheiden dürfte. Merkwürdigerweise beschränkt sich diese Struktur nicht nur auf die Linien der Seltenen Erden, sie wurde vielmehr auch bei den Linien von Cr^{+++} festgestellt, wenn dieses Ion isomorph in Aluminiumverbindungen, z.B. als Träger der Farbe im Rubin, eingelagert ist. Besonders eingehend wurden die Linien des Pr^{+++} untersucht[3]. Die Komponenten einer KHFS-Gruppe sind verschieden polarisiert, so daß man im polarisierten Licht noch Linien trennen kann, die im gewöhnlichen Licht zusammenfließen. (Dasselbe wurde auch an den Rubinlinien festgestellt.) Neben der kaum auflösbaren Mitte einer solchen Gruppe findet man in etwas größerem Abstand noch „Satelliten". Die KHFS ist gegen Gitterstörungen stark empfindlich: in ganz reinen Pr-Präparaten haben praktisch alle Linien eine solche Struktur, Verunreinigung von Pr mit Nd läßt sie durch Verschmierung verschwinden. Auch bei den Cr^{+++}-Linien wurde eine solche Abhängigkeit von der speziellen Probe gefunden. Mit steigender Temperatur (mit Rücksicht auf die Schärfe kommt nur der Bereich 4,2 bis 20° K in betracht) verschiebt sich die Gruppe als ganzes. Ein gewisser Aufschluß über die Ursache der KHFS ist vom ZEEMAN-Effekt zu erwarten, der von BROCHARD und HELLWEGE[4] gemessen wurde. Es ergab sich: die inneren Gruppen spalten als ganzes auf, jede ZEEMAN-Komponente hat die ursprüngliche KHFS. Dies entspricht einem vollständigen PASCHEN-BACK-Effekt, der bei der Kleinheit der Aufspaltung in den verwendeten starken Feldern durchaus zu erwarten ist. Die Satelliten verhalten sich aber anders: nie wurde eine Aufspaltung beobachtet, in manchen Fällen verhält sich die Linie im Feld wie eine einzelne ZEEMAN-Komponente, in andern Fällen verschiebt sie sich in anderer Weise. Durch diese Beobachtungen scheidet jedenfalls die naheliegende Erklärung aus, daß es sich bei den Satelliten um Pr^{+++}-Ionen handelt, die an nicht gleichwertigen Gitterplätzen sitzen. Eine Erklärung durch Überlagerung von Gitterschwingungen scheidet deshalb aus, weil die Frequenzen ja einer Wellenlänge entsprechen, die bereits zwischen 1 und 10 cm liegt.

---

[1] K. H. HELLWEGE u. W. SCHRÖCK-VIETOR: Z. Physik **143**, 451 (1955).
[2] A. M. u. K. H. HELLWEGE: Z. Physik **130**, 549 (1951).
[3] K. H. HELLWEGE: Ann. d. Phys. (5) **40**, 529 (1941). — A. M. u. K. H. HELLWEGE: Z. Physik **135**, 615 (1953).
[4] J. BROCHARD u. K. H. HELLWEGE: Z. Physik **135**, 620 (1953).

Es handelt sich höchstwahrscheinlich, wie auch Hellwege annimmt, um eine neue Art von Aufspaltung, die zur Unterscheidung von der „Bethe-Aufspaltung" als „Davydov-Aufspaltung" bezeichnet wird und im folgenden Abschnitt ausführlich besprochen wird. Sie bedeutet eine weitere Aufspaltung durch eine schwache Wechselwirkung der Ionen derselben Zellen oder vielleicht sogar der Ionen verschiedener Zellen. Vielleicht ist die innere Gruppe dieser zweiten, und die Satelliten der ersten Koppelung zuzuordnen.

IV. Die Termaufspaltung in unpolaren Molekülkristallen (Davydov-Aufspaltung).

Während die Bethesche Theorie die Aufspaltung der entarteten Terme *eines* Ions unter dem Einfluß des an seinem Gitterplatz herrschenden Kristallfeldes ergibt, erhält man nach A. Davydov[1] eine andere, besonders für das Verständnis der Molekülkristallspektren wichtige Aufspaltung infolge der Wechselwirkung *verschiedener* Kristallbausteine. Die hierauf beruhende „Davydov-Aufspaltung" wurde von H. Winston[2] für den Fall eines total-symmetrischen Grundzustandes gruppentheoretisch erfaßt, indem die Elektronenenergieniveaus eines Molekülkristalls nach den Darstellungen der Raumgruppe des Kristalls klassifiziert werden.

43. Raumgruppe, Faktorgruppe und Lagegruppe. Die *Raumgruppe* $\Re$ eines Kristalls besteht neben den Symmetrieoperationen der Kristallklasse (= Punktgruppe), d. h. den Drehungen, Drehspiegelungen usw., noch aus den Translationen, die, gleichzeitig mit den Drehungen oder der Spiegelung ausgeführt, zu Schraubungen oder Gleitspiegelungen führen. Die Darstellungen der 230 Raumgruppen wurden von F. Seitz[3] untersucht. Die Translationen bilden eine Untergruppe, die *Translationsgruppe* $\mathfrak{T}$ von $\Re$. Werden die Ecken des (endlichen) Kristalls durch $N_1 \mathfrak{t}_1$, $N_2 \mathfrak{t}_2$ und $N_3 \mathfrak{t}_3$ ($\mathfrak{t}_1$, $\mathfrak{t}_2$, $\mathfrak{t}_3$ = primitive Gittervektoren der Elementarzelle) beschrieben, so ist $N_1 N_2 N_3$ die Ordnung von $\mathfrak{T}$. Da $\mathfrak{T}$ eine abelsche Gruppe ist, gibt es $N_1 N_2 N_3$ eindimensionale Darstellungen von $\mathfrak{T}$, nämlich $\exp(2\pi i \mathfrak{f} \mathfrak{a})$,

wobei $\mathfrak{f} = \sum_{\nu=1}^{3} \dfrac{l_\nu}{N_\nu} \mathfrak{b}_\nu$, ($\mathfrak{b}_\nu$ = reziproke Gittervektoren, l_ν = ganz) und $\mathfrak{a}$ ein Translationsvektor der Gruppe ist[4]. Die Elemente von $\mathfrak{T}$ sind mit allen Elementen von $\Re$ vertauschbar, $\mathfrak{T}$ ist ein Normalteiler von $\Re$. Bilden wir mit den Elementen r von $\Re$ die verschiedenen „Nebenklassen" $r\mathfrak{T}$, so stellen die Gesamtheiten $\mathfrak{T}, r_1\mathfrak{T}, r_2\mathfrak{T}, \ldots$ die Elemente einer Gruppe ($\mathfrak{T}$ = Einheitselement), die „*Faktorgruppe* $\mathfrak{F}$ von $\Re$ nach $\mathfrak{T}$", dar, die zu $\Re$ homomorph und zu einer der 32 Kristallklassen (= Punktgruppen) isomorph ist. Ist F die Ordnung von $\mathfrak{F}$, so ist $N_1 N_2 N_3 F$ die Ordnung der (endlichen) Raumgruppe $\Re$. — Alle Operationen, die irgendeinen Punkt („Lage") L eines Kristalls invariant lassen, bilden die „*Lagegruppe*" (Sitegroup)[4] $\mathfrak{L}$ der Lage L. $\mathfrak{L}$ ist eine Untergruppe von $\mathfrak{F}$. Betrachten wir z. B. in Fig. 4 ($\mathfrak{F}$ isomorph zu D_4) eine Lage L auf der y-Achse ($y = A > 0$), so besteht $\mathfrak{L}$ aus den Elementen e und f. Die Gruppenelemente von $\mathfrak{F}$ transformieren die Lage L in n „zu L äquivalenten" Lagen $L_1 = L, L_2, \ldots, L_n$. In unserem Beispiel ergibt etwa die Anwendung von h (ebenso wie die von c) auf L eine äquivalente Lage auf der x-Achse; insgesamt erhält man 4 (= n) äquivalente Lagen $x = \pm A$, $y = \pm A$, $z = 0$. Ist L die Ordnung der Lagegruppe, so gilt stets $F = nL$. Äquivalente Lagen haben isomorphe Lagegruppen.

---

[1] A. Davydov: J. exp. theor. Phys. USSR. **18**, 210 (1948).
[2] H. Winston: J. Chem. Phys. **19**, 156 (1951).
[3] F. Seitz: Annals Math. **37**, 17 (1936).
[4] H. Winston u. R. S. Halford: J. Chem. Phys. **17**, 607 (1949).

Betrachten wir z.B. die Verhältnisse bei einem Benzolkristall. Im freien Zustand ist die Symmetrie der Moleküle D_{6h}. Die Raumgruppe des Kristalls ist D_{2h}^{15}; die Faktorgruppe $\mathfrak{F}$ ist also isomorph zu D_{2h}. In einer Elementarzelle befinden sich $n = 4$ Benzolmoleküle; ihre Lagegruppe hat die Symmetrie C_i.

44. Die Davydov-Aufspaltung. Die Funktionen, die zu irreduziblen Darstellungen der Lagesymmetrie $\mathfrak{L}$ der Moleküle im Kristall gehören, bestimmen sich aus den Funktionen des freien Moleküls nach der Methode von H. Bethe. Diese zur Lagesymmetrie gehörigen Terme spalten nach Davydov weiter auf. Ist in der k-ten Elementarzelle das Molekül in der α-ten Lage angeregt, so bezeichne $\Psi^p_{j,k\alpha}$ die zugehörige p-te Eigenfunktion in der j-ten Darstellung der Lagegruppe $\mathfrak{L}$. Zur $\mathfrak{k}$-Darstellung der Translationssymmetrie gehört die Linearkombination

$$\Phi^p_{j,\alpha}(\mathfrak{k}) = \sum_k e^{2\pi i \mathfrak{k}_k k\alpha}\, \Psi^p_{j,k\alpha}. \tag{44.1}$$

Beschränkt man sich mit H. Winston auf Funktionen mit $\mathfrak{k} = 0$, d.h. auf die totalsymmetrische Darstellung der Translationsgruppe, so bleibt eine Aufspaltung in Energiebänder unberücksichtigt[1]. Durch die Funktionen $\Phi^p_{j,\alpha}(0)$ wird eine im allgemeinen reduzible Darstellung der Faktorgruppe $\mathfrak{F}$ induziert, die zur Davydov-Aufspaltung Anlaß gibt. Da ein Gruppenelement f der Faktorgruppe, das nicht Element der α-ten Lagegruppe ist, die Lage α in eine andere Lage β überführt, transformiert sich $\Phi^p_{j,\alpha}(0)$ in $\Phi^q_{j,\beta}(0)$, wobei die Transformationsmatrix keine Diagonalelemente enthält. Wenn jedoch f ein Element der Lagegruppe ist, geht die Lage α in sich über, wobei die Summe über die Diagonalelemente den Charakter $\chi^{\mathfrak{L}}_j(f)$ der Lagegruppe L liefert. Damit erhält man für den Charakter der reduziblen Darstellung der Faktorgruppe

$$\chi^{\mathfrak{F}}_j(f) = \sum_\alpha \delta_\alpha(f) \cdot \chi^{\mathfrak{L}}_j(f), \tag{44.2}$$

wobei $\delta_\alpha(f) = 1$ oder 0 ist, je nachdem ob f ein Element der α-ten Lagegruppe ist oder nicht. Die Ausreduzierung in die irreduziblen Bestandteile $\Gamma^{\mathfrak{F}}_i$ von $\mathfrak{F}$ liefert nach Gl. (11.9) die gesuchte Anzahl der irreduziblen Bestandteile:

$$n_{ij} = \frac{1}{F} \sum_f \chi^{\mathfrak{F}}_j(f)\, \chi^{\mathfrak{F}}_i{}^*(f). \tag{44.3}$$

Da $\mathfrak{L}$ eine Untergruppe von Γ ist, ergibt sich bei der Ausreduzierung derjenigen Elemente f, die auch in $\mathfrak{L}$ enthalten sind, nach Gl. (11.8)

$$\chi^{\mathfrak{F}}_i(f) = \sum_k a_{ik}\, \chi^{\mathfrak{L}}_k(f). \tag{44.4}$$

Die a_{ik} stellen bekannte Größen dar, die aus den Charakterentabellen entnommen werden können. Setzt man Gl. (44.2) und (44.4) in Gl. (44.3) ein,

$$n_{ij} = \frac{1}{F} \sum_\alpha \sum_k \sum_f a_{ik}\, \chi^{\mathfrak{L}}_j(f)\, \chi^{\mathfrak{L}}_k{}^*(f)\, \delta_\alpha(f), \tag{44.5}$$

so folgt aus den Orthogonalitätsrelationen [vgl. (11.3)] der Gruppencharaktere

$$n_{ij} = a_{ij}. \tag{44.6}$$

Diese Gleichung gibt an, wieviele Energieniveaus, die zur i-ten Faktorgruppendarstellung im Kristall gehören, aus *einem* zur j-ten Darstellung von $\mathfrak{L}$ gehörigen Term entstehen (Davydov-Aufspaltung).

---

[1] L. Bouckaert, R. Smoluchowski u. E. Wigner: Phys. Rev. **50**, 58 (1936).

Die bei den Seltenen Erden auftretende Kristallhyperfeinstruktur (S. 265) läßt sich möglicherweise als Davydov-Aufspaltung erklären[1].

Die Davydovsche Theorie hat noch eine weitere Konsequenz: Die Anregung eines bestimmten Moleküls eines Kristalles stellt ohne Koppelung einen entarteten Zustand dar. Ist jedoch eine Wechselwirkungsenergie ΔE zwischen benachbarten Molekülen vorhanden, so wird die Entartung aufgehoben und die Anregung geht nach der Zeit $\hbar/\Delta E$ mit großer Wahrscheinlichkeit auf das Nachbarmolekül über, d.h. der Anregungszustand wandert im Gitter. Diese Wanderung kann formal wie die Bewegung eines materiellen Teilchens („*Exciton*")[2] behandelt werden.

D. Ergebnisse der Kristallspektroskopie.

I. Kristallspektren von Ionen mit unabgeschlossenen Schalen.

a) Die Lanthanidenreihe.

45. Allgemeiner Charakter der Spektren. Die besten Objekte zum Studium der Gesetzmäßigkeiten der Kristallspektren sind die Salze der Seltenen Erden (Lanthaniden). Dadurch, daß die für die Spektren maßgebenden $4f$-Elektronen

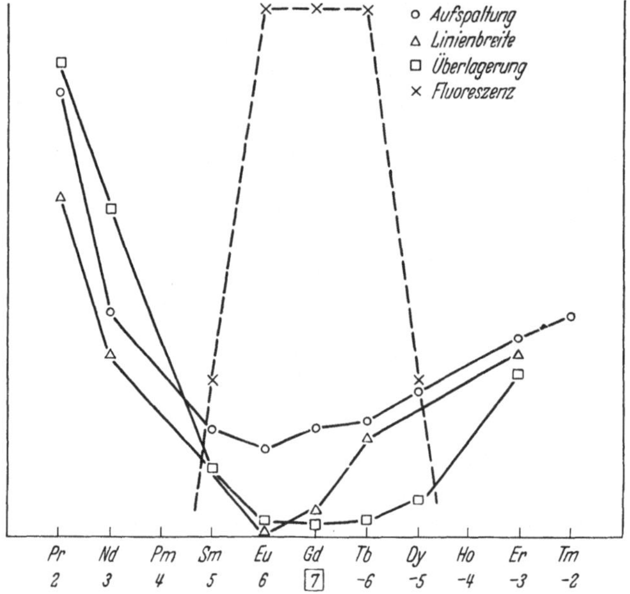

Fig. 11. Gang des Gittereinflusses in der Reihe der Seltenen Erden.

durch eine Hülle von zwei $5s$- und sechs $5p$-Elektronen vor allzugroßer Störung durch die Gitternachbarn geschützt sind, sind insbesondere bei tiefen Temperaturen die Linien kaum weniger scharf als die freier Atome und die Einwirkung der Umgebung ist so, daß die Kristallfeldaufspaltung noch als klein gegen die Multiplett-Aufspaltung angesehen werden kann; andererseits ist sie doch wieder hinreichend groß, um die Linienstruktur auflösen zu können. Die Einwirkung ist in der Mitte der Reihe am geringsten. Wenn man als Zeichen starker Störung die Größe der Aufspaltung, die Breite der Linien und das Auftreten von Schwingungskombinationslinien, als Zeichen geringer Störung die Fluorescenzfähigkeit

---

[1] H. Winston: J. Chem. Phys. **19**, 156 (1951).
[2] J. Frenkel: Phys. Rev. **37**, 17, 1276 (1931). — Phys. Z. Sowjet. **9**, 158 (1936).

der reinen Salze wertet, so erhält man durch Auftragen dieser Größen über der Ordnungszahl die von HELLWEGE[1] stammende Übersichtsfigur 11, welche den Gang der Gitterstörung schön veranschaulicht.

46. Die Lage der Terme der freien Ionen. Schon die Berechnung der aus n $4f$-Elektronen unter Berücksichtigung des PAULI-Prinzips möglichen Terme stellt eine interessante Aufgabe der Quantenmechanik dar. Die *Zahl* der möglichen Terme beträgt bei Ce^{+++} mit einem $4f$-Elektron, abgesehen von der Multiplett-, in diesem Fall Dublett-Aufspaltung, 1, nämlich ^2F. Dasselbe gilt für Yb mit 13 $4f$-Elektronen. Bei Pr^{+++} mit zwei Elektronen kann man bereits 7 Terme bilden. (Eine Anleitung zur Abzählung der nach dem PAULI-Gesetz möglichen Terme findet man z.B. in SOMMERFELDs Atombau und Spektrallinien[2]. Bei Pr^{+++} sind es die Terme $^1S_0$; $^3P_{0,1,2}$; $^1D_2$; $^3F_{2,3,4}$; $^1G_4$; $^3H_{4,5,6}$; $^1I_6$. Bei Gd^{+++} mit sieben Elektronen wird die Höchstzahl von 119 Termen, die selbst noch Multiplettstruktur aufweisen, erreicht. Die Berechnung ihrer Lage (Schwerpunkte des Multipletts) ist für die am Anfang und am Ende der Reihe stehenden Elemente von verschiedener Seite in Angriff genommen worden[3]. Man geht von Wasserstoff- oder besser HARTREE-Eigenfunktionen aus und hat nach den Regeln der Störungsrechnung die Wechselwirkung der Elektronen als Störung hinzuzunehmen; im Grund dieselbe Aufgabe wie die Berechnung der He-Terme. Dabei kommen die Energiestufen als lineare Funktionen von vier Austauschintegralen heraus, wobei das vierte als sehr klein zunächst vernachlässigt werden kann.

Da die Austauschintegrale wegen der Störung durch die nicht zur $4f$-Gruppe gehörenden Elektronen nur ungenau berechenbar sind, wurden sie durch Ausgleichsrechnung für die Terme des Pr^{+++} bestmöglich der Erfahrung angepaßt, wobei sich zunächst ein sehr gutes Bild ergab. Offenbar ist dabei aber die gelbe Liniengruppe nicht richtig zugeordnet worden. Eine Neubearbeitung unter Zuordnung entsprechend den jetzigen Kenntnissen[4] ergab die in Fig. 12 wiedergegebene Gegenüberstellung von Beobachtung und verschieden berechneten Termwerten.

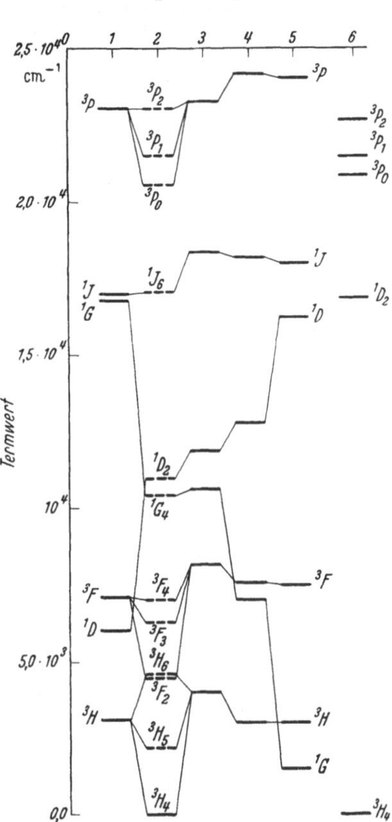

Fig. 12. Verschiedene Berechnungen der Termschwerpunkte von Pr^{+++}. 1. Nach LANGE [Ann. Physik 31, 609 (1938)]. 2. Nach SPEDDING [Phys. Rev. 58, 255 (1940)] mit Spin-Bahn Wechselwirkung. 3. Desgleichen ohne Spin-Bahn-Koppelung. 4. Nach SLATER [Phys. Rev. 34, 1293 (1929)]. 5. Desgleichen mit Nullsetzung zweier Austauschintegrale. 6. Zuordnung der beobachteten Terme von HELLWEGE [Z. Physik 130, 549 (1951)]. ($^1S_0$ liegt wahrscheinlich über ^3P.)

---

[1] K. H. HELLWEGE: Naturwiss. 34, 225 (1947).

[2] A. SOMMERFELD: Atombau und Spektrallinien, 5. u. folg. Aufl., Bd. 1, S. 491 ff. Braunschweig 1931 und später. — R. C. GIBBS, D. T. WILBER u. H. E. WHITE: Phys. Rev. 29, 790 (1927).

[3] J. C. SLATER: Phys. Rev. 34, 1293 (1929). — E. U. CONDON u. G. H. SHORTLEY: Phys. Rev. 37, 1025 (1931). — H. LANGE: Ann. d. Phys. (5) 31, 609 (1938). — H. GOBRECHT: Ann. d. Phys. (5) 28, 673 (1937). — F. H. SPEDDING: Phys. Rev. 58, 255 (1940).

[4] E. TREFFTZ: Z. Physik 130, 561 (1951).

Bei dem komplementären Element Tm haben sich bisher keine Bedenken gegen die Zuordnung ergeben, welche nach Rechnung und Messung von Go-brecht[1] eine sehr gute Übereinstimmung zeigt.

Besser der Rechnung zugänglich ist die Multiplettaufspaltung der Grundterme. Es ergibt sich für die Totalaufspaltung, d.h. den Abstand des höchsten vom Grund-Niveau beim Term größter zulässiger Multiplizität, der zugleich nach der HUNDschen Regel der Grundterm ist, die Formel

$$\Delta \nu = \frac{R\,\alpha^2\,(2L+1)\,(Z-\sigma)^4}{n^3\,l\,(l+1)\,(2l+1)} \tag{46.1}$$

(R RYDBERG-Zahl, α Feinstrukturkonstante, L Gesamtdrehimpulsquantenzahl, l Einzelbahnmomentquantenzahl, Z Ordnungs-, σ Abschirmzahl). Unterteilt man

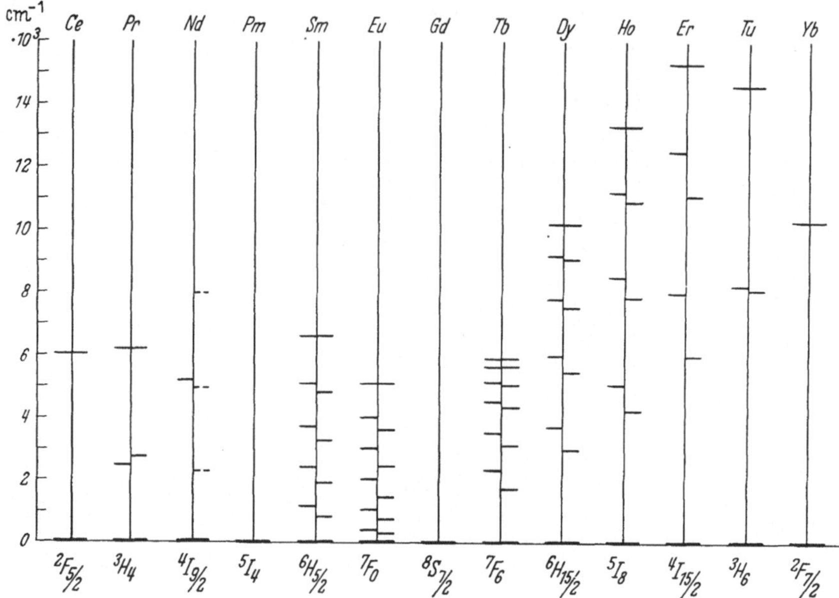

Fig. 13. *J*-Aufspaltung der Grundterme der dreiwertigen Ionen der Seltenen Erden.
(Links Rechnung, rechts Beobachtung.)

die Totalaufspaltung nach der LANDÉschen Intervallregel, so ergibt sich das in Fig. 13 wiedergegebene Aufspaltungsbild[2], das sehr gut von der Erfahrung bestätigt ist. Die geringen Abweichungen zeigen, daß die RUSSELL-SAUNDERS-Koppelung nicht mehr streng erfüllt ist, was dem Verhalten dieser Elemente durchaus entspricht. Nach dieser Formel hat GOBRECHT z.B. den einzigen bei Yb ohne Änderung der Hauptquantenzahl möglichen Übergang $^2F_{7/2}$—$^2F_{5/2}$ im Infrarot bei 930 mµ vorausberechnet und dort auch gefunden. Auch die übrigen Multiplett-Stufen, die in Fig. 13 eingezeichnet sind, hat GOBRECHT im Infrarot aufgefunden. Bei freien Atomen sind Übergänge zwischen den Niveaus eines Multipletts streng verboten, unter dem Einfluß der Felder der Nachbaratome sind aber erzwungene Dipolübergänge möglich. Die eingezeichneten experimentellen Werte sind die Schwerpunkte der im Kristallfeld aufgespaltenen Liniengruppen, die bei der geringen Auflösung im Infrarot ohnehin nicht getrennt sind. Das Dublett bei Cer ist nicht direkt in Absorption gemessen (vgl. unten).

[1] H. GOBRECHT: Ann. d. Phys. (5) **31**, 600 (1938).
[2] H. GOBRECHT: Ann. d. Phys. (5) **28**, 673 (1937); **31**, 755 (1938).

In Fig. 14 sind die nach (46.1) aus den beobachteten Aufspaltungen berechneten Abschirmzahlen σ als Funktion von Z aufgetragen, Röntgenspektren können für die Abschirmung im Bereich der $4f$-Elektronen nur rohe Werte liefern; sie passen aber gut zu den hier ermittelten. Wie gut die Theorie der Elektronenhüllen in diesem Gebiet in sich stimmt, zeigt sich z.B. darin, daß sich aus dem ganz anomalen Temperaturgang der Suzceptibilität der Europiumsalze nach der Rechnung von VAN VLECK[1] und den Messungen von FRITSCH[2] $\sigma = 33$ ergibt gegenüber von 34 aus der spektroskopisch beobachteten Grundtermaufspaltung.

47. Die Kristallfeldaufspaltungen der einzelnen Ionen. Bis hierher handelte es sich nur um die Spektroskopie der freien Ionen. Die eigentlichen kristallspektroskopischen Erscheinungen werden, da sie individuell sehr verschieden sind, bei den einzelnen Ionen behandelt. Allgemein kann man sagen, daß ohne KHFS die theoretischen Aussagen hinsichtlich der Zahl der Komponenten der Aufspaltungen gut stimmen, daß aber nur in wenigen Fällen ein ins einzelne gehender Vergleich zwischen dem berechneten und dem beobachteten Linienbild möglich ist, da insbesondere bei den Elementen mit mehr als zwei $4f$-Elektronen sich die einzelnen Multipletts durcheinanderschieben und dadurch der Zuordnung der einzelnen Liniengruppen immer größere Schwierigkeiten erwachsen. Das gilt sogar schon beim ersten Element, bei dem Kristallfeldaufspaltungen zu beobachten sind, beim Pr^{+++} mit seinen zwei $4f$-Elektronen.

Es sollen nun die einzelnen Elemente der Lanthanidenreihe durchgesprochen werden.

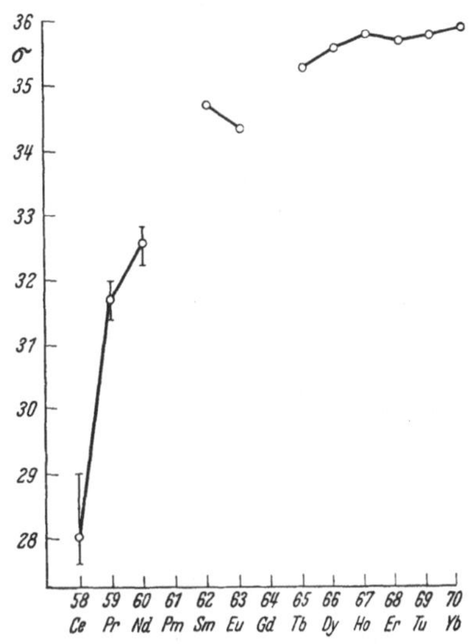

Fig. 14. Die Abschirmungszahlen der $4f$-Elektronen bei den Ionen der Seltenen Erden.

α) *Cer.* Wie bereits mehrfach erwähnt, ist bei Ce^{+++} nur der Übergang $^2F_{\frac{5}{2}} \to {}^2F_{\frac{7}{2}}$ ohne Änderung der Hauptquantenzahl möglich. In Absorption ist er nicht beobachtet, dagegen hat GOBRECHT[3] aus Fluorescenzübergängen von höheren, wegen Änderung der Quantenzahl sehr verwaschenen Termen nach $^2F_{\frac{5}{2}}$ bzw. $^2F_{\frac{7}{2}}$ die Dublettaufspaltung mit ziemlicher Fehlerbreite ermittelt (vgl. Fig. 13).

β) *Praseodym.* Die Linien des nächsteinfachen Elements sind sehr eingehend untersucht von MERZ[4], GOBRECHT[5] und HELLWEGE[6]. Beim trigonalen Praseodym-Zink- und Praseodym-Magnesium-Nitrat gelang HELLWEGE eine weitgehende Analyse der Kristallfeldaufspaltung. Der Grundterm $^3H_4$ spaltet im Feld der Symmetrie C_{3v} (trigonal) in drei einfache und drei zweifache Komponenten auf, von denen bei 58° K die drei untersten besetzt sind und denen nach HELLWEGE die Kristallquantenzahlen $\mu = \pm 1, 0$ und ± 1 zukommen (Fig. 15 und 16). Der unterste und der oberste Teilterm sind also noch entartet und können im Magnetfeld

---

[1] I. H. VAN VLECK: The theory of electric and magnetic susceptibilities, S. 251. Oxford 1932.

[2] H. FRITSCH: Ann d. Phys. (5) **39**, 31 (1941).

[3] H. GOBRECHT: Ann. d. Phys. (5) **31**, 181 (1938).

[4] A. MERZ: Ann. d. Phys. (5) **28**, 569 (1937).

[5] H. GOBRECHT: Ann. d. Phys. (5) **28**, 673 (1937).

[6] K. H. HELLWEGE u. A. M. HELLWEGE: Zahlreiche Arbeiten in Ann. d. Phys. und Z. Physik, Lit. bis 1951: Z. Physik **130**, 549 (1951); ferner Z. Physik **133**, 174 (1952); **135**, 92, 615 (1953).

noch aufspalten. Dies entspricht der Beobachtung eines linearen Zeeman-Effekts mit 0,78 Δv_{norm}, die Merz an den Übergängen nach $^3P_{0,1}$ machte. Die Richtigkeit dieser Zuordnung ist durch hervorragende Zeeman-Effekt-Aufnahmen

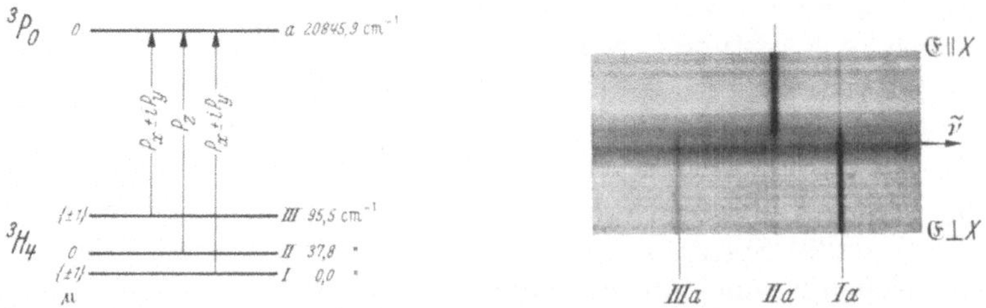

Fig. 15. Kristallfeldaufspaltung des Übergangs $^3H_4 \rightarrow {}^3P_0$ von PrMg-Nitrat bei 58° K. Die erlaubten Übergänge sind im Termschema und am Rand des Spektrums eingezeichnet. Strahlrichtung $\perp X$ (X = Kristallachse).

von Brochard und Hellwege[1], die bei 4° und 20° K mit dem großen Bellevue-Magneten gemacht wurden, erhärtet worden. Von den zahlreichen in den Original-arbeiten reproduzierten Aufnahmen sei in Fig. 17 nur das Aufspaltungsbild von $^3H_4 - {}^3P_0$ wiedergegeben. Fig. 18 enthält die Feldstärkeabhängigkeit der Lage der Komponenten (s_1 und s_2 sind Satelliten der Kristall-hyperfeinstruktur; vgl. Ziff. 42).

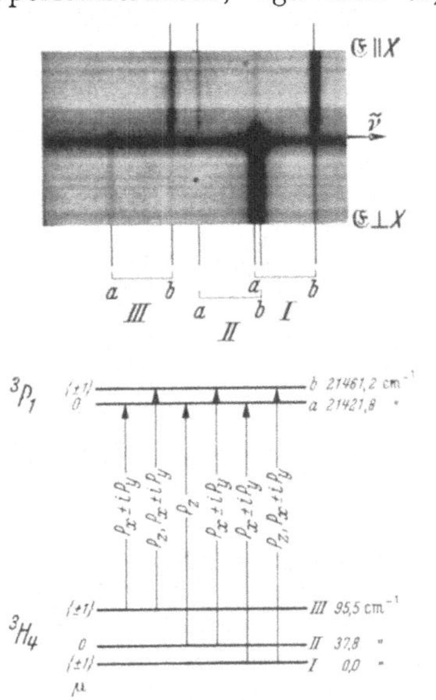

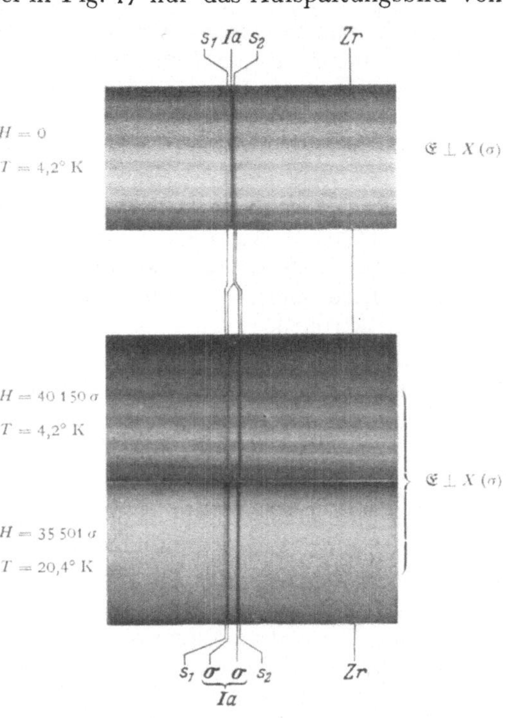

Fig. 16. Kristallfeldaufspaltung des Übergangs $^3H_4 \rightarrow {}^3P_1$ von PrMg-Nitrat bei 58° K. (Die Linien Ia und IIb sind auf der Originalplatte deutlich getrennt.)

Fig. 17. Zeeman-Effekt der Linie Ia der Gruppe $^3H_4 - {}^3P_0$ im achsenparallelen Magnetfeld ($T = 20,4°$ K). s_1 und s_2 Satelliten.

Fig. 19 enthält das Termschema für sämtliche gemessenen Linien. Die Ausmessung der Aufnahmen ergibt einen erheblichen quadratischen Anteil der Aufspaltung, der vom linearen durch Messung bei verschiedenen Feldstärken abgetrennt wurde. Ist H senkrecht zur Achse gerichtet, so ergibt sich in Übereinstimmung mit der Theorie überhaupt keine lineare Aufspaltung. Der aus

---

[1] J. Brochard u. K. H. Hellwege: Z. Physik **135**, 620 (1953).

den Aufspaltungen der Teilniveaus gemäß der Theorie des ZEEMAN-Effekts von Kristall-Linien für den Gesamtterm $^3H_4$ ermittelte g-Wert weicht etwas von dem für reine RUSSELL-SAUNDERS-Koppelung geltenden Wert ab.

Das wasserfreie $PrCl_3$, dessen Struktur röntgenographisch genau bekannt ist, wurde von SAYRRE, SANCIER und FREED[1] eingehend untersucht. Die Symmetrie am Ort des Ions ist C_{3h}. Die Auswahlregeln bei elektrischer Dipolstrahlung sind für diese Symmetrie durch das Schema (vgl. Ziff. 23)

μ_I	0	± 1	± 2	3
0			σ	π
± 1		σ	π	σ
± 2	σ	π	σ	
3	π	σ		

gekennzeichnet. Damit gelingt den Autoren nicht nur die völlige Einordnung der blauen und gelben Liniengruppen unter Zuordnung der HELLWEGEschen

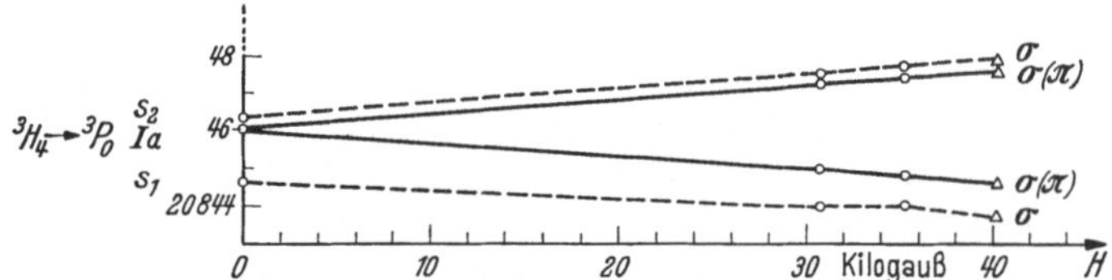

Fig. 18. Feldstärkenabhängigkeit der Lage der ZEEMAN-Komponenten.

Kristallquantenzahlen, sondern es werden auch im Infrarot noch die Übergänge nach den bisher nicht gefundenen Termen ^1G, ^3F gemessen und eingeordnet, so

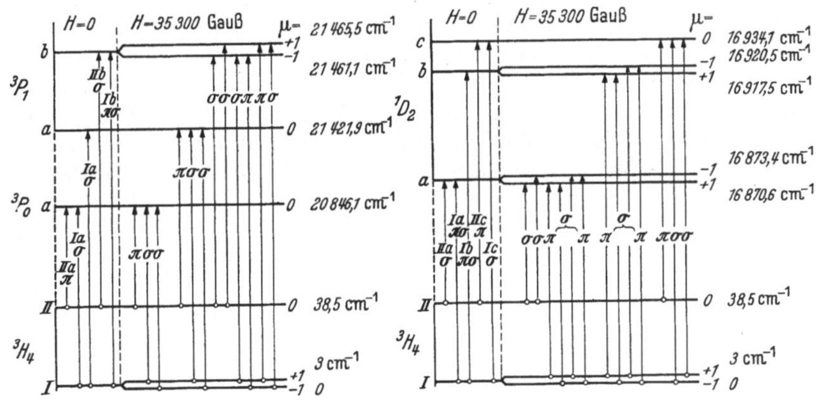

Fig. 19. ZEEMAN-Termschema der PrMg-Nitrat-Linien.

daß nur noch die Terme $^1S_0$ und $^1I_6$ fehlen. Beide fallen wahrscheinlich in einen Bereich, der durch die Absorption der anderen Bestandteile nicht mehr zugänglich ist ($^1S_0$ jedenfalls ins Ultraviolett).

Bei dem wasserfreien Praseodymfluorid tritt wieder eine Merkwürdigkeit auf: Bei 4° K treten die Auswahlregeln für die Polarisation außer Kraft, die Linien finden sich in beiden Polarisationsrichtungen gleich stark[1]. Zur

---

[1] V. E. SAYRE, KENNETH M. SANCIER u. S. FREED: J. Chem. Phys. **23**, 2060 (1955).

Erklärung wird von den Entdeckern dieses merkwürdigen Effekts angenommen, daß in der Ruhelage die Pr^{+++}-Ionen nicht in der Symmetrieebene sitzen, daß aber mit steigender Temperatur die Schwingungsmittelpunkte mehr und mehr in diese fallen. Damit hängt offenbar eine nur bei den Bromaten des Pr, Nd, Eu von HELLWEGE[1] beobachtete viel auffälligere Erscheinung zusammen, die eine Gitterumwandlung bei 80° K andeutet und die in Ziff. 58 besprochen wird.

γ) *Neodym*. Die Salze des Nd^{+++} sind von allen Seltenen Erden am meisten untersucht. Der Grund dafür liegt nicht nur in der relativen Häufigkeit und leichten Beschaffbarkeit des Elements, sondern auch in der verhältnismäßig großen Intensität der Absorptionslinien und in der durch die ungerade Elektronenzahl bedingten KRAMERS-Entartung, die ihrerseits eine große Mannigfaltigkeit linearer ZEEMAN-Aufspaltungen bewirkt. An den Nd^{+++}-Salzen wurde die Kombination mit Schwingungen von Gittermolekeln wie H_2O und NO_3 zuerst beobachtet[2], an ihnen wurde eine Temperaturabhängigkeit der Übergangswahrscheinlichkeiten selbst gefunden[3], die durch die vom Kristallfeld erzwungene Dipolstrahlung bedingt sind, wobei dieses sich mit der Temperatur etwas ändert. An den Nd-Salzen wurden auch zum ersten Mal paramagnetische Resonanzmessungen mit den ZEEMAN-Aufspaltungen des Grundterms verglichen[4]. Die Identifikation der zahlreichen aus drei f-Elektronen zu bildenden Terme ist erst spät durchgeführt. SATTEN[5] gibt bei $Nd(BrO_3)_3 \cdot 9H_2O$ für die einzelnen Liniengruppen die zugehörigen angeregten Terme an. Das Bromat hat die Symmetrie C_{3v}. In diesem Feld spaltet der Grundterm $^4I_{\frac{9}{2}}$ in fünf Komponenten 0; 115; 184; 363 und 382 cm^{-1} auf. Dies ist bereits die Höchstzahl von Teilniveaus, da in diesem Feld sich vier einfache und drei doppelte Terme ergeben, wobei aber wegen der KRAMERS-Entartung die vier einfachen Terme in zwei doppelte zusammenfließen.

Sehr eingehend sind die Nd-Salze bei Temperaturen bis herab zu 4,2° K von G. H. DIEKE und L. HEROUX[4] untersucht worden, wobei besonderes Gewicht auf die Doppelnitrate gelegt wurde. Besonders schöne ZEEMAN-Effekt-Aufnahmen zeigen, daß alle Linien in Quartette aufspalten, da sowohl der Grundterm als der angeregte Term ein Dublett gibt. Wenn die Aufspaltung in beiden gleich ist, fließt das Quartett der Linien in ein Triplett zusammen. Sowohl die Kristallfeldaufspaltung als die ZEEMAN-Aufspaltung sind von Kristall zu Kristall sehr verschieden. So ist der Abstand des nächsten Teilniveaus vom Grundterm, der beim Bromat 115 cm^{-1} beträgt, bei $Nd_2Mg_3(NO_3)_{12} \cdot 24H_2O$: 33,13 cm^{-1}, beim entsprechenden Zn-Salz 36,6, beim Sulfat 77 cm^{-1}. Die γ-Faktoren (vgl. Ziff. 36) des tiefsten Niveaus sind beim Neodym-Äthylsulfat

$$\gamma_{\parallel} = 3{,}50 \quad \text{und} \quad \gamma_{\perp} = 2{,}06.$$

Paramagnetische Resonanzmessung[6], welche die Übergänge zwischen den ZEEMAN-Niveaus selbst gibt, führt zu

$$\gamma_{\parallel} = 3{,}58 \quad \text{und} \quad \gamma_{\perp} = 2{,}09.$$

[1] A. M. u. K. H. HELLWEGE: Z. Physik **127**, 334 (1950).

[2] Unabhängig von JOOS u. EWALD z.B. auch von A. BENTON u. E. I. KINSEY: Phys. Rev. **73**, 536 (1948) festgestellt; ferner besonders K. H. HELLWEGE: Z. Physik **133**, 174 (1952).

[3] H. EWALD: Z. Physik **110**, 428 (1938).

[4] G. H. DIEKE u. L. HEROUX: Phys. Rev. **103**, 1227 (1956).

[5] R. A. SATTEN: J. Chem. Phys. **21**, 637 (1953).

[6] B. BLEANEY u. K. W. H. STEVENS: Rep. Progr. Phys. **16**, 108 (1953). — B. BLEANEY u. H. E. D. SKOVIL: Proc. Phys. Soc. Lond. **63**, 1639 (1950).

Bildet das Magnetfeld mit der Achse den Winkel ϑ, so gilt nach Gl. (36.5)

$$\gamma^2 = \gamma_\|^2 \cos^2 \vartheta + \gamma_\perp^2 \sin^2 \vartheta,$$

und bei allen möglichen Orientierungen, wie sie bei Pulvermessungen des Paramagnetismus vorkommen

$$\overline{\gamma^2} = \tfrac{1}{3}\gamma_\|^2 + \tfrac{2}{3}\gamma_\perp^2.$$

Für die verschiedenen Salze sind die Werte des untersten Teilniveaus

	$\gamma_\|$	$\gamma_\perp$	$\sqrt{\overline{\gamma^2}}$
$Nd_2Mg_3(NO_3)_{12} \cdot 24\,H_2O$	0,420	2,629	2,160
$Nd_2Zn_3(NO_3)_{12} \cdot 24\,H_2O$	0,450	2,670	2,198
$Nd(BrO_3)_3 \cdot 9\,H_2O$	2,32	2,34	2,33
$Nd(C_2H_5SO_4)_3 \cdot 9\,H_2O$	2,50	3,061	2,63

Wie man sieht, schwanken die Zahlen für $\sqrt{\overline{\gamma^2}}$ von Kristall zu Kristall viel weniger als die der $\gamma_\|$ und $\gamma_\perp$. Der g-Faktor des freien Ions, der bei höherer Temperatur gilt, bei der alle Kristallniveaus gleich besetzt sind, beträgt 0,727.

Von den folgenden Seltenen Erden sind nur bei Europium und Gadolinium besonders bemerkenswerte Einzelheiten festgestellt. Wir wollen uns daher bei den anderen Salzen der Seltenen Erden auf die wichtigsten Literaturhinweise[1] beschränken und nur diese beiden Seltenen Erden eingehender behandeln.

$\delta)$ *Europium.* Die Linien der Europiumsalze zeichnen sich durch die größte Schärfe aller Kristallinien aus. Eu liegt damit im Minimum der HELLWEGE-schen Kurve für die Wechselwirkung mit dem Gitter. Die Linien des Sulfats sind nach Messungen von JOOS und HELLWEGE[2] bei Zimmertemperatur nur 1,5 cm^{-1} breit, das entspricht der Hg-Linie 2537 Å bei einer Druckverbreiterung durch 10 Atm Argon. Bei der Temperatur der flüssigen Luft gelang es selbst mit einem 6m-Gitter nicht, eine endliche Linienbreite anzugeben; die Linien sind hier schärfer als Gaslinien, da viele Ursachen für deren Breite wie der DOPPLER-Effekt bei den Kristallinien, die bei tiefer Temperatur aufgenommen sind, stark vermindert sind. Bei den Eu^{+++}-Salzen wurde auch zum ersten Mal aus der Polarisation der Linien die Natur und Lage der für die einzelnen Linien maßgebenden Strahlungsquelle festgestellt[3]. Die Auswahlregeln der Kristallquantenzahlen lassen eindeutig über elektrische Dipol-Quadrupol- oder magnetische Dipolstrahlung entscheiden. Fig. 20b gibt für das monokline Eu$_2$Zn$_3$(NO$_3$)$_{12} \cdot$ H$_2$O die Lage der Strahler für die gelbe (eine Komponente), grüne (drei Komponenten) und blaue (fünf Komponenten) Gruppe wieder[4]. Zur Erläuterung

---

[1] Wichtige Literatur zu den Spektren der übrigen Lanthanidensalze: *Prometheum* (nur in Lösung). G. W. PARKER u. P. M. LANTZ: J. Amer. Chem. Soc. **72**, 2834 (1950). — *Samarium.* F. H. SPEDDING u. R. S. BEAR: Phys. Rev. **46**, 308, 975 (1934). — *Terbium.* H. GOBRECHT: Ann. d. Phys. (5) **28**, 673 (1937). — H. F. GEISLER u. K. H. HELLWEGE: Z. Physik **136**, 293 (1953); dort ältere Literatur. — *Dysprosium.* A. M. ROSA: Ann. d. Phys. (5) **43**, 161 (1943). — *Holmium.* H. GOBRECHT: Ann. d. Phys. (5) **28**, 673 (1937). — H. G. KAHLE: Z. Physik **145**, 347 (1956). — H. SEVERIN: Z. Physik **125**, 455 (1949). — *Erbium.* H. GOBRECHT: Ann. d. Phys. (5) **28**, 673 (1937). — H. SEVERIN: Ann. d. Phys. (6) **1**, 41 (1947). — H. G. KAHLE: Z. Physik **145**, 361 (1956). — *Thulium.* H. GOBRECHT: Ann. d. Phys. (5) **31**, 755 (1938); (6) **7**, 88 (1950). — F. H. SPEDDING: Phys. Rev. **52**, 454 (1937). — *Ytterbium.* H. GOBRECHT: Ann. d. Phys. (5) **28**, 673 (1937). — S. FREED u. R. I. MESIROW: J. Chem. Phys. **5**, 22 (1937).
[2] G. JOOS u. K. H. HELLWEGE: Ann. d. Phys. (5) **39**, 25 (1941).
[3] O. DEUTSCHBEIN: Ann. d. Phys. (5) **36**, 183 (1939).
[4] K. H. HELLWEGE u. H. SCHRÖCK-VIETOR: Z. Physik **138**, 449 (1954).

ist in Fig. 20a die Lage des Strahls $\mathfrak{S}$ zu den kristallographischen Indikatrixachsen und den optischen Achsen gezeichnet.

ε) *Gadolinium.* Bei Gd^{+++}-Ionen enthaltenden natürlichen Kristallen und bei den reinen Salzen brachten die ZEEMAN-Effekt-Messungen eine große Überraschung: Während die Linien der Salze der übrigen Seltenen Erden im Magnetfeld nur Dubletts oder Quartetts geben, spalten manche Gd-Linien in neun Komponenten mit einer Totaltrennung von $16\,\Delta\nu_{\mathrm{norm}}$ auf[1]. Auf Grund unserer heutigen Kenntnisse ist dies einfach zu verstehen: Der Grundterm von Gd^{+++} ist $^8S_{\frac{7}{2}}$. Da das elektrische Feld primär nur am Bahnmoment angreift, dieses aber für einen S-Term null ist, bleibt der Term im Kristallfeld achtfach entartet und erst das Magnetfeld macht daraus acht Teilniveaus im Abstand von $2\Delta\nu_{\mathrm{norm}}$. Die Kom

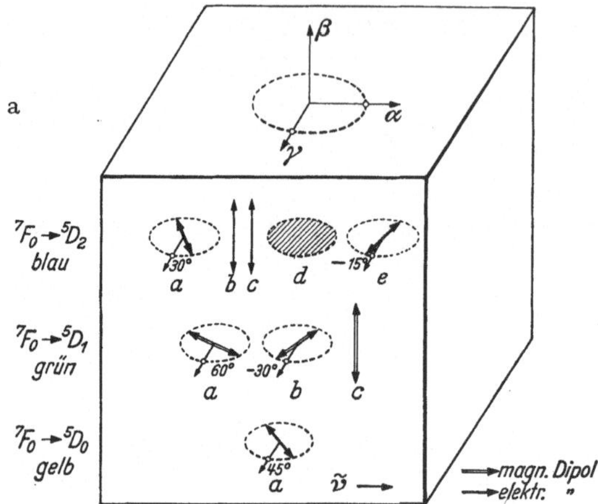

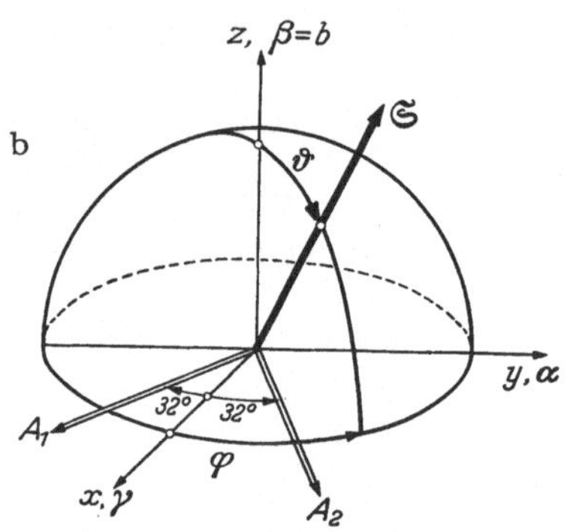

Fig. 20a u. b. a Natur und Lage der elementaren Strahler in $Eu_2Zn_3(NO_3)_{12}$. b Zur Orientierung des Koordinatensystems. α, β, γ Indicatrixachsen. Monokline Kristallachse: $b = \beta$. Optische Achsen: A_1, A_2. $\mathfrak{S}$ Strahlvektor außerhalb des Kristalls.

bination mit einem in zwei Komponenten mit Abstand $2\Delta\nu_{\mathrm{norm}}$ aufgespaltenen angeregten Term gibt die neun Komponenten mit der riesigen Aufspaltung von $16\Delta\nu_{\mathrm{norm}}$ der äußersten ZEEMAN-Komponenten.

Durch eine kleine Spin-Bahn-Wechselwirkung des Einzelelektrons (beginnende -j-Koppelung) müßte eine kleine Kristallfeldaufspaltung des Grundterms zustandekommen, deren Gesamtwert von der Größenordnung $1\ \mathrm{cm}^{-1}$, also zwei Größenordnungen kleiner als bei den anderen Salzen zu erwarten ist. Diese Aufspaltung und ihr Verhalten im Magnetfeld, die durch cm-Wellenabsorption nachzuweisen wäre, aber noch nicht gefunden wurde ist von J. DE BOER und R. VAN LIESHOUT[2], sowie KITTEL und LUTTINGER[3] berechnet worden.

b) Die Aktinidenreihe[4].

48. Die Analogie zur Lanthanidenreihe. Bekanntlich setzt bei Th in völliger Analogie zu der Reihe der Lanthaniden der Ausbau der $5f$-Schale ein. Nur drei der wieder zu erwartenden 14 Elemente (der *Aktiniden*) kommen in der Natur vor. Die schwereren können nur durch künstliche Kernumwandlungen erzeugt werden, wobei man 1956 bereits bis zu elf Elementen gekommen ist. Bis zum sechsten (Americium) sind auch die Absorptionsspektren der Salze aufgenommen

[1] F. H. SPEDDING: Phys. Rev. 38, 2080 (1931). Gd ist auch der Träger von Linien mit extrem hoher Aufspaltung, die in manchen Mineralien wie Tysonit gefunden werden.
[2] J. DE BOER u. R. V. LIESHOUT: Physica, Haag 15, 569 (1949).
[3] C. KITTEL u. J. M. LUTTINGER: Phys. Rev. 73, 162 (1948).
[4] Vgl. hierzu auch den Beitrag von HYDE und G. T. SEABORG (Transuranium elements) in Bd. XLII dieses Handbuches.

worden. Bei den minimalen Mengen von Am-Salzen, die zur Verfügung standen, wurde hier eine Mikrotechnik entwickelt, bei der das durchstrahlte Kriställchen in hundertfacher mikroskopischer Vergrößerung auf den Spektrographenspalt abgebildet wurde. Im Gegensatz zu den Lanthaniden, bei denen nur die Salze der dreiwertigen Ionen zur Untersuchung geeignet sind, kommen z.B. von Np auch vier-, fünf- und sechswertige Ionen vor, die entsprechend weniger $5f$-Elektronen besitzen. Nach den Messungen von FREED und LEITZ[1] an UCl_4, $NpCl_4$, $PuCl_3$ und $AmCl_3$ geht die Analogie zu den Lanthaniden so weit, daß auch das dem Eu^{+++} entsprechende Ion Am^{+++} wieder die schärfsten Linien gibt, die etwa dieselbe Halbwertsbreite bei Zimmertemperatur haben wie die Eu^{+++}-Linien. FREYMANN, FREYMANN und ROHMER[2] sowie SANCIER und FREED[3] führten einen Vergleich von UCl_4 mit $PrCl_3$ und UCl_3 mit $NdCl_3$ durch. Die entsprechenden Linien liegen in denselben Spektralbereichen, die Gitterionenbeeinflussung ist aber größer, was sich in folgendem zeigt: wesentlich größere Aufspaltung, stärkere Intensität der erzwungenen Dipollinien, größere Intensität der Schwingungskombinationslinien. Die stärkere Gitterbeeinflussung der Ionen zeigt sich auch beim Magnetismus, der von HOWLAND und CALVIN[4] an Np^{6+}, Np^{5+}, Np^{4+}, Pu^{4+}, Pu^{3+} und Am^{3+} studiert wurde und viel größere Abweichungen von dem des freien Ions zeigt als bei den Lanthaniden.

c) Die Eisenreihe.

49. Der Unterschied zur Lanthaniden- und Aktinidenreihe. Bei den Elementen Sc bis Cu findet der Ausbau der $3d$-Schale statt, welche im abgeschlossenen Zustand zehn Elektronen enthält. Dadurch wird die Möglichkeit von Übergängen zwischen den verschiedenen Konfigurationen der $n\ 3d$-Elektronen geschaffen, bei denen keine Änderung der Hauptquantenzahl stattfindet, was ja die Voraussetzung für einigermaßen scharfe Linien ist. Es besteht aber ein großer Unterschied gegenüber den Salzen der Seltenen Erden: Da die neutralen Atome der Eisenreihe außer den $3d$-Elektronen nur ein oder zwei $4s$-Elektronen besitzen, liegt die unvollständige Schale bei den zwei- oder dreiwertigen Ionen an der Oberfläche, ist also in ganz anderem Maß der Störung durch die Umgebung ausgesetzt wie die $4f$-Schale der Lanthaniden oder die $5f$-Schale der Aktiniden. Außerdem ist an dieser Stelle des Periodischen Systems die Multiplettaufspaltung, die ungefähr mit Z^4 geht, noch klein. Wir haben also den Fall, daß die Kristallfeldaufspaltung groß gegen die Multiplett-Aufspaltung ist. Ja man kann fragen, ob bei der Bindung der Ionen an ihre Umgebung, z. B. bei dem Komplex $Cr(H_2O)_6^{+++}$, nicht unpolare Bindungsanteile überwiegen und ob man überhaupt das Recht hat, die Umgebung durch ein elektrostatisches Feld bestimmter Symmetrie zu schematisieren. ORGEL[5] gibt dazu eine sehr einleuchtende Antwort: Für die Ermittlung der Zahl der Aufspaltungsniveaus und die Symmetrieeigenschaften der Terme zieht man die Gruppentheorie heran, welche nur die Symmetrie-Elemente der Umgebung verwendet, für die *Berechnung* des Zahlwerts der Aufspaltung benützt man empirische Parameter, die aus optischen oder magnetischen Messungen entnommen sind. Da man immer Felder von oktaedrischer Symmetrie, also kubische Kristallfelder annimmt[6], handelt es sich immer nur um einen Parameter $C = D \cdot q$, der so entsteht: Als Störungspotential dieser

---

[1] S. FREED u. F. J. LEITZ: J. Chem. Phys. **17**, 540 (1949).
[2] M. FREYMANN, R. FREYMANN u. ROHMER: C. R. Acad. Sci., Paris **230**, 1524 (1950).
[3] K. M. SANCIER u. S. FREED: J. Chem. Phys. **20**, 349 (1952).
[4] J. HOWLAND u. M. CALVIN: J. Chem. Phys. **18**, 239 (1950).
[5] L. E. ORGEL: J. Chem. Phys. **23**, 1004, 1819 (1955).
[6] W. G. PENNEY u. R. SCHLAPP: Phys. Rev. **41**, 194 (1932); **42**, 666 (1932).

Symmetrie setzt man an:

$$V = D\left(x^4 + y^4 + z^4 - \tfrac{3}{5}r^4\right). \tag{49.1}$$

(Das letzte Glied ist zur Erfüllung der LAPLACEschen Gleichung zugefügt, es verschiebt lediglich das ganze Termgebilde.) Der in die Rechnung eingehende Parameter C stellt nach den Ziff. 27 und 29 das Produkt des Proportionalitätsfaktors D mit einer Größe q dar, die gleich dem mit dem Zahlenfaktor β_L multiplizierten Integral über das mit $r^4 \cdot r^2$ multiplizierte Quadrat der radialen Eigenfunktion $R(r)$ des Termes ist:

$$q = \beta_L \int\limits_0^\infty r^4 R^2(r)\, r^2\, dr. \tag{49.2}$$

Für das System der zwei $3d$-Elektronen des V^{+++} im Komplex $[V(H_2O)_6]^{+++}$ ist die Störungsrechnung in allen Einzelheiten von ILSE und HARTMANN[1] durchgeführt worden. Die Absorptionsspektren dieses Komplexes zeigen zwei breite Banden, welche als Übergänge zwischen den Aufspaltungsniveaus des Grundterms ^3F gedeutet werden. Diese Autoren vermuten, daß alle langwelligen Absorptionsbanden der Komplexe von Elementen der Eisengruppe solchen Übergängen zuzuordnen sind, die nur unter Kombinationen mit Schwingungsenergieänderungen zulässig sind[2]. Es ist aber nicht zu verstehen, wieso bei den ersten Elementen der Reihe nur breite Banden, bei dem Dreielektronensystem Cr^{+++} aber wieder scharfe Linien auftreten. Denkbar ist, daß die Breite durch nichtpolare Anteile der Bindung bedingt ist, aber auch die Möglichkeit ist nicht ausgeschlossen, daß bei V^{+++} scharfe Linien in den Bereich der fast kontinuierlichen Absorption fallen. Da die scharfen Linien des Cr^{+++}-Ions viel mehr Einzelheiten erkennen lassen, ist dieses, und zwar hauptsächlich eingebaut in die kubischen Alaune, am meisten untersucht worden[3]. Da diese Kristalle im Rot und im Blau scharfe Linien haben und in diese Spektralbereiche die Übergänge $^4F - {}^2G$ und $^4F - {}^2H$ des freien Ions fallen, hat man in den älteren Arbeiten die Chromalaunlinien mit diesen Linien des freien Ions in Verbindung gebracht. Diese Zuordnung ist aber heute nicht mehr zu halten. Nach einer Rechnung von FINKELSTEIN und VAN VLECK[4] spaltet das Kristallfeld die Terme soweit auf, daß die Teilniveaus der Terme des freien Ions Abstände haben, die von der Größenordnung der Termabstände selbst sind, so daß in ihrer relativen Lage nichts mehr von den Termwerten des freien Ions zu erkennen ist. Das Resultat der Rechnung ist im Termschema Fig. 21 wiedergegeben. Die Bezeichnungen des Termschemas entsprechen der Gruppen-Nomenklatur. Dabei ist die charakteristische Konstante Dq aus magnetischen Messungen an anderen Ionen der Eisengruppe zu $1500\ cm^{-1}$ geschätzt. Mit diesem Zahlwert würde das „charakteristische Dublett" (vgl. unten) einem Übergang zwischen den Aufspaltungstermen des Terms ^4F zahlenmäßig entsprechen. Aus dem ZEEMAN-Effekt, der entweder keine oder doppeltnormale Aufspaltung der Linien gibt[5], schließt aber VAN VLECK[6] zwingend, daß der angeregte Term aus einem Dublett-Term hervorgegangen ist.

---

[1] F. E. ILSE u. H. HARTMANN: Z. phys. Chem. **197**, 239 (1951). — Z. Naturforsch. 6a, 751 (1951). — H. HARTMANN u. H. L. SCHLÄFER: Z. phys. Chem. **197**, 116 (1951). — Z. Naturforsch. 6a, 754, 760 (1951).

[2] F. E. ILSE u. H. HARTMANN: Z. Naturforsch. 6a, 751 (1951). — H. HARTMANN u. H. L. SCHLÄFER: Z. Naturforsch. 6a, 754, 760 (1951).

[3] H. SAUER: Ann. d. Phys. (4) **87**, 197 (1928). — F. H. SPEDDING u. C. NUTTING: J. Chem. Phys. **2**, 421 (1934). — D. L. KRAUS u. G. C. NUTTING: J. Chem. Phys. **9**, 133 (1941). — E. GOLLING: Ann. d. Phys. (6) **9**, 181 (1951).

[4] R. FINKELSTEIN u. J. H. VAN VLECK: J. Chem. Phys. **8**, 790 (1940).

[5] F. H. SPEDDING u. G. C. NUTTING: J. Chem. Phys. **3**, 369 (1935).

[6] J. H. VAN VLECK: J. Chem. Phys. **8**, 787 (1940).

Bei der Unsicherheit des Störungsparameters Dq ist eine solche Differenz von 3000 cm^{-1} (vgl. Fig. 21) durchaus möglich.

Die wesentlichen experimentellen Ergebnisse der Absorptionsmessungen im roten Teil — im blauen Teil sind die Spektren ähnlich, aber nicht so eingehend studiert — sind folgende: Alle Chromalaune haben bei 6700 Å eine scharfe Liniengruppe, an die sich nach kurzen Wellen zahlreiche immer verschwommener werdende Banden anschließen. Hinsichtlich der scharfen Liniengruppe lassen sich die Alaune in drei Klassen einteilen[1]:

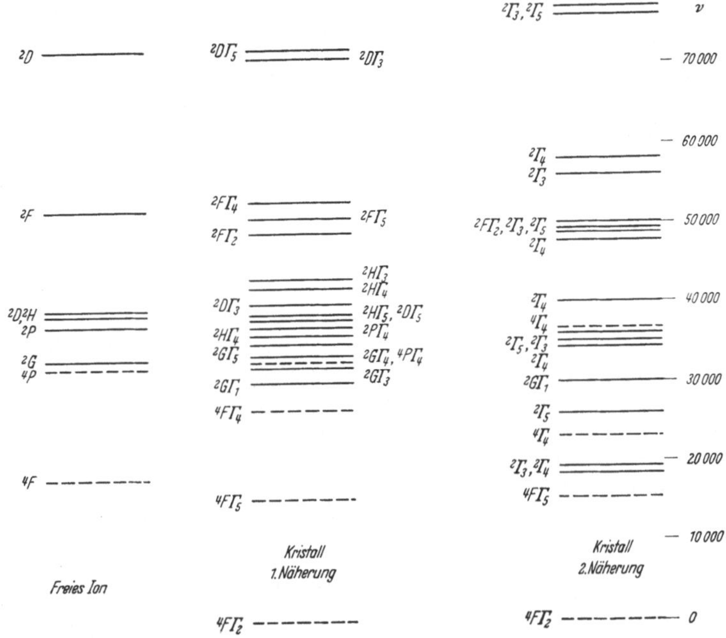

Fig. 21. Die berechneten Energieniveaus des Cr^{+++}-Ions im Chromalaun.

a) Die charakteristische Liniengruppe ist ein starkes Dublett mit einigen schwachen scharfen Begleitlinien. Hierher gehören Rb-Tl-Sulfat und -Selenat, Cs- und K-Selenat, Hydroxylamin-Sulfat und alle NH$_4$-Alaune bei höherer Temperatur.

b) Bei Methylamin und Cs-Sulfat-Alaun ist die charakteristische Gruppe noch bei 85° K völlig diffus, wird aber bei 20° K eine einzige scharfe Linie, die kein Zeichen von Auflösung in zwei Komponenten zeigt.

c) K-Sulfat und alle Tieftemperaturformen der NH$_4$-Alaune haben an dieser Stelle eine Gruppe von gleichstarken Linien, deren Zahl bis zu sechs geht. Für die Deutung ist eine Beobachtung von Golling[2] an Mischkristallen wichtig: Mit zunehmender Ersetzung von K durch NH$_4$ verschieben sich zwei Linien während vier stehen bleiben. Dies spricht dafür, daß in dieser Liniengruppe das Dublett der Klasse a und die Übergänge nach $^4\Gamma_i$ sich wahrscheinlich überdecken. Die anschließenden Banden sind Schwingungsüberlagerungen zuzuordnen. Bei Ersatz des Kristallwassers durch D$_2$O treten erhebliche Verschiebungen dieser Banden auf, wobei die Verschiebung ungefähr quadratisch mit dem Abstand von den Hauptlinien zunimmt und bis zu 30 cm^{-1} geht[3]. Mit

[1] D. L. Kraus u. G. C. Nutting: J. Chem. Phys. 9, 133 (1941).
[2] E. Golling: Ann. d. Phys. (6) 9, 181 (1951).
[3] H. Böhm: Ann. d. Phys. (5) 32, 521 (1938).

diesem Befund verträglich ist die Deutung als Schwingungen ganzer Wassermolekeln gegenüber dem Cr^{+++}-Ion. (Die Eigenschwingungen der Wassermolekeln oder der SO_4^{--}-Ionen haben viel höhere Eigenfrequenzen als sie den Abständen dieser Banden entspricht und kommen daher nicht in Frage.) Solche Schwingungen wurden von Magat[1] auch anderwärts beobachtet und als äußere Wasserschwingungen bezeichnet.

Beim Zeeman-Effekt ist noch die von Schnetzler[2] bei 85° K beobachtete Anisotropie bemerkenswert, welche der lokalen am Ort des Cr^{+++}-Iones geltenden Symmetrie folgt. Bei 20° K fanden Spedding und Nutting nichts derartiges.

Das charakteristische Dublett findet sich auch, allerdings meist viel mehr verbreitert in andern Chromkomplexen[3], so z.B. in $Cr\,(NH_3)_6^{+++}$. Wird aber z.B. eine NH_3-Molekel durch SCN ersetzt, so erhält man 4 weit getrennte annähernd äquidistante breite Banden.

Die isomorphe Einlagerung von Cr_2O_3 in das Gitter des Al_2O_3 gibt die rote Farbe des Rubin-entsprechend dem auch in Chromalaunen vorhandenen kontinuierlichen Absorptionsspektrum im Grün. Aber auch das charakteristische Dublett tritt als Rubinlinien $R_1 = 6933,7$ und $R_2 = 6919,8$ Å mit großer Schärfe auf. Dies viel untersuchte Dublett erhält man auch in Fluorescenz bei Anregung durch U.V. oder Elektronenbeschuß. Auch sein Zeeman-Effekt ist schon früher untersucht worden[4]. Jede der beiden Linien spaltet, wenn die optische Achse senkrecht zum Feld steht, in ein Quartett, wenn sie parallel zu ihm ist, in ein Triplett auf. Im großen ganzen hängt die Aufspaltung nur von der Orientierung von Feld, optischer Achse und Vektor $\mathfrak{E}$ ab. Indessen folgen feinere Veränderungen der Intensitätsverhältnisse bei Drehung um die verschiedenen ausgezeichneten Achsen der Gittersymmetrie. Obwohl die Aufspaltung feldstärkeproportional ist, liegen die vier Komponenten unsymmetrisch zur ursprünglichen Linie. Bei R_1 sind die Abstände von der feldfreien Linie z.B. $-3,15$; $-1,32$; $0,35$; $2,00\ \varDelta\nu_{\text{norm}}$. Eine theoretische Erklärung ist noch nicht gefunden.

Deutschbein[5] machte die merkwürdige Beobachtung, daß Absorptions- und Emissionslinien nicht genau zusammenfallen. Eine eingehende Nachprüfung durch Deutschbein, Joos und Teltow[6] bestätigte dies und führte zur Erklärung durch eine Kristallhyperfeinstruktur der Linien, die bei dem niedrigersymmetrischen Disthen gut meßbar ist. Die Übergänge in das eine oder andere Hyperfeinstrukturniveau sind offenbar in ihrer Häufigkeit sehr strukturempfindlich, denn die Verschiebung der beiden Spektren des Rubins, die etwa 0,1 cm^{-1} beträgt, schwankt von Probe zu Probe. Beim Disthen, wo die Aufspaltung wahrnehmbar ist, ist keine Verschiebung zwischen Absorption und Fluorescenz da.

Auch bei den Ionen Mn^{++}, Ni^{++} und Co^{++} wurden von Gielessen[7] linienhafte Absorptionen gefunden, die aus zahlreichen mehr oder weniger scharfen Streifen bestehen. Das große in dieser Arbeit steckende empirische Material harrt noch der Auswertung.

Anhang. Spektren anderer Übergangselemente. In den Salzen der homologen Elemente der 4. und 5. Reihe des Periodischen Systems wurde bisher noch keine linienhafte Absorption gefunden. Am ehesten wäre diese bei den zu Cr homologen

[1] H. Magat: Trans. Faraday Soc. **33**, 117 (1937).
[2] K. Schnetzler: Ann. d. Phys. (5) **10**, 373 (1931).
[3] G. Joos u. K. Schnetzler: Z. phys. Chem., Abt. B **20**, 1 (1933). — B. Duhm: Z. phys. Chem., Abt. B **38**, 359 (1937). (Ersatz von H durch D.)
[4] Vgl. H. Lehmann: Ann. d. Phys. (5) **19**, 99 (1934). Dort ältere Literatur.
[5] O. Deutschbein: Ann. d. Phys. (5) **14**, 712 (1932).
[6] O. Deutschbein, G. Joos u. J. Teltow: Naturwiss. **30**, 228 (1942).
[7] J. Gielessen: Ann. d. Phys. (5) **22**, 537 (1935).

Elementen Mo und W zu erwarten, die auch von GIELESSEN in verschiedenen Komplexen untersucht wurden. Es ergaben sich aber nur breite Kontinua. Der Grund liegt in der zunehmenden Größe dieser Ionen, deren Elektronenhüllen tief in die der Nachbarbausteine hineingreift. Dies zeigt sich auch in dem zunehmenden Halbleitercharakter der betreffenden Komplexe.

d) Spektren von Molekeln der Übergangselemente, die Teile des Kristallgitters sind.

50. Lanthaniden- und Aktinidenmolekeln. Es erscheint in gewissem Grad willkürlich, in einem Kristall einzelne Atom- oder Ionengruppen zu einer Molekel zusammenzufassen, da ja der ganze Kristall eine Riesenmolekel darstellt. Man wird dazu aber dann das Recht haben, wenn die Gruppe charakteristische in andern Kristallen wiederkehrende Eigenschaften hat, die nichts mehr mit denen der Bestandteile zu tun haben, also insbesondere charakteristische Eigenschwingungen und Elektronenzustände, die sich nicht aus denen der Bausteine ableiten lassen. Niemand wird die Berechtigung bestreiten, das SO_4^{--}-Ion im Kristallgitter als eine geladene Molekel anzusehen, da z.B. ihre infraroten Eigenschwingungen, wenn auch je nach Gittersymmetrie aufgespalten, in allen Sulfatkristallen wiederkehren. So gibt es auch bei den Übergangselementen Molekülionen, deren Spektren in den verschiedenen Kristallen mehr oder weniger modifiziert wiederkehren. Die schärfsten Linien sollten wieder Molekeln zeigen, bei denen eine unvollständige $4f$- oder $5f$-Schale vorhanden ist, also Molekeln der Lanthaniden- oder Aktinidenreihe. Von den Seltenen Erden sind keine Molekeln bekannt, wohl aber ist die Molekel UO_2^{++}, das Uranylradikal, seit langem eingehend untersucht worden[1].

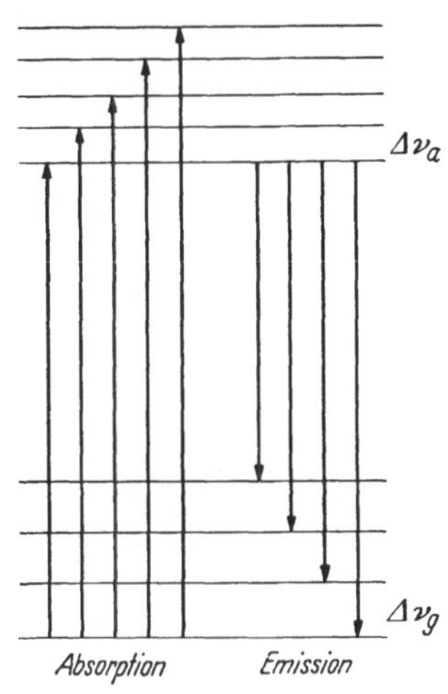

Fig. 22. Emissions- und Absorptionsbanden bei UO_2 (schematisch).

Der Molekelcharakter von UO_2^{++} tritt in seinem Spektrum durch die Folge äquidistanter Liniengruppen zutage. Außer dem Absorptionsspektrum kann man auch das Fluorescenzspektrum durch Einstrahlung von ultraviolettem Licht erhalten. Beidemal erhält man die oben erwähnten Schwingungsfolgen, und zwar ergibt nach Fig. 22 das Absorptionsspektrum durch den Abstand homologer Linien die Eigenschwingung des angeregten Zustands zu 720 cm^{-1}, die des Fluorescenzspektrums die des Grundzustands mit 860 cm^{-1}. Die Zahlen schwanken etwas in den verschiedenen Uranylsalzen. Diese Bandenstruktur ist auch noch in den wäßrigen Lösungen der Uranylsalze und in den Uranylgläsern zu erkennen. Schwieriger zu verstehen ist die Struktur der Einzelbanden, die aus 4 bis 8 Einzellinien mit Abständen der Größenordnung 50 cm^{-1} bestehen. Diese Struktur hängt von der Symmetrie des Kristallgitters ab und ist bei Einkristallen im ordentlichen und außerordentlichen Spektrum verschieden. Bei gleicher Kristallstruktur haben oft ganz verschiedene Salze dieselbe Feinstruktur. Die wahrscheinlichste Erklärung ist die, daß es sich um eine Aufspaltung des Elektronenterms im Kristallgitter handelt. Joos und DUHM[2] ersetzten bei Uralylnitrat das Kristallwasser durch D_2O und erhielten eine Verminderung der Abstände dieser

[1] E. L. NICHOLS u. H. MERRITT: Carnegie Publ. No. 298, Washington 1915. — E. L. NICHOLS u. H. L. HOWES: Phys. Rev. **8**, 364 (1916). — E. L. NICHOLS u. E. MERRITT: Phys. Rev. **9**, 113 (1917).

[2] G. JOOS u. B. DUHM: Göttinger Nachr. **2**, 123 (1936).

Linien um etwa 3 %, während die Eigenschwingung 718 cm^{-1} unverändert bleibt. Die Deutung dieses auffallend großen Einflusses des Kristallwassers durch Zuordnung der Feinstrukturlinien zu Schwingungen von Wassermolekeln gegenüber dem UO_2^{++}, ist aber nicht mehr aufrechtzuerhalten, da auch kristallwasserfreie Uranylsalze fast dieselben Aufspaltungen zeigen, wenn nur die Gittersymmetrie dieselbe ist. Die Feinstruktur muß also wohl durch eine Aufspaltung des Elektronenterms bedingt sein. Dafür spricht auch, daß bei den Spektren mit D_2O sämtliche Linien eine gemeinsame Verschiebung erfahren. Der große Unterschied $D_2O - H_2O$ ist aber bemerkenswert.

Auch das offenbar ganz ähnlich gebaute Plutonyl-Ion $(PuO_2)^{++}$ hat ein analoges Spektrum[1] mit einer Eigenfrequenz des angeregten Zustands von 710 cm^{-1} (Grundzustand aus RAMAN-Effektmessungen 865 cm^{-1}), doch ist die Feinstruktur der einzelnen Banden wesentlich verwickelter als bei UO_2 und harrt noch der Klärung.

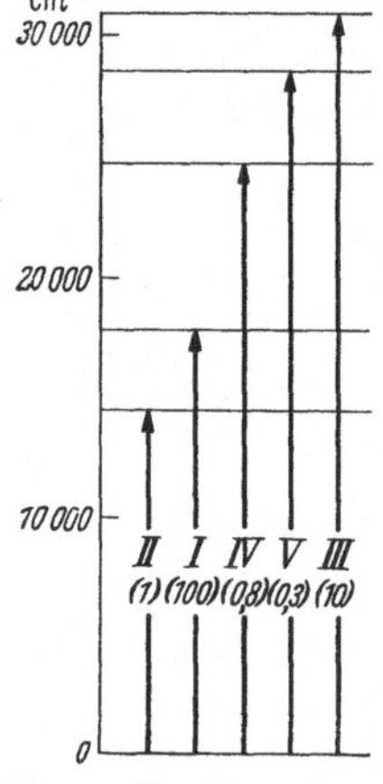

Fig. 23. Die Absorptionsbanden des MnO_4^--Ions (eingeklammerte Zahlen relative Intensitäten).

51. Spektren von Molekeln der Eisengruppe.

Seit alter Zeit sind die äquidistanten Banden einer $KMnO_4$-Lösung bekannt, die dem Molekülion MnO_4^- zuzuschreiben sind. In Wirklichkeit hat das MnO_4^- aber fünf solche Bandensysteme[2], von denen das im grünen gelegene, die violette Farbe der MnO_4^--Salze bewirkende, das stärkste ist (Fig. 23). Während bei den Seltenen Erden und auch bei den linienhaften Spektren der Ionen der Eisenreihe die Oszillatorenstärke der einzelnen Linien von etwa 10^{-4} bis 10^{-9} geht, kommt der im Grün gelegenen Permanganatabsorption die Oszillatorenstärke 1 zu. Diese hohe Oszillatorenstärke läßt bei kleinen Prismen aus $KMnO_4$ anomale Dispersion beobachten. Die Aufnahme des Kristallspektrums erfordert aber deshalb außerordentlich dünne Schichten, die technisch schwer herzustellen sind. Man lagert daher besser das MnO_4^--Ion in andere Kristalle isomorph ein, was noch den Vorteil bringt, daß man die Symmetrie des Umgebungsfelds wechseln kann. Besonders geeignet sind dazu die Perchlorate. Fig. 24 zeigt die Veränderungen des MnO_4^--Spektrums mit der Änderung des Wirtsgitters. Auffällig ist besonders die große Intensitätsänderung der zwischen den Hauptbanden liegenden Banden und die große Anisotropie.

Ein großer Teil des in der Arbeit von TELTOW enthaltenen Materials konnte von WOLFSBERG und HELMHOLZ[3] gedeutet werden. Die beiden Verfasser gehen von den Termen des Ions MnO_4^- aus, das in seinem Kerngerüst (reguläres Tetraeder) die Symmetrie T_d aufweist. Die Elektronenterme gehören nach steigender Energie geordnet zu den Darstellungen (vgl. Ziff. 17) A_1 (Grundterm), A_2, E, T_1 und T_2. Erlaubt ist der Übergang $A_1 \rightarrow T_2$; er entspricht also dem intensiven grünen Absorptionsspektrum. Beim Einbau in das monokline $KClO_4$ mit Symmetrie C_s geht A_1 in A' über, während T_2 aufspaltet in $A'' + A' + A'$. Erlaubt sind nun für $\mathfrak{E} \| a$, $\mathfrak{E} \| c$, d.h. $\mathfrak{E} \|$ zur Ebene (ac) die Übergänge $A' \rightarrow A'$, für $\mathfrak{E} \| b$ dagegen $A' \rightarrow A''$. Da es zwei A' Terme gibt, hat man für $\mathfrak{E} \| a$ und $\mathfrak{E} \| c$ Dubletts der Hauptlinien, für $\mathfrak{E} \| b$ dagegen Einfachlinien ganz in Übereinstimmung mit der Beobachtung. Die Abstandsdifferenz der Hauptlinien entspricht der Eigenfrequenz der totalsymmetrischen Schwingung des Tetraeders mit $\Delta \nu = 760$ cm^{-1}.

[1] M. KASHA: J. Chem. Phys. **17**, 349 (1949).
[2] J. TELTOW: Z. phys. Chem. Abt. B **40**, 397 (1938).
[3] M. WOLFSBERG u. L. HELMHOLZ: J. Chem. Phys. **20**, 837 (1952).

Noch mehr Einzelheiten konnten WOLFSBERG und HELMHOLZ aus dem Absorptionsgebiet II (rotes Spektrum) herausholen. Die Analyse bezieht sich auf das in wasserfreies $NaClO_4$ (Symmetrie C_{2v}) eingelagerte Molekülion. In diesem Gebiet gehen die Aufspaltungsniveaus von E und T_1 durcheinander und erzeugen mit den Tetraeder- und Gitterschwingungen zusammen ein Gewirr, dessen Ordnung TELTOW nicht gelungen war. Unter Benützung von zwei aus dem grünen Spektrum entnommenen Gitterschwingungen von 80 und 150 cm^{-1}, sowie vier Schwingungen des Tetraeders gelingt eine völlige Ordnung der Linien des Gebiets II. Am bemerkenswertesten ist die scharfe Linie 14446 cm^{-1}, welche für $\mathfrak{E} \parallel b$ (in der Arbeit von TELTOW ist b mit a zu vertauschen) und nur in dieser Polarisation erscheint. Sie ist dem schwingungslosen Übergang von A' nach dem aus E entstandenen $A''(E)$ zuzuordnen.

Analoge Banden wurden von TELTOW[1] auch bei CrO_4^{--} und MnO_4^{--} gefunden, deren Analogie auch von HELMHOLZ und WOLFSBERG klargestellt wurde. Dagegen zeigt das Vanadat-Ion VO_4^{---} nur zwei verwaschene breite Banden im Ultraviolett.

Das Bichromat-Ion $Cr_2O_7^{--}$ hat eine dem grünen MnO_4^--Absorptionsspektrum analoge kräftige Absorption von 4000 Å nach kurzen Wellen, die die bekannte gelbe Farbe verursacht. Bei tiefer Temperatur läßt sie sich in ein Bandensystem mit $\Delta\nu = 760$ cm^{-1} auflösen[2]. Viel interessanter ist aber das dem roten MnO_4^--Spektrum entsprechende schwache Kontinuum bei 5500 Å. An dessen langwelliger Seite treten bei 20° K, selbst bei 85° K noch nicht wahrnehmbare äußerst scharfe Linien auf[3], deren Lage stark von der Orientierung von $\mathfrak{E}$ zu den Achsen abhängt. Die Halbwertsbreite dieser Linien ist bei 20° K nur etwa 2 cm^{-1}. Man kann mit ziemlicher Sicherheit sagen, daß diese Linien schwingungslosen Elektronenübergängen zugehören, die der einen scharfen Linie des MnO_4^{--} entsprechen. In dem völlig unsymmetrischen $K_2Cr_2O_7$-Kristall (triklin-pediale Klasse) erfolgt eine Aufspaltung der Übergänge in zahlreiche Linien. Daß sie unter völliger Aufhebung aller Entartungen entstanden sind, beweist auch der Umstand, daß das

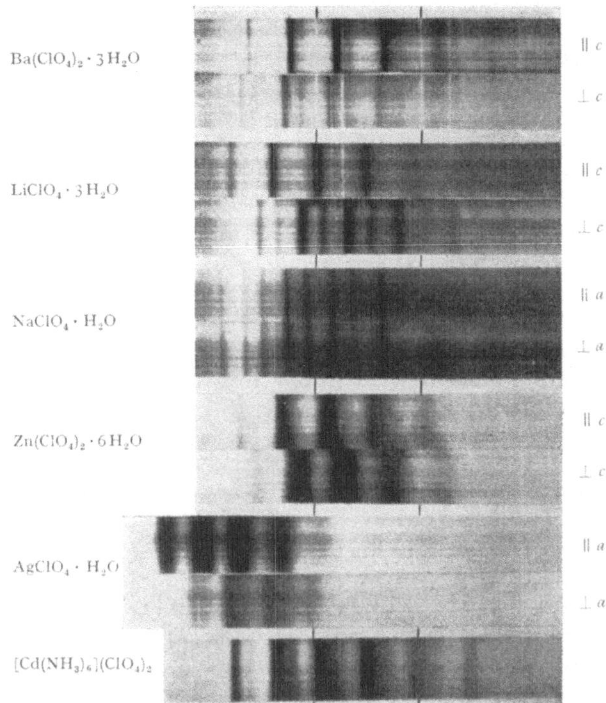

Fig. 24. Grünes Spektrum des MnO_4^--Ions in verschiedenen Wirtsgittern.

[1] J. TELTOW: Z. phys. Chem., Abt. B 43, 198 (1939).
[2] H. SCHAUMANN: Z. Physik 76, 106 (1932).
[3] Erstmals beobachtet von I. OBREIMOW u. W. J. DE HAAS, Leyden commun. 191a (1928).

Magnetfeld keine weitere Aufspaltung mehr hervorruft. Wenn zu jeder dieser Linie ein System verbreiterter Banden gehört, ist zu verstehen, daß das anschließende Spektrum durch Überlagerung als Kontinuum erscheint.

II. Kristalle organischer Moleküle.

52. Die Spektren von Benzol und Naphthalin. Bei organischen Molekeln sind ebenfalls Übergänge ohne Änderung der Hauptquantenzahl möglich, so daß die Grundbedingung für das Auftreten scharfer Linien im Kristall erfüllt ist. Da aber die Nachbarbausteine elektrisch neutral sind, kommt eine Aufspaltung der Elektronenterme durch elektrische Felder nicht in betracht. Hier kommt allein die in Ziff. 44 behandelte DAVYDOV-Aufspaltung zur Wirkung. Leider sind bereits die Spektren der freien Molekeln so verwickelt, daß es schwer fällt, diese einigermaßen zu analysieren und dann die Kristallspektren mit ihnen zu vergleichen. Schon das Spektrum des verhältnismäßig einfachen Benzolmoleküls, das im kristallinen Zustand von KRONENBERGER[1] und in Einkristallen von BROUDE, MEDVEDEW und PRIKHOTJKO[2] untersucht wurde, ist wenig geklärt. Die Banden des kristallinen Zustands, die bei 20° K sehr scharf sind, lassen sich nach einer Verschiebung um 258 cm⁻¹ nach höheren Frequenzen mit denen des Gases im allgemeinen Verlauf zur Deckung bringen, doch konnte KRONENBERGER nicht entscheiden ob die zahlreichen Linien der Feinstruktur, die besonders bei dickeren Schichten auftreten, mit denen des Gaszustands, bei dem sie auf Rotation zurückgeführt werden, identisch sind, da die Überlappung der vielen Linien kein klares Bild liefert. Eine Theorie der angeregten Zustände ist von FOX und SCHNEPP[3] ausgearbeitet worden, doch ist der Vergleich mit der Beobachtung aus den oben angeführten Gründen nur in geringem Umfang möglich. Etwas mehr läßt sich aus dem Naphthalinspektrum herausholen. Dies wurde von McCLURE und SCHNEPP im gasförmigen Zustand[4], in fester Lösung in Durol[5] und im reinen Kristall[6] untersucht. Das Naphthalinmolekül hat drei Absorptionsgebiete: I: 3200—2900 Å sehr schwach, II: 2900—2500 Å etwas stärker, und III: von 2500 Å nach kurzen Wellen sehr stark (wohl erlaubter Übergang). Die beiden ersten konnten als Übergänge $^1A_{1g} \rightarrow {}^1B_{3u}$ und $^1A_{1g} \rightarrow {}^1B_{2u}$ gedeutet werden. Wie aus der DAVYDOVschen Theorie zu erwarten, gibt es gegenüber dem Gas eine Verschiebung des Elektronenterms und eine Dublettaufspaltung als „Faktorgruppenaufspaltung". Die Verschiebung beträgt bei I 462 cm⁻¹ bei II 2213 (2290) cm⁻¹, die Aufspaltung bei I 166 cm⁻¹ bei II 173 (320) cm⁻¹. Die eingeklammerten Zahlen bei II beziehen sich auf eine andere mögliche Zuordnung.

III. Die Spektren der Alkalihalogenid-Kristalle.

a) Reinkristalle.

53. Die Deutung der Absorptionsbanden. Die Spektren der zum Kristall zusammengefügten edelgasähnlichen Ionen wie K^+ und Cl^- stellen das einfachste Beispiel von Spektren dar, bei denen die bisher immer erfüllte Bedingung nicht gilt, daß der Übergang vom Grundzustand zum angeregten Zustand ohne Änderung der Hauptquantenzahl erfolgt. Trotzdem haben diese Kristalle noch relativ schmale Absorptionsbanden, die eine atomistische Deutung als möglich erscheinen

[1] A. KRONENBERGER: Z. Physik **63**, 497 (1930).
[2] S. BROUDE, M. MEDVEDEW u. A. PRIKHOTJKO: J. exp. theor. Phys. USSR. **21**, 665 (1951).
[3] D. FOX u. O. SCHNEPP: J. Chem. Phys. **23**, 767 (1955).
[4] O. SCHNEPP u. D. S. McCLURE: J. Chem. Phys. **20**, 1375 (1954).
[5] D. S. McCLURE: J. Chem. Phys. **22**, 1668 (1954).
[6] D. S. McCLURE u. O. SCHNEPP: J. Chem. Phys. **23**, 1575 (1954).

lassen. Die experimentelle Erzeugung dieser Spektren führt auf zwei Schwierigkeiten: Da die unter Änderung der Hauptquantenzahl erfolgenden Übergänge erlaubt sind, ist die Absorption in diesen Banden so stark, daß außerordentlich dünne Schichten verwandt werden müssen. Man dampft zu diesem Zweck Schichten von etwa 20 mμ auf Spiegel auf und läßt das Licht diese Schichten zweimal durchsetzen. Weiter liegt der größte Teil dieser Spektren im Vakuumultraviolett zwischen 1000 und 2000 Å. Nur die langwelligste Bande fällt noch ins bequem zugängliche Ultraviolett. Man nahm daher anfangs, als man nur diesen Bereich untersuchte, an, daß hier eine kontinuierliche Absorption einsetze. Fig. 25 gibt für 6 Alkalihalogenide die Struktur der Banden. Wie man sieht, ist das Spektrum durch das Anion bestimmt. Dies ist auch zu erwarten: die Absorption der freien Ionen Na$^+$ usf. liegt im Gebiet von etwa 500 Å, also außerhalb des betrachteten Gebiets, wenn man wenigstens qualitativ die Absorptionsgebiete der freien Ionen auch noch im Kristall als gültig ansieht. Die Breite der Banden und die Lage ihrer Maxima hängt stark von der Temperatur ab. Nach Untersuchungen von MARTIENSSEN[1] beträgt die Halbwertsbreite der langwelligsten Bande bei 20° K nur etwa 20 Å (vgl. Fig. 26). Dies ist etwa das 10fache der durch 30 Atm H$_2$ verbreiterten Resonanzlinie des gasförmigen Hg.

HILSCH und POHL[2] haben als erste diese Bande systematisch untersucht. Sie ordneten sie einem Übergang des Elektrons vom Anion (Beispiel: Cl$^-$) zum Kation (Beispiel: K$^+$) zu. Auf Grund folgender einfacher Überlegung kamen sie zu einer bemerkenswert einfachen Beziehung zwischen der Lage dieser Bande und den Daten des Kristalls: Man entferne aus dem Gitter ein Cl$^-$- und ein K$^+$-Ion, dabei ist eine Arbeit aufzuwenden, die etwa der Gitterenergie bezogen auf ein Ionenpaar entspricht, also jedenfalls dem Kehrwert der Gitterkonstanten proportional sein wird. Man nehme nun dem Cl$^-$-Ion ein Elektron weg: aufzuwendende Arbeit gleich der Elektronenaffinität des Cl$^-$-Atoms E, man überführe es ans K$^+$-Ion: zu gewinnende Arbeit Ionisierungsspannung des K$^+$-Atoms J. Man setzte die neutralen Atome wieder ins Gitter: da keine Ladung vorhanden, kann die hierfür aufzuwendende oder dabei zu gewinnende Arbeit nur klein sein. Man hat damit die Bilanz

$$h\nu = E - J + b/a. \tag{53.1}$$

Mit einer einzigen empirischen Konstanten $b = 5 \cdot 10^{-7}$ eV cm wird diese Beziehung von 12 Alkalihalogeniden und 4 -hydriden gut erfüllt. Es sind weitere Verfeinerungen dieser Formel vorgenommen worden (vgl. z.B. das Tabellenwerk von LANDOLT-BÖRNSTEIN, 6. Aufl., Bd. I, 4 Kristalle, S. 869ff.).

Wenn trotzdem heute diese einfachen Betrachtungen durch sehr viel kompliziertere ersetzt werden, so hat dies folgende Gründe: Der Übergang des Elektrons würde neutrale Atome entstehen lassen und damit eine photochemische Veränderung des Kristalls bedeuten — welcher Art, wird unten eingehender besprochen. Bei ganz reinen Kristallen ist aber von einer solchen Veränderung nichts zu bemerken. (Streng genommen, extrapoliert man die mit steigender Reinheit zu beobachtende Abnahme solcher Veränderungen auf die Verunreinigung null.) Weiter fehlt jede Zuordnungsmöglichkeit für die kurzwelligeren Banden. Man wird daher solchen Theorien den Vorzug geben, welche die Banden als Anregungszustände des im Kristallverband befindlichen Cl$^-$-Ions deuten. Für das freie Ion Cl$^-$ kennt man keine höheren diskreten Quantenzustände, sondern nur ein der Abtrennung entsprechendes Kontinuum. Hier kommt nun im

[1] W. MARTIENSSEN: Göttinger Nachr. **1955**, 258.
[2] R. HILSCH u. R. W. POHL: Z. Physik **57**, 145 (1929).

Kristallverband etwas neues herein, was einleitend (Ziff. 1) erwähnt worden war: Da die Anregungszustände mit ihrer Elektronenhülle in die der andern Ionen

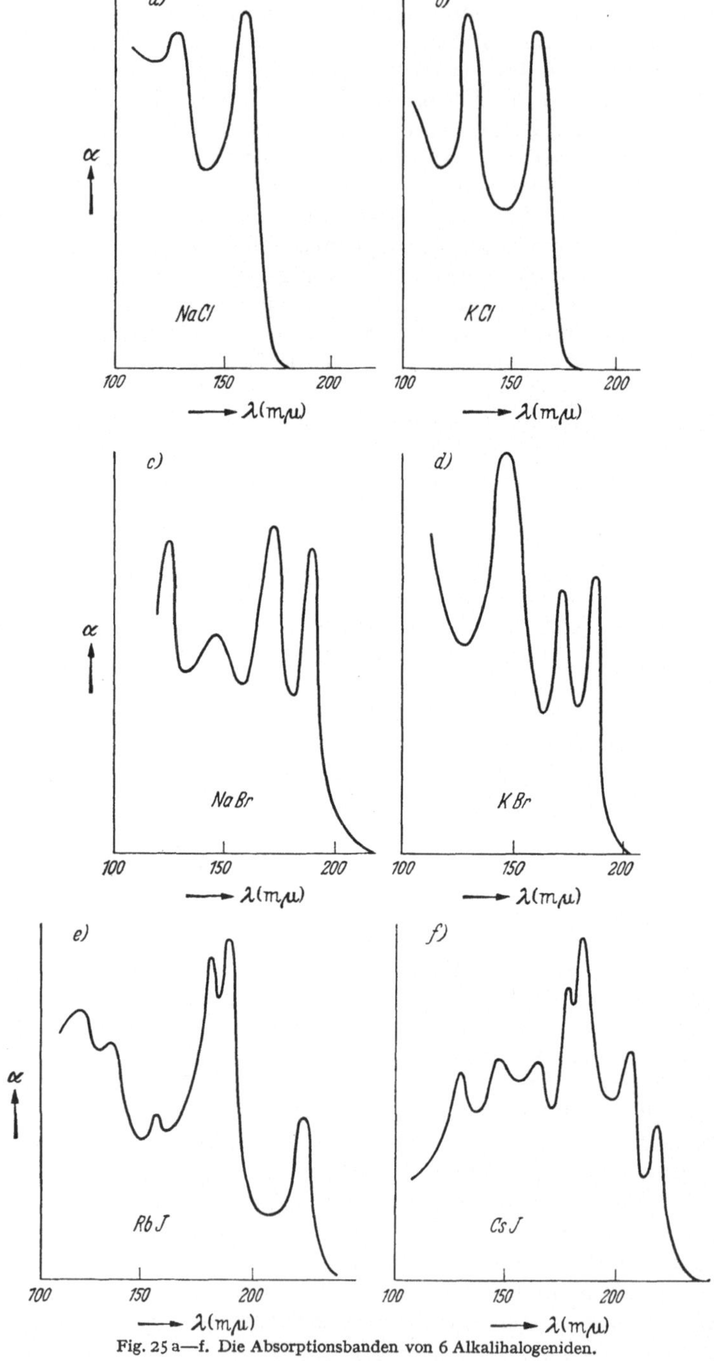

Fig. 25 a—f. Die Absorptionsbanden von 6 Alkalihalogeniden.

hineingreifen, kann man das Elektron nicht mehr streng einem bestimmten Ion zuordnen; man kommt so von der einen Seite an den Fall der Leitungselektronen heran, der bei den Metallen gewöhnlich von der andern Seite, der des freien Elektronengases, angegangen wird. Bei diesem sind wegen der Interferenz der die Elektronen führenden DE BROGLIE-Wellen nicht alle Energien möglich; es bestehen Lücken im Energiespektrum (vgl. Fig. 27). Je stärker die Wechsel-wirkung mit dem Gitter der Rest-ionen ist, desto breiter werden die Lücken und schließlich erhält man schmale Bänder, welche in der Nähe der scharfen Energie-stufen der freien Atome liegen. Durch solche schmale Bänder sind offenbar auch die Anregungs-zustände des Cl⁻-Ions im Kristall-verband gegeben, wobei die Gren-ze zwischen gebundenen und freien (Leitungs-) Elektronen flie-ßend ist. Dieses in der Theorie der Metalle und Halbleiter be-stens bewährte Bändermodell hat noch eine andere wichtige Eigen-schaft: Die Zahl der durch je zwei (wegen des Spins) Elektro-nen zu besetzenden Energiestufen ist durch das PAULI-Prinzip ge-geben. Sind alle Plätze eines Bandes besetzt, so kann ein äußeres elektrisches Feld keine Energie mehr an irgendeines der Elektronen im Band übertragen, d.h.

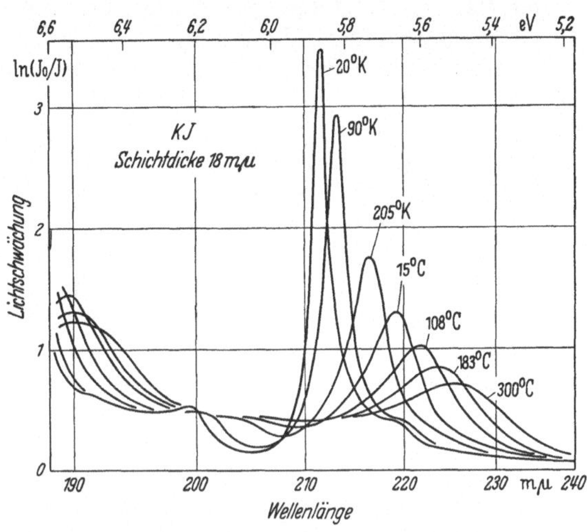

Fig. 26. Breite und Lage der langwelligsten Absorptionsbande von KJ bei verschiedenen Temperaturen.

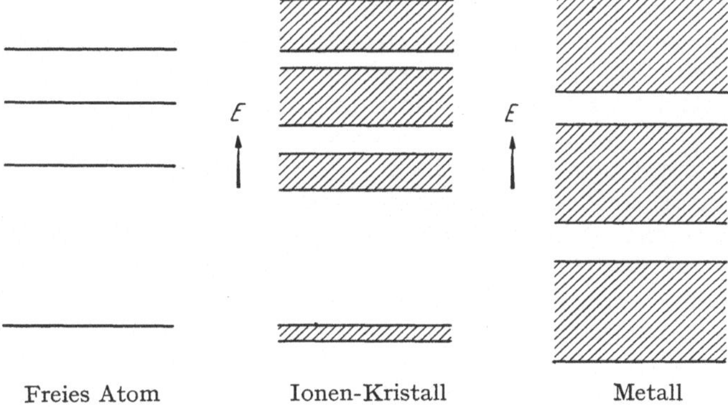

Abb. 27. Schema der Energiestufen von freiem Atom, polarem Kristall und Metall.

kein Elektron kann dem Feld als Leitungselektron folgen, der Stoff ist ein Isola-tor. Der Grundzustand der Ionen mit der abgeschlossenen Edelgasschale ent-spricht dem. Wird ein Elektron in eines der oberen Bänder gehoben, so hat nicht nur dieses darin eine gewisse Beweglichkeit, sondern es entsteht auch eine Lücke in der Füllung des unteren (Valenz-) Bandes. Bei der Bewegung der Elek-tronen im fast vollen Band läuft dann die Lücke entgegengesetzt, so daß man so rechnen kann, als ob eine positive Ladung (im folgenden kurz „Loch" genannt) beweglich geworden wäre. Wir werden später sehen, daß wir eine weitgehende Symmetrie zwischen Elektronen- und Löcher-Spektren haben.

Die wellenmechanischen Zustände eines Elektrons im schematisierten Potential eines NaCl-Kristalls berechnete zuerst Tibbs[1] für die 100-Richtung. Dies ist aber nicht der Fall der Anregung, da hierbei alle Ionen als geladen angenommen werden. Dexter[2] fügte nun als Störungsfunktion die Veränderung hinzu, welche in der Nähe des angeregten Cl⁻-Ions dadurch entstehen, daß der Rest ungeladen ist und eine veränderte Ladungswolke besitzt und daß die Nachbarionen ebenfalls in Mitleidenschaft gezogen wurden. (Dieser dritte Beitrag läßt sich durch die Dielektrizitätskonstante ausdrücken, da es sich ja um eine Veränderung der elektrostatischen Polarisation handelt.) Auf diese Weise berechnete Dexter den Absolutwert des Absorptionskoeffizienten für die erste Bande und insbesondere das Intensitätsverhältnis der beiden ersten Banden in größenordnungsmäßiger Übereinstimmung mit der Beobachtung. Trotz der großen Rechenarbeit ist man weit entfernt, die Struktur der Absorptionsbanden oder gar die einfachen Gesetzmäßigkeiten der Gl. (53.1) aus dem Bändermodell abzuleiten. Der Wert dieser Rechnungen liegt auf einem andern Gebiet: Sie zeigt, daß es mehrere Anregungszustände gibt und fügt das Gebiet organisch in die heutige Festkörperphysik ein. Daß bei voller Beherrschung der Rechnung etwas ähnliches wie Gl. (53.1) zu erwarten ist, kann man folgendermaßen überlegen: Der Verlauf des periodischen Potentials ist durch den Bau der Ionen Cl⁻ und K⁺ bedingt, der seinerseits wieder die Elektronenaffinität E und die Ionisierungsarbeit J gibt, während die Gitterkonstande a ja gerade die Periode des Potentials bestimmt. Die Rechnung von Dexter läßt sich ferner in gleicher Weise für die unten (Ziff. 54 ff.) zu besprechenden Kristalle mit Störstellen durchführen, wobei nur die Störungsfunktion eine andere Bedeutung hat.

Overhauser behandelt in einer kürzlich erschienenen Arbeit[3] das Problem ganz im Sinn der theoretischen Betrachtungen von Abschnitt B I und C IV. Ausgangspunkt aber ist bei ihm die Pohlsche Grundannahme über die Natur des angeregten Zustands (excitons), der dadurch gekennzeichnet ist, daß im NaCl-Kristall beim Na⁺-Ion ein weiteres Elektron, und zwar ein $3s$-Elektron vorhanden ist, während beim Cl⁻-Ion ein $3p$-Elektron fehlt, anders ausgedrückt ein $3p$-Loch vorliegt. Der $3p$-Zustand ist wie jeder p-Zustand 3fach entartet. Da aber jedes $3p$-Loch zu 6 Nachbarn mit $3s$-Elektron gehören kann, stellt dies eine weitere 6fache Entartung dar. Endlich können die beiden Spins von Loch und Elektron in bezug auf eine gegebene Feldrichtung noch in 4 verschiedenen Weisen orientiert sein. Insgesamt ist der Zustand also 72fach entartet. Bei der Beschränkung auf den Ausbreitungsvektor $\mathfrak{k} = 0$ (vgl. Ziff. 44) bleiben davon 30 mögliche Teilniveaus. Nach einem Satz der Störungsrechnung sind aber von dem gegen alle Symmetrieoperationen des NaCl-Gitters invarianten Grundzustand Dipolübergänge nur nach solchen Niveaus möglich, welche dieselbe Symmetrie haben wie die Störungsfunktion. Dies reduziert die Zahl der Übergänge bei völligem Fehlen einer Spin-Bahn-Koppelung auf 2, bei beginnender Spin-Bahn-Koppelung kommen 3 weitere, schwächere, hinzu. Beim CsJ, das im raumzentrierten Typ kristallisiert, gibt es so 6 statt 5 erlaubte Übergänge. Vergleicht man diese Aussagen mit Fig. 26, so haben wir bei NaCl und KCl, wo sicher die Spin-Bahn-Kopplung noch klein ist, 2 Maxima. Bei den Bromiden sind es aber nur 4, bei CsJ ist ein Maximum zuviel da. Trotzdem darf man diese teilweisen Widersprüche nicht zu schwer nehmen, da auch das experimentelle Material bei der Schwierigkeit der Erzeugung der Spektren nicht immer übereinstimmt.

[1] S. R. Tibbs: Trans. Faraday Soc. **35**, 1471 (1939).
[2] D. L. Dexter: Phys. Rev. **83**, 435 (1951).
[3] A. W. Overhauser: Phys. Rev. **101**, 1702 (1956).

b) Alkalihalogenidkristalle mit Störstellen.

54. Einteilung der Störstellen. Ein ungeheures experimentelles Material, größtenteils aus der POHLschen Schule[1] liegt über die zusätzlichen Absorptionsbanden vor, welche Alkalihalogenidkristalle mit Störstellen zeigen. Ein Teil dieser Ergebnisse ist auch theoretisch geklärt, insbesondere wissen wir heute über die Träger dieser Banden recht gut Bescheid. Es kann hier kein erschöpfender Bericht über diese Untersuchungen gegeben werden, die zu einem großen Teil ins Gebiet der Photochemie gehören, es sollen hier nur in großen Zügen die spezifisch spektroskopischen Ergebnisse besprochen werden. Ausgangspunkt ist die Tatsache, daß man diese Banden in Kristallen von nicht genau stöchiometrischer Zusammensetzung erhält, wenngleich man sie auch durch Einstrahlung energiereicher Teilchen oder Quanten erzeugen kann. Alle diese Banden gruppieren sich um zwei Arten: *F*-Banden bei Alkaliüberschuß und *V*-Banden bei Halogenüberschuß. Dabei haben wir allerdings von vornherein die Absorptionsgebiete ausgeschieden, welche durch ausgeflockte Metallteilchen, Kolloide, entstehen. Hierzu gehört die blaue Farbe mancher Steinsalzfunde. SIEDENTOPF[2] hat zuerst im Ultramikroskop die Einsprengung solcher Kolloide in Kristalle nachgewiesen.

55. *F*-Banden. Die seit langem bekannte *F*-Bande erhält man z. B. nach folgender Vorschrift: Ein aus reinstem KCl bestehender Kristall wird im Vakuum bei 415° C entgast, bei 360° wird bei einem Druck von 10^{-4} Torr K-Metall eindestilliert, darauf wird das Gefäß 28 Std auf 477° gehalten. Entscheidend für das Auftreten der *F*-Bande ist, daß man darnach die Kristalle auf Zimmertemperatur abschreckt. Langsames Abkühlen oder nachträgliches Tempern läßt Kolloide ausfallen. Eine viel höhere Konzentration von Trägern der *F*-Bande — im folgenden kurz *F*-Zentren genannt — erhält man, wenn man nach HILSCH und Mitarb.[3] im Hochvakuum auf eine tiefgekühlte Platte gleichzeitig Alkalihalogenid und Alkali aufdampft. Die *F*-Bande hat bei 20° K eine Halbwertsbreite von etwa 40 mμ. Ihre Lage (Frequenz v in $\sec^{-1}$) hängt mit der Gitterkonstante a in cm durch die empirische Formel[4] zusammen

$$v = p/a^2 \quad \text{mit} \quad p = 2\ \text{cm}^2\ \sec^{-1}. \tag{55.1}$$

Sie fällt bei KCl nach 550 mμ, gibt also eine blau-violette Färbung.

Gegen die nächstliegende Deutung der Träger der *F*-Bande, der *F*-Zentren, als in Zwischengitterplätzen sitzenden Alkali-Atomen erheben sich zahlreiche Einwände, die hier nicht alle aufgezählt werden können. In letzter Zeit kamen zwei neue hinzu: Die sorgfältigen Untersuchungen über die paramagnetische Resonanz dieser so verfärbten Kristalle, die hauptsächlich von KITTEL und Mitarbeitern[5] durchgeführt wurden, geben eine einfache Resonanzkurve, während in dem tetraedersymmetrischen Umgebungsfeld der Kernspin eine nachweisbare Aufspaltung ergeben müßte. Der bei diesen Messungen ermittelte g-Faktor beträgt 1,995 statt 2,0023 fürs freie Elektron. Versuche, diese Abweichung aus dem andern sogleich zu besprechenden Modell abzuleiten (vgl. die oben zitierten Arbeiten von KITTEL und Mitarbeitern), geben den Sinn der Abweichung richtig. Wenn auch die Möglichkeit der Deutung durch Alkali-Atome nicht restlos ausgeschlossen

---

[1] Zusammenfassende Berichte. R. W. POHL: Proc. Phys. Soc. Lond. **49**, 3 (1937). — Phys. Z. **39**, 36 (1938). — F. SEITZ: Rev. Mod. Phys. **18**, 384 (1946); **26**, 7 (1954).

[2] H. SIEDENTOPF: Phys. Z. **6**, 855 (1905).

[3] R. KAISER: Z. Physik **132**, 482 (1952).

[4] R. W. POHL: Optik und Atomphysik, 9. Aufl., S. 320. Berlin: Springer 1954.

[5] C. KITTEL u. Mitarb.: Phys. Rev. **89**, 315 (1953); **90**, 238 (1953); **91**, 1066 (1953). — A. M. PORTIS: Phys. Rev. **91**, 1071 (1953); ältere Untersuchung von C. A. HUTCHINSON: Phys. Rev. **75**, 1769 (1949).

ist, so sprechen doch sehr viel Gründe für das schon 1937 von DE BOER[1] vorgeschlagene Modell eines in einer Halogenionenlücke befindlichen Elektrons, d. h. des Ersatzes eines Halogenions durch ein Elektron. Dabei hat man sich dies so vorzustellen, daß es zwischen den 6 benachbarten Alkaliionen „aufgeteilt" wird. Dies ist auch der Sinn der vom Bändermodell ausgehenden Betrachtungsweise von DEXTER[2]. Für die Richtigkeit dieser Vorstellung sprechen vor allem auch Messungen der Dichteänderungen[3], die ein reiner Kristall erfährt, wenn in ihm durch Röntgenstrahlen — dies ist eine zweite Möglichkeit zur Erzeugung von F-Zentren — die Bande hervorgerufen wird. Restlos zwingend ist die Beobachtung der Dichteabnahme allerdings erst, wenn am gleichen Kristall außer der Dichteabnahme auch noch die Gitterkonstante mit hoher Präzision gemessen wird und keine entsprechende Änderung bei ihr gefunden wird. Besonders stark aber sprechen für die DE BOERsche Deutung zwei neue mit der F-Bande genetisch verknüpfte Banden α und β, die von DELBECQ, PRINGSHEIM und YUSTER[4] gefunden wurden. Bei einem additiv verfärbten Kristall findet sich hart neben der ersten Bande des reinen Kristalls eine neue Bande (β). Zerstört man die F-Bande durch Einstrahlung von Licht dieser Wellenlänge, so verschwindet im gleichen Maß die β-Bande, dafür entsteht etwas weiter von der Eigenabsorptionsbande entfernt, aber immer noch in enger Nachbarschaft eine neue Bande α. Diese Bande entsteht nun nach MARTIENSSEN und POHL[5] fast ausschließlich, wenn man den Kristall bei 20° K Röntgenstrahlen aussetzt. Es bietet sich zwangsläufig folgende Deutung an: Durch die Röntgenbestrahlung entstehen bei tiefer Temperatur in erster Linie Halogenionenlücken. Ein Halogenion in der Nachbarschaft dieser Lücke befindet sich in einer veränderten Umgebung[6]. Ihm wird die α-Bande zugeordnet. Wahrscheinlich wird das Halogenion auf einen Zwischengitterplatz geworfen — durch das hochenergetische Lichtquant kann eine genügend hohe lokale Erwärmung entstehen. Damit wird aber sein Absorptionsspektrum stark verändert. Wird die Lücke mit einem Elektron gefüllt, so haben wir das F-Zentrum und gleichzeitig eine Veränderung des Feldes, in dem sich die nächsten Ionen befinden; deren Absorptionsspektrum gibt die β-Bande. So lassen sich noch mehrere schwache Banden als Störstellenaggregate deuten. Ihre Bezeichnung geht aus Fig. 28a hervor, wobei die Quadrate das Fehlen eines Gitterions andeuten, die schraffierten kleinen Kreise ein Elektron bedeuten. Auf ein besonderes Zentrum sei dabei hingewiesen, auf das F'-Zentrum, das eine breite im langwelligen Gebiet gelegene Bande gibt. Bestrahlt man bei Zimmertemperatur den verfärbten Kristall mit Licht der F-Bande, so wird diese ab- und die F'-Bande aufgebaut, dabei verschwinden für jedes Lichtquant zwei F-Zentren. Dies kann nur so erklärt werden, daß ein aus einer Halogenlücke herausgeworfenes Elektron von einem andern

---

[1] J. H. DE BOER: Rec. Trav. chim. phys.-Pays-Bas **56**, 301 (1937).

[2] D. L. DEXTER: Phys. Rev. **83**, 435 (1951).

[3] I. ESTERMANN, W. J. LEIVO u. O. STERN: Phys. Rev. **75**, 627 (1949).

[4] J. DELBECQ, P. PRINGSHEIM u. PH. YUSTER: J. Chem. Phys. **19**, 574 (1951); **20**, 746 (1952).

[5] W. MARTIENSSEN: Göttinger Nachr. **1952**, 111. — W. MARTIENSSEN u. R. W. POHL: Z. Physik **133**, 153 (1952).

[6] Hier besteht ein Unterschied zwischen den Vorstellungen von MARTIENSSEN und denen von SEITZ: Aus der Dichteabnahme bei Röntgenstrahlung bei Zimmertemperatur ist zu schließen, daß es sich um die Absorption von Cl-Gitterionen in der Nähe einer Lücke handelt, während MARTIENSSEN annimmt, daß es sich um die Absorption des von seinem Gitterplatz ins Zwischengitter verdrängten Ions handelt, da man bei so tiefer Temperatur nicht wohl eine Wegdiffusion der Cl-Ionen, die bei Zimmer- und höherer Temperatur wahrscheinlich ist, annehmen kann. Möglicherweise fallen die beiden Banden so nahe zusammen, daß sie nicht getrennt werden.

F-Zentrum eingefangen wird, wodurch dieses in ein F'-Zentrum verwandelt wird, also ebenfalls für die F-Bande ausfällt.

56. *V*-Banden. MOLLWO[1] gelang es als erstem, Kristalle mit einem Überschuß von Br oder J (nicht aber Cl) dadurch herzustellen, daß er reine Kristalle bei hoher Temperatur unter großem Br_2 bzw. J_2-Druck hielt und dann abschreckte. Die dabei auftretende Bande ist breiter als die F-Bande und zeigt eine deutliche Struktur. Sie hat bei KBr ihr Hauptmaximum bei 275 mμ, und die Breite beträgt auch bei tiefer Temperatur noch 75 mμ. Die verschiedenen in der Folgezeit festgestellten, zum Teil sehr schwachen Maxima werden mit V_1, V_2, V_3, V_4 bezeichnet. V_2 ist die Hauptbande. Im Bändermodell entspricht ein neutrales Halogenatom einem Loch mit positiver Ladung, die V-Zentren sind also das Gegenstück

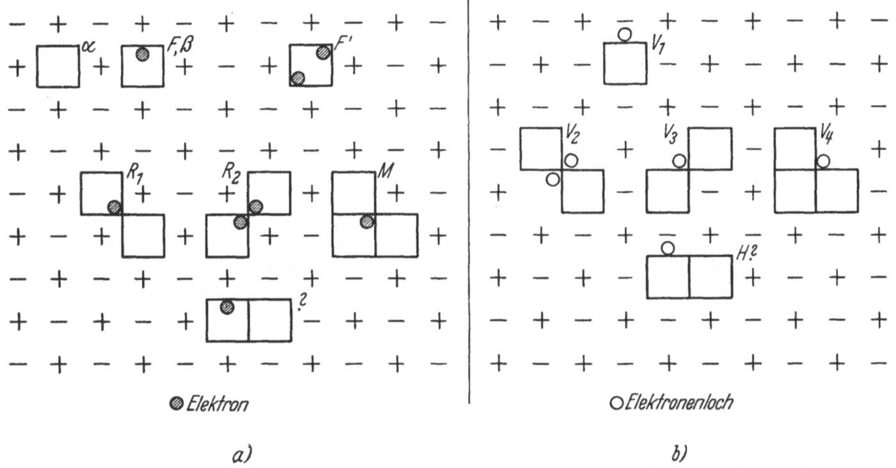

Abb. 28 a und b. Die Störstellen in Alkalihalogeniden und die ihnen zugeordneten Banden. a Halogen-, b Alkalilücken.

zu den aus Elektronen gebildeten F-Zentren. So wie die F-Zentren ein Halogenion ersetzen, so ersetzen die V-Zentren ein Alkaliion. Hier besteht allerdings ein Unterschied: Da Br-Dampf zweiatomig ist, wird hauptsächlich Br_2 ins Kristallgitter eingebaut, wie schon MOLLWO erkannte, dazu müssen aber zwei benachbarte Alkalilücken da sein. Die Zuordnung der übrigen Banden zu den vermuteten Störstellenaggregaten gibt Fig. 28 b wieder. Es bedeutete einen erheblichen Fortschritt der Erkenntnis als CASLER, PRINGSHEIM und YUSTER[2], sowie DORENDORF[3] entdeckten, daß bei Röntgenbestrahlung bei Zimmertemperatur nicht nur F-, sondern auch V-Banden, also gleichzeitig „Loch"- und „Leitungs"-Elektronen entstehen.

Man kann nun die Frage stellen, ob nicht bereits Licht von der Wellenlänge der Eigenabsorption der Kristalle dieselbe Wirkung hat wie Röntgenstrahlen. Bei sehr reinen von Baufehlern freien Kristallen ist die Wirkung minimal, so daß man wohl sagen kann, daß bei Idealkristallen keine Wirkung da ist. Die Kristalle können aber insbesondere dadurch sensibilisiert werden, daß man einen kleinen Teil der Halogenionen durch H^- ersetzt, d.h. Mischkristalle Halogenid-Hydrid herstellt. Dem H^- kommt eine der Eigenabsorption vorgelagerte Bande (U-Bande) zu. Technisch erzeugt man solche Mischkristalle, indem man additiv mit F-Zentren versehene Kristalle bei höherer Temperatur in Wasserstoff von

[1] E. MOLLWO: Ann. d. Phys. **29**, 394 (1937).
[2] R. CASLER, P. PRINGSHEIM u. PH. YUSTER: J. Chem. Phys. **18**, 1564 (1950).
[3] H. DORENDORF: Z. Physik **129**, 317 (1951).

etwa 50 Atm Druck hält, wodurch die F-Bande in die Hydrid-(U-)Bande übergeht. Einstrahlung in der U-Bande, aber auch in der Grundgitterabsorption, gibt F-Zentren. Die Hauptwirkung dürfte dabei dem kleinen H-Atom zukommen, das leicht an eine „Versetzung" („innere Oberfläche") oder die äußere Oberfläche diffundiert. Bei Einstrahlung im Bereich der Grundgitterabsorption wird wahrscheinlich über die Löcherleitung ein Elektron von H^- nach Cl^- transportiert. Strahlt man bei 20° K in die U-Bande ein, so entsteht nach Thomas[1] nicht die F-, sondern die α-Bande, von Thomas damals U'-Bande genannt. Dies spricht entschieden für die Seitzsche Auffassung der Natur der α-Bande, da ja durch diesen Prozeß sehr leicht eine Halogen-Ionenlücke entsteht, wenn das kleine neutrale Wasserstoffatom wegdiffundiert, während man keinen Grund für die Verdrängung eines Halogenions auf einen Zwischengitterplatz sieht.

IV. Die scharfen Linien von Halbleiterkristallen.

57. Die dem Kontinuum vorgelagerten Linien und ihre Deutung. E. F. Gross und S. Karryeff[2] entdeckten als erste bei Halbleiterkristallen auf der langwelligen Seite der kontinuierlichen Absorption einzelne Linien, die bei 4° K die Schärfe der Linien der Seltenen Erden aufweisen. Sie sind zu deuten als Übergänge nach erstaunlicherweise scharfen Anregungszuständen, die über dem Valenzband liegen, sich aber nicht mehr aus den Zuständen des freien Ions ableiten lassen. S. Nikitine[3] hat in einer Reihe von Arbeiten mit seinen Mitarbeitern diese überraschenden Spektren durchforscht. Besonders scharfe Linien geben die halbleitenden Verbindungen des einwertigen Kupfers, aber auch TlCl, PbJ_2, CdJ_2, CdS und HgJ_2 geben solche Linien. Um diese Spektren zu erhalten müssen außerordentlich dünne Schichten benützt werden. Sie wurden z.B. bei CuJ so erzeugt, daß das Salz bei 400° C auf eine auf 120° C vorgewärmte Glasplatte aufgedampft wurde und die Aufdampfung nach Erreichen einer eben wahrnehmbaren Schicht von etwa $0{,}1\,\mu$ abgebrochen wurde. Bei diesem Salz ist eine Dublettstruktur mit $\Delta\nu = 24$ cm^{-1} festzustellen, deren Ursache nicht geklärt ist. Das erstaunlichste ist, daß bei Cu_2O ausgesprochene Wasserstoffserien festgestellt

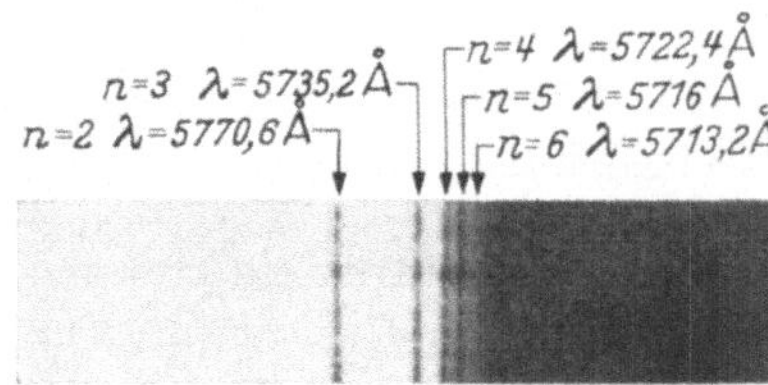

Fig. 29. „Lyman-Serie" in Cu_2O.

wurden (vgl. Fig. 29). Bei den andern Stoffen sind zu wenig Glieder beobachtbar, als daß man aus ihnen auf eine Wasserstoffserie schließen dürfte. Solche wasserstoffartige Terme wurden schon in dem 1940 erschienenen Buch von Mott und Guerney[4] angegeben. Die Deutung ist anschaulich folgende: Der Anregungszustand (das „Exciton") besteht aus einem Paar Elektron-Loch, das einige Zeit zusammenbleibt und eine Art Wasserstoffatom bildet, wobei natürlich die Masse des Lochs nicht die des Protons ist. Führt man die reduzierte Masse μ von Elektron und Loch ein und schematisiert die Umgebung durch ein Medium der Dielektrizitätskonstante ε, in welchem sich Loch und Elektron bewegen, so kommt man zu Energiestufen der Form

$$E = E_B - \frac{R\,h\,c\,\mu}{\varepsilon^2\,m} \cdot \frac{1}{n^2} \quad (n = 2, 3, 4, \ldots). \tag{57.1}$$

[1] H. Thomas: Ann. d. Phys. (5) **38**, 601 (1940).

[2] E. F. Gross u. S. Karryeff: Dokl. Akad. Nauk. S.S.S.R. **84**, 261 (1952).

[3] S. Nikitine: Helv. phys. Acta **28**, 307 (1955). — S. Nikitine u. Mitarb.: C. R. Acad. Sci. Paris **241**, 629 (1955); **242**, 1003 (1955).

[4] N. F. Mott u. R. W. Guerney: Electronic Processes in Ionic Crystals, S. 83. Oxford 1940.

Hierin bedeutet E_B den Abstand des Leitungsbandes vom Valenzband[1], m die Elektronenmasse, R die RYDBERG-Zahl.

Durch die nicht direkt angebbare Größe μ/ε^2 läßt sich die Konstante der Serie dem experimentellen Befund anpassen.

Hierher gehören wohl auch die von DELBECQ und PRINGSHEIM[2] an röntgenbestrahltem Lithiumfluorid gefundenen zum Teil sehr scharfen Absorptionslinien bzw. -banden. Sie unterscheiden sich von den oben besprochenen Excitonenspektren dadurch, daß sie keinem Kontinuum vorgelagert sind. Die im Grün gelegene Linie 5235,1 Å, welche von einer etwas verwaschenen Linie 5178,3 Å begleitet ist, ist bei 20° K so scharf, daß ihr Verhalten im Magnetfeld untersucht werden kann[3]. Es ergibt sich lediglich eine sehr kleine Blauverschiebung, die im Feld von 26700 Oersted 0,02 Å beträgt. Die Absorption im Ultraviolett hat ausgesprochenen Bandencharakter: fünf annähernd äquidistante nach kurzen Wellen immer unschärfer werdende Teilbanden ($\varDelta \nu \approx 250$ cm^{-1}) lassen das Spektrum dem grünen Teil des MnO$_4$-Spektrums auffallend ähnlich erscheinen[3]. Gegen die Erklärung, daß der Träger ein von Verunreinigungen stammender MnF$_4^-$-Komplex ist, spricht der Umstand, daß Kristalle ganz verschiedener Herkunft bei gleicher Bestrahlung dieselbe Stärke der Linien zeigen. Kommen die Linien dem reinen LiF zu, so wäre hier zum ersten Mal ein Excitonen-Bandenspektrum festgestellt. Die andern Alkalihalogenide weisen bei Röntgenbestrahlung keine derartigen Linien auf.

E. Anwendungen der Kristallspektren[4].

58. Erforschung von Feinheiten der Kristallstruktur. Die große Empfindlichkeit, welche die scharfen Linien der Übergangselemente gegenüber den Einflüssen der Umgebung ihrer Träger aufweisen, macht diese als Sonden zur Strukturuntersuchung geeignet, wobei noch Feinheiten zu Tage kommen, die mit der üblichen Röntgenstrukturanalyse nicht nachzuweisen sind. Um mit möglichst geringen Mengen von Sonden, die durch ihre Anwesenheit den zu untersuchenden Stoff beeinflussen könnten, auszukommen, verwendet man zweckmäßigerweise Stoffe, die durch Ultraviolett oder Kathodenstrahlen zur Fluorescenz angeregt werden können, also von den Seltenen Erden Europium oder Samarium, da man in Emission durch Verlängerung der Belichtungszeit auch schwache Erscheinungen noch photographieren kann. Ein eindrucksvolles Beispiel gibt TOMASCHEK[4]: Baut man Eu^{+++} in MgO ein, so erhält man je nach der Herkunft, ob z.B. das MgO aus Mg$_2$Cl oder MgSO$_4$ hergestellt wurde, verschiedene Fluorescenzspektren (vgl. Fig. 30). Die einfache Linie unterscheidet sich im wesentlichen in der Intensität; das Dublett rechts davon ändert nicht nur die Intensität, sondern verschiebt sich auch ganz erheblich. TOMASCHEK spricht hier von einem „Innenbau“, der bei der Röntgenanalyse nicht beobachtet wird. Vom heutigen Standpunkt aus wird man sagen, daß die Störstellen, auf die es ja beim festen Körper oft gerade entscheidend ankommt, bei den einzelnen Proben verschieden angeordnet sind. In diese werden die Sonden bevorzugt eingebaut. Die große Schärfe der Linien zeigt, daß die Plätze der Sonden nicht nach Zufall verteilt sind, sondern

---

[1] Daß trotz einer gewissen Breite des Valenz*bandes* eine scharfe Einsatzgrenze der kontinuierlichen Absorption und scharfe Excitonenlinien herauskommen, läßt sich durch die Erhaltungssätze des Ausbreitungsvektors $\mathfrak{k}$ begründen.

[2] CH. DELBECQ u. P. PRINGSHEIM: J. Chem. Phys. **21**, 794 (1953). — Z. Physik **136**, 573 (1954) (dort weitere Literatur).

[3] G. JOOS u. H. RITZ: Verh. dtsch. phys. Ges. **7**, 101 (1956).

[4] Zusammenfassender Bericht: R. TOMASCHEK: Ergebn. exakt. Naturw. **20**, 268 (1942).

offenbar je nach Herstellungsart etwa entlang von kristallographisch ausgezeichneten Ebenen oder Geraden geordnet sind. Dies wirft ein besonderes Licht auf die selektive Wirkung von chemisch und röntgenographisch identischen Katalysatoren je nach ihrer Herstellungsart. Es ist dabei zu bedenken, daß auch bei den TOMASCHEKschen Messungen die Substanzen nicht als Einkristalle, sondern als feine Pulver vorlagen, also von vornherein viele Gitterstörungen enthalten durften.

Ebenso aufschlußreich sind die Spektren der seltenen Erd-Ionen in mit ihnen aktivierten Phosphoren, über die in Bd. XXVI (GARLICK) berichtet wird.

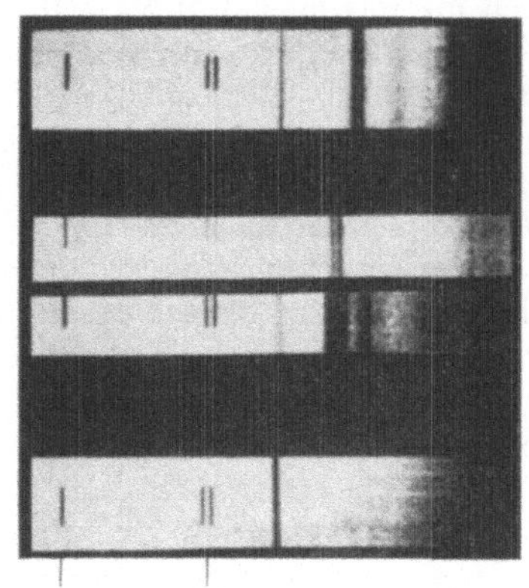

Fig. 30 a—c. Phosphorescenzspektren des Eu^{+++} in MgO verschiedener Herstellungsart (Kathodenstrahlerregung bei − 180°) a) aus $MgSO_4$, b) aus $MgCl_2$ (untere Aufnahme viel länger belichtet), c) aus Magnesia usta. Die Röntgenspektren aller 3 Proben sind identisch. In der Mitte jeder Aufnahme Hg-Vergleichspektrum.

Die Strukturänderungen des γ-$Al_2O_3 \cdot 3\,H_2O$ (Hydrargillit) bei der Entwässerung und dem Übergang in α-Al_2O_3 (Boehmit) verfolgte ASSELMEYER[1] sowohl röntgenographisch wie mit Sonden aus eingelagerten Eu^{+++}-Ionen, zum Teil auch mit den Fluorescenzspektren adsorbierter organischer Stoffe (Perilen und Coronen). Der Parallelismus mit den Veränderungen der DEBYE-Aufnahmen tritt klar hervor; zur Vorsicht gegenüber der Methode mahnt aber die dabei gemachte Beobachtung, daß der Einbau der Sonden den Umwandlungspunkt um etwa 100° herabsetzt.

Aber auch an wohl ausgebildeten Einkristallen sind sonst nicht festzustellende Umwandlungen zu beobachten. So fanden HELLWEGE und Mitarbeiter[2] bei den Bromaten von Nd und Pr zwischen 85 und 55°K eine Umwandlung, welche sich in der Durchbrechung aller bei höherer Temperatur bestehenden Auswahlregeln der Polarisation kundtut; sie ist durch den Übergang einer bei höherer Temperatur sechszähligen Achse in eine zweizählige zu erklären. Bei $Eu(BrO)_3 \cdot 9\,H_3O$ tritt mit sinkender Temperatur eine Vermehrung der Linienzahl weit über die Höchstzahl $2J + 1$ der Termaufspaltungen hinaus auf. Diese Erscheinung erklärt sich durch zwei Vorgänge: Einmal wieder durch die Verminderung der Symmetrie — der Kristall ist in Wirklichkeit nicht streng hexagonal, sondern pseudo-hexagonal —, dann aber durch die Einnahme von drei energetisch etwas verschiedenen Plätzen durch die Eu^{+++}-Ionen, so daß man in Wirklichkeit drei ineinandergeschobene Spektren hat. Bemerkenswert ist auch die kristallspektroskopische Verfolgung einer an sich bekannten Umwandlung des $BaTiO_2$ bei − 80° C durch eine japanische Forschergruppe[3]. Es wurde aus reinsten Ausgangsmaterialien unter Zusatz von Samariumnitrat ein pulverförmiger Phosphor präpariert und dieser durch Kathodenstrahlen (vgl. Ziff. 4) zum Leuchten gebracht. Bei − 80° geht $BaTiO_2$ vom orthorombischen in einen pseudokubischen rhomboedrischen Typ über. Entsprechend der erhöhten Symmetrie des fast kubischen Gitters verwandeln sich alle Dubletts des Sa^{+++}-Ions in Einzellinien.

59. Struktur der Gläser und Lösungen. Besonders tiefe Einblicke gibt die Sondenmethode bei der Glasforschung.

Zu einer Zeit, da die ZACHARIASsche Vorstellung der Glasstruktur als einer unperiodischen Gitterstruktur, noch nicht allgemein bekannt war, man vielmehr

[1] F. ASSELMEYER: Z. angew. Phys. **1**, 26 (1948).
[2] A. M. HELLWEGE u. K. H. HELLWEGE: Z. Physik **127**, 334 (1950).
[3] K. OSHIMA, S. HAYAKAWA, H. NAGANO u. M. NAGUSA: J. Chem. Phys. **24**, 903 (1956).

noch an einen feinstkristallinen Aufbau der Gläser dachte, machten Joos und SCHNETZLER[1] zur Entscheidung folgenden Versuch: Sie schmolzen reines Cr_2O_3 und B_2O_3 zusammen, wobei bekannt ist, daß sich kein Cr_2O_3 löst. In der Tat blieb das bei tiefer Temperatur relativ scharfe Spektrum des Cr_2O_3 erhalten. Wurde nun mehr und mehr Na_2O oder K_2O zugesetzt, so verschwand von 10% Na_2O oder 15% K_2O an das Linienspektrum, und es entstand ein neues ganz verwaschenes Spektrum des Glases, bei dessen Entglasung wieder die Cr_2O_3-Linien erschienen. Dies wurde als Anzeichen der statistisch schwankenden Umgebungsfelder der Cr^{+++}-Ionen gedeutet, wie sie nach der ZACHARIASschen Vorstellung vorliegen. Viel aufschlußreicher aber sind die von TOMASCHEK und DEUTSCHBEIN[2] eingebauten Eu^{+++}-Ionen, die die Unterschiede zwischen Silikat-, Phosphat- und Berylliumfluorid-Gläsern erkennen lassen, die in dieser Reihenfolge abnehmende Feldwirkung der Umgebung entsprechend der Ladungsverminderung der Ionen zeigen. Auch in organischen Gläsern kann man das Eu-Fluorescenzspektrum erzeugen. Diese geben dabei viel mehr Ähnlichkeit mit wäßrigen Lösungen als die anorganischen.

Damit kommen wir zu der Erforschung der Struktur der eigentlichen Lösungen, die hauptsächlich durch FREED und Mitarbeiter[3] durchgeführt wurde. Bezeichnend ist, daß die Linien des Eu^{+++} nicht sehr verbreitert sind, sondern eine vom Lösungsmittel abhängige Struktur, meist Dubletts, zeigen. Bei verdünnten Lösungen spielt das Anion keine Rolle, erst bei höherer Konzentration machen sich Strukturunterschiede im Sinn einer quasikristallinen Umgebung bemerkbar, wie sie den heutigen Vorstellungen vom Bau der Elektrolyte entspricht.

Herrn Professor Dr. K. H. HELLWEGE sind wir für große Hilfe bei der Literaturbeschaffung sehr zu Dank verbunden.

---

[1] G. Joos u. K. SCHNETZLER: Z. phys. Chem., Abt. B **24**, 389 (1934).

[2] R. TOMASCHEK u. O. DEUTSCHBEIN: Glastechn. Ber. **16**, 155 (1938). Vgl. auch die oben zitierte zusammenfassende Darstellung von TOMASCHEK in den Erg. exakt. Naturw. **20**, 289ff. (1942).

[3] S. FREED u. Mitarb.: J. Chem. Phys. **6**, 297, 654 (1938); **7**, 824 (1939); **8**, 291 (1940).

The Zeeman Effect[1].

By

J. C. VAN DEN BOSCH.

With 12 Figures.

I. Introduction.

1. Historical survey. In 1896 at Leiden (Holland) PIETER ZEEMAN (1865—1943) discovered the influence of an external homogeneous magnetic field on the frequencies of spectral lines emitted by a light source, placed in that magnetic field. He published his first experiments in the proceedings of the Royal Academy of Amsterdam in 1896 [2]. In this paper he points out that his work was stimulated by the experiments already done before by FARADAY. In 1845 Faraday discovered the influence of a magnetic field on the plane of polarisation of linearly polarised light[3]. At that time already FARADAY was convinced of the fact that light and magnetism were closely related. Accordingly, Maxwell describes[4] how Faraday devoted his last experiments to the study of the influence of a magnetic field on the light emitted by a light source, placed in the magnetic field. However he did not come to a result, because of the imperfection of his experimental mounting and the smallness of the effect. After him, other research workers tried to repeat his experiment, but they failed too. ZEEMAN was the first who came to a definite result[2]. Just as FARADAY he put a bunsen flame with sodium chloride between the poles of an electromagnet, which gave a field strength of 10000 oersted. It was a good idea of ZEEMAN that he did not put down the negative results of FARADAY and others to the lacking of the effect of the magnetic field, but to the smallness of the effect. So he used a ROWLAND grating with a radius of 10 ft and 14438 lines to an inch as a spectral apparatus, which gave a much higher resolving power than those used by FARADAY and others. He observed the sodium D-lines with an eyepiece and with the aid of this instrument he observed the broadening of the lines, when putting on the magnetic field. When switching off the magnetic current the lines were again as sharp as before. The experiment could be repeated indefinitely, always with the same result. Very carefully ZEEMAN interpreted his results. He remarks:

"Possibly the observed phenomena will be regarded as not being remarkable at all. One may reason this way: widening of the lines of the spectrum of an incandescent vapour is caused by increasing the density of the radiating substance and by increasing the temperature. Now, under the influence of the magnet, the outline of the flame is undoubtedly changed (as is easily seen), hence the temperature and possibly also the density of the vapour is changed. Hence one might be inclined to account for the phenomenon in this manner."

---

[1] In this article, the vector model has been used throughout. For the quantum mechanical treatment, cf. H. A. BETHE and E. E. SALPETER, Vol. XXXV of this Encyclopedia, pp. 291 to 314. See also P. KUSCH and V. W. HUGHES in Vol. XXXVII for the atomic beam method.

[2] P. ZEEMAN: Proc. Roy. Acad. Amst. **5**, 181, 242 (1896).

[3] M. FARADAY: Experimental Researches, 19th series.

[4] J. CL. MAXWELL: Collected Works **2**, 790.

The years after 1896, however, proved that the phenomenon he had observed was indeed the influence of the magnetic field on the frequencies of the sodium D-lines. ZEEMAN himself proved this already when studying the influence of the magnetic field on the D-lines as absorption lines[1]. Also he investigated the influence of a magnetic field on the band spectrum of iodine[1]. Though this spectrum is very sensitive to changes in density and temperature, no influence was observed. So the broadening of the sodium D-lines, just as that of the red lithium line 6708 Å[1] was a real, new and very remarkable effect.

It proved indeed that there is a connection between light and magnetism, as was already suggested by FARADAY. Apart from the broadening of the lines, two details were observed by ZEEMAN[1]:

1. When observing at right angles to the magnetic field, the central part of the line was linearly polarised, parallel to the magnetic field, whereas the edges of the lines were linearly polarised at right angles to the magnetic field.

2. When observing the effect parallel to the magnetic field, only the edges were visible. These edges were circularly polarised.

Just before the discovery of the magnetic splitting of spectral lines, LORENTZ had published his "theory of electrons"[2]. From this theory, the observed magnetic influence on light vibrations could completely explained. Also the fact that the edges of the lines were circularly polarised, followed from LORENTZ' theory, whereas from the direction of polarisation of the vibrations, displaced to shorter and longer wavelength, compared with the wavelength of the original line, the sign of the electron could be determined as being negative.

Finally ZEEMAN could calculate from the width of the splitting of the lines the value of e/m for the electron. ZEEMAN remarks[1]:

"The extremely remarkable conclusion of Prof. LORENTZ relating to the state of polarisation in the magnetically widened lines, I have found to be fully confirmed by experiment."

After these first experiments many other physicists investigated the magnetic splitting of spectral lines. We especially draw the attention of the reader to the work of PRESTON[3], RUNGE and PASCHEN[4], and LANDÉ[5]. The main result of these investigations was that most lines were not split into three, but into more components by the magnetic field. For this effect the theory of LORENTZ was not sufficient, and in spite of all efforts to extend the theory and to make it suitable for the anomalous Zeeman effect, a really sufficient explanation could not be given before the spin of the electron was introduced by UHLENBECK and GOUDSMIT[6] in 1925. In this case it was clear once more, that the "normal" effect (splitting of a spectral line by a magnetic field in three components with a mutual distance of one LORENTZ' unit) is an exception in atomic spectra.

Two very important rules were published during the first ten years after the discovery of the Zeeman effect, one by PRESTON[7] and one by RUNGE[8]. PRESTON'S rule was of very great importance to understand the relation between the magnetic splitting of a spectral line and its origin in the atom in question.

---

[1] See footnote 2, p. 296.

[2] H. A. LORENTZ: La théorie electromagnitique de MAXWELL. Leiden 1892. — Versuch einer Theorie der eletrischen und optischen Erscheinungen in bewegten Körpern. Leiden 1895.

[3] TH. PRESTON: Phil. Mag. 45, 325 (1895). — Nature, Lond. 59, 224 (1899).

[4] C. RUNGE, and F. PASCHEN: Phys. Z. 1, 180 (1900); 8, 232 (1907). — Astrophys. J. 15, 235 (1902); 16, 118 (1902).

[5] A. LANDÉ: Z. Physik 5, 231 (1921); 15, 189 (1923).

[6] G. E. UHLENBECK and S. GOUDSMIT: Naturwiss. 13, 953 (1925). — Nature, Lond. 107, 264 (1926).

[7] TH. PRESTON: Nature, Lond. 59, 224 (1899). — Dublin Transact. 7, 7 (1899).

[8] C. RUNGE: Phys. Z. 8, 232 (1907).

From a systematic investigation of the spectral lines from the principal and sharp series of the $^3P-^3S$ combinations in Hg, Mg, Sr, Zn, Cd and of the $^2P-^2S$ combinations in Cu, Ag, Al, Tl, Mg, Ca, Sr, Ba, Preston concluded that

1. all successive lines from one series have the same type of Zeeman effect,
2. homologous series of different elements have the same type of Zeeman effect.

This rule has been of very great importance for the use of the Zeeman effect in the analysis of spectra. For the magnetic splitting of a spectral line is a direct indication, to which series of lines in the spectrum the line in question belongs.

In spite of the contradiction which seemed to consist between Lorentz' theory and the anomalous Zeeman effect, Runge from his measurements of many spectral lines[1], established a relation between the distances of the components of the line and the normal distance of a normal triplet. The result of his measurements was, that all distances between the components of a magnetically split line and the original line could be expressed as rational multiples of a Lorentz unit. The denominators of the fractions are the same for all components of one line, the numerators are always simple integers.

This rule was expressed again when Landé introduced the g-factor in 1921[2]. He introduced this factor to express the distances between the components as simple integers. The real meaning of it became clear after the spin of the electron was introduced.

A quantum mechanical theory of the normal Zeeman effect was first given by Debye[3] and Sommerfeld[4] in 1916 for the normal effect, later on completed by Rubinowicz[5] and Bohr[6] in 1918. A theory for the anomalous effect we find with Dirac[7] and Condon and Shortley[8], without using the vector atom model.

The importance of the Zeeman effect in atomic spectroscopy, just as in microwave and radio spectroscopy, is now quite different from that at the time of the discovery of the effect. At that time, it was a large help to understand the structure of the atom; it confirmed Lorentz' theory, and our ideas on the character of light. Nowadays, we use the Zeeman effect in atomic spectra as a powerful expedient in the analysis of spectra and, in microwave- and radio spectroscopy, as a detection method. Moreover the magnetic influence on the hyperfine structure of spectral lines gives a method to determine nuclear mechanical and magnetic moments. However, not only is the Zeeman effect used, in spectroscopy, but also when studying the properties of nuclei and of solids, and in astronomy, it is an aid of invaluable importance.

2. Some experimental remarks. α) *Magnets.* In order to be able to observe the Zeeman effect in atomic spectra of as many lines as possible, it is essential to have at one's disposal a magnet which gives a homogeneous, constant and strong magnetic field in as large a region as possible. To come up to these requirements, many investigators tried to build and improve magnets. A stronger field may mean an improvement of 100% for some lines of a spectrum.

Kapitza and Strelkov[9] built an electromagnet which gave fields up to 320000 oersted. This field, however, was constant only for 0.01 sec, so only the Zeeman effect of the strongest lines of simple spectra could be observed.

---

[1] See footnote 8, p. 297.
[2] See footnote 5, p. 297.
[3] P. Debye: Göttinger Nachr., Juni **1916.**
[4] A. Sommerfeld: Phys. Z. **17**, 491 (1916).
[5] A. Rubinowicz: Phys. Z. **19**, 441, 465 (1918).
[6] N. Bohr: Kopenh. Acad. **1** and **2** (1918).
[7] P. A. M. Dirac: Quantum mechanics, 1930.
[8] E. U. Condon and G. H. Shortley: The theory of atomic spectra. 1953.
[9] P. Kapitza, P. G. Strelkov and E. Laurman: Proc. Roy. Soc. Lond., Ser. A **167**, 1 (1938).

Jacquinot and Belling[1] improved the great Bellevue magnet at Paris using supplementary coils. So they reached with this magnet a field strength of 65800 oersted, in which they investigated the Paschen-Back effect of the yellow Hg-lines.

The author[2] improved the Weiss magnet of the Zeeman laboratory in Amsterdam in two ways:

1. by introducing FeCo pieces at the end of the magnet poles, which was done already by others, for instance by Back[3].

2. by fixing supplementary coils around the iron poles of the magnet. These coils were connected in series with the coils of the magnet, also made of hollow copper tubes, so that water could run through them. At the distance most frequently used, viz. 5 mm, the improvement was about 11%, viz. from 38000 oersted to 42000 oersted.

The most important progress as to the strength of the magnetic field, was made by Bitter and Harrison[4], who built a magnet without using iron, which gave fields up to about 90000 oersted. This magnet, which was also used by the author to make Zeeman effect measurements of the spectra of several elements (U, Os, Ir, Ta), was mounted in the spectroscopy laboratory of M.I.T. at Cambridge (U.S.A.). It consists of 200 circular copper plates, connected in series. The diameter of the plates is about 20 cm, the current through the coil about 10000 A at a voltage of 170 V, giving a power of 1700 kW! The temperature of the cooling water (3600 litres per minute) is some degrees below the boiling point. The great advantage of this magnet is that the field can be varied, by regulating the current, and that the region in which the field is homogeneous (to 1%) and constant (to 0.1%) is large (25 cm^3), compared with that of the most commonly used Weiss magnets.

β) *Light sources*. Different kinds of light sources have been used in the magnetic field. Very suitable for metals is the trembler, according to Back[5]. A similar one, however in a somewhat modified form, is used with the Weiss magnet of the Zeeman laboratory of Amsterdam[6]. The voltage between tungsten and metal electrode is 90 V, the current being about 1 to 4 A. The light source is built in a vacuum box, in which the pressure can be varied from 1 to 10 cm Hg. Another filling gas, for instance argon, can also be introduced into the box. Pressure and filling gas depend upon the element under investigation.

A vacuum box was also used by Bakker[7] in the Zeeman laboratory to observe the Zeeman effect of the noble gases. A glow discharge in a noble gas between tungsten electrodes is used. This was done before by Hansen and Jacobsen[8], and later on improved by Back[9].

In order to investigate the Zeeman effect in weak fields, the Schüler tube was often used. In general this lightsource requires much more space than other light sources, and so it can only be used in weak fields. The use of an interference apparatus, such as a Fabry-Pérot etalon or a Lummer plate is necessary. Photographing the spectra with different strengths of the magnetic field is conducive to a right interpretation of the observed interference pattern.

[1] P. Jacquinot and T. Belling: C. R. Acad. Sci., Paris **201**, 778 (1935).

[2] J. C. van den Bosch: Thesis Amsterdam 1948.

[3] E. Back: Handbuch der Experimentalphysik, Bd. 22, 1. Leipzig 1929.

[4] G. R. Harrison and F. Bitter: Phys. Rev. **57**, 15 (1940).

[5] E. Back and A. Landé: Zeeman-Effekt und Multiplettstruktur. 1925.

[6] H. J. van de Vliet: Thesis, Amsterdam 1939.

[7] C. J. Bakker: Thesis, Amsterdam 1931.

[8] Hansen and Jacobsen: Kopenh. Acad. **3**, 11 (1921).

[9] E. Back: Ann. d. Physik **76**, 317 (1925).

To observe the transversal effect, Geissler tubes are not very suitable, because of the fact that the discharge cannot be maintained very easily. They are mostly used for investigation of the longitudinal effect.

As to the spectroscopic apparatus, we refer to the handbooks of optics.

In this survey we had not the intention to be complete, but in any case we gave the most important features of the Zeeman effect, since its discovery in 1896.

II. The Zeeman effect in atomic spectra.

a) Theory for weak fields.

3. Classical theory of the normal triplet. According to Lorentz' "Theory of electrons", electrons in an atom are bound quasi-elastic. The equations of motion for such an electron are

$$m\ddot{x} = -f \cdot x, \tag{3.1}$$

$$m\ddot{y} = -f \cdot y. \tag{3.2}$$

$$m\ddot{z} = -f \cdot z \tag{3.3}$$

in which m is the mass of an electron and f a constant. The solution of these equation is

$$x = a\,e^{2\pi i \nu_0 t}, \quad y = b\,e^{2\pi i \nu_0 t}, \quad z = c\,e^{2\pi i \nu_0 t}. \tag{3.4}$$

Substitution of these expressions in (3.1) to (3.3) gives

$$\nu_0 = \frac{1}{2\pi}\sqrt{\frac{f}{m}} \tag{3.5}$$

and

$$f = 4\pi^2 m \nu_0^2. \tag{3.6}$$

An external magnetic field with only a component in the z-direction acts on the electron with forces

$$k_x = \frac{e}{c} H \dot{y}, \qquad k_y = -\frac{e}{c} H \dot{x}, \qquad k_z = 0$$

in which e is the charge of the electron and c the velocity of light. Now the equations of motions become

$$m\ddot{x} = -f x + \frac{e}{c} H \dot{y}, \tag{3.7}$$

$$m\ddot{y} = -f y - \frac{e}{c} H \dot{x}, \tag{3.8}$$

$$m\ddot{z} = -f z. \tag{3.9}$$

The solution is now

$$x = a\,e^{2\pi i \nu t}, \quad y = b\,e^{2\pi i \nu t}, \quad z = c\,e^{2\pi i \nu_0 t} \tag{3.10}$$

in which ν_0 still is defined by Eq. (3.5). Substitution of (3.10) in (3.7) to (3.9) yields

$$a(\nu^2 - \nu_0^2) = -\frac{e}{2\pi m c} H b i \nu, \tag{3.11}$$

$$b(\nu_0^2 - \nu^2) = -\frac{e}{2\pi m c} H a i \nu. \tag{3.12}$$

From (3.11) and (3.12) we conclude

$$\nu_0^2 - \nu^2 = \pm \frac{e\,H\,\nu}{2\pi\,m\,c}$$

or with $\nu - \nu_0 = \varDelta\nu$ and $\nu + \nu_0 \approx 2\nu$:

$$2\nu\,\varDelta\nu = \pm \frac{e\,H\,\nu}{2\pi\,m\,c}\,;$$

so

$$\varDelta\nu = \pm \frac{e\,H}{4\pi\,m\,c}\ \mathrm{sec}^{-1}$$

and

$$\varDelta\sigma = \pm \frac{e\,H}{4\pi\,m\,c^2}\ \mathrm{cm}^{-1}$$

(3.13)

where $\varDelta\sigma$ is the change in wave number, compared with the original line.

We call $\dfrac{e\,H}{4\pi\,m\,c^2}$ a Lorentz unit. Its value in c-g-s units is $4.669 \times 10^{-5}\ \mathrm{cm}^{-1}$.

4. Polarisation of the emitted radiation. Determination of the sign of the electric charge of the electron. From (3.13) and (3.11) we see that

$$a = \mp\,b\,i.$$

(4.1)

If we want to consider the real motions of the electron under influence of the external magnetic field H_z, we have to look at the real parts of (3.10). We consider the two cases:

1. $a = +b\,i$, which corresponds to

$$\varDelta\nu = - \frac{e}{4\pi\,m\,c} \cdot H.$$

Then the motion of the electron is given by the equations

$$x = a\cos 2\pi\nu t, \quad y = a\sin 2\pi\nu t.$$

Looking along the direction of H_z these expressions give the coordinates of an electron which is moving clockwise around the magnetic field as a function of time.

2. $a = -b\,i$; this corresponds to

$$\varDelta\nu = + \frac{e}{4\pi\,m\,c} \cdot H.$$

The motion of the electron is then given by

$$x = a\cos 2\pi\nu t, \quad y = - a\sin 2\pi\nu t.$$

Looking again along the direction of the field, these expressions give the coordinates of the electron, moving counterclockwise as a function of time.

Hitherto we assumed that the charge of the electron is positive, for which has been accounted in the signs in Eqs. (3.1) to (3.3), (3.7) to (3.9).

We observe, however, that the light emitted with a lower frequency than that of the original line, is circularly polarised counterclockwise. This means that *the charge of the electron is negative.*

Both the frequency and direction of the vibration of the electron in the z-direction remain unchanged. So the emitted light with frequency ν_0 is linearly polarised in the direction of the magnetic field.

Observing the effect parallel to the magnetic field, we only see the two frequencies

$$\nu_0 + \Delta\nu = \nu_0 + \frac{e\,H}{4\pi\,m\,c},$$

$$\nu_0 - \Delta\nu = \nu_0 - \frac{e\,H}{4\pi\,m\,c},$$

These components are called the σ-components. They are circularly polarised. Observing the effect at right angles to the magnetic field, we see three lines with frequencies

$$\nu_0 + \Delta\nu = \nu_0 + \frac{e\,H}{4\pi\,m\,c},$$

$$\nu_0 = \nu_0,$$

$$\nu_0 + \Delta\nu = \nu_0 - \frac{e\,H}{4\pi\,m\,c}.$$

The component with frequency ν_0 is called the π-component. Now all vibrations are linearly polarised, those with the frequencies $\nu_0 + \Delta\nu$ and $\nu_0 - \Delta\nu$ at right angles to the magnetic field and that with the frequency ν_0 parallel to the z-axis. The intensity of the π-component is twice as big as that of each of the σ-components.

5. Determination of e/m. We found

$$\Delta\nu = \frac{e\,H}{4\pi\,m\,c}.$$

From this we conclude

$$\frac{e}{m} = \frac{4\pi\,c^2}{H}\cdot\frac{\Delta\lambda}{\lambda^2}.$$

So we can determine e/m from the difference in wavelengths of the emitted radiation. Zeeman[1] estimated this value from the broadening of the sodium D-lines, at 10^7 e.m.u. In 1897 he found from more accurate measurements 1.6×10^7 e.m.u. also from the width of the sodium D-lines and later on 2.4×10^7 e.m.u. from the Cd-line 4800 Å. It is evident that great deviations from the real value could be measured by him at that time, as he assumed that all broadened lines consisted of normal triplets.

6. The anomalous effect according to the vector model (LS-coupling). According to classical theory the magnetic moment μ, associated with a mechanical moment A of a spinning negative charge is given by

$$\mu_A = -\frac{e}{2m\,c}\,A \tag{6.1}$$

in which e is the charge and m the mass of the particle. Hence, the ratio of the magnetic and mechanical moments of an electron in an orbit is

$$\frac{\mu_l}{l} = -\frac{e}{2m\,c}. \tag{6.2}$$

Besides its orbital moment l the electron also has a spin moment s. Several experiments show that the ratio of the magnetic and mechanical moments for the spinning electron is twice that for the orbital motion, i.e.

$$\frac{\mu_s}{s} = -2\cdot\frac{e}{2m\,c}. \tag{6.3}$$

---

[1] See footnote 2, p. 296.

We presume that in case the atom has more than one valence electron, the interaction between the spin moments of these electrons on one hand and the interaction between the orbital moments on the other hand, is large compared with the interaction between spin and orbital moment of each of the electrons; that is to say we suppose that LS-coupling (or Russell-Saunders coupling) between the electrons predominates.

The magnetic moment, coupled with the total orbital moment of all electrons or of the total atom core is then

$$\boldsymbol{\mu_L} = - \frac{e}{2mc} \boldsymbol{L} \qquad (6.4)$$

whereas the magnetic moment, coupled with the total spin moment of all electrons or of the whole atom core is then

$$\boldsymbol{\mu_S} = - \frac{e}{mc} \boldsymbol{S}. \qquad (6.5)$$

In general there always remains a small interaction between the total orbital moment $\boldsymbol{L}$ and the total spin moment $\boldsymbol{S}$, with the result that both precess around their resultant $\boldsymbol{J}$. In order to explain and compute the anomalous Zeeman effect, it is important to derive the total magnetic moment $\boldsymbol{\mu_J}$ in the direction of $\boldsymbol{J}$.

The relation between $\boldsymbol{\mu_J}$ and $\boldsymbol{J}$ is given by

$$\boldsymbol{\mu_J} = - g_J \cdot \frac{e}{2mc} \boldsymbol{J}. \qquad (6.6)$$

We find an expression for g_J as a function of L, S and J as fololws:

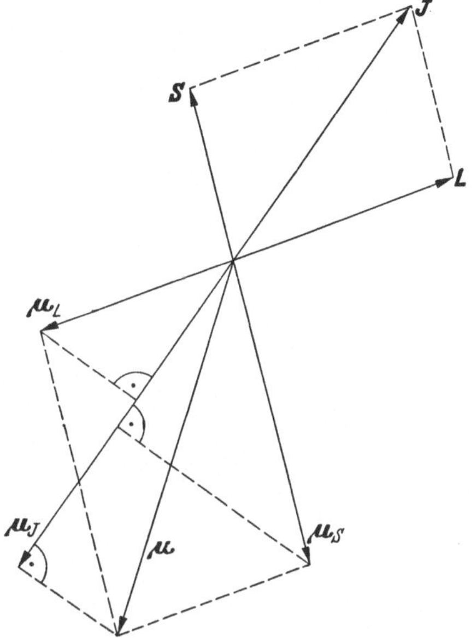

Fig. 1. Vector diagram in case of LS-coupling with mechanical and magnetic moments of electrons.

We introduce the different moments as vectors. According to wave mechanics the length of such a vector $\boldsymbol{A}$ is

$$\sqrt{A(A+1)}\ \hbar.$$

This should be considered as a very simple aid, because of the fact that the moment (for instance the total moment $\boldsymbol{J}$) as it is found from experiments always is the maximal value $J\hbar$. On the contrary the value of $\boldsymbol{J}^2$, as it follows from susceptibility measurements, is $J(J+1)\hbar^2$.

As we now consider the spin, orbital and total moment as vectors with lengths resp. $\sqrt{S(S+1)}\,\hbar$, $\sqrt{L(L+1)}\,\hbar$ and $\sqrt{J(J+1)}\,\hbar$, we are able to compute from these geometric figures, qualitative and quantitative results, which are in agreement with experiments. We then use the vector atom model.

From Fig. 1 we find

$$\mu_J = \mu_L \cos\left(\boldsymbol{L}, \boldsymbol{J}\right) + \mu_S \cos\left(\boldsymbol{S}, \boldsymbol{J}\right),$$

$$\mu_J = - \frac{e}{2mc} \cdot \left\{ \frac{L}{J} \cos\left(\boldsymbol{L}, \boldsymbol{J}\right) + 2 \frac{S}{J} \cos\left(\boldsymbol{S}, \boldsymbol{J}\right) \right\} \boldsymbol{J}, \qquad (6.7)$$

or, using Eq. (6.6) for $\boldsymbol{\mu_J}$, we obtain

$$g_J = \frac{L}{J} \cos\left(\boldsymbol{L}, \boldsymbol{J}\right) + \frac{2S}{J} \cos\left(\boldsymbol{S}, \boldsymbol{J}\right). \qquad (6.8)$$

Applying the wave mechanical cosine law to the triangle, formed by the vectors S, L and J, one easily finds

$$g_J = 1 + \frac{J(J+1) + S(S+1) - L(L+1)}{2J(J+1)}.$$

(6.9)

This factor now known as the Landé g-factor, determines the total magnetic moment of the atom, if it is in a state characterised by the quantum numbers L, S and J. It gives directly the separations of the Zeeman levels, belonging to a level with quantum numbers L, S and J. Its importance can hardly be overestimated.

The energy of an atom in an external magnetic field is determined by

1. the potential and kinetic energy of the valence electrons in the field of the nucleus or in the field of the nucleus screened by a number of electrons,
2. the spin-spin interaction energy,
3. the orbit-orbit interaction energy,
4. the spin-orbit interaction energy.

We will not go into details concerning these interaction energies. The reader may find them in all handbooks about atomic spectroscopy. The most important energy contribution, when dealing with the Zeeman effect, is the next one:

5. the energy resulting from the interaction between the total magnetic moment and an external homogeneous magnetic field. J will be quantized with respect to the magnetic field H in such a way that the maximum value of the total moment in the direction of H is $J\hbar$, the minimum value being $-J\hbar$. There are $2J+1$ values of the component of the moment in the direction of H, viz. $m_J\hbar$ with $m_J = -J, -J+1, \ldots J$, each corresponding to a different magnetic energy level.

Suppose the energy of the total atom, according to the interaction energies 1, 2, 3 and 4, mentioned above, is E_0, then the energy of the atom in the magnetic field H will be

$$E = E_0 - \mu_J \cdot H \cos(\mu_J, H)$$

(6.10)

and, according to Eq. (6.6),

$$E = E_0 + g_J \frac{e}{2mc} J H \cos(J, H)$$

or

$$E = E_0 + g_J \cdot m_J \cdot \frac{e}{2mc} \cdot \hbar H.$$

(6.11)

As the relation between term value T and energy value E is

$$T = -\frac{E}{hc}$$

we find from (6.11)

$$T = T_0 - g_J \cdot m_J \frac{eH}{4\pi mc^2} \text{ cm}^{-1}.$$

(6.12)

The separation between two neighbouring magnetic sublevels will be

$$T = g_J \frac{eH}{4\pi mc^2} \text{ cm}^{-1}.$$

(6.13)

T is determined by H and the g-factor of Landé, belonging to the energy level in question. As g_J is given in terms of L, S and J (in case of LS-coupling), it will in general be different for different levels.

We give g_J-factors in some particular cases:

1. For all S-levels ($L=0$) (except the $^1S_0$ level, which does not split up in the magnetic field) $g=2.000$.

2. For all singlet levels ($S=0$) (again with the exception of the $^1S_0$-level) $g=1.000$.

3. For all levels for which $L=S$ (^3P, ^5D etc.) $g=1.500$ with the only exception of the levels with $J=0$.

4. For a term with orbital quantum number L, $g = \dfrac{L+2S}{L+S}$ for the fine structure level with maximum J.

5. For a term with orbital quantum number L

$$\Sigma g = \begin{cases} 2(2L+1) & \text{if} \quad L < S, \\ 2S+1 & \text{if} \quad L > S. \end{cases}$$

A list of g-values in case of LS-coupling has been given in Table 8 on p. 331.

7. Selection rules. In case of electric dipole radiation there are the following selection rules for a transition between magnetic sublevels in a weak magnetic field: Allowed transitions occur if

$$\Delta m_J = 0,$$
$$\Delta m_J = +1,$$
$$\Delta m_J = -1.$$

Only $m_j = 0 \rightarrow m_j = 0$ is forbidden in case $\Delta J = 0$.

If $\Delta m_j = 0$, the radiated light is linearly polarised, parallel to the magnetic field (π-polarisation).

If $\Delta m_j = \pm 1$, the radiated light is circularly polarised in planes perpendicular to the magnetic field (σ-polarisation).

8. JJ-coupling. In case of jj-coupling of several valence electrons or of the one valence electron and the atom core, we derive the g-factor as follows. Now, the spin-orbit interaction of one electron prevailing we find for the magnetic moment of each electron

$$\boldsymbol{\mu}_j = - g_j \frac{e}{2mc} \boldsymbol{j} \tag{8.1}$$

where, according to Sect. 6,

$$g_j = 1 + \frac{j(j+1) + s(s+1) - l(l+1)}{2j(j+1)}.$$

In case the atom has two valence electrons (which example we will consider here), the total moment $\boldsymbol{J}$ is

$$\boldsymbol{J} = \boldsymbol{j}_1 + \boldsymbol{j}_2. \tag{8.2}$$

The total magnetic moment connected with $\boldsymbol{J}$ is

$$\boldsymbol{\mu}_J = - g_J \frac{e}{2mc} \cdot \boldsymbol{J}$$

with

$$g_J = g_1 \frac{J(J+1) + j_1(j_1+1) - j_2(j_2+1)}{J(J+1)} + g_2 \frac{J(J+1) + j_2(j_2+1) - j_1(j_1+1)}{J(J+1)}. \tag{8.3}$$

9. Intermediate coupling. In most atoms we have neither extreme LS-coupling nor extreme jj-coupling between the electrons, but intermediate coupling. For some cases of this intermediate coupling we shall derive the g-factor.

We consider two vectors A and B which precess around their resultant C and with which are connected the magnetic moments $g_A A$, $g_B B$ and $g_C C$ respectively. g_C may be derived from the following expression

$$g_C\,C = g_A\,A\cos(A,\,C) + g_B\,B\cos(B,\,C).\tag{9.1}$$

We give the following examples:

1. The coupling scheme may be given by

$$\{(l_1 l_2)\,s_1\}\,s_2 = (L\,s_1)\,s_2 = j'\,s_2 = J.$$

Applying (9.1) we obtain at first

$$g_{j'} = \frac{3j'(j'+1) + s_1(s_1+1) - L(L+1)}{2j'(j'+1)},$$

then, in a second step

$$g_J = 2\,\frac{s_2}{J}\cos(s_2,\,J) + g_{j'}\,\frac{j'}{J}\cos(j',\,J),$$

and therefore,

$$g_J = 2\,\frac{s_2(s_2+1)+J(J+1)-j'(j'+1)}{2J(J+1)} + g_{j'}\,\frac{j'(j'+1)+J(J+1)-s_2(s_2+1)}{2J(J+1)},\tag{9.2}$$

2. The coupling scheme may be given by

$$\{(l_1 s_1)\,s_2\}\,l_2 = (j_1 s_2)\,l_2 = j'\,l_2 = J.$$

Application of (9.1) leads in a first step to

$$g_{j_1} = \frac{3j_1(j_1+1) + s_1(s_1+1) - l_1(l_1+1)}{2j_1(j_1+1)},$$

in a second step to

$$\left.\begin{aligned}
g_{j'} &= g_{j_1}\,\frac{j_1(j_1+1)+j'(j'+1)-s_2(s_2+1)}{2j'(j'+1)} + 2\cdot\frac{s_2(s_2+1)+j'(j'+1)-j_1(j_1+1)}{2j'(j'+1)} \\
\text{and} \\
g_J &= g_{j'}\,\frac{J(J+1)+j'(j'+1)-l_2(l_2+1)}{2J(J+1)} + \frac{J(J+1)+l_2(l_2+1)-j'(j'+1)}{2J(J+1)}.
\end{aligned}\right\}\tag{9.3}$$

In an analogous way the g-factor of the resultant vector J may be derived for any given coupling scheme.

10. The calculation of Zeeman patterns. If the J- and g-values of the two combining levels are known, we can derive the Zeeman effect of the spectral lines, applying the selection rules of Sect. 7.

We give some examples.

α) *First example*: $J_1 = {}^3/_2$, $J_2 = {}^5/_2$. The value of $m_J \cdot g_J$ gives the separation in wv/cm [†] of the magnetic sublevel to the original level, so that we may draw up the following scheme for separations of magnetic sublevels concerning the levels with $J_1 = {}^3/_2$ and $J_2 = {}^5/_2$.

$$-{}^5/_2\,g_2 \qquad -{}^3/_2\,g_2 \qquad -{}^1/_2\,g_2 \qquad +{}^1/_2\,g_2 \qquad +{}^3/_2\,g_2 \qquad +{}^5/_2\,g_2$$

$$-{}^3/_2\,g_1 \qquad -{}^1/_2\,g_1 \qquad +{}^1/_2\,g_1 \qquad +{}^3/_2\,g_2$$

[†] The unit wv/cm is introduced here for the wavenumber, as this is the number of waves per cm. This unit is also called the Kayser (K). Cf. the unit of frequency: cycles per second (c/s) which is also called Hertz (Hz).

Applying the selection rules we find from this scheme:

π-components $\qquad (\pm^1/_2\,g_2\mp^1/_2\,g_1);\quad (\pm^3/_2\,g_2\mp^3/_2\,g_1)$

σ-components $\begin{cases} g_1 > g_2: \quad \underline{\pm^5/_2\,g_2\mp^3/_2\,g_1}; \quad \pm^3/_2\,g_2\mp^1/_2\,g_1; \quad \pm^1/_2\,g_2\pm^1/_2\,g_1; \\ \qquad\qquad\qquad \mp^1/_2\,g_2\pm^3/_2\,g_1\,, \\[4pt] g_1 < g_2: \quad \mp^1/_2\,g_2\pm^3/_2\,g_1; \quad \pm^1/_2\,g_2\pm^1/_2\,g_1; \quad \pm^3/_2\,g_2\mp^1/_2\,g_1; \\ \qquad\qquad\qquad \underline{\pm^5/_2\,g_2\mp^3/_2\,g_1}\,. \end{cases}$

Remark: π-components are always put between brackets; the underlined is the strongest component, both in π- and σ-components.

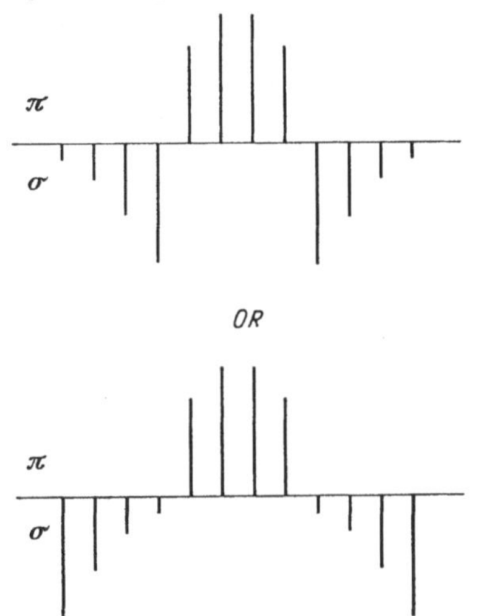

OR

Fig. 2. Zeeman pattern for $J_1=\frac{3}{2}$; $J_2=\frac{5}{2}$ in two cases: $g_1 > g_2$ (above) and $g_1 < g_2$ (below).

This Zeeman pattern is plotted in Fig. 2. As usually done, the π-components are plotted above, the σ-components below the line.

The separations of the π-components (in wv/cm units), to the position of the original line, are in the proportion of $1:3:5$ etc. There is no central π-component.

The number of π-components is $2J_1+1$, if $J_1 < J_2$.

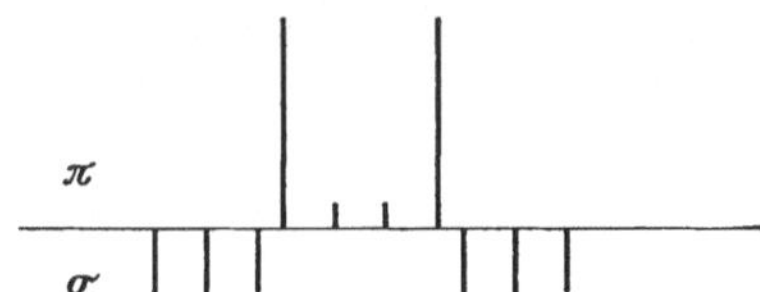

Fig. 3. Zeeman pattern for $J_1=\frac{3}{2}$; $J_2=\frac{3}{2}$.

The number of σ-components is twice as large as the number of π-components.

β) *Second example*: $J_1=^3/_2$, $J_2=^3/_2$. In the same way as before we obtain:

$$-^3/_2\,g_2 \qquad -^1/_2\,g_2 \qquad +^1/_2\,g_2 \qquad +^3/_2\,g_2$$
$$-^3/_2\,g_1 \qquad -^1/_2\,g_1 \qquad +^1/_2 g_1 \qquad +^3/_2\,g_1\,.$$

Applying the selection rules we find:

π-components $\quad (\pm^1/_2\,g_2\mp^1/_2\,g_1);\quad (\underline{\pm^3/_2\,g_2\mp^3/_2\,g_1})$

σ-components $\quad \pm^3/_2\,g_2\mp^1/_2\,g_1;\quad \underline{\pm^1/_2\,g_2\pm^1/_2\,g_1};\quad \mp^1/_2\,g_2\pm^3/_2\,g_1\,.$

Fig. 3 shows this pattern.

The separations of the π-components (in wv/cm units), to the position of the original line, are in the proportion of $1:3:5$ etc. There is no central π-component. The number of π-components is $2J+1$.

The number of σ-components is two less than twice that of the π-components.

γ) *Third example*: $J_1=2$, $J_2=3$. Now we have:

$$-3g_2 \qquad -2g_2 \qquad -g_2 \qquad 0 \qquad +g_2 \qquad +2g_2 \qquad +3g_2$$
$$-2g_1 \qquad -g_1 \qquad 0 \qquad +g_1 \qquad +2g_1$$

20*

We find from this scheme:

π-components: $\qquad \underline{(0)} \ (\pm g_2 \mp g_1) \ (\pm 2g_2 \mp 2g_1)$

σ-components: $\begin{cases} g_1 > g_2: & \pm 3g_2 \mp 2g_1; \quad \pm 2g_2 \mp g_1; \quad \pm g_2; \quad \pm g_1; \quad \mp g_2 \pm 2g_1; \\ g_1 < g_2: & \mp g_2 \pm 2g_1; \quad \pm g_1; \quad \pm g_2; \quad \pm 2g_2 \mp g_1; \quad \pm 3g_2 \mp 2g_1; \end{cases}$

Fig. 4 shows this pattern.

The separations of the π-components (in wv/cm units) to the position of the original line are in the proportion of $1:2:3$ etc. There is a central π-component.

The number of π-components is $2J_1 + 1$, if $J_1 < J_2$.

The number of σ-components is twice as large as the number of π-components.

δ) *Fourth example*: $J_1 = 2$, $J_2 = 2$.

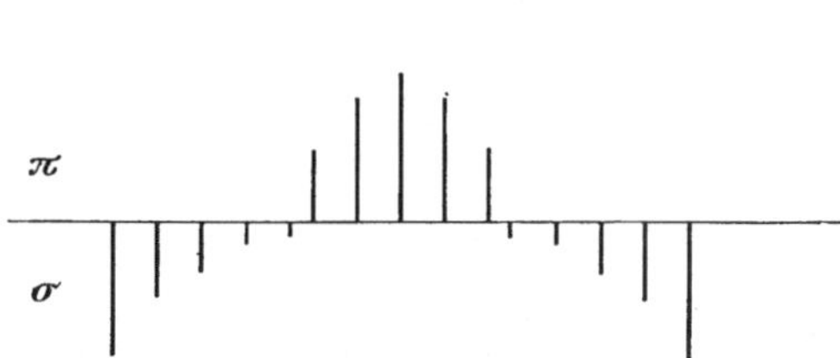

Fig. 4. Zeeman pattern for $J_1 = 2$, $J_2 = 3$ in two cases: $g_1 > g_2$ (above) and $g_1 < g_2$ (below).

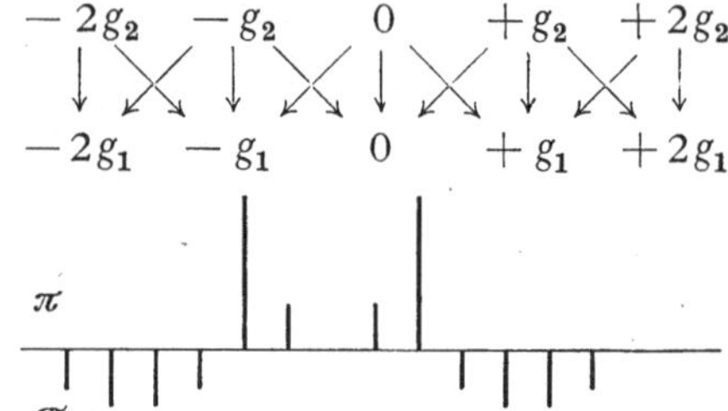

Fig. 5. Zeeman pattern for $J_1 = 2$, $J_2 = 2$.

Now this scheme gives as Zeeman pattern

π-components $\quad (\pm g_2 \mp g_1); \quad (\pm 2g_2 \mp 2g_1),$

σ-components $\quad \pm 2g_2 \mp g_1; \quad \pm g_2; \quad \pm g_1; \quad \mp g_2 \pm 2g_1.$

For this pattern see Fig. 5.

The separation of the π-components (in wv/cm units) to the position of the original line are in the proportion of $1:2:3$ etc. There is no central component.

The number of π-components is $2J$.

The number of σ-components is twice as large as that of the π-components.

The separation between a component and its nearest (both for π- and σ-components) is in all cates $(g_2 - g_1)$ wv/cm.

Finally we remark

1. transitions between singlet levels in a magnetic field give a normal triplet.

2. the Zeeman effect of a transition between levels with equal g-values is a triplet. The separation between the σ-component and the π-component depends on the value of g.

11. Wave mechanical theory of the normal triplet. We consider the motion of an electron with mass μ in the central field of the nucleus. The Hamiltonian for such an electron is:

$$\mathscr{H} = \frac{\mathbf{p}^2}{2\mu} - eV$$

when its potential energy is eV, which is a function of the distance to the nucleus.

In an external magnetic field H, for which $\boldsymbol{H} = \mathrm{rot}\,\boldsymbol{A}$, the Hamiltonian becomes

$$\mathscr{H} = \frac{1}{2\mu}\left(\boldsymbol{p} + \frac{e}{c}\,\boldsymbol{A}\right)^2 - eV, \tag{11.1}$$

and the Hamiltonian operator will be

$$\mathscr{H}_{\mathrm{op}} = -\frac{\hbar^2}{2\mu}\,\Delta + \frac{\hbar\,e}{\mu\,i\,c}\,(\boldsymbol{A}\,\mathrm{grad}) - e\,V$$

supposing that $\mathrm{div}\,\boldsymbol{A} = 0$ and neglecting terms with A^2. If we choose

$$A_x = -\tfrac{1}{2}H\,y,$$
$$A_y = \tfrac{1}{2}H\,z,$$
$$A_z = 0,$$

we have a magnetic field only along the z-axis. In this case

$$(\boldsymbol{A}\,\mathrm{grad}\,\psi) = \frac{1}{2}\,H\left(x\,\frac{\partial\psi}{\partial y} - y\,\frac{\partial\psi}{\partial x}\right) \tag{11.2}$$

or, using polar coordinates instead of rectangular coordinates, (11.2) becomes

$$(\boldsymbol{A}\,\mathrm{grad}) = \frac{1}{2}\,H\,\frac{\partial\psi}{\partial\varphi}\,.$$

The wave equation follows from its general form

$$\mathscr{H}_{\mathrm{op}}\,\psi = E\cdot\psi \tag{11.3}$$

by substituting (11.1) into (11.3):

$$\Delta\psi + \frac{2\mu}{\hbar^2}\,E_m + \frac{2\,i\,e}{\hbar\,c}\cdot\frac{1}{2}\,H\,\frac{\partial\psi}{\partial\varphi} + \frac{2\mu\,Z\,e^2}{\hbar^2\,r}\,\psi = 0\,. \tag{11.4}$$

This Schrödinger equation describing the motion of an atomic electron under the combined forces of a Coulomb field $-Z\,e^2/r$ and the magnetic field $\boldsymbol{H}$ in z-direction has a solution

$$\psi\,(r,\,\vartheta,\,\varphi) = R\,(r)\;P_l^m\,(\cos\vartheta)\;e^{\pm\,i\,m\,\varphi}$$

(m: magnetic quantum number) in which only the factor $e^{\pm\,i\,m\,\varphi}$ depends on φ, so that

$$\frac{\partial\psi}{\partial\varphi} = \pm\,i\,m\,\psi\,e^{i\,m\,\varphi}\,.$$

Eq. (11.4) therefore yields

$$\Delta\psi + \frac{2\mu}{\hbar^2}\left(E_m \pm m\,\frac{\hbar\,e}{2\mu\,c}\,H + \frac{Z\,e^2}{r}\right)\psi = 0\,. \tag{11.5}$$

The wave equation for the problem *without* external magnetic field is

$$\Delta\psi + \frac{2\mu}{\hbar^2}\left(E + \frac{Z\,e^2}{r}\right)\psi = 0\,. \tag{11.6}$$

Eqs. (11.5) and (11.6) have the same solution, if

$$E_m = E \pm m\,\frac{\hbar\,e}{2\mu\,c}\cdot H \tag{11.7}$$

in which $m = -l,\,-l+1,\,\ldots l-1,\,l$, with l being the orbital quantum number.

From (11.7) there follows:

1. For a given l-value the wave functions in the cases with and without magnetic field are the same. Only the eigenvalues differ for different values of m.

2. The energy separation between two levels with magnetic quantum numbers m and $m+1$ is $\mu_B H$, if $\mu_B = \dfrac{e\hbar}{2mc}$ is a Bohr magneton which is again the classical result for the normal triplet.

Fig. 6 shows energy level diagrams for two transitions in a magnetic field, both in case $J_1 = 2$ and $J_1 = 1$; Fig. 6a gives the transition, if $g_1 = g_2$, and Fig. 6b, if $g_1 > g_2$.

b) Theory for strong fields.

12. Paschen-Back effect for LS-coupling. In 1912 Paschen and Back[1,2] observed that several lines of the Li spectrum were normal triplets in a magnetic field. This was in contrast with Preston's rule[3]. According to this rule, these lines of a spectrum homologous with Na should be narrow doublets. Paschen and Back found that the lines were normal triplets, if the strength of the magnetic field was sufficiently high. Apparently the Zeeman effect pattern of a spectral line depends on the strength of the magnetic field. They also found this effect in the oxygen triplet at a wavelength of 3947 Å[1]. They used for the Li lines a magnetic field of 35000 oersted, and in the case of the O lines, a field strength of 41000 oersted. Whereas the Li line 6700 Å,—for which Zeeman[4] proved that

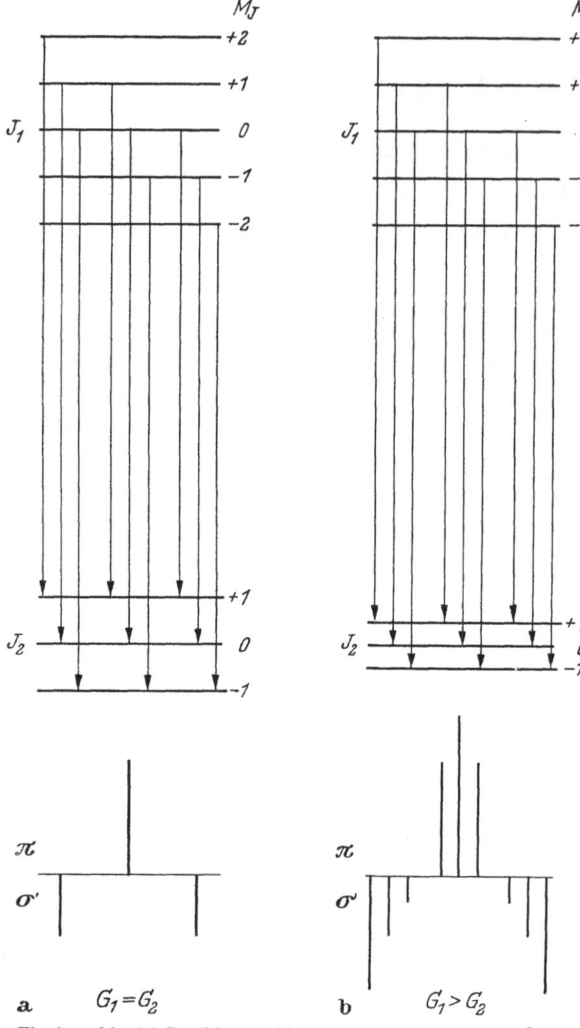

Fig. 6a and b. (a) Possible transitions in a magnetic field for $J_1 = 2$, $J_2 = 1$ in case $g_1 = g_2$. (b) Possible transitions in a magnetic field for $J_1 = 2$, $J_2 = 1$ in case $g_1 > g_2$.

it was a narrow doublet without magnetic field, corresponding to the Na D lines, with a wavelength distance of 0.13 Å—already was a normal triplet in a field of 45000 oersted, this was not the case with the oxygen triplet, not even in a field of 41000 oersted. A magnetic field of about 200000 oersted would be required for these oxygen lines to show up as a normal triplet in a magnetic field. Obviously in a field of 41000 oersted we have to do with an intermediate state between a weak and strong field pattern, and it can be said that when

[1] F. Paschen and E. Back: Ann. der Phys. **39**, 929 (1912).
[2] F. Paschen and E. Back: Ann. der Phys. **40**, 960 (1913).
[3] See footnote 7, p. 297.
[4] P. Zeeman: Phys. Z. **14**, 405 (1913).

increasing the strength of the magnetic field, the Zeeman pattern of a group of lines, belonging to one multiplet, is completely changed. This is called the Paschen Back effect. If the magnetic field is not yet strong enough to change the splitting of a group of lines in a normal triplet, we call it a *partial* Paschen Back effect. The discovery of this effect preceded the quantum theory of atomic spectra, and it was not before the introduction of the spin of the electron in 1925 that an explanation of it could be given.

From the above it will be clear that only in a few cases the Paschen Back effect, or even a partial Paschen Back effect, could be observed, as the required field strength even in the case of the oxygen triplet (where the multiplet distances are relatively small), is of the order of magnitude of about 200000 oersted. In Chap. 1, Sect. 2, we saw that the magnetic field strength, suitable for experimental work, remains below 100000 oersted. A few examples of experimental results may follow here:

WOLTJER[1] and POPOW[2] showed that there was a beginning Paschen Back effect of the Na D lines in a magnetic field of 40000 oersted. This revealed itself in an asymmetrical Zeeman pattern of the two lines.

KENT[3] investigated five lines of the Li spectrum, which were all normal triplets in a field of 36000 oersted. OLDENBURG[4], just as FÖRSTERLING and HANSEN[5] investigated the H_α line very accurately and also found an effect. Recently, KIESS and SHORTLEY[6] reported the partial Paschen Back effect of some oxygen and nitrogen lines in fields up to 90000 oersted.

In the preceding section it was assumed that the magnetic field was a "weak" field, that is to say that the distance between the magnetic sublevels is small compared with the multiplet distance. In other words the spin-orbit coupling of the electron is stronger than the coupling between the total angular momentum of the electrons and the external magnetic field. Then J is still a vector and J a quantum number, and the precession frequency of this vector around the direction of the magnetic field is small compared with that of spin and orbital momentum around the total angular momentum J. The precession frequency of J around the magnetic field increases when the field strength of the magnetic field is increased and so does the energy distance between succeeding magnetic sublevels.

In the following we assume the field strength of the magnetic field to be high enough that the interaction between spin moment and magnetic field on the one hand, and orbital moment and magnetic field on the other hand, is strong compared with the mutual interaction of spin and orbital moment. Then J is no more a vector, nor is J a quantum number. With L and S quantized independently with respect to the external magnetic field H, there are $2L+1$ and $2S+1$ states resp. for L and S. The total energy is determined by the sum of the energies of spin and orbital moment in the magnetic field. To this energy we have to add a small correction energy which is due to a remaining interaction of S and L. The energy of the orbital moment in the magnetic field is:

$$E_{L,H} = -\,\mu_L\,\boldsymbol{H}\cos(\mu_L,\boldsymbol{H})\,,$$

and of the spin moment:

$$E_{S,H} = -\,\mu_S\,\boldsymbol{H}\cos(\mu_S,\boldsymbol{H})\,. \tag{12.1}$$

[1] J. WOLTJER: Thesis, Amsterdam 1914.
[2] S. POPOW: Phys. Z. **15**, 756 (1919).
[3] N. A. KENT: Astrophys. J. **40**, 343 (1914).
[4] O. OLDENBURG: Ann. der Phys. **67**, 253 (1922).
[5] K. FÖSTERLING and G. HANSEN: Z. Physik **18**, 26 (1923).
[6] C. C. KIESS and G. H. SHORTLEY: J. Res. Nat. Bur. Stand. **42**, 183 (1949).

With $\mu_l = -\dfrac{e}{2mc}\,\boldsymbol{L}$ and $\mu_s = -2\cdot\dfrac{e}{2mc}\cdot\boldsymbol{S}$ we find for the total energy

$$E = E_0 + \frac{e\,\hbar}{2mc}\,H\,(M_L + 2M_S). \tag{12.2}$$

The extra energy, corresponding to the interaction of $\boldsymbol{S}$ and $\boldsymbol{L}$, can be written as:

$$\Delta E_{LS} = hc\,A\,L\,S\,\overline{\cos(\boldsymbol{LS})}$$

in which we have to deal with the mean value of $\cos(\boldsymbol{L}, \boldsymbol{S})$, because of the fact that the angle between $\boldsymbol{L}$ and $\boldsymbol{S}$ is changing every moment, in contrast to the

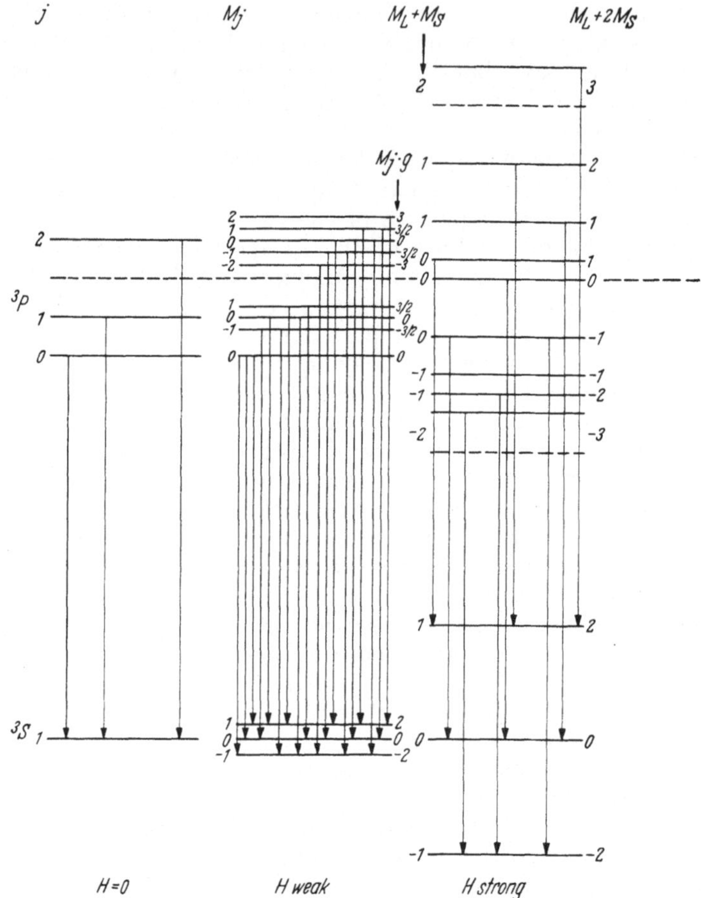

Fig. 7. The multiplet $^3P_{012}-^3S_1$ without, in a "weak", and in a "strong" external magnetic field.

case of a weak field where this angle remains constant. For this mean value we can write:

$$\overline{\cos(\boldsymbol{L}, \boldsymbol{H})} = \cos(\boldsymbol{L}, \boldsymbol{H})\cos(\boldsymbol{S}, \boldsymbol{H})$$

so that

$$\Delta E_{LS} = hc\,A\,M_L\,M_S$$

and with (12.2)

$$E = E_0 + \mu_B H(M_L + 2M_S) + hc\,A\,M_L M_S \tag{12.3}$$

and

$$T = T_0 - \frac{e\,H}{4\pi mc^2}\,(M_L + 2M_S) - A\,M_L M_S. \tag{12.4}$$

In the case of a strong magnetic field there are the following selection rules.

$$\text{For } \pi\text{- as well as } \sigma\text{-polarisation:} \quad \Delta M_S = 0,$$
$$\text{for } \pi\text{-polarisation:} \quad \Delta M_L = 0,$$
$$\text{for } \sigma\text{-polarisation:} \quad \Delta M_L = +1 \quad \text{and} \quad \Delta M_L = -1.$$

In this section we have assumed that the magnetic field is not strong enough that the coupling between the spin moments of the electrons and that between the orbital moments of the electrons will be broken down by the field. In case of *one* valence electron we may of course write the formulae (12.2) to (12.4) using small characters m_l and m_s.

As an example let us discuss the transition $^3P_{012} - {}^3S_1$.

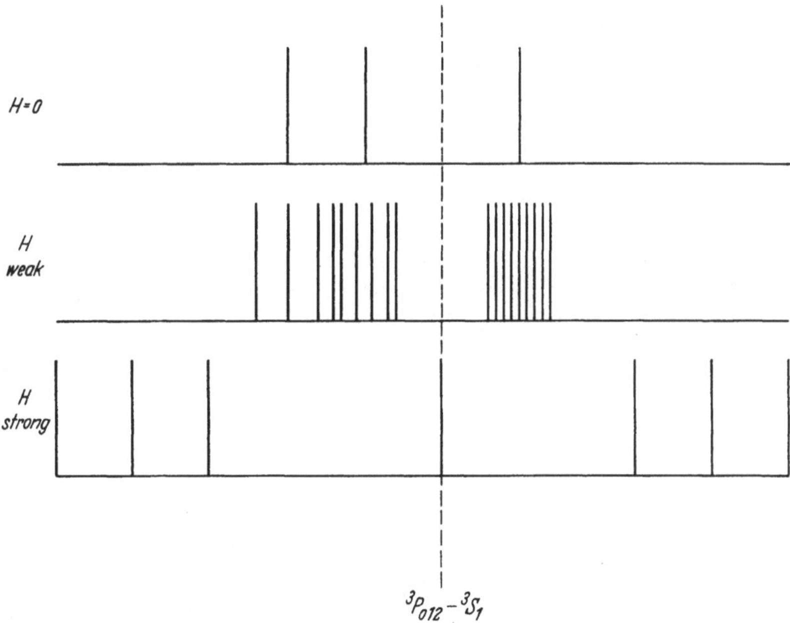

Fig. 8. Relative positions of the transitions belonging to the same multiplet of Fig. 7, without, in a "weak", and in a "strong" magnetic field.

In order to find the energy distances of the levels in a strong field to the center of gravity of the multiplets, we have to compute the values of $(M_L + 2M_S)$ and $M_L M_S$ for each level.

If σ_0 is the distance in wv/cm between the centers of gravity of the ^3P-term and the $^3S_1$ level, then the wave numbers of the other transitions in the Paschen Back effect of this example are:

$$\Delta M_L = +1 \begin{cases} \sigma = \sigma_0 + \dfrac{e}{4\pi m c^2} + A, \\[2mm] \sigma = \sigma_0 + \dfrac{e}{4\pi m c^2}, \\[2mm] \sigma = \sigma_0 + \dfrac{e}{4\pi m c^2} - A, \end{cases} \qquad \Delta M_L = -1 \begin{cases} \sigma = \sigma_0 - \dfrac{e}{4\pi m c^2} - A, \\[2mm] \sigma = \sigma_0 - \dfrac{e}{4\pi m c^2}, \\[2mm] \sigma = \sigma_0 - \dfrac{e}{4\pi m c^2} + A. \end{cases}$$

$$\Delta M_L = 0 \qquad \sigma = \sigma_0,$$

See Fig. 7.

On both sides of the central line (σ_0) there are two groups of three lines (σ-components). The distance between the middle one of each of these groups and the central line is equal to a normal unit (Lorentz unit). Fig. 8 shows the positions of the lines of this triplet relative to σ_0 and in dependence of the strength of the magnetic field.

13. Intermediate field strengths. The theory for intermediate field strengths has among others been given by C. G. DARWIN[1] for any values of L and S. These formulae have been applied in more detail to doublet and triplet terms by K. DARWIN[2]. From this theory we can link together the corresponding levels in the weak and strong field cases. We give here some essential rules for this correspondence:

1. Conservation of angular momentum. The sum of the projections of the mechanical moments on the direction of H does not change. M values keep constant.

2. No two levels with the same M will cross.
These rules may also be applied in case of jj-coupling.

14. Complete Paschen-Back effect. Theoretically we can apply a very strong magnetic field which breaks even down the mutual interaction of the spin momens and the mutual interaction of orbital moments in case of LS-coupling, and the interaction of spin and orbital moment of each electron in case of jj-coupling. L and S no longer have any meaning (nor $\boldsymbol{j}_1$ and $\boldsymbol{j}_2$), nor are L and S (or j_1 and j_2) quantum numbers. Each spin moment and each orbital moment is quantized with respect to the magnetic field. Theory shows that the distance of the energy levels from the centre of gravity of the Coulomb interaction between electrons and nucleus is given by

$$T = (2m_{s_1} + 2m_{s_2} + m_{l_1} + m_{l_2}) L \quad \text{cm}^{-1} \tag{14.1}$$

for two electrons. Again we have to add small corrections because of small interactions still remaining between spin moments, between orbital moments and between spin and orbital moments. These corrections can be written in the form

$$a\, m_{s_1} m_{s_2}, \qquad a'\, m_{l_1} m_{l_2}, \qquad a''\, m_{s_1} m_{l_1}, \qquad a'''\, m_{s_2} m_{l_2}.$$

The selection rules are

for π- as well as σ-components $\quad \begin{cases} \Delta m_{s_1} = 0 \quad \Delta m_{s_2} = 0, \\ \Delta m_{l_1} = 0, \end{cases}$

moreover for π-components $\quad \Delta m_{l_2} = 0,$ $\left.\vphantom{\begin{cases}a\\b\end{cases}}\right\}$ or interchanged.

and for σ-components $\quad \Delta m_{l_2} = +1, \Delta m_{l_2} = -1$

Apart from the small correction terms we shall always find a normal triplet.

The state of an electron when a very strong magnetic field is applied is apparently determined by the projections m_s and m_l of $\boldsymbol{s}$ and $\boldsymbol{l}$ on the direction of the external magnetic field. The consequence of this is that we can compute from all possible m_s- and m_l-values the S and L values that are present in the fieldfree case, in other words we can derive from the possible states of the electrons in a very strong field — where all interactions are broken down —the energy levels when there is no field applied.

---

[1] C. G. DARWIN: Proc. Roy. Soc. Lond., Ser. A **115**, 1 (1928).

[2] K. DARWIN: Proc. Roy. Soc. Lond., Ser. A **118**, 264 (1928).

15. The energy levels of a p^2 configuration. This is extremely important when we have to deal with equivalent electrons, viz. when they have the same n and l values. For a p-electron we have:

$$m_l = 1 \quad \text{and} \quad m_s = +\,{}^1/_2\,,$$
$$m_l = 0\,, \qquad\quad m_s = -\,{}^1/_2\,,$$
$$m_l = -\,1\,.$$

The following combinations of values of m_l and m_s are possible:

$$(1,\ \ +{}^1/_2)\,, \qquad (1,\ \ -{}^1/_2)\,,$$
$$(0,\ \ +{}^1/_2)\,, \qquad (0,\ \ -{}^1/_2)\,,$$
$$(-\,1,\ \ +{}^1/_2)\,, \quad (-\,1,\ \ -{}^1/_2)\,.$$

Taking into account the Pauli principle (no two electrons in the same atom can have the same set of quantum numbers, n, l, m_l and m_s), out of these combinations we take two at a time. A simple calculation gives the following result:

Possible energy levels for a p^2-configuration are ${}^1S_0 - {}^3P_{012} - {}^1D_2$.

In case of two non-equivalent p-electrons (we need not consider the Pauli principle) the result is:

Possible energy levels for a pp-configuration are ${}^1S_0 - {}^1P_1 - {}^1D_2 - {}^3S_1 - {}^3P_{012} - {}^3D_{123}$.

In an analogous way we can derive the possible energy states of the atoms, if jj-coupling between the electrons is predominant. Then the Pauli principle says that no two electrons in the same atom can have the same set of quantum numbers, n, l, j, m_j. In both cases of coupling the number of levels and their M values are the same, only the relative position of the levels is different.

III. Rules for g-factors.

16. g-Permanence rule. PAULI[1] has given two rules for g-factors. If no magnetic field is applied, the g-factor is given by the ratio of the total magnetic moment in the direction of $\boldsymbol{J}$ and the total mechanical moment

$$g\,\boldsymbol{J} = \mu_J \tag{16.1}$$

with μ_J in units $\dfrac{e\,\hbar}{2\,m\,c}$ and J in units $\hbar$.

In a weak external magnetic field, the g-factor will be defined as the ratio of the magnetic moment in the direction of the external field and the mechanical moment in the direction of this field

$$g\,\boldsymbol{M}_J = M_J\,\boldsymbol{g} \tag{16.2}$$

$\boldsymbol{J}$ is still a vector and J a quantum number.

In a strong magnetic field, when $\boldsymbol{L}$ and $\boldsymbol{S}$ are quantized separately with respect to the magnetic field, we find:

$$g\,(M_S + M_L) = 2M_S + M_L\,. \tag{16.3}$$

We now consider the terms of a multiplet in weak and strong fields. Then in both cases there are terms with the same total magnetic quantum number M. As an illustration we give the ${}^3D_{123}$-multiplet. In this example we find the M

---

[1] W. PAULI: Z. Physik **16**, 155 (1923).

values: -3 (1), -2 (1), -1 (3), 0 (3), $+1$ (3), $+2$ (2), $+3$. (The number between brackets means the number of magnetic sublevels in the multiplet with that magnetic quantum number.) These values of M also exist in the weak field case:

$$^3D_3: \quad M_J = -3, \; -2, \; -1, \; 0, \; +1, \; +2, \; +3,$$
$$^3D_2: \quad M_J = \qquad -2, \; -1, \; 0, \; +1, \; +2,$$
$$^3D_1: \quad M_J = \qquad\qquad -1, \; 0, \; +1,$$

Applying the above formulae (16.2) and (16.3) we find that the sum of the g-factors for a definite M value is independent of the field strength (see Table 1). *This is the so-called g-permanence rule.*

Table 1.

M	g (weak field)		Σg	(g strong field)				Σg
3	$(^3D_3)$	1.333	1.333	$(M=1$	$M=2)$		1.333	1.333
2	$(^3D_3)$ $(^3D_2)$	1.333 1.167	2.500	$(M=1$ $(M=0$	$M=1)$ $M=2)$		1.500 1.000	2.500
1	$(^3D_3)$ $(^3D_2)$ $(^3D_1)$	1.333 1.167 0.500	3.000	$(M=1$ $(M=0$ $(M=-1$	$M=0)$ $M=1)$ $M=2)$		2.000 1.000 0.000	3.000
0		%						
-1	$(^3D_3)$ $(^3D_2)$ $(^3D_1)$	1.333 1.167 0.500	3.000	$(M=1$ $(M=0$ $(M=-1$	$M=-2)$ $M=-1)$ $M=0)$		0.000 1.000 2.000	3.000
-2	$(^3D_3)$ $(^3D_2)$	1.333 1.167	2.500	$(M=0$ $(M=-1$	$M=-2)$ $M=-1)$		1.000 1.500	2.500
-3	$(^3D_3)$	1.333	1.333	$(M=-1$	$M=-2)$		1.333	1.333

The g-permanence rule is also valid in still higher fields, where no coupling between spin and orbital moments exists.

17. The g-sum rule. There is another rule for g-factors, given by Pauli[1]. This rule is a most important aid in the analysis of atomic spectra. Among the levels of one electron configuration there are often some with the same J-value. For instance in an sd-configuration there are two levels with $J=2$, viz. the $^1D_2$ and the $^3D_2$ level in case of LS-coupling. This number of levels with the same J-value is independent of the coupling scheme. Only the relative position of the levels has changed and so have the g-values. Applying the formulae for the g-values in extreme LS-coupling and extreme jj-coupling, we find for the two levels with $J=2$ in an s-d configuration:

	LS	jj	
$^3D_2$	$g=1.167$	$j_1=\tfrac{1}{2}, j_2=\tfrac{3}{2}, J=2$	$g=1.100$
$^1D_2$	$g=1.000$	$j_1=\tfrac{1}{2}, j_2=\tfrac{5}{2}, J=2$	$g=1.067$
	$\Sigma g=2.167$		$\Sigma g=2.167$

---

[1] See footnote 1, p. 315.

Evidently the sum of both g-values is the same in both kinds of coupling. In general:

The sum of the g-values for the levels in one electron configuration with the same value of the total angular momentum, is independent of the coupling-scheme.

It is easy to verify this rule applying the formulae, given in Sects. 6 and 8 for g-values in the same electron configuration, assuming first LS-coupling and second jj-coupling.

It often occurs in complex spectra that this rule does not hold well because of configuration perturbation. Not only is the position of the levels different from that calculated from LS or jj-coupling—or any other intermediate coupling scheme—but the sum of g-values belonging to the levels in one configuration with the same J has also changed. Applying the same rule to the levels with the same J, it is found that it holds for both perturbating configurations together.

It also happens that the g-sum rule can be verified for a group of levels lying far apart from other levels with the same J-values, in the same configuration. For instance in the Os I spectrum[1], several levels with $J=2$ (belonging to the $5d^7 6s$-configuration) are known. A group of them, viz. the $^5F_2$, $^3F_2$, $^5P_2$ and $^3P_2$ levels lie low in the atom. Other known levels lie high and the unknown levels

Table 2. Os I.

Configuration	g_{LS}	g_{obs}
$5d^7 6s\ ^5F_2$	1.00	1.46
$5d^7 6s\ ^3F_2$	0.67	0.94
$5d^7 6s\ ^5P_2$	1.83	1.01
$5d^7 6s\ ^3P_2$	1.50	1.62
	$\Sigma g = 5.00$	$\Sigma g = 5.03$

still higher in the atom, so that they will not influence each other. For the four levels mentioned the g-sum rule could be verified, though each of them has a g-value which is much different from the corresponding LS-value. From this it will be clear that it has no sense at all to designate these levels as $^5F_2$, $^3F_2$, $^5P_2$ and $^3P_2$ levels, though they certainly belong to the $5d^7 6s$-configuration (Table 2).

The g-sum rule for the two configurations $5d^3 6s^2$ and $5d^4 6s$ in Ta I, was verified by VAN DEN BERG, KLINKENBERG and VAN DEN BOSCH[2].

In many other spectra g-sum rules have been determined. It would go too far to mention all of them here.

IV. Intensity rules in the Zeeman effect.

18. Intensity measurements for Zeeman effect components of spectral lines were made by BURGER and DORGELO[3] and by ORNSTEIN and BURGER[4]. Intensity formulas were derived theoretically by KRONIG and GOUDSMIT[5], HÖNL[6], SOMMERFELD and HEISENBERG[7], and VAN VLECK[8] (see also CONDON and SHORTLEY[9]).

---

[1] J. C. VAN DEN BOSCH: Unpublished material. See also Lunds Univ. Årsskr., Proc. Rydberg Cent. Conf. Atomic Spectr. 1955.

[2] G. J. VAN DEN BERG, P. F. A. KLINKENBERG and J. C. VAN DEN BOSCH: Physica, **18**, 221 (1952).

[3] H. C. BURGER and B. DORGELO: Z. Physik **23**, 258 (1924).

[4] L. S. ORNSTEIN and H. C. BURGER: Z. Physik **28**, 135 (1924); **29**, 29 (1924).

[5] R. DE L. KRONIG and S. GOUDSMIT: Naturwiss. **13**, 90 (1925). — Z. Physik **31**, 885 (1925).

[6] H. HÖNL: Z. Physik **31**, 340 (1925).

[7] A. SOMMERFELD and W. HEISENBERG: Z. Physik **11**, 131 (1922).

[8] J. H. VAN VLECK: Quantum Principles and Line Spectra. 1926.

[9] See footnote 8, p. 298.

From this the intensity rules can be summarized as follows:

1. The intensities of the components of one line are symmetrical relative to the position of the original line.

2. *Sum rule.* The sum of intensities of combinations of a level characterised by the magnetic quantum number M with the levels $M-1$, M and $M+1$, is independent of M.

3. As the light, emitted by the atom outside the magnetic field, is unpolarised, so should in total also be the light emitted in a magnetic field in every direction. This means that the sum of the intensities of all π-components (light linearly polarised parallel to the direction of the magnetic field) is equal to the sum of the intensities of all σ-components (light circularly polarised in planes perpendicular to the magnetic field).

One has to take account of the fact that observing the Zeeman effect in a direction perpendicular to the magnetic field, only half of the intensity of the σ-components is observed. Observing the effect in a direction parallel to the magnetic field, one finds the other half. Hence,

$$I_\pi = I_\sigma.$$

The intensity formulas are:

$$\Delta J = 0; \quad J \to J \quad \begin{cases} M \to M+1 & I = \tfrac{1}{2}c\,(J-M)(J+M+1)\,, \\ M \to M & I = c\,M^2\,, \\ M \to M-1 & I = \tfrac{1}{2}c\,(J+M)(J-M+1)\,, \end{cases} \tag{18.1}$$

$$\Delta J = \pm 1;\ J \to J-1 \quad \begin{cases} M \to M+1 & I = \tfrac{1}{2}c'\,(J-M)(J-M-1)\,, \\ M \to M & I = c'\,(J^2 - M^2)\,, \\ M \to M-1 & I = \tfrac{1}{2}c'\,(J+M)(J+M-1)\,, \end{cases} \tag{18.2}$$

In case of a transition $J+1 \to J$, one has to replace J by $J+1$ in Eqs. (18.2). c and c' are arbitrary constants. When observing perpendicularly to the magnetic field, the intensities of the σ-components have to be divided by 2. The intensities are independent of the coupling scheme.

From (18.1) and (18.2) it is easy to derive the rules summarized above:

1. $\Delta J = 0$;

$$I_{+M \to M+1} = I_{-M \to -M-1} = \tfrac{1}{2}c\,(J-M)(J+M+1)\,,$$
$$I_{M \to M} = I_{-M \to -M} = c\,M^2\,,$$
$$I_{M \to M-1} = I_{-M \to -M+1} = \tfrac{1}{2}c\,(J+M)(J-M+1)\,,$$

$\Delta J = \pm 1$;

$$I_{M \to M+1} = I_{-M \to -M-1} = \tfrac{1}{2}c'\,(J-M)(J-M-1)\,,$$
$$I_{M \to M} = I_{-M \to -M} = c'\,(J^2 - M^2)\,,$$
$$I_{M \to M-1} = I_{-M \to -M+1} = \tfrac{1}{2}c'\,(J+M)(J+M-1)\,.$$

2. $\Delta J = 0$;

$$I_{M \to M+1} + I_{M \to M} + I_{M \to M-1} = \tfrac{1}{2}c\,(J-M)(J+M+1) +$$
$$+ c\,M^2 + \tfrac{1}{2}c\,(J+M)(J-M+1)$$
$$= c\,J(J+1), \quad \text{independent of } M,$$

$\Delta J = \pm 1$;
$$I_{M \to M+1} + I_{M \to M} + I_{M \to M-1} = \tfrac{1}{2}c'\,(J-M)(J-M-1) +$$
$$+ c'\,(J^2 - M^2) + \tfrac{1}{2}c'\,(J+M) \times$$
$$\times (J+M-1) = c'\,J(2J-1),$$
$$\text{independent of } M.$$

3. $\varDelta J = 0$ and $\varDelta J = \pm 1$. Calculation shows:

$$\sum_{M=-J}^{M=+J} I_{M \to M} = \sum_{M=-J}^{M=J-1} I_{M \to M+1} + \sum_{M=-J+1}^{M=J} I_{M \to M-1}.$$

Example.

We give numerical examples for some transitions.

$\underline{J_1 = 2 \to J_2 = 2}$ (we put $c = 1$); observation perpendicular to the magnetic field.

$$
\begin{array}{lccccccl}
I_\pi(M \to M) & 4 & 1 & 0 & 1 & 4 & \sum I_\pi = 10 \\
I_\sigma(M \to M + 1) & 1 & {}^3/_2 & {}^3/_2 & 1 & & \left. \right\} \sum I_\sigma = 10, \\
I_\sigma(M \to M - 1) & 1 & {}^3/_2 & {}^3/_2 & 1 & &
\end{array}
$$

$\underline{J_1 = 2 \to J_2 = 1}$ (We put $c' = 1$).

$$
\begin{array}{lcccl}
I_\pi(M \to M) & 3 & 4 & 3 & \sum I_\pi = 10 \\
I_\sigma(M \to M + 1) & 3 & {}^3/_2 & {}^1/_2 & \left. \right\} \sum I_\sigma = 10, \\
I_\sigma(M \to M - 1) & 3 & {}^3/_2 & {}^1/_2 &
\end{array}
$$

$\underline{J_1 = {}^3/_2 \to J_2 = {}^3/_2}$.

$$
\begin{array}{lccccl}
I_\pi(M \to M) & {}^9/_4 & {}^1/_4 & {}^1/_4 & {}^9/_4 & \sum I_\pi = 5 \\
I_\sigma(M \to M + 1) & {}^3/_4 & 1 & {}^3/_4 & & \left. \right\} \sum I_\sigma = 5, \\
I_\sigma(M \to M - 1) & {}^3/_4 & 1 & {}^3/_4 & &
\end{array}
$$

$\underline{J_1 = {}^3/_2 \to J_2 = {}^1/_2}$.

$$
\begin{array}{lcccl}
I_\pi(M \to M) & 2 & 2 & \sum I_\pi = 4 \\
I_\sigma(M \to M + 1) & {}^3/_2 & {}^1/_2 & \left. \right\} \sum I_\sigma = 4, \\
I_\sigma(M \to M - 1) & {}^3/_2 & {}^1/_2 &
\end{array}
$$

In the following Table 3 the sum rule is verified for each of these cases (intensities of σ-components have to be multiplied by 2).

Table 3.

$J=2$ / $J=2$			M_J			$\sum I$
M_J	2	1	0	-1	-2	
2	4	2				6
1	2	1	3			6
0		3	0	3		6
-1			3	1	2	6
-2				2	4	6
$\sum I$	6	6	6	6	6	

$J=2$ / $J=1$			M_J			$\sum I$
M_J	2	1	0	-1	-2	
1	6	3	1			10
0		3	4	3		10
-1			1	3	6	10
$\sum I$	6	6	6	6	6	

$J={}^3/_2$ / $J={}^3/_2$		M_J			$\sum I$
M_J	${}^3/_2$	${}^1/_2$	$-{}^1/_2$	$-{}^3/_2$	
$+{}^3/_2$	${}^9/_4$	${}^3/_2$			${}^{15}/_4$
$+{}^1/_2$	${}^3/_2$	${}^1/_4$	2		${}^{15}/_4$
$-{}^1/_2$		2	${}^1/_4$	${}^3/_2$	${}^{15}/_4$
$-{}^3/_2$			${}^3/_2$	${}^9/_4$	${}^{15}/_4$
$\sum I$	${}^{15}/_4$	${}^{15}/_4$	${}^{15}/_4$	${}^{15}/_4$	

$J={}^3/_2$ / $J={}^1/_2$		M_J			$\sum I$
M_J	${}^3/_2$	${}^1/_2$	$-{}^1/_2$	$-{}^3/_2$	
${}^1/_2$	3	2	1		6
$-{}^1/_2$		1	2	3	6
$\sum I$	3	3	3	3	

For $\Delta J = 0$ the intensities of π-components differ very much, the intensities of σ-components are not very different. If $\Delta J = \pm 1$ the intensities of π-components are not very different, the intensities of σ-components differ widely. So it is difficult to detect the weakest π-components in case of $\Delta J = 0$ and the weakest σ-component in case of $\Delta J = \pm 1$ (see also Figs. 2, 3, 4, 5, 6).

V. The Zeeman effect and hyperfine structure of spectral lines in atomic spectra.

19. Introduction. Since spectroscopists have had at their disposal interference apparatus with high resolving power to detect very fine structures of spectral lines[1,2,3], they have learnt that many lines in an atomic spectrum have what we still call hyperfine structure (hfs). Up to that time these lines were considered as transitions between single energy levels. From experiments made as far back as 1909, for instance by Janicki[4], we know that we have to consider these levels as consisting of a number of levels with a very small mutual energy distance. Not before 1924, when Pauli[5] and independently Russell[6] introduced a mechanical moment of the nucleus connected with a magnetic moment, could an explanation of this effect be given. Pauli predicted that an external magnetic field would have influence on the hfs of spectral lines, so that it should be possible to detect a Zeeman effect and Paschen Back effect of the hfs lines. After the introduction of the "spin" of the electron in 1925 by Uhlenbeck and Goudsmit and the explanation of the fine structure of spectral lines associated with it, Back and Goudsmit[7] succeeded in drawing up a level scheme for the hfs components of several lines in the neutral Bi spectrum. They accepted a new quantum vector and a new quantum number, which had to be added to the series of quantum numbers already known in the atom. This new quantum number I had to be connected with the nucleus, and determined the mechanical moment of it. In agreement with the prediction Back and Goudsmit[7] in 1927 observed the influence of an external magnetic field on the hfs components of the Bi I line λ 4722 Å (Back Goudsmit effect). They could determine the spin of the nucleus of the atom. It was the first case of a mechanical nuclear moment determined by atomic spectroscopy. As energy distances between levels of a hyperfine multiplet arising from the interaction between the spin of the nucleus and the total angular momentum of the electron, are small (compared with the distances in fine structure multiplets), it is very difficult to detect the Zeeman effect of hfs components of a spectral line. In that case, the magnetic field should not break down the coupling between nuclear spin and angular momentum of the electrons. Such a magnetic field would have a strength of the order of magnitude of 100 oersted. At this magnetic field strength the energy distance between magnetic sublevels is of the order of magnitude of 10^{-3} cm^{-1}. In general it is not possible to detect such distances.

In the following section we shall give the theory of the Zeeman effect and Paschen Back effect of hfs levels according to the vector atom model.

20. Zeeman effect. The magnetic moment of the nucleus μ_I is connected with the mechanical moment I according to

$$\mu_I = g_I I \frac{e}{2Mc} \tag{20.1}$$

in which g_I is a factor for the nucleus formally corresponding to the Landé factor of an electron, and M is the mass of a proton. This magnetic moment is quantized to the magnetic field H at the place of the nucleus, caused by the movement of the electron. This gives rise to a coupling of the vectors I and J to a resultant F, determining the total angular momentum of the whole atom. I and J have a

[1] A. A. Michelson: Phil. Mag. **31**, 338 (1891).
[2] C. Fabry and A. Perot: Ann. de chim. et phys. **12**, 459 (1897).
[3] O. Lummer and E. Gehrcke: Ann. der Phys. **10**, 457 (1903).
[4] L. Janicki: Ann. der Phys. **29**, 1833 (1909).
[5] W. Pauli: Naturwiss. **12**, 741 (1924).
[6] W. F. Meggers and K. Burns: J. Opt. Soc. Amer. **14**, 449 (1927).
[7] S. Goudsmit and E. Back: Z. Physik **43**, 321 (1927); **47**, 174 (1928).

precession movement around F. The possible values of the quantum number F are:

$$F = I + J, \quad I + J - 1, \quad I + J - 2, \dots |I - J|;$$

so there are $2I + 1$ values of F, if $I < J$

$\qquad\qquad 2J + 1$ values of F, if $J < I$.

Coupled with the mechanical moment F there is a magnetic moment

$$\mu_F = - g_F F. \tag{20.2}$$

In a similar way as we found for μ_J from Fig. 1, we derive

$$\mu_F = \mu_J \cos(J, F) + \mu_I \cos(I, F). \tag{20.3}$$

Measuring angular momenta in units of $\hbar$ again, and rewriting Eqs. (6.6) and (20.1), we obtain

$$\mu_J = - g_J J \frac{e\hbar}{2mc}$$

and

$$\mu_I = g_I \frac{e\hbar}{2mc} I \cdot \frac{1}{1838}$$

(in which g_J is the Landé factor for electrons) Eq. (20.3), becomes:

$$\mu_F = g_J J \mu_B \cos(J, F) + g_I \frac{\mu_B}{1838} I \cos(I, F) \tag{20.4}$$

(in which μ_B is a Bohr magneton). Introducing the formulae for a "wave mechanical" cosine, Eqs. (20.2) and (20.4) together give:

$$g_F = g_J \frac{F(F+1) + J(J+1) - I(I+1)}{2F(F+1)} - \frac{g_I}{1838} \frac{F(F+1) + I(I+1) - J(J+1)}{2F(F+1)}$$

or, as g_J and g_I are of the same order of magnitude:

$$g_F \approx g_J \frac{F(F+1) + J(J+1) - I(I+1)}{2F(F+1)}. \tag{20.5}$$

In an external magnetic field, the energy perturbation of the total atom due to this field is

$$E_{F,H} = - \mu_F H \cos(F, H)$$

or, with (20.4) and (20.5)

$$E_{F,H} = \frac{e\hbar}{2mc} \cdot H M_F g_F \tag{20.6}$$

where $M_F = -F, -F+1 \dots + F$.

From this formula and Eq. (20.5) we see that the mutual energy distances of the magnetic sublevels, corresponding to the possible values of M_F, is of the same order of magnitude as those of the magnetic sublevels of the fine structure levels. The total number of levels in a magnetic field is

$$\sum_{F=I-J}^{F=I+J} (2F + 1) = (2I + 1)(2J + 1).$$

In Fig. 9, as an example, there has been shown the hfs of a $^2S_{1/2}$ term in case of $I = \frac{1}{2}$. Without a magnetic field applied there is a hfs splitting into two levels with $F = 0$ and $F = 1$. An ordinary Zeeman effect of this hfs will be obtained if we apply a magnetic field weak enough to keep the Zeeman splitting a small compared with the hfs splitting A, i. e. a magnetic field that does not break down the coupling between the angular moments I and J. In order to observe the splitting of each separate hfs level still keeping its individuality, $E_{F,H}$ of Eq. (20.6) has to be small compared with the hfs distance A. Now this distance is proportional to $\mu_I H_J$, and $E_{F,H}$ is proportional to $\mu_B H$, so that

$$\mu_B H \ll \mu_I H_J \quad \text{or} \quad H \ll \frac{\mu_I}{\mu_B} H_J,$$

i.e.

$$H \ll 10^{-3} H_J.$$

In the alkali metals the field H_J is of the order of magnitude of 10^5 to 10^6 oersted, for the valence electron in the ground state, so that for these elements in order to observe an ordinary Zeeman effect of hfs components in the ground state,

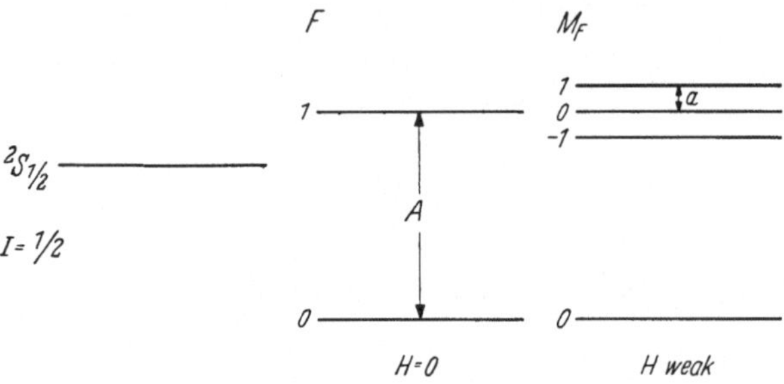

Fig. 9. Hfs of a $^2S_{1/2}$ level with $I = \tfrac{1}{2}$ without, and in a "weak" magnetic field.

we have to use magnetic fields that would be small compared with 100 to 1000 oersted. With such weak fields however, the distances become too small to observe them with any interference spectroscope.

21. Back Goudsmit effect. Increasing the strength of the external magnetic field causes the uncoupling of the vector I and J, and already at a field strength of a few thousands oersted they are both separately quantized with the magnetic field. F no longer is a quantum vector, nor is F a quantum number. Both I and J precess independently around H, I with a frequency small compared with the precession frequency of J. The quantization of J with respect to the direction of the magnetic field gives rise to $2J+1$ levels, characterised by the magnetic quantum number M_J, and lying symmetrically with respect to the original electron level. The distance of these levels from the undisturbed electron level is given by

$$E_{J,H} = g_J \cdot M_J H \cdot \frac{e\,\hbar}{2\,m\,c}, \tag{21.1}$$

with M_J varying from $-J$ to $+J$.

To this we have to add the energy resulting from the interaction of I with the magnetic field which amounts to

$$E_{I,H} = g_I \cdot \frac{e\,\hbar}{2\,m\,c} M_I H \cdot \frac{1}{1838} \tag{21.2}$$

so that this energy correction is about 2000 times smaller than $E_{J,H}$. To Eqs. (21.1) and (21.2) we still have to add the interaction energy of I and J, so that the total energy distance from the original electron level is given by (cf. Sect. 12):

$$E_H = g_J \frac{e\,\hbar}{2\,m\,c} M_J H - \frac{g_I}{1838} \cdot \frac{e\,\hbar}{2\,m\,c} M_I H + B M_I M_J,$$

or, as the second term is small compared to the first and the third term:

$$E_H = g_J \frac{e\,\hbar}{2\,m\,c} M_J H + B\,M_I M_J. \qquad (21.3)$$

The last term in this formula which is independent of the magnetic field strength accounts for the interaction between $\boldsymbol{I}$ and $\boldsymbol{J}$. Now, for each level M_J, there are $2I+1$ levels, each characterised by its M_I value, and the total number of levels

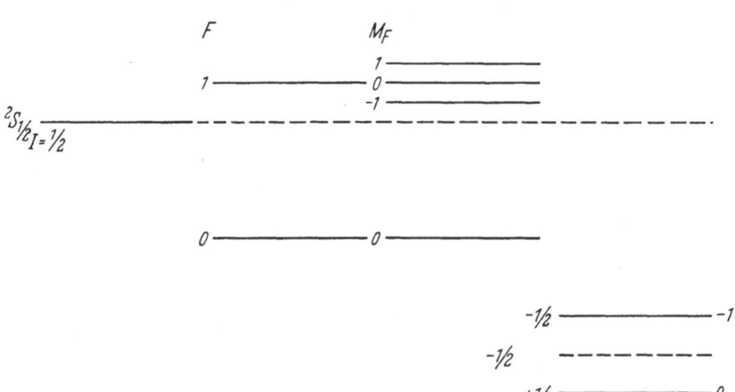

Fig. 10. Hfs of a $^2S_{1/2}$ level with $I = \tfrac{1}{2}$ without, in a "weak", and in a "strong" magnetic field.

is again $(2I+1)(2J+1)$. Again angular momentum is conserved: the sum of the projections of the mechanical moments on the direction of $\boldsymbol{H}$ undergoes no alterations. In the weak field case, $M = M_F$, in the strong field, $M = M_I + M_J$. In Fig. 10 the left-hand side of which agrees with Fig. 9, the strong field case has been added on the right.

22. Determination of the nuclear spin from the Back Goudsmit effect. The Back Goudsmit effect of hfs lines yields a beautiful way to determine the mechanical moment of the nucleus. In the weak field case, the selection rules are

$$\Delta M_F = 0 \quad \text{for } \pi\text{-components},$$

$$\Delta M_F = \pm 1 \text{ for } \sigma\text{-components},$$

so that a relatively large number of lines may be expected in the Zeeman effect of a hfs line, supposed it can be observed. The number of components decreases with increasing field strength.

In a strong magnetic field the selection rules are (cf. the Paschen Back effect for fine structure lines, Sect. 12):

$$\Delta M_I = 0 \quad \text{for all components},$$

$$\Delta M_J = 0 \quad \text{for } \pi\text{-components},$$

$$\Delta M_J = \pm 1 \text{ for } \sigma\text{-components}.$$

A magnetic field, which is a weak field for the magnetic splitting of a fine structure line, is a strong field for hfs. Each of the $2J+1$ magnetic sublevels of a fine

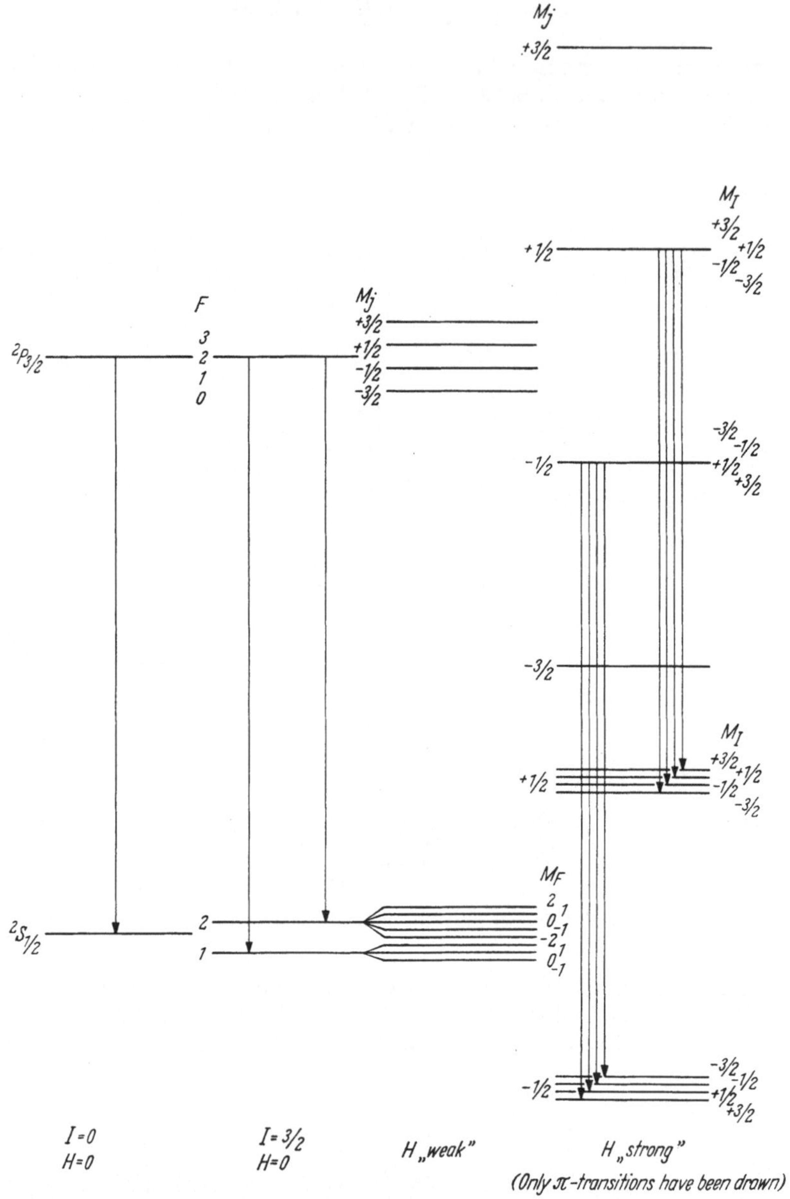

Fig. 11. Zeeman effect and Back-Goudsmit effect of the transition $^2P_{3/2}-^2S_{1/2}$ with $I=\frac{3}{2}$. In the strong field case only π-transitions have been drawn.

structure level will split up into $2I+1$ levels, due to the interaction of $\boldsymbol{I}$ and $\boldsymbol{J}$, which is accounted for by the term BM_IM_J in Eq. (21.3). The application of the selection rule, $\varDelta M_I=0$ for all components, leads to the result that each of the Zeeman components of a fine structure line splits up into $2I+1$ components.

Fig. 11 shows the energy level scheme for the Na D line λ 5890 Å in case of $H=0$, weak field, intermediate field strength and strong field. Only those transitions have been drawn for which $\Delta M_I = 0$ (π-components). Moreover it is assumed that the hfs splitting of the $3s\,^2P_{3/2}$ level is negligibly small compared with that of the $3s\,^2S_{1/2}$ level. BACK and GOUDSMIT succeeded in resolving the structure of the components of the Bi I line $\lambda\,4722$ Å. In each transition they found 10 components, so that in this case $2I+1=10$, and the nuclear spin of Bi^{209} becomes $I=9/2$. Later on their results for a group of components were reproduced and published by ZEEMAN, BACK and GOUDSMIT[1].

There are only a few examples where the influence of a magnetic field on the hfs of spectral lines has been shown experimentally in weak, intermediate and strong fields. One of them is the beautiful experiment by JACKSON and KUHN[2] on the Zeeman effect of the hfs of the Na D line $\lambda\,5890$ Å. These authors used an atomic beam light source and photographed the lines in absorption. They showed that for the components of this line there was a real Zeeman effect in magnetic fields up to a magnetic field strength of about 1600 oersted, whereas there is a transition to the Back Goudsmit effect in stronger fields, the latter being completed at a field strength of about 3000 oersted.

VI. The use of the Zeeman effect in the analysis of atomic spectra.

23. Introduction. The analysis of atomic spectra makes it its object to determine the classification of as many spectral lines as possible, emitted by an atom or ion. There are several aids and appliances which spectroscopists have at their disposal for this aim. For instance there is the observation of spectra of one element under different circumstances (arc, spark, furnace, electric discharge in gases at reduced pressures, electrodeless discharge, hollow-cathode discharge), and arising from this, the determination of wavelengths and associated wave numbers in vacuum; study of the character of spectral lines (intensity, sharpness, cathode enhancement in vertical arc); interval investigation of wave numbers, influence of external fields on a spectrum as a whole and on each spectral line in particular, especially that of a magnetic field. It has been known for long that an external magnetic field has an influence on the relative intensity of spectral lines of ions, in such a way that the intensity of these lines is increased in a magnetic field[3-5] whereas those of the lines of the neutral atom are decreased. This effect depends among other things on the strength of the magnetic field and the ionisation potential of the atom. Hence it is difficult to observe spectra of neutral atoms with a low ionisation potential (e.g. the rare earth elements) in a strong magnetic field. For this effect KLINKENBERG[6] has given an explanation, as far as the influence of an external magnetic field on the spectrum as a whole is concerned. However, the influence of a magnetic field on a spectral line in particular seems to be even more important because of the fact that from the magnetic splitting of a spectral line, the J- and g-values of the combining levels can be calculated, in case the line splits up into all its components.

Each energy level is characterised by its angular momentum and its g-factor and in some cases the combination of J- and g-values can be a direct indication which levels, written in LS-symbols, are associated with the spectral line in

[1] P. ZEEMAN, E. BACK and S. GOUDSMIT: Z. Physik **66**, 7 (1930).
[2] D. JACKSON amd H. KUHN: Proc. Roy. Soc. Lond., Ser. A **167**, 205 (1938).
[3] T. L. DE BRUIN and P. F. A. KLINKENBERG: Proc. Roy. Acad. Amst. **43**, 581 (1940).
[4] G. R. HARRISON: Rep. Progr. Phys. **8**, 212 (1941).
[5] G. RAOULT: Ann. de Phys. **4**, 369 (1949).
[6] P. F. A. KLINKENBERG: Physica, Haag **16**, 185 (1950).

question. However, in many atoms the coupling between electrons is far from ideal LS-coupling, so that it has no sense attributing a symbol, as we do in the case of ideal LS-coupling, to the levels involved in such an atom. Then only J- and g-value of the level remain, which can be used to reach a better interpretation of the spectrum. It is just the g-value that gives an indication as to the kind of coupling we have to do with in the atom.

The Zeeman effect of a spectral line gives an unambiguous determination of the angular momentum and g-factor of the energy levels involved.

In the following sections we shall discuss how these values can be determined from the Zeeman effect of a spectral line, in the case of a completely resolved pattern as well as in that of an unresolved one.

24. Determination of J- and g-values of energy levels from the completely resolved Zeeman pattern of a spectral line. α) *J-values.* $J_1 \neq J_2$. If the smallest J-value is J, the number of π-components is always $(2J+1)$, in case the J-values are integers (odd multiplicity) as well as in case the J-values are half-integers (even multiplicity).

Counting the number of π-components gives directly both J-values of the combining terms, involved in a spectral line.

Since in many cases π- and σ-components overlap, it is recommended to photograph both kinds of polarisations separately, which of course increases the exposure time because of the absorption by the calcite rhomb.

$J_1 = J_2 = J$. Now the number of π-components is—as the transition $M_J = 0 \rightarrow M_J = 0$ is forbidden—, $2J$, so that in this case the J-value of both combining energy levels again follows from counting the Zeeman effect components.

β) *g-values.* In Sect. 6 we saw that the energy distance between magnetic sublevels, characterised by the magnetic quantum numbers M_J, is given by

$$E = M_J g \frac{e \hbar}{2mc} H$$

if H is the magnetic field strength and g the Landé factor of the level. For a transition between two magnetic sublevels we find an energy difference with that of the unsplit line

$$\Delta E = (M_{J_1} g_1 - M_{J_2} g_2) \frac{e \hbar}{2mc} H$$

and

$$\Delta \nu = (M_{J_1} g_1 - M_{J_2} g_2) \frac{e}{4\pi mc} H.$$

Since the corresponding wavelength difference is

$$|\Delta \lambda| = \lambda^2 \frac{\Delta \nu}{c}$$

or

$$\left| \frac{\Delta \lambda}{\lambda^2} \right| = \frac{\Delta \nu}{c},$$

in this scale we obtain

$$\frac{\Delta \lambda}{\lambda^2} = (M_{J_1} g_1 - M_{J_2} g_2) \frac{e}{4\pi mc^2} H \quad \text{cm}^{-1} \tag{24.1}$$

with always

$$M_{J_2} - M_{J_1} = 0 \quad \text{or} \quad \pm 1.$$

For the Zeeman effect of each component the value of $M_{J_1} g_1 - M_{J_2} g_2$ is given in Lorentz units.

Form the number of π-components both J_1 and J_2 are known, and so are the values of M_{J_1} and M_{J_2} in Eq. (24.1). $\Delta\lambda$ follows from measuring on the plate the distances of the components from the central line.

These distances can be measured on the plate with an accuracy of 0.001 mm, from which—with the aid of the plate factor and the wavelength—the value of $\frac{\Delta\lambda}{\lambda^2}$ is computed. Dividing this by $\frac{eH}{4\pi m c^2}$ gives the value of $M_{J_1}g_1 - M_{J_2}g_2$ for each component. In order to compute g_1 and g_2 from the expression $M_{J_1}g_1 - M_{J_2}g_2$, we need two equations. In addition to this, all other $M_{J_1}g_1 - M_{J_2}g_2$ values of the other components may be used, provided they are known accurately enough.

In order to ascribe the right values of M_{J_1} and M_{J_2} to the different components, one has to observe the intensity rules very carefully. If $J_1 \neq J_2$, the highest values of M_J are associated to the strongest σ-components, whereas, if $J_1 = J_2$, the weakest σ-components have the highest values of M_J. Concerning the π-components this consideration is of no importance because only the $g_2 - g_1$ value can be computed from the mutual distances of these components.

The following example will make this clear. For the Zeeman effect of the Ge III line $\lambda\,4178.96$ Å [1], a Zeeman pattern was found, corresponding to Fig. 6b, only with this difference that the stronger σ-components lie nearest to the central line. The distances between the components were measured with a comparator to an accuracy of 0.001 mm. Table 4 shows the calculation that led to the definite Zeeman effect of the line, expressed in Lorentz units. Column 1 gives the readings of the comparator in 0.001 mm (d); column 2 the distances between symmetrical components on the plate, again in mm (Δd). Then these values are multiplied by half the plate factor. The result we find in column 3, so that then the distances $\Delta\lambda$ between all components and the place of the fieldless line are known in Å. Next these values are divided by λ^2, so that in column 4 we find the values of $\Delta\lambda/\lambda^2$ in Å^{-1} units. Column 5 at last shows what we get when dividing the values of column 4 by $\Delta\lambda/\lambda^2$ for one Lorentz unit. For if we substitute $M_{J_1}g_1 - M_{J_2}g_2 = 1$ in Eqs. (24.1), we find this unit value, $(\Delta\lambda/\lambda^2)_{\text{L.U.}}$ and so column 5 gives $M_{J_1}g_1 - M_{J_2}g_2$ in Lorentz units for each component.

Table 4.

d (mm)	Δd (mm)	$\Delta\lambda(Å)$	$\frac{\Delta\lambda}{\lambda^2}$ (Å^{-1})$\times 10^7$	$\frac{\Delta\lambda}{\lambda^2}\Big/\Big(\frac{\Delta\lambda}{\lambda^2}\Big)_{\text{L.U.}}$
23.452	0.514	0.62708	0.35908	1.972
23.386	0.385	0.46970	0.26896	1.477
23.324	0.258	0.31476	0.18024	0.990
23.259	0.128	0.15616	0.08942	0.491
23.196	0.000	0.00000	0.00000	0
23.131				
23.066				
23.001				
22.938				

From this calculation the Zeeman effect of this line is given as

Ge III $\lambda\,4178.96$ Å $(\underline{0})$ (0.49) $\underline{0.99}$ 1.48 1.97

The underlined components are the strongest. We can not give more than two digits in these figures, i.e. not determine the Zeeman effect more accurately than in hundredths of a Lorentz unit.

---

[1] J. C. van den Bosch and P. F. A. Klinkenberg: Proc. Roy. Acad. Amst. **44**, 556 (1941).

We further conclude from this Zeeman effect:

1. the number of π-components is 3, so the smallest J-value is 1, the largest one is 2, and $g_{J=1} > g_{J=2}$.

2. the strongest σ-component is the 0.99 component, so that M_{J_1} and M_{J_2} for this component are 1 and 2, (or -1 and -2); consequently those of the 1.48 component are 0 and 1 (or 0 and -1) and those for the 1.97 component -1 and 0 (or 1 and 0).

3. From the distances between successive components (π or σ) there follows that $g_1 - g_2 = 0.49$. This result combined with the distances of all separate σ-components (for which now M_{J_1} and M_{J_2} are known) to the central line leads to the definite g_1 and g_2-values:

$$g_{J=1} = 1.97, \quad g_{J=2} = 1.48.$$

From these considerations it follows that the line is a transition

$$1(1.97) \to 2(1.48).$$

From the above mentioned calculation it follows as a matter of fact that we can predict the value of $\Delta\lambda/\lambda^2$ if $M_{J_1}g_1 - M_{J_2}g_2 = 1$. For this purpose we have to use a spectral line of which we know (for instance from theoretical consideration) the Zeeman effect.

Of course, if we know

$$\left(\frac{\Delta\lambda}{\lambda^2}\right)_{1\,\mathrm{L.U.}} = \frac{e}{4\pi m c^2} H \quad \mathrm{cm^{-1}}$$

we can compute the strength of the magnetic field. However, to compute the value of $\Delta\lambda/\lambda^2$ for a component of an arbitrary line, it is sufficient to know $(\Delta\lambda/\lambda^2)_{1\,\mathrm{L.U.}}$. Table 5 gives a survey of most commonly used spectral lines with their Zeeman effect in L.U., for calculating the magnetic field necessary to make $\Delta\lambda/\lambda^2$ equal 1 L.U. Besides those mentioned here, CuI lines are often used for this purpose.

Table 5.

El.	λ (Å)	Combination	Zeeman effect	
			π	σ
ZnI	4810.53	$^3P_2 - {}^3S_1$	0 0.50	1.00 1.50 2.00
	4722.16	$^3P_1 - {}^3S_1$	0.50	1.50 2.00
	4680.14	$^3P_0 - {}^3S_1$	0	2.00
CrI	4289.89	$^7S_3 - {}^7P_2$	0 0.33 0.67	1.34 1.67 1.99 2.33 2.65
	4274.96	$^7S_3 - {}^7P_3$	0.22	1.94
	4254.50	$^7S_3 - {}^7P_4$	0 0.25 0.50 0.75	1.00 1.25 1.50 1.75 2.00 2.24
	3605.48	$^7S_3 - {}^7P_2$	0 0.34 0.67	1.35 1.68 2.01 2.33 2.66
	3593.64	$^7S_3 - {}^7P_3$	0.21	1.97
	3578.84	$^7S_3 - {}^7P_4$	0 0.25 0.50 0.75	1.00 1.25 1.50 1.74 1.99 2.24 2.49
AlI	3961.54	$^2P_{\frac{3}{2}} - {}^2S_{\frac{1}{2}}$	0.33	1.00 1.67
	3944.03	$^2P_{\frac{1}{2}} - {}^2S_{\frac{1}{2}}$	0.67	1.33

The importance of the determination of J- and g-values of energy levels can hardly be overestimated. It can lead to the detection of real intervals in the spectrum. Moreover, if one level with a definite J-value in a configuration, could not yet be inserted in the analysis, there is the possibility that it may be found by applying the g-sum rule.

If it is found that a high J-value is combined with a low g-value, then this points to the presence of electrons with high quantum numbers in the atom under investigation. We find this in the spectra of the rare earth elements and, for instance, the detection of such electron configurations in uranium resulted in considering this element in some sense as a rare earth element too[1].

25. Unresolved Zeeman pattern. There are not only the completely resolved spectral lines in a magnetic field which have their importance for the analysis of atomic spectra. From unresolved Zeeman patterns we too can draw conclusions which may lead to the determination of g-values of the combining levels. We distinguish these patterns in *pseudo quartets* and *pseudo triplets*. In case the line shows up as a pseudo quartet, one can decide from this that both J-values of the combining levels are equal. But on the other hand, when a line is a pseudo triplet in a magnetic field, it is not allowed to decide that the two J-values are not equal. Viz. if the g-values of both combining levels are equal or nearly equal, the groups on both sides of the central line cannot be observed separately (cf., for instance, the Se II lines $\lambda\,4630.56$, $\lambda\,4516.25$ and $\lambda\,4467.60$ Å[2].

In this connection it is advisable to mention whether an unresolved Zeeman pattern is sharp or diffuse.

For unresolved Zeeman pattern SHENSTONE and BLAIR[3] using theoretical formulas by HÖNL[4] gathered the following formulas for the centres of gravity of π- and σ-components, with respect to the place of the unsplit line.

$$\underline{J_1 \neq J_2.} \qquad B_\sigma = \tfrac{1}{2}(g_2 - g_1)\,J_1 + g_2 \quad \text{if} \quad J_2 = J_1 + 1\,,$$
$$B_\pi = 0\,,$$

$$\underline{J_1 = J_2 = J.} \quad B_\sigma = \tfrac{1}{2}(g_1 + g_2)\,,$$
$$B_\pi = (g_1 - g_2)\,\{1 + c\,(J - 1)\} \quad \text{if } J \text{ is an integer}\,,$$
$$B_\pi = (g_1 - g_2)\,\{\tfrac{1}{2} + c'(J - \tfrac{1}{2})\} \quad \text{if } J \text{ is half integer}.$$

All values are given in Lorentz units. In these formulae

$$c = \frac{3J + 2}{2(2J + 1)}\,,$$

$$c' = \frac{12J^2 + 16J + 3}{16J(J + 1)}\,.$$

CATALAN and POGGIO[5] computed the centres of gravity of π- and σ-components for several J-values. VAN DE VLIET[6] recalculated these values. His results are given in Table 6 from which Table 7 can easily be deduced.

We give a survey of the cases to which these formulas apply. It is necessary to know both J-values.

1. $\underline{J_1 \neq J_2.}$ One of both g-values has to be known. Then using Table 6 or 7 the other one can be computed.

2. $\underline{J_1 = J_2.}$ If the line is observed as a pseudo quartet, so that the distance of the centres of gravity of the groups of π-components on both sides can be

---

[1] J. C. VAN DEN BOSCH: Thesis Amsterdam 1948. — C. C. KIESS, C. J. HUMPHREYS and D. D. LAUN: Bur. Stand. J. Res. **37**, 57 (1946). — PH. SCHUURMANS: Physica, Haag **11**, 419 (1946). — Thesis Amsterdam 1946. — PH. SCHUURMANS, J. C. VAN DEN BOSCH and N. DIJKWEL: Physica, Haag **13**, 117 (1947).—

[2] See footnote 2, p. 299.

[3] A. G. SHENSTONE and H. A. BLAIR: Phil. Mag. **8**, 765 (1929).

[4] See footnote 6, p. 317.

[5] M. A. CATALAN and F. POGGIO: Pieter Zeeman Jubilee Vol. p. 387. Nijhoff 1935.

[6] See footnote 6, p. 299.

measured, both g-values can be computed from Table 6 or 7. In such a case a Zeeman effect of an unresolved pattern, as far as the determination of g-values is concerned, is as useful as that of a completely resolved line, though the results are less accurate.

Table 6.

$J_1 \neq J_2$ J's are integers				$J_1 \neq J_2$ J's are half integers			
J_1	J_2	B_π	B_σ	J_1	J_2	B_π	B_σ
0	1		g_2	$\tfrac{1}{2}$	$1\tfrac{1}{2}$		$1.25g_2 - 0.25g_1$
1	2		$1.50g_2 - 0.50g_1$	$1\tfrac{1}{2}$	$2\tfrac{1}{2}$		$1.75g_2 - 0.75g_1$
2	3	In all cases:	$2.00g_2 - 1.00g_1$	$2\tfrac{1}{2}$	$3\tfrac{1}{2}$	In all cases:	$2.25g_2 - 1.25g_1$
3	4	0	$2.50g_2 - 1.50g_1$	$3\tfrac{1}{2}$	$4\tfrac{1}{2}$	0	$2.75g_2 - 1.75g_1$
4	5		$3.00g_2 - 2.00g_1$	$4\tfrac{1}{2}$	$5\tfrac{1}{2}$		$3.25g_2 - 2.25g_1$
5	6		$3.50g_2 - 2.50g_1$	$5\tfrac{1}{2}$	$6\tfrac{1}{2}$		$3.75g_2 - 2.75g_1$
6	7		$4.00g_2 - 3.00g_1$				

$J_1 = J_2$ J's are integers			$J_1 = J_2$ J's are half integers		
J	B_π	B_σ	J	B_π	B_σ
1	$1.00\,(g_1 - g_2)$		$\tfrac{1}{2}$	$0.500\,(g_1 - g_2)$	
2	$1.80\,(g_1 - g_2)$		$1\tfrac{1}{2}$	$1.400\,(g_1 - g_2)$	
3	$2.57\,(g_1 - g_2)$	In all cases:	$2\tfrac{1}{2}$	$2.186\,(g_1 - g_2)$	In all cases:
4	$3.33\,(g_1 - g_2)$	$0.50\,(g_1 + g_2)$	$3\tfrac{1}{2}$	$2.952\,(g_1 - g_2)$	$0.50\,(g_1 + g_2)$
5	$4.09\,(g_1 - g_2)$		$4\tfrac{1}{2}$	$3.712\,(g_1 - g_2)$	
6	$4.85\,(g_1 - g_2)$		$5\tfrac{1}{2}$	$4.469\,(g_1 - g_2)$	
7	$5.60\,(g_1 - g_2)$		$6\tfrac{1}{2}$	$5.223\,(g_1 - g_2)$	

[By courtesy of H. J. van de Vliet from Thesis (Amsterdam 1939).]

Table 7.

$J_1 \neq J_2$ J's are integers				$J_1 \neq J_2$ J's are half integers			
J_1	J_2	g_1	g_2	J_1	J_2	g_1	g_2
0	1	0	$1.00\,B_\sigma$	$\tfrac{1}{2}$	$1\tfrac{1}{2}$	$5g_2 - 4B_\sigma$	$0.800\,B_\sigma + 0.200g_1$
1	2	$3.00g_2 - 2.00\,B_\sigma$	$0.33g_1 + 0.67\,B_\sigma$	$1\tfrac{1}{2}$	$2\tfrac{1}{2}$	$2.333g_2 - 1.333\,B_\sigma$	$0.571\,B_\sigma + 0.429g_1$
2	3	$2.00g_2 - 1.00\,B_\sigma$	$0.50g_1 + 0.50\,B_\sigma$	$2\tfrac{1}{2}$	$3\tfrac{1}{2}$	$1.800g_2 - 0.800\,B_\sigma$	$0.444\,B_\sigma + 0.556g_1$
3	4	$1.67g_2 - 0.67\,B_\sigma$	$0.60g_1 + 0.40\,B_\sigma$	$3\tfrac{1}{2}$	$4\tfrac{1}{2}$	$1.571g_2 - 0.571\,B_\sigma$	$0.364\,B_\sigma + 0.636g_1$
4	5	$1.50g_2 - 0.50\,B_\sigma$	$0.67g_1 + 0.33\,B_\sigma$	$4\tfrac{1}{2}$	$5\tfrac{1}{2}$	$1.444g_2 - 0.444\,B_\sigma$	$0.308\,B_\sigma + 0.692g_1$
5	6	$1.40g_2 - 0.40\,B_\sigma$	$0.71g_1 + 0.29\,B_\sigma$	$5\tfrac{1}{2}$	$6\tfrac{1}{2}$	$1.364g_2 - 0.364\,B_\sigma$	$0.267\,B_\sigma + 0.733g_1$
6	7	$1.33g_2 - 0.33\,B_\sigma$	$0.75g_1 + 0.25\,B_\sigma$				

$J_1 = J_2$ J's are integers			$J_1 = J_2$ J's are half integers		
J	g_1	g_2	J	g_1	g_2
1	$B_\sigma + 0.500\,B_\pi$	$B_\sigma - 0.500\,B_\pi$	$\tfrac{1}{2}$	$B_\sigma + B_\pi$	$B_\sigma - B_\pi$
2	$B_\sigma + 0.278\,B_\pi$	$B_\sigma - 0.278\,B_\pi$	$1\tfrac{1}{2}$	$B_\sigma + 0.357\,B_\pi$	$B_\sigma - 0.357\,B_\pi$
3	$B_\sigma + 0.195\,B_\pi$	$B_\sigma - 0.195\,B_\pi$	$2\tfrac{1}{2}$	$B_\sigma + 0.229\,B_\pi$	$B_\sigma - 0.229\,B_\pi$
4	$B_\sigma + 0.150\,B_\pi$	$B_\sigma - 0.150\,B_\pi$	$3\tfrac{1}{2}$	$B_\sigma + 0.169\,B_\pi$	$B_\sigma - 0.169\,B_\pi$
5	$B_\sigma + 0.122\,B_\pi$	$B_\sigma - 0.122\,B_\pi$	$4\tfrac{1}{2}$	$B_\sigma + 0.135\,B_\pi$	$B_\sigma - 0.135\,B_\pi$
6	$B_\sigma + 0.103\,B_\pi$	$B_\sigma - 0.103\,B_\pi$	$5\tfrac{1}{2}$	$B_\sigma + 0.112\,B_\pi$	$B_\sigma - 0.112\,B_\pi$
7	$B_\sigma + 0.089\,B_\pi$	$B_\sigma - 0.089\,B_\pi$	$6\tfrac{1}{2}$	$B_\sigma + 0.096\,B_\pi$	$B_\sigma - 0.096\,B_\pi$

[By courtesy of H. J. van de Vliet from Thesis (Amsterdam 1939).]

26. Separation of ionisation stages. We have already mentioned that an external magnetic field influences the ionisation stage of an atom. The Zeeman effect of a spectral line is a good aid to determine to which ionisation stage of

Table 6.

	^2S	^2P	^2D	^2F	^2G	^2H	^2I	^2K
Singlets	For all singlet levels $g = 1.000$ (except $^1S_0 = \%$)							
Doublets 2.000	J: 1/2 g: 2.000	J: 1/2, 3/2 g: 2/3, 4/3	J: 3/2, 5/2 g: 4/5, 6/5	J: 5/2, 7/2 g: 6/7, 8/7	J: 7/2, 9/2 g: 8/9, 10/9	J: 9/2, 11/2 g: 10/11, 12/11	J: 11/2, 13/2 g: 12/13, 14/13	J: 13/2, 15/2 g: 14/15, 16/15

	^3S	^3P	^3D	^3F	^3G	^3H	^3I	^3K
Triplets 2.000	J: 1 g: 2.000	J: 0, 1, 2 g: 0/0, 3/2, 3/2	J: 1, 2, 3 g: 1/2, 7/6, 4/3	J: 2, 3, 4 g: 2/3, 13/12, 5/4	J: 3, 4, 5 g: 3/4, 21/20, 6/5	J: 4, 5, 6 g: 4/5, 31/30, 7/6	J: 5, 6, 7 g: 5/6, 43/42, 8/7	J: 6, 7, 8 g: 6/7, 57/56, 9/8

	^4S	^4P	^4D	^4F	^4G	^4H	^4I	^4K
Quartets 2.000	J: 3/2 g: 2.000	J: 1/2, 3/2, 5/2 g: 8/3, 26/15, 8/5	J: 1/2, 3/2, 5/2, 7/2 g: 0, 6/5, 48/35, 10/7	J: 3/2, 5/2, 7/2, 9/2 g: 2/5, 36/35, 78/63, 4/3	J: 5/2, 7/2, 9/2, 11/2 g: 4/7, 62/63, 116/99, 14/11	J: 7/2, 9/2, 11/2, 13/2 g: 2/3, 32/33, 162/143, 16/13	J: 9/2, 11/2, 13/2, 15/2 g: 8/11, 138/143, 72/65, 6/5	J: 11/2, 13/2, 15/2, 17/2 g: 10/13, 188/195, 278/255, 20/17

	^5S	^5P	^5D	^5F	^5G	^5H	^5I	^5K
Quintets 2.000	J: 2 g: 2.000	J: 1, 2, 3 g: 5/2, 11/6, 5/3	J: 0, 1, 2, 3, 4 g: 0/0, 3/2, 3/2, 3/2, 3/2	J: 1, 2, 3, 4, 5 g: 0, 1, 5/4, 27/20, 7/5	J: 2, 3, 4, 5, 6 g: 1/3, 11/12, 23/20, 19/15, 4/3	J: 3, 4, 5, 6, 7 g: 1/2, 9/10, 11/10, 17/14, 9/7	J: 4, 5, 6, 7, 8 g: 3/5, 9/10, 15/14, 33/28, 5/4	J: 5, 6, 7, 8, 9 g: 2/3, 19/21, 59/56, 83/72, 11/9

	^6S	^6P	^6D	^6F	^6G	^6H	^6I	^6K
Sextets 2.000	J: 5/2 g: 2.000	J: 3/2, 5/2, 7/2 g: 12/5, 66/35, 12/7	J: 1/2, 3/2, 5/2, 7/2, 9/2 g: 10/3, 28/15, 58/35, 100/63, 14/9	J: 1/2, 3/2, 5/2, 7/2, 9/2, 11/2 g: −2/3, 16/15, 46/35, 88/63, 142/99, 16/11	J: 3/2, 5/2, 7/2, 9/2, 11/2, 13/2 g: 0, 6/7, 8/7, 14/11, 192/143, 18/13	J: 5/2, 7/2, 9/2, 11/2, 13/2, 15/2 g: 2/7, 52/63, 106/99, 192/143, 50/39, 4/3	J: 7/2, 9/2, 11/2, 13/2, 15/2, 17/2 g: 4/9, 82/99, 148/143, 226/195, 316/255, 22/17	J: 9/2, 11/2, 13/2, 15/2, 17/2, 19/2 g: 6/11, 130/143, 66/65, 96/85, 390/323, 24/19

an atom a spectral line observed belongs. A neutral atom and its succeeding ions have alternating multiplicity. Is that one of the atom even, then the multiplicity of the twice, four times (an so on) ionised atom is also even. Since with this changing multiplicity an alternating picture is associated as far as the π-components are concerned (viz. in case of even multiplicity there is no central component, and in case of odd multiplicity there is one) one can decide to a certain extent for a completely resolved pattern, whether a special line belongs to the atom, or to which one of its succeeding ions.

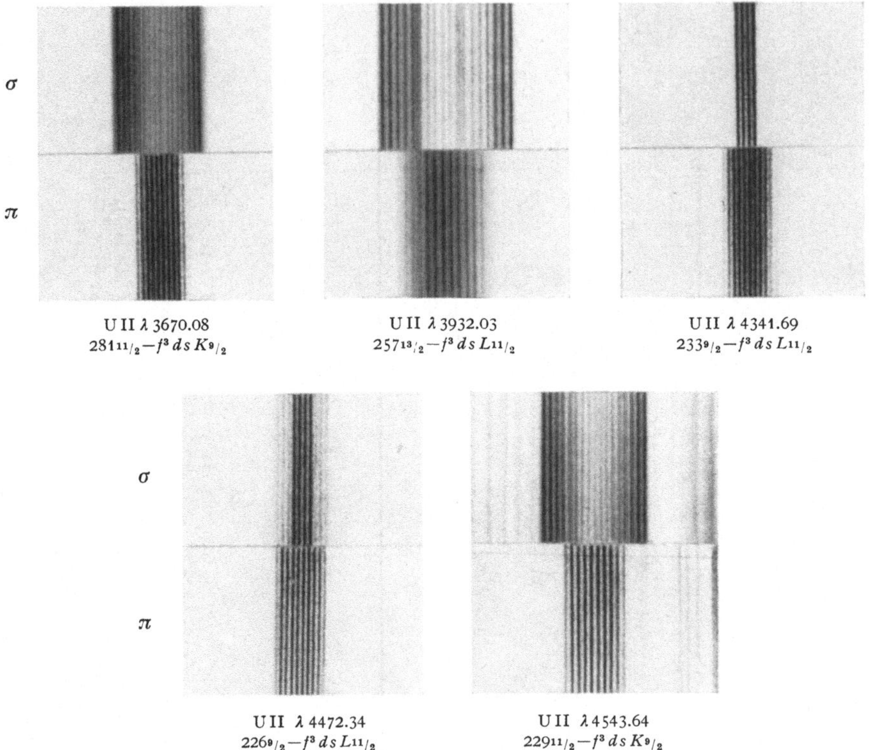

Fig. 12. Zeeman effect patterns of some spectral lines of the U II spectrum. The σ-groups of the lines 4341.69 Å and 4472.34 Å overlap, so there are four very strong components in the middle.

Combining this with the fact that the centre of gravity of the intensity of the spectrum of higher ionisation stages is shifted to the ultraviolet, the above consideration will lead to an unambiguous determination, to which ionisation stage a spectral line belongs. Moreover, the intensity of a spectral line of an ion is increased by a magnetic field, which is often associated with an increased haziness of the line, because of internal Stark effect, so that for lines unresolved in the magnetic field, the ionisation stage also can be determined.

In this way about 120 lines of the Os spectrum, for which no separation in ionisation stages has been known hitherto, could be detected as belonging to the Os II spectrum by means of indicating the multiplicity and magnetic enhancement. On this a preliminary analysis of the spectrum could be built up[1].

Table 8 gives g-values for a number of levels in case of LS-coupling.

---

[1] See footnote 1, p. 317.

Activité optique naturelle.

Par

J. P. MATHIEU.

Avec 71 Figures.

A. Historique et lois générales.

1. Phénomènes, lois et définitions relatifs à l'activité optique naturelle. α) *L'activité optique naturelle*[1], appelée également *pouvoir rotatoire naturel* ou *polarisation rotatoire*, fut découverte par ARAGO[2] qui, examinant entre polariseurs croisés des lames de quartz perpendiculaires à l'axe, remarqua que les couleurs observées en lumière blanche ne changent pas lorsqu'on fait tourner la lame dans son plan, contrairement aux couleurs données, dans ces conditions, par les lames biréfringentes parallèles à l'axe. BIOT[3] rencontra le phénomène dans des substances organiques isotropes: liquides et vapeurs (essence de térébenthine), solutions (acide tartrique, sucres), et trouva que certaines substances, comme le sucre de canne, ont le pouvoir rotatoire à la fois à l'état cristallin et en solution, tandis que d'autres, comme le quartz, perdent cette propriété par fusion ou dissolution. On dit que les premières ont un *pouvoir rotatoire moléculaire*, les secondes un *pouvoir rotatoire de structure*.

BIOT[4] énonça les lois suivantes:

1. Certains corps isotropes transparents, traversés par un faisceau parallèle de lumière monochromatique polarisé rectilignement, font tourner d'un certain angle ϱ la direction de la vibration lumineuse autour de celle du rayon. La même loi est valable pour certains cristaux cubiques et aussi pour certains cristaux uniaxes ou biaxes, traversés par la lumière suivant la direction d'un axe optique.

2. La grandeur de l'angle de rotation ϱ est proportionnelle à l'épaisseur traversée. Cette loi permet de diviser les substances optiquement actives en *dextrogyres* et *lévogyres*. Pour les premières, la rotation qui amène la vibration incidente sur la vibration émergente s'effectue, pour l'observateur recevant la lumière, dans le sens des aiguilles d'une montre et on la compte positivement; pour les secondes, la rotation se fait en sens inverse du précédent et on la compte négativement.

3. La rotation produite ne dépend pas de l'azimut de la vibration incidente, ni du sens dans lequel la lumière se propage suivant une direction donnée.

4. La rotation dépend de la longueur d'onde: c'est le phénomène de la *dispersion rotatoire*. Pour un corps incolore, la rotation croît généralement en valeur absolue depuis le rouge jusqu'au violet (Fig. 1). Depuis lors, on a étendu les mesures de rotation dans l'ultraviolet et dans l'infrarouge.

---

[1] L'adjectif *naturel* distingue le phénomène étudié de la polarisation rotatoire magnétique (effet FARADAY).

[2] D. F. ARAGO: Mém. Cl. Sci. Math. Phys. Inst. France **1**, 115 (1811).

[3] J. B. BIOT: Mém. Acad. Sci., Paris **2**, 114 (1817).

[4] J. B. BIOT: Mém. Acad. Sci., Paris **15**, 93 (1838).

Biot définit le *pouvoir rotatoire spécifique* $[\varrho]$ d'un fluide pour une longueur d'onde λ et une température T, au moyen de la relation:

$$[\varrho]_\lambda^T = \frac{\varrho}{l\,d}, \tag{1.1}$$

ϱ désignant la rotation observée, exprimée en degrés et fractions décimales; l, l'épaisseur traversée en décimètres; d, la densité du fluide par rapport à l'eau à 4° C. Pour une solution fluide d'une substance active dans un solvant inactif, la relation (1.1) est remplacée par la suivante:

$$[\varrho]_\lambda^T = \frac{\varrho}{l\,c}, \tag{1.2}$$

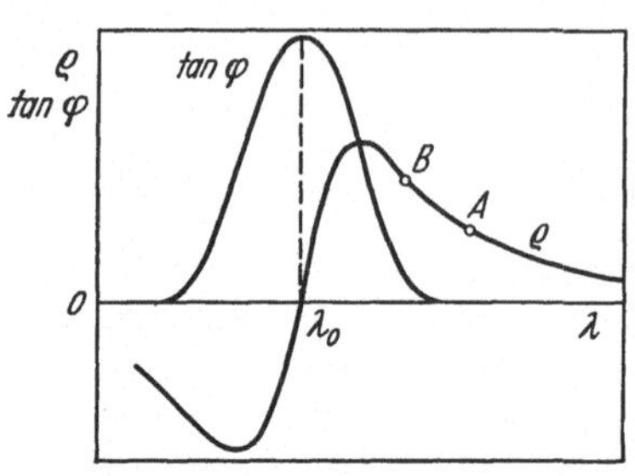

Fig. 1. Dispersion du pouvoir rotatoire ϱ et de l'ellipticité $\tan\varphi$.

c désignant le nombre de grammes du composé optiquement actif contenu dans un cm³ du mélange.

Lorsqu'on connaît la masse moléculaire M du composé actif, son *pouvoir rotatoire moléculaire* $[M]$ est défini par la formule:

$$[M] = \frac{M}{100}\,[\varrho].$$

β) Pasteur[1] put séparer, à partir du racémate double de sodium et d'ammonium, deux sortes de cristaux qui donnent des solutions respectivement dextrogyres et lévogyres, mais dont la valeur absolue du pouvoir rotatoire spécifique est la même. Les développements de cette découverte conduisirent à la généralisation suivante: à tout composé optiquement actif doit en correspondre un autre, appelé *isomère optique, inverse optique, antipode optique* ou *énantiomorphe*, dont les propriétés chimiques, sauf à l'égard de composés eux-mêmes optiquement actifs, sont les mêmes. Les pouvoirs rotatoires des deux isomères optiques sont égaux, mais de signes contraires.

A plusieurs reprises, on a prétendu que diverses propriétés physico-chimiques (point de fusion, solubilité, absorption de la lumière, réactivité chimique, valeur absolue du pouvoir rotatoire) n'étaient pas les mêmes pour les isomères optiques d'un même composé. Il semble bien que toutes ces différences soient dues simplement à des erreurs expérimentales ou à une purification imparfaite et que rien ne permette actuellement d'affirmer que toutes les propriétés des antipodes optiques n'ont pas la même valeur absolue. On trouvera le détail de ces controverses en [5].

γ) Cotton[2] établit les relations entre la dispersion rotatoire et l'absorption sélective de certaines radiations par des solutions de corps colorés optiquement actifs (tartrates de chrome et de cuivre). Soit λ_0 (Fig. 1) la longueur d'onde du maximum d'absorption: la rotation varie au voisinage de λ_0 comme le montre la courbe ϱ. On a donné parfois au phénomène qui vient d'être décrit le nom de *dispersion rotatoire anomale*; mais on ne doit pas le considérer comme une anomalie, car il existe chez tous les composés actifs, dans une partie au moins de leurs bandes d'absorption. On appelle *bandes actives* les bandes dans lesquelles la courbe de dispersion rotatoire a une forme analogue à celle de la Fig. 1.

Les bandes actives sont le siège d'un second phénomène[2]: la vibration rectiligne qui tombe sur la substance étudiée n'est plus rectiligne à la sortie, comme dans les régions de transparence, mais devient elliptique. Le pouvoir rotatoire est alors mesuré par l'angle ϱ que fait le grand axe de l'ellipse avec la direction

---

[1] L. Pasteur: Ann. Chim. Phys., Paris **28**, 56 (1850).
[2] A. Cotton: Ann. Chim. Phys., Paris **8**, 347 (1896).

de la vibration rectiligne incidente (Fig. 2). L'ellipticité, mesurée par la valeur du rapport $\frac{B}{A} = \tan \varphi$ entre le petit axe et le grand axe de l'ellipse, a une valeur nulle dans les régions de transparence et passe par un maximum à l'intérieur de la bande d'absorption (Fig. 1). Suivant une convention analogue à celle que l'on fait pour les rotations, on compte positivement l'ellipticité lorsque l'observateur voit l'ellipse parcourue dans le sens des aiguilles d'une montre.

Les phénomènes de dispersion rotatoire anomale et d'ellipticité, dont l'ensemble constitue l'*effet* COTTON, montrent que la dispersion rotatoire est sous la dépendance de l'absorption. Les relations que montre la Fig. 1 sont très générales. Dans le spectre visible et ultraviolet, on les rencontre en étudiant non seulement la dispersion rotatoire des substances liquides, gazeuses ou dissoutes (Figs. 43, 44) mais encore celle des cristaux ayant uniquement un pouvoir rotatoire de structure (Fig. 61) et aussi celle des arrangements hélicoïdaux de molécules qui caractérisent la phase cholestérique des substances mésomorphes (Sect. 60). Dans le spectre hertzien, on a pu les produire en plaçant sur le parcours de l'onde électromagnétique un ensemble de résonateurs métalliques constitués soit par quatre boules identiques assemblées aux sommets d'un tétraèdre irrégulier[1], soit plus récemment

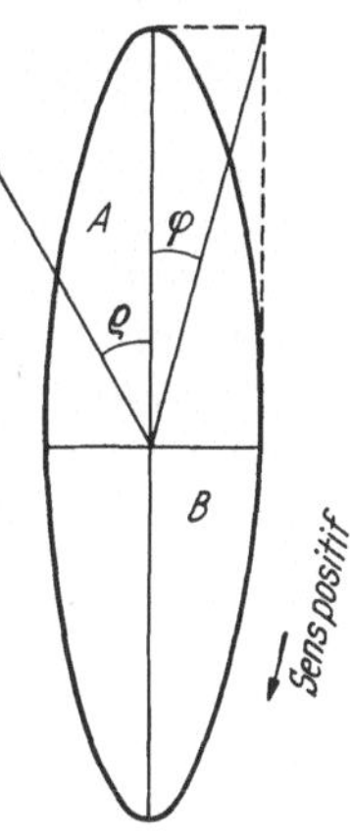

Fig. 2. Eléments d'une vibration elliptique.

par des hélices placées suivant l'axe de guides d'onde cylindriques fonctionnant dans le mode TE_{II}[2].

2. Biréfringence circulaire et dichroïsme circulaire. α) La Fig. 3 rappelle comment se propage un rayon de lumière circulaire gauche monochromatique de fréquence v. L'extrémité du vecteur lumineux $\boldsymbol{E}$ décrit une hélice, dont le pas est égal à la longueur d'onde de la radiation, soit $\lambda_0 = c/v$

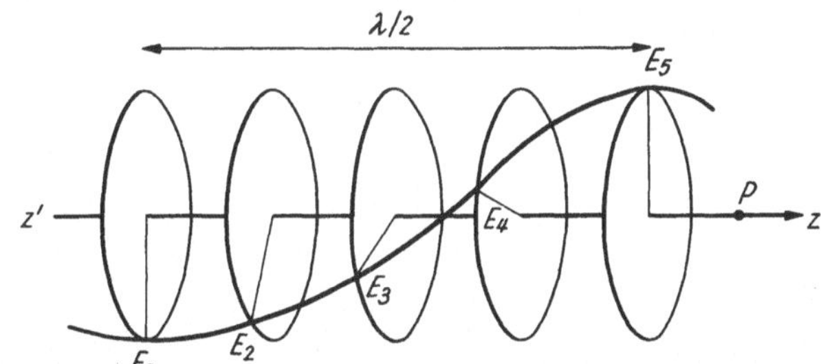

Fig. 3. Propagation d'un rayon de lumière circulaire gauche.

dans le vide (c = célérité de la lumière dans le vide) et $\lambda = c'/v$ dans un milieu matériel où la célérité $c' = c/n$ (n = indice de réfraction). L'observateur placé en P voit l'hélice tourner en sens inverse des aiguilles d'une montre. FRESNEL[3] a montré que la rotation ϱ d'une vibration rectiligne produite par un milieu optiquement actif est due à l'existence de deux indices de réfraction différents pour les circulaires droits et gauches, soient n_d et n_g. Décomposons à l'entrée dans le milieu la vibration lumineuse $OR = 2A$, dont la direction est Ox, en deux vibrations circulaires OD et OG (Fig. 4a). A l'entrée dans le milieu actif, elles sont en phase (Fig. 4b). Si les deux vibrations ne se propagent pas avec la même vitesse, c'est-à-dire avec le même indice, la résultante des vecteurs tournants OD et OG n'a plus la direction Ox à la sortie du milieu actif, car les phases de ces vecteurs ne sont plus égales. En effet, soit δ l'angle de phase du vecteur OD après traversée d'une

[1] K. F. LINDMAN: Ann. Phys., Lpz. **77**, 337 (1925).
[2] R. SERVANT, P. LOUDETTE et A. CHARRU: C. R. Acad. Sci., Paris **240**, 1978 (1955).
[3] A. FRESNEL: Ann. Chim. Phys., Paris **28**, 147 (1825).

épaisseur l du milieu. Le chemin optique est $n_d\,l$ et la phase correspondante:

$$\delta = \frac{2\pi\,n_d\,l}{\lambda_0}\,.$$

L'angle de phase γ du vecteur OG au même instant est différent de δ:

$$\gamma = \frac{2\pi\,n_g\,l}{\lambda_0}\,.$$

Par suite (Fig. 4c), la résultante des deux circulaires à la sortie du milieu est encore une vibration rectiligne, mais dirigée suivant une droite Ox', qui fait avec Ox l'angle:

$$\varrho = \frac{\gamma - \delta}{2} = \frac{\pi\,l}{\lambda_0}\,(n_g - n_d)\,. \tag{2.1}$$

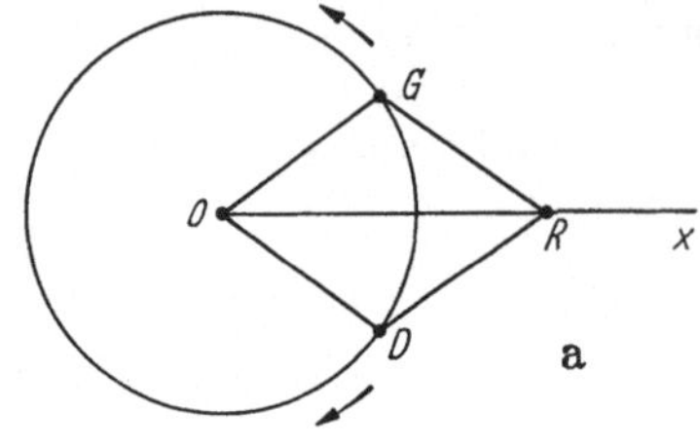

L'angle de rotation donné par cette formule est positif lorsque $\gamma > \delta$ ou $n_g > n_d$. Comme γ est compté dans le sens des aiguilles d'une montre, on retrouve pour ϱ la convention de signe donnée à la Sect. 1. La rotation se fait

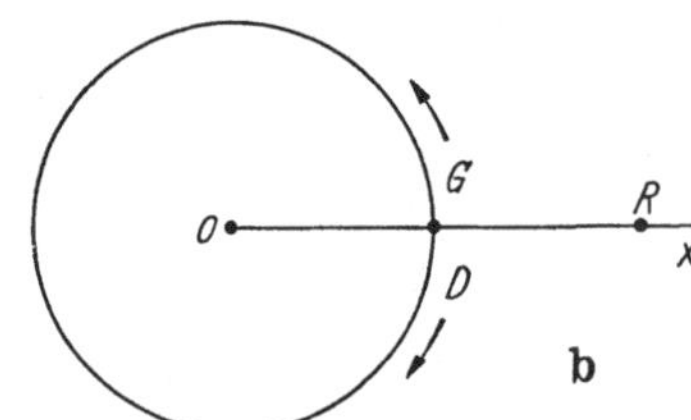
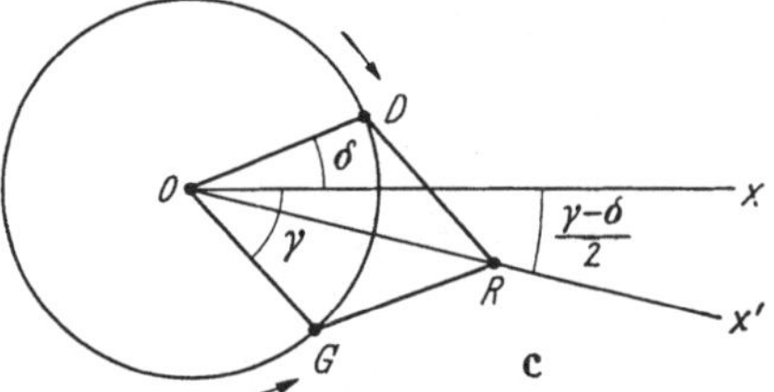

Fig. 4a—c. a) Décomposition d'une vibration rectiligne en deux vibrations circulaires de même amplitude. b) Vibration rectiligne à l'entrée d'un milieu optiquement actif. c) La même à la sortie.

donc dans le sens de la vibration circulaire de plus faible indice, c'est-à-dire de la plus rapide. La différence $n_g - n_d$ mesure la *biréfringence circulaire*; elle est toujours faible. Par exemple, pour une substance de densité $d = 1$, de pouvoir rotatoire spécifique $[\varrho] = 100°$ (valeur que l'expérience montre d'une grandeur peu fréquente), la formule (1.1) donne $\varrho = 10°$ par cm ou 0,17 radians. On calcule alors, à l'aide de la formule (2.1) pour $\lambda = 0,5\,\mu$: $n_g - n_d = 2,8 \cdot 10^{-6}$, soit une valeur de l'ordre du millionième de l'indice moyen de la substance.

β) Cotton[1] montra que l'on peut rendre compte de la transformation d'une vibration rectiligne en vibration elliptique, qui se produit dans le passage à travers un milieu actif absorbant, en admettant l'inégale absorption des rayons circulaires droits et gauches par ce milieu: d'où le nom de *dichroïsme circulaire*, donné à ce phénomène. Les lois de l'absorption lumineuse enseignent que l'amplitude d'une vibration de forme quelconque est réduite exponentiellement par la traversée du milieu. Si A est l'amplitude commune aux circulaires à l'entrée dans le milieu, leurs amplitudes respectives à la sortie peuvent s'écrire:

$$OD = A\,e^{-\frac{2\pi}{\lambda_0}\chi_d\,l} \quad \text{et} \quad OG = A\,e^{-\frac{2\pi}{\lambda_0}\chi_g\,l}\,, \tag{2.2}$$

χ_d et χ_g désignant les indices d'absorption du milieu pour les rayons circulaires droits et gauches de longueur d'onde λ. Si $\chi_g > \chi_d$, $OD > OG$. D'après ce que

---

[1] A. Cotton: Ann. Chim. Phys., Paris **8**, 347 (1896).

l'on a vu au sujet de la composition des circulaires d'amplitudes inégales, le rapport des axes de l'ellipse résultante, défini à la fin de la Sect. 1, a pour valeur:

$$\tan \varphi = \frac{D - G}{D + G} = \frac{1 - e^{-\frac{2\pi l}{\lambda_0}(\chi_g - \chi_d)}}{1 + e^{-\frac{2\pi l}{\lambda_0}(\chi_g - \chi_d)}}. \tag{2.3}$$

La différence $\chi_g - \chi_d$ étant toujours faible, on peut écrire:

$$\varphi = \frac{\pi l}{\lambda_0}(\chi_g - \chi_d). \tag{2.4}$$

Au lieu de l'indice d'absorption χ, on utilise souvent, pour mesurer l'absorption des solutions, le coefficient d'extinction moléculaire ε, défini par la relation:

$$I = I_0\, 10^{-\varepsilon c l}, \tag{2.5}$$

c désignant la concentration moléculaire de la solution, I_0 l'intensité de la lumière à l'entrée, I son intensité à la sortie. Comme l'intensité lumineuse est proportionnelle au carré de l'amplitude de la vibration, on tire de (2.2) et de (2.5) la relation

$$e^{-\frac{4\pi}{\lambda_0}\chi l} = 10^{-\varepsilon c l}$$

ou

$$\chi = 2{,}3\,\frac{c\lambda_0}{4\pi}\,\varepsilon.$$

La formule (2.4) devient

$$\varphi = 2{,}3\,\frac{c l}{4}\,(\varepsilon_g - \varepsilon_d), \tag{2.6}$$

ε_g et ε_d désignant les coefficients d'extinction moléculaire respectifs des rayons circulaires gauches et droits. Le dichroïsme circulaire moléculaire a pour expression $\varDelta = \frac{100}{c l}\,\varphi$, d'où $\varDelta = \frac{2300}{4}\,(\varepsilon_g - \varepsilon_d)$ en radians et $\varDelta = 3300\,(\varepsilon_g - \varepsilon_d)$ en degrés.

L'ellipse est décrite dans le sens positif (sens de rotation des aiguilles d'une montre) lorsque $\chi_g > \chi_d$; c'est-à-dire que le sens de parcours de l'ellipse est celui de la vibration circulaire la moins absorbée. Quant à l'orientation du grand axe de l'ellipse, elle serait celle de la vibration rectiligne incidente, si le milieu ne possédait pas de biréfringence circulaire pour la longueur d'onde considérée. L'existence du pouvoir rotatoire modifie inégalement la phase des circulaires OD et OG comme dans le cas où elles ont des amplitudes égales; le grand axe de l'ellipse fait alors avec la vibration rectiligne incidente un angle ϱ donné par la formule (2.1).

La différence $\chi_g - \chi_d$ mesure le dichroïsme circulaire; il est toujours faible. L'expérience montre, en effet, que si l'on veut que l'intensité de la lumière qui a traversé le milieu actif absorbant soit suffisante pour permettre des pointés polarimétriques à l'aide d'un analyseur à pénombres, les angles φ mesurés ne dépassent jamais quelques degrés. Prenons $\varphi = 5° = 0{,}087$ radians (valeur exceptionnellement forte); $l = 1$ cm; $\lambda = 0{,}5\,\mu$; la formule (2.4) donne $\chi_g - \chi_d = 1{,}4 \cdot 10^{-6}$. Le dichroïsme circulaire est donc du même ordre de grandeur que la biréfringence circulaire.

γ) On peut montrer la biréfringence circulaire[1] à l'aide d'un prisme de quartz de 60° d'angle, dont l'axe est normal au plan bissecteur. Un rayon de lumière

---

[1] A. Cornu: C. R. Acad. Sci., Paris **92**, 1365 (1881).

tombant au minimum de déviation donne deux rayons émergents (Fig. 5), l'un polarisé circulairement à droite, l'autre à gauche. Pour la raie D, la différence des indices du quartz droit correspondant à ces deux rayons est $n_g - n_d = 7 \cdot 10^{-5}$. On calcule que l'écart angulaire des rayons émergents est de 27'', valeur plus

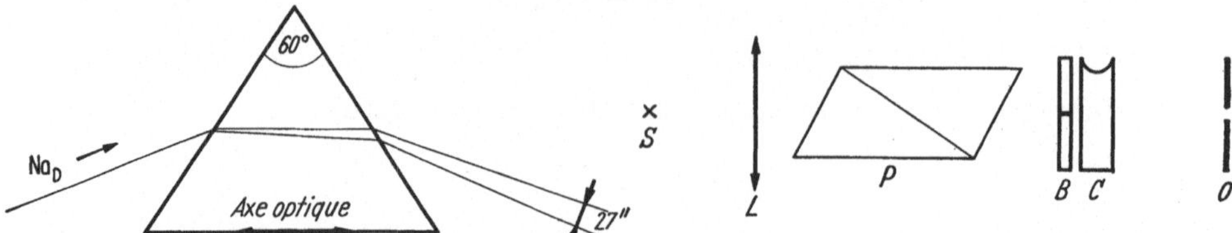

Fig. 5. Observation de la biréfringence circulaire.   Fig. 6. Observation du dichroïsme circulaire.

de trois fois supérieure à l'écart angulaire des raies D_1 et D_2; il est donc facile à constater.

δ) Le dichroïsme circulaire peut être mis en évidence sur des solutions de tartrates complexes de chrome colorés en vert, qui présentent une différence des indices d'absorption particulièrement forte $(\chi_g - \chi_d = 1{,}8 \cdot 10^{-6})$ dans une bande située dans le jaune orangé. Le montage est celui de la Fig. 6.

3. Conditions de symétrie[1,2]. α) Le pouvoir rotatoire étant indépendant de la direction de propagation de la lumière, l'hypothèse la plus simple que l'on

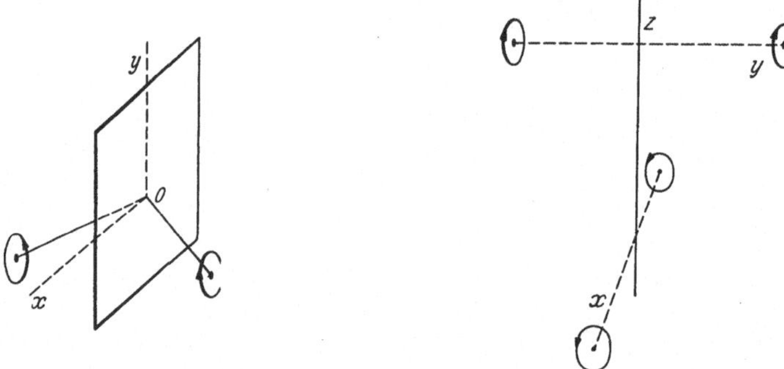

Fig. 7. Pouvoir rotatoire dans un milieu possédant un plan de symétrie.   Fig. 8. Pouvoir rotatoire dans un milieu possédant un axe de symétrie inverse d'ordre 4.

puisse faire sur sa variation dans un milieu anisotrope est de le représenter par une fonction quadratique des cosinus directeurs s_x, s_y, s_z de la direction de propagation considérée:

$$\varrho = a_{xx} s_x^2 + a_{yy} s_y^2 + a_{zz} s_z^2 + 2 a_{xy} s_x s_y + 2 a_{yz} s_y s_z + 2 a_{zx} s_z s_x. \qquad (3.1)$$

1. Il ne peut y avoir de centre de symétrie. Car l'inversion, qui change les signes de s_x, s_y, s_z à la fois, transforme un rayon circulaire droit en gauche et devrait changer le signe de ϱ. Cela exige que ϱ s'annule, car c'est une fonction paire des s.

2. S'il existe un plan de symétrie, xOy par exemple, ϱ doit avoir la même valeur et des signes contraires dans deux directions symétriques par rapport au

---

[1] W. Voigt: Ann. Phys., Lpz. **18**, 649 (1905).
[2] H. Chipart: Theorie gyrostatique de la lumière. Paris 1904.

plan (Fig. 7). Donc $\varrho \to -\varrho$ pour $s_z \to -s_z$, d'où:

$$\varrho = 2 s_z (a_{yz} s_y + a_{zx} s_x). \tag{3.2}$$

3. S'il existe un axe de symétrie inverse d'ordre 4, soit Oz (Fig. 8) ϱ doit changer de signe lorsque $s_x \to s_y$, $s_y \to -s_x$, $s_z \to -s_z$, d'où $a_{xx} = -a_{yy}$ et:

$$\varrho = a_{xx} (s_x^2 - s_y^2) + 2 a_{xy} s_x s_y. \tag{3.3}$$

4. S'il existe un axe de symétrie direct, le signe de ϱ ne change pas dans deux directions symétriques. Pour un axe binaire, Oz par exemple, $\varrho \to \varrho$ lorsque $s_x \to -s_x$ et $s_y \to -s_y$:

$$\varrho = a_{xx} s_x^2 + a_{yy} s_y^2 + a_{zz} s_z^2 + 2 a_{xy} s_x s_y. \tag{3.4}$$

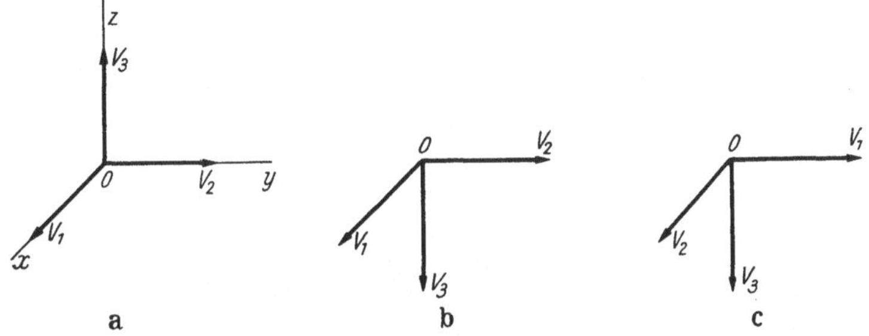

Fig. 9. Symétrie du produit mixte de trois vecteurs.

Un axe d'ordre supérieur à 2, Oz par exemple, se comporte comme un axe de révolution à l'égard du pouvoir rotatoire; d'où $a_{xx} = a_{yy}$, $a_{xy} = a_{yz} = a_{zx} = 0$ et

$$\varrho = a_{xx} (s_x^2 + s_y^2) + a_{zz} s_z^2. \tag{3.5}$$

β) On peut représenter la symétrie caractéristique de l'activité optique naturelle en un point 0 d'un milieu isotrope par le produit scalaire d'un vecteur polaire V par un vecteur axial V' ayant leur origine en 0, ou, ce qui revient au même, par le triple produit mixte de trois vecteurs polaires V_1, V_2, V_3

$$[V_1 \times V_2] \cdot V_3. \tag{3.6}$$

On voit, en effet, que le signe de cette expression (Fig. 9a) change par une inversion (Fig. 9b), mais demeure le même si l'on change à la fois (Fig. 9c) le sens de la rotation qui amène V_1 sur V_2 et celui de V_3. Or la troisième loi de Biot (Sect. 1) montre que si l'on change simultanément le sens de rotation et le sens de propagation d'un rayon lumineux polarisé circulairement, sa vitesse de propagation doit demeurer la même. On trouve des expressions analogues à (3.6) dans les théories moléculaires de l'activité optique (Sects. 5 et 24).

4. Propagation des ondes électromagnétiques dans les milieux isotropes optiquement actifs. α) Dans un milieu non conducteur, les équations de Maxwell s'écrivent:

$$\operatorname{rot} E + \frac{1}{c} \dot{B} = 0, \tag{4.1}$$

$$\operatorname{rot} H - \frac{1}{c} \dot{D} = 0, \tag{4.2}$$

$$\operatorname{div} D = 0, \tag{4.3}$$

$$\operatorname{div} B = 0, \tag{4.4}$$

(E = champ électrique; H = champ magnétique; D = déplacement électrique, B = induction magnétique; c = vitesse de la lumière dans le vide). Les vecteurs E, H et le vecteur unité s porté par la normale à l'onde ont les orientations mutuelles représentées par la Fig. 10.

On peut poser

$$D = E + 4\pi P, \qquad (4.5)$$

$$B = H + 4\pi M. \qquad (4.6)$$

P et M sont deux vecteurs, le premier polaire, le second axial, qui représentent respectivement la polarisation électrique et la polarisation magnétique du milieu matériel. Tenant compte de ces définitions, on peut remplacer (4.2) par:

$$\operatorname{rot} B - \frac{1}{c}\, E = \frac{4\pi}{c}\, (P - c\operatorname{rot} M). \qquad (4.7)$$

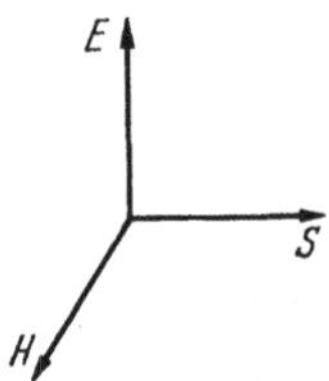

Fig. 10. Orientation des vecteurs caractéristiques d'une onde électromagnétique.

Les polarisations sont produites par les champs respectifs. Si le milieu est diamagnétique et dépourvu d'activité optique, la polarisation magnétique peut être négligée et l'on peut écrire en outre

$$P = K E. \qquad (4.8)$$

K est la susceptibilité électrique du milieu matériel; d'où:

$$D = (1 + 4\pi K)\, E = \varepsilon E. \qquad (4.9)$$

ε est la constante diélectrique du milieu. C'est une grandeur scalaire dans un milieu isotrope. Les vecteurs P et D sont alors parallèles à E. De façon analogue à (4.8) on peut écrire

$$B = \mu H, \qquad (4.10)$$

où μ est la perméabilité électrique du milieu; mais dans les milieux transparents μ est très voisin de l'unité. Dans ces conditions, les équations (4.1) et (4.2) deviennent,

$$\operatorname{rot} E + \frac{1}{c}\, \dot{H} = 0, \qquad (4.11)$$

$$\operatorname{rot} H - \frac{\varepsilon}{c}\, \dot{E} = 0. \qquad (4.12)$$

On en tire l'équation de propagation d'une onde électromagnétique dans le milieu

$$\varepsilon \ddot{E} = c^2 \Delta E, \qquad (4.13)$$

Δ désignant le laplacien.

La vitesse de propagation a pour valeur

$$c' = \frac{c}{\sqrt{\varepsilon}}. \qquad (4.14)$$

β) Pour une onde plane monochromatique, le champ électrique est représenté par

$$E = E^0\, e^{-2\pi i\, \frac{\Re(r \cdot s)}{\lambda_0}} \cos 2\pi \nu t = E^0\, e^{-2\pi i\, \frac{\Re(r \cdot s)}{\lambda_0}} \left(\frac{e^{2\pi i \nu t} + e^{-2\pi i \nu t}}{2} \right), \qquad (4.15)$$

où ν est la fréquence, λ_0 la longueur d'onde dans le vide, s un vecteur unité porté par la normale à l'onde, r un vecteur joignant un point quelconque du plan d'onde à une origine arbitraire. $\Re$ est l'indice complexe:

$$\Re = n - i\,\chi, \tag{4.16}$$

où n est l'indice de réfraction, χ l'indice d'absorption.

Les équations de MAXWELL (4.1) et (4.2) donnent:

$$\operatorname{rot} \boldsymbol{E} = \frac{2\pi i\,\Re}{\lambda}\,\boldsymbol{E} \times \boldsymbol{s}, \tag{4.17}$$

$$\operatorname{rot} \boldsymbol{H} = \frac{2\pi i\,\Re}{\lambda}\,\boldsymbol{H} \times \boldsymbol{s}, \tag{4.18}$$

$$\Re\,\boldsymbol{E} \times \boldsymbol{s} = -\boldsymbol{H}, \tag{4.19}$$

$$\Re\,\boldsymbol{H} \times \boldsymbol{s} = \boldsymbol{D}, \tag{4.20}$$

d'où

$$\boldsymbol{D} = -\Re^2\,(\boldsymbol{E} \times \boldsymbol{s}) \times \boldsymbol{s} = \Re^2\,[\boldsymbol{E} - \boldsymbol{s}(\boldsymbol{E} \cdot \boldsymbol{s})]. \tag{4.21}$$

γ) Dans un milieu non conducteur qui possède le pouvoir rotatoire, $\boldsymbol{P}$ et $\boldsymbol{D}$ ne peuvent être parallèles à $\boldsymbol{E}$ car, par raison de symétrie, la rotation ne pourrait alors se produire. Il faut modifier la relation (4.9) en respectant le caractère linéaire de l'équation de propagation (4.13).

L'examen de la Fig. 9 montre que la symétrie du pouvoir rotatoire exige que si l'on change à la fois le sens de rotation et le sens de propagation du rayon, il conserve le même indice de réfraction, de façon que l'angle de rotation (1.1) demeure le même pour l'observateur.

La modification la plus simple consiste à ajouter à l'expression (4.9) de $\boldsymbol{D}$ une fonction linéaire des dérivées partielles de $\boldsymbol{E}$ par rapport aux coordonnées. Comme cette fonction doit en outre respecter la condition de symétrie et être indépendante de l'orientation des axes de coordonnées, car le pouvoir rotatoire d'un milieu isotrope ne dépend pas de la direction de propagation, on trouve que les dérivées partielles qu'elle contient doivent être d'ordre impair et que si l'on se borne aux dérivées du premier ordre, les relations (4.9) et (4.13) doivent être remplacées respectivement par:

$$\boldsymbol{D} = \varepsilon\,\boldsymbol{E} + j\,\operatorname{rot}\boldsymbol{E} \tag{4.22}$$

et par:

$$\frac{\partial^2}{\partial t^2}\,[\varepsilon\,\boldsymbol{E} + j\,\operatorname{rot}\boldsymbol{E}] = c^2\,\varDelta\,\boldsymbol{E} \tag{4.23}$$

où j est un paramètre. Cette dernière équation a été proposée par CAUCHY[1] avant la théorie de MAXWELL.

Ce type d'équation conduit à la biréfringence circulaire. Pour le voir, raisonnons sur une onde plane se propageant suivant l'axe Oz

$$\operatorname{rot}_x \boldsymbol{E} = -\frac{\partial E_y}{\partial z} = \frac{2\pi\,\Re\,i}{\lambda_0}\,E_y, \quad \operatorname{rot}_y \boldsymbol{E} = \frac{\partial E_x}{\partial z} = -\frac{2\pi\,\Re\,i}{\lambda_0}\,E_x, \quad \operatorname{rot}_z \boldsymbol{E} = 0.$$

Le déplacement (4.21) a pour composantes:

$$D_x = \varepsilon E_x + i\,\frac{2\pi\,\Re\,j}{\lambda_0}\,E_y, \quad D_y = \varepsilon E_y - i\,\frac{2\pi\,\Re\,j}{\lambda_0}\,E_x, \quad D_z = 0. \tag{4.24}$$

Le facteur i est dû à ce que la valeur de $\boldsymbol{D}$ en un point dépend des dérivées de $\boldsymbol{E}$ en même temps que de $\boldsymbol{E}$ lui-même.

---

[1] A. CAUCHY: C. R. Acad. Sci., Paris **25**, 331 (1847).

D'autre part, les relations (4.19) et (4.20) s'écrivent:

$$\Re E_x = H_y, \qquad \Re E_y = - H_x,$$
$$\Re H_x = - D_y, \qquad \Re H_y = D_x,$$

d'où

$$D_x = \Re^2 E_x, \qquad D_y = \Re^2 E_y,$$

ce qui, transporté dans (4.13), donne les équations:

$$E_x (\Re^2 - \varepsilon) = i \, \frac{2\pi \Re j}{\lambda_0} \, E_y,$$

$$E_y (\Re^2 - \varepsilon) = - i \, \frac{2\pi \Re j}{\lambda_0} \, E_x$$

d'où l'on tire

$$E_x = \pm \, i \, E_y$$

avec la condition

$$\Re^2 = \varepsilon \pm \frac{2\pi \Re j}{\lambda_0} . \tag{4.25}$$

ε représente, d'après la relation de MAXWELL, le carré n^2 de l'indice de réfraction du milieu pour la lumière naturelle. Le second terme est lié à la biréfringence circulaire et doit être petit (Sect. 2α).

On peut donc écrire

$$\Re = \sqrt{\varepsilon} \pm \frac{\pi j}{\lambda_0} . \tag{4.26}$$

Aux deux solutions (4.25), qui représentent deux ondes planes polarisées circulairement de sens contraires, correspondent deux indices différents donnés par (4.26). Dans les régions de transparence $\chi = 0$, $\Re = n$, et l'on tire de (4.26) l'expression de la biréfringence circulaire:

$$n_g - n_d = \frac{2\pi j}{\lambda_0} \tag{4.27}$$

ainsi que celle du pouvoir rotatoire par unité de longueur, d'après (2.1):

$$\varrho = \frac{\pi}{\lambda_0} (n_g - n_d) = \frac{2\pi^2 j}{\lambda_0^2} . \tag{4.28}$$

5. Propagation des ondes électromagnétiques dans les milieux anisotropes optiquement actifs. Dans un milieu diélectrique anisotrope, comme le sont les cristaux (à l'exception des cristaux cubiques) mais qui ne possède pas de pouvoir rotatoire, le pouvoir inducteur spécifique ε est un tenseur symétrique du deuxième ordre; le déplacement D est une fonction vectorielle linéaire du champ E et la relation (4.9) doit être remplacée par:

$$D = [\varepsilon] \, E \tag{5.1}$$

ou

$$D_x = \sum_y \varepsilon_{xy} E_y . \tag{5.2}$$

Les coefficients du tenseur $[\varepsilon]$ sont réels. Cette dernière condition n'existe plus dans un cristal doué d'activité optique. La relation (5.2) est remplacée par une relation analogue à (4.23), pour la propagation d'ondes planes:

$$D_x = \sum_y \varepsilon_{xy} E_y + \sum_y i \, G_{xy} \, E_x . \tag{5.3}$$

Les termes additionnels G, supposés petits, doivent être conservatifs, c'est-à-dire qu'ils ne doivent pas contribuer à la densité d'énergie électrique. On a:

$$D \cdot E = \sum_x \sum_y \varepsilon_{xy} E_x E_y + i \sum_x \sum_y G_{xy} E_x E_y .$$

Pour que le second terme s'annule pour toute valeur de E, il faut que:

$$G_{xx} = G_{yy} = G_{zz} = 0, \quad G_{xy} + G_{yx} = 0 \quad \text{etc.}$$

Les G_{xy} sont donc les composantes purement imaginaires d'un tenseur anti-symétrique. Un tel tenseur peut être remplacé par un vecteur axial G, dont les composantes sont:

$$G_x = G_{yz} = -G_{zy}, \quad G_y = G_{zx} = -G_{xz}, \quad G_z = G_{xy} = -G_{yx}.$$

G est le *vecteur de gyration*. La formule (5.3) peut alors s'écrire:

$$D_x = \sum_y \varepsilon_{xy} E_y + i\,(E \times G)_x. \tag{5.4}$$

Les conditions de symétrie du milieu qui possède le pouvoir rotatoire (Sect. 3) sont bien respectées par la forme de cette relation. Par exemple, s'il existe un centre de symétrie, l'inversion change les signes des composantes de D, de E et de $E \times G$, mais aussi celui de la phase représentée par i, de sorte que le dernier terme de (5.4) ne changeant pas de signe, doit s'annuler, d'où $G = 0$.

B. Méthodes de mesure.

Dans le cas le plus général, la mesure de l'activité optique d'un fluide comprend celle du pouvoir rotatoire et celle du dichroïsme circulaire. L'angle de rotation ϱ et l'angle φ défini par la formule (2.6) sont tous deux petits si le corps est absorbant; ils ne dépassent jamais quelques degrés. Dans les régions de transparence, au contraire, où le pouvoir rotatoire existe seul, les angles ϱ peuvent atteindre plusieurs tours. Les méthodes de mesure sont différentes dans les deux cas.

D'autre part, des précautions spéciales doivent être prises pour la mesure de l'activité optique des cristaux.

I. Mesure du pouvoir rotatoire.

6. Généralités. Dans tous les appareils de mesure, appelés *polarimètres* (ou *spectropolarimètres* lorsqu'ils permettent les mesures pour diverses longueurs d'onde), la substance est mise sous forme d'un cylindre droit traversé suivant son axe par un faisceau lumineux polarisé rectilignement, qui passe ensuite à travers un dispositif analyseur.

Dans la région visible du spectre, on utilise des polariseurs en spath, prismes de NICOL, ou, mieux, prismes à champ normal de GLAZEBROOK ou de GLAN[1]. Ce dernier, comme le prisme de FOUCAULT, ou encore le prisme de GLAZEBROOK collé à la glycérine, peut servir dans l'ultraviolet jusque vers 0,24 µ. Pour des longueurs d'onde inférieures, on peut utiliser soit le prisme de ROCHON en quartz, en éliminant les rayons extraordinaires[2], soit la réflexion[3]. Dans l'infrarouge, on polarise la lumière par réflexion ou par transmission à travers des lames de sélénium[4] ou de chlorure d'argent.

Le récepteur dépend de la région spectrale étudiée. Dans le visible, on emploie l'œil, la plaque photographique ou la cellule photoélectrique. Ces deux derniers récepteurs ont été employés dans l'ultraviolet. Dans l'infrarouge, on a utilisé des cellules photorésistantes ou des piles thermoélectriques.

---

[1] Voir l'article de M. FRANÇON, tome XXIV de cette Encyclopédie p. 430.
[2] W. KUHN: Ber. dtsch. chem. Ges. **62**, 1727 (1929).
[3] R. SERVANT: Ann. Phys., Paris **12**, 397 (1939).
[4] A. ELLIOTT, E. AMBROSE et R. TEMPLE: J. Opt. Soc. Amer. **38**, 212 (1948).

La nature du récepteur intervient dans le choix de la méthode de mesure. Les courants des piles thermoélectriques et des cellules photoélectriques peuvent être rendus proportionnels aux flux lumineux Φ qu'elles recoivent et la plus petite différence $\Delta\Phi$ de flux perceptible est indépendante de la valeur absolue de Φ. Dans les mesures oculaires, il faut considérer l'éclairement E de l'image rétinienne, c'est-à-dire le flux lumineux reçu par l'unité de surface de la rétine; si ΔE désigne la plus petite différence d'éclairement perceptible, l'expérience montre que c'est le rapport $\Delta E/E$ qui est, dans de larges limites, indépendant de la valeur absolue de E. Pour la plaque photographique enfin, le problème est complexe: la densité optique, pour une durée d'exposition déterminée, est reliée à l'éclairement par la courbe de noircissement; mais le plus petite variation d'éclairement perceptible dépend également de la méthode qui sert à apprécier les différences de densité optique.

a) Méthodes fondées sur l'appréciation d'une variation de flux.

7. Considérations générales. Le flux lumineux transmis par un analyseur qui laisse passer une vibration rectiligne faisant un angle ϑ avec la direction de la

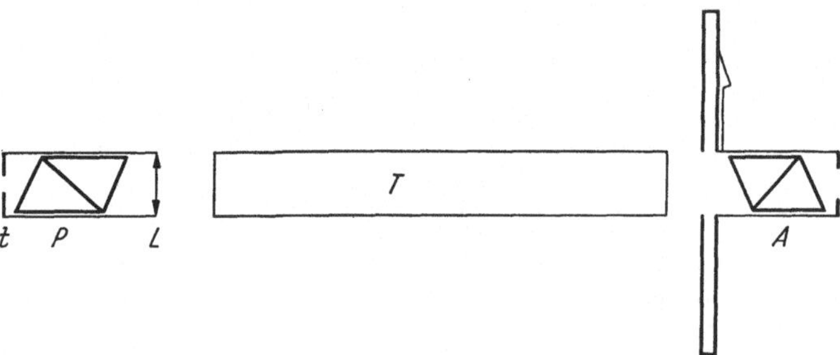

Fig. 11. Polarimètre visuel à extinction (Mitscherlich).

vibration qu'il reçoit, varie avec ϑ suivant la loi de Malus:

$$\Phi = \Phi_0 \cos^2 \vartheta.$$

Une petite rotation $\Delta\vartheta$ de la vibration produit une variation absolue de flux:

$$\Delta\Phi = \Phi_0 \sin 2\vartheta \cdot \Delta\vartheta, \tag{7.1}$$

qui est maximum pour $\vartheta = \pi/4$ et une variation relative

$$\frac{\Delta\Phi}{\Phi} = 2 \tan \vartheta \cdot \Delta\vartheta, \tag{7.2}$$

qui est maximum pour $\vartheta = \pi/2$.

D'après ce qu'on a vu à la section précédente, la valeur $\vartheta = \pi/2$ convient aux mesures visuelles (on l'emploie aussi dans les mesures photographiques); la valeur $\vartheta = \pi/4$ aux récepteurs thermiques et électroniques.

8. Méthodes d'extinction. α) *Les polarimètres visuels* à extinction dérivent de l'appareil de Mitscherlich[1], représenté schématiquement par la Fig. 11. L'image de la source se projette sur le trou t placé au foyer de la lentille L. Le polariseur P et l'analyseur A sont des prismes de Nicol. On mesure l'angle ϱ par la différence des azimuts de l'analyseur produisant l'extinction avant et après introduction du liquide dans le tube T. Pour distinguer une rotation ϱ d'une rotation $\pi - \varrho$

---

[1] Mitscherlich: Lehrbuch der Chemie, p. 361 (1844).

en sens contraire, on fait la mesure pour deux épaisseurs de substance l_1 et l_2 et on applique la formule (1.1). L'incertitude est de plusieurs dizièmes de degré. La précision peut être beaucoup accrue par l'emploi d'une source lumineuse très intense qui permet de pointer la frange de LIPPICH[1]. La méthode est cependant abandonnée pour les mesures visuelles, au profit des méthodes de pénombre (Sect. 10).

β) *Dans l'ultraviolet*, la méthode d'extinction a été appliquée[2] en employant comme analyseur et polariseur des prismes à champ normal[3], en photographiant la trace du faisceau monochromatique pour une suite d'azimuts connus de l'ana-

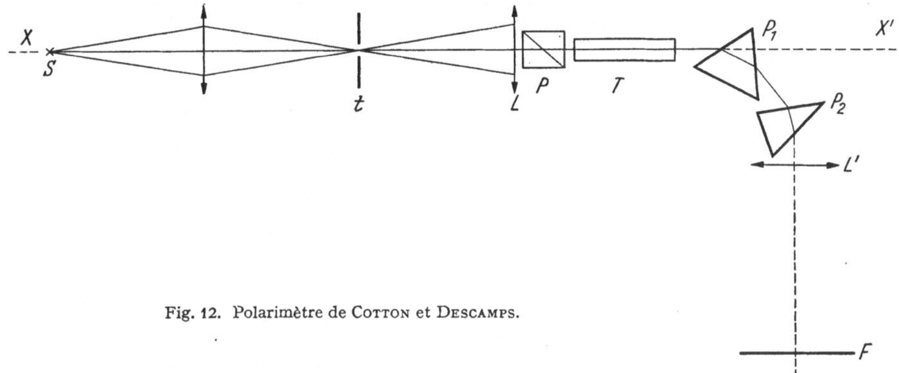

Fig. 12. Polarimètre de COTTON et DESCAMPS.

lyseur et en déterminant à l'aide d'un microphotomètre le minimum de densité optique. Dans les meilleures conditions, l'incertitude ne dépasse pas 0,01°. On peut appliquer la même méthode, en utilisant un rayonnement complexe et en cherchant à éteindre, par rotation de l'analyseur[4], les diverses raies du spectre qu'il donne.

Dans l'appareil de COTTON et DESCAMPS[5] la source S est un arc à vapeur de mercure (Fig. 12). Deux prismes de spath P_1 et P_2, d'axe optique parallèle à l'arête, servent à la fois d'analyseur et de système dispersif, donnant une déviation moyenne de 90°. Les prismes et l'objectif de chambre L' ont un mouvement oscillatoire autour de la direction XX'. Les images du trou t données par les diverses radiations se déplacent sur la surface cylindrique d'axe XX' d'un film photographique, qui coupe en F le plan de la figure. On détermine au microphotomètre la position d'extinction pour chaque raie du spectre. Les mesures s'étendent jusqu'à 2537 Å; leur précision atteint 0,02°.

γ) *Dans l'infrarouge*, on a employé[6] jusqu'à 1,2 μ des polariseurs de spath et la méthode fondée sur la relation (7.1). Entre 1 et 25 μ on a utilisé des polariseurs au sélénium associés à un monochromateur à prisme; le détecteur était une pile thermoélectrique; on mesurait ϱ par le maximum de courant, obtenu lorsque la section principale de l'analyseur est parallèle à la vibration qu'il reçoit[7].

δ) *Dans le domaine des ondes hertziennes*[8], on a utilisé comme polariseurs des empilements de feuilles d'étain séparées par des feuilles de papier. L'ensemble

[1] Voir l'article de M. FRANÇON, tome XXIV de cette Encyclopédie.
[2] G. BRUHAT et M. PAUTHENIER: Rev. Opt. 6, 163 (1927).
[3] Voir l'article de M. FRANÇON.
[4] R. SERVANT: Rev. Opt. 28, 558 (1949).
[5] R. DESCAMPS: Rev. Opt. 5, 481 (1926).
[6] R. FREYMANN: Ann. Phys., Paris 20, 329 (1933).
[7] H. S. GUTOWSKY: J. Chem. Phys. 19, 438 (1951).
[8] J. C. BOSE: Proc. Roy. Soc. Lond. 63, 146 (1898).

est dichroïque. L'absorption est complète pour les vibrations parallèles au plan des feuilles.

ε) *Dans le cas des substances mésomorphes à l'état cholestérique*[1], dont les rotations peuvent atteindre des dizaines de milliers de degrés par millimètre, on introduit la substance entre une lame plan-parallèle de verre L et la face convexe d'une lentille plan convexe L' de grand rayon de courbure R (Fig. 13). Entre un polariseur P et un analyseur A faisant entre eux l'angle ϑ, on voit en lumière monochromatique des anneaux noirs de rayon r, correspondant aux rotations

$\vartheta + (2n+1)\dfrac{\pi}{2}$. On évite la mesure de la faible épaisseur $e = \dfrac{r^2}{2R}$ en mesurant les rayons de deux anneaux consécutifs r_n et r_{n+1}. On a:

$$\varrho = \pm\frac{2\pi R}{r_{1+n}^2 - r_n^2}. \qquad (8.1)$$

Le sens de ϱ est celui dans lequel il faut tourner l'analyseur pour observer une augmentation du diamètre des anneaux.

9. Emploi des spectres cannelés. α) Lorsqu'une substance est assez active et assez transparente pour qu'on puisse l'étudier sous une épaisseur telle que l'angle de rotation soit au moins de quelques dizaines de degrés et varie rapidement avec la longueur d'onde, on peut utiliser la production des spectres cannelés pour mesurer le pouvoir rotatoire, en utilisant, dans le montage de la Fig. 11, une source S intense donnant un spectre continu. Soit ϑ l'angle que font entre elles les sections principales de P et de A. Les radiations pour lesquelles la condition (Fig. 14):

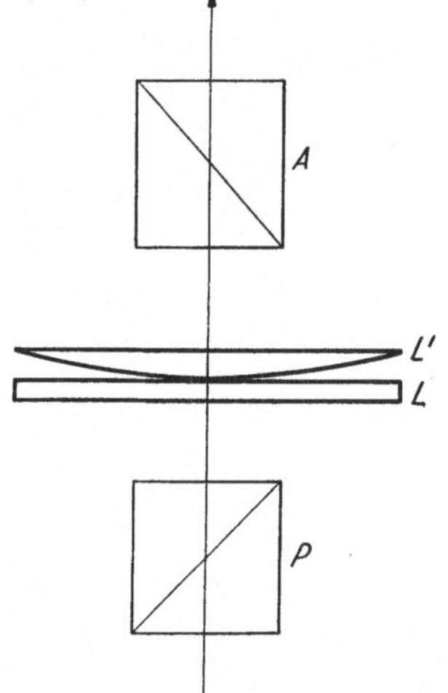

Fig. 13. Mesure du pouvoir rotatoire des substances cholestériques (Stumpf).

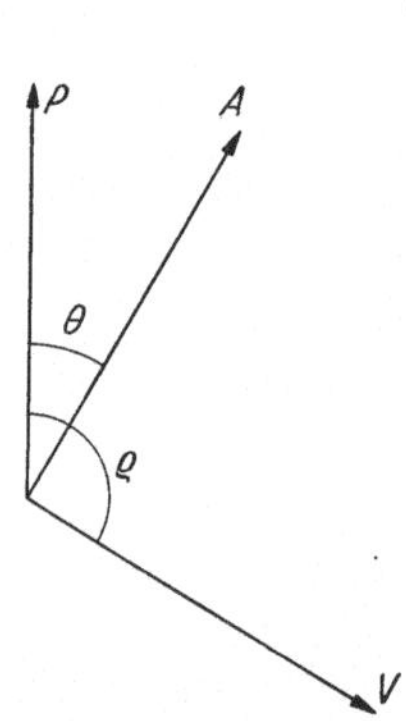

Fig. 14. Substance optiquement active entre polariseur et analyseur.

$$\varrho - \vartheta = (2n+1)\frac{\pi}{2} \qquad (n \text{ entier}) \qquad (9.1)$$

est réalisée, manquent dans le rayonnement sortant de A.

Si l'on concentre le faisceau lumineux sur la fente d'un spectroscope, on peut déterminer les longueurs d'onde des bandes obscures ou *cannelures* correspondant à la relation (9.1). Si l'on a déterminé la rotation pour une longueur d'onde connue, l'ordre n des cannelures s'en déduit. En tournant l'analyseur de façon à faire varier l'angle ϑ, on déplace les cannelures et l'on peut tracer point par point la courbe de dispersion rotatoire. Si l'on est dans le domaine spectral correspondant à la branche AB de cette courbe (Fig. 1) le sens dans lequel il faut tourner l'analyseur pour déplacer les cannelures vers le violet est celui du pouvoir rotatoire. D'une cannelure à la suivante, la rotation varie de 180°. On ne peut guère déterminer leur position à mieux que $\frac{1}{200}$ de leur intervalle; l'incertitude absolue est supérieure à 1°. La méthode, due à Fizeau et Foucault[2], a été appliquée

---

[1] O. Lehmann: Ann. Phys., Lpz. **2**, 649 (1900). — F. Stumpf: Ann. Phys., Lpz. **37**, 351 (1912).

[2] H. Fizeau et L. Foucault: C. R. Acad. Sci., Paris **21**, 1155 (1845).

à la mesure de la dispersion rotatoire des cristaux[1], des liquides[2], des substances mésomorphes[3].

La méthode a été utilisée dans l'ultraviolet, en photographiant les sannelures[4,5]. Elle a été également employée dans l'infrarouge, en envo yantour le récepteur (thermopile) une radiation de longueur d'onde connue et en faisant varier l'angle ϑ de façon à réaliser la condition (9.1)[6] ou la condition $\varrho - \vartheta = n\pi$[7].

β) *Cas du quartz.* On a employé un compensateur de SENARMONT. Il se compose de deux coins de quartz D et G (Fig. 15) de même angle et de rotations contraires. Traversé suivant AB par un faisceau de lumière parallèle polarisé, il fait tourner la vibration d'un angle croissant de A en A'. Si l'on reçoit la lumière sur un analyseur, puis sur la fente d'un spectrographe parallèle à AA', on observe un spectre traversé par des cannelures noires, obliques par suite de la dispersion. Pour chaque longueur d'onde, la distance de deux cannelures est inversement proportionnelle à la rotation produite par le quartz. La méthode a été appliquée dans l'infrarouge[8] et dans l'ultraviolet[9] jusque 1500 Å.

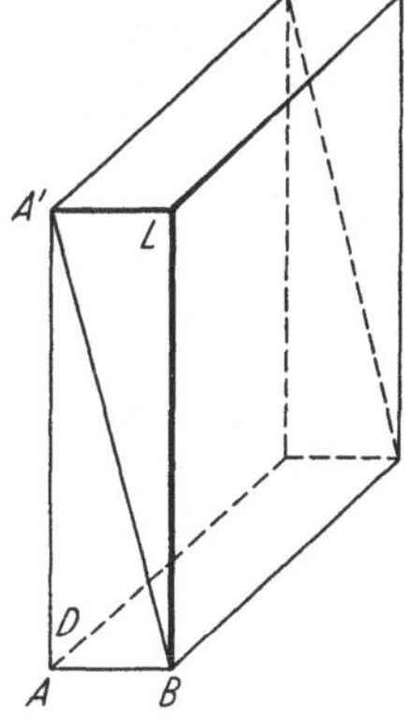
Fig. 15. Compensateur de SENARMONT.

b) Méthodes différentielles.

Elles remplacent la mesure d'un maximum ou d'un minimum de réponse du récepteur, soit par l'appréciation simultanée ou successive de l'égalité de deux flux ou de deux éclairements, soit par la mesure de leur différence. Les conditions d'emploi de ces méthodes, comme des précédentes, dépendent des propriétés du récepteur.

10. Polarimètres visuels à pénombre. α) Le faisceau issu du polariseur est divisé en deux parties, dans lesquelles la direction des vibrations rectilignes V_1 et V_2 (Fig. 16) diffère d'un angle 2β petit (quelques degrés) appelé *angle de pénombre.* L'œil observe (Fig. 19) l'image d'un diaphragme limitant une section droite du faisceau; le champ lui apparait divisé, par une ligne de séparation fine, en deux *plages*, dont il apprécie l'égalité d'éclairement. Celle-ci existe lorsque la section principale de l'analyseur est parallèle ou perpendiculaire à la bissectrice V de l'angle de V_1 et V_2, ce qu'on obtient

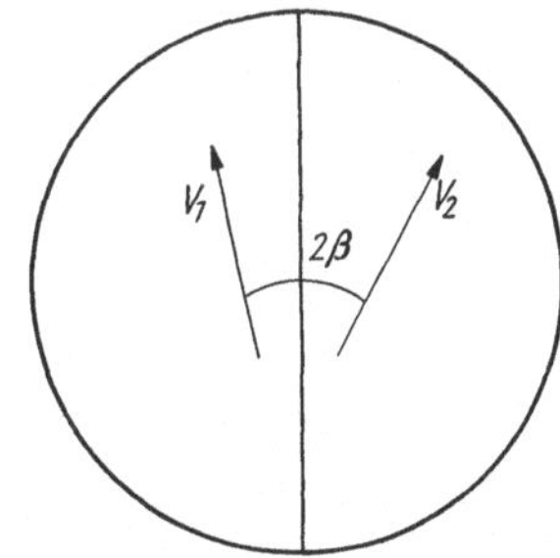
Fig. 16. Principe des polarimètres à pénombre (LAURENT).

en tournant l'analyseur. L'éclairement commun aux deux plages a pour valeur: dans la première disposition $E = E_0 \cos^2\beta$; dans la seconde $E = E_0 \sin^2\beta$. Lorsqu'on tourne l'analyseur d'un petit angle $\Delta\vartheta$, les éclairements varient très peu dans la première disposition, que l'on doit éviter; dans la seconde, ils deviennent respectivement $E_0 \sin^2(\beta + \Delta\vartheta)$ et $E_0 \sin^2(\beta - \Delta\vartheta)$. C'est à leur différence relative, égale si β est petit, à:

$$\frac{\Delta E}{E} = \frac{2\sin 2\Delta\vartheta}{\tan\beta} \approx \frac{4\Delta\vartheta}{\beta} \tag{10.1}$$

[1] BROCH: Dove's Rep. Physik **7**, 113 (1846).
[2] J. RABINOVITCH: Rev. Opt. **17**, 161 (1938).
[3] J. P. MATHIEU: Bull. Soc. franç. Minér. **61**, 174 (1938).
[4] J. SORET et E. SARRAZIN: Arch. de Genève **8**, 97 (1882).
[5] J. DUCLAUX et P. JEANTET: J. Phys. Radium **7**, 200 (1926).
[6] J. CARVALLO: Ann. Chim. Phys., Paris **26**, 113 (1892).
[7] T. M. LOWRY et C. SNOW: Proc. Roy. Soc. Lond., Ser. A **127**, 271 (1930).
[8] A. HUSSEL: Ann. Phys., Lpz. **43**, 498 (1891).
[9] R. SERVANT: Ann. Phys., Paris **12**, 397 (1939).

séparation R des deux plages. Le viseur L, avantageusement constitué par une sorte de lunette de GALILÉE pour réduire sa longueur, permet à l'œil de mettre au point sur R et donne de D_1 une image D_1' servant d'anneau oculaire.

$\delta)$ Dans les cristaux, le pouvoir rotatoire est généralement superposé à la biréfringence. Celle-ci disparaît dans les directions des axes optiques, et c'est dans ces directions que l'on peut définir et mesurer l'activité optique. Pour cela, LONGCHAMBON[1] modifie le montage précédent (Fig. 20). La lame cristalline C, traversée par un faisceau parallèle de lumière monochromatique, est suivie d'une

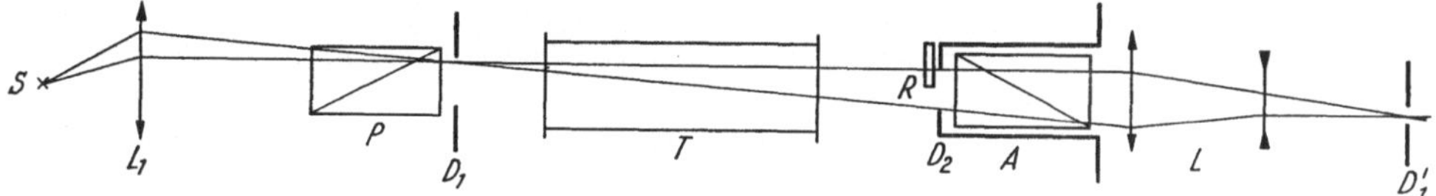

Fig. 19. Polarimètre à pénombre.

lentille L_2 dont la longueur focale est de l'ordre de 7 mètres. Dans son plan focal se forme une figure d'interférences (Sect. 51), sur le pôle de laquelle on place le très petit diaphragme D.

Fig. 20. Mesure du pouvoir rotatoire des cristaux (LONGCHAMBON).

11. Polarimètres photographiques à pénombre. On a remplacé parfois l'œil par la plaque photographique, en formant sur la couche sensible une image réelle du diaphragme D_2 de la Fig. 19. Sur une série de clichés correspondant à des angles ϱ différents, on détermine celui où l'égalité de noircissement des plages est réalisée. Dans le cas où cette égalité est appréciée au moyen de l'œil, le temps de pose doit être choisi de façon qu'on se trouve avant le début de la partie recti-ligne de la courbe de noircissement de l'émulsion.

Dans un certain nombre de spectropolarimètres, la lumière totale d'une source donnant un spectre discontinu (Hg, Fe), traverse le corps optiquement actif. L'image du diaphragme D_2 se forme sur la fente d'un spectrographe. On observe un spectre de raies divisé en deux parties correspondant aux deux plages. On photographie le spectre pour une série d'azimuts connus du polariseur et on détermine la longueur d'onde de la raie spectrale pour laquelle le noircissement est le même dans les deux moitiés du spectre. Avec des polariseurs de spath collés à la glycérine, il est possible de faire des mesures jusque $0,25~\mu$ environ[2]. Avec des prismes biréfringents de quartz, KUHN[3] est parvenu jusque $0,185~\mu$.

12. Polarimètres photoélectriques différentiels. La relation (7.1) montre que l'angle de pénombre 2β correspondant au maximum de sensibilité d'un montage photoélectrique est égal à $\pi/2$. Pour l'obtenir, on a utilisé le montage de la Fig. 11, mais en employant comme analyseur un prisme biréfringent, qui partage le flux incident entre deux faisceaux polarisés à angle droit. Les deux flux $\Phi_1 = \Phi_0 \sin^2 \vartheta$ et $\Phi_2 = \Phi_0 \cos^2 \vartheta$ (ϑ désignant l'angle du polariseur et de l'analyseur)

[1] L. LONGCHAMBON: Bull. Soc. franç. Minéral. **45**, 161 (1922).

[2] S. LANDAU: Z. Physik **9**, 417 (1908). — T. M. LOWRY et W. R. C. COODE-ADAMS: Phil. Trans. Roy. Soc. Lond., Ser. A **226**, 391 (1927).

[3] W. KUHN: Ber. dtsch. Chem. Ges. **62**, 1727 (1929).

sont égaux à $\Phi_0/2$ pour $\vartheta = \pi/4$. Une petite rotation $\Delta\vartheta$ produit une différence de flux:

$$\Delta\Phi = \Phi_1 - \Phi_2 = 4\Phi \cdot \Delta\vartheta, \tag{12.1}$$

que l'on mesure en envoyant alternativement les deux faisceaux sur la cellule[1].

Dans certains cas, les conditions d'emploi de la cellule conduisaient au courant de saturation pour $2\beta = \pi/2$; on a alors utilisé des valeurs de 2β beaucoup plus faibles (10 à 50°) produites par un biquartz (Sect. 10β) dont on peut substituer l'une des moitiés à l'autre devant la cellule[2].

13. Polarimètres à récepteurs thermoélectriques différentiels. La méthode d'Ingersoll[3] présente sur celles de la (Sect. 8) l'avantage de s'affranchir des variations du flux Φ_0 émis par la source. Elle est analogue à la première des méthodes de la Sect. 12: l'analyseur est un prisme biréfringent, les directions des vibrations dans les deux faisceaux qu'il fournit faisant des angles $\beta = \pi/4$ avec la vibration V donnée par le polariseur. Les flux égaux Φ transportés par ces faisceaux tombent sur deux bolomètres placés dans deux branches d'un pont de Wheatstone, qui demeure équilibré. Si la vibration V tourne d'un petit angle $\Delta\vartheta$, les flux prennent des valeurs Φ_1 et Φ_2, dont la différence est donnée par la formule (12.1). La déviation du galvanomètre est proportionnelle à $\Phi_1 - \Phi_2$. On mesure Φ en cachant l'une des images. La méthode a été utilisée par Meyer[4] avec des polariseurs par réflexion, jusque 2,5 μ.

Pour mesurer de grandes rotations données par du quartz, Dongier[5] a utilisé également un analyseur biréfringent. On obtient deux spectres cannelés correspondant à des angles ϑ (Fig. 14) qui diffèrent de $\pi/2$ et dans lesquels les cannelures alternent. On fait défiler les mêmes radiations des deux spectres sur les deux éléments d'une pile ou d'un bolomètre différentiels: là où la différence des flux est nulle, on a $\varrho = (2n+1)\dfrac{\pi}{4}$.

II. Mesure du dichroïsme circulaire.

14. Conditions des mesures. Tandis que, dans les régions de transparence, le pouvoir rotatoire existe seul, le dichroïsme circulaire (Sect. 2), qui apparaît dans certaines régions d'absorption, est toujours accompagné de pouvoir rotatoire: la vibration que reçoit l'analyseur possède une ellipticité φ et son grand axe fait un angle ϱ avec la vibration rectiligne que produit le polariseur. Lorsque les angles φ et ϱ sont quelconques, leur mesure demande l'emploi des méthodes générales d'analyse des vibrations elliptiques. Mais, dans le cas actuel, le problème est simplifié, car la symétrie du pouvoir rotatoire fait que les angles ϱ et φ sont les mêmes, quel que soit l'azimut de la vibration rectiligne incidente; on peut donc, en tournant le polariseur, donner à l'ellipse une orientation déterminée sans changer la valeur de φ.

15. Emploi de compensateurs. α) *Compensateur de* Babinet. Dans le cas exceptionnel des substances cholestériques[6] où un dichroïsme circulaire très élevé ne s'accompagne pas d'une absorption importante, on place après la substance, entre polariseur et analyseur croisés à 45° de la verticale, un compensateur de

[1] G. Bruhat et G. Chatelain: Rev. Opt. **12**, 1 (1933).
[2] G. Bruhat et A. Guinier: Rev. Opt. **12**, 396 (1933).
[3] L. R. Ingersoll: Phys. Rev. **23**, 489 (1906); **9**, 257 (1917).
[4] U. Meyer: Ann. Phys., Lpz. **30**, 607 (1909).
[5] R. Dongier: J. Phys. Radium **7**, 637 (1898).
[6] J. P. Mathieu: Bull. Soc. franç. Minér. **61**, 174 (1938).

Les principaux organes de l'appareil de mesure sont représentés sur la Fig. 21[1]. Q est une lame quart d'onde que l'on peut enlever du faisceau ou remettre en place sans modifier son orientation. En l'absence de substance active, on l'oriente de façon que sa direction principale de grand indice soit parallèle à la vibration donnée par le polariseur; puis on l'enlève. La substance étant mise en place,

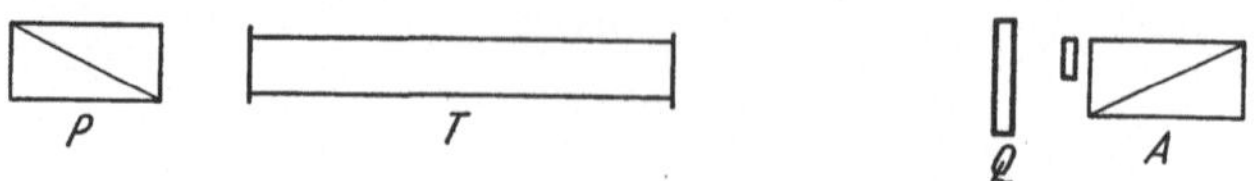

Fig. 21. Polarimètre ellipsomètre visuel (Bruhat).

on rétablit l'égalité des plages de l'analyseur en tournant le polariseur d'un angle $-\varrho$. On replace la lame Q et on rétablit l'égalité des plages, de nouveau détruite, en tournant l'analyseur d'un angle égal à $-\varphi$.

β) *Emploi d'un analyseur à pénombre sensible à l'ellipticité et non à la rotation* Un tel appareil (Fig. 22) peut être constitué par un polariseur à pénombre P donnant deux vibrations rectilignes faisant entre elles l'angle 2β (par exemple[2]

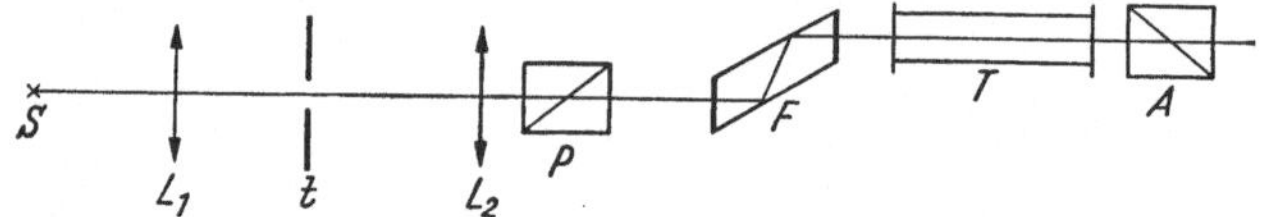

Fig. 22. Mesure du dichroïsme circulaire (Kuhn et Braun).

celui de la Fig. 17) suivi d'une lame quart d'onde ou d'un parallélépipède de Fresnel F dont les directions principales sont les bissectrices de l'angle 2β. Les deux faisceaux transportent des vibrations elliptiques de même orientation, ayant des ellipticités égales en valeur absolue et de signes contraires. Si l'égalité des plages est d'abord réalisée, la section principale de l'analyseur A étant perpendiculaire au grand axe des ellipses, elle ne change pas lorsque la substance produit une rotation ϱ: l'éclairement est seulement augmenté. Mais le dichroïsme circulaire augmente de φ l'une des ellipticités, diminue l'autre de la même quantité: l'égalité des plages est détruite. On peut la rétablir en tournant le polariseur d'un angle $-\varphi$.

C. Activité optique des milieux isotropes.

I. Principes de la stéréochimie des molécules dissymétriques.

17. Activité optique des fluides et dissymétrie moléculaire. Lorsque le milieu qui possède l'activité optique est homogène et isotrope (vapeurs, liquides, solutions liquides) le pouvoir rotatoire est le même dans toutes les directions. Il ne peut donc exister dans un tel milieu, non seulement de centre de symétrie, mais encore de plan de symétrie ou d'axe de symétrie inverse.

Dans les fluides ou les solutions, les molécules ou les ions sont en mouvements constants. Dans un fluide parfait, ils sont distribués et orientés au hasard. Les molécules ou les ions d'un fluide qui possède l'activité optique doivent eux-mêmes être dépourvus d'éléments de symétrie de seconde espèce (centre, plan ou axe inverse) car c'est à cette condition que l'ensemble n'en possédera pas. On aboutit ainsi à la conclusion énoncée par Pasteur[3]: *l'activité optique naturelle d'un fluide révèle la dissymétrie de ses molécules.*

[1] G. Bruhat: Rev. Opt. **8**, 412 (1929).
[2] W. Kuhn et K. Braun: Z. phys. Chem. Abt. B **8**, 445 (1930).
[3] L. Pasteur: Ann. Chim. Phys., Paris **24**, 442 (1848).

Le nom de «dissymétrie» est réservé, dans l'étude de l'activité optique, à cette sorte de symétrie qui ne comporte que des éléments de symétrie de première espèce (axes directs); le terme «asymétrie» désignant, comme à l'ordinaire, l'absence de tout élément de symétrie.

Les molécules de deux isomères optiques doivent avoir entre elles les relations géométriques de deux figures énantiomorphes (Fig. 23) se déduisant l'une de l'autre par une réflexion, opération de deuxième espèce, sur un plan ou par une inversion par rapport à un point, mais ne peuvent être amenées en superposition par une opération de première espèce (translation et rotation).

Ces conceptions exigent que l'on considère la molécule comme un objet dans l'espace, non comme écrite sur le papier. Les isomères optiques sont des *stéréoisomères*. Ce fut l'étude du pouvoir rotatoire qui provoqua la naissance de la *stéréochimie*[1].

La réciproque de la loi de PASTEUR: *l'arrangement dissymétrique des atomes dans une molécule entraîne l'activité optique du composé*, n'est pas exigée par les lois de symétrie, car les effets peuvent être plus symétriques que les causes. Les méthodes de localisation des atomes dans les molécules d'un fluide (diffrac-

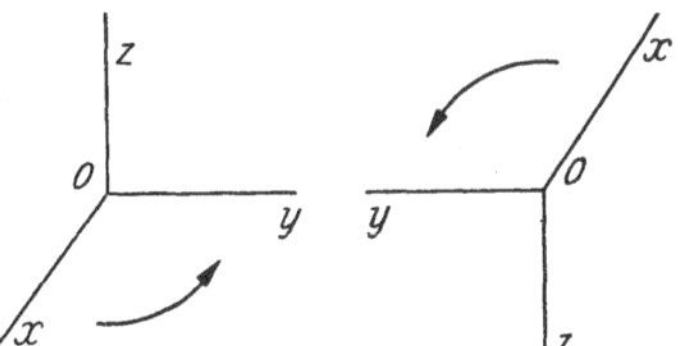

Fig. 23. Figures énantiomorphes.

tion des rayons X ou des électrons) n'ont encore jamais permis de prévoir l'existence du pouvoir rotatoire. Mais des raisonnements, déduits d'hypothèses hardies sur les valences dirigées de divers éléments et confirmées par les déductions de la chimie quantique, ont permis de très nombreuses prévisions de ce genre.

18. Asymétrie moléculaire. L'atome de carbone asymétrique. α) Les méthodes d'hybridation de la mécanique quantique[2], qui consistent à former des combinaisons linéaires orthogonales des fonctions d'onde relatives aux orbitales stables d'un atome, de façon qu'elles aient la plus grande extension radiale possible, ont permis de prévoir la direction des covalences formées par les atomes de la plupart des éléments. Les prévisions de cette stéréochimie théorique ont été soumises à des vérifications multiples. La prévision de l'activité optique des dérivés d'un élément A se fait en fixant sur le squelette formé par ses liaisons des atomes ou des groupements atomiques en nombre et en nature tels que la molécule ou l'ion formés soient dissymétriques.

β) Dans le cas du carbone, l'hybridation fait intervenir une orbitale $2s$ et trois orbitales $2p$, pour former quatre liaisons de même force, dirigées vers les sommets d'un tétraèdre régulier. Mais bien avant les développements modernes de la théorie de la valence, VAN'T HOFF[3] et LE BEL[4] avaient simultanément deviné la disposition tétraédrique régulière des valences de l'atome de carbone, précisément afin d'interpréter l'isomérie optique. Lorsque quatre atomes ou groupes différents sont liés à l'atome de carbone, on trouve deux dispositions énantiomorphes (Fig. 24). Dans ce cas, la molécule n'admet aucun élément de symétrie (groupe C_1) et on parle à bon droit d'un *atome de carbone asymétrique*[3]; un tel atome sera désigné dans ce qui suit par C*.

---

[1] V. MEYER: Ber. dtsch. Chem. Ges. **23**, 567 (1890). — J. H. VAN'T HOFF: Stéréochimie. Paris 1892.

[2] L. PAULING: J. Amer. Chem. Soc. **53**, 1367 (1931). — The Nature of the Chemical Bond. Ithaca 1944.

[3] J. H. VAN'T HOFF: Arch. néerlandaises Sci. **9**, 445 (1875).

[4] A. LE BEL: Bull. Soc. chim. France **22**, 337 (1874).

Exemples: acide fluorochlorobromacétique[1] $C^*F \cdot Cl \cdot Br \cdot CO_2H$ (Fig. 25); acide chloroiodométhanesulfonique[2] $C^*H \cdot Cl \cdot I \cdot SO_3H$; méthyléthylcarbinol $C^* \cdot H \cdot OH \cdot CH_3 \cdot C_2H_5$.

γ) Lorsqu'il existe dans la molécule plusieurs atomes de carbone, chacun d'eux doit être représenté par un tétraèdre. Dans le cas simple de l'acide fluoro-

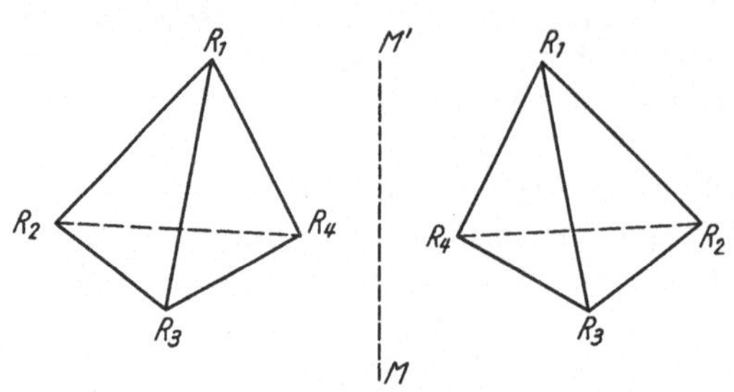

Fig. 24. Atome de carbone asymétrique. MM' est la trace du miroir par rapport auquel les deux molécules sont symétriquement placées.

Fig. 25. Liaison mobile.

chlorobromacétique, par exemple, on aura autant de molécules isomères que de positions relatives des deux tétraèdres obtenues par rotation autour de la liaison des deux atomes C. Le nombre d'isomères est restreint, comme le montre l'expérience, si l'on admet soit que la rotation se produit librement (liaison mobile), soit que différents isomères sont en équilibre réciproque, soit que l'une des formes moléculaires est plus stable que les autres. Diverses méthodes physico-chimiques peuvent fournir des renseignements à ce sujet: mesure des moments dipolaires[3], des spectres de vibration[4].

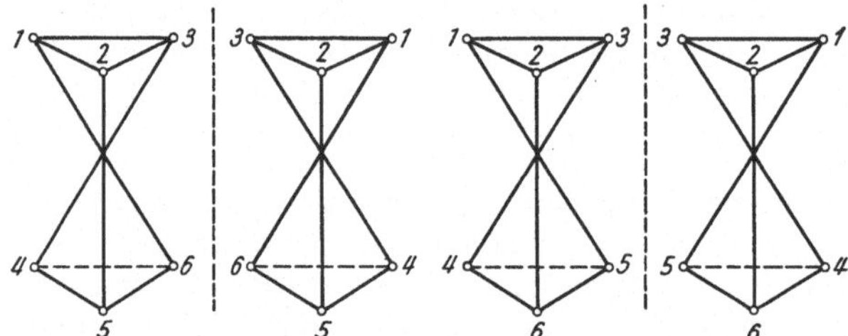

Fig. 26. Antipodes optiques d'une molécule contenant deux atomes de carbone asymétrique.

δ) Lorsqu'une molécule contient deux atomes de carbone asymétriques liés entre eux, en admettant une seule orientation privilégiée ou une liaison mobile, on est conduit à prévoir deux paires d'antipodes optiques, ainsi qu'on le vérifie sur les schémas de la Fig. 26. Si les deux groupements sont égaux, c'est-à-dire si dans la figure, on avait 1, 2, 3 = 4, 5, 6, on n'aurait plus que trois isomères, deux antipodes optiques devenant identiques car la molécule acquiert un plan ou un centre de symétrie: c'est le cas des acides tartriques (Fig. 27).

Mais dans ce cas, les mesures de dispersion rotatoire[5] conduisent à admettre un équilibre entre plusieurs formes (Sect. 59).

A mesure que croît le nombre d'atomes de carbone enchaînés par une liaison simple, celui des isomères possibles — et en particulier des isomères optiques — croît énormément.

ε) Pour représenter commodément les combinaisons organiques les plus simples, en tenant compte de leur configuration spatiale et sans avoir à des-

[1] Swarts: Bull. Acad. Belg. **31**, 28 (1896).

[2] W. J. Pope et Read: J. Chem. Soc., Lond. **105**, 811 (1914).

[3] K. L. Wolf: The Structure of Molecules. London 1932. — Trans Faraday Soc. **26**, 418 (1930).

[4] S. Mizushima, Y. Morino et T. Shimanouchi: J. Chem. Phys. **56**, 324 (1952).

[5] R. Lucas: Ann. Phys., Paris **9**, 381 (1928).

β) Composés à structure octaédrique. Chez les éléments pour lesquels les orbitales d ont des énergies peu différentes des orbitales s et p, on peut combiner une orbitale s, trois p et deux d pour former six orbitales dont les liaisons sont dirigées vers les sommets d'un octaèdre régulier (Fig. 29, II). Ce cas est celui des métaux M des séries de transition (séries du fer, du palladium et du platine). C'est le cobalt trivalent qui a donné le plus grand nombre de composés de ce genre, appelés souvent complexes de Werner. Cependant, on ne connait pas d'atome asymétrique octaédrique au sens défini plus haut, c'est-à-dire à six atomes ou groupements atomiques tous différents. Mais il existe des complexes dépourvus de tout élément de symétrie. Les composés du type $[MA_2RR']$ cis (Fig. 29, III) où A désigne un groupe bivalent, R et R' des groupes monovalents, sont asymétriques. Exemples: $[Coen_2NH_3Cl]^{++}$ [1] (en = éthylène-diamine), $[Ru(C_2O_4)_2(C_5H_5N)(NO)]^-$ [2].

Il en est de même des composés $[MAR_2R_2']$ cis-cis (Fig. 29, IV). Exemples: $[Coen(NH_3)_2Cl_2]^+$ [25].

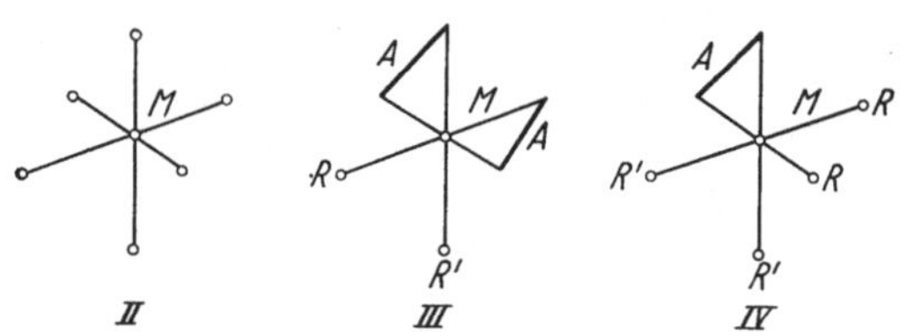

$$II \qquad\qquad III \qquad\qquad IV$$

Fig. 29. Complexes octaédriques asymétriques.

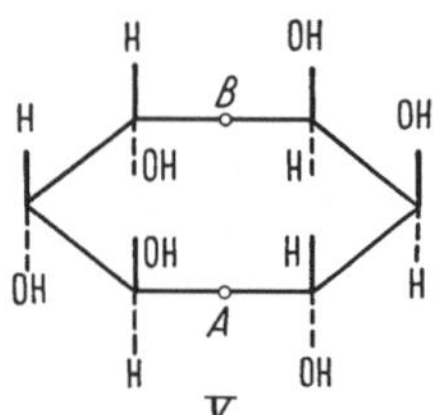

$$V$$

Fig. 30. Inositol actif (l'axe binaire passe par les points A et B).

20. Dissymétrie moléculaire sans asymétrie. Quelle que soit l'importance de la notion d'atome asymétrique, surtout dans le domaine de la chimie organique, elle s'est montrée trop étroite pour rendre compte de tous les cas d'activité optique. Au contraire, la notion de dissymétrie moléculaire (absence de plan, de centre, d'axe de symétrie alterne) sert toujours de principe directeur.

α) L'exemple de l'acide tartrique racémique (Sect. 18) montre bien que la présence d'atomes de carbone asymétriques dans une molécule ne suffit pas à lui conférer dans tous les cas l'activité optique, ainsi que l'avait vu le Bel [3].

β) C'est en partant de ces principes qu'il faut étudier l'isomérie optique des composés organiques à chaînes fermées, dans lesquels il peut y avoir énantiomorphie sans carbone asymétrique. Exemple: inositols, une forme active, de symétrie C_2 (Fig. 30).

Pour les systèmes polycycliques, voir [5].

γ) *Allènes.* Selon une remarque de van't Hoff, les composés $R_1R_2C:C:CR_1R_2$, dérivés de l'allène $H_2C:C:CH_2$ sont asymétriques (Fig. 31). Exemple:

$$\begin{array}{c} C_6H_5 \\ C_{10}H_7 \end{array}\!\!\!>\!C\!=\!C\!=\!C\!<\!\!\!\begin{array}{c} C_6H_5 \\ CO_2CH_2CO_2H \end{array} \qquad (V)$$

ester glycolique de l'acide diphényl 1—3 — naphtyl 1 allène carbonique 3 [4].

δ) *Spiranes.* Ce sont des composés bicycliques, dont les deux cycles, qui ont un atome de carbone en commun, ne sont pas coplanaires, grâce à la disposition tétraédrique des valences de l'atome de carbone.

---

[1] A. Werner: Ber. dtsch. Chem. Ges. **44**, 3275 (1911).
[2] R. Charonnat: C. R. Acad. Sci., Paris **178**, 1279, 1423 (1924).
[3] A. le Bel: Bull. Soc. chim. France **22**, 337 (1874).
[4] Kohler: J. Amer. Chem. Soc. **57**, 1743 (1935).

Le premier exemple de ce genre fut donné par un composé carboné: la céto-dilactone de l'acide benzophénone tetracarboxylique[1]:

$$H_2OC \langle \text{—} \rangle \overset{O\text{—}OC}{\underset{CO\text{—}O}{C}} \langle \text{—} \rangle CO_2H \qquad (VI)$$

On connait des composés du même type pour les éléments suivants: N, B, Be, Cu, Zn. Leur existence s'accorde avec la disposition tétraédrique des valences, prévue par la théorie des liaisons dirigées. Des expériences de dédoublement qui avaient paru conduire à la même conclusion pour les composés de Pt et Pd, contrairement aux prévisions de la théorie de la valence, ne résistent pas à l'examen. Pour l'histoire de ce débat, voir [19].

ε) On peut regarder comme intermédiaires entre allènes et spiranes des composés dans lesquels un cycle est uni à une chaîne latérale par l'intermédiaire d'une double liaison.

Exemple: acide méthyl 1 cyclohexylidène 4 acétique[2]:

$$\underset{H_3C}{\overset{H}{>}}C\underset{C\ C}{\overset{C\ C}{<}}C=CHCO_2H \qquad (VII)$$

ζ) *Dissymétrie par empêchement stérique.* Dans la molécule de diphényle H_5C_6—C_6H_5 les cycles sont alignés (Fig. 32, VIII) car les dérivés disubstitués en para ont tous un moment électrique nul. Mais s'il existe au moins deux groupements substituants R_1 et R_2, dans les positions indiquées par la figure, et si leur encombrement est assez grand, les deux cycles ne peuvent plus être coplanaires; dans toute position qu'ils prennent l'un par rapport à l'autre autour de la liaison qui les unit, il existe une configuration énantiomorphe.

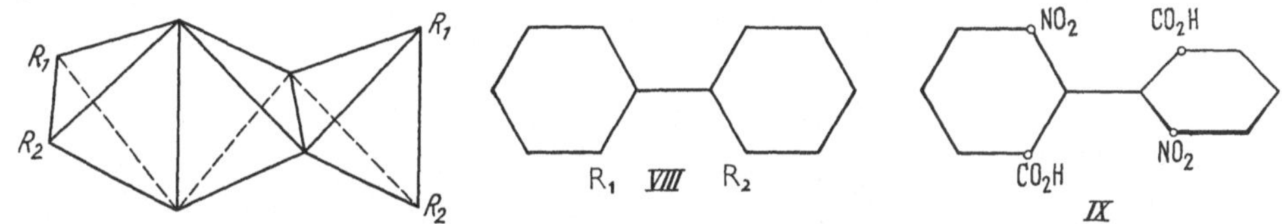

Fig. 31. Allène.   Fig. 32a et b. a) Diphényle et b) acide dinitro 2-2'-diphénique.

Exemple: acide 2 2' dinitrodiphénique (Fig. 32, IX)[3]. Pour un exposé détaillé, voir [7].

η) Les cas de dissymétrie sont plus nombreux que ceux d'asymétrie chez les complexes de WERNER. On connait des ions complexes dissymétriques des types suivants (où R_1', R', désignent des atomes ou des groupes monovalents, A un groupe bivalent: en = éthylènediamine, dip = dipyridyle).

1. Type [MA_3], symétrie D_3 (Fig. 33, X). Exemples [Coen_3]Cl_3[4] où Co peut être remplacé par Cr, Rh, Ir; [Fedip_3]Cl_2[5]; [Co(C_2O_4)_3]K_3[6] où Co peut être remplacé par Cr, Rh, Ir et même Al; $\left[As\left(\overset{O}{\underset{O}{>}}C_6H_4\right)_3\right]H$[7]; $\{Co[(OH)_2Co(NH_3)_4]_3\}Cl_6$[8].

---

[1] MILLS et NODDER: J. Chem. Soc., Lond. **117**, 1407 (1920).
[2] PERKIN, POPE et WALLACH: J. Chem. Soc., Lond. **95**, 1789 (1909).
[3] CHRISTIE et KENNER: J. Chem. Soc., Lond. **121**, 614 (1922).
[4] A. WERNER: Ber. dtsch. Chem. Ges. **45**, 121 (1912).
[5] A. WERNER: Ber. dtsch. Chem. Ges. **45**, 433 (1912).
[6] F. M. JAGER: Rec. Trav. chim. Pays-Bas **38**, 180 (1919).
[7] A. ROSENHEIM et W. PLATO: Ber. dtsch. Chem. Ges. **58**, 2000 (1925).
[8] A. WERNER: C. R. Acad. Sci., Paris **159**, 426 (1914).

2. Type [MA_2A'], symétrie C_2 (Fig. 34, XI). Exemple [$Coen_2CO_3$]Cl[1].

3. Type [MA_2R_2]cis, symetrie C_2 (Fig. 34, XII). Exemples: [$Coen_2(NH_3)_2$]Cl, [$Ir(C_2O_4)_2Cl_2$]K_3.

Lorsque les groupes A sont eux-mêmes dissymétriques, les isomères optiques deviennent plus nombreux; voir [*19*].

Il existe des composés contenant deux atomes à structure octaédrique (complexes à deux noyaux). Ils sont reliés par un ou plusieurs groupements atomiques formant ponts entre les deux noyaux. La série la mieux étudiée est celle des complexes à deux ponts:

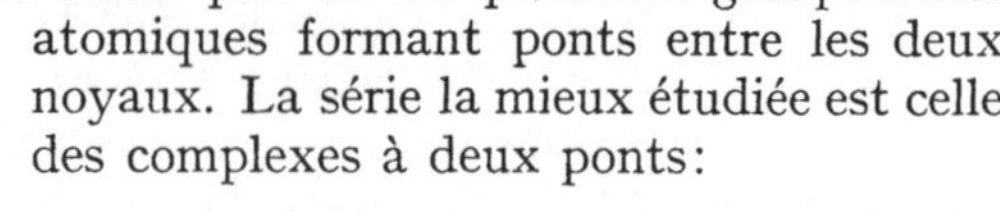

$$\left[en_2Co \!<\!\genfrac{}{}{0pt}{}{R}{R'}\!>\! Coen_2 \right] . \qquad \text{(XIII)}$$

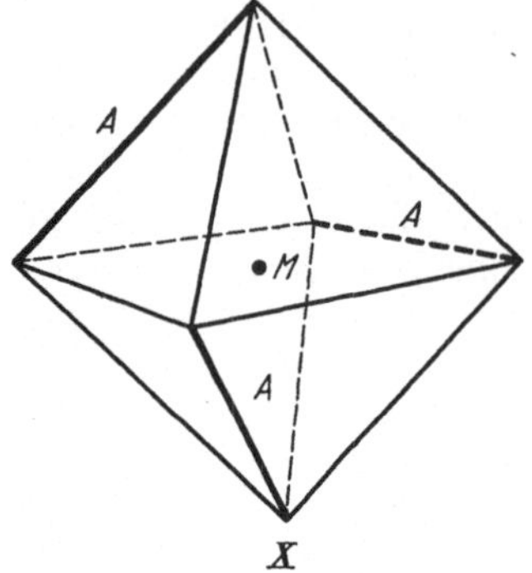

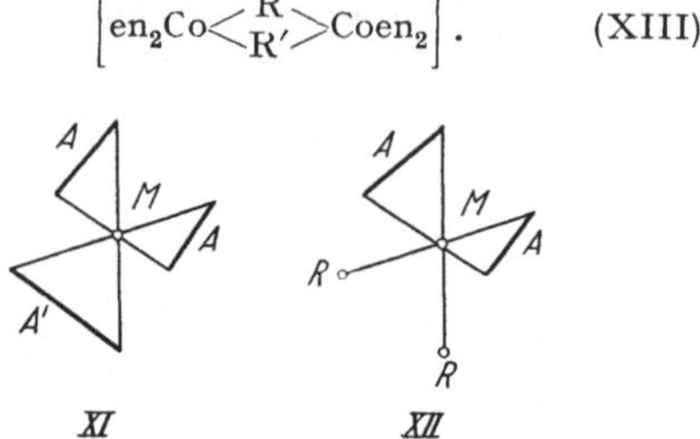

Fig. 33. Complexe octaédrique de Werner de symétrie D_3.   Fig. 34. Complexes octaédriques dissymétriques.

Si les deux atomes de cobalt sont trivalents, les configurations des deux groupements $>Coen_2$ peuvent être identiques ou énantiomorphes. Il doit alors exister trois formes isomères, deux énantiomorphes et une méso, comme pour l'acide tartrique (Fig. 27). Werner[2] a isolé celles du composé:

$$Br_4 \left[en_2Co \!<\!\genfrac{}{}{0pt}{}{NH_2}{NO_2}\!>\! Coen_2 \right] . \qquad \text{(XIV)}$$

Pour plus de détails voir [*19*], [*22*].

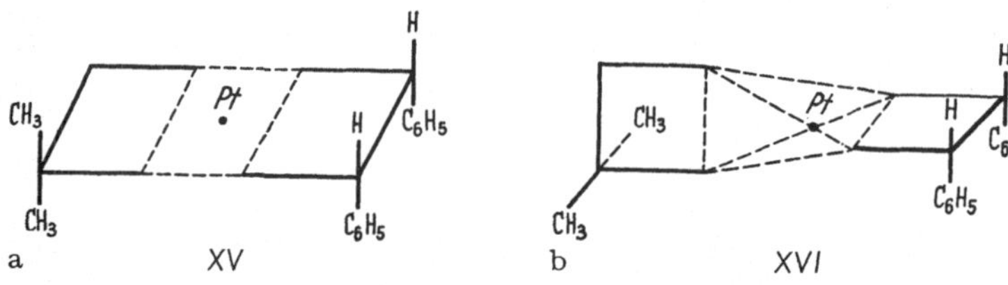

Fig. 35a et b. Configuration a) plane ou b) tétraédrique des complexes du platine.

$\vartheta)$ En combinant l'orbitale $6s$ avec trois orbitales $6p$ et une orbitale $5d$ de l'ion Pt^{++}, on peut former quatre liaisons fortes, dirigées suivant les sommets d'un carré. Le dédoublement du chlorure de platine II isobutylène-diamine-méso-stilbènediamine[3] s'accorde avec cette configuration, car l'ion est asymétrique (Fig. 35, XV), tandis que le spirane formé par une configuration tétraédrique serait indédoublable (Fig. 35, XVI).

Les mêmes prévisions, valables pour Ni^{++} n'ont pas reçu, dans ce cas, de confirmation expérimentale par l'activité optique.

21. Configuration des molécules dissymétriques et activité optique. α) Déterminer la *configuration absolue* d'un composé optiquement actif, c'est rattacher

<hr>

[1] A. Werner: Ber dtsch. Chem. Ges. **44**, 3279 (1911).

[2] A. Werner: Ber. dtsch. Chem. Ges. **46**, 3674 (1913).

[3] W. H. Mills et T. H. Quibell: J. Chem. Soc., Lond. **1935**, 839.

le signe du pouvoir rotatoire de l'un de ses antipodes optiques à l'un des deux schémas énantiomorphes de structure moléculaire. Ce problème a donné lieu à des études théoriques (Sect. 41) et a reçu dans un cas une solution expérimentale (Sect. 57).

D'autre part, si l'on sait établir une relation expérimentale entre deux composés A et B optiquement actifs, on pourra, dans certains cas, déterminer les schémas de structure des énantiomorphes qui se correspondent dans le phénomène étudié et déterminer les *configurations relatives* des molécules A et B, même si l'on ne connaît pas leurs configurations absolues.

Du point de vue de la représentation stéréochimique, la comparaison des configurations n'a de signification que si elle porte sur des composés qui ne sont pas trop différents. On comprend, par exemple, ce qu'on entend lorsqu'on dit que les composés suivants ont la même configuration relative

CHO	CO$_2$H	CO$_2$H	CO$_2$H	CO$_2$H
H—C*—OH	H—C*—OH	H—C*—OH	H—C*—OH	H—C*—OH
CH$_2$OH	CH$_2$OH	CH$_2$NH$_2$	CH$_3$	OH—CH
				CO$_2$H
(XVII)	(XVIII)	(XIX)	(XX)	(XXI)
Aldéhyde	acide	$d(+)$ isosérine	acide	acide
$d(+)$glycérique	$(d-)$glycérique		$d(-)$ lactique	$d(+)$ tartrique

car la permanence du groupe $\overset{\text{C}}{\underset{\text{C}}{\text{H—C*—OH}}}$ permet de faire correspondre sans ambiguité les schémas de FISCHER de tous ces composés. Mais on ne sait déjà plus comment disposer par exemple le schéma

$$\begin{matrix} \text{CO}_2\text{H} \\ \text{H—C*—NH}_2 \\ \text{CH}_3 \\ d(-)\ \text{alanine} \end{matrix} \qquad\qquad \text{(XXII)}$$

par rapport au schéma XVII pour comparer leurs configurations. De même, la comparaison entre tous les complexes MA$_3$ de la Fig. 33 a un sens stéréochimique; mais la corrélation entre Fig. 29, IV et Fig. 33 n'en pas.

β) Du point de vue expérimental, il existe essentiellement deux méthodes permettant de relier l'un à l'autre les isomères actifs de deux composés:

1. Les transformations chimiques. On trouvera en [21] une mise au point récente de ce sujet immense. Nous devons nous borner à de brèves indications. Signalons simplement que l'on sait transformer par voie chimique les composés (XVII) à (XXI) les uns dans les autres. Par convention, on rapporte leurs configurations relatives à celle de l'aldéhyde glycérique, dont on désigne arbitrairement par d le schéma (XVII) de FISCHER. De même, on peut rattacher la configuration des acides aminés et de leurs dérivés à celle de l'alanine (XXII). Mais on a pu passer, par des réactions, de l'alanine à l'acide lactique (XX), lui même relié à l'aldéhyde glycérique et attribuer ainsi le schéma (XXII) à la série d.

Comme on connaît la configuration absolue de l'acide tartrique (Sect. 57) celle de l'aldéhyde glycérique est fixée aujourd'hui: le schéma (XVII) n'est plus arbitraire. L'affiliation chimique est rendue difficile par le fait que les réactions qui portent sur des groupes directement liés à l'atome de carbone asymétrique peuvent entrainer le passage d'une configuration à son inverse. On en a la preuve par le fait que certains cycles de réactions qui ramènent au composé initial donnent l'antipode optique du corps dont on est parti. C'est *l'inversion de* Walden, phénomène qui a donné lieu à de nombreux travaux, mais est encore imparfaitement éclairci [5], [21]. Il empêche, par exemple, de rattacher la configuration de l'acide malique (XXIII) à celle de l'acide monochloracétique (XXIII bis)

$$
\begin{array}{ccc}
CO_2H & & CO_2H \\
| & & | \\
H-C*-OH & \text{et} & H-C*-Cl \\
| & & | \\
CH_2CO_2H & & CH_2CO_2H \\
(XXIII) & & (XXIII\ bis)
\end{array}
$$

car en partant de l'antipode dextrogyre de ce dernier composé, on obtient l'acide malique dextrogyre ou lévogyre, suivant qu'on fait agir sur lui l'hydroxyde d'argent ou la potasse.

2. *La formation de racémiques actifs.* Lorsqu'on fait cristalliser certains mélanges à parties égales de deux antipodes optiques fondus ou dissous, il peut arriver qu'on obtienne, non pas un mélange des cristaux des inverses optiques, mais des cristaux d'un composé *racémique*, dont la maille contient des nombres égaux de molécules d et g. Soient alors deux racémiques de composés isomorphes, que nous représenterons par $A_d + A_g$ et par $B_g + B_d$. Un *racémique actif* $A_d + B_g$ ou $A_g + B_d$ est formé par l'union deux à deux de molécules actives, d'espèces différentes, dont les configurations sont nécessairement inverses l'une de l'autre[1]. La méthode est limitée, mais elle est rigoureuse et elle peut s'appliquer à des cas où il est impossible de passer de A à B par des réactions chimiques, par exemple[2] à la comparaison des complexes $[MA_3]$ (Sect. 20η).

3. *Autres méthodes.* On a parfois admis que dans des séries de composés de structure voisine, ce sont les antipodes de même configuration qui donnent la combinaison diastéréo-isomére la moins soluble avec un même composé optiquement actif. La règle a été appliquée en chimie organique[3] et minérale[4]. D'autre part, les cristaux de quartz optiquement actif adsorbent de préférence l'un des antipodes optiques[5]. On a cherché[6] à relier par des raisonnements stériques la configuration de cet antipode à celle du quartz (Sect. 57). Les deux méthodes précédentes, qui sont l'une et l'autre fondées sur des considérations géométriques, sont d'une valeur incertaine[7].

γ) En ce qui concerne l'activité optique, on constate que le signe de $[\varrho]_D$ n'est pas le même pour tous les termes des séries de configurations relatives déterminées par l'une des deux méthodes précédentes, ainsi que l'indiquent les signes $+$ et $-$ qui accompagnent les schémas (XVII) à (XXI)[8]. *On ne peut donc nullement déduire de l'identité du signe du pouvoir rotatoire la similitude des configurations.*

On comprendra plus tard qu'une comparaison des activités optiques fondée sur le signe de $[\varrho]$ pour une même radiation n'a pas de signification claire pour des corps incolores dont la dispersion rotatoire peut n'être pas simple (Figs. 37 et

---

[1] M. Delepine: Bull. Soc. chim. France **29**, 656 (1921).

[2] M. Delepine: Bull. Soc. chim. France **1**, 1256 (1934).

[3] Ch. Winther: Ber. dtsch. Chem Ges. **28**, 3013 (1895).

[4] A. Werner: Bull. Soc. chim. France **11** (1912).

[5] G. M. Schwab, F. Rost et L. Rudolph: Kolloid. Z. **68**, 157 (1934). — R. Tsuchida M. Kobayashi et A. Nakamura: Bull. Soc. Chim. Japan **11**, 38 (1936).

[6] E. U. Condon, W. Altar et H. Eyring: J. Chem. Phys. **5**, 753 (1937).

[7] W. Kuhn: Z. Elektrochem. **56**, 506 (1952).

[8] La convention qui attache la lettre d ou l à la configuration, le signe $+$ ou $-$ à la rotation, n'est pas universellement adoptée.

44) et pour des corps colorés dont la rotation peut changer plusieurs fois de signe dans le spectre visible (Figs. 39 et 56). Il serait plus rationnel de comparer le signe de l'effet CotTON dans les bandes d'absorption correspondantes que possè‑ dent des composés ayant des spectres analogues, c'est-à-dire ayant des structures comparables [*12*].

Ce n'est donc que dans les séries très limitées de composés voisins, en tenant compte de règles semi-empiriques que l'on étudiera plus loin (Sect. 42 à 46), qu'une comparaison des configurations fondée sur l'activité optique est digne de confiance.

II. Théories générales de l'activité optique des molécules.

22. Théorie des électrons. Les théories moléculaires se sont développées après que LORENTZ, partant de l'existence et des propriétés mécaniques et électrodynamiques classiques des électrons, eût donné une interprétation de la propagation des ondes électromagnétiques dans la matière. Il faut alors remplacer les vecteurs «macroscopiques» $\boldsymbol{E}$ et $\boldsymbol{H}$ des équations de MAXWELL, par les champs «microscopiques» $\boldsymbol{e}$ et $\boldsymbol{h}$ qui règnent entre les particules chargées électriquement, ou même à l'intérieur de celles-ci, et ne pas faire intervenir les vecteurs $\boldsymbol{D}$ et $\boldsymbol{B}$, qui n'ont qu'une signification macroscopique, car ils expriment, d'après (4.9) et (4.10), des propriétés globales de la matière. Les équations de LORENTZ[1] pour un milieu non conducteur sont

$$\operatorname{rot} \boldsymbol{e} + \frac{1}{c}\,\dot{\boldsymbol{h}} = 0,$$
$$\operatorname{rot} \boldsymbol{h} - \frac{1}{c}\,\dot{\boldsymbol{e}} = 0, \qquad (22.1)$$
$$\operatorname{div} \boldsymbol{e} = \operatorname{div} \boldsymbol{h} = 0.$$

Fig. 36. Moment de dipôle électrique.

Une particule matérielle, de masse m et de charge e, est soumise de la part du champ électromagnétique à une force

$$\boldsymbol{F} = e\,\boldsymbol{e} + \frac{e}{c}\,\boldsymbol{v} \times \boldsymbol{h}, \qquad (22,2)$$

$\boldsymbol{v}$ désignant sa vitesse. La premier terme représente la force de COULOMB, le second la force de LAPLACE. Cette expression peut se déduire de l'application des équations de LAGRANGE au mouvement de la particule, dont l'énergie potentielle, en présence de l'onde électromagnétique, a pour expression

$$V = -\frac{e}{m\,c}\,\boldsymbol{\pi} \cdot \boldsymbol{A}. \qquad (22.3)$$

$\boldsymbol{\pi} = m\boldsymbol{v}$ est la quantité de mouvement de la particule, $\boldsymbol{A}$ le potentiel vecteur de l'onde à l'endroit où se trouve la particule. Lorsque la force $\boldsymbol{F}$ déplace une particule depuis sa position d'équilibre en A, définie par un vecteur $\boldsymbol{r}^0$, jusqu'en B ($\overline{BA} = \boldsymbol{r}$), cela revient (Fig. 36) à superposer à l'état initial ($-e$ en A) un dipôle électrique induit ($+e$ en A, $-e$ en B) dont le moment a pour valeur

$$\boldsymbol{p} = e\,\boldsymbol{r}. \qquad (22.4)$$

D'autre part, un électron qui décrit une trajectoire fermée équivaut à un dipôle magnétique dont le moment moyen a pour valeur

$$\boldsymbol{m}_0 = \frac{e}{2c}\,\overline{\boldsymbol{r}^0 \times \dot{\boldsymbol{r}}^0} = \frac{e}{2mc}\,\overline{\boldsymbol{r}^0 \times \boldsymbol{\pi}}. \qquad (22.5)$$

---

[1] H. A. LORENTZ: The Theory of electrons. Leipzig 1916.

On tient compte du déplacement r produit par l'onde, en remplaçant r^0 par $r^0 + r$. Si le moment m_0 est nul, le moment magnétique induit peut se mettre sous la forme[1]

$$m = \frac{e}{c}\,\overline{r^0 \times \dot{r}}. \tag{22.6}$$

Il faut maintenant établir les relations qui existent entre les équations (4.1) à (4.4) et les équations (22.1) à (22.6). On peut démontrer[2] les formules suivantes

$$E = \bar{e}, \qquad B = \bar{h}, \tag{22.7}$$

$$P = N\,p, \qquad M = N\,m \tag{22.8}$$

d'où

$$D = \bar{e} + 4\pi N\,\bar{p}, \tag{22.9}$$

$$H = \bar{h} - 4\pi N\,\bar{m}. \tag{22.10}$$

23. Théorie électronique de Drude. α) Lorentz ne considérait pas explicitement que les électrons font partie d'atomes ou de molécules; il admettait simplement que, dans un milieu diélectrique, ils sont rappelés vers des positions d'équilibre par des forces $/r$ isotropes et proportionnelles à l'écart r. L'équation du mouvement pris par un électron soumis à l'action de l'onde, supposée réduite à celle du champ E, est

$$m\,\ddot{r} + /r = e\,E. \tag{23.1}$$

Pour une onde monochromatique de fréquence ν

$$E = E_0 \cos 2\pi \nu t = \frac{E_0}{2}\,(e^{2\pi i\nu t} + e^{-2\pi i\nu t}). \tag{23.2}$$

La solution de l'équation (23.1) est un déplacement sinusoïdal

$$r = \frac{e\,E_0}{4\pi^2 m\,(\nu_k^2 - \nu^2)}\cos(2\pi\nu t + \varphi),$$

où $\nu_k^2 = 4\pi^2\,/m$ est la fréquence propre du mouvement libre. Le moment électrique correspondant,

$$p = e\,r$$

peut s'écrire

$$p = \alpha\,E,$$

en posant

$$\alpha = \alpha_0\,e^{i\varphi} = \frac{e^2}{4\pi^2 m\,(\nu_k^2 - \nu^2)}\,e^{i\varphi}. \tag{23.3}$$

α est la *polarisabilité* de la particule. On calcule ensuite la polarisation $P = N\,p$ et le déplacement

$$D = 1 + 4\pi P = \varepsilon\,E \tag{23.4}$$

avec

$$\varepsilon = 1 + \frac{N e^2}{\pi m\,(\nu_k^2 - \nu^2)}\,e^{i\varphi}. \tag{23.5}$$

β) Drude[3] admettait que les électrons d'une molécule active pouvaient se déplacer suivant des trajectoires en hélice, et parvenait[4] à ajouter au second

---

[1] M. Born: Optik. Berlin 1933.

[2] J. H. van Vleck: Electric and magnetic susceptibilities. Oxford 1932.

[3] P. Drude: Göttinger Nachr. **1892**, 400.

[4] W. Kuhn [Z. phys. Chem., Abt. B **20**, 325 (1933)] a montré que le raisonnement de Drude contient une erreur qui annule l'activité optique du modèle sur lequel il a raisonné.

membre de l'équation (23.1) un terme $\gamma_k \operatorname{rot} \boldsymbol{E}$, γ_k désignant un coefficient de proportionnalité. On obtient alors la relation (4.22) avec

$$j_k = \frac{N e^2 \gamma_k}{\pi m (v_k^2 - v^2)},\tag{23.6}$$

et d'après (4.28) la valeur du pouvoir rotatoire

$$\varrho = \frac{2\pi N e^2}{m \lambda_0^2} \frac{\gamma_k}{v_k^2 - v^2}.\tag{23.7}$$

Les équations de DRUDE sont les premières qui aient fait intervenir les fréquences propres. Mais les paramètres γ_k sont indéterminés et l'addition du terme $\gamma_k \operatorname{rot} \boldsymbol{E}$ est arbitraire.

24. Théorie moléculaire classique. Un grand progrès fut réalisé dans l'étude de la propagation de la lumière à travers un milieu matériel, lorsqu'on groupa les particules de LORENTZ en molécules, généralement anisotropes, considérées comme les unités élémentaires du milieu.

α) La molécule, en dehors de l'action de tout champ électrique ou magnétique, peut posséder un moment de dipôle électrique permanent $\boldsymbol{p}_0$ (molécule polaire) ou un moment magnétique permanent $\boldsymbol{m}_0$ (molécule paramagnétique). Ces moments n'interviennent pas dans l'activité optique. Par contre, il importe de connaître les valeurs des moments, électrique et magnétique, induits dans la molécule par le déplacement des particules sous l'action d'un champ électromagnétique. Soit n le nombre de particules contenues dans une molécule, $\boldsymbol{r}_k^0$ le rayon vecteur qui définit la position moyenne de la particule d'indice k dans la molécule soustraite au champ. L'ensemble de ces rayons vecteurs définit une configuration moyenne d'équilibre de la molécule, sur laquelle nous raisonnerons dans ce qui suit, comme si elle était statique. Le moment de dipôle électrique permanent est

$$\boldsymbol{p}_0 = \sum_{k=1}^n e_k \boldsymbol{r}_k^0.\tag{24.1}$$

Le moment de dipôle magnétique permanent est donné par une généralisation de la formule (22.5)

$$\boldsymbol{m}_0 = \frac{1}{2c} \sum_{k=1}^n e_k \overline{\boldsymbol{r}_k^0 \times \dot{\boldsymbol{r}}_k^0}.\tag{24.2}$$

Si $\boldsymbol{r}_k$ est le vecteur qui définit le déplacement de la particule k produit par le champ de l'onde, le moment électrique de dipôle induit par l'onde a pour expression

$$\boldsymbol{p} = \sum_k \boldsymbol{p}_k = \sum_k e_k \boldsymbol{r}_k.\tag{24.3}$$

Pour une molécule diamagnétique, le moment $\boldsymbol{m}_0$ est nul; le moment magnétique induit est donné par une formule analogue à (22.6)

$$\boldsymbol{m} = \frac{1}{c} \sum_k e_k (\boldsymbol{r}_k^0 \times \dot{\boldsymbol{r}}_k).\tag{24.4}$$

β) D'autre part, on tient compte de l'anisotropie moléculaire[1] en posant, comme dans le cas des cristaux (Sect. 5), que le moment $\boldsymbol{p}$ est une fonction linéaire

---

[1] P. LANGEVIN: Radium **7**, 249 (1910).

vectorielle du champ E, la *polarisation moléculaire* α étant un tenseur du second ordre, asymétrique dans le cas général.

$$p = [\alpha]\, E \tag{24.5}$$

ou

$$p_j = \sum_h \alpha_{jh}\, E_h. \tag{24.6}$$

Les coefficients α_{jh} sont en général des grandeurs complexes, comme la polarisabilité de la relation (23.3). Deux coefficients tels que α_{jh} et α_{hj} sont des quantités complexes conjuguées: cela résulte de ce que l'énergie de polarisation de la molécule doit avoir une valeur réelle. Le tenseur $[\alpha]$ peut alors se décomposer en une partie symétrique $[\alpha]^s$, dont les coefficients sont réels, et une partie antisymétrique $[\alpha]^a$, dont les coefficients sont purement imaginaires, et qui est équivalente à un vecteur axial $\boldsymbol{d}$. On peut écrire une relation analogue à (5.4):

$$p_j = \sum_h \alpha_{jh}\, E_h + i\,(\boldsymbol{d} \times \boldsymbol{E}_j). \tag{24.7}$$

γ) Le passage de la formule (24.6) à la formule (24.7) se fait dans les théories de Born[1], Oseen[2], de Mallemann [*18*], sans recourir à des hypothèses trop particulières ou arbitraires. Ces théories partent toutes des deux propositions suivantes:

1. La longueur d'onde des radiations est grande devant les dimensions des molécules, mais n'est pas regardée comme infinie.

2. Les particules constituant la molécule sont couplées entre elles et leurs mouvements suivent les lois de la mécanique classique.

Nous suivrons essentiellement le mode d'exposition de Born [*2*].

25. Equations des mouvements internes d'une molécule. Mouvements libres. L'énergie potentielle V du système des forces qui s'exercent entre les n particules constituant la molécule est supposée nulle à l'équilibre. Si les déplacements $r_k(x_k, y_k, z_k)$ des particules à partir de leurs positions d'équilibre définies par les rayons vecteurs r_k^0 sont assez petits, V est une fonction quadratique de ces déplacements.

$$2V = \sum_{kl}\sum_{xy} f_{xy}^{kl}\, x_k\, y_l. \tag{25.1}$$

Les forces étant supposées former un système conservatif, les coefficients f, au nombre de $3n^2$, satisfont à des relations du genre

$$f_{xy}^{kl} = f_{yx}^{lk}. \tag{25.2}$$

L'énergie cinétique T a pour expression

$$2T = \sum_k\sum_x m_k\, \dot{x}_k^2. \tag{25.3}$$

Les équations de Lagrange relatives aux mouvements libres des particules sont de la forme

$$m_k\, \ddot{x}_k + \sum_l\sum_y f_{xy}^{kl}\, y_l = 0. \tag{25.4}$$

Ces équations sont au nombre de $3\,n$. Elles admettent des solutions harmoniques de la forme

$$x_k = x_k^0\, e^{2\pi i\nu t}.$$

---

[1] M. Born: Ann. Phys., Lpz. **55**, 177 (1918).
[2] C. W. Oseen: Ann. Phys., Lpz. **48**, 1 (1915).

En substituant ces solutions dans (25.4) et en posant

$$u_{kx} = \sqrt{m_k}\, x_k, \qquad f'^{kl}_{xy} = \frac{1}{\sqrt{m_k m_l}}\, f^{kl}_{xy}, \qquad \omega = 2\pi\nu, \tag{25.5}$$

ces équations s'écrivent

$$-\omega^2 u^0_{kx} + \sum_l \sum_y f'^{kl}_{xy} u^0_{yl} = 0. \tag{25.6}$$

La condition de compatibilité des $3n$ équations fournit $3n$ valeurs de ω^2, dont les racines positives représentent les *pulsations fondamentales* des $3n$ modes de mouvement harmoniques appelés *oscillations fondamentales*, auxquels prennent part toutes les particules.

On peut introduire $3n$ coordonnées q_j appelées *coordonnées normales* ($j = 1, 2, \ldots, 3n$), qui permettent d'écrire T et V comme des sommes de carrés

$$2V = \sum_j \omega^2_j q^2_j,$$

$$2T = \sum_j \dot{q}^2_j$$

et les équations (25.4) sont remplacées par

$$\ddot{q}_j + \omega^2_j q_j = 0. \tag{25.7}$$

Les q sont reliées aux u par les transformations linéaires suivantes

$$u_{kx} = \sum_j a^j_{kx} q_j, \tag{25.8}$$

$$q_j = \sum_k \sum_x a^j_{kx} u_{kx} \tag{25.9}$$

les coefficients a^j_{kx} étant les éléments d'une matrice de $3n$ lignes et $3n$ colonnes satisfaisant aux conditions d'orthogonalité suivantes

$$\left.\begin{aligned} \sum_{k=1}^{n} (a^i_{kx} a^i_{kx} + a^i_{ky} a^i_{ky} + a^i_{kz} a^i_{kz}) &= 0, \\ \sum_{j=1}^{3n} a^j_{kx} a^j_{ly} &= 0. \end{aligned}\right\} \tag{25.10}$$

26. Mouvements contraints. Sous l'action d'une force extérieure $\boldsymbol{F}_k\,(F_{kx}, F_{ky}, F_{kz})$, les équations telles que (25.4), sont remplacées par

$$m_k \ddot{x}_k + \sum_l \sum_y f^{kl}_{xy} y_l = F_{kx}. \tag{26.1}$$

Ces équations peuvent se transformer. En posant

$$\mathfrak{F}_j = \sum_l \sum_y a^j_{ly} \frac{F_{ly}}{\sqrt{m_l}},$$

les équations (25.6) sont remplacées par

$$\ddot{q}_j + \omega^2_j q_j = \mathfrak{F}_j.$$

En utilisant les formules (25.5) et (25.8) pour revenir des coordonnées q aux coordonnées cartésiennes, il vient

$$x_k = \frac{1}{\sqrt{m_k}} \sum_j \frac{a^j_{kx}}{\omega^2_j - \omega^2} \sum_l \sum_y a^j_{ly} \frac{F_{yl}}{\sqrt{m_l}}. \tag{26.2}$$

Posons

$$K_{xy}^{kl} = \frac{1}{\sqrt{m_k m_l}} \sum_j \frac{a_{kx}^j\, a_{ly}^j}{\omega_j^2 - \omega^2}.$$ (26.3)

Pour des indices k et l déterminés, les coefficients K_{xy}^{kl} sont les composantes d'un tenseur, généralement asymétrique. On a

$$K_{xy}^{kl} = K_{yx}^{lk}.$$ (26.4)

Finalement

$$x_k - \sum_l \sum_y K_{xy}^{kl} F_{yl} = 0.$$

La force que le champ électrique d'une onde monochromatique plane exerce sur la particule k a pour composante

$$F_{kx} = e_k E_{kx} = e_k E_x^0 \, e^{-\frac{2\pi i}{\lambda}\, r_k^0 s} \left(\frac{e^{2\pi i v t} + e^{-2\pi i v t}}{2} \right).$$ (26.5)

$\lambda = \lambda_0/n$ désigne la longueur d'onde dans le milieu d'indice n. Posons

$$x_k = x_k^0 \, e^{2\pi i v t} = X_k \, e_{\cdot}^{-\frac{2\pi i}{\lambda}\, r_k^0 \cdot s} \left(\frac{e^{2\pi i v t} + e^{-2\pi i v t}}{2} \right).$$

X_k est la composante suivant $0x$ de *l'amplitude relative* de la particule, en phase avec E_{kx}. Les équations (26.1) deviennent

$$X_k = \sum_l \sum_y K_{xy}^{kl} e_l E_y^0 \, e^{\frac{2\pi i}{\lambda}\, (r_k^0 - r^0)\, s}.$$ (26.6)

27. Calcul du moment électrique induit. *Le moment électrique relatif* induit dans la molécule, dont l'amplitude a, par définition, pour composante suivant $0x$

$$p_x'^0 = \sum_k e_k X_k$$ (27.1)

a pour expression

$$p_x'^0 = \sum_{kl} \sum_y K_{xy}^{kl} e_k e_l E_y^0 \, e^{\frac{2\pi i}{\lambda}\, (r_k^0 - r_l^0)\, s}.$$

Introduisons maintenant la seconde hypothèse fondamentale de la théorie moléculaire. On peut développer l'expression du champ $\boldsymbol{E}$ contenue dans (26.5) en fonction des puissances croissantes du rapport $\dfrac{r_k^0 \cdot \boldsymbol{s}}{\lambda}$ et, comme ce rapport est toujours petit (de l'ordre de 10^{-3} à 10^{-4} pour les radiations visibles), limiter le développement à ses deux premiers termes. On obtient ainsi, pour le moment $\boldsymbol{p}'$

$$p_x' = \sum_{kl} \sum_y K_{xy}^{kl} e_k e_l E_y \left[1 + \frac{2\pi i}{\lambda} (r_k^0 - r_l^0) \cdot \boldsymbol{s} \right] = \sum_y \alpha_{xy} E_y + \sum_{yz} \alpha_{xy,z} E_y s_z$$ (27.2)

en posant

$$\alpha_{xy} = \sum_{kl} K_{xy}^{kl} e_k e_l$$

et

$$\alpha_{xy,z} = \frac{2\pi i}{\lambda} \sum_{kl} K_{xy}^{kl} e_k e_l (z_k^0 - z_l^0).$$

En tenant compte de (26.4) on peut écrire

$$\alpha_{xy} = \alpha_{yx} = \tfrac{1}{2} \sum_{kl} (K^{kl}_{xy} + K^{kl}_{yx})\, e_k\, e_l = \sum_{kl} S^{kl}_{xy}\, e_k \cdot e_l, \tag{27.3}$$

$$\left.\begin{aligned}
\alpha_{xy,z} = -\,\alpha_{yx,z} &= -\frac{2\pi i}{\lambda} \sum_{kl} \frac{(K^{kl}_{xy} - K^{kl}_{yx})}{2} (z^0_k - z^0_l)\, e_k\, e_l \\
&= \frac{2\pi i}{\lambda} \sum_{kl} A^{kl}_{xy} (z^0_k - z^0_l)\, e_k\, e_l.
\end{aligned}\right\} \tag{27.4}$$

S et A désignent respectivement la partie symétrique et la partie antisymétrique du tenseur des coefficients K. Les expressions (27.3) donnent les composantes du tenseur de polarisabilité ordinaire, dont dépend la réfraction. Ce sont les expressions (27.4), qui s'annulent pour $\lambda = \infty$, dont dépend l'activité optique. On peut remplacer dans (27.4) le terme A^{kl}_{xy} par la composante A^{kl}_z d'un vecteur axial $\boldsymbol{A}^{kl}$; d'autre part, on remarque que tous les termes tels que $\alpha_{xx,z}$, dont les deux premiers indices sont égaux, s'annulent, puisque A est un tenseur antisymétrique. Les relations (27.4) définissent donc seulement neuf coefficients distincts, que l'on peut écrire, par exemple

$$\alpha_{xy,z} = -\,\alpha_{yx,z} = -\,i\,g_{zz}. \tag{27.5}$$

Les neuf quantités $g_{ij}\,(i, j = x, y, z)$ sont les composantes d'un tenseur $[g]$ asymétrique, le *tenseur de gyration* [2] ou *tenseur rotatoire*[1].

Le second terme de la formule (27.2), qui définit la partie antisymétrique p'_{xa} du moment induit, dont dépend l'activité optique, s'écrit, en tenant compte des remarques précédentes

$$\left.\begin{aligned}
p'_{xa} = \sum_{yz} \alpha_{xy,z}\, E_y\, s_z &= -\,i\,E_y (g_{zx}\, s_x + g_{zy}\, s_y + g_{zz}\, s_z) + \\
&\quad + i\,E_z (g_{yx}\, s_x + g_{yy}\, s_y + g_{yz}\, s_z).
\end{aligned}\right\} \tag{27.6}$$

Les parenthèses représentent les composantes δ_z et δ_y d'un vecteur $\boldsymbol{\delta}$ que l'on obtient en appliquant le tenseur $[g]$ au vecteur $\boldsymbol{s}$. On a donc

$$p'_{ax} = i\,(\delta_z\, E_y - \delta_y\, E_z) = i\,(\boldsymbol{\delta} \times \boldsymbol{E})_x$$

et

$$p'_x = \sum_y \alpha_{xy}\, E_y + i\,(\boldsymbol{\delta} \times \boldsymbol{E})_x. \tag{27.7}$$

On retrouve ainsi la formule (24.7).

28. Expression classique du pouvoir rotatoire des gaz et des liquides. $\alpha)$ Dans un gaz parfait, les molécules peuvent prendre, indépendamment les unes des autres et avec une égale probabilité, toutes les orientations possibles. On peut écrire pour la polarisation induite

$$\boldsymbol{P} = N\,\overline{\boldsymbol{p}}', \tag{28.1}$$

$\overline{\boldsymbol{p}}'$ désignant la valeur du moment électrique moyen induit dans une molécule par le champ $\boldsymbol{E}$. La valeur de $\boldsymbol{E}$ et celles des composantes des tenseurs $[\alpha]$ et $[g]$-étant des constantes pour des molécules données, le calcul des moyennes se ramène à celui de combinaisons de cosinus directeurs [2], [20]. On trouve

$$\overline{\boldsymbol{p}}' = \overline{\boldsymbol{p}}'_s + \overline{\boldsymbol{p}}'_a = \alpha\,\boldsymbol{E} + i\,g\,(\boldsymbol{s} \times \boldsymbol{E}). \tag{28.2}$$

---

[1] R. DE MALLEMANN: Trans. Faraday Soc. **26**, 281 (1930).

α est la *polarisabilité moyenne de la molécule:*

$$\alpha = \tfrac{1}{3}\left(\alpha_x + \alpha_y + \alpha_z\right),\tag{28.3}$$

g est la *constante de gyration*

$$g = \tfrac{1}{3}\left(g_{xx} + g_{yy} + g_{zz}\right).\tag{28.4}$$

D'après (27.4) et (27.5) et en tenant compte de l'expression d'un produit scalaire, on peut écrire

$$g = \frac{2\pi}{3\lambda}\sum_{kl} \boldsymbol{A}^{kl}\cdot\left(\boldsymbol{r}_k^0 - \boldsymbol{r}_l^0\right) e_k\,e_l.\tag{28.5}$$

Le signe Σ porte sur le produit scalaire du vecteur axial $\boldsymbol{A}^{kl}$ par le vecteur polaire $(\boldsymbol{r}_k^0 - \boldsymbol{r}_l^0)$. La constante g a donc la dissymétrie caractéristique des paramètres rotatoires (Sect. 3).

Les formules (28.1), (28.2) et (4.5) donnent

$$\boldsymbol{D} = \varepsilon\,\boldsymbol{E} + i\,\Gamma(\mathbf{s}\times\boldsymbol{E})$$

avec

$$\varepsilon - 1 = 4\pi N\alpha\tag{28.6}$$

et

$$\Gamma = 4\pi N g.\tag{28.7}$$

Pour une onde plane, on a (Sect. 4) dans les régions de transparence où $\mathfrak{N} = n$

$$\mathrm{rot}\,\boldsymbol{E} = -\frac{2\pi i n}{\lambda_0}\,\mathbf{s}\times\boldsymbol{E}.$$

On retrouve la relation (4.22) en posant

$$j = \frac{\Gamma\lambda_0}{2\pi n} = \frac{2 N g\lambda_0}{n}.$$

D'où l'expression (4.27) de la biréfringence circulaire et celle (4.28) du pouvoir rotatoire par unité de longueur

$$\varrho = \frac{\pi}{\lambda_0}\left(n_g - n_d\right) = \frac{4\pi^2 N}{n\lambda_0}\,g.\tag{28.8}$$

β) Dans un milieu condensé isotrope, le champ agissant $\boldsymbol{E}'$ sur une molécule n'est pas égal au champ $\boldsymbol{E}$ de l'onde; il faut tenir compte du champ créé par les molécules. Celui-ci se compose, suivant Lorentz, du *champ de polarisation* produit par les molécules extérieures à une sphère de diamètre grand devant les dimensions moléculaires et centrée sur la molécule considérée et du *champ moléculaire* produit par les molécules intérieures à cette sphère.

Born [2] néglige le champ moléculaire et évalue le champ de polarisation suivant la méthode de Lorentz

$$\boldsymbol{E}' = \boldsymbol{E} + \frac{4\pi}{3}\,\boldsymbol{P} = \frac{\varepsilon + 2}{3}\,\boldsymbol{E}\tag{28.9}$$

car

$$\boldsymbol{P} = \frac{\varepsilon - 1}{4\pi}\,\boldsymbol{E}.\tag{28.10}$$

En portant la valeur de $\boldsymbol{E}'$ dans (28.2), on trouve, au lieu de (28.8), avec $\varepsilon = n^2$

$$\varrho = \frac{4\pi^2 N}{n\lambda_0}\,\frac{n^2 + 2}{3}\,g.\tag{28.11}$$

Kooy[1] remarque que la relation (28.11), fondée sur l'hypothèse d'une polarisation uniforme du milieu, n'est valable que si l'on néglige l'activité optique. En posant, au lieu de (28.2)

$$\boldsymbol{p}'\left(1 - \frac{4\pi}{3}\alpha\right) = \alpha\,\boldsymbol{E} + i\,g\cdot\boldsymbol{s}\times\left(\boldsymbol{E} + \frac{4\pi}{3}\,\boldsymbol{P}\right)$$

et au lieu de (28.10)

$$\boldsymbol{P} = \frac{N\alpha}{1 - \dfrac{4\pi}{3}N\alpha}\,\boldsymbol{E}$$

on trouve, au lieu de (28.8)

$$\varrho = \frac{4\pi^2 N}{n\,\lambda_0}\left(\frac{n^2 + 2}{3}\right)^2 g. \tag{28.12}$$

De Mallemann[2] n'admet pas une valeur nulle pour le champ moléculaire et pose

$$\boldsymbol{E}' = \boldsymbol{E} + 4\pi\sigma\boldsymbol{P},$$

σ étant un coefficient qui caractérise la forme des molécules et leur distribution. On trouve

$$\varrho = \frac{4\pi^2 N}{n\,\lambda_0}\,\sigma^2\left(n^2 + \frac{1-\sigma}{\sigma}\right)^2 g = \frac{(n^2 - 1)^2}{4\,n\,\lambda_0\,N\,\alpha^2}\,g. \tag{28.13}$$

γ) Nous avons seulement tenu compte de la polarisation électrique du milieu pour établir la formule (28.12). Si l'on ne confond plus le champ $\boldsymbol{H}$ avec l'induction $\boldsymbol{B}$, mais que l'on tienne compte de la valeur du moment magnétique induit $\boldsymbol{m}'$, on trouve[3] pour le pouvoir rotatoire une valeur deux fois plus grande que ne l'indique la formule (28.11), mais Kooy[4] a montré que cette formule est correcte et qu'en considérant, comme l'a fait Born, le moment electrique relatif donné par (27.1), on tient compte, par là même, du rayonnement dû au moment dipolaire magnétique de la molécule.

29. Théorie moléculaire quantique. α) La théorie quantique de l'activité optique naturelle a été faite par Rosenfeld[5], en partant de la mécanique des matrices de Heisenberg. Nous l'exposerons dans le langage équivalent de la mécanique ondulatoire[6]. C'est la théorie de Dirac qui permet l'étude la plus précise des interactions entre rayonnement et matière. Mais nous utiliserons, comme l'a fait Condon, une extension de la méthode plus simple de Schrödinger, fondée sur le principe de correspondance et qui regarde l'action de l'onde électromagnétique sur une molécule comme une perturbation dépendant du temps.

La molécule non perturbée est dans l'état stationnaire a, où l'équation d'onde est

$$H_0\,\Psi_a^0 = E_a\,\Psi_a^0$$

H_0 désignant le hamiltonien. La fonction propre solution de cette équation est

$$\Psi_a^0 = \psi_a^0\,e^{-i\frac{2\pi E_a}{h}t}. \tag{29.1}$$

En présence de l'onde électromagnétique, le hamiltonien devient

$$H = H_0 + H_1$$

---

[1] J. M. Kooy: Diss. Leiden 1936.

[2] R. de Mallemann: Ann. Phys., Paris **2**, 137 (1924); **4**, 456 (1925).

[3] M. Born: Optik. Berlin 1933.

[4] J. M. Kooy: Diss. Leiden 1936.

[5] L. Rosenfeld: Z. Physik **52**, 161 (1928).

[6] G. Temple: Trans. Faraday Soc. **26**, 272 (1930).

H_1 désignant le hamiltonien perturbateur, fonction du temps. La solution de l'équation d'onde perturbée peut se développer suivant le système des fonctions propres, soit:

$$\Psi_a = \Psi_a^0 + \sum_b c_b(t)\,\Psi_b^0 \tag{29.2}$$

les fonctions Ψ_b^0 ayant des expressions analogues à (29.1) et les coefficients $c_b(t)$ étant nuls au temps $t = 0$, pris pour origine de la perturbation. La méthode de variation des constantes donne

$$\frac{dc_b}{dt} = \frac{2\pi i}{h}\,(\Psi_b^{0*}|H_1|\Psi_a^0) = \frac{2\pi i}{h}\,(\psi_b^{0*}|H_1|\psi_a^0)\,e^{2\pi i\nu_{ba}t} \tag{29.3}$$

en tenant compte de (29.1) et en posant $\nu_{ba} = \dfrac{E_b - E_a}{h}$. $(\Psi_b^{0*}|H_1|\Psi_a^0)$ désigne l'élément de matrice du hamiltonien correspondant à la transition $a \to b$.

Il faut alors expliciter l'expression de H_1

$$H_1 = -\sum_k \frac{e}{m_k c}\,\boldsymbol{A}_k \cdot \boldsymbol{\pi}_k \tag{29.4}$$

$\boldsymbol{A}_k$ désignant la valeur du potentiel vecteur de l'onde électromagnétique à l'endroit occupé par la particule d'indice k. Pour une onde plane monochromatique de fréquence ν

$$\boldsymbol{A}_k = \boldsymbol{A}_0\,e^{-\frac{2\pi i}{\lambda}\,\boldsymbol{r}_k \cdot \boldsymbol{s}} = \tfrac{1}{2}\left[\boldsymbol{A}_0^0\,e^{2\pi i\nu t} + \boldsymbol{A}_0^{0*}\,e^{-2\pi i\nu t}\right]e^{-\frac{2\pi i}{\lambda}\,\boldsymbol{r}_k \cdot \boldsymbol{s}} \tag{29.5}$$

$\boldsymbol{A}_0$ désignant la valeur de $\boldsymbol{A}$ à l'origine des coordonnées prise dans la molécule, $\boldsymbol{A}_0^0$ l'amplitude complexe correspondante.

Comme dans la théorie classique (Sect. 27) on peut se borner à considérer les deux premiers termes du développement de (29.5) soit:

$$\boldsymbol{A}_k = \boldsymbol{A}_0\left[1 - \frac{2\pi i}{\lambda}\,(\boldsymbol{r}_k \cdot \boldsymbol{s})\right].$$

En introduisant cette expression dans (29.4), il vient:

$$H_1 = H_1' + H_1'' = -\sum_k \frac{e_k}{m_k c}\left[\boldsymbol{A}_0 \cdot \boldsymbol{\pi}_k - \frac{2\pi i}{\lambda}\,(\boldsymbol{A}_0 \cdot \boldsymbol{\pi}_k)(\boldsymbol{r}_k \cdot \boldsymbol{s}_k)\right].$$

Le terme H_1'' peut se transformer, suivant les règles de l'algèbre vectorielle

$$H_1'' = \frac{2\pi i}{\lambda c}\sum_k \frac{e_k}{m_k}\,[\boldsymbol{\pi}_k \cdot (\boldsymbol{r} \cdot \boldsymbol{s}) \cdot \boldsymbol{A}_0] = \sum_k \frac{e_k}{m_k c}\,\frac{2\pi i}{\lambda}\,[\vartheta]_k\,\boldsymbol{s} \cdot \boldsymbol{A}_0,$$

où $[\vartheta]_k$ désigne un tenseur, généralement asymétrique. En décomposant ce tenseur en une partie symétrique $[\vartheta]_k^s$ et une partie antisymétrique $[\vartheta]_k^a$, H_1'' devient la somme de deux termes. Le premier se rapporte au moment de quadrupole électrique de la molécule, qui n'a pas d'intérêt pour les phénomènes qui nous concernent. En ne conservant que le second terme, on peut écrire [20], en tenant compte de (24.4)

$$H_1'' = \frac{2\pi i}{c\lambda}\sum_k \frac{e_k}{m_k}\,[\vartheta]_k^a\,\boldsymbol{s} \cdot \boldsymbol{A}_0 = -\frac{2\pi i}{\lambda}\,\boldsymbol{m} \cdot (\boldsymbol{s} \times \boldsymbol{A}_0) = \boldsymbol{m}\,(\text{rot}\,\boldsymbol{A})_0. \tag{29.6}$$

Ce terme d'interaction fait donc intervenir le moment de dipôle magnétique $\boldsymbol{m}$ de la molécule perturbée.

L'équation (29.3) s'écrit

$$\frac{dc_b}{dt} = \frac{2\pi i}{h}\left[\left(\psi_b^0 * \Big| \sum_k \frac{e_k}{m_k}\,\boldsymbol{\pi}_k \Big| \psi_a^0\right) \cdot \frac{\boldsymbol{A}_0}{c} + (\psi_b^0|\boldsymbol{m}|\psi_a^o) \cdot (\text{rot}\,\boldsymbol{A})_0\right]e^{2\pi i\nu_{ba}t}. \tag{29.7}$$

Mais on a

$$\left(\psi_b^{0\,*}\left|\sum_k \frac{e_k}{m_k}\,\pi_k\right|\psi_b^0\right) = -\,2\pi\,i\,\nu_{ba}\left(\psi_b^{0\,*}\left|\sum_k e_k\,r_k\right|\psi_a^0\right) = -\,2\pi\,i\,\nu_{ba}\,(\psi_b^0{}^{*}\,|\,p\,|\,\psi_a^0).$$

En tenant compte de cette expression, et de la variation de A_0 avec le temps, donnée par (29.5), l'intégration de (29.7) donne

$$\left.\begin{aligned}
c_b &= \left[\frac{\pi\,i\,\nu_{ba}}{c\,h^2}\,(b\,|\,p\,|\,a)\cdot A_0^0 + (b\,|\,m\,|\,a)\cdot(\text{rot}\,A^0)_0\right]\frac{e^{2\pi i(\nu_{ba}+\nu)t}}{\nu_{ba}+\nu} + \\
&\quad + \left[\frac{\pi\,i\,\nu_{ba}}{c\,h^2}\,(b\,|\,p\,|\,a)\,A_0^{0\,*} + (b\,|\,m\,|\,a)\cdot(\text{rot}\,A^{0\,*})_0\right]\frac{e^{2\pi i(\nu_{ba}-\nu)t}}{\nu_{ba}-\nu}.
\end{aligned}\right\} \tag{29.8}$$

La constante d'intégration peut être négligée, car elle ne contient pas de terme dépendant du champ perturbateur. On pose

$$(b\,|\,p\,|\,a) = \int \psi_b^0{}^{*}\,p\,\psi_a^0\,d\tau$$
$$(b\,|\,m\,|\,a) = \int \psi_b^0{}^{*}\,m\,\psi_a^0\,d\tau.$$

Cette expression permet d'obtenir celle de la fonction d'onde (29.2) de la molécule perturbée. On calcule alors l'émission induite de rayonnement par la molécule à partir des éléments diagonaux des matrices qui représentent les moments dipolaires, électrique et magnétique, car dans les phénomènes optiques que nous étudions, le système retombe à l'état initial a. Ces composantes diagonales sont des grandeurs réelles, ce que nous désignerons par Re.

β) Le moment électrique s'écrit, compte tenu de (29.1) et en négligeant les termes en c_b^2

$$\text{Re}\,(\psi_a^{*}\,|\,p\,|\,\psi_a) = (\psi_a^0\,|\,p\,|\,\psi_a^0) + 2\,\text{Re}\sum_b c_b\,(\psi_a^0\,|\,p\,|\,\psi_b^0)\,e^{-2\pi i\nu_{ba}t}.$$

Le second terme représente le moment de dipôle induit. En exprimant les c_b au moyen de (29.8), et en tenant compte des relations entre A, E et H, soit:

$$E = -\frac{1}{c}\dot{A} \quad \text{et} \quad H = \text{rot}\,A,$$

on trouve l'expression de l'amplitude du moment induit:

$$\left.\begin{aligned}
p_a' = \frac{1}{\pi h}\sum_b \frac{1}{\nu_{ba}^2-\nu^2}\,\text{Re}\,\big[&2\pi\nu_{ba}\,(a\,|\,p\,|\,b)\,(b\,|\,p\,|\,a)\,E^0 + \\
&+ i\,\frac{\nu_{ba}^2}{\nu^2}\,(a\,|\,p\,|\,b)\,(b\,|\,p\,|\,a)\,\dot{E}^0 + \\
&+ 2\pi\nu_{ba}\,(a\,|\,p\,|\,b)\,(b\,|\,m\,|\,a)\,H^0 + \\
&+ i\,(a\,|\,p\,|\,b)\,(b\,|\,m\,|\,a)\,\dot{H}^0\big].
\end{aligned}\right\} \tag{29.9}$$

Le second terme est nul, car $(a\,|\,p\,|\,b)\,(b\,|\,p\,|\,a)$ étant réel, son produit par i est purement imaginaire. Le premier contribue seulement à la réfraction; après un calcul de moyennes analogue à celui de la Sect. 28α, il donne l'expression de la polarisabilité moléculaire moyenne α, analogue à (23.3)

$$\alpha = \frac{2}{3h}\sum_b \frac{\nu_{ba}}{\nu_{ba}^2-\nu^2}\,(b\,|\,p\,|\,a)^2 = \frac{2}{3h}\sum_b \frac{\nu_{ba}}{\nu_{ba}^2-\nu^2}\,S_{ba}. \tag{29.10}$$

$S_{ba} = (b\,|\,p\,|\,a)^2$ mesure l'intensité de la raie d'absorption correspondant à la transition $a \to b$ [4].

Les deux derniers termes ont la dissymétrie nécessaire pour intervenir dans l'expression de l'activité optique et il reste finalement, Im désignant une quantité purement imaginaire

$$\boldsymbol{p}_a' = \frac{1}{\pi h} \sum_b \frac{1}{v_{ba}^2 - v^2} \left\{ 2\pi v_{ba} \operatorname{Re}\left[(a\,|\boldsymbol{p}|\,b)\,(b\,|\boldsymbol{m}|\,a)\,\boldsymbol{H}^0\right] - \operatorname{Im}\left[(a\,|\boldsymbol{p}|\,b)\,(b\,|\boldsymbol{m}|\,a)\,\dot{\boldsymbol{H}}^0\right]\right\}. \tag{29.11}$$

γ) Le moment magnétique s'obtient de façon analogue au moment électrique. On trouve pour la partie qui peut contribuer au pouvoir rotatoire

$$\boldsymbol{m}_a' = \frac{1}{\pi h} \sum_b \frac{1}{v_{ba}^2 - v^2} \left\{ 2\pi v_{ba} \operatorname{Re}\left[(a\,|\boldsymbol{m}|\,b)\,(b\,|\boldsymbol{p}|\,a)\,\boldsymbol{E}^0\right] - \operatorname{Im}\left[(a\,|\boldsymbol{m}|\,b)\,(b\,|\boldsymbol{p}|\,a)\,\dot{\boldsymbol{E}}^0\right]\right\}. \tag{29.12}$$

30. Expression quantique du pouvoir rotatoire des gaz et des liquides. α) En phase vapeur ou liquide, ou en solution, il faut évaluer les valeurs moyennes $\overline{\boldsymbol{p}}'$ et $\overline{\boldsymbol{m}}'$ des moments induits, en tenant compte de ce que les vecteurs $\boldsymbol{p}$ et $\boldsymbol{m}$ ont des directions fixes dans la molécule, mais que toute orientation de celle-ci par rapport aux champs $\boldsymbol{E}$ et $\boldsymbol{H}$ est également probable. On trouve [4], [20][1]

$$\overline{\boldsymbol{p}}_a' = \overline{\beta}_a \boldsymbol{H} - \overline{\gamma}_a \dot{\boldsymbol{H}},$$
$$\overline{\boldsymbol{m}}_a' = \overline{\beta}_a \boldsymbol{E} + \overline{\gamma}_a \dot{\boldsymbol{E}}$$

où

$$\overline{\beta}_a = \frac{2}{3h} \sum_b \frac{v_{ba}}{v_{ba}^2 - v^2} \operatorname{Re}\left[(a\,|\boldsymbol{p}|\,b)\,(b\,|\boldsymbol{m}|\,a)\right], \tag{30.1}$$

$$\overline{\gamma}_a = \frac{c}{3\pi h} \sum_b \frac{1}{v_{ba}^2 - v^2} \operatorname{Im}\left[(a\,|\boldsymbol{p}|\,b)\,(b\,|\boldsymbol{m}|\,a)\right]. \tag{30.2}$$

On montre [4] que le coefficient $\overline{\beta}_a$ ne contribue qu'au deuxième ordre à l'activité optique naturelle. L'expression des vecteurs $\boldsymbol{D}$ et $\boldsymbol{B}$ est alors

$$\left.\begin{aligned} \boldsymbol{D} &= \boldsymbol{E} + 4\pi N \overline{\boldsymbol{p}}_a' = \varepsilon\,\boldsymbol{E} - \frac{j}{c}\,\dot{\boldsymbol{H}}, \\[4pt] \boldsymbol{B} &= \boldsymbol{H} + 4\pi N \overline{\boldsymbol{m}}_a' = \boldsymbol{H} + \frac{j}{c}\,\dot{\boldsymbol{E}} \\[6pt] \varepsilon &= 1 + 4\pi N \alpha_a, \\ j &= 4\pi N \gamma_a. \end{aligned}\right\} \tag{30.3}$$

avec

En tenant compte de (4.1) on retrouve pour $\boldsymbol{D}$ une formule analogue à (4.22). L'expression (4.28) du pouvoir rotatoire en radians par unité de longueur est donc:

$$\varrho = \frac{16\pi^2 N}{3hc} \sum_a \varPi_a \sum_b \frac{v^2}{v_{ba}^2 - v^2} \operatorname{Im}\left[(a\,|\boldsymbol{p}|\,b)\,(b\,|\boldsymbol{m}|\,a)\right]. \tag{30.4}$$

$\varPi_a$ désigne la probabilité que la molécule a de se trouver dans l'état initial a

$$\operatorname{Im}\left[(a\,|\boldsymbol{p}|\,b)\,(b\,|\boldsymbol{m}|\,a)\right] = R_{ba} \tag{30.5}$$

est la *force rotatoire* de la transition $(a \to b)$.

β) La formule (30.4) a été établie en admettant que le champ électrique agissant sur la molécule se confond avec celui de l'onde lumineuse. Si les actions

---

[1] Voir aussi J. KIRKWOOD: J. Chem. Phys. **5**, 479 (1937).

intermoléculaires ne sont pas négligeables, une meilleure approximation consiste à évaluer le champ de polarisation comme on l'a vu à la Sect. 28β: le second membre de la formule (30.4) sera, par exemple, multiplié par le facteur $\dfrac{n^2+2}{3}$.

III. Dispersion rotatoire et dichroïsme circulaire.

31. Dispersion rotatoire dans les régions de transparence.   Règles de sommation. α) La formule (23.7) de DRUDE donne l'expression du pouvoir rotatoire en fonction de la fréquence de l'onde.   S'il existe plusieurs catégories d'électrons, ayant chacun sa fréquence propre et son coefficient γ_k, on a

$$\varrho = \frac{2\pi N e^2}{m\,\lambda_0^2} \sum_k \frac{\gamma_k}{\nu_k^2 - \nu^2}\,. \tag{31.1}$$

Les fréquences ν_k sont les fréquences des vibrations élastiques des électrons.

β) La théorie de BORN donne également la variation du pouvoir rotatoire avec la fréquence, lorsqu'on explicite, dans (27.3) et (27.4), les valeurs des coefficients K tirés de (26.3).   Posons

$$\mathfrak{L}_x^j = \sum_k \frac{e_k}{\sqrt{m_k}}\, a_{kx}^j\,. \tag{31.2}$$

La formule (27.3) devient

$$\alpha_{xy} = \sum_j \frac{\mathfrak{L}_x^j \mathfrak{L}_y^j}{\omega_j^2 - \omega^2}\,. \tag{31.3}$$

De même, d'après (27.4), on a

$$A_z^{kl} = \frac{1}{2\sqrt{m_k m_l}} \sum_j \frac{a_{kx}^j a_{ly}^j - a_{ky}^j a_{lx}^j}{\omega_j^2 - \omega^2}$$

ou encore

$$A_z^{kl} = \frac{1}{2\sqrt{m_k m_l}} \sum_j \frac{(\boldsymbol{a}_l^j \times \boldsymbol{a}_k^j)_z}{\omega_j^2 - \omega^2}\,,$$

$(\boldsymbol{a}_l^j \times \boldsymbol{a}_k^j)_z$ désignant la composante suivant l'axe z du produit vectoriel entre parenthèses.   Par suite, d'après (27.5), on a,

$$g_{zz} = \frac{\pi\,n}{\lambda_0} \sum_j \sum_{kl} \frac{e_k e_l}{\sqrt{m_k m_l}} \frac{(\boldsymbol{a}_l^j \times \boldsymbol{a}_k^j)_z}{\omega_j^2 - \omega^2}\,. \tag{31.4}$$

La constante de gyration (28.5) devient

$$g = \frac{2\pi\,n}{6\,\lambda_0} \sum_j \frac{1}{\omega_j^2 - \omega^2} \sum_{kl} \frac{e_k e_l}{\sqrt{m_k m_l}}\, (\boldsymbol{r}_k^0 - \boldsymbol{r}_l^0)\cdot(\boldsymbol{a}_l^j \times \boldsymbol{a}_k^j)\,,$$

ou, après transformations [2], [20] et en remplaçant ω^2 par $4\pi^2\nu^2$ et λ par c/ν

$$g = \frac{n\,\nu}{6\pi\,c} \sum_j \frac{\mathfrak{L}^j \cdot \mathfrak{R}^j}{\nu_j^2 - \nu^2} \tag{31.5}$$

avec

$$\mathfrak{R}^j = \sum_k \frac{e_k}{\sqrt{m_k}}\, (\boldsymbol{a}_k^j \times \boldsymbol{r}_k^0)\,. \tag{31.6}$$

D'où l'expression du pouvoir rotatoire (28.8)

$$\varrho = \frac{2\pi N}{3\,c^2} \sum_j \mathfrak{L}^j \mathfrak{R}^j \frac{\nu^2}{\nu_j^2 - \nu^2}\,. \tag{31.7}$$

Les fréquences ν_j sont les fréquences mécaniques des oscillations fondamentales de la molécule.

γ) La théorie quantique conduit à la formule (30.4)

$$\varrho = \frac{16\,\pi^2\,N}{3\,h\,c} \sum_a \Pi_a \sum_b R_{ba} \frac{\nu^2}{\nu_{ba}^2 - \nu^2}\,. \tag{31.8}$$

Les fréquences ν_{ba} sont ici des fréquences optiques, absorbées ou émises lorsque la molécule subit une variation d'énergie.

δ) Des définitions (31.2) et (31.6), on tire

$$\sum_j \mathfrak{L}^j \cdot \mathfrak{R}^j = \sum_{kl} \frac{e_k\,e_l}{\sqrt{m_k\,m_l}} \left(\boldsymbol{r}_k \cdot \sum_i \boldsymbol{a}_l^j \times \boldsymbol{a}_k^j\right).$$

La condition d'orthogonalité (25.10) annule le produit vectoriel, d'où la *règle de sommation*

$$\sum_j \mathfrak{L}^j \cdot \mathfrak{R}^j = 0\,. \tag{31.9}$$

Les forces rotatoires (30.5) obéissent à une règle de sommation analogue

$$\sum_b R_{ba} = \mathrm{Im}\left[\sum_b (a\,|\boldsymbol{p}|\,b)\cdot(b\,|\boldsymbol{m}|\,a)\right] = \mathrm{Im}\,(a\,|\boldsymbol{p}\cdot\boldsymbol{m}|\,a) = 0\,, \tag{31.10}$$

car les éléments diagonaux du produit de la matrice $\boldsymbol{p}$ par la matrice $\boldsymbol{m}$ sont des quantités réelles.

Les formules (31.7) ou (31.8) montrent que le pouvoir rotatoire tend vers zéro lorsque la fréquence ν devient très basse, car le numérateur s'annule, tandis que le dénominateur conserve une valeur finie. D'autre part, pour les fréquences très élevées, les fréquences caractéristiques deviennent négligeables devant ν. Les relations (31.9) ou (31.10) montrent qu'alors la valeur limite vers laquelle tend le pouvoir rotatoire est nulle. Ainsi, le pouvoir rotatoire naturel doit s'annuler aux deux extrémités du spectre électromagnétique.

32. Théorie classique du dichroïsme circulaire. α) Drude[1], reprenant les idées appliquées par Helmholtz à l'étude de la dispersion de réfraction dans les bandes d'absorption, modifie l'équation (23.1) en ajoutant un terme d'amortissement

$$m\,\ddot{\boldsymbol{r}} + 2\pi\,m\,\nu'\,\dot{\boldsymbol{r}} + 2\pi\,m\,\nu_k\,\boldsymbol{r} = e\,(\boldsymbol{E} + \gamma_k\,\mathrm{rot}\,\boldsymbol{E})\,. \tag{32.1}$$

$2\pi\,m\,\nu'$ désigne le coefficient de la force amortissante et ν_k la fréquence propre du mouvement non amorti. La solution de régime permanent de cette équation s'écrit

$$\boldsymbol{r} = \boldsymbol{r}^0\,e^{2\pi i\nu t}\,,$$

$\boldsymbol{r}^0$ désignant une amplitude généralement complexe. En substituant cette valeur dans (32.1) et en menant le calcul comme à la Sect. 23, on retrouve la relation (4.22) en posant au lieu de (23.5) et de (23.6)

$$\varepsilon = 1 + \frac{N e^2}{\pi m} \cdot \frac{1}{\nu_k^2 - \nu^2 + i\,\nu\,\nu'}\,,$$

$$j_k = \frac{N e^2\,\gamma_k}{\pi\,m}\,\frac{1}{\nu_k^2 - \nu^2 + i\nu\nu'}\,.$$

---

[1] P. Drude: Lehrbuch der Optik, 3e édit., p. 392. Leipzig 1912.

Dans les régions d'absorption, la relation $n = \sqrt{\varepsilon}$ est remplacée par (4.16)

$$n - i\chi = \sqrt{\varepsilon},$$

d'où

$$n^2 - \varkappa^2 = 1 + \frac{Ne^2}{\pi m} \frac{\nu_k^2 - \nu^2}{(\nu_k^2 - \nu^2)^2 + \nu^2 \nu'^2}, \tag{32.2}$$

$$2n\chi = \frac{Ne^2}{\pi m} \frac{\nu \nu'}{(\nu_k^2 - \nu^2)^2 + \nu^2 \nu'^2}. \tag{32.3}$$

De même, la relation (4.27) fait place à

$$n_g - n_d - i(\chi_g - \chi_d) = \frac{2\pi j_k}{\lambda_0},$$

d'où

$$n_g - n_d = \frac{Ne^2 \gamma_k}{2\pi^2 mc} \frac{\nu(\nu_k^2 - \nu^2)}{(\nu_k^2 - \nu^2)^2 + \nu^2 \nu'^2}, \tag{32.4}$$

$$\chi_g - \chi_d = \frac{Ne^2 \gamma_k}{2\pi^2 mc} \frac{\nu' \nu^2}{(\nu_k^2 - \nu^2)^2 + \nu^2 \nu'^2}. \tag{32.5}$$

Les courbes qui représentent ces fonctions ont l'allure de la Fig. 1. On tire des formules précédentes

$$\frac{n_g - n_d}{\chi_g - \chi_d} = \frac{\nu_k^2 - \nu^2}{\nu \nu'}. \tag{32.6}$$

Ce rapport est égal au rapport de la rotation ϱ à l'ellipticité φ pour une fréquence donnée, d'après (2.1) et (2.4). Comme ν' est une constante positive, on voit que $\varrho/\varphi > 0$ pour $\nu < \nu_k$. Ce résultat est connu sous le nom de *règle de* NATANSON[1]: Du côté de la bande d'absorption situé vers les basses fréquences, les ellipses dues au dichroïsme circulaire sont décrites dans le sens de la rotation produite par la substance active.

β) On a cherché à améliorer les formules (32.2) à (32.5) en les multipliant par un facteur F_k appelé *intensité de l'oscillateur* correspondant[2]. La valeur de F_k peut se déduire de la mesure de l'absorption de la bande à laquelle il se rapporte. Si celle-ci est assez étroite pour que l'on puisse remplacer $\nu_k^2 - \nu^2$ par $2\nu(\nu_k - \nu)$, on trouve

$$F_k = \frac{mc}{N\pi e} \int_0^\infty \mu \, d\nu \tag{32.7}$$

μ désignant le coefficient d'absorption relié à l'indice d'absorption par

$$\mu = \frac{4\pi \chi \nu}{c}.$$

Il est intéressant de considérer le *facteur d'anisotropie* [13] ou mieux *facteur de dissymétrie* [17] défini par:

$$G = \frac{\chi_g - \chi_d}{\chi}. \tag{32.8}$$

Lorsqu'on peut poser:

$$n^2 - \varkappa^2 - 1 \approx 2(n-1) \quad \text{et} \quad 2n\chi \approx 2\chi,$$

---

[1] L. NATANSON: J. Phys. Radium **8**, 321 (1909).

[2] M. BORN: Optik, Sect. 91. Berlin 1933. — L. DE BROGLIE: Le principe de correspondance, Chap. 5. Paris 1938.

on a

$$G = \frac{n_g - n_d}{n - 1} = \frac{\gamma \nu}{c \pi}. \tag{32.9}$$

Le facteur de dissymétrie n'est pas constant dans une bande d'absorption. Soit G_k sa valeur pour $\nu = \nu_k$. On a

$$G = G_k \frac{\nu}{\nu_k}. \tag{32.10}$$

En introduisant les grandeurs F et G, l'expression de l'absorption devient

$$\mu = \frac{2 N e^2}{m c} F \frac{\nu'}{4 (\nu_k - \nu)^2 + \nu'^2}, \tag{32.11}$$

celle du pouvoir rotatoire

$$\varrho = \frac{1}{2 \pi} \frac{G_k}{\nu_k} \frac{\nu^2 (\nu_k^2 - \nu^2)}{(\nu_k^2 - \nu^2)^2 + \nu^2 \nu'^2} \int \mu \, d\nu \tag{32.12}$$

et celle du dichroïsme circulaire

$$\chi_g - \chi_d = \frac{N e^2}{2 \pi m} \frac{F_k G_k}{\nu_k} \frac{\nu' \nu^2}{(\nu_k^2 - \nu^2)^2 + \nu^2 \nu'^2}. \tag{32.13}$$

γ) Au lieu de représenter les courbes d'absorption des liquides par résonance accompagnée d'amortissement, on les considère actuellement comme l'enveloppe d'un système de bandes de vibration-rotation non résolu, correspondant à une transition électronique déterminée; on les représente par des courbes en cloche[1], symétriques dans l'échelle des fréquences, par exemple, soit

$$\mu = \mu_k \, e^{-\left(\frac{\nu_k - \nu}{\vartheta}\right)^2}. \tag{32.14}$$

Pour $\mu = \frac{\mu_k}{2}$, on a $\nu_k - \nu = \frac{\nu'}{2}$, où ν' est la demi-largeur de la bande représentée par (32.11). Par suite

$$\vartheta = \frac{\nu'}{2 \sqrt{\log 2}}.$$

Le dichroïsme circulaire est alors représenté par la formule

$$\chi_g - \chi_d = (\chi_g - \chi_d)_k \frac{\nu}{\nu_k} \, e^{-\left(\frac{\nu_k - \nu}{\vartheta}\right)^2}. \tag{32.15}$$

Quant au pouvoir rotatoire, en admettant qu'une formule du genre (32.11) le représente pour chacune des bandes du système, de fréquence ν_1, et en faisant quelques autres approximations, on trouve [12], [20],

$$\varrho = \frac{1}{2 \sqrt{\pi}} \frac{\nu}{\nu_k} G_k \mu_k \left[e^{-\left(\frac{\nu_k - \nu}{\vartheta}\right)^2} \int_0^{\frac{\nu_k - \nu}{\vartheta}} e^{x^2} \, dx - \frac{\vartheta}{2 (\nu_k + \nu)} \right], \tag{32.16}$$

avec

$$x = \frac{\nu_k - \nu_1}{\vartheta}.$$

Il existe en [12] et [17] des tables de valeurs du premier membre de la parenthèse.

---

[1] W. Kuhn et E. Braun: Z. phys. Chem., Abt. B **8**, 281 (1930).

Lowry et Hudson[1] ont proposé une formule semblable à (32.14), mais symétrique en longueurs d'onde. On en tire des expressions de l'activité optique analogues à (32.15) et (32.16), pour le calcul desquelles on a dressé des tables [17].

33. Théorie quantique du dichroïsme circulaire[2] **[4].** Lorsque la fréquence v de l'onde est voisine de la fréquence d'absorption v_{ba}, le deuxième terme de la formule (29.8) a une importance prépondérante. On sait que $|c_b|^2$ représente la probabilité de la transition induite $a \to b$. En ne conservant dans l'élévation au carré que le terme du premier degré en $(\mathrm{rot}\, A^0)$ dont dépend l'activité optique et en supposant une onde plane, pour laquelle $[\mathrm{rot}\, A^{0*}] = \dfrac{2\pi i v}{c}\, A^{0*} \times s$, il vient pour l'amplitude de la probabilité de transition :

$$\frac{\pi^2 v_{ba}^2}{c^2 h^4}\left[(b\,|\boldsymbol{p}|\,a)\, A_0^{0*}\right]^2 + \frac{2\pi v\, v_{ba}}{c^2 h^4}\,\mathrm{Re}\left[A_0^0\,(a\,|\boldsymbol{p}|\,b)\cdot(b\,|\boldsymbol{m}|\,a)\,(A_0^{0*}\times s)\right].$$

La moyenne de cette expression, pour un gaz parfait, a pour valeur

$$\frac{\pi^2 v_{ba}^2}{c^2 h^4}\cdot\frac{1}{3}\,A_0^{0\,2}\,(b\,|\boldsymbol{p}|\,a)^2 + \frac{2\pi^2 v\, v_{ba}}{c^2 h^4}\cdot\frac{1}{3}\,A_0^0\,(A_0^{0*}\times s)\,\mathrm{Im}\left[(a\,|\boldsymbol{p}|\,b)\cdot(b\,|\boldsymbol{m}|\,a)\right]. \quad (33.1)$$

Pour une onde polarisée circulairement

$$A_0^0\,(A_0^{0*}\times s) = \mp\, i\, A^2$$

et l'expression (33.1) devient

$$\frac{\pi^2 v_{ba}^2}{3 c^2 h^4}\,A^2\left[(b\,|\boldsymbol{p}|\,a)^2 \mp 2\,\frac{v}{v_{ba}}\,\mathrm{Im}\left\{(a\,|\boldsymbol{p}|\,b)\cdot(b\,|\boldsymbol{m}|\,a)\right\}\right]. \quad (33.2)$$

Le premier terme se rapporte à la lumière naturelle : c'est de lui que dépend le coefficient d'absorption μ. La différence $\mu_g - \mu_d$ des coefficients d'absorption pour les rayons circulaires gauches et droits est donnée par le double du second terme. Le facteur de dissymétrie (32.8) a pour expression

$$G = \frac{\mu_g - \mu_d}{\mu} = 4\,\frac{v}{v_{ba}}\,\frac{\mathrm{Im}\left\{(a\,|\boldsymbol{p}|\,b)\cdot(b\,|\boldsymbol{m}|\,a)\right\}}{(b\,|\boldsymbol{p}|\,a)^2} = 4\,\frac{v}{v_{ba}}\,\frac{R_{ba}}{S_{ba}}, \quad (33.3)$$

en tenant compte des définitions (29.10) et (30.5).

La théorie quantique n'a encore rien apporté en ce qui concerne la forme des courbes $\varphi = f(v)$: on a simplement admis que les formules (32.4) et (32.5) sont valables.

34. Fréquences caractéristiques du dichroïsme circulaire. α) A chaque formule représentant le dichroïsme circulaire $\chi_g - \chi_d$ d'une bande d'absorption correspond une formule donnant le pouvoir rotatoire ϱ. Il doit donc être possible de faire la synthèse des courbes de dispersion rotatoire, en essayant de les représenter par la superposition de plusieurs courbes théoriques.

Pour des fréquences très éloignées de la fréquence caractéristique v_k d'une bande d'absorption, la formule (32.12) se ramène à

$$\varrho_R = \frac{G_k}{2\pi v_k}\,\frac{v^2}{v_k^2 - v^2}\int \mu\, dv \quad (34.1)$$

On peut montrer que l'expression (32.16) prend la même forme ; il en est de même de la formule de Lowry et Hudson ; nous venons de voir qu'il en est de même dans la théorie quantique. Ainsi, dans les régions de transparence, la contribution d'une bande au pouvoir rotatoire ne dépend pas des hypothèses

[1] T. M. Lowry et H. Hudson: Phil. Trans. Roy. Soc. Lond., Ser. A **132**, 117 (1933).
[2] E. U. Condon, W. Altar et H. Eyring: J. Chem. Phys., **5**, 753 (1937).

faites sur la répartition des intensités à l'intérieur de la bande, mais seulement de l'intensité totale et du facteur de dissymétrie.

β) On cherche souvent à représenter les mesures de pouvoir rotatoire, faites dans les régions de transparence, par une formule de Drude où les longueurs d'onde remplacent les fréquences:

$$\varrho = \frac{2\pi N e^2}{m c^2} \sum_k \frac{\gamma_k \lambda_k^2}{\lambda_0^2 - \lambda_k^2} = \sum_k \frac{C_k}{\lambda_0^2 - \lambda_k^2}. \tag{34.2}$$

Les constantes λ_k et C_k servent alors de paramètres. Lorsqu'un seul terme suffit à une représentation exacte, la dispersion est dite *simple* [17]; dans le cas

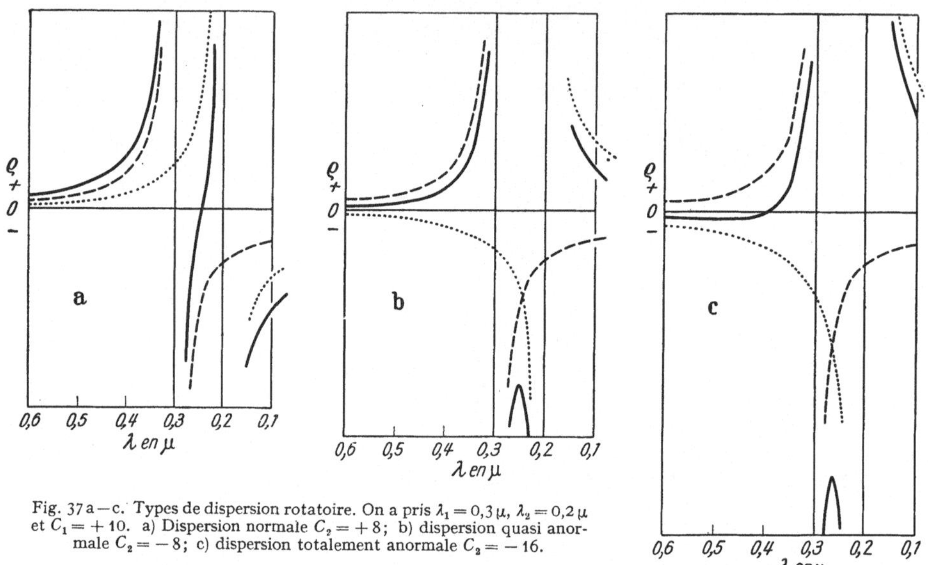

Fig. 37 a—c. Types de dispersion rotatoire. On a pris $\lambda_1 = 0,3\,\mu$, $\lambda_2 = 0,2\,\mu$ et $C_1 = +10$. a) Dispersion normale $C_2 = +8$; b) dispersion quasi anormale $C_2 = -8$; c) dispersion totalement anormale $C_2 = -16$.

contraire, on la dit *complexe* et on essaie une formule à deux termes. En posant $\lambda_1 > \lambda_2$, on distingue (Fig. 37) la dispersion *normale* lorsque C_1 et C_2 ont le même signe, et, lorsqu'ils ont des signes contraires, les dispersions *quasi anormale* ($|C_1| > |C_2|$) et *totalement anormale* ($|C_1| < |C_2|$).

Lorsque les longueurs d'onde du domaine étudié sont beaucoup plus grandes que les longueurs d'onde caractéristiques, la formule (34.2) se ramène à la formule de Biot (Sect. 1)

$$\varrho = \frac{C}{\lambda_0^2}. \tag{34.3}$$

γ) A titre d'exemple, indiquons que pour les composés tels que les alcools[1] ou les acides[2] dont les bandes d'absorption sont éloignés dans l'ultraviolet, une formule du genre (34.2) à un seul terme suffit généralement à représenter la dispersion. Par contre, pour l'acide tartrique, en solution dans l'eau ou à l'état fondu, une formule à deux termes est nécessaire:

$$\varrho = -\frac{C_1}{\lambda_0^2 - (0,233)^2} + \frac{C_2}{\lambda_0^2 - (0,1749)^2}. \tag{34.4}$$

[1] T. M. Lowry et Dickson: J. Chem. Soc., Lond. **103**, 1067 (1913).
[2] P. A. Levene et A. Rothen: Rotatory dispersion. In Gilman, Organic Chemistry, tome II. New York 1938.

C_1 et C_2 varient avec la concentration de la solution, mais $|C_2| > |C_1|$. La dispersion est donc totalement anormale dans le spectre visible. Voir [17].

δ) On constate fréquemment que les valeurs des fréquences ν_k, déterminées à partir des mesures d'activité optique, sont sensiblement différentes de celles qui interviennent dans la représentation des mesures de l'absorption ou de la dispersion de réfraction qui lui est liée [au moyen de formules du genre (32.7) et 32.4) par exemple]. Cela se produit dans divers cas.

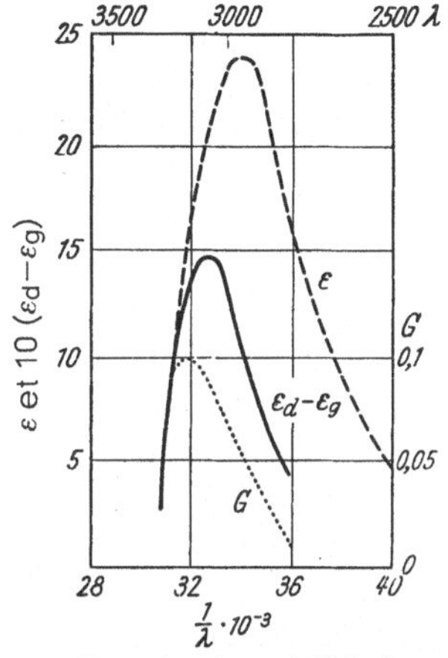

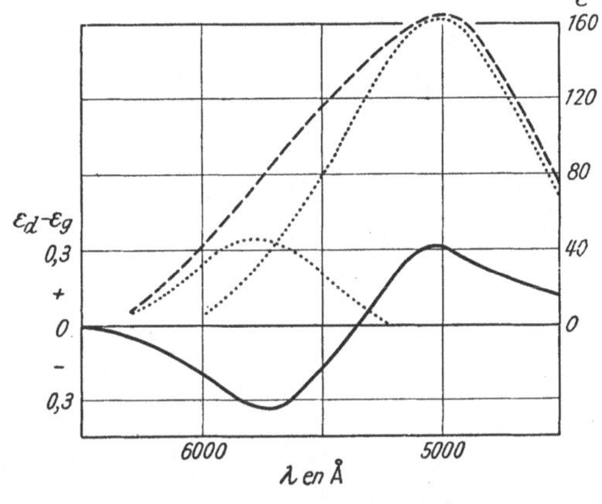

Fig. 38. Absorption (———) dichroïsme circulaire (———) et facteur de dissymétrie (···) de la bande $C = 0$ du camphre

Fig. 39. Absorption (———) et dichroïsme circulaire (———) de la bande d'absorption visible du complexe $Cl[Coen_2Cl \cdot NCS]$.

1. Lorsqu'on cherche à représenter n et ϱ par des formules de DRUDE à un terme. Exemples de valeurs de λ_k en Å:

	réfraction	rotation
Quartz . . .	800	1878
Limonène . .	998	1200

Ces différences viennent de ce que la représentation adoptée est trop grossière [4]. Elle confond plusieurs termes en un seul, ce qui revient à poser dans (29.10)

$$\frac{\nu_r S_r}{\nu_r^2 - \nu^2} = \sum_b \frac{\nu_{ba} S_{ba}}{\nu_{ba}^2 - \nu^2},$$

et dans (30.2)

$$\frac{R_r}{\nu_r'^2 - \nu^2} = \sum_b \frac{R_{ba}}{\nu_{ba}^2 - \nu^2}.$$

ν_r et ν_r' sont des *fréquences réduites*. Lorsque ν est très inférieure à ν_r ou à ν_r' et aux ν_{ba}, on peut développer les expressions précédentes en série suivant les puissances croissantes de ν^2/ν_r^2 et ν^2/ν_{ba}^2 et ne conserver que les deux premiers termes; cela conduit aux conditions

$$\nu_r^2 = \left(\sum \frac{S_{ba}}{\nu_{ba}}\right) : \left(\sum \frac{S_{ba}}{\nu_{ba}^3}\right) \quad \text{et} \quad \nu_r'^2 = \left(\sum \frac{R_{ba}}{\nu_{ba}^2}\right) : \left(\sum \frac{R_{ba}}{\nu_{ba}^4}\right).$$

Bien que les ν_{ba} soient les mêmes, ν_r et ν_r' peuvent donc être différentes.

2. L'existence dans l'ultraviolet d'une absorption générale croissant vers les hautes fréquences et superposée aux bandes d'absorption sélective peut faire

attribuer une valeur ν_k trop élevée pour le maximum d'absorption apparent [*17*]. D'autre part, si la représentation de la rotation obéit effectivement à une formule du genre (32.16) et qu'on essaie de lui adapter une formule de DRUDE, on est conduit à donner une valeur trop faible au paramètre ν_k de cette dernière formule[1].

3. Dans certains cas enfin, une bande d'absorption en apparence unique est composée de plusieurs parties où le dichroïsme circulaire a des valeurs très inégales (Fig. 38), parfois même des signes opposés (Fig. 39).

IV. Théories simplifiées de l'activité optique.

35. Bandes d'absorption actives et inactives. Données empiriques. α) Les théories générales exposées dans les sections précédentes permettent en principe

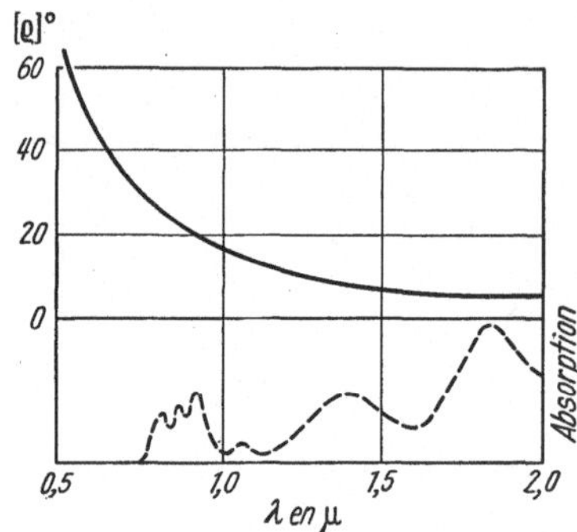

Fig. 40. Absorption (— — —) et dispersion rotatoire (———) du pinène β.

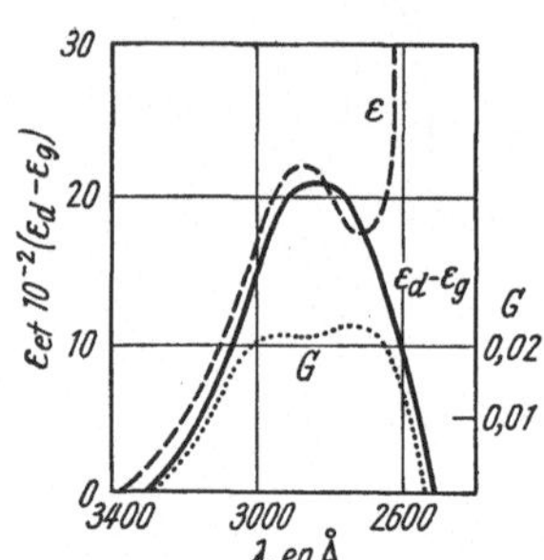

Fig. 41. Absorption (— — —), dichroïsme circulaire (———) et facteur de dissymétrie (·····) du d diméthylamine de l'acide α-azidopropionique $CH_3CHN_3CON(CH_3)_2$.

de calculer le pouvoir rotatoire moléculaire d'un fluide pour une fréquence quelconque, en faisant la somme des contributions apportées par l'ensemble des bandes d'absorption de la substance; celles-ci peuvent se déterminer à partir des mesures de dichroïsme circulaire (Sect. 31). Au premier examen, la méthode paraît impraticable, car les sommes $\sum_j$ et $\sum_b$ qui interviennent dans les formules (31.7) et (31.8) s'étendent à toutes les fréquences propres de la molécule. Or, pour des raisons pratiques, on n'a pas encore pu faire de mesures au-dessous de 2000 Å, dans une région où les bandes d'absorption sont nombreuses et intenses.

β) Cependant le problème se simplifie dans une certaine mesure, l'expérience enseignant que toutes les bandes d'absorption ne sont pas actives et permettant d'énoncer les règles suivantes:

1. Les bandes d'absorption infrarouges sont inactives. La Fig. 40 représente, par exemple, les mesures relatives au β-pinène[2,3]. Cette règle s'interprète en remarquant que dans la formule (31.4) intervient le facteur $\sqrt{m_k\,m_l}$ qui est de l'ordre de grandeur de la masse d'une des particules qui constituent la molécule. La contribution apportée au pouvoir rotatoire par les bandes de vibration-rotation, qui font intervenir la masse des noyaux atomiques, doit donc être plusieurs milliers de fois plus faible que celle des bandes électroniques (car les charges e et les distances r sont du même ordre de grandeur dans les deux cas). La formule (31.8) conduit à la même conclusion, car la masse intervient au dénominateur de m dans (30.5).

---

[1] W. KUHN et E. BRAUN: Z. phys. Chem., Abt. B **8**, 281 (1930).
[2] L. R. INGERSOLL: Phys. Rev. **9**, 257 (1917).
[3] R. FREYMANN: Ann. Phys., Paris **20**, 329 (1933).

2. *La structure de vibration des bandes électroniques a peu ou point d'influence sur l'activité optique*[1]. Le plus souvent les mesures d'absorption faites sur les bandes des corps à l'état liquide ne montrent pas de structure. Dans plusieurs cas, on trouve une valeur du facteur de dissymétrie G qui satisfait à la relation (32.10) (exemple: Fig. 41). Dans d'autres cas, G varie en grandeur et même en signe (exemples: Figs. 39 et 40); on admet alors que l'on a affaire à plusieurs transitions électroniques. Dans quelques cas, enfin, comme celui des nitrites, les mesures d'absorption montrent une structure de vibration (Fig. 42).

3. *Les bandes d'absorption électroniques situées dans l'ultraviolet extrême n'apportent souvent qu'une faible contribution au pouvoir rotatoire dans le spectre visible*. La valeur de ϱ décroît en effet lorsque ν^2 est beaucoup plus grand que les termes ν_j^2 ou ν_{ba}^2 des formules (31.7) et (31.8). De plus, les règles de sommation (31.9) ou (31.10) laissent prévoir que les nombreuses bandes de l'ultraviolet lointain ne doivent pas toutes apporter des contributions de même signe: celles-ci tendent donc à se compenser, au moins partiellement, à mesure que la fréquence ν croît.

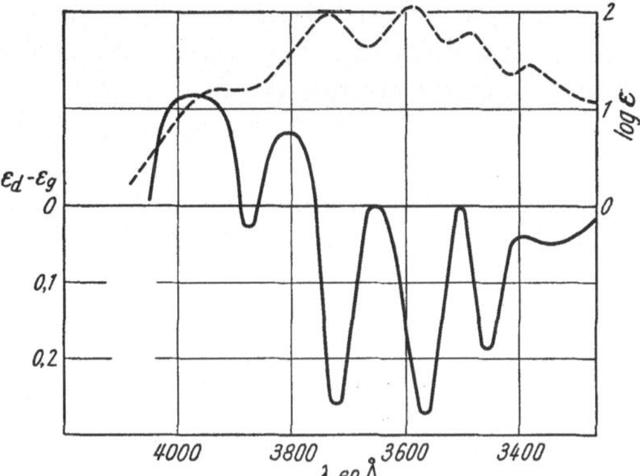

Fig. 42. Absorption (— — —) et dichroïsme circulaire (————) du nitrite d'octyle $CH_2CHONOC_6H_{13}$.

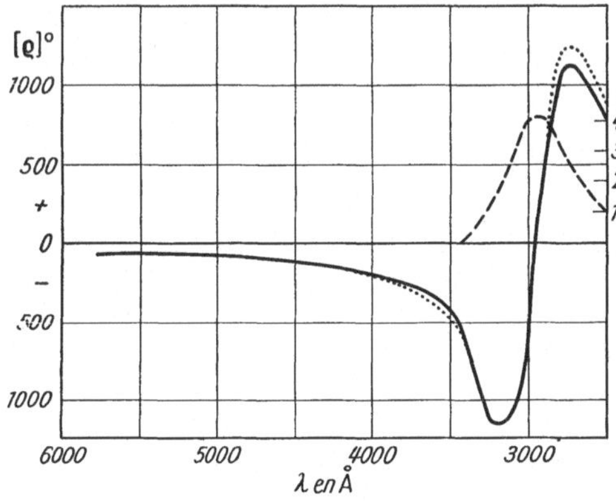

Fig. 43. Absorption (— — —) dispersion rotatoire mesurée (————) et calculée par la formule de Lowry [17] (·······) pour le tétracétyle μ arabinose.

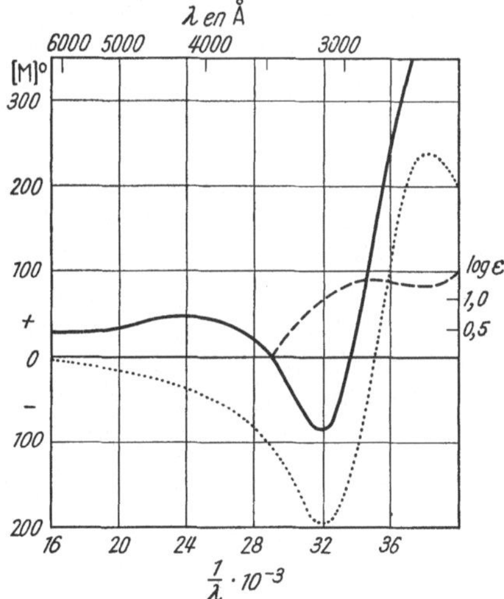

Fig. 44. Absorption (— — —), dispersion rotatoire mesurée (————) et calculée par la formule (32.16) (·······) pour l'α azidopropionate de méthyle.

Par exemple, la Fig. 43 montre que la faible bande d'absorption du tétracétyle-μ-arabinose, dont le maximum est à 2850 Å, est responsable de la totalité du pouvoir rotatoire dans le spectre visible. La Fig. 44 montre que la bande de l'α-azidopropionate de méthyle située vers 2800 Å, qui représente environ 10^{-5} de l'absorption totale, contribue pour 40% au pouvoir rotatoire dans le jaune. On trouvera d'autres exemples en [17].

γ) On peut conclure de ce qui précède que les particules, dont les formules (31.7) et (31.8) ne précisent pas la nature, sont des électrons; en outre, qu'il

---

[1] W. Kuhn et H. Lehmann: Z. phys. Chem., Abt. B **18**, 32 (1932).

suffit, pour obtenir un ordre de grandeur satisfaisant du pouvoir rotatoire dans le spectre visible, de considérer les fréquences d'absorption électroniques les plus proches.

La plupart des bandes d'absorption moléculaires sont dues aux transitions d'un seul électron, que l'on peut fréquemment localiser dans un atome ou dans un groupe d'atomes déterminé de la molécule. A ce *chromophore* correspond souvent une bande d'absorption de fréquence assez bien définie. Par exemple, la bande des Figs. 41 et 44 est commune à tous les dérivés de l'acide azidopropionique (chromophore N_3); celle de la Fig. 43 se retrouve chez tous les aldéhydes et cétones (chromophore C=O).

δ) Dans l'acétone, dont la molécule possède deux plans de symétrie, le facteur de dissymétrie de cette bande est nul. Il ne l'est plus dans les cétones dissymétriques: on dit que *l'action de voisinage* (vicinal effect [17], Vizinalwirkung [12]) exercée sur l'électron du groupe C=O, auquel est due cette bande, par les autres atomes de la molécule, dont la disposition produit la dissymétrie nécessaire à l'activité optique, crée dans le groupe C=O une *dissymétrie induite* (induced dissymetry [17]). Cette façon de s'exprimer conduit à considérer l'action de voisinage comme une perturbation exercée sur le chromophore par son entourage.

L'ensemble des données qui précèdent est important, car il a, jusqu'à présent, fourni la base de toutes les tentatives faites pour pousser jusqu'au bout l'application des théories générales à des cas particuliers concrets.

36. Théorie monoélectronique[1]. α) La théorie quantique peut aborder l'étude directe des transitions d'un électron chromophore et de leur action sur le pouvoir rotatoire. Pour cela, on est parti de l'approximation du champ self-consistent de Hartree[2], dans laquelle chaque électron d'un atome est placé dans un champ de force électrostatique central, créé par la distribution moyenne de charges due aux noyaux et aux autres électrons. La fonction d'onde relative à l'atome se décompose alors en un produit de fonctions d'onde qui se rapportent chacune à un seul électron. C'est une telle fonction d'onde qui caractérise l'électron chromophore non perturbé, c'est-à-dire soumis au seul champ de l'atome auquel il appartient. Dans cet état, l'électron n'apporte pas de contribution au paramètre γ donné par (30.2). Les éléments de matrice qui entrent dans cette formule ont pour expression,

$$\left.\begin{aligned} (b\,|\boldsymbol{p}_0|\,a) &= \int \psi_{0b}^* \,\boldsymbol{p}_0\, \psi_{0a}\, dv\,, \\ (b\,|\boldsymbol{m}_0|\,a) &= \int \psi_{0b}^*\, \boldsymbol{m}_0\, \psi_{0a}\, dv \end{aligned}\right\} \tag{36.1}$$

où $\boldsymbol{p}_0$ et $\boldsymbol{m}_0$ désignent les amplitudes des moments et ψ_a^0 et ψ_b^0 celles des fonctions d'onde monoélectroniques non perturbées. La perturbation que l'on considère est créée par le champ électrostatique dérivant d'un potentiel V dû aux autres atomes de la molécule. Le terme perturbateur H_1 du hamiltonien est égal à eV, e désignant la charge électronique. Lorsqu'il n'y a pas dégénérescence, les fonctions d'onde perturbées ont la forme[3]

$$\left.\begin{aligned} \psi_a &= \psi_{0a} + \sum_j c_{aj}\,\psi_{0j}\,, \\ \psi_b &= \psi_{0b} + \sum_k c_{bk}\,\psi_{0k} \end{aligned}\right\} \tag{36.2}$$

---

[1] E. U. Condon, W. Altar et H. Eyring: J. Chem. Phys. **5**, 753 (1937).

[2] D. R. Hartree: Proc. Cambridge Phil. Soc. **24**, 89, 111, 426 (1928).

[3] La perturbation considérée ici n'est plus celle qui est produite par l'onde lumineuse, comme à la Sect. 29, mais celle qui résulte des actions mutuelles des groupes en lesquels on divise la molécule.

avec

$$c_{aj} = -\frac{(a\,|\,H_1\,|\,j)}{E_{0a}-E_{0j}} = \frac{\int \psi_{0a}^* e\,V\,\psi_{0b}\,dv}{E_{0a}-E_{0j}} \qquad (36.3)$$

H_1 désignant le terme du premier ordre du hamiltonien perturbateur.

Les éléments de matrice perturbés qui entrent dans (36.1) se calculent au moyen des relations

$$\left.\begin{aligned}
(b\,|\,\boldsymbol{p_0}\,|\,a) = \int \psi_b^* \,\boldsymbol{p_0}\,\psi_a\,dv = \int \psi_{0b}^*\,\boldsymbol{p_0}\,\psi_{0a}\,dv + \sum_j c_{aj} \int \psi_{0b}^*\,\boldsymbol{p_0}\,\psi_{0j}\,dv \\
+ \sum_k c_{bk} \int \psi_{0k}^*\,\boldsymbol{p_0}\,\psi_{0a}\,dv + \sum_{jk} c_{aj}c_{bk} \int \psi_{0k}^*\,\boldsymbol{p_0}\,\psi_{0j}\,dv.
\end{aligned}\right\} \quad (36.4)$$

Une expression analogue existe pour $\boldsymbol{m}$. Par suite, dans la formule (30.2), on peut mettre le terme entre accolades sous la forme

$$\left.\begin{aligned}
(b\,|\,\boldsymbol{p_0}\,|\,a)\cdot(a\,|\,\boldsymbol{m_0}\,|\,b) + \sum_k c_{bk}[(b\,|\,\boldsymbol{p_0}\,|\,a)\cdot(a\,|\,\boldsymbol{m_0}\,|\,k) + (k\,|\,\boldsymbol{p_0}\,|\,a)\cdot(a\,|\,\boldsymbol{m_0}\,|\,b)] + \\
+ \sum_j c_{aj}\,[(b\,|\,\boldsymbol{p_0}\,|\,a)\cdot(j\,|\,\boldsymbol{m_0}\,|\,b) + (b\,|\,\boldsymbol{p_0}\,|\,j)\,(a\,|\,\boldsymbol{m_0}\,|\,b)].
\end{aligned}\right\} \quad (36.5)$$

Les termes écrits, qui contiennent un seul coefficient c, sont les termes du premier ordre; le termes omis contiennent les produits de plusieurs coefficients c et sont appelés d'ordre supérieur.

Le premier terme de (36.5) est nul; les coefficients c_{aj}, c_{bk} se calculent au moyen de la relation (36.3). Les potentiels scalaires s'ajoutant algébriquement, ce calcul conduit à évaluer des intégrales du genre

$$\int \psi^*\,V_l\,\psi\,dv, \qquad (36.6)$$

l étant un indice affecté au groupe atomique qui produit le potentiel V_l. Les coefficients c_{aj} s'expriment donc par une somme de termes qui dépendent chacun de l'électron considéré et d'un groupe voisin; les actions de voisinage du premier ordre sont donc additives, comme les potentiels. Les coefficients c étant petits, les actions de voisinage décroissent régulièrement à mesure que leur ordre croît, à moins que certains termes de (36.5) ne soient nuls par raison de symétrie.

β) La méthode a été appliquée[1] à un électron de masse m dont l'énergie potentielle, dans l'état non perturbé, est

$$2\,V_0 = f_1\,x_1^2 + f_2\,x_2^2 + f_3\,x_3^2, \qquad (36.7)$$

les x désignant les coordonnées cartésiennes, les f des constantes:

$$4\pi^2 v_t^2 = \frac{f_t}{m} \quad (t = 1,\,2,\,3).$$

On reconnaît un oscillateur harmonique (anisotrope pour éviter la dégénérescence). Les états d'un tel oscillateur sont caractérisés par trois nombres quantiques, soit:

$$a\,(n_1,\,n_2,\,n_3) \quad \text{et} \quad b\,(n_1',\,n_2',\,n_3').$$

Les amplitudes des fonctions d'onde peuvent s'écrire

$$\psi^0\,(n_1,\,n_2,\,n_3) = \psi_{n_1}(x_1)\cdot\psi_{n_2}(x_2)\cdot\psi_{n_3}(x_3)$$

où les facteurs ont l'expression connue

$$\psi_n(x) = \frac{1}{\sqrt{\pi\,2^n\,n!}}\,\mathrm{e}^{-\frac{x^2}{2q^2}}\,H_n\!\left(\frac{x}{q}\right)$$

---

[1] E. U. Condon, W. Altar et H. Eyring: J. Chem. Phys. **5**, 753 (1937).

où H_n désigne le polynome d'Hermite de degré n et

$$q = \frac{1}{2\pi}\sqrt{\frac{h}{m\nu}}, \tag{36.8}$$

Les règles de sélection de l'oscillateur harmonique donnent l'expression des éléments de la matrice p à partir de ceux de la matrice élongation r. On trouve par exemple, que les seuls éléments non nuls de la matrice r_1 sont ceux pour lesquels $n_1' = n_1 + 1$ et qu'on a

$$(a\,|\,p_1|\,b) = (n_1, n_2, n_3\,|\,e\,r_1|\,n_1 \pm 1, n_2, n_3) = i\,e\,\sqrt{\frac{(n)_1}{2}}\,q_1\,\delta_{n_2 n_2'}\,\delta_{n_3 n_3'} \tag{36.9}$$

avec $\delta_{n n'} = 0$ pour $n \neq n'$ et $\delta = 1$ pour $n = n'$; i est un vecteur unitaire dirigé suivant l'axe de coordonnées 1; (n) désigne le plus grand des deux nombres quantiques n ou n'.

Les éléments de la matrice m s'expriment, d'après (22.5), à partir de ceux de la matrice du moment cinétique $L = r \times \pi$. On trouve, par exemple, que les éléments de la matrice L ne sont différents de zéro que pour $n_2' = n_2 \pm 1$ et $n_3' = n_3 \pm 1$ simultanément et que

$$\left.\begin{aligned}
(b\,|\,m_1|\,a) &= \left(n_1', n_2', n_3'\,\left|\,\frac{e}{2mc}\,L_1\,\right|\,n_1, n_2, n_3\right) \\
&= \frac{i\,e\,h}{8\pi m c}\,i\left(\pm\frac{1}{q_3^2} - \pm\frac{1}{q_2^2}\right)\sqrt{(n_2)\,(n_3)}\,\delta_{n_1 n_1'}.
\end{aligned}\right\} \tag{36.10}$$

Les règles de sélection pour p et m sont donc telles que le produit scalaire de (30.2) est nul pour toutes les transitions permises: l'électron non perturbé n'a pas d'activité optique. Par exemple, si l'on part de l'état fondamental (000), l'élément de matrice $(000|p_1|100)$ n'est pas nul, tandis que $(100|m_1|000)$ l'est.

On ajoute alors à l'énergie potentielle un terme perturbateur tel que

$$2V = 2V_0 + 2A\,x_1 x_2 x_3. \tag{36.11}$$

Cette forme particulière de perturbation permet un calcul facile des intégrales (36.3). Les fonctions d'onde perturbées se calculent au moyen de (29.2) et les éléments de matrices perturbés au moyen de formules analogues à (36.1). On obtient, par exemple, pour l'élement perturbé $[100|m_1|000]$ l'expression suivante

$$[100\,|\,m_1|\,000] = -\,i\,\frac{A\,q_1(q_2^2 - q_3^2)\,e}{16\sqrt{2}\,\pi m c}\,\frac{2\nu_1}{(\nu_1 + \nu_3)^2 - \nu_1^2}\,i,$$

ce qui, combiné avec la valeur de l'élement de matrice non perturbé

$$(000\,|\,p_1|\,100) = \frac{e\,q_1}{\sqrt{2}}\,i$$

et les valeurs (36.7) des q donne

$$(000\,|\,p_1|\,100)\,[100\,|\,m_1|\,000] = -\,i\,\frac{A\,h^2 e^2}{256\pi^5 m^3 c}\left(\frac{1}{\nu_2} - \frac{1}{\nu_3}\right)\frac{1}{(\nu_2 + \nu_3)^2 - \nu_1^2}. \tag{36.12}$$

Les transitions $(000) \to (010)$ et $(000) \to (001)$ sont également rendues actives par la perturbation du premier ordre de m. Les transitions $(000) \to (110)$, $(000) \to (101)$ et $(000) \to (011)$ le sont par suite de la perturbation de p. Lorsqu'on en tient compte, on obtient pour le paramètre rotatoire γ_{000} dans (30.2) l'expression

$$
\begin{aligned}
\gamma_{000} &= \frac{c}{3\pi h}\sum \frac{\mathrm{Im}(000\,|\boldsymbol{p}|\,100)\,[100\,|\boldsymbol{m}|\,000]}{\nu_1^2 - \nu^2} \\
&= \frac{A\,h\,e^2\,c}{512\,\pi^6\,m^3}\left\{\left(\frac{1}{\nu_2} - \frac{1}{\nu_3}\right)\frac{1}{(\nu_2+\nu_3)^2 - \nu_1^2}\left[\frac{1}{(\nu_2+\nu_3)^2 - \nu^2} - \frac{1}{\nu_1^2 - \nu^2}\right] + \right. \\
&\quad + \left(\frac{1}{\nu_3} - \frac{1}{\nu_1}\right)\frac{1}{(\nu_3+\nu_1)^2 - \nu_2^2}\left[\frac{1}{(\nu_3+\nu_1)^2 - \nu^2} - \frac{1}{\nu_2^2 - \nu^2}\right] + \\
&\quad \left. + \left(\frac{1}{\nu_1} - \frac{1}{\nu_2}\right)\frac{1}{(\nu_1+\nu_2)^2 - \nu_3^2}\left[\frac{1}{(\nu_1+\nu_2)^2 - \nu^2} - \frac{1}{\nu_3^2 - \nu^2}\right]\right\}.
\end{aligned}
\qquad (36.13)
$$

On obtient une expression de même forme pour γ lorsque l'état initial est caractérisé par trois nombres quantiques quelconques. Lorsqu'on y remplace $h\nu_j$ par $\frac{1}{2}kX_j^2$ où X est l'amplitude du mouvement harmonique de l'électron, on trouve pour γ une expression qui ne contient plus h, et on déduit du principe de correspondance que l'activité optique du modèle étudié existe également dans la théorie classique[1].

Lorsqu'on calcule la contribution apportée à γ_{000} par la transition $(000) \to (011)$, on trouve pour la force rotatoire $[000\,|\boldsymbol{p}_1|\,011]\,{\scriptstyle\bullet}\,(011\,|\boldsymbol{m}_1|\,000)$ l'expression (36.13) au signe près. Or cette transition ne possède un moment électrique que par suite de la perturbation; elle correspond donc à une faible absorption, tandis que la transition $(000) \to (100)$ possède un moment électrique sans être perturbée et produit par suite une forte absorption. Le facteur de dissymétrie (32.9) de la première doit donc être beaucoup plus grand que celui de la seconde.

37. Théories de couplage. L'échec du modèle électronique de DRUDE (cf. note 4 p. 362) a pu faire croire, jusqu'à l'avènement de la théorie monoélectronique, que les théories de l'activité optique devaient être fondées sur un modèle moléculaire contenant des oscillateurs couplés (Sect. 24).

Les théories classiques, inaptes à traiter la mécanique des électrons, considèrent comme unités de structure des atomes ou des groupes d'atomes auxquels on attribue des propriétés d'ensemble (masse, volume, charge, fréquences propres, etc.). Cette méthode, qui a donné des résultats valables dans l'étude de divers phénomènes optiques[2], présente l'avantage de se rattacher au point de vue de la stéréochimie.

Les forces de couplage utilisées par tous les auteurs se ramènent à des actions de dipoles électriques. Dans ces conditions, le nombre minimum d'oscillateurs couplés nécessaire pour former un modèle moléculaire doué d'activité optique dépend de l'anisotropie des oscillateurs. Si les oscillateurs sont linéaires, les moments électriques induits ne peuvent avoir qu'une direction, et deux oscillateurs couplés peuvent fournir une contribution qui n'est pas nulle à l'expression (28.5), pourvu que leurs directions privilégiées ne soient pas coplanaires: il suffit donc, dans ce cas, de tenir compte des actions mutuelles du premier ordre. Si les oscillateurs sont isotropes, le moment électrique induit est parallèle au champ et il faut considérer des actions du troisième ordre pour obtenir une contribution à l'activité optique; ce résultat ne peut s'obtenir que si la molécule contient au moins quatre oscillateurs différents non coplanaires, en accord avec les représentations stéréochimiques fondées sur une représentation sphérique des atomes.

---

[1] Cette conclusion ne s'applique que si X a une valeur finie, car l'activité optique disparaît en théorie classique si l'amplitude est infinement petite. Voir E. GORIN, J. WALTER et H. EYRING: J. Chem. Phys. **7**, 327 (1939).

[2] R. DE MALLEMANN: Rev. gén. Sci. **38**, 453 (1927). — Trans. Faraday Soc. **26**, 281 (1930).

Un modèle contenant des oscillateurs couplés anisotropes a été utilisé par
Kuhn. Un modèle formé d'oscillateurs isotropes a été étudié par Born.

38. Modèle moléculaire de Kuhn [4], [12]. α) Le modèle moléculaire de Kuhn,
le plus simple qui puisse posséder l'activité optique, est représenté sur la Fig. 45.
En théorie classique, il se compose de deux oscillateurs linéaires harmoniques,
de masses m_1 et m_2, de charges e_1 et e_2, dont les positions d'équilibre O_1 et O_2 sont
séparées par la distance d sur l'axe des z, et qui peuvent se déplacer respectivement
suivant les directions x et y. Les oscillateurs isolés auraient une énergie potentielle

$$V = \frac{f_1}{2} x_1^2 + \frac{f_2}{2} y_2^2$$

et une énergie cinétique

$$T = \frac{m_1}{2} \dot{x}_1^2 + \frac{m_1}{2} \dot{y}_2^2.$$

On suppose que leurs mouvements sont couplés de
telle sorte que l'énergie potentielle s'écrit

$$V = \frac{f_1}{2} x_1^2 + f_{12} x_1 y_2 + \frac{f_2}{2} y_2^2, \tag{38.1}$$

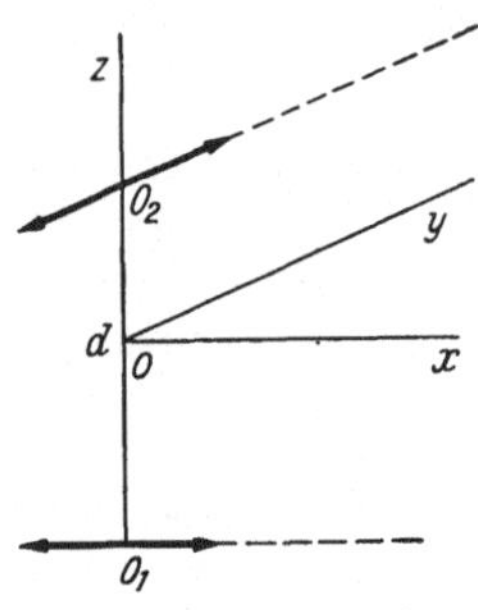

Fig. 45. Modèle moléculaire
de Kuhn.

l'expression de l'énergie cinétique demeurant la même.

On met T et V sous une forme ne contenant que des carrés, en remplaçant
les coordonnées x et y par des coordonnées normales q_1 et q_2 (Sect. 25) au moyen
de la transformation orthogonale

$$\left.\begin{aligned}
\sqrt{m_1}\, x_1 &= q_1 \cos\beta + q_2 \sin\beta . \\
\sqrt{m_2}\, y_2 &= -q_1 \sin\beta + q_2 \cos\beta,
\end{aligned}\right\} \tag{38.2}$$

β étant défini par

$$\left(\frac{f_1}{m_1} - \frac{f_2}{m_2}\right) \sin 2\beta + \frac{f_{12}}{\sqrt{m_1 m_2}} \cos 2\beta = 0. \tag{38.3}$$

On a alors

$$V = 2\pi^2 \left(\nu_1^2 q_1^2 + \nu_2^2 q_2^2\right),$$
$$T = \tfrac{1}{2}\left(\dot{q}_1^2 + \dot{q}_2^2\right),$$

où ν_1 et ν_2 représentent les fréquences fondamentales du modèle

$$\left.\begin{aligned}\nu_1^2 \\ \nu_2^2\end{aligned}\right\} = \frac{1}{8\pi^2}\left[\frac{f_1}{m_1} + \frac{f_2}{m_2} \pm \sqrt{\left(\frac{f_1}{m_1} - \frac{f_2}{m_2}\right)^2 + \frac{4 f_{12}^2}{m_1 m_2}}\right].$$

Le moment électrique produit par le mouvement des oscillateurs a pour
expression

$$\left.\begin{aligned}
\boldsymbol{p}' &= e_1 \boldsymbol{x}_1 + e_2 \boldsymbol{y}_2 \\
&= \left(\boldsymbol{i}\,\frac{e_1}{\sqrt{m_1}} \cos\beta - \boldsymbol{j}\,\frac{e_2}{\sqrt{m_2}} \sin\beta\right) q_1 + \left(\boldsymbol{i}\,\frac{e_1}{\sqrt{m_1}} \sin\beta + \boldsymbol{j}\,\frac{e_2}{\sqrt{m_2}} \cos\beta\right) q_2,
\end{aligned}\right\} \tag{38.4}$$

$\boldsymbol{i}$ et $\boldsymbol{j}$ désignant des vecteurs unitaires portés respectivement par Ox et Oy.
Le moment magnétique créé dans les mêmes conditions est, d'après (24.4)

$$\begin{aligned}
\boldsymbol{m}' &= \frac{1}{2c}\left[-e_1\left(\frac{\boldsymbol{d}}{2} \times \dot{\boldsymbol{x}}_1\right) + e_2\left(\frac{\boldsymbol{d}}{2} \times \dot{\boldsymbol{y}}_2\right)\right] \\
&= -\frac{d}{4c}\left\{\left[\boldsymbol{j}\,\frac{e_1}{\sqrt{m_1}} \cos\beta - \boldsymbol{i}\,\frac{e_2}{\sqrt{m_2}} \sin\beta\right]\dot{q}_1 - \left[\boldsymbol{j}\,\frac{e_1}{\sqrt{m_1}} \sin\beta + \boldsymbol{i}\,\frac{e_2}{\sqrt{m_2}} \cos\beta\right]\dot{q}_2\right\}.
\end{aligned}$$

On peut poursuivre le calcul du pouvoir rotatoire par la méthode de la Sect.27 (voir aussi [2], [20]). Nous le ferons en appliquant la théorie quantique [4], [16].

β) Dans cette théorie, chacun des niveaux d'énergie du modèle est défini par deux nombres quantiques, n_1 et n_2. Les règles de sélection de l'oscillateur harmonique ne permettent que les transitions d'un niveau $(a) = (n_1, n_2)$ à l'un des quatre suivants

$$(b_1) = (n_1 + 1, n_2); \quad (b_1') = (n_1 - 1, n_2); \quad (b_2) = (n_1, n_2 + 1); \quad (b_2') = (n_1, n_2 - 1).$$

L'expression des éléments de matrice des coordonnées normales q et de leurs dérivées $\dot{q}$ se déduit de celle de l'oscillateur harmonique. On a, par exemple:

$$\left.\begin{aligned}
(a \,|\, q_1 |\, b_1) &= (n_1, n_2 \,|\, q_1 |\, n_1 + 1, n_2) = \sqrt{\frac{h\,(n_1 + 1)}{8\,\pi^2 \nu_1}}\; \delta_{n_2 n_2'}, \\
(b_1 \,|\, \dot{q}_1 |\, a) &= (n_1 + 1, n_2 \,|\, \dot{q}_1 |\, n_1, n_2) = \pm\, i \sqrt{\frac{h\,(n_1 + 1)\,\nu_1}{2}}\; \delta_{n_2 n_2'}
\end{aligned}\right\} \tag{38.5}$$

$\delta_{n_2 n_2'}$ vaut 0 ou 1, selon que $n_2 \neq n_2'$ ou $n_2 = n_2'$. Des formules du même genre, où n_1 remplace $n_1 + 1$, se rapportent aux transitions $(a) \rightleftharpoons (b_1')$. Le remplacement de q_1 et $\dot{q}_1$ par q_2 et $\dot{q}_2$ annule les éléments de matrice (38.5).

Mais on a quatre formules analogues aux précédentes, relatives aux transitions $(a) \rightleftharpoons (b_2)$ et $(a) \rightleftharpoons (b_2')$, en remplaçant partout l'indice 1 par 2.

De ce qui précède, on tire:

$$\left.\begin{aligned}
(a \,|\, \boldsymbol{p} |\, b_1) &= \sqrt{\frac{h\,(n_1 + 1)}{8\,\pi^2 \nu_1}} \left(\boldsymbol{i}\, \frac{e_1}{\sqrt{m_1}} \cos\beta - \boldsymbol{j}\, \frac{e_2}{\sqrt{m_2}} \sin\beta \right), \\
(b_1 \,|\, \boldsymbol{m} |\, a) &= -\frac{i\,d}{4\,c} \sqrt{\frac{h\,(n_1 + 1)\,\nu_1}{2}} \left(\boldsymbol{j}\, \frac{e_1}{\sqrt{m_1}} \cos\beta - \boldsymbol{i}\, \frac{e_2}{\sqrt{m_2}} \sin\beta \right), \\
(a \,|\, \boldsymbol{p} |\, b_1') &= \sqrt{\frac{h\,n_1}{8\,\pi^2 \nu_1}} \left(\boldsymbol{i}\, \frac{e_1}{\sqrt{m_1}} \cos\beta - \boldsymbol{j}\, \frac{e_2}{\sqrt{m_2}} \sin\beta \right), \\
(b_1' \,|\, \boldsymbol{m} |\, a) &= -\frac{i\,d}{4\,c} \sqrt{\frac{h\,n_1\,\nu_1}{2}} \left(\boldsymbol{j}\, \frac{e_1}{\sqrt{m_1}} \cos\beta - \boldsymbol{i}\, \frac{e_2}{\sqrt{m_2}} \sin\beta \right)
\end{aligned}\right\} \tag{38.6}$$

et des formules analogues pour les transitions $(a) \rightleftharpoons (b_2)$ et $(a) \rightleftharpoons (b_2')$. On calcule alors les forces rotatoires moyennes au moyen de la formule (30.5), en se rappelant que $\boldsymbol{i} \cdot \boldsymbol{j} = 0$ et $\boldsymbol{i} \cdot \boldsymbol{i} = \boldsymbol{j} \cdot \boldsymbol{j} = 1$:

$$R_{b_1 a} = (n_1 + 1)\, \frac{h}{2\pi}\, \frac{d}{4\,c}\, \frac{e_1 e_2}{\sqrt{m_1 m_2}}\, \sin\beta \cos\beta,$$

$$R_{b_1' a} = -\, n_1\, \frac{h}{2\pi}\, \frac{d}{4\,c}\, \frac{e_1 e_2}{\sqrt{m_1 m_2}}\, \sin\beta \cos\beta.$$

En prenant pour valeurs totales des forces rotatoires

$$\overline{R}_1 = \overline{R}_{b_1 a} + \overline{R}_{b_1' a} \quad \text{et} \quad \overline{R}_2 = \overline{R}_{b_2 a} + \overline{R}_{b_2' a}, \tag{38.7}$$

on trouve, en accord avec la règle de sommation (31.10):

$$\overline{R}_1 = -\,\overline{R}_2 = \frac{h}{2\pi}\, \frac{d}{4\,c}\, \frac{e_1 e_2}{\sqrt{m_1 m_2}}\, \sin\beta \cos\beta. \tag{38.8}$$

En introduisant ces valeurs dans la formule (30.4) et en admettant que la probabilité pour que le modèle se trouve dans l'état (a) est égale à l'unité, on obtient la formule donnée par Kuhn[1] lorsque la distance d est petite devant la longueur d'onde λ de l'onde lumineuse

$$\bar{\varrho} = \frac{2\pi N d}{3 c^2} \frac{e_1 e_2}{\sqrt{m_1 m_2}} \sin\beta \cos\beta \, \nu^2 \left[\frac{1}{\nu_1^2 - \nu^2} - \frac{1}{\nu_2^2 - \nu^2} \right]. \tag{38.9}$$

La valeur finie de d est essentielle à l'existence du pouvoir rotatoire, de même que l'existence du couplage, car d'après (38.3), si $f_{12} = 0$, $\sin 2\beta = 0$.

La contribution de la fréquence propre ν_1 à la biréfringence circulaire a pour valeur

$$(n_g - n_d)_{\nu_1} = \frac{\lambda_0 \varrho_1}{\pi} = \frac{2 N d}{3 \mathbf{c}} \frac{e_1 e_2}{\sqrt{m_1 m_2}} \sin\beta \cos\beta \frac{\nu}{\nu_1^2 - \nu^2}. \tag{38.10}$$

D'autre part, en introduisant les expressions (38.4) dans la formule (29.10), compte tenu de (38.8), on obtient la part de polarisabilité due à la fréquence propre ν_1 et la contribution à la réfraction

$$(n - 1)_{\nu_1} = 2\pi N \alpha_1 = \frac{N}{6\pi} \left(\frac{e_1^2}{m_1} \cos^2\beta + \frac{e_2^2}{m_2} \sin^2\beta \right) \frac{1}{\nu_1^2 - \nu^2}. \tag{38.11}$$

Dans les régions d'absorption, Kuhn exprime le dichroïsme circulaire en ajoutant dans les équations du mouvement écrites à l'aide des coordonnées normales, un coefficient d'amortissement, ce qui conduit à une expression analogue à (32.12).

Le facteur de dissymétrie (32.9) a pour expression, d'après (38.10) et (38.11)

$$G_{\nu_1} = \frac{(n_g - n_d)_{\nu_1}}{(n - 1)_{\nu_1}} = \frac{4\pi\nu}{c} d \frac{\dfrac{e_1 e_2}{\sqrt{m_1 m_2}} \sin\beta \cos\beta}{\dfrac{e_1^2}{m_1} \cos^2\beta + \dfrac{e_2^2}{m_2} \sin^2\beta}$$

On introduit les intensités des oscillateurs (32.7) en posant

$$\frac{e_1^2}{m_1} = F_1 \frac{e^2}{m}, \qquad \frac{e_2^2}{m_2} = F_2 \frac{e^2}{m}, \tag{38.12}$$

e et m représentant respectivement la charge et la masse de l'électron:

$$G_{\nu_1} = \frac{4\pi\nu}{c} d \frac{\sqrt{F_1 F_2} \sin\beta \cos\beta}{F_1 \cos^2\beta + F_2 \sin^2\beta}. \tag{38.13}$$

La formule (32.10) est valable ici encore.

δ) Examinons quelques conséquences des formules précédentes:

1. Le facteur de dissymétrie (38.13) a une valeur maximum

$$G_{\nu_1 m} = \frac{2\pi\nu d}{c} \tag{38.14}$$

lorsque

$$\sqrt{F_1} \cos\beta = \sqrt{F_2} \sin\beta.$$

Cette dernière relation exprime, d'après (38.4), que le moment électrique produit par l'oscillateur 1 dans l'oscillation fondamentale q_1 est égal à celui que produit l'oscillateur 2.

2. On tire de (29.10), (38.8), (38.11) et (38.13):

$$G_{\nu_1} S_{\nu_1} = - G_{\nu_2} S_{\nu_2}. \tag{38.15}$$

---

[1] Référence [13], formule 47. Le facteur $\tfrac{1}{3}$ de la formule (38.9) vient du calcul de moyennes.

Seules les bandes d'absorption faibles peuvent posséder de grands facteurs de dissymétrie. Ainsi s'expliquent les faits expérimentaux signalés à la Sect. 35. Voir aussi l'article de Kuhn dans [12].

3. La relation permet de calculer une valeur minimum pour la distance d qui sépare dans la molécule les oscillateurs couplés. On trouve parfois des valeurs acceptables (azidopropionate de méthyle, $d = 2,9$ Å) mais d'autres beaucoup trop grandes (camphre, $d = 45$ Å). Kuhn et Bein[1] ont modifié le modèle de la Fig. 45, en supposant que les deux oscillateurs font un angle égal à $\pi - \eta$, voisin de π (Fig. 46) et ont obtenu des valeurs de d plus normales (camphre, $d = 0,6$ Å).

ε) En ce qui concerne la nature des forces qui donnent naissance au terme $f_{12} x_1 y_2$ de la formule (38.1), Kuhn, après avoir essayé de montrer qu'elles ne peuvent pas être de nature électrodynamique [13], a admis qu'elles sont dues aux actions mutuelles des dipôles électriques induits, lorsqu'il a cherché à appliquer son modèle moléculaire à l'étude des spiranes[2], des complexes de Werner[2] et de l'acide lactique[3].

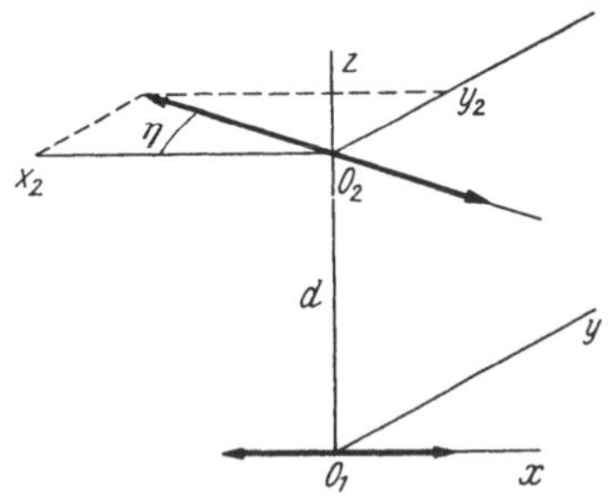

39. Modèle moléculaire de Born[4]. Il se compose de quatre oscillateurs isotropes disposés aux sommets d'un tétraèdre irrégulier. Pour appliquer une méthode de perturbation, on met la formule (25.1) donnant l'énergie potentielle sous la forme

$$2V = 2V_0 + 2V_1$$
$$= \sum_k \sum_{xy} f_{xy}^k x_k y_k + \sum_{kl}' \sum_{xy} f_{xy}^{kl} x_k y_l, \qquad (39.1)$$

z
y_2
η
x_2
O_2
d
y
O_1
x

Fig. 46. Modification du modèle de la Fig. 45.

l'accent indiquant que les cas $k = l$ sont exclus.

Le premier terme se rapporte aux oscillateurs non couplés; le second est dû aux interactions des oscillateurs et constitue la fonction perturbatrice. Les équations du mouvement de l'oscillateur k non perturbé, analogues à (25.4), admettent une solution harmonique, dont on désigne le carré de la pulsation par Ω_k^{0j}; il y a en principe trois fréquences $(j = 1, 2, 3)$, que l'on doit confondre si l'oscillateur est isotrope. En introduisant des coordonnées normales q_k^i comme en (25.9)

$$u_{kx} = \sqrt{m_k}\, x_k = \sum_{j=1}^{3} a_{kx}^j q_k^j,$$

les équations du mouvement non perturbé s'écrivent

$$(\Omega - \Omega_k^{0j})\, q_k^j = 0. \qquad (39.2)$$

La fonction perturbatrice a pour expression

$$\sum_{j, i = 1}^{3} \sum_{kl} a_{ji}^{kl}\, q_k^i q_l^i$$

avec

$$a_{ji}^{kl} = \sum_{xy} f_{xy}^{kl}\, a_{kx}^i a_{ly}^j \quad (i \neq j), \qquad a_{ij}^{kl} = 0.$$

Les équations du mouvement perturbé sont

$$(\Omega - \Omega_k^{0j})\, q_k^j = \sum_{j=1}^{3} \sum_l a_{ji}^{kl}\, q_l^i. \qquad (39.3)$$

[1] W. Kuhn et K. Bein: Z. phys. Chem., Abt. B **22**, 406 (1933).
[2] W. Kuhn et K. Bein: Z. phys. Chem., Abt. B **24**, 335 (1934).
[3] W. Kuhn: Z. phys. Chem., Abt. B **31**, 23 (1935).
[4] M. Born: Proc. Roy. Soc., Lond., Ser. A **150**, 84 (1935).

En développant Ω et les q en fonction des puissances croissantes d'un paramètre λ et en égalant séparément les coefficients des puissances successives dans les deux membres, on trouve des équations qui expriment des perturbations d'ordre croissant. Il est nécessaire de pousser l'approximation jusqu'au troisième ordre, de calculer les valeurs de q correspondantes, puis, au moyen des relations (25.8), les amplitudes u_{kx} dans les divers ordres d'approximation, et les expressions de la constante de gyration g correspondantes (pour le détail des calculs, voir [20]).

On admet que les coefficients f_{xy}^{kl} de (39.1) sont dus aux actions de dipole, c'est-à-dire que

$$f_{xy}^{kl} = \frac{e_k e_l}{r_{kl}^3} \left(3 \frac{x_{kl} y_{kl}}{r_{kl}^2} - \delta_{xy} \right) \tag{39.4}$$

avec $\delta_{xy} = 1$ pour $x = y$ et $\delta_{xy} = 0$ pour $x \neq y$. On obtient finalement pour la valeur de la constante de gyration g^l correspondant au mouvement de la particule l et en négligeant l'anisotropie des résonateurs, la formule:

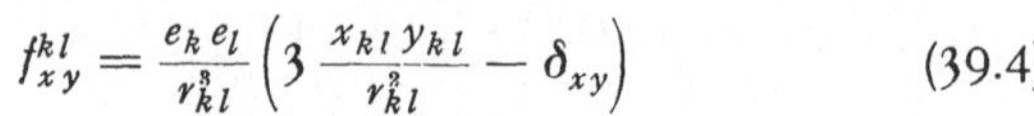

Fig. 47. Modèle moléculaire de Born.

$$g^l = 27 \frac{e^8}{m^4} F_l \sum_{k < m < p} \frac{F_k F_m F_p}{(\Omega_l^0 - \Omega_p^0)(\Omega_l^0 - \Omega_m^0)(\Omega_l^0 - \Omega_k^0)} \Delta(k\,m\,p\,l)\, G_{k\,m\,p}^l \tag{39.5}$$

dans laquelle les F représentent les intensités des oscillateurs données par (38.12) et où $\Delta(k m p l)$ et $G_{k m p}^l$ sont uniquement fonctions de la forme et des dimensions du tétraèdre dont les sommets sont occupés par les quatre résonateurs k, m, p et l.

Born a appliqué les formules précédentes au cas d'un modèle moléculaire asymétrique, formé de quatre résonateurs de deux espèces différentes I et II, disposés aux sommets d'un tétraèdre irrégulier comme le montre la Fig. 47.

40. Théories de polarisabilité globale. α) Elles constituent la forme la plus simplifiée des théories de l'activité optique fondées sur le couplage des oscillateurs. On décompose la molécule en groupements atomiques assez symétriques pour n'avoir pas isolément d'activité optique. On leur associe un tenseur de polarisabilité $[\alpha]_k$ (Sect. 24) dont la trace est la polarisabilité moyenne

$$\alpha_k = \frac{\alpha_{1k} + \alpha_{2k} + \alpha_{3k}}{3} \tag{40.1}$$

et dont l'anisotropie est definie par

$$\delta_k^2 = \frac{(\alpha_{1k} - \alpha_{2k})^2 + (\alpha_{2k} - \alpha_{3k})^2 + (\alpha_{3k} - \alpha_{1k})^2}{2\,\alpha_k^2}. \tag{40.2}$$

On peut écrire les relations telles que (23.3) ou (29.10) sous la forme

$$\alpha_k = \sum_j \alpha_k^j$$

α_k^j désignant la contribution de la bande j à la polarisabilité du groupe k. En considérant la polarisabilité globale moyenne α_k au lieu des α_k^j, ces theories s'interdisent la synthèse de la courbe de dispersion rotatoire (Sect. 34) et donnent une approximation plus grossière que les théories examinées précédemment, comme le montre la critique suivante: la réfractivité de la molécule est, d'après (28.6), proportionnelle à la polarisabilité α de la molécule.

$$n^2 - 1 = C\,\alpha = C \sum \alpha_k. \tag{40.3}$$

L'existence de l'effet Cotton dans une bande d'absorption due à un groupe chromophore exigerait que le terme α_k relatif à ce groupe s'annule. Or cette conclusion est en désaccord avec les données de l'expérience: par exemple[1], α_{CO} est loin de s'annuler dans la bande d'absorption de la cyclohexanone située vers 2900 Å, qui exerce cependant une grande influence sur la rotation des cétones actives (cf. Fig. 38).

Les forces de couplage considérées par tous les auteurs se ramènent, en dernière analyse, à des actions de dipôle. L'ordre d'approximation auquel il faut parvenir pour expliquer l'activité optique dépend, ici comme dans les autres théories de couplage (Sect. 37), des hypothèses faites sur l'anisotropie des groupes atomiques.

β) De Mallemann [18], [23] a appliqué la théorie classique à divers modèles moléculaires formés d'atomes ou de groupements isotropes ou anisotropes. Boys[2] a étudié un modèle en tétraèdre asymétrique du type $CR_1R_2R_3R_4$ (Sect. 18) dans lequel les groupements R sont supposés isotropes, et trouve pour le pouvoir rotatoire d'un gaz parfait formé de ces molécules l'expression

$$\varrho = \pm \frac{N(n^2+2)(n^2+5)\,\alpha_1\alpha_2\alpha_3\alpha_4}{\lambda^2}\,I \qquad (40.4)$$

I désignant un facteur de structure d'expression compliquée, mais qui n'est fonction que des constantes géométriques du tétraèdre. Pour le détail de ces théories, voir [9], [20].

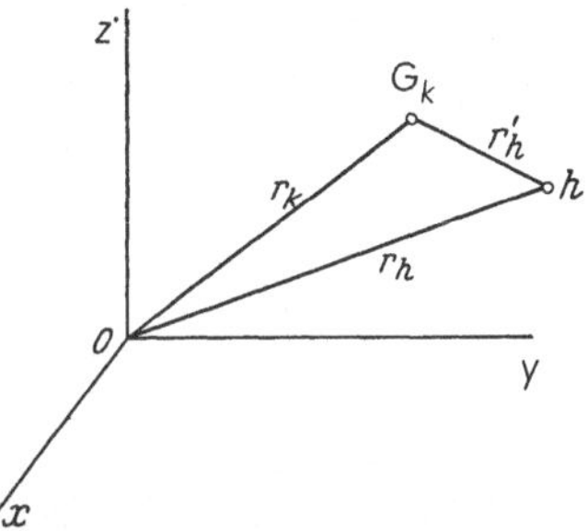

Fig. 48. Méthode de Kirkwood.

γ) Kirkwood[3] a transformé la formule générale (30.2) de la théorie quantique de façon à y faire intervenir les propriétés de groupes atomiques pris pour unité de structure. On répartit les n électrons entre n' groupes pris pour constituants de la molécule, ce qui revient à négliger les phénomènes d'échanges électroniques, et on localise dans des groupes déterminés les transitions $a \to b$ entre niveaux d'énergie électronique de la molécule. On peut alors écrire le moment électrique de la molécule

$$\boldsymbol{p} = \sum_{h=1}^{n} \boldsymbol{p}_h = \sum_{k=1}^{n'} \boldsymbol{p}_k = \sum_{k=1}^{n'} e \sum_{h=1}^{s} \boldsymbol{r}'_h \qquad (40.5)$$

$\boldsymbol{r}'_h$ désignant la distance de l'électron d'indice h au centre de gravité G_k (Fig. 48) du groupe d'indice k qui contient s électrons. Le moment magnétique de la molécule a pour expression

$$\boldsymbol{m} = \frac{e}{2\,m\,c}\sum_{h=1}^{n}\overline{\boldsymbol{\pi}}_h \times \boldsymbol{r}_h = \frac{e}{2\,m\,c}\left(\sum_{k=1}^{n'}\overline{\boldsymbol{\pi}}_k \times \boldsymbol{r}_k + \sum_{k=1}^{n'}\boldsymbol{m}_k\right) \qquad (40.6)$$

$\boldsymbol{\pi}_k$ désignant la quantité de mouvement totale des électrons du groupe k et $\boldsymbol{m}_k$ le moment magnétique du groupe k rapporté au point G_k.

Les formules (40.5) et (40.6) permettent de transformer l'expression (30.2) en la suivante:

$$\left.\begin{aligned}\gamma = \frac{e}{6\pi h m}\sum_b \frac{1}{v_{ba}^2 - v^2}\,\mathrm{Im}\Big\{&\sum_k \sum_l (a\,|\,\boldsymbol{p}_k\,|\,b)\cdot[\boldsymbol{r}_l \times (b\,|\,\boldsymbol{m}_l\,|\,a)] + \\ &+ \sum_l \sum_{l \neq k}(a\,|\,\boldsymbol{p}_k\,|\,b)\cdot(b\,|\,\boldsymbol{m}_l\,|\,a) + \sum_k (a\,|\,\boldsymbol{p}_k\,|\,b)(b\,|\,\boldsymbol{m}_k\,|\,a)\Big\},\end{aligned}\right\} \quad (40.7)$$

[1] T. M. Lowry et C. B. Allsopp: Proc. Roy. Soc. Lond., Ser. A **146**, 313 (1934).
[2] S. F. Boys: Proc. Roy. Soc. Lond., Ser. A **144**, 655 (1934).
[3] J. Kirkwood: J. Chem. Phys. **5**, 479 (1937); **7**, 139 (1939).

l désignant un groupe différent de k. Kirkwood ne conserve que le premier terme de l'expression (40.7) qui doit, en effet, être plus grand que les autres. On le transforme en remarquant que

$$(b \,|\, \boldsymbol{m}_l \,|\, a) = \frac{2\pi i m}{e}\, \nu_{ba}\, (b \,|\, \boldsymbol{p}_l \,|\, a)$$

et que Im $(ix) = \mathrm{Re}(x)$; on obtient, en remaniant les sommes

$$\gamma = \frac{1}{6h} \sum_b \sum_k \sum_{l \neq k} \frac{\nu_{ba}}{\nu_{ba}^2 - \nu^2}\, \mathrm{Re}\left\{ \boldsymbol{r}_{kl}\left[(a \,|\, \boldsymbol{p}_k \,|\, b) \times (b \,|\, \boldsymbol{p}_l \,|\, a) \right] \right\}. \tag{40.8}$$

$\boldsymbol{r}_{kl}$ est le vecteur qui va de G_k à G_l. Pour effectuer le calcul des éléments de matrice de la formule (40.8), on néglige en première approximation les interactions des groupes atomiques; on en tient compte ensuite, en les considérant comme une perturbation de l'état précédent. On choisit comme terme perturbateur du hamiltonien une somme d'expressions qui représentent les énergies potentielles V_{kl} d'interaction des groupes pris deux à deux. La formule (36.3) s'écrit

$$c_{ab} = - \frac{(a \,|\, H_1 \,|\, b)}{E_b - E_a} = - \frac{1}{h} \sum_{l > k} \frac{(a_k\, a_l \,|\, V_{kl} \,|\, b_k\, b_l)}{\nu_{a_k b_k} + \nu_{a_l b_l}}, \tag{40.9}$$

car l'élément de matrice $(a \,|\, V_{kl} \,|\, b)$ contient les variations des nombres quantiques des groupes k et l. On calcule les fonctions d'onde perturbées et les éléments de matrice $\boldsymbol{p}$ au moyen de la formule (36.4). On trouve finalement, en supposant que l'état initial a est l'état fondamental 0 pour chacun des groupes

$$\gamma_0 = - \frac{2}{3h^2} \sum_{k \neq l} \sum_{b_k} \sum_{b_l} \frac{\nu_{0 b k}\, \nu_{0 b l}}{(\nu_{0 b_k}^2 - \nu^2)(\nu_{0 b_l}^2 - \nu^2)}\, \boldsymbol{r}_{kl}\left[(0 \,|\, \boldsymbol{p}_k \,|\, b_k) \cdot (0 \,|\, \boldsymbol{p}_l \,|\, b_l) \right] (00 \,|\, V_{kl} \,|\, b_k\, b_l). \tag{40.10}$$

On admet alors que les seules interactions qui interviennent dans V_{kl} sont celles des dipôles électriques, ce qui donne

$$(00 \,|\, V_{kl} \,|\, b_k\, b_l) = (0 \,|\, \boldsymbol{p}_k \,|\, b_k) \cdot T_{kl} \cdot (0 \,|\, \boldsymbol{p}_l \,|\, b_l)$$

avec

$$T_{kl} = \frac{1}{r_{kl}^2}\left[1 - \frac{3\, \boldsymbol{r}_{kl} \cdot \boldsymbol{r}_{kl}}{r_{kl}^2} \right].$$

La formule (40.10) devient

$$\left. \begin{aligned} \gamma_0 = &- \frac{2}{3h^2} \sum_{k \neq l} \sum_{b_k} \sum_{b_l} \frac{\nu_{0 b_k}\, \nu_{0 b_l}}{(\nu_{0 b_k}^2 - \nu^2)(\nu_{0 b_l}^2 - \nu^2)} \times \\ &\times \boldsymbol{r}_{kl}\left[(0 \,|\, \boldsymbol{p}_k \,|\, b_k)(0 \,|\, \boldsymbol{p}_l \,|\, b_l) \right]\left[(0 \,|\, \boldsymbol{p}_k \,|\, b_k)\, T_{kl}\, (0 \,|\, \boldsymbol{p}_l \,|\, b_l) \right]. \end{aligned} \right\} \tag{40.11}$$

On attribue à chaque groupe atomique un tenseur de polarisabilité ayant trois axes principaux définis par trois vecteurs unités $\boldsymbol{b}_t (t = 1, 2, 3)$; les composantes des moments $\boldsymbol{p}_k \ldots$ peuvent s'écrire $\boldsymbol{p}_{tk} = p_{tk}\, \boldsymbol{b}_{tk}$ et comme elles sont reliées aux coefficients α_{tk} du tenseur de polarisabilité par une formule analogue à (29.10)

$$\alpha_{tk} = \frac{2}{h} \sum_{b_k} \frac{\nu_{0 b_k}}{\nu_{0 b_k}^2 - \nu^2}\, (a_k \,|\, \boldsymbol{p}_{tk} \,|\, b_k)^2,$$

la formule (40.11) s'écrit

$$\gamma_0 = - \frac{1}{6} \sum_{l \neq k} \sum_{t, u = 1}^{3} \alpha_{tk}\, \alpha_{ul}\, (\boldsymbol{b}_{tk} \cdot T_{kl} \cdot \boldsymbol{b}_{ul})\, \boldsymbol{r}_{kl} \cdot (\boldsymbol{b}_{tk} \times \boldsymbol{b}_{ul}). \tag{40.12}$$

Lorsque le tenseur de polarisabilité des groupes k et l est de révolution autour de la direction $\boldsymbol{b}_1$ par exemple, on peut faire coïncider les directions $\boldsymbol{b}_{2k}$ et $\boldsymbol{b}_{2l}$; la somme de (40.12) relative à ces groupes se simplifie et l'on a, en introduisant l'anisotropie δ donnée par (40.2)

$$\gamma_0 = - \frac{1}{6} \sum_{l \neq k} \alpha_k\, \delta_k\, \alpha_l\, \delta_l\, (\boldsymbol{b}_k \cdot T_{kl} \cdot \boldsymbol{b}_l)\, \boldsymbol{r}_{lk}\, (\boldsymbol{b}_k \times \boldsymbol{b}_l). \tag{40.13}$$

Cette formule montre que le terme (k, l) de la somme qui entre dans γ_0 s'annule si l'un des groupes est isotrope ou si les axes $\boldsymbol{b}_1$ de ces groupes sont parallèles entre eux, comme l'exige la symétrie, et en outre, si les axes $\boldsymbol{b}_1$ sont tous deux perpendiculaires à $\boldsymbol{r}_{kl}$, ce qui vient de la forme particulière de V_{kl}. En particulier, pour $\delta_k = \delta_l = 1$, on retrouve les oscillateurs linéaires de KUHN et le modèle de la Fig. 45 serait dépourvu d'activité optique; effectivement, lorsque KUHN a couplé ses oscillateurs par actions de dipôle, il a dû placer hors de l'axe Oz leurs positions d'équilibre (Sect. 41γ).

La formule (40.10) présente une grande analogie avec la formule (31.5) de la théorie classique. Les moments électriques des oscillateurs, leurs distances mutuelles y interviennent; le facteur V_{kl} exprime le couplage qui est une des conditions fondamentales de la théorie classique. Celle-ci apparaît donc comme un cas particulier de la théorie quantique, puisque la formule (40.10) se déduit de la théorie quantique générale par un ensemble d'hypothèses supplémentaires.

41. Configuration absolue des molécules dissymétriques. Les théories modernes du pouvoir rotatoire ont été appliquées[1] à la recherche des configurations absolues (Sect. 21).

α) La molécule de méthyléthylcarbinol $C^* \cdot H \cdot OH \cdot CH_3 \cdot C_2H_5$, $[\varrho]_D = 13{,}9°$, a fait l'objet de divers calculs dont on trouvera le détail en [*20*].

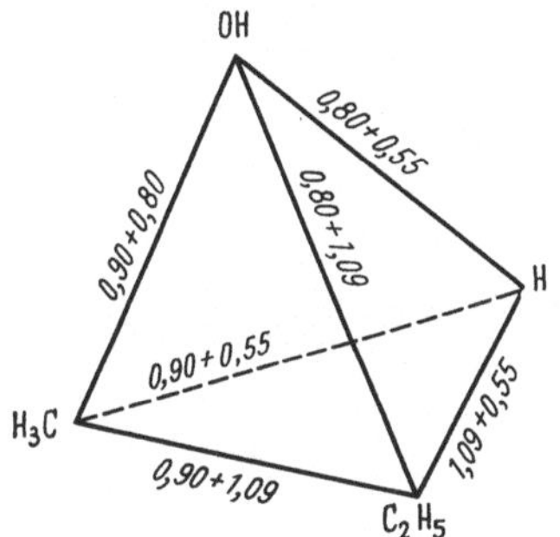

Fig. 49. Modèle de la molécule de méthyléthylcarbinol (BOYS).

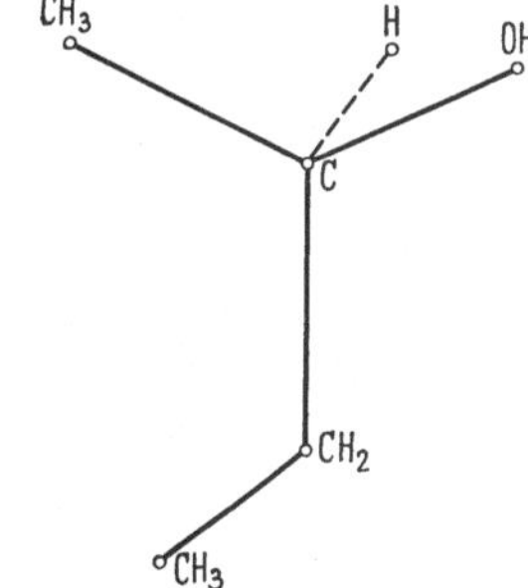

Fig. 50. Modèle de la molécule de méthyléthylcarbinol (KIRKWOOD).

BOYS[2] suppose isotropes et de forme sphérique les quatre groupes liés à l'atome C^*. Le facteur de structure I de la formule (40.4) s'exprime au moyen des rayons des groupes; son signe, donc celui de ϱ, est positif pour le schéma de FISCHER de la Fig. 49. On calcule $[\varrho]_D = 9{,}3°$.

KIRKWOOD[3] part du modèle moléculaire de la Fig. 50 et admet qu'il est également probable que l'axe de la liaison $CH_2 - CH_3$ soit coplanaire avec l'une des directions $C^* - CH_3$ (cas a) ou $C^* - OH$ (cas b). Les seules interactions qui interviennent dans la formule (40.13) sont celles du groupe $CH_2 - CH_3$ avec

[1] W. KUHN: Z. Elektrochem. **56**, 506 (1952).
[2] S. F. BOYS: Proc. Roy. Soc. Lond., Ser. A **144**, 655 (1934).
[3] J. G. KIRKWOOD: J. Chem. Phys. **5**, 479 (1937).

C*—OH dans le cas a, avec C*—CH_3 dans le cas b, avec C*—H dans les deux cas. Mais cette dernière change de signe en passant d'un cas à l'autre et disparaît lorsqu'on calcule la moyenne arithmétique γ_0 des facteurs γ_{a0} et γ_{b0} donnés par (40.13). Si le vecteur b relatif au groupe CH_2—CH_3 est orienté comme l'indique la Fig. 51a, on trouve $[\varrho]_D = +21,9°$; s'il l'est comme sur la Fig. 51b, $[\varrho]_D = +9,5°$ La configuration absolue est la même que celle trouvée par Boys; mais elle serait inverse[1] si la direction CH_2—CH_3 était coplanaire avec C*—H.

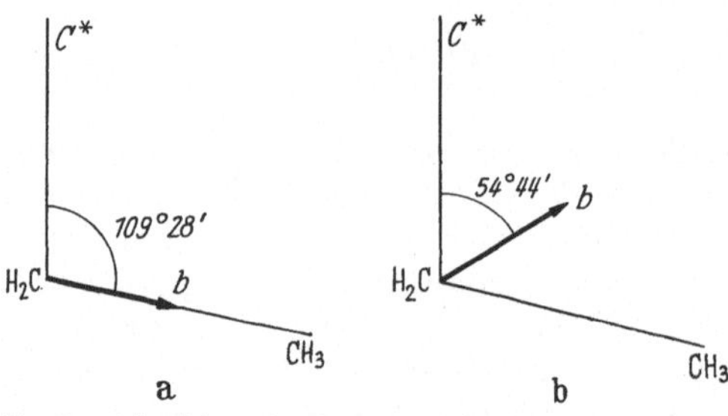

Fig. 51 a et b. Orientation du groupe C_2H_5 dans la molécule de méthyléthylcarbinol.

Kuhn[2] admet que la rotation du méthyléthylcarbinol dans le visible est déterminée par le signe de la contribution des bandes d'absorption dues au groupe OH. L'absorption de ce groupe est rendue anisotrope par le champ électrostatique dû aux autres atomes de la molécule et possède trois fréquences propres. C'est la plus faible, c'est à dire la plus proche du visible, qui fixerait le signe de la

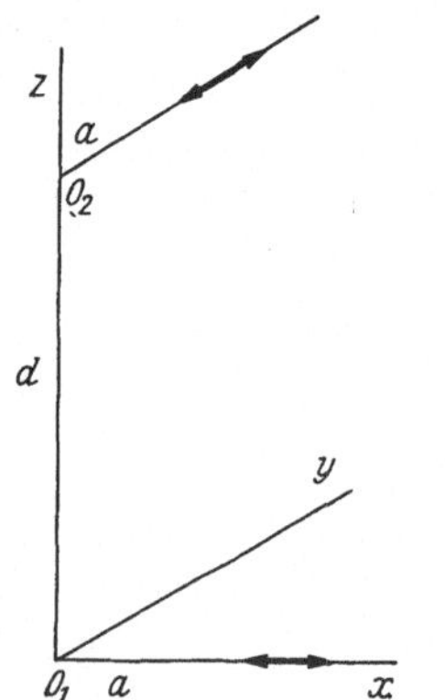

Fig. 52. Modèle moléculaire de spirane (Kuhn et Bein).

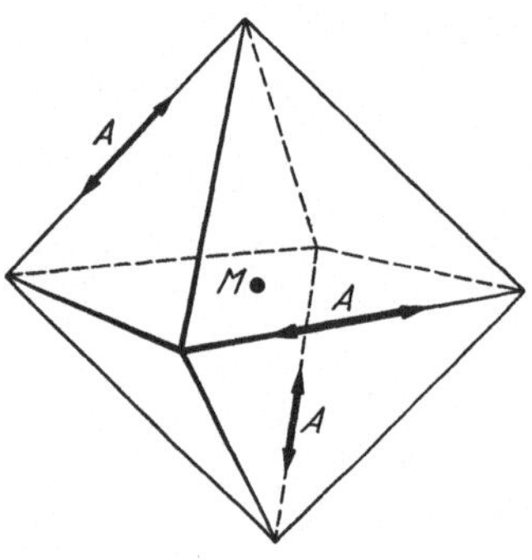

Fig. 53. Modèle moléculaire de complexe MA_3 (Kuhn et Bein).

rotation. Des considérations stériques incertaines font penser que la configuration absolue du composé dextrogyre est inverse de celle de la Fig. 50.

Condon, Altar et Eyring[3], puis Gorin, Walter et Eyring[4] ont appliqué la théorie monoélectronique (Sect. 36) en admettant que la transition qui donne naissance à la bande d'absorption dont dépend essentiellement le pouvoir rotatoire est due à un électron de l'atome O qui n'intervient pas dans les liaisons avec les atomes C* et H. L'état initial $2p$ est regardé comme non perturbé. L'état final est formé d'états $2p$, $3s$, $3p$, $3d$, combinés par la perturbation due aux atomes avoisinants. Le calcul des fonctions d'onde conduit pour l'antipode de la Fig. 49, à des valeurs de $[\varrho]_D$ comprises entre $-85°$ et $+6°$, suivant les hypothèses faites sur la nature de la perturbation (méthode de Hartree ou actions dipolaires) et sur la disposition relative des atomes.

---

[1] Au sujet des isomères de rotation du méthyléthylcarbinol, voir H. J. Bernstein et E. E. Pedersen: J. Chem. Phys. **17**, 885 (1949).

[2] W. Kuhn: Z. phys. Chem., Abt. B **31**, 23 (1935).

[3] E. U. Condon, W. Altar et H. Eyring: J. Chem. Phys. **5**, 753 (1937).

[4] E. Gorin, J. Walter et H. Eyring: J. Chem. Phys. **6**, 824 (1938).

β) La méthode de KIRKWOOD a été appliquée[1] à l' epoxy 2,3 butane $(CH_3CH)_2O$ et au dichloro 1,2 propane $C*HCl \cdot CH_3 \cdot CH_2Cl$. Dans les deux cas, la configuration de FISCHER désignée conventionnellement par d correspond aux composés dextrogyres.

γ) La configuration des spiranes a été étudié par KUHN et BEIN[2] sur un schéma moléculaire qui diffère du modèle de la figure 45 en ce que les positions d'équilibre des deux oscillateurs ne sont plus sur l'axe Oz, mais à une distance a de celui-ci (Fig. 52). Le même problème a été étudié par BORN[3] sur un modèle obtenu en posant $r_2 = r_3$ dans celui de la Fig. 47. Les deux auteurs parviennent à des conclusions opposées pour la configuration absolue.

δ) La configuration absolue des complexes minéraux du type MA_3 (Sect. 20η) a été étudiée par KUHN et BEIN[2] sur un modèle où l'atome M est constitué par un oscillateur isotrope et les groupes A par des oscillateurs linéaires orientés suivant les arêtes de l'octaèdre (Fig. 53). On trouve que l'édifice de la Fig. 53 doit avoir un effet COTTON positif dans la bande d'absorption électronique de plus faible fréquence.

ε) Des incertitudes et des contradictions qui précèdent, on retiendra la difficulté du problème, l'extrême sensibilité de l'activité optique (en grandeur et en signe) à l'égard des distances interatomiques et de l'orientation des liaisons. On verra à la Sect. 57 que la configuration absolue de l'acide tartrique a été établie de façon sûre par voie expérimentale.

V. Interprétation théorique de règles empiriques.

On sait que le pouvoir rotatoire d'une substance varie beaucoup avec les modifications de la constitution chimique et des conditions expérimentales de mesure. Pour des raisons de commodité, on peut distinguer, parmi les facteurs de cette variation, ceux qui sont intérieurs à la molécule (substitutions chimiques ou stériques) de ceux qui lui sont extérieurs (température, état physique et, dans le cas des solutions, concentration et nature du solvant). Cette division est un peu arbitraire, car les changements du pouvoir rotatoire doivent se ramener, en dernière analyse, à des modifications intramoléculaires.

Nous examinerons dans ses grandes lignes le parti qu'on a pu tirer des théories récentes pour relier, au moins qualitativement, les variations du pouvoir rotatoire aux modifications de la structure moléculaire et pour mettre de l'ordre dans le très grand nombre d'observations empiriques qui attendent une classification.

42. Actions de voisinage et nature des liaisons chimiques. La dissymétrie induite dans un chromophore donné n'est déterminée que si les groupes qui entourent ce chromophore forment une configuration bien définie, c'est-à-dire si les liaisons de valence qui les unissent à lui ont une direction fixe dans les conditions expérimentales d'étude. Ce n'est pas le cas des électrovalences des sels dissous. Par suite, si un chromophore, tel qu'un ion métallique simple, est relié par électrovalence à un reste dissymétrique, tel qu'un ion organique optiquement actif, la dispersion rotatoire du sel dissocié par dissolution ne dépend pas de la nature du métal et les bandes d'absorption du chromophore métallique ne montrent pas d'effet COTTON. Exemple: le phénylsuccinate de chrome $[CO_2CH(C_6H_5)CH_2CO_2]_3Cr_2$.

[1] W. W. WOOD, W. FICKETT et J. G. KIRKWOOD: J. Chem. Phys. **20**, 561 (1952).
[2] W. KUHN et K. BEIN: Z. phys. Chem., Abt. B **24**, 335 (1934).
[3] M. BORN: Proc. Roy. Soc. Lond., Ser. A **150**, 84 (1935).

Par contre, les liaisons de covalence et les liaisons de coordinence, qui sont dirigées et résistent également à la dissociation électrolytique, ne diffèrent pas dans leur aptitude à transmettre les actions de voisinage[1]. Par exemple, l'effet Cotton existe dans les bandes d'absorption du chromophore Co, aussi bien chez les trioxalates complexes $Co(C_2O_4)_3^{---}$, où les liaisons $Co=C_2O_4$ sont des covalences, que chez les sels de triéthylènediamine $Co(C_2N_2H_8)_3^{+++}$ où les liaisons $CO=C_2N_2H_8$, sont des coordinences.

43. Actions de voisinage et libre rotation. Supposons que, tout en étant maintenu par une liaison rigide à une distance déterminée d'un chromophore, un groupe atomique puisse tourner autour de la direction de cette liaison et cherchons les modifications qui résultent de cette rotation pour l'action de voisinage.

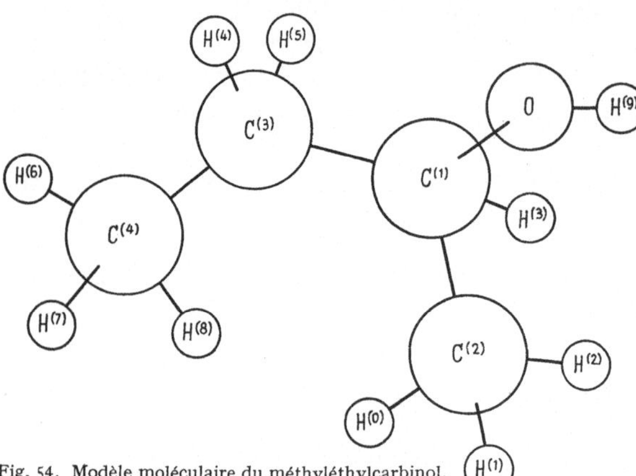

Fig. 54. Modèle moléculaire du méthyléthylcarbinol.

Considérons le modèle de la molécule de méthyléthylcarbinol représenté par la Fig. 54 et examinons l'action de voisinage du groupe éthyle sur l'atome d'hydrogène $H^{(3)}$ lié à l'atome de carbone asymétrique $C^{(1)}$. Si les atomes $H^{(3)}$, $C^{(1)}$, $C^{(3)}$ et $C^{(4)}$ ne sont pas coplanaires, cette action est du premier ordre. Mais si le groupe éthyle peut tourner autour de la liaison $C^{(1)}-C^{(3)}$, de façon à acquérir une symétrie moyenne de révolution autour de cette direction ou à prendre avec une égale probabilité des positions symétriques par rapport au plan $H^{(3)}C^{(1)}C^{(3)}$, l'action du premier ordre s'annule et la contribution à la rotation n'est plus due qu'aux actions de voisinage d'ordre supérieur.

De ce qui précède et de l'ordre de grandeur des contributions des divers ordres à la rotation (Sect. 37), on conclut que les influences capables de restreindre la libre rotation doivent accroître l'ordre de grandeur du pouvoir rotatoire. Parmi ces influences, on peut citer: a) La formation de cycles; on trouvera en [9] et en [10] des tableaux probants à cet égard. b) L'empèchement stérique, auquel les dérivés du diphényle, par exemple, doivent leur activité optique (Sect. 20ζ). c) La cristallisation, qui n'admet dans le réseau que des molécules ayant la même configuration, ou du moins un très petit nombre de configurations différentes (voir le Chap. D).

44. Influence des distances intramoléculaires. Quelle que soit la nature des forces interatomiques admises par les diverses théories (Sects. 39 et 40), ces forces décroissent rapidement quand la distance croît. Il en est, par suite, de même des actions de voisinage d'un ordre donné.

En ce qui concerne les actions du premier ordre, on déduit de la remarque précédente que *l'effet* Cotton *d'une bande donnée est d'autant plus grand que le chromophore est plus voisin du centre de dissymétrie de la molécule.* Les seules vérifications directes que l'on ait de cette proposition se trouvent dans le domaine des complexes minéraux: l'effet Cotton des bandes d'absorption dues à l'atome métallique dans des composés tels que

$$Cl[Co(NO_2)_2en\ d\ pn] \quad et \quad Br_2[Coen_2\ d\ valine]^2$$

[1] J. P. Mathieu: C. R. Acad. Sci., Paris **215**, 325 (1942).
[2] en = éthylènediamine, pn = propylènediamine.

est cinq à quinze fois plus grand lorsque cet atome est lui-même un centre de dissymétrie que lorsqu'il est seulement relié à une molécule optiquement active d'amine ou d'acide aminé[1]. L'effet COTTON apparaît dans les bandes d'absorption dues au métal dans les lactates ou les tartrates complexes, dont les atomes de carbone asymétriques sont voisins de l'atome metallique, mais il n'apparaît pas dans les oxyméthylènecamphorates complexes, où les atomes de carbone asymétriques sont plus éloignés de l'atome métallique chromophore[2].

Dans le domaine des composés organiques, les mesures de rotation dans le spectre visible, région de transparence, ont fourni de la règle précédente une vérification indirecte, que TSCHUGAEFF[3] a énoncé sous la forme suivante: plus un substituant inactif se trouve proche d'un groupe asymétrique, plus son influence sur la rotation est marquée. Cette règle a été vérifiée en étudiant l'action de la place occupée par le groupe phényle dans divers acides (de l'acide benzoïque à l'acide phénylpropionique) sur la rotation des esters menthyliques de ces acides.

Pour ce qui est des actions de voisinage d'ordre supérieur, l'influence de la distance est analogue à celle que nous venons d'étudier. Reprenons l'exemple de la Sect. 43, où le groupe C_2H_5 de la molécule de méthyléthylcarbinol tournait librement autour de la liaison $C^{(1)}-C^{(3)}$. Les actions de voisinage sont alors d'ordre supérieur au premier; les plus importantes sont celles qui font intervenir les atomes $C^{(2)}$, $C^{(3)}$, O et $H^{(3)}$ liés à l'atome de carbone asymétrique $C^{(1)}$. Mais l'ensemble de ces atomes admet un plan de symétrie; par suite, les actions de voisinage efficaces sur l'activité optique font intervenir le radical méthyle du groupe C_2H_5, qui est éloigné du centre d'asymétrie et dont l'action de voisinage est faible. Dans les conditions de libre rotation du groupe C_2H_5 — conditions réalisées par élévation de température — le pouvoir rotatoire du méthyléthylcarbinol doit donc être beaucoup inférieur à celui d'un composé tel que $CHClI(SO_3)^-$ dans lequel les atomes directement liés au centre de dissymétrie sont tous différents. On a effectivement, pour ce dernier composé, $[M] = 36°$. Pour d'autres exemples, voir [10].

45. Influence de substitutions. Il faut tenir compte à la fois de la fonction de chromophore et de l'action de voisinage d'un groupe pour estimer les effets produits par sa substitution sur l'activité optique. Le meilleur moyen d'apprécier soit le premier, soit le second de ces effets est d'étudier l'effet COTTON des bandes appartenant soit au groupe considéré, soit à ses voisins. En fait, pour des raisons techniques, on n'a jamais étudié directement que les bandes d'absorption d'un seul chromophore, et l'on a déduit l'effet des autres sur l'activité optique en mesurant la part de dispersion rotatoire qui leur revient. Dans ces conditions, on a pu découvrir diverses règles relatives aux substitutions, principalement chez les composés organiques.

1. On effectue *une seule substitution*, qui ne porte pas sur le chromophore dont on étudie les bandes. On constate alors que l'effet COTTON du chromophore non modifié varie assez peu, tandis que la part de dispersion rotatoire attribuable au groupe substitué peut subir des changements considérables. Voici deux exemples:

a) Passage de l'ester méthylique de l'acide α-azidopropionique $CH_3 \cdot CHN_3 \cdot CO_2 \cdot CH_3$ à la diméthylamide de cet acide $CH_3CHN_3CON(CH_3)_2$[4]. Le maximum de la bande d'absorption étudiée, due au radical N_3, passe, dans cette substitution, de 2850 à 2900 Å. La part de la rotation due à cette bande — calculée au

---

[1] J. P. MATHIEU: Ann. Phys., Paris **19**, 335 (1944).
[2] F. M. JAEGER: Bull. Soc. chim. France **33**, 853 (1923).
[3] L. TSCHUGAEFF: Ber. dtsch. Chem. Ges. **31**, 1775 (1898).
[4] W. KUHN et E. BRAUN: Z. phys. Chem., Abt. B **8**, 281 (1930).

moyen de l'équation (32.16) — est représentée par les courbes *a* (ester) et *b* (amide) de la Fig. 55. En retranchant ces courbes de la courbe expérimentale de dispersion rotatoire, on obtient les courbes *a'* et *b'*, qui représentent les parts de rotation dues respectivement aux groupes $-CO_2CH_3$ et $-CON(CH_3)_2$. On voit que la substitution modifie profondément la rotation du groupe substitué, et assez peu celle du groupe N_3.

b) Passage du complexe $Cl_2[Coen_2ClNH_3]$ à $Cl[Coen_2ClNO_2]$[1]. Le maximum principal de dichroïsme circulaire se déplace depuis 5550 Å jusqu'à 5450 Å pour la bande due à Co et l'effet Cotton reste du même signe et du même ordre de grandeur (Fig. 56). Les courbes de dispersion rotatoire de la figure indiquent

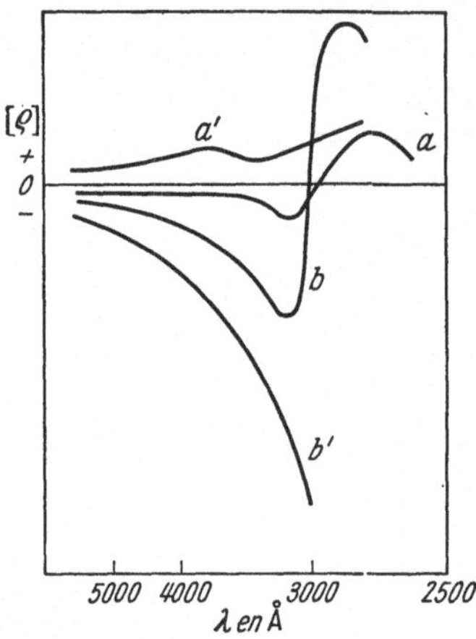

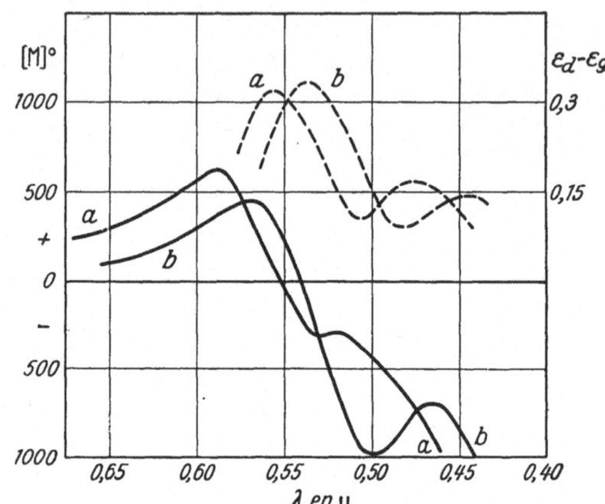

Fig. 55a et b. Règle de voisinage a) α azidopropionate de méthyle. b) Diméthylamide de l'acide α azidopropionique.

Fig. 56a et b. Règle de voisinage a) $Cl_2[Coen_2 ClNH_3]$; b) $Cl[Coen_2 ClNO_2]$. Rotation (———) et dichroïsme circulaire (------)

une contribution des bandes ultraviolettes qui est positive pour le premier composé, négative pour le second et change par suite de signe lorsqu'on substitue NO_2 à NH_3.

Les faits observés se traduisent en termes théoriques par la *règle de voisinage* (Vizinalregel) de Kuhn et Freudenberg: lorsqu'on fait subir une modification faible à la nature chimique d'un substituant dans une molécule optiquement active, sa dissymétrie induite propre peut changer beaucoup, son action de voisinage varie peu. Il convient de définir ce qu'on entend par «modification chimique faible». Il s'agit évidemment de substitutions qui conservent le type de structure moléculaire; de plus, Kuhn et Freudenberg ajoutent que les deux groupes sur lesquels porte la substitution doivent être suffisamment différents des autres groupes de la molécule. Ces conditions sont vagues; l'invariance approximative de l'effet de voisinage de deux substituants doit pouvoir s'exprimer de façon plus précise dans le langage des diverses théories. Dans les théories de couplage, elle se traduit par l'analogie d'orientation des axes des dipôles ou de ceux des ellipsoïdes de polarisabilité; dans la théorie monoélectronique, par une dépendance analogue des potentiels atomiques envers la direction et la distance[2]. On trouvera

---

[1] J. P. Mathieu: Bull. Soc. chim. France **3**, 463 (1936).

[2] L'exemple le plus parfait de similitude d'actions de voisinage est fourni par celles des isotopes, dont la formule électronique est la même. Il s'ensuit que si, parmi quatre groupes différents attachés à un atome de carbone asymétrique, il se trouve deux isotopes, tout se passe à peu près, du point de vue des actions de voisinage, comme si l'ensemble admettait un plan de symétrie. On s'explique ainsi que les tentatives faites pour créer la dissymétrie moléculaire en remplaçant H par D n'aient conduit qu'à une activité optique douteuse ou nulle [voir D. Duveen et A. Willemart: Rev. Sci. **78**, 279 (1940)].

en [12] et en [13] des exemples d'application de la règle de voisinage à des problèmes de stéréochimie.

2. On effectue à la fois *deux substitutions*, dont l'une porte sur le chromophore dont on étudie les bandes. Les relations sont alors plus compliquées que dans le cas précédent. On trouve cependant la régularité suivante, énoncée par HUDSON, puis par FREUDENBERG sous le nom de *règle de déplacement* (Verschiebungsregel): une même substitution, faite sur des composés actifs de constitution et de configuration analogues, produit des différences de rotation de même sens et du même ordre de grandeur. Cette règle est issue de la remarque que le passage d'un acide actif de constitution quelconque à l'amide s'accompagne d'un fort accroissenent dextrogyre de la rotation pour la raie D. On la justifie théoriquement de la façon suivante. Considérons, par exemple, l'acide azidopropionique $CH_3CHN_3CO_2H$ et l'acide chloropropionique $CH_3CHClCO_2H$. Ils peuvent être regardés comme analogues dans leur constitution, car le radical N_3 est très voisin des halogènes par ses propriétés chimiques et électrochimiques. Le remplacement de N_3 par Cl modifie toutefois beaucoup l'emplacement de la bande ultraviolette la plus proche, dont le maximum passe de 2900 à 2300 Å, change le signe de l'effet COTTON et en décuple la grandeur. Appelons ϱ_{N_3} et ϱ_{Cl} les contributions apportées à la rotation pour une longueur d'onde donnée par les groupes N_3 et Cl. La règle du 1. enseigne que les actions de voisinage des groupes N_3 et Cl sont analogues. Par suite, si l'on écrit la rotation de l'ester méthylique de l'acide azidopropionique $\varrho_{N_3} + \varrho_e + \varrho'$ (ϱ_e désignant la

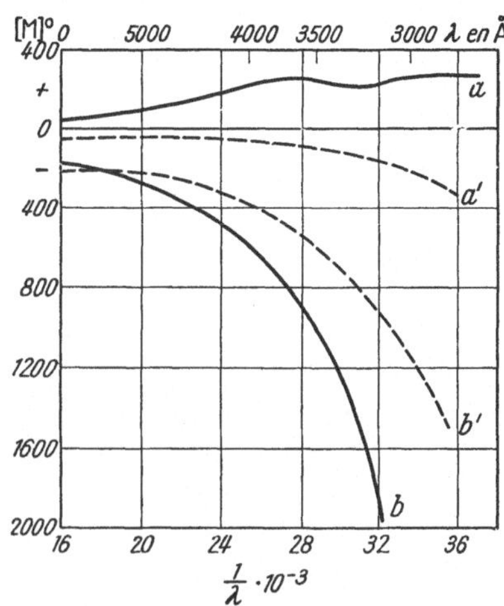

Fig. 57. Règle de déplacement, a) rotation de CO_2CH_3 dans $CH_3CHN_3CO_2CH_3$; b) rotation de $CON(CH_3)_2$ dans $CH_3CHN_3CON(CH_3)_2$; a') rotation de CO_2CH_3 dans $CH_3CHClCO_2CH_3$; b') rotation de $CON(CH_3)_2$ dans $CH_3CHClCON(CH_3)_2$.

contribution du groupe CO_2CH_3 et ϱ' celle du reste de la molécule), on a pour la rotation de l'ester méthylique de l'acide chloropropionique l'expression $\varrho_{Cl} + \varrho_e + \varrho'$. Les rotations des diméthylamides des mêmes acides peuvent s'écrire $\varrho_{N_3} + \varrho_a + \varrho'$ et $\varrho_{Cl} + \varrho_a + \varrho'$, ϱ_a désignant la contribution du groupe $CON(CH_3)_2$. Par suite, on a

$$\varrho_{N_3} + \varrho_e + \varrho' - (\varrho_{N_3} + \varrho_a + \varrho') = \varrho_{Cl} + \varrho_e + \varrho' - (\varrho_{Cl} + \varrho_a + \varrho')$$

c'est-à-dire

$$\varrho\,[CH_3CHN_3CO_2CH_3] - \varrho\,[CH_3CHN_3CON(CH_3)_2]$$

$$= \varrho\,[CH_3CHClCO_2CH_3] - \varrho\,[CH_3CHClCON(CH_3)_2].$$

Ce résultat, que l'expérience vérifie (Fig. 57), est conforme à la règle de déplacement. D'autres exemples de cette règle se rencontrent dans le domaine des complexes minéraux; ainsi, on a

$$\varrho\,[Cl_3(Cren_3)] - \varrho\,[Cl(Cren_2Cl_2)] = \varrho\,[Cl_3(Coen_3)] - \varrho\,[Cl(Coen_2Cl_2)].$$

La règle de déplacement a servi à comparer les configurations des molécules actives. Voir à ce sujet[1] et [13].

---

[1] E. GORIN, W. KAUZMANN et J. WALTER: J. Chem. Phys. **7**, 327 (1939).

tourage immédiat sur une molécule active donnée M soit égal à $\dfrac{4\pi \boldsymbol{P}}{3}$, avec, comme en (28.1): $\boldsymbol{P} = N\overline{\boldsymbol{p}}$, N désignant le nombre des molécules actives contenues dans l'unité de volume, $\overline{\boldsymbol{p}}$ leur moment de dipôle moyen. La formule $\overline{\boldsymbol{p}} = \alpha \boldsymbol{E}$ doit être remplacée par la suivante dans le cas de molécules polaires (théorie de DEBYE)

$$\overline{\boldsymbol{p}} = \left(\alpha + \frac{p_0^2}{3kT}\right)\boldsymbol{E} = \beta\,\boldsymbol{E}, \tag{47.1}$$

p_0 désignant le moment de dipôle permanent, k la constante de BOLTZMANN, T la température absolue. Le champ $\boldsymbol{E}$ est supposé dû à la molécule active M considérée, que l'on assimile à une sphère rigide portant un dipôle électrique en son centre. Sous son influence, les molécules voisines se polarisent dans sa direction et leurs dipôles tendent à s'orienter dans cette même direction. Dans ces conditions, le champ $\overline{\boldsymbol{F}}$ est parallèle à l'axe du dipôle de la molécule M. Or dans cette direction, le champ $\boldsymbol{E}$ exercé par la molécule M sur une molécule voisine a pour expression

$$\boldsymbol{E} = \frac{2\boldsymbol{p}}{R^3},$$

R désignant la distance minimum d'approche des molécules. Par suite

$$\overline{\boldsymbol{F}} = \frac{4\pi}{3} N\overline{\boldsymbol{p}} = \frac{4\pi}{3} N\beta \frac{2\boldsymbol{p}}{R^3}.$$

D'autre part, une généralisation de la formule de LORENTZ-LORENZ

$$\frac{\varepsilon - 1}{\varepsilon + 2} = \frac{4\pi N}{3}\alpha,$$

au cas des molécules polaires donne

$$\frac{\varepsilon - 1}{\varepsilon + 2} = \frac{4\pi N}{3}\beta,$$

d'où, enfin,

$$\overline{\boldsymbol{F}} = \frac{2\boldsymbol{p}}{R^3}\frac{\varepsilon - 1}{\varepsilon + 2}. \tag{47.2}$$

La formule (47.2) s'étend au cas d'une solution. On a alors, d'une part,

$$\boldsymbol{P} = N_1\overline{\boldsymbol{p}}_1 + N_2\overline{\boldsymbol{p}}_2,$$

d'autre part,

$$\frac{\varepsilon - 1}{\varepsilon + 2} = \frac{4\pi}{3}(N_1\beta_1 + N_2\beta_2), \tag{47.3}$$

l'indice 1 se rapportant à l'un des constituants de la solution, l'indice 2 à l'autre. Supposons que la distance R soit la même pour les molécules du solvant que pour celles du soluté actif. On a

$$\overline{\boldsymbol{F}} = \frac{4\pi}{3}(N_1\beta_1 + N_2\beta_2)\frac{2\boldsymbol{p}}{R^3} \tag{47.4}$$

et, compte tenu de (47.3), on retrouve la formule (47.2), que BECKMANN et COHEN obtiennent par des raisonnements statistiques.

Cherchons maintenant quels peuvent être les effets du champ $\overline{\boldsymbol{F}}$ sur une molécule active M. On a suggéré que ce champ pouvait déformer la molécule. En admettant la proportionnalité des déformations à l'intensité du champ et la proportionalité du paramètre γ (Sect. 30) — donc de $[\varrho]$ — aux déformations, on trouve aussitôt que la rotativité doit être une fonction linéaire du champ

$$\Omega = \Omega_0 + \Omega_1 F. \tag{47.5}$$

Cette formule, qu'il est possible de soumettre à l'expérience[1], compte tenu de (47.2), ne représente pas complètement l'action de solvant. On peut, en effet, imaginer d'autres modes d'influence du champ $\overline{F}$; il se peut, par exemple, que les dipôles des molécules voisines de M, qui s'orientent, comme on l'a vu, sous l'action du dipôle global de M, se trouvent alors en position favorable pour exercer une action de voisinage sur certains des principaux groupes chromophores de la molécule M: il en résultera des modifications du pouvoir rotatoire. C'est sans doute ainsi que doivent s'interpréter les fortes variations de l'effet Cotton d'une bande donnée avec la nature du solvant, qui se produisent, par exemple, pour le camphre cyané[2].

Dans certains cas, enfin, la molécule active forme avec les solvants des combinaisons définies, dont on peut déterminer la formule où que l'on peut même isoler. Exemple: solutions d'iodacétate de β-octyle dans la pyridine[3]; formation de complexes par l'acide tartrique, le mannitol, etc., dans les solutions d'acide borique et de borates.

48. Influence de la température sur l'activité optique. L'influence de la température sur les actions intramoléculaires peut être mise en relation avec l'effet de la libre rotation discuté à la Sect. 43. Considérons une molécule organique de composition un peu compliquée, par exemple la molécule de méthyléthylcarbinol. Nous avons vu à la Sect. 41 que l'on peut imaginer pour cette molécule plusieurs structures, qui se distinguent par la situation du groupe C_2H_5 relativement aux autres groupes. De l'une à l'autre de ces structures, le pouvoir rotatoire calculé change, comme on l'a vu, par suite des variations de l'action mutuelle des groupes; pour la même raison, les valeurs de l'énergie interne de la molécule correspondant aux diverses structures sont généralement différentes. Tant que l'énergie d'agitation thermique est faible devant ces différences, la forme dont l'énergie est minimum représente une forme d'équilibre stable pour le groupe C_2H_5. Plus exactement, si W_1 et W_2 sont les énergies internes de deux formes moléculaires possibles, le rapport des nombres de molécules de ces formes est égal à

$$\frac{n_1}{n_2} = e^{-\frac{W_1 - W_2}{kT}},$$

k désignant la constante de Boltzmann, T la température absolue. On voit que, tant que la température reste assez basse, les groupes à liaison mobile effectuent simplement des librations autour de leur position d'équilibre stable. La température croissant, le passage à une forme d'énergie interne plus élevée devient plus probable. Pour des températures suffisamment élevées enfin, toutes les formes moléculaires possibles tendent à devenir également probables, les librations devenant alors des rotations autour des liaisons mobiles[4].

D'après ce qu'on a vu à la Sect. 43, on doit s'attendre à ce que le pouvoir rotatoire des molécules dont la structure n'est pas rigidement fixée diminue lorsque la température s'élève. Cette prévision est vérifiée par l'expérience: tandis que les composés cycliques montrent une variation thermique du pouvoir

[1] C. O. Beckmann et K. Cohen: J. Chem. Phys. **4**, 784 (1936); **6**, 163 (1938).
[2] J. P. Mathieu et M. Ronayette: Rev. Opt. **19**, 1 (1940).
[3] H. G. Rule et R. K. S. Mitchell: J. Chem. Soc., Lond. **3**, 202 (1926).
[4] K. L. Wolf: Molekülstruktur, Vorträge. Leipzig 1931.

rotatoire tantôt positive et tantôt négative, presque tous les composés aliphatiques ont une rotation qui décroît par accroissement de température [9].

Le rôle de la température dans les actions intermoléculaires se manifeste par son influence sur les effets des solvants. Quel que soit, en effet, le mécanisme de ces actions, nous avons vu plus haut qu'il doit faire intervenir des orientations mutuelles des molécules du solvant et du soluté. Une augmentation de l'énergie d'agitation thermique, en contrariant ces orientations privilégiées, doit réduire l'action spécifique du solvant. Il en résulte que les rotativités d'un composé dans divers solvants doivent tendre vers une valeur limite commune, lorsque la température croît. Cette prévision a été vérifiée dans le cas du tartrate d'éthyle, dont la rotation a été étudiée dans quarante-sept solvants [10].

D. Activité optique des milieux anisotropes.

49. Anisotropie naturelle et artificielle. Pouvoir rotatoire de structure. Une première catégorie de milieux anisotropes est formée par des phases cristallines ou mésomorphes de certains composés; leur anisotropie est spontanée. En second lieu, on peut rendre anisotrope un fluide en le soumettant à un champ magnétique ou électrique; cette anisotropie artificielle disparaît en même temps que le champ imposé.

L'étude de l'activité optique de tels milieux est moins simple que celle des milieux isotropes. Les bases de la théorie phénoménologique ont été étudiées à la Sect. 5 et quelques particularités des procédés de mesure ont été indiquées à la Sect. 10. L'interprétation théorique des résultats est compliquée par l'intervention de plusieurs phénomènes. Lorsqu'une variation de l'activité optique naturelle avec la direction d'observation se manifeste dans un milieu artificiellement anisotrope, elle est due à ce que les molécules sont orientées par le champ et à ce que la rotation produite par l'une d'elles varie avec l'orientation de cette molécule par rapport à l'onde lumineuse. En d'autres termes, l'activité optique d'une molécule individuelle est anisotrope.

Dans les milieux cristallisés, l'activité optique se présente sous deux formes (Sect. 1). Les cristaux qui perdent leur pouvoir rotatoire par fusion ou dissolution ont uniquement un *pouvoir rotatoire de structure*, dû à la dissymétrie de l'édifice cristallin. D'autre part, une substance qui possède le *pouvoir rotatoire moléculaire* et dont les molécules sont par suite dissymétriques, donne des cristaux dans lesquels l'arrangement des molécules est lui-même dissymétrique: en effet, la présence, dans la structure cristalline, d'un élément de symétrie de deuxième espèce entraînerait la présence, en nombres égaux, de molécules antipodes optiques et produirait l'inactivité optique. Il en résulte que, sauf dans des cas exceptionnels, un pouvoir rotatoire de structure accompagne le pouvoir rotatoire moléculaire. Les variations de l'activité optique avec la direction peuvent alors aussi bien être attribuées à l'influence de la structure qu'à l'anisotropie du pouvoir rotatoire de la molécule. D'ailleurs, lorsque la maille cristalline contient plusieurs molécules, la rotation moléculaire résulte de la composition des rotations dues à celles-ci.

Nous étudierons d'abord la théorie atomique générale de l'activité optique cristalline, applicable aux cristaux qui possèdent un pouvoir rotatoire de structure pur ou mélé de pouvoir rotatoire moléculaire. Nous examinerons ensuite les preuves de l'anisotropie moléculaire de l'activité optique.

I. Activité optique des cristaux.

50. Propagation de la lumière dans les cristaux optiquement actifs. Biréfringence elliptique [2], [3], [24]. α) Dans un cristal transparent dépourvu d'activité optique, rapportons le tenseur $[\varepsilon]$ de (5.1) à ses axes principaux; il vient:

$$D_x = \varepsilon_x E_x, \quad D_y = \varepsilon_y E_y, \quad D_z = \varepsilon_z E_z. \tag{50.1}$$

On tire de la formule (4.21), la relation suivante:

$$D_x = n^2 \left[E_x - s_x (\boldsymbol{E} \cdot \boldsymbol{s}) \right] = n^2 \left[\frac{D_x}{\varepsilon_x} - s_x (\boldsymbol{E} \cdot \boldsymbol{s}) \right] \tag{50.2}$$

ou

$$D_x = - \frac{s_x (\boldsymbol{E} \cdot \boldsymbol{s})}{\dfrac{1}{n^2} - \dfrac{1}{\varepsilon_x}}$$

et des relations analogues. En multipliant ces relations par s_x, s_y, s_z respectivement et en les ajoutant, on obtient, puisque $\boldsymbol{D} \cdot \boldsymbol{s} = 0$ et que $\boldsymbol{E} \cdot \boldsymbol{s} \neq 0$:

$$\frac{s_x^2}{\dfrac{1}{n^2} - \dfrac{1}{\varepsilon_x}} + \frac{s_y^2}{\dfrac{1}{n^2} - \dfrac{1}{\varepsilon_y}} + \frac{s_z^2}{\dfrac{1}{n^2} - \dfrac{1}{\varepsilon_z}} = 0. \tag{50.3}$$

Cette équation définit la *surface des vitesses normales*; elle peut s'écrire:

$$n^4 (\varepsilon_x s_x^2 + \varepsilon_y s_y^2 + \varepsilon_z s_z^2) - n^2 \left[\varepsilon_y \varepsilon_z (s_y^2 + s_z^2) + \varepsilon_z \varepsilon_x (s_z^2 + s_x^2) + \varepsilon_x \varepsilon_y (s_x^2 + s_y^2) \right] + \varepsilon_x \varepsilon_y \varepsilon_z = 0,$$

ou encore, en désignant ses racines par n_0' et n_0'':

$$(n^2 - n_0'^2)(n^2 - n_0''^2) = 0. \tag{50.4}$$

β) Dans un cristal doué de pouvoir rotatoire, les relations (50.1) sont remplacées par les suivantes, de même que (24.6) par (24.7)

$$\left. \begin{aligned} D_x &= \varepsilon_x E_x + i (\boldsymbol{G} \times \boldsymbol{E})_x = \varepsilon_x E_x - i G_z E_y + i G_y E_z, \\ D_y &= \varepsilon_y E_y + i (\boldsymbol{G} \times \boldsymbol{E})_y = + i G_z E_x + \varepsilon_y E_y - i G_x E_z, \\ D_z &= \varepsilon_z E_z + i (\boldsymbol{G} \times \boldsymbol{E})_z = - i G_y E_x + i G_x E_y + \varepsilon_z E_z. \end{aligned} \right\} \tag{50.5}$$

$\boldsymbol{G}$ est le *vecteur de gyration*, déja introduit à la Sect. 5. En égalant les expressions (50.2) et (50.5) de D_x, on trouve l'équation:

$$E_x \left[\varepsilon_x - (1 - s_x^2) n^2 \right] + E_y (n^2 s_x s_y - i G_z) + E_z (n^2 s_x s_z + i G_y) = 0.$$

Les expressions de D_y et D_z donnent des équations analogues. La condition de compatibilité de ce système d'équations linéaires conduit à remplacer l'équation (50.4) par:

$$(n^2 - n_0'^2)(n^2 - n_0''^2) = G^2, \tag{50.6}$$

avec:

$$G^2 = \frac{\varepsilon_x G_x^2 + \varepsilon_y G_y^2 + \varepsilon_z G_z^2 - n^2 (\boldsymbol{s} \times \boldsymbol{G})^2}{\varepsilon_x s_x^2 + \varepsilon_y s_y^2 + \varepsilon_z s_z^2} \tag{50.7}$$

G est le *paramètre scalaire de gyration*. Il prend en principe deux valeurs, G' et G'', selon que l'on remplace, dans (50.7), n par n' ou n''. Supposons $n_0' > n_0''$; les racines de (50.6) s'écrivent:

$$\left. \begin{aligned} n'^2 \\ n''^2 \end{aligned} \right\} = \frac{1}{2} \left[n_0'^2 + n_0''^2 \pm \sqrt{(n_0'^2 - n_0''^2)^2 + 4 G^2} \right] \tag{50.8}$$

L'expérience montre que G, dont dépend le pouvoir rotatoire, est toujours petit devant la différence $n_0' - n_0''$ des indices principaux. Tant que la direction de propagation des ondes dans le cristal n'est pas très proche de celle des axes optiques, l'équation (50.4) est approximativement valable et la forme des deux nappes de la surface des vitesses normales ne subit que des altérations minimes. Au voisinage immédiat des axes optiques, il n'en est plus de même: les deux nappes sont séparées, au lieu de posséder un ombilic dans les cristaux biaxes (Fig. 58a) ou d'être tangentes dans les cristaux uniaxes (Fig. 58b).

γ) Pour déterminer la forme des vibrations qui se propagent dans un cristal optiquement actif, on tire de (50.2):

$$E_x = \frac{D_x}{n^2} + s_x(\boldsymbol{E}\cdot\boldsymbol{s})$$

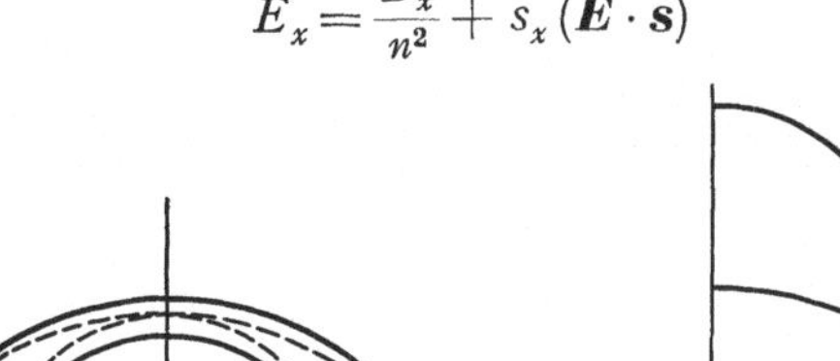
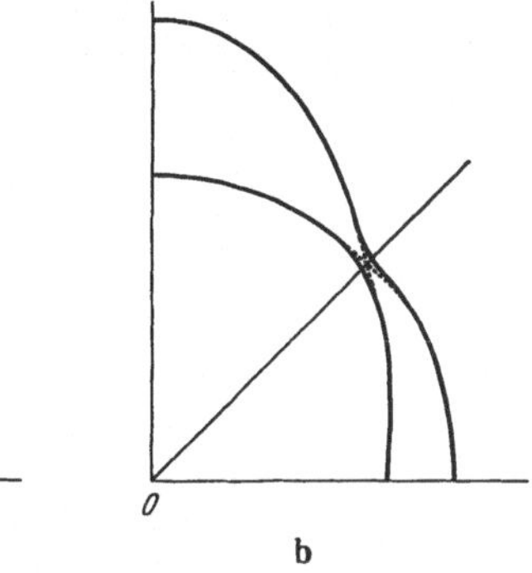

Fig. 58a et b. Section de la surface des vitesses normales par le plan des axes optiques dans un cristal optiquement actif:
a) uniaxe; b) biaxe.

et on porte cette expression dans le terme de (50.5) indépendant de G. Il vient

$$D_x = \varepsilon_x\left[\frac{D_x}{n^2} - s_x(\boldsymbol{E}\cdot\boldsymbol{s}) - \frac{i}{\varepsilon_x}(\boldsymbol{E}\times\boldsymbol{G})_x\right], \tag{50.9}$$

ou, en posant dans le dernier terme $\varepsilon_x = \varepsilon_y = \varepsilon_z = \bar{n}^2$, ce qui revient à négliger la biréfringence et en y substituant $\boldsymbol{D}/n^2$ à $\boldsymbol{E}$:

$$D_x = \varepsilon_x\left[\frac{D_x}{n^2} - s_x(\boldsymbol{E}\cdot\boldsymbol{s}) - \frac{i}{n^4}(\boldsymbol{D}\times\boldsymbol{G})_x\right]. \tag{50.10}$$

On montre[1] que ces équations admettent deux solutions:

$$\left.\begin{array}{ll} n = n', & D = D' - ikD'', \\ n = n'', & D = D'' - ikD', \end{array}\right\} \tag{50.11}$$

avec $|D'| = |D''| = D$ et

$$k = \frac{1}{2G}\left[\sqrt{(n_0'^2 - n_0''^2) - 4G^2} - (n_0'^2 - n_0''^2)\right]. \tag{50.12}$$

En prenant pour D' et D'' des expressions correspondant à des ondes planes monochromatiques:

$$D = D_0\, e^{2\pi i\left(\nu t - \frac{\boldsymbol{r}\cdot\boldsymbol{s}}{\lambda}\right)} = D_0\, e^{i\varphi},$$

en choisissant l'axe des x suivant D', l'axe des y suivant D'', et en prenant les parties réelles des solutions (50.11), il vient:

$$n = n', \quad D_x = D\cos\varphi, \quad D_y = kD\sin\varphi, \quad D_z = 0,$$
$$n = n'', \quad D_x = kD\sin\varphi, \quad D_y = D\cos\varphi, \quad D_z = 0.$$

---

[1] M. Born: Z. Physik **8**, 414 (1922).

Ces équations représentent les vibrations qui se propagent sans déformation dans toute direction prise dans le cristal. Ce sont deux vibrations elliptiques dont le rapport k des axes est le même et qui sont décrites en sens inverses l'une de l'autre. Les grands axes de ces vibrations sont croisés; ils ont les orientations des vibrations rectilignes qui se propageraient dans la même direction sans déformation, avec les indices n_0' et n_0'', si le cristal était biréfringent, mais dépourvu de pouvoir rotatoire. Ce résultat est celui que Airy[1] avait pris pour hypothèse.

L'existence des vibrations elliptiques privilégiées a pu être montrée expérimentalement[2].

δ) Dans la direction des axes optiques, $n_0' = n_0'' = \bar{n}$, $k = 1$: les vibrations sont circulaires. Leurs indices n_d et n_g sont donnés par (50.8) et le pouvoir rotatoire par unité de longueur.

$$\varrho = \frac{\pi}{\lambda}\,(n_d - n_g) = \frac{\pi G}{\lambda\, n^2}\,. \tag{50.13}$$

ε) On peut donner aux formules qui précèdent une expression simplifiée. Dans la formule (50.12), remplaçons G par son expression tirée de (50.7), $n_0' + n_0''$ par $2\bar{n}$, et $n_0' - n_0''$ par $\frac{\varphi_0 \lambda_0}{2\pi}$ (φ_0 est la biréfringence ordinaire dans la direction de propagation de l'onde); il vient

$$k = \sqrt{\left(\frac{\varphi_0}{2\varrho}\right)^2 + 1} + \frac{\varphi_0}{2\varrho}\,. \tag{50.14}$$

Les mêmes approximations donnent, à partir de la formule (50.8), pour la différence de phase par unité de longueur entre les vibrations elliptiques privilégiées, l'expression:

$$\varphi = \frac{2\pi}{\lambda}\,(n' - n'') = \sqrt{\varphi_0^2 + 4\varrho^2}\,. \tag{50.15}$$

Ces formules avaient été obtenues jadis par Gouy[3]. Elles reposent sur l'hypothèse que φ_0 n'est pas modifiée par le pouvoir rotatoire et que ϱ ne dépend pas de la direction de propagation. Si l'on envoie dans une direction quelconque du cristal une onde polarisée rectilignement, la vibration qui sort est généralement elliptique. La mesure de l'azimut et de l'ellipticité de cette vibration permet de calculer les valeurs de k et de φ et de soumettre les formules de Gouy à l'expérience. Les mesures pour le quartz[4] sont d'accord avec ces formules, tant que la direction de propagation de l'onde ne fait pas avec l'axe un angle supérieur à 10°; au-delà, on observe pour k des valeurs plus petites que celles calculées. Pour des ondes se propageant normalement à l'axe, par exemple, $\varphi_0 \gg \varrho$ et la formule (50.14) devient $k = \varphi_0/\varrho$, soit 0,0039 pour le jaune, tandis que les mesures[5] donnent 0,0019.

51. Interférences en lumière polarisée convergente dans les cristaux optiquement actifs. α) *Cristaux uniaxes.* Lorsqu'on observe entre polariseur et analyseur croisés la figure d'interférences donnée en lumière convergente par une lame uniaxe, de quartz par exemple, d'épaisseur e perpendiculaire à l'axe, les rayons qui parviennent dans la partie centrale du champ sont peu inclinés sur l'axe: la biréfringence est négligeable et l'effet de la lame se réduit pratiquement à une rotation ϱe de la vibration rectiligne incidente. Le bord du champ

[1] G. B. Airy: Cambridge Phil. Trans. **4**, 79, 199 (1831).
[2] M. Croullebois: Ann. Chim. Phys., Paris **28**, 433 (1873).
[3] G. Gouy: J. Phys., Paris **4**, 149 (1885).
[4] F. Beaulard: J. Phys., Paris **2**, 393 (1893).
[5] W. Voigt: Ann. Phys., Lpz. **18**, 645 (1905).

correspond à des rayons écartés de l'axe, pour lesquels la biréfringence l'emporte. On observe donc sur ce bord les anneaux colorés et la croix noire des cristaux uniaxes dépourvus de pouvoir rotatoire, mais la croix s'estompe vers le centre où l'on observe une tache de teinte uniforme (Fig. 59).

La théorie du phénomène repose sur les bases suivantes: d'un point de la lame correspondant à la traversée par des ondes dont la direction de propagation est inclinée sur l'axe optique, sortent deux vibrations elliptiques dont les axes x et y coïncident avec les directions principales d'une lame normale à la direction considérée, en l'absence de pouvoir rotatoire. Si la lame est gauche, la vibration

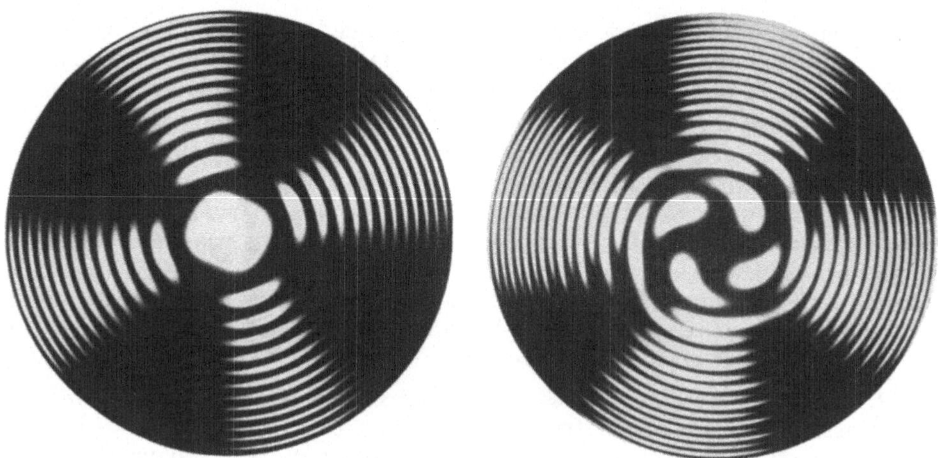

Fig. 59. Figure d'interférence donnée en lumière convergente par une lame de quartz perpendiculaire à l'axe optique.

Fig. 60. Spirales d'Airy.

droite a un retard de phase φ sur la vibration gauche. Le calcul[1] donne pour l'intensité transmise par l'analyseur:

$$I = \frac{I_0}{(1 + k^2)^2} \left[4k^2 + (1 - k^2) \sin^2 2i \right] \sin^2 \frac{\varphi}{2}, \tag{51.1}$$

I_0 désignant l'intensité de la vibration rectiligne incidente, i l'angle que fait la section de l'analyseur avec l'axe x au point de la lame considéré. L'intensité I s'annule le long des circonférences pour lesquelles $\varphi = 2K\pi (K = 0, 1, 2 \ldots)$. Si, de plus, k est assez petit, I est presque nul pour $\sin 2i = 0$: une croix sombre apparaît.

$\beta)$ *Spirales d'*Airy[1]. En examinant en lumière convergente, entre polariseur et analyseur croisés, l'ensemble de deux lames de quartz perpendiculaires à l'axe, de même épaisseur, l'une droite, l'autre gauche, on observe sur le bord du champ l'anneau et la croix noire donnés par une lame. Le centre du champ est noir, puisque la rotation est nulle au total. Il en part quatre courbes noires en forme de spirales, comme le montre la Fig. 60[2]. Leur sens de rotation à partir du centre est celui de la lame traversée la première.

Pour faire la théorie du phénomène, on remarque que les directions x et y des vibrations principales considérées ci-dessus en α s'échangent dans la seconde lame, où la vibration elliptique gauche prend le retard φ. Le calcul montre que

---

[1] G. B. Airy: Cambridge Phil. Trans. **4**, 111 (1831).

[2] Les clichés des Figs. 59 et 60 sont dûs à l'obligeance de M. Françon.

l'intensité transmise par l'analyseur a pour expression, avec les mêmes notations que dans la formule (51.1):

$$I = I_0(1 - k^2)^2 \sin\frac{\varphi}{2}\left[2k\cos 2i\sin\frac{\varphi}{2} - (1 + k^2)\sin 2i\cos\frac{\varphi}{2}\right]^2. \qquad (51.2)$$

L'intensité I s'annule: au centre, pour $k = 1$; le long des circonférences pour lesquelles $\varphi = 2K\pi$; enfin le long des spirales, lieux des points pour lesquels s'annule la quantité entre crochets et que définit l'équation:

$$\tan 2i = \frac{2k}{1 + k^2}\tan\frac{\varphi}{2}. \qquad (51.3)$$

On peut observer les spirales avec une seule lame, en faisant subir au faisceau qui l'a traversée une réflexion normale qui lui fait traverser la lame en sens inverse. Ce mode d'observation, qui se réalise aisément avec l'appareil de Nörremberg, sert à vérifier qu'une lame est bien perpendiculaire à l'axe optique.

γ) *Cristaux biaxes*. L'observation en lumière convergente, entre nicols croisés, d'une lame biaxe normale à l'un des axes optiques montre un système d'anneaux obscurs traversé par une ligne neutre obscure diamétrale. Pour les mêmes raisons que dans le cas des uniaxes, l'intensité n'est pas nulle au centre du champ. On trouvera des détails et des figures en[1] et en [3]. La biréfringence φ_0 croît beaucoup plus vite, lorsqu'on s'écarte de l'axe, dans les biaxes que dans les uniaxes, ce qui rend nécessaires les précautions prises pour la mesure du pouvoir rotatoire de ces derniers (Sect. 10δ).

52. Surface de gyration. Classes de cristaux optiquement actifs. Revenons à l'expression (50.7) du paramètre scalaire de gyration. L'étude de ses variations avec la direction de propagation s de l'onde se simplifie lorsqu'on néglige la biréfringence dans cette expression, ce qui revient à poser $\varepsilon_x = \varepsilon_y = \varepsilon_z = \bar{n}^2$. On a alors [2]

$$G = s \cdot G. \qquad (52.1)$$

Tableau 1.

Système	Classe		Expression de G	Exemples
Triclinique	1	C_1	(52.2)	bitartrate de strontium
Monoclinique	2	C_2	$g_{xx}s_x^2 + g_{yy}s_y^2 + g_{zz}s_z^2 + 2g_{xy}s_xs_y$	acide tartrique, saccharose[2]
	m	C_s	$2s_z(g_{yz}s_y + g_{zx}s_x)$	
Orthorhombique	222	D_2	$g_{xx}s_x^2 + g_{yy}s_y^2 + g_{zz}s_z^2$	acide iodique, sel de Seignette
	mm	C_{2v}	$2g_{xy}s_xs_y$	
Rhomboédrique	3	C_3	$g_{xx}(s_x^2 + s_y^2) + g_{zz}s_z^2$	periodate de sodium
	32	D_3	(comme C_3)	quartz α, camphre, cinabre
Quaternaire	4	C_4	(comme C_3)	antimoniotartrate de baryum
	42	D_4	(comme C_3)	sulfate d'ethylènediamine
	$\bar{4}$	S_4	$g_{xx}(s_x^2 - s_y^2) + 2g_{xy}s_xs_y$	
	$\bar{4}2m$	S_{4u}	$g_{xx}(s_x^2 - s_y^2)$	
Hexagonal	6	C_6	(comme C_3)	sulfate de lithium et de potassium
	62	D_6	(comme C_3)	quartz β
Cubique	23	T	$g_{xx} = g_{yy} = g_{zz}$	chlorate de sodium
	43	O	(comme T)	

---

[1] L. Longchambon: Bull. Soc. franç. Minér. **45**, 161 (1922).

[2] Sommerfeldt: Neues Jb. Mineral., Geol. u. Paläont. **58**, 1 (1908), avait placé dans cette classe le «Mesityloxydoxalsäuremethylester». D'après Rogers [Nature, Lond. **171**, 929 (1953)], ce composé appartient en réalité à la classe C_{2h} dépourvue de pouvoir rotatoire.

On peut écrire, comme en (27.6), l'expression des composantes de G

$$G_x = g_{xx} s_x + g_{xy} s_y + g_{xz} s_z, \quad \text{etc.}$$

d'où

$$G = g_{xx} s_x^2 + g_{yy} s_y^2 + g_{zz} s_z^2 + 2\bar{g}_{yz} s_y s_z + 2\bar{g}_{zx} s_y s_x + 2\bar{g}_{xy} s_x s_y \tag{52.2}$$

où l'on a posé:

$$\bar{g}_{xy} = \tfrac{1}{2}(g_{xy} + g_{yx}), \dots \tag{52.3}$$

L'extrémité d'un segment de longueur égale à $1/\sqrt{G}$, porté par la direction de propagation, décrit une surface du second degré, appelée *surface de gyration*. Les considérations de symétrie de la Sect. 3 concernant l'expression (3.1) de ϱ analogue à (52.2), sont valables pour la surface de gyration. En les appliquant aux groupes d'éléments de symétrie des classes cristallines, on trouve qu'il existe 15 classes dans lesquelles l'activité optique naturelle peut exister. Le tableau suivant donne ces classes[1] les systèmes cristallins auxquels elles appartiennent, l'expression de G dans chacune d'elles, ainsi que des exemples d'espèces cristallines.

53. Résultats expérimentaux. Nous nous bornerons à quelques exemples importants. Pour les méthodes de mesure, voir la Sect. 10δ.

α) Le pouvoir rotatoire des cristaux orthorhombiques est le même suivant les directions des deux axes optiques. Il n'en est pas de même pour les cristaux monocliniques et tricliniques, car les axes optiques ne se correspondent par aucune opération de symétrie; leurs directions, équivalentes pour la biréfringence, ne le sont pas nécessairement pour l'activité optique. Par exemple, le saccharose $C_{12}H_{22}O_{11}$ a un pouvoir rotatoire de 5,4 degré/mm suivant l'axe a' et $-1,6$ degré par mm suivant l'axe a.

β) Le quartz α appartient au groupe de symétrie D_3^4. Son pouvoir rotatoire, dû seulement à la structure, a fait l'objet de mesures précises et étendues. Les nombres suivants donnent quelques valeurs de ϱ suivant l'axe:

λ en μ	0,1555	0,1690	0,2178	0,2669	0,3694	0,4358	0,5889	0,671
ϱ en degré/mm	713	517	226	130	60	41,55	21,75	16,53

λ en μ	1,34	1,90	2,21	3,04	3,86	4,05	4,65	6,84	9,71
ϱ en degré/mm	3,9	1,83	1,3	0,59	0,38	0,43	1,08	1,28	1,31

Entre 0,23 et 3,0 μ, on représente [17] les mesures à mieux que 0,01 degré/mm au moyen de la formule[2]:

$$\varrho = \frac{9{,}5639}{\lambda^2 - 0{,}0127493} - \frac{2{,}3113}{\lambda^2 - 0{,}000974} - 0{,}1905.$$

Entre 0,8 et 0,2 μ on a proposé la formule plus simple[3]

$$\varrho = (n_0 + 2)\,\frac{1{,}656}{\lambda_0^2 - 0{,}01325},$$

où n_0 est l'indice ordinaire, λ_0 la longueur d'onde dans le vide. La dispersion rotatoire est donc, en première approximation, commandée par une longueur

---

[1] Elles sont désignées par le symbole de Mauguin normalisé et par celui de Schoenflies. Une direction privilégiée est prise pour axe des z.

[2] T. M. Lowry et Coode-Adams: Phil. Trans. Roy. Soc. Lond., Ser. A **226**, 391 (1927).

[3] R. Servant: J. Phys. Radium **3**, 90 (1942).

d'onde caractéristique un peu inférieure à 0,12 µ (voir la Sect. 34). Dans l'infrarouge, des mesures[1] poursuivies jusqu'à 9,7 µ ont paru montrer une anomalie de la dispersion rotatoire, avec un minimum vers 3,8 µ, dans une région d'absorption; mais on a attribué ce résultat[2] à une purification spectrale insuffisante du rayonnement.

Pour le pouvoir rotatoire normalement à l'axe, voir la Sect. 56.

γ) Le cinabre HgS appartient au même groupe de symétrie que le quartz α et n'a qu'un pouvoir rotatoire de structure. Entre 0,719 et 0,598 µ, le pouvoir rotatoire en degré/mm est représenté[3] par la formule:

$$\varrho = \frac{19,13\,\lambda^2}{(\lambda^2 - 0,243)^2}\,.$$

δ) Le sulfate de nickel hexahydraté $NiSO_4$, $6H_2O$ appartient[4] au groupe D_4^4. La Fig. 61 montre les résultats des mesures d'activité optique entre 0,25 et 2 µ[5].

Pour l'acide tartrique, voir la Sect. 59γ; pour le sel de Seignette, la Sect. 64δ.

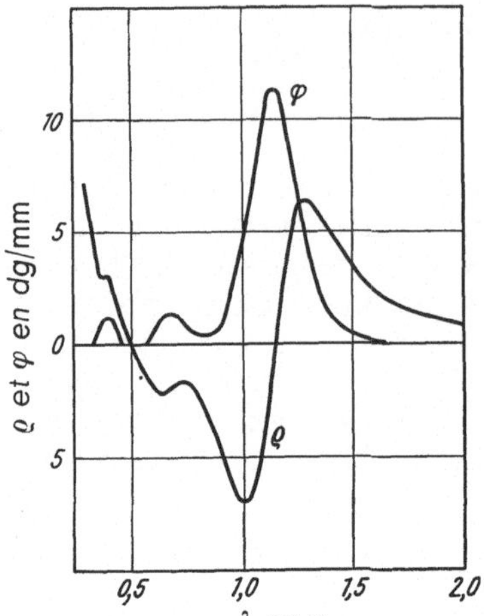

Fig. 61. Dispersion de la rotation ϱ et de l'ellipticité φ pour les cristaux de $NiSO_4$, $6H_2O$.

II. Théorie atomique du pouvoir rotatoire cristallin.

La théorie quantique générale de l'activité optique des cristaux n'a pas encore été faite. La théorie atomique a été développée par Born[6] sur les mêmes bases que la théorie classique relative aux molécules (Sect. 24): elle s'applique au cas où la longueur d'onde est grande devant les dimensions de la maille cristalline, mais non infinie; elle suppose que les atomes composant la maille sont couplés entre eux et que leurs mouvements suivent les lois de la mécanique classique.

54. Dynamique des réseaux cristallins. L'équation (25.4) est valable pour un cristal. Mais il est commode de modifier les notations. On distingue par un numéro d'ordre l, l' … les mailles dont se compose le cristal; chacune d'elles contient un motif formé de n atomes. Chaque atome est caractérisé par deux indices: son déplacement s'écrit $u_k^l(x_k^l, y_k^l, z_k^l)$ et l'équation (25.4) s'écrit:

$$m_k \ddot{x}_k^l + \mathsf{S} \sum_{l'} \sum_{k'} \sum_y f_{xy}^{kl,k'l'} \, y_{k'l'} = 0. \tag{54.1}$$

Le champ de force étant périodique dans le milieu cristallin, les atomes homologues accomplissent, à la phase près, les mêmes oscillations harmoniques, qui forment des ondes élastiques planes progressives. La solution des équations (54.1)

[1] H. S. Gutowsky: J. Chem. Phys. **19**, 438 (1951).

[2] C. D. West: J. Chem. Phys. **22**, 749 (1954).

[3] S. Chandrasekhar: Proc. Indian Acad. Sci. **36**, 697 (1953). Cet auteur a essayé de représenter la dispersion rotatoire de divers cristaux, en particulier du quartz [Proc. Indian Acad. Sci. **35**, 103 (1952)] par des formules quadratiques.

[4] C. A. Beevers et H. Lipson: Z. Kristallogr. **83**, 123 (1932).

[5] L. R. Ingersoll, P. Rudnick, F. G. Slack et N. Underwood: Phys. Rev. **57**, 1145 (1940). — J. P. Mathieu et G. Vuldy: C. R. Acad. Sci., Paris **222**, 223 (1946). Au sujet de la variation du pouvoir rotatoire avec la température, voir [*16*].

[6] M. Born: Z. Physik **8**, 390 (1922). — Atomtheorie des festen Zustandes, p. 604. Leipzig 1923. — M. Born et M. Göppert-Mayer: In: Geiger-Scheels Handbuch der Physik, Bd. XXIV/2, p. 657. Berlin 1933.

peut s'écrire:

$$x_k^l = U_{kx}\,e^{-2\pi i\nu t}\,e^{\frac{2\pi i}{\Lambda}r_l^k\cdot s}\,.$$ (54.2)

En la substituant dans (54.1) on obtient, avec $\omega = 2\pi\nu$:

$$\omega^2 m_k U_{kx}\,e^{\frac{2\pi i}{\Lambda}r_k^l\cdot s} + S\sum_{l'}\sum_{k'}\sum_{y} f_{xy}^{kl,k'l'}\,U_{k'y}\,e^{\frac{2\pi i}{\Lambda}r_{k'}^{l'}\cdot s}$$ (54.3)

ou, sous une forme analogue à (25.6)

$$\omega^2 m_k U_{kx} + \sum_{k'}\sum_{y} \begin{bmatrix} k\,k' \\ x\,y \end{bmatrix} U_{k'y} = 0$$ (54.4)

en représentant, pour abréger, les séries de FOURIER triples par:

$$\begin{bmatrix} k\,k' \\ x\,y \end{bmatrix} = S_l\, f_{xy}^{kl,k'l'}\,e^{\frac{2\pi i}{\Lambda}r_{kk'}^l\cdot s}$$ (54.5)

avec

$$r_{kk'}^l = r_k^l - r_{k'}^l\,.$$

Pour un cristal fini, contenant N mailles, on peut introduire, comme on l'a fait à la Sect. 25 pour une molécule, des coordonnées normales q, au nombre de $3nN$, qui effectuent, indépendamment les unes des autres, des oscillations fondamentales harmoniques, dont les fréquences (fréquences fondamentales) sont fournies par la condition de compatibilité des équations (54.4).

Lorsque la longeur d'onde Λ des ondes élastiques tend vers l'infini, tous les atomes homologues oscillent avec la même amplitude et la même phase. Il reste $3n$ oscillations fondamentales distinctes, les *oscillations principales*. Trois des fréquences fondamentales tendent alors vers une valeur nulle: ce sont les *fréquences acoustiques*. Les $3n-3$ autres tendent vers des valeurs finies: ce sont les *fréquences optiques*. Plusieurs de ces valeurs peuvent être confondues par dégénérescence, si le cristal possède une symétrie suffisante.

Pour des longueurs d'onde finies, mais grandes, on peut développer les coefficients (54.5) en fonction des puissances croissantes de $\tau = \dfrac{2\pi}{\Lambda}$, soit

$$\begin{bmatrix} k\,k' \\ x\,y \end{bmatrix} = \begin{bmatrix} 0 \\ k\,k' \\ x\,y \end{bmatrix} + \begin{bmatrix} 1 \\ k\,k' \\ x\,y \end{bmatrix} + \cdots ,$$ (54.6)

où

$$\begin{bmatrix} 0 \\ k\,k' \\ x\,y \end{bmatrix} = S_l\, f_{xy}^{kl,k'l'}\,,$$ (54.7)

$$\begin{bmatrix} 1 \\ k\,k' \\ x\,y \end{bmatrix} = -i\,S_l\, f_{xy}^{kl,k'l'}\,r_{kk'}^l\cdot s\,.$$ (54.8)

On développe les solutions des équations (54.4) suivant les puissances de τ:

$$\omega^2 = \omega^{(0)2} + \tau\,\omega^{(1)2} + \cdots ,$$ (54.9)

$$U_k = U_k^{(0)} + \tau\,U_k^{(1)} + \cdots$$ (54.10)

et l'on obtient des équations d'approximations successives.

L'équation de première approximation ($\tau = 0$, $\Lambda = \infty$) s'écrit

$$\omega^{(0)2} m_k U_{kx}^{(0)} + \sum_{k'}\sum_{y} \begin{bmatrix} 0 \\ k\,k' \\ x\,y \end{bmatrix} U_{k'y}^{(0)} = 0\,.$$ (54.11)

On peut alors définir le mouvement harmonique commun à tous les atomes d'espèce k dans l'oscillation principale d'indice j au moyen de vecteurs propres

a_k^j orthonormés, comme on l'a fait à la Sect. 25 pour une molécule. On obtient ainsi les solutions des équations (54.11) et on a comme en (25.8)

$$U_k^{(0)} = \sum_j a_k^j q_j. \tag{54.12}$$

L'équation de seconde approximation est:

$$\omega^{(0)2} m_k U_{kx}^{(1)} + \sum_{k'} \sum_y \begin{bmatrix} 0 \\ k\ k' \\ x\ y \end{bmatrix} U_{k'y}^{(1)} = -\omega^{(1)2} m_k U_{kx}^{(0)} - \sum_{k'} \sum_y \begin{bmatrix} 1 \\ k\ k' \\ x\ y \end{bmatrix} U_{k'y}^{(0)}. \tag{54.13}$$

En y introduisant les valeurs (54.12) et en tenant compte des conditions d'orthogonalité et de normalisation des a_k^j pour effectuer les sommes, on obtient:

$$\omega^{(1)2} q_j + \sum_{j'} \sum_{kk'} \sum_{xy} \begin{bmatrix} 1 \\ k\ k' \\ x\ y \end{bmatrix} a_{kx}^j a_{k'y}^{j'} q_j = 0, \tag{54.14}$$

ou

$$\omega^{(1)2} q_j + \sum_{j'} i\,(\mathbf{s} \cdot \mathfrak{R}_{jj'}), \tag{54.15}$$

en posant

$$\mathfrak{R}_{jj} = -\mathbf{S} \sum_{l} \sum_{kk'} r_{kk'}^l \sum_{xy} f_{xy}^{kl,k'l'} a_{kx}^j a_{k'y}^{j'}. \tag{54.16}$$

55. Expression du pouvoir rotatoire. α) Le champ électrique d'une onde monochromatique plane de fréquence ν exerce sur l'atome k, l une force dont 'expression est analogue à (26.5):

$$F_{kx}^l = e_k E_{kx}^l = e_k E_{0x} e^{-\frac{2\pi i}{\lambda} r_k^l \cdot \mathbf{s}} \left(\frac{e^{2\pi i \nu t} + e^{-2\pi i \nu t}}{2} \right). \tag{55.1}$$

Les équations du mouvement contraint, analogues à (26.1), admettent des solutions analogues à (54.2) avec $\Lambda = \lambda$ et s'écrivent

$$\omega^2 m_k U_{kx} + \sum_{k'} \sum_y \begin{bmatrix} k\ k' \\ x\ y \end{bmatrix} U_{k'y} = e_k E_{0x}. \tag{55.2}$$

Pour des longueurs d'onde lumineuses grandes devant les dimensions de la maille, on cherche une solution de ces équations sous forme d'un développement suivant les puissances de $\tau = 2\pi/\lambda$ analogue à (54.10). On obtient ainsi des équations correspondant aux ordres d'approximation successifs, que l'on peut résoudre si l'on connait les solutions des équations homogènes (54.4).

L'équation de première approximation dérive de (54.11)

$$\omega^2 m_k U_{kx}^{(0)} + \sum_{k'} \sum_y \begin{bmatrix} 0 \\ k\ k' \\ x\ y \end{bmatrix} U_{k'y}^{(0)} = e_k E_{0x}. \tag{55.3}$$

Sa solution s'obtient comme en (26.2):

$$U_k^{(0)} = \frac{1}{\sqrt{m_k}} \sum_j \frac{a_k^j}{\omega_j^{(0)2} - \omega^2} \sum_{k'} \frac{e_{k'}}{\sqrt{m_{k'}}} a_{k'}^j E_0. \tag{55.4}$$

L'équation de deuxième approximation est:

$$\omega^2 m_k U_{kx}^{(1)} + \sum_{k'} \sum_y \begin{bmatrix} 0 \\ k\ k' \\ x\ y \end{bmatrix} U_{k'y}^{(1)} = -\sum_{k'} \sum_y \begin{bmatrix} 1 \\ k\ k' \\ x\ y \end{bmatrix} U_{k'y}^{(0)}. \tag{55.5}$$

En la traitant comme (55.3) et en tenant compte de la valeur de $U_k^{(0)}$ donnée par (55.4), on obtient pour solution:

$$U_k^{(1)} = \frac{i}{\sqrt{m_k}} \sum_{jj'} \frac{a_k^j\,(\mathbf{s} \cdot \mathfrak{R}_{jj'})}{(\omega_j^{(0)2} - \omega^2)\,(\omega_{j'}^{(0)2} - \omega^2)} \sum_{k'} \frac{e_{k'} a_{k'}^{j'}}{\sqrt{m_{k'}}} E_0. \tag{55.6}$$

β) La polarisation $\boldsymbol{P}$, moment de dipôle électrique induit dans l'unité de volume, a pour expression:

$$\boldsymbol{P} = \frac{1}{\varDelta} \sum_k e_k \, \boldsymbol{U}_k \,,$$

$\varDelta$ désignant le volume de la maille du cristal. Aux deux valeurs (55.4) et (55.6) de $\boldsymbol{U}_k$ correspondent deux approximations successives pour $\boldsymbol{P}$, soit:

$$\boldsymbol{P}^{(0)} = \varDelta \sum_j \frac{\mathfrak{L}^j \, (\mathfrak{L}^j \cdot \boldsymbol{E}_0)}{\omega_j^{(0)^2} - \omega^2} \,, \tag{55.7}$$

$$\boldsymbol{P}^{(1)} = i \, \frac{2\pi \varDelta}{\lambda} \sum_{jj'} \frac{(\boldsymbol{s} \cdot \mathfrak{R}_{jj'}) \, \mathfrak{L}^j \, (\mathfrak{L}^{j'} \cdot \boldsymbol{E}_0)}{(\omega_j^{(0)^2} - \omega^2) \, (\omega_{j'}^{(0)^2} - \omega^2)} \,, \tag{55.8}$$

en posant

$$\mathfrak{L}^j = \sum_k \frac{e_k}{\sqrt{m_k}} \, \boldsymbol{a}_k^j \,. \tag{55.9}$$

La partie $\boldsymbol{P}^{(0)}$ de la polarisation est en relation avec la biréfringence. Le tenseur $[\varepsilon]$ du pouvoir inducteur spécifique a des composantes définies par (23.4):

$$D_x = E_x + 4\pi \, P_x^{(0)} = \varepsilon_{xx} E_x + \varepsilon_{xy} E_y + \varepsilon_{xz} E_z, \tag{55.10}$$

d'où

$$\varepsilon_{xx} = 1 + 4\pi \varDelta \sum_j \frac{\mathfrak{L}_{jx}^2}{\omega_j^{(0)^2} - \omega^2} \,, \dots \tag{55.11}$$

$$\varepsilon_{xy} = 4\pi \varDelta \sum_j \frac{\mathfrak{L}_{jx} \, \mathfrak{L}_{jy}}{\omega_j^{(0)^2} - \omega^2} \,, \dots . \tag{55.12}$$

On peut toujours rapporter ce tenseur symétrique à ses axes principaux. Seules subsistent les composantes diagonales, qui ont l'expression (55.11).

La partie $\boldsymbol{P}^{(1)}$ de la polarisation est liée à l'activité optique. On peut l'écrire:

$$\boldsymbol{P}^{(1)} = i \, (\boldsymbol{E} \times \boldsymbol{G}), \tag{55.13}$$

où $\boldsymbol{G}$ désigne le vecteur de gyration (Sect. 50β).

$$\boldsymbol{G} = i \, \frac{2\pi \varDelta}{\lambda} \sum_{jj'} \frac{(\boldsymbol{s} \cdot \mathfrak{R}_{jj}) \, (\mathfrak{L}^j \times \mathfrak{L}^{j'})}{(\omega_j^{(0)^2} - \omega^2) \, (\omega_{j'}^{(0)^2} - \omega^2)} \,. \tag{55.14}$$

C'est une fonction vectorielle linéaire de la direction $\boldsymbol{s}$ de propagation de l'onde. Les composantes du tenseur g (52.3) ont pour expressions:

$$g_{xx} = \frac{\tau}{2} \sum_{jj'} \frac{\mathfrak{R}_{jj'x} \, [\mathfrak{L}^j \times \mathfrak{L}^{j'}]_x}{(\omega_j^{(0)^2} - \omega^2) \, (\omega_{j'}^{(0)^2} - \omega^2)} \,, \dots \tag{55.15}$$

$$\bar{g}_{xy} = \frac{\tau}{2} \sum_{jj'} \frac{\mathfrak{R}_{jj'y} \, [\mathfrak{L}^j \times \mathfrak{L}^{j'}]_z + \mathfrak{R}_{jj'z} \, [\mathfrak{L}^j \times \mathfrak{L}^{j'}]_y}{(\omega_j^{(0)^2} - \omega^2) \, (\omega_{j'}^{(0)^2} - \omega^2)} \dots \tag{55.16}$$

γ) La variation du pouvoir rotatoire avec la longueur d'onde est donnée par les formules (55.15) et (55.16). Il existe une différence entre la dispersion rotatoire des molécules et des cristaux[1]. L'expression (31.5) de la constante de gyration d'un milieu isotrope peut s'écrire:

$$g = \frac{n\,\omega}{3c} \sum_j \frac{\mathfrak{L}^j \cdot \mathfrak{R}^j}{\omega_j^2 - \omega^2} \,.$$

____

[1] M. Born: Z. Physik **8**, 414 (1922).

Elle ne contient pas, au dénominateur, de produits tels qu'il en existe dans (55.15). Les fréquences multiples ($\omega_j^{(0)} = \omega_{j'}^{(0)}$) peuvent donc donner, dans la dispersion rotatoire des cristaux, des termes contenant des carrés ($\omega_j^{(0)2} - \omega^2)^2$ au dénominateur. Il en résulte que le pouvoir rotatoire pourrait être de même sens de part et d'autre d'une bande d'absorption, au lieu de changer de signe, comme cela se produit nécessairement dans un fluide (Fig. 1). Toutefois, ce phénomène n'a pas encore été observé dans les cas, peu nombreux il est vrai, où l'on a fait des mesures de dispersion rotatoire assez étendues.

Chandrasekhar[1] remarque que le modèle de Kuhn (Sect. 38) conduit à une formule de dispersion contenant des termes quadratiques au dénominateur, lorsqu'on admet que les oscillateurs couplés ont des fréquences très voisines. Posons, dans la formule (38.9):

$$v_0 = \frac{v_1^2 + v_2^2}{2} \qquad \varepsilon = v_2^2 - v_1^2.$$

On peut alors écrire:

$$\varrho = \frac{a\,\varepsilon\,v^2}{(v_0^2 - v_2^2)^2}.$$

Pour cet auteur, seules les formules de ce type sont valables, car elles satisfont automatiquement à la règle de sommation, tandis que les formules du type (31.7) fondés sur l'expérience (par exemple dans le cas du quartz) et comprenant un nombre très limité de termes n'y satisfont pas. Cet argument n'est pas décisif, car la règle de sommation fait intervenir la totalité des fréquences et non celles qui sont directement accessibles aux mesures.

On remarquera, enfin, que les raisons qui rendent négligeables la contribution des bandes infrarouges à l'activité optique des molécules (Sect. 35), ne sont pas valables dans le cas des cristaux. On a vu, en effet (Sect. 53), que des bandes infrarouges du quartz, par exemple, sont actives. Mais la théorie n'a pas été développée sur ce point.

56. Calcul du pouvoir rotatoire de structure. Bien que la théorie générale soit valable pour les cristaux qui possèdent le pouvoir rotatoire moléculaire, comme pour ceux qui n'ont que le pouvoir rotatoire de structure, c'est à des cristaux de cette dernière catégorie qu'on l'a appliquée. La composition de leur motif cristallin est, en effet, souvent beaucoup plus simple que celle des cristaux à molécules optiquement actives; de plus certains d'entre eux sont cubiques.

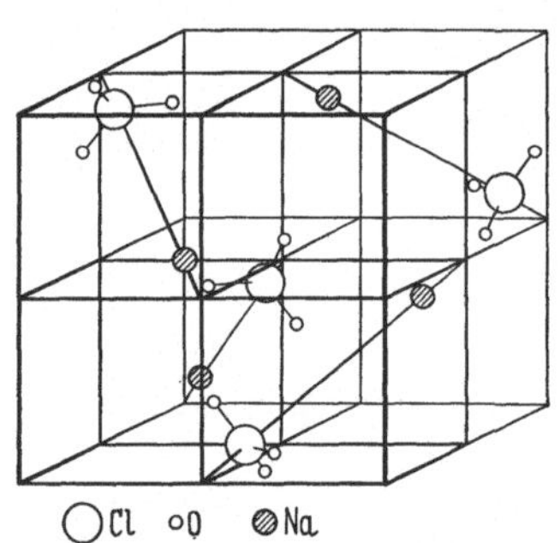

Fig. 62. Structure cristalline de NaClO₃.

α) Hermann[2] a appliqué la théorie générale (Sect. 55) au chlorate de sodium et au bromate de sodium, dont la structure cristalline est connue (Fig. 62). A chacune des quatre molécules $NaClO_3$ ou $NaBrO_3$ de la maille, on associe, pour simplifier, un seul oscillateur, constitué par l'ensemble de N_e électrons de l'anion, rassemblés en leur centre de gravité. On suppose que ce dernier coïncide avec l'atome Cl ou Br; cette hypothèse, qui simplifie considérablement les calculs déjà compliqués, est critiquable aujourd'hui que l'on sait que les ions ClO_3^- et BrO_3^- ne sont pas plans. L'oscillateur est supposé soumis à une force de rappel quasi-élastique, à laquelle correspond une fréquence propre ω_1. On admet que

---

[1] S. Chandrasekhar: Proc. Indian Acad. Sci. A **36**, 118 (1952).
[2] C. Hermann: Z. Physik **16** 103 (1923).

les actions exercées sur lui par les autres oscillateurs de charge $-N_e e$, par les anions de charge $-(N_e-1)e$ et par les cations de charge $+e$, sont des forces de COULOMB. Les sommes (54.7) et (54.8) se calculent au moyen des séries d'EWALD. L'équation (54.11) du mouvement, en première approximation, permet de calculer les valeurs de $N_e=9$ et $\omega_1=2,2\cdot10^{16}$ (pour $NaClO_3$) ainsi que les vecteurs propres a_k^i, les moments $\mathfrak{L}^j$ et les fréquences ω_j. L'expression du vecteur gyration s'obtient par la formule (55.14) et enfin celle de la dispersion rotatoire. L'accord est bon pour le bromate ($\varrho_{\mathrm{calc}}=3,14$ degré/mm, $\varrho_{\mathrm{obs}}=2,62$, pour $\lambda=0,546\,\mu$); pour le chlorate, les nombres calculés sont environ deux fois trop faibles ($\varrho_{\mathrm{calc}}=1,45$ degré/mm, $\varrho_{\mathrm{obs}}=3,13$, pour $\lambda=0,589\,\mu$).

RAMACHANDRAN[1] applique aux mêmes cristaux une théorie de polarisabilité (cf. Sect. 40). En supposant qu'une onde plane polarisée ($E=E_x$) se propage dans la direction z, le moment de dipôle induit dans l'atome k est

$$p_k = [\alpha]_k \left(E_x + \frac{4\pi P_x}{3}\right) e^{\frac{2\pi i}{\lambda} z_k}, \tag{56.1}$$

en admettant que le champ de polarisation a l'expression donnée par LORENTZ. De plus, on tient compte du champ moléculaire E_m produit sur l'atome consideréré par les dipôles voisins $p_{k'}$. Les phases des mouvements des divers atomes k' étant en général différentes de celle de l'atome k, on peut écrire

$$E_m = E_m' + i E_m''.$$

Les composantes du moment induit dans l'atome k sont donc

$$p_{kx} = \alpha_{xx}\left(E_x + \frac{4\pi P_x}{3} + E_{mx}\right) + \alpha_{xy}E_{my} + \alpha_{xz}E_{mz}, \ldots. \tag{56.2}$$

Les composantes de la polarisation P peuvent s'ecrire:

$$P_x = \frac{1}{\varDelta}\sum_k p_{kx} = \left(E_x + \frac{4\pi P_x}{3}\right)(A_x' + i A_x''), \ldots. \tag{56.3}$$

Ayant calculé ces expressions, on tire de (55.13) les valeurs de certaines composantes du tenseur de gyration.

Dans le cas des cristaux cubiques, on a $g_{xx}=g_{yy}=g_{zz}=g$; $\bar{g}_{xy}=\bar{g}_{yz}=\bar{g}_{zx}=0$. L'application a été faite au chlorate de sodium (Fig. 62). On admet que seuls les groupes O_3 sont polarisables et à partir de la réfraction, on calcule par la méthode de BRAGG[2] les composantes du tenseur de polarisabilité de la maille suivant les axes du cube. Dans le calcul du champ E_m, on fait intervenir les 12 groupes O_3 les plus voisins de l'un d'eux. On calcule le moment p_k par (56.2), puis g par (55.15). On trouve $g/n=9,806\cdot10^{-6}$ pour $\lambda=0,589\,\mu$, d'où $\varrho=3,0$ degré/mm, en bon accord avec la valeur mesurée.

β) En ce qui concerne les cristaux anisotropes, on a jusqu'ici seulement calculé le pouvoir rotatoire du quartz β, hexagonal, stable au-dessus de 570° C, dont la structure est plus simple que celle du quartz α, stable à la température ordinaire.

HYLLERAAS[3] considère la polarisation du quartz β comme due uniquement à celle des atomes O. Le calcul de ϱ (dont le détail n'est pas donné) ne fait intervenir aucun paramètre arbitraire. On trouve pour $\lambda=0,518\,\mu$, $\varrho=70$ degré/mm environ, alors que la valeur mesurée est 32 degré/mm à 600° C. Mais le calcul

---

[1] G. N. RAMACHANDRAN: Proc. Indian Acad. Sci. **38**, 217, 309 (1951).
[2] W. L. BRAGG: Proc. Roy. Soc. Lond., Ser. A **105**, 370 (1924).
[3] E. A. HYLLERAAS: Z. Physik **44**, 871 (1927).

montre que la valeur de ϱ est très sensible aux variations de la distance x des atomes O à l'axe sénaire, qui intervient comme paramètre dans l'interprétation des diagrammes de rayons X.

Ramachandran[1] applique les principes exposés ci-dessus pour les cristaux cubiques à un modèle hypothétique de cristal tétragonal à structure en spirale, puis au quartz β[2], en ne tenant compte que de la polarisabilité des atomes O, dont l'anisotropie est estimée assez arbitrairement. Il considère le champ moléculaire $\boldsymbol{E}_m$ produit par 33 atomes O entourant dans un rayon de 5 Å celui que l'on considère et il calcule au moyen de la formule (56.2) le moment induit dans chacun des atomes de la maille. On trouve dans la direction de l'axe optique

$$\frac{g_{zz}}{n_{,}} = -\,7{,}72 \cdot 10^{-5}, \quad \text{d'où} \quad \varrho_{\|} = -\,23{,}6 \text{ degré/mm} \quad \text{pour} \quad \lambda = 0{,}589\,\mu \quad \text{(valeur}$$

mesurée $-25{,}2$ degré/mm) et normalement à cet axe $\varrho_{\perp} = +\,11{,}5$ degré/mm. Cette dernière quantité n'a pas été mesurée, mais, pour le quartz α, on a une valeur voisine du rapport $\dfrac{\varrho_{\perp}}{\varrho_{\|}} = -\,0{,}50.$

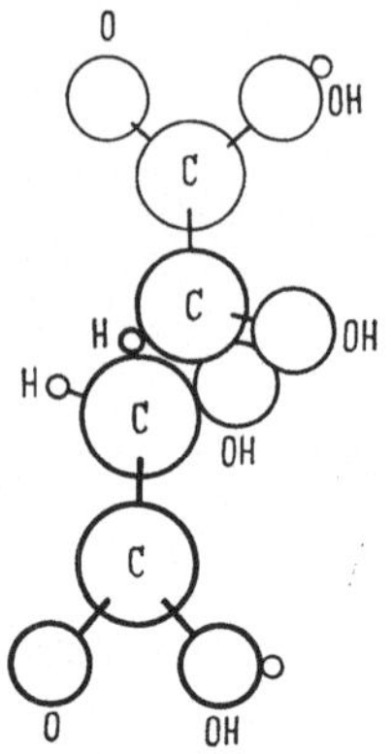

Chandrasekhar[3] a étendu la théorie d'oscillateurs couplés (Sect. 55γ) au cas du quartz α, en admettant que chaque groupe SiO_2 peut être remplacé par un oscillateur linéaire et que leur couplage donne naissance à deux fréquences propres[4].

On calcule $\varrho_D = 21{,}7$ degré/mm et $\dfrac{\varrho_{\perp}}{\varrho_{\|}} \neq -\,0{,}5.$

Fig. 63. Configuration absolue de l'acide tartrique dextrogyre.

57. Configuration absolue des cristaux optiquement actifs. α) La disposition relative des atomes dans la maille de nombreux cristaux optiquement actifs est connue, grâce à la diffraction des rayons X. Cependant les diagrammes obtenus pour deux antipodes sont identiques en général, car ils ne diffèrent que par le signe de differences de phase dues à des différences de marche, ce qui n'a pas d'influence sur les intensités. Les réflexions sur des plans $k\,h\,l$ et $\bar{k}\,\bar{h}\,\bar{l}$ sont généralement équivalentes[5]. Elles peuvent cependant cesser de l'être, si l'on utilise des rayons X de fréquences voisines des fréquences propres des électrons K de l'un des atomes. Des variations de phase des ondes diffractées par cet atome se produisent alors et ne changent pas de signe lorsqu'on passe d'une molécule à son antipode, entraînant des variations de l'intensité réfléchie par des plans de mêmes indices. En étudiant le diagramme de Weissenberg donné par le tartrate de sodium et de rubidium sur les rayons K_{α} de Zr, qui excitent l'atome Rb, on a pu[6] distinguer les deux isomères optiques et montrer que la configuration absolue de l'ion tartrique est celle qui correspond au schéma de Fischer (Fig. 63).

β) Une méthode moins sure est analogue à celle qui cherche à relier l'anisotropie de la polarisabilité au signe de la biréfringence. On sait que, dans les structures en couches ou en chaines, l'indice de réfraction pour les vibrations rectilignes a une valeur maximum lorsque le vecteur électrique est parallèle

---

[1] G. N. Ramachandran: Proc. Indian Acad. Sci. **33**, 222 (1951).

[2] G. N. Ramachandran: Proc. Indian Acad. Sci. **34**, 127 (1951).

[3] S. Chandrasekhar: Proc. Indian Acad. Sci. A **37**, 468 (1953).

[4] Cette hypothèse est exagérément simple et les raisonnements invoqués par l'auteur pour la justifier sont sujets à caution.

[5] G. Friedel: C. R. Acad. Sci., Paris **157**, 1533 (1913). — F. M. Jaeger et H. Haga: Proc. Akad. van Wet., Amsterdam **17** (1914).

[6] A. F. Peerdeman, A. J. van Bommel et J. M. Bijvoet: Nature, Lond. **168**, 271 (1951). — Proc. Akad van Wet., Amsterdam B **54**, 16 (1951).

aux couches ou aux chaînes. Par analogie, on admet que lorsqu'un rayon de lumière circulaire traverse une structure en spirale, son action sur les atomes est plus forte et son indice de réfraction plus élevé si la spirale décrite par la vibration (Fig. 3) tourne dans le même sens que les spirales atomiques. C'est la conclusion obtenue pour le quartz par des théories plus approfondies[1].

γ) Nous ne citerons que pour mémoire des considérations tirées du développement des faces cristallines et qui ont été reconnues sans valeur[2].

III. Pouvoir rotatoire moléculaire et pouvoir rotatoire de structure.

58. Anisotropie du pouvoir rotatoire d'une molécule. L'activité optique naturelle d'une molécule considérée isolément varie avec l'orientation de cette molécule par rapport à l'onde lumineuse. C'est cette hypothèse, implicitement contenue dans les théories que nous avons étudiées, qui rend nécessaire les calculs de moyenne pour obtenir l'expression du pouvoir rotatoire d'un milieu fluide.

Si l'on considère une molécule ayant une orientation déterminée dans le trièdre de propagation d'une onde lumineuse, les composantes des moments qui interviennent dans l'expression des propriétés optiques sont normales au vecteur d'onde. On trouve [*16*] qu'elles se réduisent essentiellement à

$$p'_a = \alpha\, E - \gamma_a\, \dot{H}, \tag{58.1}$$

$$m'_a = \gamma'_a\, \dot{E} \tag{58.2}$$

avec

$$\gamma_a = \frac{1}{\pi\, h\, \dot{H}^2} \sum_b \frac{1}{v_{ab}^2 - v^2}\, \mathrm{Im}\,[(a\,|\boldsymbol{p}|\,b) \cdot \boldsymbol{H}\,(b\,|\boldsymbol{m}|\,a)\,\dot{\boldsymbol{H}}], \tag{58.3}$$

$$\gamma'_a = -\frac{1}{\pi\, h\, \dot{E}^2} \sum_b \frac{1}{v_{ab}^2 - v^2}\, \mathrm{Im}\,[(a\,|\boldsymbol{m}|\,b) \cdot \dot{\boldsymbol{E}}\,(b\,|\boldsymbol{p}|\,a)\,\boldsymbol{E}]. \tag{58.4}$$

Si le milieu est formé de molécules dans l'état a, fixes et toutes parallèles entre elles, on a des relations analogues à (30.3)

$$D = E + 4\pi N\, p'_a = \varepsilon\, E - \frac{j}{c}\, \dot{H}, \tag{58.5}$$

$$B = H + 4\pi N\, m'_a = H + \frac{j'}{c}\, \dot{E}, \tag{58.6}$$

j et j' étant liés de façon évidente à γ_a et γ'_a.

Le pouvoir rotatoire est donné par une expression du même genre que (31.8), mais où la force rotatoire a une expression différente de (30.5)

$$R_{ab} = \frac{3}{2}\, \mathrm{Im}\left[\frac{1}{\dot{H}^2}\,(a\,|\boldsymbol{p}|\,b) \cdot \dot{\boldsymbol{H}}\,(b\,|\boldsymbol{m}|\,a)\,\dot{\boldsymbol{H}} - \frac{1}{\dot{E}^2}\,(a\,|\boldsymbol{m}|\,b) \cdot \dot{\boldsymbol{E}}\,(b\,|\boldsymbol{p}|\,a)\,\dot{\boldsymbol{E}}\right]. \tag{58.7}$$

On peut définir une *surface de gyration* (Sect. 52) liée à la molécule; on l'obtient en portant, sur toutes les directions issues d'un point fixe de la molécule, une longueur fonction du pouvoir rotatoire que possède la molécule lorsque la lumière se propage dans la direction correspondante. Cette surface existe, car on peut

---

[1] E. A. Hylleraas: Z. Physik **44**, 871 (1927). — G. N. Ramachandran: Proc. Indian Acad. Sci. **34**, 127 (1951).

[2] J. Waser: J. Chem. Phys. **17**, 498 (1949). — F. E. Turner et K. Lonsdale: J. Chem. Phys. **18**, 156 (1950).

montrer [16] que la force rotatoire R_{ab} ne dépend pas de l'orientation de la vibration lumineuse incidente dans le plan d'onde. La surface de gyration est définie par l'équation

$$R_{ab} = \tfrac{2}{3} \operatorname{Im} \left[(a\,|\boldsymbol{p}|\,b) \cdot (b\,|\boldsymbol{m}|\,a) - (a\,|\boldsymbol{p}|\,b)\,\boldsymbol{s}\,(b\,|\boldsymbol{m}|\,a) \cdot \boldsymbol{s} \right], \qquad (58.8)$$

où $\boldsymbol{s}$ désigne le vecteur unitaire porté par la direction de propagation de l'onde lumineuse.

Par exemple, le pouvoir rotatoire ϱ du modèle moléculaire de Kuhn dans une direction $\boldsymbol{s}$, définie en coordonnées polaires dans le trièdre de la Fig. 45 par les angles ϑ et φ a pour expression

$$\varrho = \frac{3}{2}\,\bar{\varrho} \left[1 - \frac{1}{2}\sin^2 \vartheta \left(1 - \frac{\dfrac{e_1^2}{m_1}\cos^2 \beta + \dfrac{e_2^2}{m_2}\sin^2 \beta}{\dfrac{2\,e_1 e_2}{\sqrt{m_1 m_2}}\sin \beta \cos \beta}\sin 2\varphi \right) \right]. \qquad (58.9)$$

$\bar{\varrho}$ désigne le pouvoir rotatoire moyen donné par (38.9).

On remarque que si les oscillateurs de Kuhn ne sont pas couplés ($\sin 2\beta = 0$) le pouvoir rotatoire d'une molécule individuelle ne s'annule pas, tandis que le pouvoir rotatoire moyen disparaît d'après (38.9).

59. Pouvoir rotatoire moléculaire et pouvoir rotatoire cristallin. α) Tous les cristaux provenant de composés qui possèdent le pouvoir rotatoire moléculaire possèdent aussi le pouvoir rotatoire cristallin[1].

Pour comparer le pouvoir rotatoire cristallin d'une substance moléculairement active à son pouvoir rotatoire à l'état liquide ou dissous, il faut ramener les valeurs mesurées dans les deux cas à la même quantité de substance active. Pour cela, on définit le pouvoir rotatoire spécifique cristallin par une relation analogue à (1.1) $[\varrho]_c = 100\,\varrho/d$, où ϱ est la rotation en degré/mm et d la densité du cristal, et on compare $[\varrho]_c$ au pouvoir rotatoire spécifique $[\varrho]_f$ à l'état fluide. On trouve que le rapport $[\varrho]_c/[\varrho]_f$ est très variable, comme le montrent les nombres suivants relatifs à la raie D [3].

Sulfate de strychnine	$+25$
Camphre de Matico	$+\ 6{,}14$
Tartrate de rubidium	$-14{,}8$
Saccharose, axe a'	$+\ 4{,}9$
Saccharose, axe a	$-\ 1{,}45$

Toutefois, on ne peut, dans tous les cas précédents, faire la part de l'anisotropie du pouvoir rotatoire moléculaire et celle du pouvoir rotatoire de structure, car ou bien la structure cristalline est encore inconnue, ou bien la maille contient plusieurs molécules. Longchambon[2], remarquant que les cristaux de rhodium III trioxalate de potassium actifs et les ions $(Rh\,o\,x_3)$ ont tous deux la symétrie ternaire, avait pensé que les ions étaient tous parallèles entre eux et que les différences de dispersion rotatoire de ce composé à l'état solide et dissous montraient l'anisotropie de la rotation moléculaire. Mais les recherches récentes ont montré qu'il y a six molécules dans la maille cristalline; il doit donc exister un pouvoir rotatoire de structure.

β) Pour un grand nombre de composés incolores, Longchambon[3] a trouvé que la dispersion rotatoire, mesurée dans le spectre visible, est la même pour

[1] L. Longchambon: Bull. Soc. franç. Minér. **45**, 161 (1922).
[2] L. Longchambon: C. R. Acad. Sci., Paris **178**, 1828 (1924).
[3] L. Longchambon: Bull. Soc. Franç. Minér. **45**, 161 (1922).

la substance cristallisée ou fluide. Les nombres suivants donnent les valeurs du rapport $\Delta = [\varrho]_{436}/[\varrho]_{579}$ pour divers composés:

	Tartrate d'ammonium	Saccharose axe a axe a'	Camphre des laurinèes
Δ cristal	1,80	1,90　1,88	2,7
Δ solution	1,81	1,87	2,92

Ce résultat montre simplement que, loin des bandes d'absorption, on doit pouvoir représenter le pouvoir rotatoire du corps cristallisé ou dissous par des formules de DRUDE à un terme (Sect. 34β) dans lesquelles la longueur d'onde propre est à peu près la même. L'identité des dispersions rotatoires cesse, en effet, lorsqu'on étudie l'activité optique au voisinage de bandes d'absorption[1,2].

γ) L'étude de l'activité optique de l'acide tartrique cristallisé et dissous fournit la preuve de l'anisotropie du pouvoir rotatoire moléculaire[3]. La maille de l'acide actif contient deux molécules qui se déduisent l'une de l'autre par une rotation hélicoïdale autour d'un axe binaire. Toutes les molécules sont donc parallèles ou antiparallèles dans le cristal. Il en résulte que la détermination du pouvoir rotatoire de l'acide cristallisé fournit la rotation produite par une molécule individuelle dans la direction de l'un ou de l'autre des axes optiques. La rotation, mesurée par LONGCHAMBON[4], peut être représentée par la formule

$$[\varrho]_c = \frac{-183,65}{\lambda^2 - (0,233)^2} \, . \tag{59.1}$$

La rotation en solution dans l'eau est représentée par la formule:

$$[\varrho]_f = -\frac{C_1}{\lambda^2 - (0,233)^2} + \frac{C_2}{\lambda^2 - (0,1749)^2} \, . \tag{59.2}$$

Les coefficients C_1 et C_2 sont fonctions du titre τ de la solution. L'étude de leurs variations conduit à admettre[5] l'existence de trois formes moléculaires, l'une α lévogyre, les autres β et γ dextrogyres. LONGCHAMBON suppose que la forme α constitue le cristal, ce qui est rendu vraisemblable par l'identité des longueurs d'onde caractéristiques en (59.1) et dans le premier terme de (59.2). Les mesures du pouvoir rotatoire des solutions entre 0,64 et 0,26 μ sont bien représentées lorsqu'on pose:

$$C_1 = 17,83 + 0,0105\,\tau, \quad C_2 = 24,12 - 0,0395\,\tau. \tag{59.3}$$

Soit $[\varrho_1]$ la rotation spécifique de la forme lévogyre, $[\varrho_2]$ celle de l'ensemble des constituants dextrogyres. On a:

$$[\varrho]_f = x\,[\varrho_1] + (1 - x)\,[\varrho_2], \tag{59.4}$$

x étant la proportion de la forme α. Des relations (59.2), (59.3) et (59.4) on tire:

$$C_1\,x = 24,24, \quad C_2\,(1 - x) = 91,20 \, .$$

---

[1] F. M. JAEGER, J. TERBERG et P. TERPSTRA: Proc. Akad. van Wet., Amsterdam **40**, 574 (1937).

[2] E. DROUARD et J. P. MATHIEU: C. R. Acad. Sci., Paris **236**, 2395 (1953).

[3] M. LEVY: J. Phys. Radium **11**, 80 (1950).

[4] L. LONGCHAMBON: C. R. Acad. Sci., Paris **178**, 951 (1924).

[5] R. LUCAS: Ann. Phys., Paris **9**, 381 (1928).

Le pouvoir rotatoire spécifique de l'acide α pur ($x=1$) en solution peut donc s'écrire:

$$[\varrho_1] = - \frac{24,24}{\lambda^2 - (0,233)^2} \cdot$$

En comparant cette expression avec (59.1), on voit que:

$$\frac{[\varrho]_c}{[\varrho_1]} = 7,57.$$

Le pouvoir rotatoire de la molécule dans la direction d'un des axes optiques du cristal est donc beaucoup plus grand que sa valeur moyenne.

60. Activité optique des substances cholestériques.

α) On sait qu'un assez grand nombre de composés organiques sont capables de donner des phases intermédiaires entre le cristal et le liquide et sont appelés pour cette raison *cristaux liquides* [14] ou mieux *substances mésomorphes* [6]. Certains d'entre eux, dont la molécule est dissymétrique, montrent en phase liquide un pouvoir rotatoire de l'ordre de grandeur ordinaire; mais à l'état mésomorphe, lorsqu'on les observe en couche mince, entre deux lames de verre, ils peuvent présenter un pouvoir rotatoire considérable, de plusieurs dizaines de milliers de degrés par mm[1]. Ce type de composés comprend de nombreux esters du cholestérol (esters des acides aliphatiques, du formiate au myristate; benzoate; cinnamate, etc. ...), d'où le nom général *d'état cholestérique* qu'on donne à ces phases à pouvoir rotatoire élevé; mais elles peuvent être formées par d'autres composés, le p-cyanobenzylidène-amino-p-cinnamate d'amyle par exemple. Elles se produisent également par l'addition d'une certaine quantité d'une substance cholestérique, ou même d'une substance à pouvoir rotatoire moléculaire (saccharose) à la phase nématique d'un corps tel que le p-azoxyphenetol [6].

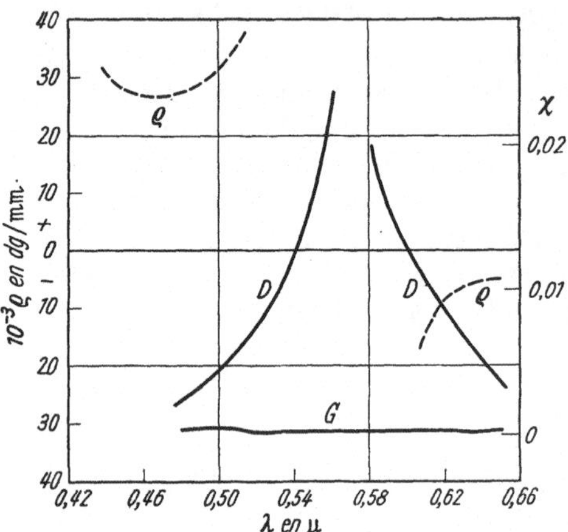

Fig. 64. Activité optique du cyanobenzylidèneamino-cinnamate d'amyle. Les courbes en traits pleins représentent l'absorption des rayons circulaires droit (D) et gauche (G).

Le pouvoir rotatoire observé est dû presque entièrement à la structure. Lorsqu'on fait passer le composé de l'état solide à l'état mésomorphe par élévation de température, de façon qu'il forme une couche mince entre deux lames de verre, un petit mouvement d'une des lames provoque, en même temps que l'apparition du pouvoir rotatoire élevé, la formation d'une structure stratifiée, révélée par l'observation des *plans* de Grandjean[2], limitant les strates, dont l'épaisseur l est de l'ordre de 0,2 μ. Cette structure, analogue à celle des photographies en couleurs de Lippmann, a pour conséquence la réflexion sélective d'une longueur d'onde $\lambda = 2nl$ (n étant l'indice de la substance) souvent située dans le spectre visible: les préparations ont alors des couleurs très vives par transmission et par réflexion. La longueur d'onde de réflexion sélective varie avec la température; elle augmente, en général, quand la température s'abaisse.

L'une des deux ondes polarisées circulairement, droite et gauche, en lesquelles se décompose une onde incidente de lumière naturelle (Sect. 2) est réfléchie sélectivement (Fig. 64). Les corps droits réfléchissent une vibration circulaire

---

[1] Pour la mesure du pouvoir rotatoire, voir en particulier la Sect. 8 ε.
[2] F. Grandjean: C. R. Acad. Sci., Paris **172**, 71 (1921).

droite et celle-ci reste droite après réflexion; c'est l'inverse dans le cas des corps gauches[1]. Les substances cholestériques stratifiées possèdent donc un dichroïsme circulaire considérable: les préparations dont l'épaisseur dépasse quelques dizaines de microns transmettent de la lumière circulaire. Toutefois, les relations générales entre dispersion rotatoire et dichroïsme circulaire (Fig. 1) ont été vérifiées[2] pour des épaisseurs très faibles.

β) Pour interpréter par la structure les propriétés des lames stratifiées des substances cholestériques, on admet que les directions d'allongement, ou axes, de leurs molécules, qui sont nettement anisodiamétriques, sont parallèles aux plans des strates, mais disposées en empilements hélicoïdaux formant des groupes (Fig. 65). Chaque groupe ayant suivant l'axe une longueur d égale à la distance des plans de GRANDJEAN comprendrait 100 â 300 molécules.

Les molécules des substances cholestériques ont toutes de forts moments électriques permanents. Leurs actions électrostatiques mutuelles peuvent peut-être rendre compte de la structure, car on peut voir [16] que le champ électrostatique E', créé sur l'une d'elles par l'ensemble des autres molécules d'un groupe hélicoidal, est dirigé suivant l'axe de la molécule considérée, ce qui montre que cet arrangement est en équilibre. S'il y a dans l'unité de longueur, comptée sur l'axe du groupe, n molécules de moment p, le champ E' est donné par la formule:

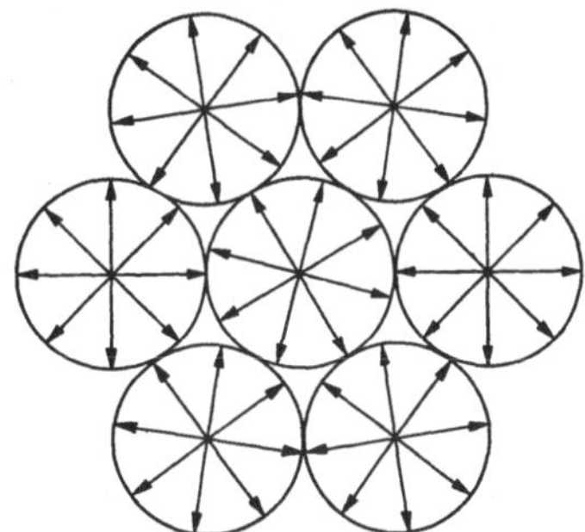

Fig. 65. Modèle de LÉVY pour les empilements hélicoïdaux de molécules d'une substance cholestérique.

$$E' = \frac{2\,p\,n^3}{d^3}\,\varphi\,(\eta) \tag{60.1}$$

où l'on a posé $\eta = \dfrac{\alpha\,d}{n}$; $\varphi\,(\eta)$ se calcule au moyen du développement en série suivant:

$$\varphi\,(\eta) = 1{,}2021 + \frac{\eta^2}{2}\log\eta - \frac{3\,\eta^2}{4} - \frac{\eta^4}{288} - \frac{\eta^6}{86\,400} - \cdots.$$

On ne possède pas, jusqu'à présent, de théorie moléculaire du pouvoir rotatoire des substances cholestériques. La théorie de BORN[3], qui ramène l'activité optique observée à celle des molécules, ne semble pas pouvoir rendre compte des phénomènes.

OSEEN[4] a développé une théorie macroscopique, en admettant que l'énergie potentielle et l'énergie libre d'un volume de la substance de densité constante D, où les actions intermoléculaires décroissent vite avec la distance, sont de la forme:

$$I = \iiint D^2 \{K_1\,\boldsymbol{L}\,\mathrm{rot}\,\boldsymbol{L} + K_{11}\,(\boldsymbol{L}\,\mathrm{rot}\,\boldsymbol{L})^2 + K_{22}\,(\mathrm{div}\,\boldsymbol{L})^2 + \\ + K_{33}\,[(\boldsymbol{L}\cdot\mathrm{grad})\,\boldsymbol{L}]^2 + 2K_{12}\,\mathrm{div}\,\boldsymbol{L}\,(\boldsymbol{L}\cdot\mathrm{rot}\,\boldsymbol{L})\}\,dx\,dy\,dz. \tag{60.2}$$

K_1, K_{11} sont des constantes à température donnée et $\boldsymbol{L}$ désigne un vecteur unitaire porté par l'axe d'une molécule, dont l'orientation est supposée une fonction continue des coordonnées x, y, z. La structure d'équilibre du milieu est déterminée par la condition $\delta I = 0$.

---

[1] F. GIESEL: Phys. Z. **11**, 192 (1910).

[2] J. P. MATHIEU: Bull. Soc. franç. Minér. **61**, 174 (1938).

[3] M. BORN: Sitzgsber. preuß. Akad. Wiss. Berlin **1916**, 614.

[4] C. W. OSEEN: Die anisotropen Flüssigkeiten. Berlin: Gebrüder Bornträger 1929. — Trans. Faraday Soc. **29**, 883 (1933).

Si les forces intermoléculaires sont de nature purement électrostatiques, $K_1 = K_{12} = 0$ et le pouvoir rotatoire disparaît. Il faut tenir compte de forces de nature magnétique pour que ces coefficients ne soient pas nuls. L'hypothèse la plus simple qui permette de rendre compte de la structure cholestérique â plans est $K_{12} = 0$, $K_1 \neq 0$. On peut alors satisfaire à la condition $\delta I = 0$ en posant:

$$L_x = \cos \alpha z, \quad L_y = \sin \alpha z, \quad L_z = 0. \tag{60.3}$$

(α étant une constante fonction des K). Les axes moléculaires ont alors une distribution hélicoïdale autour de la direction Oz.

Admettons que la polarisabilité de chaque molécule soit représentable par un ellipsoïde de révolution autour de $\boldsymbol{L}$. La polarisation $\boldsymbol{P}$ du milieu peut alors s'écrire

$$\boldsymbol{P} = (\varepsilon_0 - \varepsilon_1)\,\boldsymbol{L}\,(\boldsymbol{L} \cdot \boldsymbol{E}) + \varepsilon_1\,\boldsymbol{E}, \tag{60.4}$$

où ε_0 et ε_1 sont les constantes diélectriques principales, la première dans la direction $\boldsymbol{L}$. Les équations de Maxwell (4.1) et (4.2) prennent la forme

$$\left.\begin{aligned} &\operatorname{rot} \boldsymbol{E} - \frac{1}{c}\,\dot{\boldsymbol{H}} = 0, \\[2mm] &\operatorname{rot} \boldsymbol{H} - \frac{\varepsilon_0 - \varepsilon_1}{c}\,(\boldsymbol{L} \cdot \dot{\boldsymbol{E}})\,\boldsymbol{L} - \frac{\varepsilon_1}{c}\,\dot{\boldsymbol{E}} = 0. \end{aligned}\right\} \tag{60.5}$$

Ces équations permettent de retrouver les propriétés principales des substances cholestériques stratifiées. On montre que le domaine spectral de réflexion sélective s'étend aux longueurs d'onde λ pour lesquelles

$$\frac{2\pi \sqrt{\varepsilon_0}}{\alpha} > \lambda > \frac{2\pi \sqrt{\varepsilon_1}}{\alpha}. \tag{60.6}$$

A l'état liquide, les orientations des axes moléculaires sont indéterminées, l'angle α est infini. Lorsque la phase cholestérique apparaît, α diminue et la zône de réflexion sélective se déplace de l'ultraviolet vers le visible. La distance l des plans de Grandjean est reliée à α:

$$l = \frac{\lambda}{2n} = \frac{\lambda}{2\sqrt{\varepsilon}} \approx \frac{\pi}{\alpha}. \tag{60.7}$$

L'angle que font entre eux les axes moléculaires extrêmes des groupes hélicoïdaux de molécules est donc égal à π.

En résolvant les équations de Maxwell, on trouve que le pouvoir rotatoire a pour expression

$$\varrho = \frac{9\,\delta^2}{4\pi}\,\frac{\omega^3}{c^3 \sqrt{\varepsilon}}\,\frac{1}{\alpha^2 - \dfrac{\omega^2}{c^2}\,\varepsilon}, \tag{60.8}$$

où $2\varepsilon = \varepsilon_0 + \varepsilon_1$, $2\delta = \varepsilon_1 - \varepsilon_0$ et $\omega = \dfrac{2\pi c}{\lambda}$. Le calcul suppose que δ est petit devant ε.

61. Action d'un champ électrostatique ou magnétostatique sur l'activité optique naturelle d'un milieu isotrope. α) L'application d'un champ extérieur, électrique $\boldsymbol{E}_0$ ou magnétique $\boldsymbol{H}_0$, à un milieu isotrope, y crée une direction privilégiée, qui, dans les cas les plus intéressants, est parallèle ou perpendiculaire à la direction de propagation de la lumière.

Un champ magnétique parallèle à la direction d'observation crée du pouvoir rotatoire magnétique. Un champ électrique ou magnétique perpendiculaire à la direction d'observation crée de la biréfringence.

β) Dans les deux cas, la mesure de l'activité optique est moins simple qu'on l'a vu à la Sect. 10. Dans le premier cas, on mesurera les rotations ϱ_+ et ϱ_- pour deux champs égaux et de sens inverses [*16*]. On en tire les valeurs de la rotation naturelle ϱ_n (en présence du champ) et de la rotation magnétique ϱ_m

$$2\varrho_n = \varrho_+ + \varrho_-, \quad 2\varrho_m = \varrho_+ - \varrho_- . \tag{61.1}$$

Une vibration rectiligne incidente donne une vibration elliptique dont on détermine l'orientation α et l'ellipticité γ. Dans le deuxième cas, DE MALLEMANN[1] admet la validité de la relation de GOUY (50.15) dans le milieu biréfringent. Les grandeurs α et γ sont reliées à φ_0 et à ϱ_n. La rotation ϱ_n peut être mesurée hors du champ. On a donc un système de deux équations contenant une seule inconnue φ_0. Si ces équations sont compatibles, elles indiquent que la rotation ϱ_n n'est pas influencée par le champ. Dans le cas contraire, c'est-à-dire si les écarts entre valeurs calculées et observées sont systématiques et supérieurs aux erreurs d'expériences, c'est qu'il y a une variation du pouvoir rotatoire due à l'anisotropie moléculaire.

L'ellipticité γ et l'orientation α du grand axe ont des expressions générales compliquées[1].

L'ellipticité γ passe par un maximum γ_m pour un azimut α_{0m} de la vibration incidente donné par

$$\tan 2\alpha_{0m} = \frac{\sin 2\gamma_{45}}{\sin 2\gamma_0}, \tag{61.2}$$

où γ_0 et γ_{45} désignent les ellipticités mesurées pour $\alpha_0 = 0$ et $\alpha_0 = 45°$. On a alors:

$$\sin 2\gamma_m = \frac{2\varphi_0}{\varphi}\sin\frac{\varphi}{2}\sqrt{\cos^2\frac{\varphi}{2} + \frac{4\varrho^2}{\varphi^2}\sin^2\frac{\varphi}{2}} . \tag{61.3}$$

γ) Au point de vue théorique, le champ imposé induit dans la molécule un moment électrique ou magnétique. Lorsque les molécules ne possèdent pas de moment électrique ou magnétique permanent (Sect. 24), c'est leur moment induit qui tend à s'orienter dans le champ. Si les molécules sont polaires ou paramagnétiques, aux actions d'orientation précédentes s'ajoutent celles du champ sur les moments permanents, qui sont toujours plus fortes que les premières.

Dans tous les cas, pour calculer l'expression de l'activité optique du milieu, on procédera de la façon suivante. D'après ce qu'on a vu aux Sects. 29 et 30, la partie du moment électrique (29.11), induit par l'onde lumineuse qui est à l'origine du pouvoir rotatoire, a pour expression, pour une transition $a \to b$,

$$\boldsymbol{p}'_{ab} = -\frac{1}{\pi h}\frac{1}{v_{ba}^2 - v^2}\,\mathrm{Im}\left[(a\,|\boldsymbol{p}|\,b)\,(b\,|\boldsymbol{m}|\,a)\cdot\dot{\boldsymbol{H}}\right] . \tag{61.4}$$

Considérons un trièdre $Oxyz$ ayant son origine en un point 0 de la molécule et ses axes parallèles á ceux de l'ellipsoïde de polarisabilité de la molécule. Les vecteurs $(a\,|\boldsymbol{p}|\,b)$ et $(b\,|\boldsymbol{m}|\,a)$ sont fixes dans ce trièdre et font entre eux l'angle δ_{ab}. On peut écrire:

$$\left.\begin{aligned}
(a\,|\boldsymbol{p}|\,b) &= (a\,|\boldsymbol{p}|\,b)\,(\sin\xi\cos\eta\,\boldsymbol{i} + \sin\xi\sin\eta\,\boldsymbol{j} + \cos\xi\,\boldsymbol{k}), \\
(b\,|\boldsymbol{m}|\,a) &= (b\,|\boldsymbol{m}|\,a)\,(\sin\zeta\cos\chi\,\boldsymbol{i} + \sin\zeta\sin\chi\,\boldsymbol{j} + \cos\zeta\,\boldsymbol{k}).
\end{aligned}\right\} \tag{61.5}$$

---

[1] R. DE MALLEMANN: Ann. Phys., Paris **22**, 192 (1924).

i, j, k étant trois vecteurs unitaires portés par Ox, Oy et Oz respectivement. Les angles ξ, η, ζ, χ sont unis par la relation:

$$\sin\zeta \sin\xi \cos(\eta - \chi) + \cos\zeta \cos\xi = \cos\delta_{ab}. \tag{61.6}$$

Il faut calculer la valeur moyenne de la composante de $\boldsymbol{p}'_{ab}$ dirigée suivant $\boldsymbol{H}$. Le résultat du calcul dépend des conditions imposées pour le champ extérieur et de la nature des molécules.

62. Théorie de l'action d'un champ extérieur longitudinal sur l'activité optique (molécules non polaires). Considérons [16] deux trièdres trirectangles ayant leur origine en un point 0 de la molécule. Le premier $Oxyz$ a été défini ci-dessus. Le second $OXYZ$ est celui de la Fig. 66, OZ étant la direction du champ appliqué $\boldsymbol{E_0}$. L'orientation de $Oxyz$ par rapport à $OXYZ$ peut être défini par les angles d'Euler, ϑ, φ et ψ. On calcule le moment électrique $\boldsymbol{p}_i$ induit par le champ $\boldsymbol{E_0}$ au moyen des angles précédents et des valeurs des polarisabilités principales α_x, α_y et α_z, puis l'énergie potentielle de ce moment dans le champ $\boldsymbol{E_0}$:

$$U = -\tfrac{1}{2}\boldsymbol{p}_i \cdot \boldsymbol{E_0} = -\tfrac{1}{2}(\alpha_x \sin^2\vartheta \, \sin^2\varphi + \alpha_y \sin^2\vartheta \cos^2\varphi + \alpha_z \cos^2\vartheta)\, E_0^2. \tag{62.1}$$

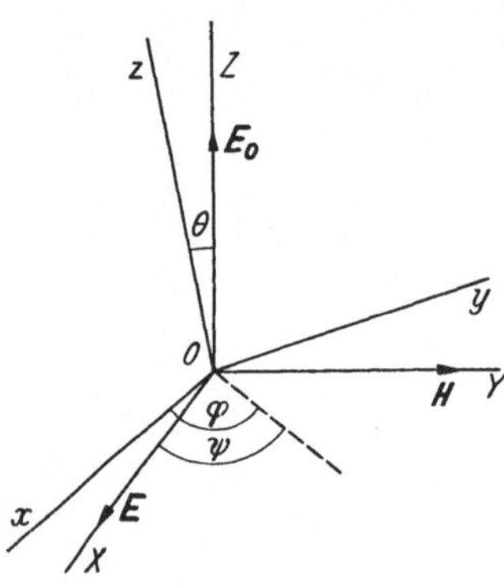

Fig. 66. Action d'un champ extérieur longitudinal sur l'activité optique.

Le moment $\boldsymbol{p}'_{ab}$ donné par (61.4) a pour composante suivant $\boldsymbol{H}$, c'est-à-dire suivant OY

$$\boldsymbol{p}'^{(y)}_{ab} = \boldsymbol{p}'_{ab}\left[\sin\xi \cos\eta (\cos\varphi \sin\psi + \sin\varphi \cos\psi \cos\vartheta) - \sin\xi \sin\eta (\sin\varphi \sin\psi - \cos\varphi \cos\psi \cos\vartheta) - \cos\xi \cos\psi \sin\vartheta\right]. \tag{62.2}$$

La valeur moyenne de cette expression est donnée par la formule

$$\overline{\boldsymbol{p}}'^{(y)}_{ab} = \frac{\displaystyle\int\limits_{\vartheta=0}^{\pi}\int\limits_{\varphi=0}^{\pi}\int\limits_{\psi=0}^{2\pi} e^{-\frac{U}{kT}}\, p'^{(y)}_{ab} \sin\vartheta \, d\vartheta \, d\varphi \, d\psi}{\displaystyle\int\limits_{\vartheta=0}^{\pi}\int\limits_{\varphi=0}^{\pi}\int\limits_{\psi=0}^{2\pi} e^{-\frac{U}{kT}} \sin\vartheta \, d\vartheta \, d\varphi \, d\psi}. \tag{62.3}$$

Le calcul se fait en tenant compte de ce que U ne dépend pas de ψ et de ce que, dans tous les cas qui se présentent pratiquement, $\dfrac{U}{kT}$ est petit devant l'unité. Il conduit au résultat suivant:

$$\overline{p}'^{(y)}_{ab} = \dot{H}\left\{\frac{1}{3}(a\,|\boldsymbol{p}|\,b)\cdot(b\,|\boldsymbol{m}|\,a) + \frac{1}{30}\alpha_{ab}\frac{E_0^2}{kT}\left[(a\,|\boldsymbol{p}|\,b)(b\,|\boldsymbol{m}|\,a) - \sum_j \delta^j_{ab}\,\boldsymbol{b}_j\cdot(a\,|\boldsymbol{p}|\,b)\,\boldsymbol{b}_j(b\,|\boldsymbol{m}|\,a)\right]\right\} \tag{62.4}$$

où

$$\alpha_{ab} = \frac{\alpha_x + \alpha_y + \alpha_z}{3},$$

$$\delta^j_{ab} = \frac{\alpha_j}{\alpha_{ab}} \qquad (j = x, y, z)$$

et où les $\boldsymbol{b}_j$ sont des vecteurs unitaires portés par les axes de l'ellipsoïde de polarisabilité. Par suite, en se reportant à la définition (30.5) de la force rotatoire R_{ab}

obtenue à partir du moment (29.11), et en posant:

$$R_{ab}^0 = \mathrm{Im}\,[(a\,|\,\boldsymbol{p}\,|\,b)\cdot(b\,|\,\boldsymbol{m}\,|\,a)], \qquad (62.5)$$

on a:

$$R_{ab} = R_{ab}^0 + \frac{1}{30}\,\alpha_{ab}\,\frac{E_0^2}{kT}\,\mathrm{Im}\,\Big[(a\,|\,\boldsymbol{p}\,|\,b)\cdot(b\,|\,\boldsymbol{m}\,|\,a) - \sum_j \delta_{ab}^j\,\boldsymbol{b}_j\,(a\,|\,\boldsymbol{p}\,|\,b)\,\boldsymbol{b}_j\,(b\,|\,\boldsymbol{m}\,|\,a)\Big]. \qquad (62.6)$$

Si $\delta_{ab}^j = 1$, c'est-à-dire si les molécules sont isotropes, le second terme entre crochets de (62.6) est égal au premier et $R_{ab} = R_{ab}^0$. Le champ $\boldsymbol{E}_0$ ne produit pas alors de variations du pouvoir rotatoire.

On tire de (62.6) pour la variation relative du pouvoir rotatoire dû à la transition $a \rightarrow b$:

$$\left(\frac{\varDelta\varrho}{\varrho}\right)_{ab} = \frac{R_{ab} - R_{ab}^0}{R_{ab}^0} = \frac{E_0^2}{30\,kT}\,\alpha_{ab}\left(1 - \frac{\sum\limits_j \mathrm{Im}\,[\delta_{ab}^j\,\boldsymbol{b}_j\,(a\,|\,\boldsymbol{p}\,|\,b)\,\boldsymbol{b}_j\,(b\,|\,\boldsymbol{m}\,|\,a)]}{\mathrm{Im}\,[(a\,|\,\boldsymbol{p}\,|\,b)\,(b\,|\,\boldsymbol{m}\,|\,a)]}\right). \qquad (62.7)$$

Le facteur entre parenthèses est compris entre 0 et 1. En le prenant égal à $\tfrac{1}{2}$, en évaluant α_{ab} à partir de la réfractivité moléculaire, on trouve pour un indice $n = 1,5$, une masse molaire 150 et une masse spécifique 1,5:

$$\left(\frac{\varDelta\varrho}{\varrho}\right)_{ab} = 1,4\cdot 10^{-10}\,\frac{E_0^2}{T},$$

soit une valeur voisine de $0,5\cdot 10^{-8}$ pour $T = 300°\,\mathrm{K}$ et $E_0 = 30000\,\mathrm{V/cm}$. La variation est impossible à déceler et le reste aux températures les plus basses utilisables actuellement dans ces mesures.

63. Théorie de l'action d'un champ extérieur sur l'activité optique (molécules polaires). Les mêmes calculs s'appliquent à l'action d'un champ électrique $\boldsymbol{E}_0$ sur un milieu formé de molécules polaires ou d'un champ magnétique $\boldsymbol{H}_0$ sur des molécules paramagnétiques.

α) Champ longitudinal. Outre le trièdre $OXYZ$ lié à l'onde lumineuse, il faut considérer ici deux trièdres liés à la molécule. Le premier $Ox_0y_0z_0$ est tel que son axe Oz_0 soit parallèle au moment permanent $\boldsymbol{p}_0$ (ou $\boldsymbol{m}_0$) et que le plan x_0Oz_0 contienne le champ imposé $\boldsymbol{E}_0$ (ou $\boldsymbol{H}_0$). Le second $Oxyz_0$ a ses axes Ox et Oy arbitrairement choisis dans le plan x_0Oy_0. L'orientation du moment $\boldsymbol{p}_0$ (ou $\boldsymbol{m}_0$) dans le trièdre $OXYZ$ est définie par les angles ϑ et φ (Fig. 67). La relation (61.5) s'applique, en remplaçant $\boldsymbol{i}, \boldsymbol{j}$ et $\boldsymbol{k}$ par $\boldsymbol{i}_0, \boldsymbol{j}_0$ et $\boldsymbol{k}_0$ relatifs au trièdre $Ox_0y_0z_0$. On a, en outre:

$$\dot{\boldsymbol{H}} = \dot{H}\,(\boldsymbol{j}_0 \sin\alpha + \boldsymbol{k}_0 \cos\alpha). \qquad (63.1)$$

On considère seulement l'action du champ $\boldsymbol{E}_0$ sur le moment permanent, négligeant le moment induit par ce champ. La probabilité d'orientation de $\boldsymbol{p}$ dans le champ $\boldsymbol{E}_0$ ne dépend pas de la rotation du trièdre $Oxyz_0$ autour de Oz_0 et l'énergie de $\boldsymbol{p}$ dans ce champ ne dépend que de ϑ. Le calcul de moyennes donne

$$\overline{\boldsymbol{p}_{ab}^{(y)}} = \tfrac{1}{2}\,\boldsymbol{p}_{ab}'\,[\cos\delta_{ab}\,(1 + \overline{\cos^2\vartheta}) + \cos\xi\,\cos\zeta\,(1 - 3\,\overline{\cos^2\vartheta})]. \qquad (63.2)$$

Le calcul de $\overline{\cos^2\vartheta}$ a été fait, dans le cas de divers modèles de molécules polaires dans un champ électrique, par van Vleck[1]; dans celui de molécules paramagnétiques dans un champ magnétique, par Levy [16]. Dans ce dernier cas, on trouve:

$$\overline{\cos^2\vartheta} = \frac{j+1}{j} - \frac{1}{j}\,\mathrm{Cot}\,\frac{m_0 H_0}{2j\,kT}\left(\frac{2j+1}{2j}\,\mathrm{Cot}\,\frac{2j+1}{2j}\,\frac{m_0 H_0}{kT} - \frac{1}{2j}\,\mathrm{Cot}\,\frac{m_0 H_0}{2j\,kT}\right), \qquad (63.3)$$

---

[1] J. H. van Vleck: Electric and Magnetic Susceptibilities. Oxford 1932.

j étant le nombre quantique «interne» de l'état a, m_0 le module du moment magnétique permanent.

La force rotatoire de la transition $a \rightarrow b$ a pour expression

$$R_{ab}^L = \tfrac{3}{4}\left[(1 + \overline{\cos^2 \vartheta})\, R_{ab}^0 + (1 - 3\,\overline{\cos^2 \vartheta})\, K_{ab}\right] \tag{63.4}$$

où R_{ab}^0 est donnée par (62.5) et

$$K_{ab} = \frac{1}{m^2}\, \mathrm{Im}\,\left[\boldsymbol{m}_0 \cdot (a\,|\,\boldsymbol{p}\,|\,b)\,\boldsymbol{m}_0 \cdot (b\,|\,\boldsymbol{m}\,|\,a)\right].$$

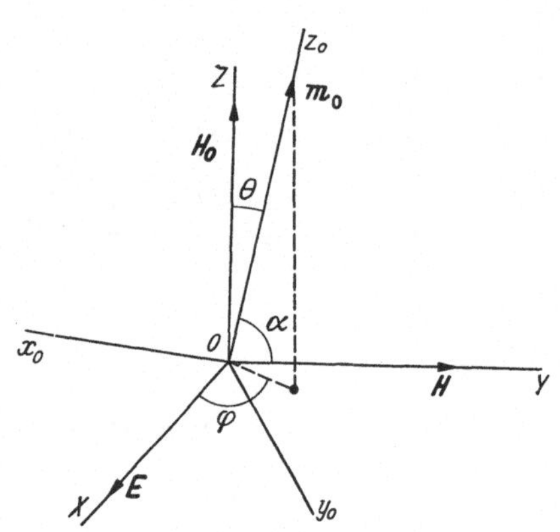

Fig. 67. Action d'un champ longitudinal sur l'orientation de molécules polaires.

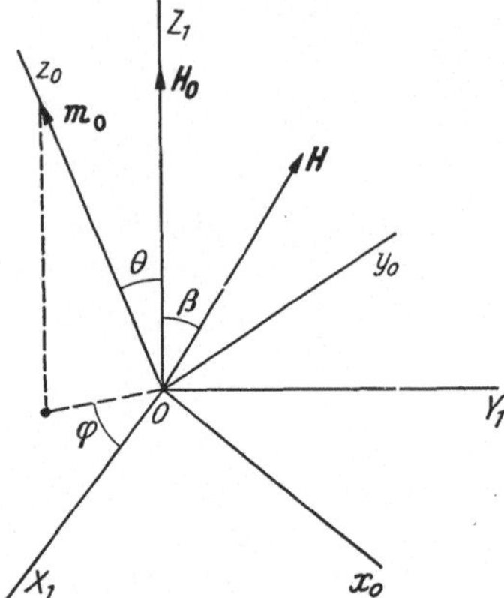

Fig. 68. Action d'un champ transversal sur l'orientation de molécules polaires.

Pour les faibles valeurs de $\dfrac{m_0 H_0}{kT}$, on peut développer $\cos^2 \vartheta$ en série; on trouve:

$$R_{ab} = R_{ab}^0 + \frac{1}{120}\,\frac{(2j+3)(2j-1)}{j(j+1)}\,(R_{ab}^0 - 3K_{ab})\left(\frac{m_0 H_0}{kT}\right)^2. \tag{63.5}$$

Cette dernière formule permet de calculer la variation $\left(\dfrac{\varDelta\varrho}{\varrho}\right)_{ab}$ comme on l'a fait pour les molécules dépourvues de moment, en prenant le cas classique où $j \rightarrow +\infty$. En admettant que:

$$1 - \frac{3K_{ab}}{R_{ab}^0} = \frac{1}{2}, \qquad m_0 = 5\,\frac{he}{4\pi m_e c}$$

($m_e =$ masse de l'électron), il vient

$$\left(\frac{\varDelta\varrho}{\varrho}\right)_{ab} = 1{,}88 \cdot 10^{-9}\,\frac{H_0^2}{T^2}. \tag{63.6}$$

Pour $H_0 = 30000$ oersteds et $T = 300°$ K, $\left(\dfrac{\varDelta\varrho}{\varrho}\right)_{ab} = 1{,}88 \cdot 10^{-5}$, variation inappréciable. Mais cette variation devient de l'ordre de $\frac{1}{1000}$ pour $T = 41°$ K. Lorsque $\dfrac{m_0 H_0}{kT}$ augmente indéfiniment, $\cos^2 \vartheta$ tend vers l'unité et R_{ab}^L vers la valeur $\tfrac{3}{2}(R_{ab}^0 - K_{ab})$.

$\beta)$ *Champ transversal* [16]. On conserve les trièdres moléculaires $Oxyz$ et $Ox_0 y_0 z_0$ précédemment définis, mais dans le trièdre $OX_1 Y_1 Z_1$ lié à l'onde lumineuse (Fig. 68) OX_1 est maintenant dirigé suivant la normale à l'onde, et OZ_1 suivant le champ appliqué $\boldsymbol{E}_0$ (ou $\boldsymbol{H}_0$). Le champ $\boldsymbol{H}$ de l'onde est dans le plan $Y_1 O Z_1$ et fait l'angle β avec OZ_1. Les relations (61.5) et (63.1) sont encore valables, l'angle α que fait $\boldsymbol{p}_0$ (ou $\boldsymbol{m}_0$) avec $\boldsymbol{H}$ satisfaisant à la relation

$$\cos \alpha = \cos \vartheta \cos \beta - \sin \vartheta \sin \beta \sin \varphi.$$

Le calcul donne, au lieu de la formule (62.2), la suivante:

$$\left. \begin{aligned} \overline{p_a'^{(y)}} = \tfrac{1}{2} p_a' \big[&\cos \delta_{ab}(1 + \overline{\cos^2 \vartheta}) + \cos \xi \cos \zeta (1 - 3\,\overline{\cos^2 \vartheta}) + \\ &+ \cos^2 \beta \,(\cos \delta_{ab} - 3 \cos \xi \cos \zeta)\,(1 - 3\,\overline{\cos^2 \vartheta})\big]. \end{aligned} \right\} \tag{63.7}$$

La valeur de $\overline{\cos^2 \vartheta}$ est donnée par (63.3), d'où

$$\left. \begin{aligned} R_{ab}^T = \tfrac{3}{4}\big[(1 + \overline{\cos^2 \vartheta}) + \cos^2 \beta\,(1 - 3\,\overline{\cos^2 \vartheta})\big] \times \\ \times R_{ab}^0 + \tfrac{3}{4}(1 - 3\cos^2 \beta)\,(1 - 3\,\overline{\cos^2 \vartheta})\,K_{ab} \end{aligned} \right\} \tag{63.8}$$

ou, en remplaçant β par l'angle ω que fait le champ appliqué avec le champ électrique de l'onde incidente:

$$\left. \begin{aligned} R_{ab}^T = \tfrac{3}{4}\big[&R_{ab}^0(1 + \overline{\cos^2 \vartheta}) + K_{ab}(1 - 3\,\overline{\cos^2 \vartheta}) + \\ &+ \sin^2 \omega\,(1 - 3\,\overline{\cos^2 \vartheta})\,(R_{ab} - 3 K_{ab})\big]. \end{aligned} \right\} \tag{63.9}$$

En comparant (63.4) avec (63.9), on voit que la force rotatoire en champ transversal est la même qu'en champ longitudinal (pour une même valeur du champ appliqué) lorsque la vibration lumineuse incidente est parallèle aux lignes de force ($\omega = 0$). La variation du pouvoir rotatoire produite par le champ est du même ordre de grandeur dans les deux cas.

64. Résultats expérimentaux. α) Les essais de MALLEMANN[1] ont porté sur l'action d'un champ électrique transversal. Le pinène, le tartrate d'éthyle, le camphre en solution dans l'hexane n'ont pas montré d'écarts à la formule de GOUY. Dans le cas de la carvone, on a observé des écarts systématiques, assez peu supérieurs aux erreurs d'expérience: l'orientation moléculaire permet à l'anisotropie du pouvoir rotatoire de se manifester.

On a vu à la Sect. 62 que les effets prévus sont si faibles à la température ordinaire qu'il n'est guère possible de les observer.

β) M. LÉVY [*16*] a mis en évidence une action d'un champ magnétique longitudinal sur le pouvoir rotatoire de la phase cholestérique que l'oléate de cholestéryle prend spontanément par refroidissement de la phase liquide, entre 35 et 42° C. A l'état liquide les valeurs de la rotation observées pour des champs égaux et de sens contraires s'écartent également de la valeur ϱ_n^0 hors du champ. La différence

$$\Delta \varrho_n = \tfrac{1}{2}(\varrho_+ + \varrho_- - 2\varrho_n^0)$$

est nulle. Il n'en est plus de même à l'état cholestérique, comme le montre la courbe I de la Fig. 69.

La théorie est la suivante. Le champ magnétique, parallèle à l'axe Oz des empilements hélicoïdaux de molécules (Sect. 60) exerce son action de la même façon sur chacune de ces molécules. Admettons que l'axe $\boldsymbol{L}$ d'une molécule fasse avec la direction du champ un angle ϑ. Les relations (60.3) donnant les composantes de $\boldsymbol{L}$ hors du champ sont remplacées par les suivantes:

$$L_x = \sin \vartheta \cos \alpha\,z, \qquad L_y = \sin \vartheta \sin \alpha\,z, \qquad L_z = \cos \vartheta,$$

---

[1] R. DE MALLEMANN: Ann. Phys., Paris **22**, 192 (1924).

et la polarisation $\boldsymbol{P}$ s'écrit, au lieu de (60.4):

$$\boldsymbol{P} = (\varepsilon_0 - \varepsilon_1)\sin^2\vartheta\,\boldsymbol{L}\,(\boldsymbol{L}\cdot\boldsymbol{E}) + \varepsilon_1\,\boldsymbol{E}.$$

On trouve que la variation du pouvoir rotatoire (60.8) produite par le champ a pour expression

$$\Delta\varrho = -\frac{9}{16\pi}\,\frac{\omega^3}{c^3}\,\frac{\alpha^3\delta^3}{\varepsilon\sqrt{\varepsilon}}\,\frac{1}{\left(\alpha^2 - \dfrac{\omega^2}{c^2}\,\varepsilon\right)^2}\,\overline{\cos^2\vartheta}.$$

Le calcul de $\overline{\cos^2\vartheta}$ se fait en partant de l'expression de l'énergie U d'une molécule soumise à la fois au champ magnétique appliqué $\boldsymbol{H}_0$ et au champ électrostatique $\boldsymbol{E}'$ donné par (60.1). On a

$$U = -(k_0\cos^2\vartheta + k_1\sin^2\vartheta)H_0^2 + \left. + (\boldsymbol{m}_0\cdot\boldsymbol{E}')\sin\vartheta\cos\varphi, \right\} \quad (64.1)$$

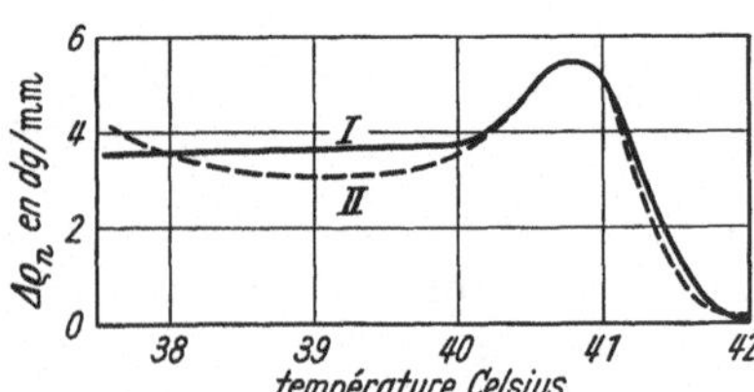

Fig. 69. Variation du pouvoir rotatoire de l'oléate de cholestéryle sous l'action d'un champ magnétique, I courbe expérimentale, II courbe déduite de (64.2) pour $H_0 = 24\,000$ oersteds.

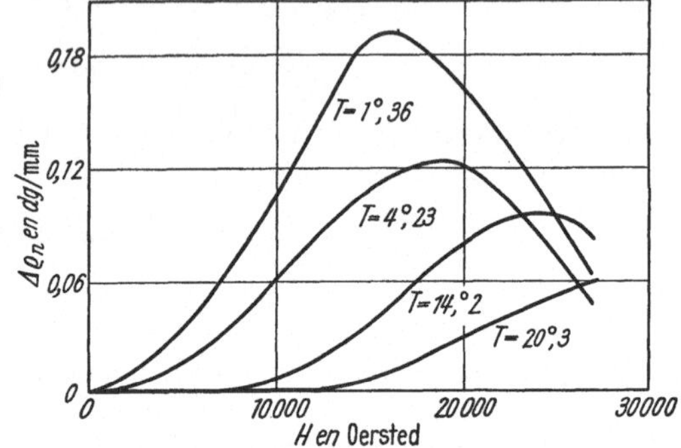

Fig. 70. Variation du pouvoir rotatoire des cristaux de $NiSO_4$, $6\,H_2O$ sous l'action d'un champ magnétique longitudinal ($\lambda = 0{,}546\,\mu$).

k_0 et k_1 désignant les susceptibilités magnétiques de la molécule suivant la direction de $\boldsymbol{L}$ et une direction perpendiculaire, $\boldsymbol{m}_0$ le moment magnétique permanent, φ l'angle que fait avec $\boldsymbol{E}'$ la projection de $\boldsymbol{L}$ sur le plan xOy. L'évaluation de $\overline{\cos^2\vartheta}$ est laborieuse; on trouve

$$\Delta\varrho_n = \frac{-9}{16\pi}\,\frac{\omega^3}{c^3}\,\frac{\alpha^3\delta^3}{\varepsilon\sqrt{\varepsilon}}\,\frac{k_0 - k_1}{\left(\alpha^2 - \dfrac{\omega^2}{c^2}\,\varepsilon\right)^2}\,\frac{kT}{m_0^2 e^2} \times$$
$$\times H^2\left(3 + \frac{8\,k^2\,T^2}{m_0^2 e^2} - \frac{7\,kT}{m_0 e}\,\mathrm{Cot}\,\frac{m_0 e}{kT} - \mathrm{Cot}^2\,\frac{m_0 e}{kT}\right). \quad\left.\right\} \quad (64.2)$$

La courbe II de la Fig. 69 montre la variation de $\Delta\varrho_n$ en fonction de T. On voit que l'accord avec la courbe expérimentale I est satisfaisant. Le maximum de $\Delta\varrho_n$ correspond à une structure des groupes hélicoïdaux pour laquelle le champ $\boldsymbol{E}'$ qui agit sur chaque molécule s'annule; le champ magnétique peut alors exercer plus facilement son action.

On a pu également expliquer l'action d'un champ magnétique transversal et prévoir une action orientatrice considérable d'un champ électrique.

γ) Lévy [16] a observé qu'au-dessous de 20° K, un champ magnétique, longitudinal ou transversal, fait varier le pouvoir rotatoire du sulfate de nickel cristallisé $NiSO_4$, $6\,H_2O$ (Sect. 53 δ). La Fig. 70 représente la variation $\Delta\varrho_n$ pour la longueur d'onde 0,546 μ, en fonction de l'intensité du champ longitudinal, â diverses températures.

L'action du champ magnétique peut être double dans ce cas: elle peut orienter les ions Ni^{++} paramagnétiques (Sect. 63 α); elle peut déformer le contenu de la maille cristalline.

L'action orientatrice du champ longitudinal se traduit par la formule (63.4). On exprime $\overline{\cos^2\vartheta}$ à partir des formules utilisées pour calculer l'aimantation dans la théorie quantique du paramagnétisme[1], soit:

$$\overline{\cos^2\vartheta} = \frac{1}{m_0^2} \frac{\sum_i \left(\frac{\partial W_i}{\partial H}\right)^2 e^{-\frac{W_i}{kT}}}{\sum_i e^{-\frac{W_i}{kT}}}. \tag{64.3}$$

m_0 est le moment magnétique déduit de la mesure de la susceptibilité à température assez élevée ($m_0 = 2{,}03 \cdot 10^{-20}$); W_i désigne l'énergie d'un niveau de l'ion paramagnétique et les sommes s'étendent à l'ensemble des niveaux. Sous l'action du champ cristallin, le niveau fondamental $3F_4$ des ions Ni^{++} se décompose. On trouve que les trois niveaux les plus bas résultant de cette decomposition sont voisins de quelques cm^{-1} et éloignés des autres par plusieurs milliers de cm^{-1}. A basse température, on ne tient compte que de ces trois niveaux dans la formule (64.3). On trouve ainsi, pour les champs faibles, la formule approchée:

$$\overline{\cos^2\vartheta} = \Gamma H^2 \tag{64.4}$$

avec

$$\Gamma = \frac{\frac{m_0^2}{\varepsilon^2} \operatorname{Cos} \frac{\varepsilon}{kT}}{\operatorname{Cos} \frac{\varepsilon}{kT} + \frac{1}{2} e^{\frac{\delta}{kT}}}, \tag{64.5}$$

où 2ε est l'écart des deux niveaux extrêmes et δ l'écart entre leur moyenne et le troisième.

D'après (63.4), la variation du pouvoir rotatoire due à l'orientation peut s'écrire:

$$\Delta\varrho_n = \Phi \overline{\cos^2\vartheta} \tag{64.6}$$

où Φ dépend de la longueur d'onde et, en principe, de la température qui exerce une influence sur la dispersion rotatoire. Or, les courbes de la Fig. 70 ne varient pas, en fonction du champ, comme $\overline{\cos^2\vartheta}$, d'après (64.6).

Il faut donc tenir compte de la déformation de la maille sous l'action du champ. Comme cette action ne dépend pas du sens du champ on peut essayer d'écrire:

$$\Delta\varrho_n = \Phi\left[\overline{\cos^2\vartheta} - Q^2 H^2\right], \tag{64.7}$$

Q pouvant dépendre de la température. Dans les champs assez élevés, on peut négliger l'exponentielle de la formule (64.5) et écrire:

$$\Delta\varrho_n = \Phi\left[\frac{m_0^2 H^2}{m_0^2 H^2 + \varepsilon^2} - Q^2 H^2\right]. \tag{64.8}$$

Comme le montre l'expérience, $\Delta\varrho_n$ a une valeur maximum pour une certaine valeur du champ H. Ces deux valeurs permettent de calculer Q et Φ (pour une longueur d'onde donnée) à partir des données expérimentales. Dans les champs faibles, compte tenu de (64.4), on a:

$$\lim \frac{\Delta\varrho_n}{H^2} = \Phi\left[\Gamma(T) - Q^2(T)\right] \tag{64.9}$$

---

[1] J. H. van Vleck: Electric and magnetic susceptibilities, p. 181. Oxford 1932.

d'où l'on tire Γ. D'autre part, Γ est calculable par la formule (64.5). On trouve par exemple, pour $\lambda = 0{,}546\,\mu$ et $T = 1{,}36^\circ\,$K, $\Gamma_{\text{exp}} = 4{,}25 \cdot 10^{-10}$, $\Gamma_{\text{cal}} = 4{,}22 \cdot 10^{-10}$.

Des calculs analogues permettent de représenter correctement l'influence d'un champ magnétique transversal sur le pouvoir rotatoire, du moins pour des valeurs du champ qui ne sont pas trop élevées.

δ) Lévy [16] a également mis en évidence les variations considérables du pouvoir rotatoire que subit le sel de Seignette, spontanément entre les températures où il est ferroélectrique (Fig. 71) et, dans cet état, sous l'influence d'un champ électrique longitudinal. La théorie des deux phénomènes est analogue, lorsqu'on admet que les modifications du champ interne sont responsables de la variation thermique de la rotation.

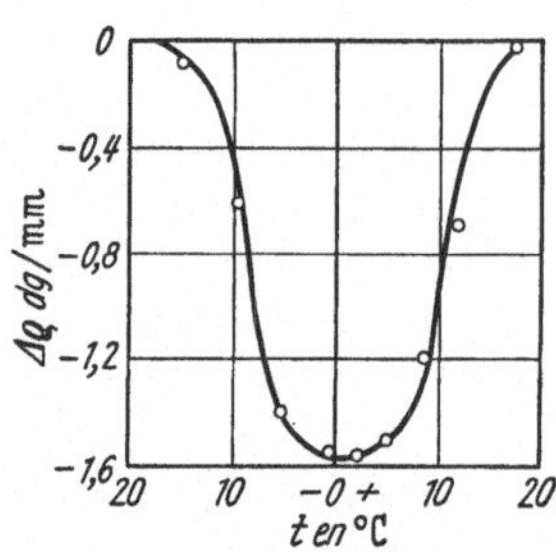

Fig. 71. Variation du pouvoir rotatoire due à la polarisation spontanée du sel de Seignette dans le domaine où il est Seignette-électrique.

La maille du sel de Seignette $\mathrm{NaKC_4H_4O_6}$, $4\,\mathrm{H_2O}$ (groupe 222) contient quatre molécules, deux à deux antiparallèles. Ce cas est donc plus compliqué que celui de l'acide tartrique (Sect. 59 δ), car il existe du pouvoir rotatoire de structure. On a $\varrho = 1{,}35$ degré/mm pour la raie D.

Admettons que la ferroélectricité soit liée à une variation de l'orientation des molécules de la maille. Pour l'une des paires de molécules antiparallèles, la force rotatoire devient, avec les notations de la Sect. 63,

$$R_1 = R_1^0 + (3\,K_1^0 - R_1^0)\,[2 - 3\,(\overline{\cos^2 \vartheta_1} + \overline{\cos^2 \vartheta_2})], \qquad (64.10)$$

où ϑ_1 et ϑ_2 désignent les angles que font avec le dipôle électrique de la molécule le vecteur $\boldsymbol{k}$ porté par l'axe optique et le vecteur $\boldsymbol{j}$ normal à $\boldsymbol{k}$. La vibration incidente est dirigée suivant $\boldsymbol{j}$. On a une équation analogue à (64.10) pour la force rotatoire R_2 de la seconde paire de molécules. Si l'on peut représenter le pouvoir rotatoire, comme celui de l'acide tartrique, par une formule de Drude à un seul terme, on a:

$$\varrho = 96\,\frac{\pi N}{h c}\,\frac{v^2}{v_0^2 - v^2}\,[R_1 + R_2 + R_3].$$

R_3 correspond au pouvoir rotatoire de structure.

La variation de pouvoir rotatoire produite par les champs interne et externe est:

$$\Delta \varrho = \frac{192\,\pi N}{h c}\,\frac{v^2}{v_0^2 - v^2}\,[3\,(K_1^0 + K_2^0) - (R_1^0 + R_2^0)]\left[1 - \frac{3}{2}\,(\overline{\cos^2 \vartheta_1} + \overline{\cos^2 \vartheta_2})\right], \quad (64.11)$$

en négligeant les variations du pouvoir rotatoire de structure. Le calcul de $\overline{\cos^2 \vartheta_1}$ se fait à l'aide de la formule:

$$\overline{\cos^2 \vartheta_1} = \frac{\displaystyle\int_0^\pi \cos^2 \vartheta\; e^{-a_1 \cos \vartheta}\, \sin \vartheta\, d\vartheta}{\displaystyle\int_0^\pi e^{-a_1 \cos \vartheta}\, \sin \vartheta\, d\vartheta},$$

avec

$$a_0 = \frac{m_0\,\boldsymbol{F_1}}{k\,T},$$

où $\boldsymbol{F_1}$ est le champ agissant. On a une formule analogue pour $\overline{\cos^2 \vartheta_2}$. On trouve finalement:

$$\Delta \varrho = \Delta \varrho_\infty \left[1 - \frac{3}{2 a_1}\left(\mathrm{Cot}\, a_1 - \frac{1}{a_1}\right) - \frac{3}{2 a_2}\left(\mathrm{Cot}\, a_2 - \frac{1}{a_2}\right)\right], \qquad (64.12)$$

avec

$$\varDelta \varrho_\infty = - \frac{384\,\pi\,N}{h\,c}\,\frac{v^2}{v_0^2 - v^2}\,[3\,(K_1^0 + K_2^0) - (R_1^0 + R_2^0)].$$

On pose $\boldsymbol{F} = \boldsymbol{E} + \beta\,\boldsymbol{P}$ où $\boldsymbol{P}$ désigne la polarisation; la théorie de MASON[1] permet d'évaluer $\boldsymbol{F}_1$ et $\boldsymbol{F}_2$, d'où a_1 et a_2 qui sont fonctions de la température T et du champ E. L'expérience donne la valeur de la variation $\varDelta \varrho_\infty$ due au champ spontané et au champ électrique imposé à saturation. L'équation (64.12) permet de calculer les variations $\varDelta \varrho$ produites par la température (polarisation spontanée) et par le champ. La courbe de la Fig. 71, tracée d'après ces calculs, passe bien par les points expérimentaux.

Ouvrages généraux.

[1] BORN, M.: Atomtheorie des festen Zustandes. Leipzig: J. B. Teubner 1923. — Ouvrage d'importance historique, contenant la théorie moléculaire classique des réseaux cristallins. Le chapitre III est consacré aux propriétés optiques.

[2] BORN, M.: Optik. Berlin: Springer 1933. — Ouvrage classique, donnant un résumé de l'ensemble des travaux théoriques de l'auteur sur l'activité optique.

[3] BRUHAT, G.: Traité de Polarimétrie, Paris: Rev. Opt. 1930. — Contient une étude détaillée des méthodes de mesure, des appareils et des applications pratiques.

[4] CONDON, E. U.: Theories of optical rotatory power. Rev. Mod. Phys. 9, 432—457 (1937). — Résumé général des théories quantiques et particulièrement de celle de l'auteur.

[5] DELEPINE, M.: Isomérie optique. Dans V. GRIGNARD, Traité de Chimie organique, Tome I, p. 843—1008. Paris: Masson & Cie. 1935. — Exposé historique très complet de la stéréochimie des composés optiquement actifs et importante bibliographie.

[6] FRIEDEL, G.: Les états mésomorphes de la matière. Ann. Phys. Paris 18, 273—474 (1922). — Ce mémoire a apporté un ordre essentiel dans le classement des propriétés des cristaux liquides.

[7] HILLEMANN, H.: Molekulare Asymmetrie. Angew. Chem. 50, 435—447 (1937). — Exposé sur la stéréochimie des composés organiques optiquement actifs, avec une bibliographie nombreuse.

[8] JAEGER, F. M.: Spatial arrangements of atomic systems and optical activity. New York: Mc. Graw Hill 1930. — Consacré principalement à l'étude des complexes minéraux et des travaux de l'auteur.

[9] KAUZMANN, W., J. E. WALTER et H. EYRING: Theories of optical rotatory power. Chem Rev. 26, 339—407 (1940). — Exposé des théories modernes dégagées de leur appareil mathématique.

[10] KAUZMANN, W., et H. EYRING: The effect of the rotation of groups about bonds on optical rotatory power. J. Chem. Phys. 9, 41—53 (1941). — Discussion de nombreux facteurs physicochimiques qui influencent le pouvoir rotatoire des fluides.

[11] KORTÜM, G.: Die Dispersion der optischen Drehung amorpher Systeme. Stuttgart: Ferdinand Enke 1932. — Contient le résumé des théories classiques et de nombreux résultats expérimentaux.

[12] KUHN, W.: Theorie und Grundgesetze der optischen Aktivität. Dans: K. FREUDENBERG, Stereochemie, p. 317—434. Leipzig: Franz Deuticke 1932. — Exposé d'ensemble, sans développements mathématiques compliqués, des théories classiques et des importants travaux des auteurs.

[13] KUHN, W., et K. FREUDENBERG: Natürliche Drehung der Polarisationsebene des Lichtes. In Handbuch und Jahrbuch der chemischen Physik, Bd. VIII/3, p. 1—142. Leipzig: Akademische Verlagsgesellschaft 1932. — Ouvrage général, contenant un exposé détaillé de la théorie de KUHN et de ses applications.

[14] LEHMANN, O.: Flüssige Kristalle. Leipzig: Wilhelm Engelman 1904. — Die neue Welt der flüssigen Kristalle. Leipzig 1911. — Ouvrages d'importance historique, contenant la description de nombreuses observations originales.

[15] LEVENE, P. A., et A. ROTHEN: Rotatory dispersion. Dans: GILMAN, Organic Chemistry, Tome II, p. 1779—1849. New York: Wiley 1938. — Résume les travaux expérimentaux importants des auteurs sur la dispersion rotatoire des composés organiques.

[16] LEVY, M.: L'anisotropie moléculaire du pouvoir rotatoire naturel. Ann. Phys., Paris 5, 153—310 (1950). — Mémoire fondamental pour l'étude théorique et expérimentale de l'action des champs électrique et magnétique sur l'activité optique.

---

[1] W. P. MASON: Phys. Rev. 72, 854 (1947).

[17] Lowry, T. M.: Optical rotatory power. London: Longmans 1935. — Exposé détaillé de l'historique et des méthodes de mesure, ainsi que de nombreux résultats expérimentaux.

[18] Mallemann, R. de: La propagation de la lumière et la structure des molécules. Rev. gén. Sci. 38, 453—479 (1927). — Discussion des bases de la théorie moléculaire du pouvoir rotatoire.

[19] Mathieu, J. P.: Aperçus sur la stéréochimie des complexes minéraux, Bull. Soc. chim. France 5, 725—805 (1938). — Etude critique et bibliographie.

[20] Mathieu, J. P.: Les théories moléculaires du pouvoir rotatoire naturel, Paris, C. N. R. S., 1946. — Exposé détaillé des théories classiques et quantiques du pouvoir rotatoire des fluides.

[21] Mills, J. A., et W. Klyne: The correlation of configurations. Dans: W. Klyne, Progress in Stereochemistry, p. 177—222. London: Butterworth 1954. — Exposé détaillé des méthodes chimiques de comparaison des configurations des composés optiquement actifs.

[22] Nyholm, R. S.: The Sterochemistry of complex compounds. Dans: W. Klyne, Progress in Stereochemistry, p. 322—360. London: Butterworth 1954. — Exposé fondé sur la théorie des valences dirigées; consacre une discussion aux complexes optiquement actifs.

[23] Optical Rotatory Power, A general discussion. Trans. Faraday Soc. 26, 266—461. (1930). — Contient environ 25 mémoires et des discussions sur la théorie classique du pouvoir rotatoire, les méthodes de mesures, les aspects physico-chimiques de l'activité optique des fluides.

[24] Szivessy, G.: Kristalloptik. Dans: Geiger et Scheel, Handbuch der Physik, tome XX. Berlin: Springer 1928. — Les chapitres III (p. 804—861) et IVb (p. 900—904) sont consacrés à l'étude détaillée des phénomènes classiques des cristaux optiquement actifs et contiennent une importante bibliographie.

[25] Werner, A.: Sur les composés métalliques à dissymétrie moléculaire. Bull. Soc. chim. France 2, 1—24 (1912). — Mémoire d'importance historique, résumant les travaux de l'auteur dans ce domaine.

On trouve des données numériques sur le pouvoir rotatoire dans les Tables de Constantes suivantes:

Bureau of Standards, Publication Nr. 118, 1924.

International Critical Tables, Tome VII. 1930.

Landolt-Börnstein: Physikalisch-Chemische Tabellen, depuis 1894.

Tables Annuelles Internationales de Constantes. de 1910 à 1937. Enfin, les Tables de Constantes Sélectionnées de l'Union Internationale de Chimie publient une série de fascicules consacrés au pouvoir rotatoire des principaux groupes de composés chimiques. Le fascicule 1 (Steroides) est paru. Paris: Masson & Cie. 1956.

Subject Index.

(English-German.)

Where English and German spelling of a word is identical the German version is omitted.

Table des matières

pour la contribution écrite en français:

J. P. MATHIEU: Activité optique naturelle.